Coviability of Social and Ecological Systems: Reconnecting Mankind to the Biosphere in an Era of Global Change

Olivier Barrière • Mohamed Behnassi
Gilbert David • Vincent Douzal
Mireille Fargette • Thérèse Libourel
Maud Loireau • Laurence Pascal
Catherine Prost • Voyner Ravena-Cañete
Frédérique Seyler • Serge Morand
Editors

Coviability of Social and Ecological Systems: Reconnecting Mankind to the Biosphere in an Era of Global Change

Vol. 1: The Foundations of a New Paradigm

Springer

IRD Institut de Recherche pour le Développement FRANCE
French National Research Institute for Sustainable Development

Université Montpellier

Editors
Olivier Barrière
IRD - French National Research Institute for Sustainable Development
UMR ESPACE-DEV
Montpellier, France

Gilbert David
IRD - French National Research Institute for Sustainable Development
UMR ESPACE-DEV
Montpellier, France

Mireille Fargette
IRD - French National Research Institute for Sustainable Development
UMR ESPACE-DEV
Montpellier, France

Maud Loireau
IRD - French National Research Institute for Sustainable Development
UMR ESPACE-DEV
Montpellier, France

Catherine Prost
The Federal University of Bahia
Salvador, Bahia, Brazil

Frédérique Seyler
IRD - French National Research Institute for Sustainable Development
UMR ESPACE-DEV
Montpellier, France

Mohamed Behnassi
Center for Research on Environment, Human Security and Governance (CERES)
Ibn Zohr University of Agadir
Agadir, Morocco

Vincent Douzal
IRSTEA
National Research Institute of Science and Technology for Environment and Agriculture
Montpellier, France

Thérèse Libourel
University of Montpellier
UMR ESPACE-DEV
Montpellier, France

Laurence Pascal
University of Montpellier
Montpellier, France

Voyner Ravena-Cañete
The Federal University of Pará
Belém, Brazil

Serge Morand
CNRS – The French National Center for Scientific Research
Paris, France

CIRAD – The French Agricultural Research Centre for International Development
Paris, France

Kasetsart University
Bangkok, Thailand

ISBN 978-3-319-78496-0 ISBN 978-3-319-78497-7 (eBook)
https://doi.org/10.1007/978-3-319-78497-7

Library of Congress Control Number: 2018950090

This Springer imprint is published by the registered company Springer Nature Switzerland AG.
The registered company address is: Gewerbestrasse 11, 6330 Cham, Switzerland

Foreword: Coviability as a Premise of a New Shared Model for Sustainable Development by Jean-Marc Châtaignier

> *« Rien n'est solitaire, tout est solidaire. L'homme est solidaire avec la planète, la planète est solidaire avec le soleil, le soleil est solidaire avec l'étoile, l'étoile est solidaire avec la nébuleuse, la nébuleuse, groupe stellaire, est solidaire avec l'infini. Ôtez un terme de cette formule, le polynôme se désorganise, l'équation chancelle, la création n'a plus de sens dans le cosmos et la démocratie n'a plus de sens sur la terre. Donc, solidarité de tout avec tout, et de chacun avec chaque chose. La solidarité des hommes est le corollaire invincible de la solidarité des univers. Le lien démocratique est de même nature que le rayon solaire ».*

"Nothing is isolated, everything is connected. Man is connected to the planet, the planet is connected to the sun, the sun is connected to the star, the star is connected to the nebula, the nebula, the stellar group, are connected to the infinity. Remove an element from this formula and the polynomial becomes disorganized, the equation staggers; creation loses meaning in the cosmos, and democracy loses meaning on earth. Thus, there should be solidarity

---

Until September 2017 Jean-Marc Châtaigner was Executive Vice-President of the French Research Institute for Development (*l'Institut de Recherche pour le Développement*, IRD), which is an internationally recognized multidisciplinary organization, and whose partners are mainly countries located in the intertropical zone. IRD is a public institution under the simultaneous authority of the Minister of Higher Education and Scientific Research, and the Minister of Foreign Affairs and International Development. Through its network and its structured presence in about fifty countries, this institution represents an original approach to research, expertise, training, and knowledge sharing for the benefit of countries and territories that consider science and innovation the primary levers of their development. Jean-Marc Châtaigner is a former student of the National School of Administration (ENA – graduating class of Jean Monnet), he graduated from the Institute of Political Studies (IEP) of Bordeaux. Since 1990, he has held influential positions concerning development in France, Africa, and the United Nations. He was the Ambassador of France in Madagascar between 2009 and 2012, and Executive Vice-President of Globalization, Development, and Partnerships (DGM) between 2012 and 2014 at the French Ministry of Foreign Affairs and International Development. He wrote several articles on issues concerning international security, African politics, development, and governance. He has also published "L'ONU en Sierra Leone Les méandres d'une négociation" *(2005)* and "États et sociétés fragiles. Entre conflits, reconstruction et développement" (with Hervé Magro 2007), both published by Karthala. His latest book (ed.) is called "Fragilités et résilience, les nouvelles frontières de la mondialisation" (Karthala 2014).

between everyone and everything. Human solidarity is the invincible corollary for universal solidarity. The democratic connection is similar in nature to a solar ray". (Victor Hugo, *Proses philosophiques* (1860–1865))

We live in a permeable, unstable world where successive crises combine with increasingly complex threats. Our individual and collective social rules no longer seem effective against this "new international disorder." Humanity can no longer afford to focus on adaptation only. The unprecedented population boom that we have almost finished experiencing (1 billion people in 1800, probably between 10 and 11 billion in 2100) questions the survival of our species in an environment whose existence we have dramatically changed. With the new threats engendered by our industrial economic development, such as global warming and the most massive extinction of terrestrial and marine species since 66 million years ago, we have to invest time rebuilding our relationship with our planet to create conditions of a new *modus operandi*.

This volume investigates the coviability of ecological and social systems. Being the result of about twelve years of multidisciplinary research around the globe, it flourishes today to clarify the possibilities of a harmonious, peaceful, fair, and balanced life. It provides holistic markers of the changes and the evolutions noticed in the relationship between human and the world of animals, plants, and minerals, and the aquatic world This innovative perspective is based on the cutting edge of research in various[1] and complementary domains.

The concept of viability originates from the mathematical patterns of control theory developed in the early 1990s by J-P Aubin.[2] This concept transcended several disciplines before developing and maturing by research programs on fragility and resilience. Emerging from the meeting point of the economic, social, and environmental pillars of sustainability, also related to the polysemous notion of sustainability, the concept of coviability is to participate in the elaboration of a new corpus of humanistic, social, and economic principles to take over social and liberal ideologies that are collapsing or in major crisis. It is implied in the book that the concept of coviability as choosing "the environment as a monitor for our development" becomes tangible and real, as stated by Johan Rockström, Director of the Resilience Research Center at the University of Stockholm.

[1]Such as mathematics, geography, law, anthropology, biology, agronomy, ecology, political science, computer science, modeling, human and social sciences, economics, philosophy, hydrology, physics, chemistry, oceanography, electrical engineering, botany, entomology, ethnology, etc.

[2]See particularly Aubin, JP (1996) "A mathematical metaphor of the precautionary principle" Nature Sciences Societies, 4, 2, 146-154.

The Limits of the Anthropocene, The Danger Facing Humanity

The winner of a Nobel Prize in chemistry, Paul Crutzen, suggests in a famous article published in 2002 by *Nature*[3] that 200 years ago the human condition mutated; it moved from an era called the Holocene to a new geological era, the Anthropocene, where humans developed the desire to construct and control their environment. If there is no rapid transmutation of this evolution, it will condemn humanity in an irreversible manner. The limits of this system are visible in the repeated ruptures and fractures that it undergoes with increasing intensity.

In "Le dérèglement du monde, quand nos civilisations s'épuisent," which was published in 2009 (ed. Grasset), the Franco-Lebanese scholar Amin Maalouf also points out that the disorders present in all domains, be it intellectual, financial, climatic, geopolitical, ethical, are a sign that humanity reached the "threshold of moral incompetence."

Johan Rockström reaches the same conclusions in his environmental research on the "Nine Planetary boundaries."[4] With his coauthors, Rockström calls for a new model for the development of humanity in its planetary environment: "here is an urgent need for a new paradigm that integrates the continued development of human societies and the maintenance of the Earth system (ES) in a resilient and accommodating state" (Steffen et al. 2009, 736).

His most recent collective article, published in *Science*[5] in 2015, warns us against the extremely risky and critical situation of four of these nine limits. They are climate change, total change of the biosphere, biogeochemical flows, and the profound changes in terrestrial ecosystems.

Despite these disturbing conclusions, we cannot fail to notice the slow awakening of consciousness, of which the environmentalist candidate in the French presidential election of 1974, René Dumont[6], was probably one of the most emblematic precursors. The movement is marked by the regular organization of environmental

[3]Crutzen P. (2002), « *Geology of mankind* », *Nature* 415, 23.

[4]Climate change; the change of the biosphere (loss of biodiversity and species extinction); the exhaustion of stratospheric ozone; ocean acidification; biogeochemical fluxes (phosphorus and nitrogen cycles); changes in terrestrial ecosystems (e.g., deforestation); the use of fresh water; atmospheric change to aerosols (microscopic particles in the atmosphere that affect the climate and living organisms); the introduction of new entities (e.g., organic pollutants, radioactive materials, nanomaterials, and microplastics) (Johan Rockström et al. (2009) « A safe operating space for humanity » Nature 461, 472-475 (24 September 2009) | doi:10.1038/461472a; Published online 23 September 2009).

[5]Will Steffen, Katherine Richardson, Johan Rockström, Sarah E. Cornell, Ingo Fetzer, Elena M. Bennett, Reinette Biggs, Stephen R. Carpenter, Wim de Vries, Cynthia A. de Wit, Carl Folke, Dieter Gerten, Jens Heinks, Georgina M. Mace, Linn M. Persson, Veerabhadran Ramanathan, Belinda Reyers, Sverker Sörlin, « *Planetary boundaries: Guiding human development on a changing plane*t », *Science,* 13 February 2015, Vol. 347, Issue 6223.

[6]https://en.wikipedia.org/wiki/René_Dumont

summits, the first taking place at Rio in 1992. Since then, annual conferences of parties are held to investigate climate change, preserve biodiversity, or combat desertification.

The state of our planet enjoins humans to stop their productive development and reproductive frenzy. We are not only facing an ecological crisis but an existential crisis as well.

A Collective Consciousness of the "Human-Nature" Unity Converted into Institutional and Legal Frameworks

Being aware of an inevitable change, of our organizations, and our behaviors is translated into different social levels.

At the global level, this change was marked by an alteration of focus from the Millennium Development Goals, which is a referential framework of public development aid adopted at the Millennium Summit in New York in 2000, whose priority was the fight against extreme poverty; the aid was also given to the Sustainable Development Goals which was adopted by the UN General Assembly for the period 2015/2030, in an integrated, processed, and universal framework. For the first time, the human sustainability agenda is not limited to the citizens of the southern hemisphere only. Shared from Ouagadougou to Seattle, via Nouméa and La Paz, we are all responsible for bearing it. If the concept of coviability is not quoted as such in the 2030 calendar, it still establishes the philosophy and framework for renewed collective action.

Meanwhile, the actors of civil or transnational private societies (NGOs, companies, foundations, universities, think tanks, trade unions, etc.) are developing an increasing influence throughout the intergovernmental processes by means of consultation and lobbying processes. The upsurge of civil societies on the international scene of relations and regulations is unsettling the traditional diplomatic relations between States, especially as certain private actors do not hesitate to discard procedures that the States are supposed to respect. This gives them a feeling of power which the said actors sometimes use to excess. Nowadays, we face fierce competition between private actors and States. States lose the control of diplomatic action leading to a certain decline of sovereignty; it will in its turn serve as a ground for the development of nationalist and populist ideas, nostalgic of an often fantasized glorious past.

At the national level, an increasing number of laws revolving around energy, biodiversity, and the management of nonrenewable resource recognize the principles of interdependence and the necessity to protect the common public goods. The precautionary principle arises from a philosophical context to affirm itself in the international and European juridical norms, and to inscribe itself through the Environmental Charter in the French constitution in 2005.

The promotion of the institutional and legislative frameworks mentioned above requires appropriate conditions for fruitful dialogue and thinking along with skillful instruction in order to facilitate their development and application. Several chapters in this book tackle the environmental right from different perspectives; as they constitute the cradle of this progressing reflection. I particularly esteem the efforts of the researchers who investigated this issue: Olivier Barrière, Mohamed Benhassi, Thérèse Libourel, Aline Treillard, and Jessica Makowiak.

The Consciousness of Our Civilization of an Ecological Action Based on the Cutting Edge of a Multidisciplinary and Global Research

The complex and occasionally conflictual contemporary context in which we have to survive requires an integrated scientific interpretation capable of reaching private and public decisions. Through a collective and coordinated method, this scientific clarification should help in addressing the set of new challenges our planet faces, challenges related to living with others and living harmoniously within ecosystems.

Scientists contribute to anticipating and measuring risks through the production and selection of indicators shaped by data that result from observations of the earth. They offer solutions concerned with adaptation or reduction while keeping up with changes in order to appraise and follow public decisions. The case studies presented in this book illustrate the cross-sectoral articulation and the applications of this process, be it in Amazonian farming borders, protected marine areas in Brazil, national parks in South Africa, or the Mediterranean pastoral ecosystems.

Managing migration, mitigating climatic change, conserving biodiversity, removing carbon from soils, building a genuine transitory energy, and managing global health crises are also themes involving the concept of coviability.

Calling for a Coalition for Coviability

This volume thus invites us to step beyond models marked by competition or simple coexistence. It offers the opportunity to adopt a new collective, holistic, essential, and unique model: "the coviability of social and ecological systems."

This volume addresses our collective understanding both in the actual sense of the word, as it calls for a "Right understanding," and in its etymological sense of intergere which denotes "recreation of links." We simply need to reconnect to the ecological system in which we are born and to which we are connected. At the dawn of a new era, we have the responsibility of combining our efforts and intentions to follow this conceptual change in order to solitarily thrive by means of scientific guidance.

To conclude, I quote once more Amin Maalouf who suggests that « les *seuls vrais combats qui méritent d'être menés par notre espèce au cours des prochains siècles seront scientifiques et éthiques* » ["*In the coming centuries, the only real battles worth fighting for by our species will be scientific and ethical*."]

Jean-Marc Châtaignier

French Ambassador, Special Envoy for the Sahel since October 2017
Until October 2017, Vice-President of the french Institute of Research for Sustainable Development (IRD), France

Foreword: The Coviability New Humanistic Paradigm by Philippe AUGE

"How to unite nature and society, human and non-human, individuals and groups, in a new assembly where they would no longer present themselves to us as comprising substances, processes and simulators, but as the established expressions of relations between multiple entities whose ontological status and capacity for action vary according to the positions they occupy in relation to each other?" Such is the question of Philippe Descola, in *L'écologie des autres. L'anthropologie et la question de la nature. 2011* (The ecology of others. Anthropology and the question of nature).

The collective work "Coviability of Social and Ecological Systems: Reconnecting Man to the Biosphere in an Era of Global Change" aims to answer this question through more than fifty articles from various scientific fields: ecology, health, mathematics, computer science, geography, human and social sciences, anthropology, environmental law, and philosophy. Under the leadership of the publishing collective (Olivier Barrière, Mohamed Behnassi, Gilbert David, Vincent Douzal, Mireille Fargette, Thérèse Libourel, Maud Loireau, Laurence Pascal, Catherine Prost, Voyner Ravena-Cañete, Frédérique Seyler, Serge Morand), the authors from different corners of the world have exchanged their points of view from the first writings and ideas emanating from theoretician colleagues (Aubin[1], Bourgine[2]) or from environmental colleagues, modellers, and economists (coviability of fisheries systems IRD and SHS[3] project). About twenty faculty members and researchers from four research units at our university are involved and have strong links with their international colleagues to make sense of the coviability paradigm.

[1] Aubin, J.P. (1991), Viability theory, Birkhäuser, Boston.

[2] Bourgine Paul (1996) Models of autonomous agents and their co-evolutionary interactions, Thinking the Spirit: from the Sciences of Cognition to a Cognitive Philosophy, V. Rialle and D. Fisette (eds), University Presses of Grenoble, pp. 421-443

[3] Le Fur, Curry, Laloe, Durand, Chaboud (1999) Co-viability of Fisheries Systems, Nature Sciences Societies, vol. 7, No. 2, 19-32

The intended ambition is to start from the concept of viability introduced by Aubin to reach an innovative paradigm, politically and ideologically neutral. But a paradigm from the world of research can contribute to the prospect of the ecological future of the planet.

In a way, the contribution of researchers, in addition to analyses, diagnosis and technical solutions, a sociopolitical and legal framework is defined, making humans responsible for their choices towards their habitat.

Reconnecting Human to the Biosphere

At the end of the war, in 1945, Vladimir I. Vernadsky wrote in *American Scientist*,[4] in a premonitory way:

> In the thick of life today, intense and complex as it is, a person practically forgets that he, and all of mankind, from which he is inseparable, are inseparably connected with the biosphere [...] Man is an element which cannot be separated from the biosphere. And this inseparability is only now beginning to become precisely clear to us. In reality, no living organism exists in a free state on Earth. All of these organisms are inseparably and continuously connected.

However, our will and our understanding can regulate the course of phenomena in the future. The stakes are high: regulate threats and find a harmonious compromise to ensure the global viability of the Earth system.

As a legal expert, I measure all research and proposals in social sciences that can bring a compromise between the human and the nonhuman. Suggestions are presented in this book and there is no doubt that the training courses and laboratories of our university will contribute to these challenges related to the law, protection, and management of the environment in these various facets.

The University in Montpellier was born and flourished by the merit of shared teaching by experts from diverse backgrounds and diverse cultures. They have met in Montpellier since the twelfth century to create in this city, a university that overtime has always been able to promote strong scientific, humanistic, and cultural links with all research centers in and beyond Europe.

The academic history of the city has been demonstrated in the medical field since 1180 with the edict of the lord of Montpellier Guilhem VII liberalizing the teaching of medicine. Forty years later, in 1220, the Pope's legate, Cardinal Conrad d'Urach officially founded the School of Medicine, which is the oldest in the Western world, still in practice. The University of Montpellier was officially created in 1289 by Pope Nicholas IV bringing together the teaching of medicine, law, literature, and theology.

[4]Published in *American Scientist* in January 1945 "La Biosphère et la Noosphère."

In the sixteenth century, the city of Montpellier became a high-level intellectual center and confirmed its position as a European crossroads of law and medicine. At that time, it attracted many scholars and scientists who shared humanist values, including François Rabelais, Guillaume Rondelet, and Pierre Richer de Belleval.

The University of Montpellier was awarded original subjects: anatomy, botany, biology, etc. Close to medicine, the study of medicinal plants spread more and more in Montpellier with the creation of the Jardin des Plantes, in 1593 by Henri IV, the oldest official botanical garden in France, forty years before Paris. The first masters of Parisian botany were trained in Montpellier, notably by Pierre Magnol. The city was considered the capital of botany until the eighteenth century.

The scientific landscape of Montpellier changed at the beginning of the nineteenth century with the creation of the Faculty of Sciences in 1809. It was endowed with seven master chairs: transcendental mathematics, astronomy, physics, chemistry, zoology, botany, and mineralogy.

Divided into three universities after 1968, the University of Montpellier was merged in January 2015 with UM1 and UM2. Today, with 16 training and research units (*Unités de formation et de recherche* -UFR), schools and institutes, 77 research structures grouped into 9 scientific departments, more than 45,000 students, and 4,500 civil servant workers, it is the largest university in the Occitanie region and the 6th largest in France.

The list of scholars, masters, and students of our university who, over the centuries, have distinguished themselves by their decisive contributions in mathematics, medicine, botany, law, chemistry, etc. is long.

The initiative of this book is therefore part of this tradition.

Our most famous scientific and humanist student was François Rabelais. I will paraphrase it[5]: "Do as you please but do not forget the crux of the matter" by reading the text of this book to find some of the spirit that will make our future more coviable.

Coviability is the business of each and everyone of us.

Our university has just been labeled "I-site" in February 2017 through the MUSE project, promoting transdiscipline and aiming to become a recognized center of excellence around major societal challenges: contributing to food security; managing sustainably the natural resources and ecosystems; and improving the treatment of emerging infectious diseases, chronic diseases, or cancers. There is no doubt that researchers from the University of Montpellier have worked and will continue to work to make our world more coviable. Our university is aiming to create added value from interdisciplinary work and interactions with international researchers.

[5]Rabelais, François, and Pierre Grimal. *Gargantua*. A. Colin, 1964. Prologue de Gargantua « A l'exemple d'icelluy vous convient estre saiges, pour fleurer, sentir et estimer ces beaulx livres de haulte gresse , legiers au prochaz et hardiz à la rencontre; puis, par curieuse leçon et meditation frequente, rompre l'os et sugcer la sustantificque mouelle – c'est à dire ce que j'entends par ces symboles Pythagoricques – avecques espoir certain d'être faictz escors et preux à ladicte lecture ».

Philippe AUGE, professor of Public Law, was President of the University of Montpellier 1 from 2009 to 2014. Since January 2015, he is President of the University of Montpellier (the University of Montpellier results from the UM1 and UM2 merger).

Foreword: Preserving Ecological and Human Viability by João Carlos Salles Pires da Silva

In a more sustainable world, the concept of coviability is emerging as a concept to respond to humanity's current global concern about its future. Coviability forces real challenges to the future of humanity, and demands attention to all types of inter-relationships existing on a global scale, so that life oriented by the logic of "development" of a few does not interfere negatively on the way of life of other social groups, leading to uneven development and increasing exclusion. In fact, despite official discourse on environmental protection, the logic of development that is promoted and perpetuated reveals a lack of recognition for much of the rights of Nature..

Coviability invites us therefore to rethink the relationship between society and Nature, in order to preserve ecological and human sustainability in its various dimensions—economic, social, cultural—thus demanding a huge effort in order to build a dialogue between the empirical knowledges and the different areas of science, in search for the analysis and solutions to the problems.

Nearly 25 years after Rio 92, and despite considerable efforts by researchers to define and evaluate the concept of sustainable development, there are still serious limitations in the effective implementation of actions; environmental problems are increasing and directly affect societies and their various socio-cultural groups. In this sense, consideration and actions based on coviability invite us to listen to other representations and relationships with Nature, in order to inspire us on the building of more harmonious ways of life. Therefore, the participation and dialogue between researchers from the most diverse corners of the world become important so as to enable diverse viewpoints to be considered, from those of industrialized countries to those of the many traditional populations, with their symbolic representations and specific production practices.

In line with the academic tradition of the Federal University of Bahia in its various fields of research, and based on its dialogue with various segments of society that sustain and legitimize our Institution, I take this opportunity to reiterate my support to the members of UFBA, congratulating those who enthusiastically engaged themselves in this project.

Rector of the Federal University of Bahia João Carlos Salles Pires da Silva
Salvador, Brazil

Preface: Coviability, the First Step in a Long-Overdue Need for Coherence in Our Living with Nature by Daniel W. Bromley

An enduring challenge to students of ethics is the question—"Ought I to do what I can do?" The question cuts to the heart of modernity, for it challenges the alleged autonomy of the individual embedded in a social context. Indeed, my question challenges the fundamental idea of what it means to be an individual. In the limit, the autonomous individual is a fiction—an artifact of the thrilling hedonism of the Enlightenment. That stark philosophical break with tradition, the ideational "creation" of the individual, has brought mixed results to the human community—and to the physical world we so uneasily inhabit. That fateful project of conjured individuation now stands before the Alter of History as an existential threat to the future of humanity. The acquired arrogance of the artificial yet autonomous individual has not freed humans from the historic bondage of superstition, original sin, and obligatory guilt. It has rather transformed that bondage into fealty to other superstitions, other sins, and serial penance so that the next conquest might then come clearly into view—and within reach. We may now purchase "carbon credits" before, not after, our next gratuitous airplane trip.

In the Prelude to my 2006 book *Sufficient Reason: Volitional Pragmatism and the Meaning of Economic Institutions* (Princeton University Press), I offered the following comment on the presumed autonomy of the individual:

It may be supposed that the most fundamental of human needs concerns food, water, and staying warm. This supposition would be mistaken. The most funda-

Born in 1940, Daniel W. Bromley' research fields include the institutional foundations of the economy property rights; the economics of natural resources and the environment; and economic development. He has been editor of the journal "Land Economics" since 1974. Since 2009, Bromley is Visiting Professor at the Faculty of Agriculture and Horticulture of the Humboldt University of Berlin, Germany. In 2011, he was honored with the Reimar Lüst Prize for International Transfer of Science and Culture awarded jointly by the German Alexander von Humboldt-Stiftung and the Fritz Thyssen-Stiftung. Daniel Bromley is a Fellow of the Association of Environmental and Resource Economists, the American Agricultural Economics Association, and is listed in Who's Who in Economics. At last, he is the recipient of the Veblen-Commons Award from the Association for Evolutionary Economics.

mental human need concerns what to believe. Believing is precedential to eating and drinking (and staying warm) for the simple reason that even the seemingly basic acts of eating and drinking require a concept about surviving and thereby experiencing the future. This attribution of value to the future is what renders survival a conceptual rather than a physical matter. Without the idea of the future, and without the attribution of value to the future, eating is not an obvious or compelling activity. Eating requires the will to live.

With the future driving actions in the present, believing becomes the predicate for all action. What should I eat? What should I drink? How might I stay warm? From this one may further suppose that believing is an individual enterprise. This supposition, too, would be mistaken.

As social beings we tend toward, indeed we are defined by, social beliefs. The essence of socialization is precisely the stabilization of beliefs. And stabilized beliefs define for us what is normal, natural, correct, right.

It could not be otherwise.

And from this spare beginning one can begin to make out the ground beneath the social arrangements—the institutions—that define our very being as social creatures…

There is no such thing as the individual—we only assume that to be the case. And of course it is this presumptive deceit that now stands indicted in our collective inability, our arrogant unwillingness, to live within—and at peace with—our natural surroundings. I stress here the problematic individual because too often the problem as laid at the feet of some abstraction called "society" or "communities" or "capitalism." These labels are mere bait to throw us off the scent of the deeper problem. Societies or communities or capitalism are not volitional agents. And of course when the active agents are finally located, they will appeal to their rights or their freedom. They will never admit to self-interest.

The project before us offers new and nicely creative ways to remind us of how hard it is to rein in the alleged autonomy of the individual so that we can, collectively and individually, exist and thrive within a biosphere that has its own ideas. The environmental ethics community has tried, for several decades now, and with indifferent success, to convince the arrogant individual that we must be nice to nature. Only our abiding "false consciousness" impedes progress on that front. Not the false consciousness of Marx, but the false consciousness of our own gilded creation myth.

The coviability approach brings us a wealth of insightful scholarship on an age-old challenge. By stressing the mutual viability of the two distinct domains, we gain a new perspective on how hard it will always be to bring the desired consilience (with apologies to E.O. Wilson). It will be hard precisely because the agents in each system operate on contrary logics. The biosphere operates on the logic of function, while the human system operates on the logic of purpose—an end in view. Humans deny the centrality of function because we see the biosphere as an entirety—an "other"—rather than as interconnected thrusting entities. The biosphere cares little for our fickle and self-serving purposes. Therein lies the eternal conflict.

Coviability wants to help us see those conflicts and modify our purposes accordingly. I like to think of the coviability project as the first step in a long-overdue need for coherence in our living with nature. Distinct moving parts cohere when they fail to clang and clash and resist. They become one out of many.

Professor of Applied Economics (Emeritus) Daniel W. Bromley
University of Wisconsin-Madison, Editor *Land Economics*

Acknowledgments

This contributed volume is the outcome of a long process of teamwork, seminars, and an international conference on Human and Environmental Security in the Era of Global Risks held in 2015.[1] The content and approach of this publication are mainly based on the empirical research undertaken by the majority of coeditors and chapter authors, highly aware to local and global ecological and social challenges.

From 24 to 28 November 2014, a seminar held in the Cévennes (Valleraugue, located in the south of France) made it possible to draft the first editorial roadmap by addressing the following theme: "The negotiated territorial regulation for the coviability of social and ecological systems: between concepts and transdisciplinary experiences." From this starting point, the idea to edit this volume emerged and subsequently urged on by my partner Serge Morand who is a bioecologist. The team was quickly formed thanks to existing partnerships with Mohamed Behnassi (Specialist in Environmental Law and Governance, Morocco), Voyner Ravena-Cañete (Anthropologist, Brazil), and Catherine Prost (Geographer, Brazil). The team was further extended to include Laurence Pascal (Biologist, France), Thérèse Libourel (Computing Scientist, France), Maud Loireau (Geographer, France), Gilbert David (Geographer, France), Mireille Fargette (Bio-Geographer, France), Vincent Douzal (Mathematician, France), and Frédérique Seyler (Hydrologist, France). This international Editorial Board was supported by the UMR ESPACE-DEV, French National Research Institute for Sustainable Development (IRD),

[1]International Conference on "Human and Environmental Security in the Era of Global Risks (HES2015)," organized on November 25–27, 2015, in Agadir by the Environment and Human Security Program (EHSP), North-South Center for Social Sciences (NRCS), Morocco (now the Center for Research on Environment, Human Security and Governance (CERES)), the French Institute of Research for Development (IRD), UMR ESPACE-DEV, France, and the Project 4C/IKI-Morocco of the Deutsche Gesellschaft für Internationale Zusammenarbeit (GIZ) GmbH, Germany, in collaboration with the Interdisciplinary Research Centre for Environmental, Urban Planning and Development Law (CRIDEAU), University of Limoges, France, the Universidade Federal do Pará (UFPA) and the Universidade Federal da Bahia (UFBA), Brazil, and the CLIMED Project (ANR: 2012-2015/CIRAD), France.

France, which managed to involve other partners – such as the Universities of Montpellier, Guyana, Reunion, and the West Indies, Antilles. This research unit has even integrated a new transversal research area within its program, labeled "Coviability," a move which helped support the entire editorial process.

This interdisciplinary work focuses on the links between humans and nonhumans (the natural environment) and the interrelations between culture and nature. The core theme of this publication is timely and complex that the objective of founding a new paradigm was a real challenge in scientific, human, and editorial terms. The important number of coeditors, six women and five men, reflects the need to involve a variety of disciplines and institutions in this transversal and interdisciplinary research enterprise.

Such an editorial team has managed to ensure the depth, relevance, and accuracy of the analysis both theoretically and empirically. The open-minded spirit of all coeditors enabled relevant scientific approaches to address the question of coviability. Based on this, I wholeheartedly thank all the members of the Editorial Board for their trust, perseverance, and valuable contribution making the whole publishing process a true success.

I also pay tribute to the IRD and the University of Montpellier, two French leading institutions. Without their support, it would not have been possible to complete this project.

I also sincerely thank the key participation of the Center for Research on Environment, Human Security and Governance (CERES, Morocco) in translating the majority of the chapters from French to English and in preparing the final book manuscripts. The work of the team, composed of Laila Sougri, Abdessamad Nafissi, and Bourhim Rabiaa, made a valuable contribution to the publishing process. All translated works subsequently underwent English language-proofing by Babel TREE under the coordination of Catherine Collier (www.babeltree.fr).

Finally, the achievement of this substantial publication (over 1200 pages, 43 chapters) in two volumes is the result of the outstanding participation of over a hundred researchers from various disciplines and research institutions around the world. On behalf of the Editorial Team, my thanks and recognition are extended to all researchers, the contributions of whom have served the relevance, the originality, and value of this timely publication.

Montpellier, France Olivier Barrière

General Introduction

> Today's planetary challenge is to profoundly revise existential paradigms. (Dalaï-Lama and Stril-Rever 2016)

Introduction

The planet Earth has entered a new geological era. The International Geological Congress held in Cape Town, South Africa, on August 29th, 2016, voted for the "official" switch from the Holocene Epoch (10,000 years ago) to the "Human Epoch" (Crutzen 2002; Monastersky 2015), or Anthropocene Epoch.[1] Human activity imprints the geology of the planet Earth and is measured by indicators found in lake sediments, Antarctic ice cores, corals or tree rings. Evidence of anthropogenic impacts refers to the rise in sea level, global temperature, the atmospheric concentration of greenhouse gases (including carbon dioxide CO_2 and methane CH_4), biological changes (extinctions of species, global species migrations, loss of biodiversity), changes in the oceans, and so forth. It also refers to the physical changes of the continents (erosion of soils and shorelines, plundering of land, sediment production, occupation and modification of areas through urbanization, intensive agriculture, deforestation, land degradation, and so forth): "sufficient evidence has emerged of stratigraphically significant change (both elapsed and imminent) in recognition of the Anthropocene Epoch—currently a vivid yet informal metaphor of global environmental change—as a new geological epoch

[1] Stratigraphic recognition is set to be validated by the International Union of Geological Sciences (IUGS) during the next three years. The definition of the "Anthropocene Epoch" can apply three levels: (a) the new geological time interval (Rudwick 2005); (b) the Earth system which has left the Holocene Epoch: "the Anthropocene Epoch is not about being able to detect human influence in stratigraphy, but reflects a change in the Earth system" (Zalasiewicz 2014) ; and (c) the human footprint: the consideration of transformations made directly to landscapes, the extinction of species, the nitrogen cycle, etc. (Syvitski 2012).

to be considered for formalization by international discussion. The Anthropocene Epoch may be defined by a GSSP (Global Stratotype Section and Point) in sediments or ice cores or simply by a numerical value" (Zalasiewicz et al. 2008, 2010, 2014). However, from the mid-twentieth century, other markers appeared, such as micro-plastics, pesticides, concrete, or radioactive elements (Vince 2014, Bonneuil and Fressoz 2016).

One of the most striking features of the Anthropocene Epoch is the ability of human to occupy and transform the planet at a rapid rate, with a greater intensity and uncertain consequences. This is accompanied by the notion of global environmental crisis which is leading us to rethink modernity (Hamilton et al. 2015). Climate change represents the most obvious manifestation of this transformation. As highlighted in the 5th report of the Intergovernmental Panel on Climate Change (IPCC 2014): "Human influence on the climate system is clear, and recent anthropogenic emissions of greenhouse gases are the highest in history. Recent climate changes have had widespread impacts on human and natural systems. (. . .) Warming of the climate system is unequivocal, and since the 1950s, many of the observed changes are unprecedented over decades to millennia. The atmosphere and oceans have warmed, the amounts of snow and ice have diminished, and sea levels have risen."

In 1950, 2.5 billion men and women inhabited the planet. Forty years later, by the time of the 1992 Earth Summit in Rio de Janeiro, which adopted the United Nations Framework Convention on Climate Change (UNFCCC),[2] the world population had doubled to reach almost 5.5 billion human beings. When the Paris Agreement (COP21) was signed on December 12th, 2015, there were between 7.3 and 7.4 billion people living on our planet and GHG emission had never been higher (Blunden and Arndt 2016).

Realization of a Planetary Ecological Emergency

The pressures that human exerts on the planet contribute toward defining an ecological emergency that penetrates collective consciousness and is a cause for concern among a large number of players throughout the world. While international environmental law has adopted treaties and multilateral agreements for more than forty years now, the watchword of "ecological emergency" is recent. In addition, in 1995 for example in Chicago, the Parliament of the World's Religions[3] warned:

[2]Adopted by 154 States in Rio de Janeiro (Brazil) during the 1992 Earth Summit. It came into force on 21 March, 1994.

[3]The Parliament of the World's Religions "aims to establish a global interfaith dialogue. Its origin dates back to September 1893 when it first met in Chicago. It re-emerged in 1993 with a meeting in Chicago, and then every five years: 1999 in Cape Town, 2004 in Barcelona, 2009 in Melbourne, and 2015 in Salt Lake City. This Parliament of Religions is not an institution but a process of dialogue, open to the religious and philosophical convictions of the whole world. A longside representatives of religions and believers, humanists are involved in the preparations and participate in events. It

"the world is in agony, a general and dramatic agony: the planet is gradually being destroyed; its ecosystem is endangered. Anarchy and violence are threatening our societies. We must respect the entire community of living beings: humans, animals, and plants. We must turn our attentions to protecting the planet, its atmosphere, soil and water" (Küng and Kuschel 1995).

This realization is universally shared thanks to a variety of initiatives, such as the "Summit of Consciences" held on July 21th, 2015, in Paris which brought together more than forty moral and religious figures from around the world in response to the international campaign "The climate, why do I care?" and to launch a "Call to Conscience for the climate."[4]

The Paris Political Agreement, adopted within the framework of the UNFCCC (COP21 December 12th, 2015), and which brought together 195 States, acknowledges that "climate change represents an immediate and a potentially irreversible threat to human societies and the planet." The relationship between mankind and the planet Earth defines a Man-Nature unity which was emphasized by the World Charter for Nature as early as 1982 (adopted by the General Assembly of the United Nations on October 28th, 1982): "Humanity is part of Nature"; and the latter has "shaped human culture."[5]

The symbiosis described by the West[6] comes within the following definition of nature: "the set of conditions of human nature itself, its global renaissance or extinction constraints, the hotel which provides it with a roof, warmth and food. Moreover, it takes it all away as soon as it is taken advantage of. It conditions human nature which, from now on, conditions it in return. Nature behaves like a subject" (Serres 1992, 64). This nature/subject[7] paradigm leads to the idea of a community of living beings, humans, and non-humans, which is crystallized in the planet Earth personified by "Gaia" (Lovelock 1979) and through a draft Universal Declaration of the rights of "Mother Earth" (2010)[8]: "recognizing that Mother Earth

---

is a place of meeting for people of faith and convictions, face to face, heart to heart. Every meeting brings together some 10,000 people of all religious and "cultural traditions.""

[4]http://www.whydoicare.org/en/summit-of-the-consciences-for-the-climate

[5]It is important to emphasize that this nature/culture division, which determines reflection and the position taken regarding the condition of Man as a being integrated into nature, is anchored in a dichotomy conceived by Western rationale. Strathem (2014) points out that such dichotomy does not exist in any other human group. This also implies that the idea of reconnecting Man with the biosphere is a Western invention. Such an affirmation underlines the imposing and hierarchical character which models the reconnection itself, which only occurs because the Western world considers the biosphere to be disconnected from humans; in reality, it thinks of it in a dichotomous manner.

[6]Humanity has various ways of thinking, of being and of seeing the world, so it would be preferable to talk about humanities. These sociocultural differences mean that by definition, discourse cannot be universalist; see Strathern, 2014.

[7]On nature as a subject of rights, see: Rühs and Jones, 2016; Laastad, 2016; Hermitte, 2011; Regarding perspectivism, see: Viveiros De Castro, 2009; 2014, and Perez, 2010.

[8]From the «World People's Conference on Climate Change and the Rights of Mother Earth», held in Cochabamba, Bolivia, between April 19 and 22, 2010.

is an indivisible community of diverse and interdependent beings with whom we share a common destiny and to whom we must relate in ways that benefit Mother Earth."

Politicians, members of religious communities, and citizens are joining forces around the challenging issue of Man's relationship to his habitat. Awareness of this issue is affecting ethics and is taking shape owing to political decisions which are transformed into laws on the basis of scientific alerts. In this way, the Intergovernmental Scientific and Political Platform on Biodiversity and Ecosystem Services (IPBES[9]) stresses the risks linked to the disappearance of pollinators for biodiversity and global agriculture and advocates a "transformation of society's links with nature" by making sure that human populations are aware of the values of pollination and by establishing "links between human populations and pollinators" (IPBES 2016: 25). These relationships go so far as to define a "biocultural diversity," which IPBES presents as the links between cultural and biological diversities.[10]

The linking of culture and biodiversity is not only ideological (see Maffi 2005, Maffi and Woodley 2010) or limited to certain societies. If we accept to disengage from the so-called Western naturalist ontology (Thomas 2011, Kholer 2011), bio-cultural diversity[11] could be conceivable in societies all over the world. This notion, widely promoted by the International Union for the Conservation of Nature (IUCN 2010) with regard to traditional societies that are erroneously presupposed "by nature" to be respectful of the environment, can break free from the sole context of indigenous or so-called traditional populations; owing to the fact that "bio-cultural diversity is not external to us: it is the total sum of nature and culture..." (Maffi and Woodley 2010), which Christophe Grenier qualifies as geodiversity (1998, 2003). Such respect, which seems to be "derived from nature," is in fact the product of the longstanding companionship between nature and culture, a biocultural symbiosis.

The Biosphere: The Focus of Life

This "sum of nature and culture" derives from reciprocal interactions that unite living beings within the biosphere. In 1926, the notion of biosphere was defined in a biogeological and ecological sense by Vladimir Ivanovitch Vernadski, who hypothesized that life is a geological force that transforms the Earth. He was the first to scientifically consider the impact of human activity on the climate, at a time when natural resources were thought to possess inexhaustible regeneration capacities. The

[9]Intergovernmental Science-Policy Platform on Biodiversity and Ecosystem Services.

[10]"A certain number of cultural practices based on indigenous and local knowledge contribute towards maintaining an abundance and diversity of pollinators and a "biocultural diversity" the value of which is appreciated (in this assessment, biological and cultural diversity and the links between them will be referred to as "biocultural diversity")" (IPBES 2016, 11).

[11]See the introductory chapter which explains this concept.

concept of biosphere was born from a multidisciplinary approach that namely brings together naturalists, geologists, and biologists.

For Vernadski, the biosphere is "the unique region of earth's crust which is occupied by life. It is only in the biosphere, a thin outer layer of our planet, that life is to be found. All living organisms are found here and are always separated from surrounding raw material by a clear and insurmountable limit. (...) All life, all living matter may be considered as an indivisible whole in the mechanism of the biosphere" (Vernadski 2002 (1929), T19 and T22). The author clearly stresses the limitations of the development of life owing to the finite dimensions of the planet and the physicochemical constitution of its environment (Vernadski 2002/1929)[12].

Human Development: How and in Which Direction?

By definition, the human being is an integral part of the biosphere. From an anthropological point of view, the intertwining of Man and the biosphere reflects a mental representation of the relationship between social systems and ecological systems; a connection which maintains the more or less clear distinction between human and nonhuman. This representation depends on rational thought, ethics (values), a cultural pattern leading to the paradigm of the Man-biosphere relationship. In this time of global change, which includes environmental changes, the ecological stake becomes essential. The reasoning behind the objective of "viable/sustainable development" is confronted with two unavoidable obstacles: that of the "development" model and that of its temporal perspective: how far and until when?

The question of the development model becomes paramount since it directly affects the relationship between Man and the planet and, consequently, Man's place *vis-à-vis* the biosphere. Owing to its responsibility for the Anthropocene Epoch, humanity finds itself in an ecological emergency that requires adaptations, transitions, and ruptures. For the past forty years, the prospect of "sustainable development" has been perceived as a recurring leitmotiv and the only statement of a new relationship with our Planet. However, are we convinced of the necessity to leave the Anthropocene Epoch for another era,[13] unifying Man in the biosphere from which, "by nature," he cannot exclude himself?

---

[12]For the author, the elements of the biosphere are divided into three groups: living matter (autotrophic species and heterotrophic organisms); biogenic matter (fossil fuels, peat, humus); and bioinert matter (including water, low atmosphere, sedimentary rocks). A protective ozone shield serves as an upper limit to the biosphere. The laws governing biosphere organizational processes are mainly the biogenic migration of chemical elements and the evolution of species (Vernadski, 2002/1929).

[13]It could be the noosphere (sphere of human thought, Vernadsky 1945, Teihard de Chardin 1956; Grenier 2000), but in this context we remain in a very anthropocentric logic.

The grip of humanity on the planet is pushing mankind toward an obligation of "planetary stewardship" in order to ensure his survival. Does the human viability stake therefore relate to human's own "development," or rather to his capacity to adapt to himself and to the transformations of his environment (Toussaint et al. 2012)? The basis of life lies in a limitless continuity, that of "persisting" (whilst remaining sustainable). Consequently, should planetary stewardship consist in inexorably pursuing demographic, economic, technological growth, etc.?[14] The need to break with consumerist rationale is obligatory in order to permit the emergence of a social resilience and adaptive capacities through an economic and ethical re-founding of societies. Societies subjected to globalization suffer from ecological constraints which define the limits and fragilities of the planet, developed by human and for human justifying a stewardship on this scale: "the advent of the Anthropocene Epoch, the time interval in which human activities now rival global geophysical processes, suggests that we need to fundamentally alter our relationship with the planet we inhabit. (. . .). The Anthropocene Epoch is a reminder that the Holocene Epoch, during which complex human societies have developed, has been a stable, accommodating environment and is the only state of the Earth System that we know for sure can support contemporary society. The need to achieve effective planetary stewardship is urgent. As we go further into the Anthropocene Epoch, we risk driving the Earth System onto a trajectory towards more hostile states from which we cannot easily return" (Steffen et al. 2011).[15]

The ecological stake becomes an issue of unity based on the interdependence of human systems with ecological systems, and owing to this, an issue of reciprocity. Rethinking humanity in terms of the biosphere is a way of reconnecting with it (as presented below in Part 1), but this remains dependent on a paradigmatic relationship (as presented below in Part 2). The objective of this work, the outline of which we will present (in Part 3), consists in bringing to light the definition of a concept and the formalization of a paradigm of socioecological unity.

Reconnecting Human to the Biosphere

"We are the first generation with the knowledge of how our activities influence the Earth System, and thus the first generation with the power and the responsibility to change our relationship with the planet" (Steffen et al. 2011).

When we speak of reconnecting human to the biosphere, it implies that a disconnection has occurred. By directly taking action, by becoming a player in the

[14] See "Is there a limit to the evolutions of man?" (Toussaint 2012).

[15] And for the poet Friedrich Hölderlin (1968), "there are two ideal states: extreme simplicity, where owing to natural organization alone, without our involvement, our needs are in accord with themselves, and extreme culture, where the same result is attained thanks to the organization that we are capable of giving ourselves."

geological transformation of the planet Earth, humanity believed itself to be free of its relationship with other living beings. By becoming the master of his own destiny, *Homo sapiens* has "moved away" from his origins, from the original matrix which saw his creation. The beginning of the Anthropocene Epoch dates from this break, from this "disconnection."

Becoming stewards of the Earth system by conforming to a new social contract for "global sustainability" (Carl Folke et al. 2011) would define humanity's survival objective. Achieving this means rethinking our relationship with the biosphere, repositioning mankind in his role and his place on the planet. It is certain that reconnecting human to the biosphere requires a balanced management (responsible) of the planet's resources and implies considering the limits which it can tolerate (Jansson et al. 1994; Rockstrom et al. 2009). However, achieving this planetary stewardship objective is not necessarily based on the paradigm of the economy which confers the status of "natural capital" to nature. A status which was established by IUCN during the "World Conservation Congress" held in Hawaii, Honolulu, from September 1–10, 2016.[16] A "Natural Capital Protocol"[17] is being developed internationally (Natural Capital Coalition 2016). The nature paradigm, seen as "our most important bank account," is being promoted by the Capital Natural Project[18] from a human well-being perspective, by integrating the value that nature represents for society.[19] This approach is supported by Stanford University, the University of Minnesota, The Nature Conservancy, and the World Wildlife Fund.

The biosphere functions through interactions and interconnections. The latter, between local and global levels, contribute toward the human footprint on the planet,

[16]"Understanding that the aim of natural capital approaches is to make the value of nature more visible in decision-making by governments, businesses, financial institutions and society, and to drive better outcomes for biodiversity, the environment, and human well-being (. . .) Acknowledging Council Decision C/84/16 outlining a roadmap to develop an IUCN policy on Natural Capital" (Declaration of the congress entitled "Planet at the crossroads" with the objective of "securing systems that support nature so that humanity and the community of life as a whole continues to thrive on Earth. This is our collective challenge for the next 15 years and this is the challenge that the 2016 IUCN Congress is proposing to the world to meet." IUCN, September 1st to 10th, 2016.

[17]The Natural Capital Protocol is a standardized framework which will permit companies from all over the world to understand and quantify their impacts and dependencies on natural capital, launched on July 13th, 2016, in London by two consortiums, WBCSD and UICN. The WBCSD consortium comprises Accenture, ARCADIS, The B Team, Climate Disclosure Standard Board, Conservation International, Deloitte, eCountability Ltd, eftec, ERM, GIST Advisory, Imperial College, Integrated Sustainability Services, Natural Capital Project, PwC, Sustain Value, Synergiz, The Nature Conservancy, World Resources Institute, and WWF US. The IUCN consortium comprises the Dutch Comittee of IUCN, Ernst and Young (EY), Trucost, the University of Cambridge (Institute of Sustainability Leadership (CISL), True Price, Industrial Ecology Research Services (IERS), and the UNO Food and Agriculture Organisation (FAO).

[18]Available online: http://www.naturalcapitalproject.org/library/

[19]"We work to integrate the value nature provides to society into all major decisions. Our ultimate objective is to improve the well-being of all people and nature by motivating greater and more targeted natural capital investments." Available online: www.naturalcapitalproject.org

but should we necessarily reason in terms of accountability for natural capital (Carl Folke et al. 2011)? Freeing ourselves of this rationale of capital equates to reasoning differently in terms of common heritage and interconnections, and even solidarity. Thinking about the links of human societies with the biosphere, as well as inter-societal links, equates to thinking about coexistence of the whole which is categorized according to "ecumene," as described by Berque (1987, 1996). It remains impossible to reconnect human to the biosphere on the basis that the planet belongs to human, as if it were an object, a source of capital. We shall investigate this point in the second part of the present introduction. For the time being, we will concentrate on the notions of connection, of "reconnection" (a) and of "reconciliation" (b), which will lead us on to the notion of harmony (c).

How Can We "Reconnect"?

- Means of "mediation"

Connecting consists in "making contact" or being connected by close relationships. The nerve center of the connection is situated in what generates the link. Connecting leads to creating or recreating links, establishing or re-establishing a link by means of a mediating system. Indeed, the act of connecting or connecting oneself (defined by "reliance" in French) requires drivers such as mediators between a person, a group, and a system of which they are a part, or between subsystems (Bolle 2003, 102).[20] The mediating systems involved in this mediation can be: systems of being (identity, sociocultural patterns, sociocognitive representations, behavioral systems); social institutions (institutional systems); practices (ways of working, action systems), habitus (provisional systems for practices, ways of thinking), and standards (endogenous and state regulatory systems).

Normative mediation may be formalized by an "ecological pact" advocated by Nicolas Hulot (2006) and Lester R. Brown (2007). The program of actions embodied in the pact translates a "natural contract" which presupposes a symbiotic and reciprocal relationship (see Serres 1992: 67).

Mediation systems characterize human's relationship to the biosphere by the paradigms that underpin them, and reflect the values and statuses (subject, object) accorded to the components of the planet. Creating a relationship will depend on the distance established between the human and the nonhuman. The cursor of the distance between the two will depend on cultural conditioning: "each individual is caught up in a particular socio-cultural system that makes him see and understand reality in a certain way, a way which is specific to the society he belongs to" (Berque 1996: 39). The mediation of the link depends on an ethical dimension (values and

[20] "Social players are both linked (they have direct links among themselves), and connected by one or more mediating systems (whether a social institution or a cultural system of signs or collective representations)" (Bolle 2003: 103).

virtues), a cultural dimension (this relates specifically to the human), and a legal dimension (regulation). The latter originates from the "parasitic" relationship of human toward the biosphere and recognizes the desirable conclusion of a contract of symbiosis and reciprocity. [21]

Reconnecting through mediation constitutes the very challenge of a "socio-ecological" pact, where coviability is defined by an intertwined relationship because of the interconnections that underpin the uniqueness of the whole (human as part of the biosphere). But this uniqueness would almost certainly be accompanied by an intrinsic pact which would be part of the common world or biosphere.

- Internalizing nature rather than externalizing our actions: merging in a common world[22]

The common world defines a "collective" which must neither combine nature and societies nor juxtapose them because the two paralyze each other (nature versus culture). Rather, they should meet and exist together within the biosphere. The composition of this world, the result of a shared and co-constructed history with all living beings, is jointly realized: we are therefore moving away from a subject/object oppositional relationship to a human/nonhuman couple. If the collective means "the whole but not two separated" (Latour 2004, 95), the properties of the couple are interrelated, resulting in defining the anthroposystem as the result of a long biological than biocultural evolution because the human is a part of nature[23] and nature is a part of the human. By transcending the subject/object relationship, we admit that the nonhuman is not an object, and we consequently free ourselves of the anthropomorphic vision of the notion of environment itself. The "biosphere world" therefore does not include objects of nature and human subjects but combines all of the biotic and abiotic elements that constitute the same planet.

- The bind of "usefulness"

Humanity is linked to the biosphere because the latter renders its existence possible. What enables the human species to exist, live, and develop is found in its environment, the biosphere. More than just a simple habitat, the planet Earth defines a "human environment" (Berque 1987) associating the physical, the chemical, the biological, and so forth. This sphere of the human being defines a relationship of usefulness, that is, a means to exist with a view to a given purpose (Barrière 2016; Barrière 2015). Usefulness concerns this aptitude to ensure the viability of humanity.

---

[21]"The parasite takes everything and gives nothing. The host gives everything and takes nothing. The law of control and property is reduced to parasitism. Conversely, the law of symbiosis is defined by reciprocity: man has to give back to nature as much as it gives him; nature has become a subject of rights" (Serres 1992: 67).

[22]The unification of realities (Latour 2004).

[23]« Formerly man was a part of nature; now he is the exploiter of nature » (White 1967, 1205).

However, even if the notion of usefulness should not be understood here in economic and utilitarian terms, the notion of "service" is generally understood in these terms because it relates to a political structure aiming to integrate nature into the market logic. In this way, the biosphere complies with human ends. The term "service," first mentioned in the 1970s[24] for pedagogical reasons, became a key concept in biodiversity protection policies with the Millennium Assessment (2005) and the TEEB initiative (The Economics of Ecosystems and Biodiversity)[25] (TEEB 2010). It is also at the center of the Capital Natural Project (op.cit.), the objective of which is to protect and restore nature by developing payments for ecosystem services.

The ecosystem service is defined by the usefulness of ecological functions for human, and consequently, the expected guarantee that these functions are preserved. This usefulness demonstrates a "need" for the biosphere (natural resources, energy flows, different components such as water, air, and so forth) and the biosphere, which is highly artificial, depends on anthropogenic practices and actions (their presence, type, orientation, intensity, etc). Therefore, usefulness works for both partners, human and biosphere, which actually form one: the biosphere being a whole which is defined within an anthroposystem (see below). Usefulness can therefore be seen as "socioecological": an obligatory, integrating, or even subservient relationship which reintegrates human into the biosphere.

"Reconciliation": Building a Socioecological Unity for a Human-Oriented Future

Re-reaching an understanding consists in envisaging a perpetuation of humanity. Reconciliation initiates a process that looks to re-found the existential paradigm. Initially, it could be a question of "... working toward reconnecting humans to nature (which) could make it possible to give every citizen the feeling of being part of a complex and dynamic living whole" (Fleury and Prevot-Julliard 2012: 11), the biosphere. Because the socioecological entity that we are defining is based on interdependence, it should reconcile humans and nonhumans, but should firstly also achieve a (re)conciliation between humans among themselves.

However, the Western view of human as the "Master of nature" has required the domination of human by human. Technology is in part responsible. Consequently, the technological revolution has permitted this increasing domination of nature by

[24] de Groot 1987 et Westman 1977, cités par Teillac-Deschamps et Clavel, 2012 : 311

[25] "The Economics of Ecosystems and Biodiversity (TEEB) is a global initiative focused on "making nature's values visible." Its principal objective is to mainstream the values of biodiversity and ecosystem services into decision-making at all levels. It aims to achieve this goal by following a structured approach to valuation that helps decision-makers recognize the wide range of benefits provided by ecosystems and biodiversity, demonstrate their values in economic terms and, where appropriate, capture those values in decision-making." Online available: http://www.teebweb.org

human while concealing a revolution of power relationships through technicality (Porto-Gonçalves 2006). Capitalism, owing to its invasive character (Santos 1996), extends and intensifies these processes of domination and ecological destruction, but also sociocultural processes in a time when globalization reigns. Re-founding our relationships with nature therefore requires re-founding relationships within and between societies.

This grail of unity reflects the intertwining of human societies in the sphere of life, the biosphere on a global scale. In this case, the stake exceeds the "ecology of reconciliation" (Rosenzweig 2003), which promotes the articulation of human activities to avoid disturbing biodiversity (Fleury and Prévot-Julliard 2012: 11). In this representation, biodiversity is only the human "environment," not the biosphere.

The passage to a condition of "reconciliation," links with the nonhuman by this "disconnection," owing to the passage to the Anthropocene Epoch, dissociates Man from the biosphere by objectivizing the latter. What surrounds Man becomes his "environment." Within a same habitat of existence between all of the living beings present on the planet Earth, this "disconnection" is consequently formalized by the distance that societies (especially of Western allegiance) set by distinguishing Man from the systemic biosphere: the latter becoming the "environment" or "nature." As a consequence, the human system is differentiated from the biosphere system via the elements it shares—culture and what is external to it. This results in anthropocentrism which clearly expresses international environmental law, as well as national rights, with some exceptions[26].

Why, Then, Are We Now Discussing a "Reconnection" or Even a "Reconciliation"?

The 2015 UN Plan of Action for Humanity addresses this issue as part of a desire to "transform our world": "We are determined to protect the planet from degradation, including through sustainable consumption and production, sustainably managing its natural resources and taking urgent action on climate change, so that it can support the needs of the present and future generations."[27]

Given the magnitude of climate change and the deterioration of biodiversity, humanity is gradually becoming aware of its uncertain future and of the value of the Planet, its sole habitat: "Man is both a creature and a creator of his environment (. . .) we have reached a time in history when we must orient our

[26] Such as the Ecuadorian Constitution of September 29th, 2008, art. 71, and the Bolivian Constitution of February 7th, 2009, which recognize a right of nature.

[27] Sustainable Development Program for 2030, Resolution of the General Assembly of the United Nations, September 25th, 2015 Annex: Transforming our world: the 2030 Agenda for sustainable development, Preamble, August 12, 2015 (A/69/L.85), "Draft outcome document of the United Nations summit for the adoption of the post-2015 development agenda."

actions (. . .) by thinking more about their repercussions on the environment" (UN World Conference on the environment in Stockholm in 1972). As from 1980, the World Conservation Strategy affirmed that "humanity is putting its survival at risk if it does nothing to safeguard the fertility and productivity of the Earth" (IUCN, UNEP, WWF, 1980). The World Charter of Nature was adopted (resolution 37/7 of the United Nations GA, 1980), affirming that humanity is a part of nature and that life depends on the uninterrupted functioning of natural systems. In 1987 the World Commission on the Environment published a report entitled "Our Common Future" highlighting global interdependence. In June 1992, the Rio de Janeiro Earth Summit recognized the urgent need to significantly modify our production and consumption modes[28] and entrusted the States with the responsibility of adopting a sustainable development model (Principle 8 of the Rio Declaration). In 2012, the Rio + 20 Earth Summit led to developing the idea of "the future we want" (UN Resolution 66/228) by "abandoning unsustainable modes of consumption and production in favor of sustainable modes as well as protecting and managing the natural resources that underlie economic and social development (which) are both the primary objectives and prerequisites of sustainable development." In 2009, April 22nd was declared International Mother Earth Day (resolution 63/278 of 22 April 2009), but without reference being made to the biosphere.

The Harmony of Human Within the Biosphere (or "Human-Nature" Harmony)

The aspiration to achieve harmony between human and nature in the biosphere[29] is reiterated[30] in the Resolution of the General Assembly of the United Nations adopted on December 21, 2009 (64/196) and the Resolution on harmony with nature of December 20, 2010 (65/164).[31]

This undefined harmony also generates an ambiguity insofar as, on the one hand, human belongs to nature (Nature Charter 1980, op. cit.), and on the other, anthropocentrism characterizes our relationship to nature (see the first principle of the Rio Declaration, 1992, "human beings are at the center of the concerns of ... sustainable development" op. cit.). Indeed, the Western naturalistic representation clearly divides nature and human beings. The Millennium Ecosystem Assessment (MA 2005) provides a clear demonstration of this separation between nature and

[28]Report of the United Nations Conference on the Environment and Development, Rio de Janeiro, 3–14 June 1992, vol. I, resolutions adopted by the Conference, resolution 1, annex I.

[29]It is always the term "nature" that is used.

[30]Principle 1 of the Rio Declaration states: "Human beings are at the center of sustainable development concerns. They have a right to a healthy and productive life in harmony with nature."

[31]See also the report of the Secretary-General of the United Nations on harmony with nature, delivered on August 19, 2010 (A / 65/314).

societies by the utilitarian relationship of ecosystems *for* human, as they provide "services" to human (twenty-four of which are identified). The four main services provided by ecosystems (supporting, provisioning, regulating, cultural) contribute toward human well-being in terms of safety, freedom of choice and actions, basic material for a good life, and health and social relations.

To be "in harmony" would consequently equate to linking societies and the rest of the biosphere with the objective of ensuring that they correspond to one another by means of a specific combination of elements connected in a relationship of interdependency. The new United Nations Sustainable Development Program for 2030, which is "human-centered," aims to "transform our world" to "a world where humanity lives in harmony with nature and where wild fauna and flora and other living species are protected" (UN 2015: 4). Firstly, the human/nature dichotomy clearly resonates with the eighth Millennium Development Goals (MDG 2000, see the UN 2015 report), of which a single objective was directly related to a "sustainable environment" that had to be "ensured." Secondly, this resonance continues via the 17 Sustainable Development Goals (with 169 targets). Commitment focuses on the human being through eliminating poverty all over the world, but also through "radical changes in the way in which our societies produce and consume goods and services," giving rise to Goal 12 ("establishing sustainable consumption and production modes through a significant reduction of wastes").[32] With respect to the climate, we retain the resolution of providing "a decisive response to the threat posed by climate change and environmental degradation" (UN 2015: 9), which is reflected in Objective 13 ("urgent actions to fight climate change and its impacts"). Furthermore, the importance given to the sustainable management of natural resources is justified by economic (promoted growth[33]) and social[34] development, leading to Objectives 6 ("sustainable management of water resources"),[35] 14 ("conserving and sustainably exploiting oceans, seas and marine resources as part of sustainable development"), and 15 ("preserving and

[32]Particularly "(...) a rational ecological management of chemicals and all wastes throughout their life cycle . . . " (UN 2015, 24).

[33]"Progressively improving, until 2030, the efficiency of the use of the world's resources, both from a consumption and a production point of view, and ensuring that economic growth no longer leads to environmental degradation (...)" (8.4) (UN 2015, 21). Economic growth is very important for urban development: 54% of the world's population lives in cities, a figure which will reach 66% by 2050. See UN 2014 Report "Perspectives on the urbanization of the world population," division of the United Nations population. This is why objective 11.6 focuses on reducing the "negative environmental impact of cities per capita, including by paying particular attention to air quality and namely (municipal) the management of waste" (UN 2015, 24).

[34]"We are therefore committed to ensuring the conservation and the reasonable usage of the seas and oceans, of freshwater resources, forests, mountains and drylands, and to protect biological diversity, ecosystems and wild flora and fauna" (UN 2015, 10).

[35]"By 2030, improve water quality by reducing pollution, eliminate waste dumping, reduce emissions of chemical products and hazardous materials to a minimum, decrease by half the proportion of untreated wastewater and significantly increase, on a global scale, recycling and reuse without posing a hazard to water" (6.3) (ONU 2015, 20).

restoring terrestrial ecosystems, while ensuring their sustainable use, managing forests sustainably, fighting desertification, stopping and reversing the process of land degradation, and stopping the loss of biodiversity").

Change the world. The recent appeal by the United Nations General Assembly to change the world through these Sustainable Development Goals (SDGs) remains resolutely based on a consumerist rationale, a growing pursuit of human development. Are these "arrangements" of human life on Earth sufficient to limit the impacts of human on the biosphere? Does this UN framework place us within a perspective of the harmony of humankind within the biosphere?

Alongside the SDGs, and to go even further, the ecological challenge lies in the paradigms of the position of the human being in the biosphere.

Continuity and Discontinuity Between the Human and Nonhuman: Animism, Totemism, Analogy, and Naturalism

The place of living beings within the biosphere or the cosmos questions the human/biosphere distribution or a continuity between both of them. A clarification is necessary in order to anchor socioecological coviability, which is based on a systemic dimension.

Religare, linking and re-linking human to the cosmos, refers to finding "re-linking" for humanity, a link to the surrounding world (Bolle 2003). The focus is on the relationship and not on the being[36]; the relationship goes beyond the player: "for the survival of humanity, it is necessary for each and every one of us to recognize the need to connect with ourselves, with each other, and with the Earth (- . . .)" (Morin 2004: 248).

Everything is linked owing to the unicity of the world. The term "holism" (Smuts1926) finds its origin in ancient times when Greek medicine approached systems in their globality according to the concept that the whole is superior to the sum of all its parts.

The link between human and the biosphere is both material and mental. It is developed within the paradigm of the place of human within Life and the living world (his way of perceiving and considering the world). Consequently, the anthropological approach requires entering into cultural diversity through mental worlds and sociocognitive systems, in order to be able to understand the manner in which people act, do things, exist, and transmit. The relationship to the invisible, questions of death and existence are wholly part of anthropological sciences (set of sciences that study human from different perspectives[37]). Consequently,

[36] See for Dessalles JL., Gaucherel C. and Gouyon PH., 2016, "the struggle for existence is not that of beings but the messages that pass through them and of which they are the ephemeral hosts."

[37] See the French Association of Anthropologists: http://www.afa.msh-paris.fr/?page_id=32

Table 1 Ontological matrix of the properties of the world's living beings (Descola 2005)

	Continuity physicality	Discontinuity physicality
Continuity interiority	Totemism	Animism
Discontinuity interiority	Naturalism	Analogism

studying human in society, and especially in his relationships with the whole of the living world he is a part of, the biosphere, cannot overlook the joint existence of metaphysics and its diversified foundations, beliefs (in transcendent powers, in one or more divinities, in spirits . . .) and acts (ritual practices), "which aim to establish specific relationships between men and other beings or extra-human powers" (Bonte and Izard 1991: 619). The interdependent system of values and practices defines moral, ethical, and religious frameworks, with their paradigms. Because of the lack of associated skills, the religious factor is not included in this work. Our purely scientific approach does not aim to "impose a higher representation of reality, intended to be authoritative, but to open up perceptions to enable them to be seen differently by people" (Auray and Bulle 2014).

As a consequence, we will refer to an approach of nature in anthropology which presents examples of the relationships between humans and nonhumans. Concerning the connection of human to the biosphere, anthropological dualism, the soul and the body, continuity and discontinuity, interiority and physicality within and between humans, plants, and animals all participate in the paradigm of coviability.

Man is distinguished from what is not human on two levels: materiality (carnal, corporeal, organic, physical: "morphological and physiological characteristics that are intrinsic to identity," Descola 2005: 169), and immateriality (incorporeal[38] and more precisely, "mind, soul or consciousness – intentionality, subjectivity, reflexivity, affects, the ability to signify or to dream," ibid., 168). The two aspects of the living entity are consequently reflected both by a "physicality," any material device, including the body, allowing humans to act on the world, and an "interiority," which is invisible and which manifests itself through its effects (Descola 2005).

Differences in human conceptions of nature are defined in the internal-external relationship which participates in living beings' manner of existing. We are referring back (Table 1) to the categories of the nature/culture relationship which Philippe Descola (2005) presents in the form of ontologies based on a classification of the physical and/or internal continuity or discontinuity between human and nonhuman entities (see summary in Table 1).

Of the four defined ontological types, only naturalism creates a rupture between human societies and nature (between human and nonhuman). It creates a distinction between the universality of physical laws (and thus the continuity of physicalities)

[38]Which may be practices, representations, expressions, knowledge, and know-how (UNESCO Convention for safeguarding intangible cultural heritage, 17 October 2003).

and the uniqueness of humanity (a discontinuity of internalities because culture is only human)[39].

Animism expresses a continuity of minds (plants and animals are "disguised humans" owing to the metamorphosis capacity that allows humans to be incorporated into an animal or a plant) with a discontinuity of physical forms (every species has its own physical characteristics—feathers, hair, scales, or bark).

Totemism translates an indistinction between human and nature owing to a resemblance of interiorities and physicalities, an "intimate relationship" between the human and nonhuman.

Analogism defines an ontology in which everything is unique: everything which exists is broken up by "small deviations" resulting in "a dense network of analogies (of correspondence) linking the intrinsic properties of the entities identified" (Descola 2005).

These ontologies stem from forms of aggregation from humans to nonhumans. They participate in the diversity of paradigms that do not authorize the human to be considered in the singular form.

Cultural diversity, especially ways of being, thinking, doing, and the geodiversity[40] of local knowledge, clearly lead to a diversity of paradigms. However, this diversity of human ontologies in their relationships with other beings in the biosphere is confronted with a kind of "technocratic anthropocentrism." Indeed, this anthropocentrism promotes technical reasoning above human reality because human "no longer possesses the feeling that nature is a valid reference or that it offers him a living refuge. He sees nature without prior assumptions, objectively, in the form of space and matter for a work into which everything is thrown no matter what the result is" (Guardini 1956: 68).

Our era of planetary change is confronted with the globalization of the technocratic paradigm as a deviation from the naturalistic westernization of the world: "life is being abandoned to circumstances conditioned by technique, understood as the primary means of interpreting existence" (Pope Francis 2015,110).

---

[39]The debate between Viveiros de Castro and Descola (Latour 2009) deserves to be mentioned here because it problematizes the framework proposed by a Western structuralist anthropology. According to Viveiros de Castro, based on Latour's account (2009, 2) "perspectivism, in his view, should not be regarded as a simple category within Descola's typology, but rather as a bomb with the potential to explode the whole implicit philosophy so dominant in most ethnographers interpretations of their material." Latour continues: "As Viveiros explained, perspectivism has become something of a fashion in Amazonian circles, but this fashion conceals a much more troublesome concept, that of 'multi-naturalism'. Whereas hard and soft scientists alike agree on the notion that there is only one nature but many cultures, Viveiros wants to push Amazonian thought (which is not, he insists, the '*pensée sauvage*' that Lévi-Strauss implied, but a fully domesticated and highly elaborated philosophy) to try to see what the whole world would look like if all of its inhabitants had the same culture but many different natures" (Latour 2009, 2).

[40]Geodiversity, according to Grenier (2003, 2014 and in part xx of this book), can be defined as a diversity of geographies or anthropogenic footprints, that is accompanied by cultural diversity (of languages, environments, etc.) and which maintains biological diversity resulting in biocultural diversity.

"Reconnecting" human to the biosphere, in various forms, is the subject of declarations, of objectives set by the international community and philosophical postulates. The incantation to "reconcile Man with all living entities," which is becoming recurrent in global discourse, is based on the acceptance of a universal responsibility,[41] human being clearly responsible for his impacts on the Planet.

The basis of this acceptance lies not only in ethical, philosophical, or moral convictions, but also in scientific evidence demonstrating the joint and inextricable relationships between the notions of human, nonhuman, and ecology. Socioecological coviability stems from the interdependence of societies and the biosphere. This dependence is reciprocal. There is no doubt that the planet Earth could have done without humans. If this had been the case, it wouldn't have constituted the living environment of humanity, and the analysis would stop there. Human wouldn't exist without the Earth and the Earth would have been quite different without this particular inhabitant. The only known habitat in the solar system for humans, planet Earth, is irremediably tied to a humanity that has rapidly transformed it into an anthroposystem. Indeed, the time during which human has been present on Earth is insignificant compared to the planet's 4.5 billion years of existence.

Socioecological interdependence (human-biosphere) generates this universal paradigm which consists in a mass of links and relationships between living organisms belonging to the biosphere; the objective is to analyze the organization of these links. An interdependency and links on which almost one hundred researchers are working in this book in order to contribute toward the emergence of a concept which defines and provides the basis for this paradigm. It is here, in the scientific world and not in the economic, financial, or political world that a view on the relationship between human and his environment is being developed.

Although the term "environment" is used throughout this book, the scientific analysis is objective and moves away from anthropocentric preconfigurations: there is therefore human, as part of the biosphere, and what surrounds human does not constitute an object but the complement of a unique system to which human belongs. The term "viability" allows neutrality to be maintained, leading to a demonstrative and undemanding analysis. Consequently, there is no "sustainable development" flag being brandished here; but a demonstration, through socio-ecological "viability," of a coviability free from duality. Indeed, it is not a question of two elements, a socio-system and a cohabiting ecosystem, but rather of a single matrix of interactions and interconnections within a single system. The difficulty of the approach lies in the fact that the "social" and the "ecological" became increasingly intertwined over the course of evolution. The challenge of this book therefore lies in involving science in the emergence of the paradigm serving as a basis for this socio-ecosystem.

[41] See Dalaï Lama and Stril-Rever, 2016.

The Book's Outline: The Emergence of a Concept/Paradigm, the Coviability of Social and Ecological Systems

The challenge of this first work on the question of coviability is based on the premise of the interrelations between human and his environment. The challenge is to imagine a different relationship to the world, not only leaving behind anthropocentrism and a consumerist rationale, but also ethnocentrism ("of one's own culture"). The expected result is to objectively consider this relationship by freeing ourselves of the dominant factor of sociocognitive references which are particular to each person. The editors of this book are from three different continents and eleven different disciplines.[42] This is a first step in developing the perspective of the new socioecological coviability paradigm.

Associating the two systems (social and ecological) stems from a naturalistic rationale based on a material continuity between the human and the biosphere, but with culture separating them. Indeed, the latter, according to the West's naturalist paradigm, provokes a discontinuity which breaks the continuum of the link between human and the biosphere.

However, coviability defines socioecological interdependency by a socioecosystem owing to the interweaving of culture and the biosphere. Interconnection forms a single set which cannot be dissociated owing to the fact that the socioecosystem is defined within the biosphere and not in the interweaving of two worlds whose distinction is based on their opposition. This super-system finds its foundation in a single existence, a single biosphere. Consequently, what is not human in appearance still participates in the anthroposystem. Otherwise, we would have been in a situation of an exo-planet as in the case of Kepler-452b for example (Jenkins et al. 2015), which is free from human beings.

The social system is not an ecosystem, yet the biosphere combines the two because on Planet Earth, one does not exist without the other; and further still, one contributes toward the other. Indeed, if the biosphere system conditions human, we now condition the biosphere (Serres 1992, who uses the term "nature"). This is where coviability lies, a joint viability for a common reality. However, the idea of coviability goes beyond the idea of sharing, which results from the different ontological paradigms of the relationship between human and the biosphere: the "socioecosystem" constitutes a single system.

This book addresses the complexity of the concept coviability by establishing it as a paradigm on the basis of an analysis and related perspectives concerning the relationship between human and the biosphere. From the beginning of the project, the process was collective. Disciplinary departitioning is a leitmotiv that structures the work to permit a more in-depth examination of this complexity. It is not easy to step out of one's biotope, context, equations, test tubes and lab benches, terrains,

[42]Environmental law, legal anthropology, bio-ecology, geography, agro-computing, anthropology, computer science, agro-bio-ecology, agro-economics, biology, hydro-geography.

etc. of one's own research topic. Breaking out of one's prism to address fundamental issues on planetary ecology at a regional or local level was a first challenge. The second challenge consisted in a genuine quest for a grail, the coviability of social and ecological systems, repositioning human as a key element of the biosphere, having "the capacity both to destroy and integrate."

This book compiles forty-three chapters of works and experiments, of analyzes and scientific approaches that allow us to take a first step which can be summarized by the definition of contours, the laying of milestones, data clearance. The last chapter of the book reviews the results achieved: the concept of coviability established as a paradigm by the scientific world with a view to providing an alternative approach. The latter opens up another perspective in order to: (a) renew the statutory and regulatory frameworks, thereby avoiding restricting ourselves to a financial rationale (nature as capital); (b) free ourselves of the "sustainable development," obligation which restricts the ontology of the relationship between humankind and the biosphere.

The book develops the possibility of a paradigm shift by mobilizing one hundred or so researchers from a wide range of disciplines (more precisely: law, legal anthropology, economics, sociology, philosophy, mathematics, biology, ecology, botany, agro-bio-ecology, hydrology, computer science, modeling, anthropology, ethnoecology, geography, electrical engineering, chemistry, oceanography, architecture and urban planning, geomatics, and so forth). It adopts a process of consilience (coordinating disparate results from various sources), based on interdisciplinarity to endeavor to adopt a transdisciplinary profile which imposes deciphering humaan's actions and relationships in the biosphere. From the foreword, the tone is given with respect to the stake of a new relationship between humanity and its habitat. The preface, written by a distinguished economist from the University of Wisconsin-Madison, highlights the conflict stemming from the contradiction of logics between humanity, which has set itself ultimate objectives and the biosphere, which is positioned in a function logic. A realignment is necessary to permit human's viability in the biosphere.

The introductory chapter clearly sets out the issue: Formalizing a paradigm, the biosphere in the face of the relationship between human and nonhuman. Examples of socioecological coviability facilitate an initial understanding of the contours of the book's contribution. It is structured according to four parts, spread over two volumes,[43] that present three phases: conceptual, applied, and reflexive. From the outset, the need for the conceptualization of the coviability paradigm was clear and its disciplinary bases were scrutinized in order to use them within an interdisciplinary perspective. The first part (in vol.1), *Towards the Theoretical Foundation of Coviability: In Search of Multi-disciplinarity*, is devoted to the foundations of the coviability paradigm.

---

[43]Volume 1: *The foundations of a new paradigm*; Volume 2: *Coviability questioned by a diversity of situations*

The conceptual explanation leads on to more empirical analysis includes an "applied" dimension that initially focuses on the governance of coviability by norms, policies, and actors (Part 2, vol.1). The third part (Part 3 integrated in Vol.1) establishes coviability as a challenge for the future. Faced with an ecological emergency, the challenge for humanity consists in taking account of its own complexity, which is deeply rooted in and an integral part of the biosphere. Two chapters summarizing the definition of the purpose of the work. An overview analysis leads us to an ontology of the coviability paradigm. Transdisciplinarity is then used to cover the determinants of the coviability paradigm. Of course, no discipline takes precedence over or imposes itself on the others, as the reader will have guessed from the outset of the work.

To introduce the second volume which is the last part of the book, Edna Maria Ramos Castro, an Amazonian and Brazilian sociologist, highlights the opposition of profit and preservation that forces human society to make choices; choices that necessitate breaks in lifestyles and economic systems, in order to re-establish the link between human, their own nature, and the biosphere. Volume 2 presents a variety of situations that questioned coviability. Two-thirds of the cases studied reveal a need, a search for coviability while existing cases illustrating this paradigm are becoming few and far between.

As the challenge of the book exceeds a single disciplinary field, requiring a multitude of contributions, twelve scientific editors from twelve different disciplines were called on to complete this editorial project. The very paradigm of coviability is, in essence, the product of this alchemy, this transcendence of fields, scientific approaches, and languages.

If the preface stresses the challenge of overcoming the contradiction which opposes Man and the biosphere, the conclusion poses coviability as a categorical imperative in the face of an ecological emergency: a necessary and unconditional action, based on the principle of universality and the "immediate principle of legislation" (Kant 1785).

References

Auray N, Bulle S (2014) L'anthropologie entre les lignes. Entretien avec Tim Ingold, La Vie des idées. Online available: http://www.laviedesidees.fr/L-anthropologie-entre-les-lignes.htm

Barrière O (2015) Repenser le droit de l'environnement dans une conception renouvelée du développement durable : prospective d'un "droit de la coviabilité" des systèmes sociaux et écologiques. In: Blaise S, David C, Victor D (ed) Le développement durable en Océanie: vers une éthique nouvelle ? PUP; PUAM, Aix-en-Provence, pp 215–242

Barrière O (2016, in press) Human relationship to the land from a legal perspective as a human and environmental security challenge. In: Behnassi M, McGlade K (eds) Environmental change and human security in the Middle East and Africa, Chap.14, Springer

Berque A (1987) Ecoumène. Introduction à l'étude des milieux humains, Belin

Berque A (1996) Etre humain sur la terre, Gallimard

Blunde J, Arndt DS (dir) (2016) State of the climate in 2015. American Meteorolgical Society

Bolle de Bal M (2003) Reliance, déliance, liance: émergence de trois notions sociologiques, Sociétés, 2–80. De Boeck supérieur, pp 99–131
Bonneuil C, Fressoz J-B (2016) L'événement Anthropocène. La Terre l'histoire et nous, Seuil
Bonte P, Izard M (1991) Dictionnaire de l'ethnologie et de l'anthropologie, Puf
Brown LR (2007) Le Plan B : pour un pacte écologique mondial, Hachettte/Calmann-Lévy/english edition, 2006, Plan B 2.0: Rescuing a Planet under Stress and a Civilisation in Trouble, Earth Policy Institute
Crutzen P (2002) Geology of mankind. Nature 415(6867):23
Dalaï-Lama, Stril-Rever S (2016) Nouvelle réalité. L'âge de la responsabilité universelle, Les Arènes
Descola P (2005) Par delà nature et culture, Gallimard
Dessalles JL, Gaucherel C, Gouyon PH (2016) Le fil de la vie. La face immatérielle du vivant, Odile Jacob
Fleury C, Prévot-Julliard A-C (dir) (2012) L'exigence de la réconciliation, Biodiversité et société. Fayard/Museum National d'Histoire Naturelle
Folke C, Jansson A, Rockström J, Per O, Carpenter SR, Chapin SF, Crépin A-S, Daily G, Danell K, Ebbesson J et al (2011) Reconnecting to the biosphere. Ambio 40(7):719–738
Grenier C (1998) Lieux, réseaux et modernité: les îles et la diversité terrestre. In: Joel Bonnemaison A, le voyage inachevé, pp 233–240
Grenier E (2000) Vladimir Vernadski, de la biosphère à la noosphère. Fusion 83:4–10
Grenier C (2003) Garder l'espace: la notion de géodiversité et la conservation du patrimoine littoral, Océanis, fascicule 28-1/2: Environnement, politiques publiques et dynamique des activités littorales, pp 233–251
Grenier C (2014) Géodiversité et mondialisation. Les fondements géographiques de la diversité terrestre et de son érosion, volume scientifique d'Habilitation à Diriger des Recherches, Université de La Rochelle.
Guardini R (1952) La fin des temps modernes, Seuil
Hamilton C, Bonneuil C, Gemenne F (2015) The Anthropocene and the global environmental crisis. Routledge, Rethinking modernity in a new epoch
Hermitte M-A (2011) La nature, sujet de droit? Annales Histoire Sciences Sociales 1:173–212. Online: http://www.cairn.info/revue-annales-2011-1-page-173.htm
Hölderlin F (1968) Hypérion ou l'ermite en Grèce, trad. P. Jaccottet (fragment Thalia) in Œuvres. Gallimard, Paris, p 113
Hulot N (2006) Pour un pacte écologique, Calmann-Lévy
IPBES (2016) Résumé à l'intention des décideurs du rapport d'évaluation de la Plateforme intergouvernementale scientifique et politique sur la biodiversité et les services écosystémiques concernant les pollinisateurs, la pollinisation et la production alimentaire. Potts SG, Imperatriz-Fonseca VL, Ngo HT, Biesmeijer JC, Breeze TD, Dicks LV, Garibaldi LA, Hill R, Settele J, Vanbergen AJ, Aizen MA, Cunningham SA, Eardley C, Freitas BM, Gallai N, Kevan PG, Kovács-Hostyánszki A, Kwapong PK, Li J, Li X, Martins DJ, Nates-Parra G, Pettis JS et Viana BF (sous la dir. de). Rapport ONU. Online available: http://www.ipbes.net/sites/default/files/downloads/Pollination_Summary%20for%20policymakers_FR.pdf
Jansson AM, Hammer M, Folke C, Costanza R (eds) (1994) Investing in natural capital: The ecological economics approach to sustainability. Island Press, Washington, DC
Jenkins JM et al (2015) Discovery and validation of kepler-452b: a 1.6-r⊕ super earth exoplanet in the habitable zone of a g2 star. Astron J 150:56 (19pp). https://doi.org/10.1088/0004-6256/150/2/56
Kant E (1785) Fondements de la métaphysique des mœurs. Online: http://classiques.uqac.ca/classiques/kant_emmanuel/fondements_meta_moeurs/fondem_meta_moeurs.pdf
Kohler F (2011) Diversité culturelle et diversité biologique: une approche critique fondée sur l'exemple brésilien. Nat Sci Sociétés 2 19:113–124. URL: www.cairn.info/revue-natures-sciences-societes-2011-2-page-113.htm
Küng H, Kuschel KJ (1995) Manifeste pour une éthique planétaire: la Déclaration du Parlement des religions du monde (Parlement des religions du monde), Cerf / (1993) Global Ethic: The Declaration of the Parliament of the World's Religions, Bloomsbury Academic

Laastad SG (2016) Nature as a subject of rights, a discourse analysis on Ecuador's constitutional rights of nature. Master Thesis in Human Geography Department of Sociology and Human Geography, University of Oslo. On line Avalaible: https://www.duo.uio.no/handle/10852/51892
Latour B (2004) Politiques de la nature. Comment faire entrer les sciences en démocratie, La découverte
Latour B (2009) Perspectivism: 'type' or 'bomb'. Anthropol Today 25(2):1–2
Lovelock J 1990 (1979) La terre est un être vivant. L'hypothèse Gaïa, Flammarion
MA Millennium Ecosystem Assessment (2005) Ecosystems and human well-being: synthesis. Island Press, Washington, DC
Maffi L (2005) Linguistic, cultural and biological diversity. Annu Rev Anthropol 29:599–617. https://doi.org/10.1146/ annurev.anthro.34.081804.120437
Maffi L, Woodley E (2010) Biocultural diversity conservation, a global sourcebook, Earthscan
Monastersky R (2015) The human age. Nature 519:147
Morin E (2004) La Méthode, t4, L'Ethique, Ed du Seuil
Natural Capital Coalition (2016) Natural Capital Protocol. Online available at: www.naturalcapitalcoalition.org/protocol
ONU (2015) Transformer notre monde : le programme de développement durable à l'horizon 2030. Résolution adoptée par l'Assemblée générale le 25 septembre 2015, A/RES/70/1
Pérez P (2010) Des Américains très inquiétants, ou de quelques conséquences de la théorie du perspectivisme. Séminaire conjoint "Transversalités philosophiques", ERRAPHIS (CNRS et ERASMUS Mundus) et CAS-LISST (EHESS, CNRS), Université de Toulouse, France. Online: halshs-00735671
Porto-Gonçalves CW (2006) A globalização da natureza e a natureza da globalização. Rio de Janeiro, Civilização brasileira
Rockstrom J, Steffen W, Noone K, Persson A, Chapin SF III, Lambin EF, Lenton TM, Scheffer M et al (2009) A safe operating space for humanity. Nature 461:472–475
Rosenzweig Michael L (2003) Win-win ecology: how the Earth's species can survive in the midst of human enterprise. Oxford University Press
Rudwick M (2005) Bursting the limits of time: the reconstruction of geohistory in the age of revolution. University of Chicago Press, Chicago
Rühs N, Jones A (2016) The implementation of Earth Jurisprudence through substantive constitutional rights of nature. Sustainability 8:174. https://doi.org/10.3390/su8020174
Santos M (1996) A natureza do espaço. técnica, tempo, razão e emoção. Hucitec, São Paulo
Serres M (1992) Le contrat naturel, Flammarion
Smuts JC (1926) Holism and evolution. Macmillan, London
Steffen W, Persson A, Deutsch L, Zalasiewicz J, Williams M, Richardson K, Crumley C, Crutzen P, Folke C, Gordon L, Molina M, Ramanathan V, Rockstrom J, Scheffer M, Joachim SH, Svedin U (2011) The anthropocene: from global change to planetary stewardship. Ambio 40:739–761. https://doi.org/10.1007/s13280-011-0185-x
Strathern AM (2014) Sem natureza, sem cultura: o caso Hagen. In: O efeito etnográfico e outros ensaios: Marilyn Strathern. Cosac Naify, São Paulo, pp 23–75
Syvitski J (2012) The anthropocene: an epoch of our making. Global Chang (78):12–15
TEEB (2010) The economics of ecosystems and biodiversity ecological and economic foundations, Pushpam Kumar (ed) Earthscan, London/Washington DC
Teihard de Chardin P (1956) La place de l'homme dans la nature, Le groupe zoologique humain, Seuil, Paris/Reprint: 2007, Albin Michel
Teillac-Deschamps P, Clavel J (2012) Les services écosystémiques, un outil politique et non écologique. In: L'exigence de la réconciliation, Biodiversité et société, Fleury C, Prévot-Julliard AC (dir) Fayard/Museum National d'Histoire Naturelle, pp 311–323
Thomas F (2011) Cosmologies, diversité bioculturelle et préservation de l'environnement. Natures Sciences Sociétés 2 19:129–132
Toussaint J-F (2012) Une limite aux évolutions de l'homme ? In: L'homme peut-il s'adapter à lui-même ? Toussaint J-F, Swynghedauw B, Bœuf G (dir) Quae editions, pp 21–28

Toussaint J-F, Swynghedauw B, Bœuf G (dir) (2012) L'homme peut-il s'adapter à lui-même ? Quae editions
UICN (2010) La diversité bio-culturelle conservée par les peuples autochtones et les communautés locales, exemples et analyses, Document d'accompagnement à la Note d'information UICN/CEESP n°10-2010, 71p
UICN, PNUE, WWF (1980) Stratégie mondiale de la conservation: la conservation des ressources vivantes au service du développement durable
Vernadski VI (2002)/1929. La Biosphère, 2e édition, Paris, Librairie Félix Alcan, 323p. – Rééd.: Paris, Seuil, coll. « Points/Science »
Vince G (2014) Anthopocene: a journey of the heart of the planet we made. Chatto & Windus
Viveiros de Castro E (2009) Métaphysiques cannibales, Paris, Presses Universitaires de France, « Métaphysiques ». https://doi.org/10.3917/puf.castro.2009.01. Online: http://www.cairn.info/metaphysiques-cannibales%2D%2D9782130578116.htm
Viveiros De Castro E (2014) Perspectivisme et multinaturalisme en Amérique indigène. J Anthropol, pp 161–181. URL: http://jda.revues.org/4512
White L Jr (1967) The historical roots of our ecologic crisis. Science 155(3767):1203–1207
Zalasiewicza J, Williamsa M, Smithb A, Barryc TL, Coec AL, Bownd PR, Brenchleye P, Cantrillf D, Galeg A, Gibbardh P, Gregoryi FJ, Hounslowj MW, Kerrk AC, Pearsonk P, Knoxl R, Powelll J, Watersl C, Marshallm J, Oatesn M, Rawsono P, Stone P (2008) Are we now living in the Anthropocene? GSA Today 18(2):4–8. Online: http://www.mindfully.org/Heritage/2008/Anthropocene-Zalasiewicz-GSA1feb08.htm
Zalasiewicza J, Williams M, Steffen W, Crutzen P (2010) The new world of Anthropocene. The Anthropocene, following the lost world of the Holocene, holds challenges for both science and society. Environ Sci Technol 44:2228–2231. https://doi.org/10.1021/es903118j
Zalasiewicza J et al (2014) When did the Anthropocene begin? A mid-twentieth century boundary level is stratigraphically optimal. Quater Int. https://doi.org/10.1016/j.quaint.2014.11.045

Contents of Volume 1

Contents of Volume 2

Contributors

Jean-Pierre Aubin VIMADES, (Viabilité, Marchés, Automatique et Décision); Paris Antenna, Paris, France

Philippe Augé Université de Montpellier, Montpellier, France

Catherine Barrière PACIM Association («Passeurs de cultures, Passeurs d'Images»/links between cultures), Montpellier, France

Olivier Barrière IRD - French National Research Institute for Sustainable Development, UMR ESPACE-DEV, Montpellier, France

Mohamed Behnassi Center for Research on Environment, Human Security and Governance (CERES), Ibn Zohr University of Agadir, Agadir, Morocco

Nadia Belaidi Department of Human & Environment, UMR CNRS 7206 Eco-Anthropology and Ethnobiology, Paris, France

Abdel Samed Bernoussi GAT, FST Tanger, Tanger, Morocco

Aliénor Bertrand CNRS, Faculty of Law and Political Science, UMR 5815, Dynamiques du droit, University of Montpellier, Montpellier, France

Philippe Billet University Jean Moulin – Lyon 3, Lyon, France

Institute of Environmental Law (CNRS, UMR 5600, EVS-IDE), University of Lyon, Labex IMU, Lyon, France

Daniel W. Bromley University of Wisconsin-Madison, Madison, WI, USA

Jean-Marc Châtaignier French Institute of Research for Sustainable Development (IRD), Paris, France

Espérance Cillaurren IRD - French National Research Institute for Sustainable Development, UMR ESPACE-DEV, Montpellier, France

Christian Corniaux Center for International Cooperation in Agricultural Research for Development (CIRAD), Dakar, Senegal

Fernando Raúl Coronato CONICET/CENPAT – Patagonia National Center, Puerto Madryn, Chubut, Argentina

Anne Coudrain IRD - French National Research Institute for Sustainable Development, UMR ESPACE-DEV, Montpellier, France

João Carlos Salles Pires da Silva Federal University of Bahia, Salvador, Brazil

Gilbert David IRD - French National Research Institute for Sustainable Development, UMR ESPACE-DEV, Montpellier, France

Vincent Douzal IRSTEA, National Research Institute of Science and Technology for Environment and Agriculture, Montpellier, France

Marie-Hélène Durand French Institute of Research for Sustainable Development (IRD), UMR GRED, Montpellier, France

Sébastien Girardin Conservatoire d'Espaces Naturels du Languedoc-Roussillon (CEN L-R), Immeuble le Thèbes, Montpellier, France

Abdelhaq El Jai IMAGES, UMR ESPACE-DEV, UPVD, Université de Perpignan Via Domitia, Perpignan, France

Marie Claude Simon El Jai IMAGES, UMR ESPACE-DEV, UPVD, Université de Perpignan Via Domitia, Perpignan, France

Samira El Yacoubi IMAGES, UMR ESPACE-DEV, UPVD, Université de Perpignan Via Domitia, Perpignan, France

Auguste Eyene Essono Ecole Normale Supérieure de Libreville, Libreville, Gabon

Mireille Fargette IRD - French National Research Institute for Sustainable Development, UMR ESPACE-DEV, Montpellier, France

Adama Faye Animal Production Scientist, Université Cheikh Anta Diop, Dakar, Senegal

Laurent Gavotte Evolutionary Sciences Institute of Montpellier (ISEM 5554), University of Montpellier, Montpellier, France

Christophe Grenier Nantes University, UMR CNRS 6554 LETG-Géolittomer, Nantes, France

Hermes Morales Grosskopf IPA – Institute for Livestock and Agriculture Planning, Montevideo, Uruguay

Bernard Hubert Agropolis International, Montpellier, France

Jacques Lepart Conservatoire d'Espaces Naturels du Languedoc-Roussillon (CEN L-R), Immeuble le Thèbes, Montpellier, France

Jessica Huron Conservatoire d'Espaces Naturels du Languedoc-Roussillon (CEN L-R), Immeuble le Thèbes, Montpellier, France

Nicolas Lescureux National Centre for Scientific Research (CNRS), UMR5175 Centre d'Ecologie Fonctionnelle et Evolutive (CNRS-UM-UM3-EPHE), Montpellier Cedex 5, France

Christian Lévêque French Research Institute for Sustainable Development (IRD), Marseille Cedex 02, France

Thérèse Libourel Université Montpellier, UMR ESPACE-DEV, Montpellier, France

Maud Loireau IRD - French National Research Institute for Sustainable Development, UMR ESPACE-DEV, Montpellier, France

Ruijun Long International Center for Tibetan Plateau Ecosystem ManagementI (CTPEM), Lanzhou University, Gansu Province, China

Jessica Makowiak CRIDEAU-OMIJ (EA 3177), Université de Limoges, Limoges, France

Morgan Mangeas IRD - French National Research Institute for Sustainable Development, UMR ESPACE-DEV, Montpellier, France

Serge Morand CNRS-ISEM, University of Montpellier, Montpellier, France

Catherine Moulia Evolutionary Sciences Institute of Montpellier (ISEM 5554), University of Montpellier, Montpellier, France

Laurence Pascal University of Montpellier, Montpellier, France

Catherine Prost The Federal University of Bahia, Salvador, Bahia, Brazil

Voyner Ravena-Cañete The Federal University of Pará, Belém, Brazil

Catherine Sabinot IRD - French National Research Institute for Sustainable Development, UMR ESPACE-DEV, Noumea, Nouvelle-Calédonie, France

Mohamed Taher Sraïri IAV Hassan II – Hassan II Institute of Agronomy and Veterinary Medicine, Rabat, Morocco

Jean François Tourrand DVM, Center for International Cooperation in Agricultural Research for Development (CIRAD), Montpellier, France

Aline Treillard CRIDEAU-OMIJ (EA 3177), Université de Limoges, Limoges, France

About the Editors

Olivier Barrière, PhD is an environmental jurist and a researcher at the IRD (French Research Institute for Development), which develops a legal socioecological approach through the anthropology of law. He got an HDR (accreditation to supervise research) in May 2012 on the topic "Elements of a Legal Socio-ecology: The Right Facing the Ecological Emergency."

For 20 years, his work has focused on the relationship which bond human beings to their environment, within the limits of a legal regulation which is faced with progressing global and environmental changes. He thus works on the local law concerned with the viability of systems by promoting innovative concepts such as "land tenure-environment," the "coviability of social and ecological systems," and "negotiated right" which creates a relationship

---

The book is led by researchers from several French research institutions, the Research Institute for Development (IRD), the National Centre for Scientific Research (CNRS), the National Research Institute of Science and Technology for Environment and Agriculture (IRSTEA), Brazilian universities (Federal University of Pará and Federal University of Bahia), and the Moroccan Ibn Zuhr University of Agadir.

Editing is carried by IRD which focused its research for over 65 years on the relationship between man and its environment, in Africa, Mediterranean, Latin America, Asia, and the French tropical overseas territories, and particularly by the Research Unit "Space and Development" which is placed in a perspective of coviability of social and ecological systems in the territories at local, regional, and global levels, particularly in the tropics.

between international law and endogenous rights. His working areas are Africa (Morocco, Senegal, Mali, Tchad, Rwanda), French Guyana, Nouvelle Calédonie, and France (Causses-Cévennes).

For several years, O. Barrière formalizes a network of researchers and experts in the field of coviability by bringing together a variety of disciplines through meetings and seminars. Within its research institution (IRD), he leads an interdisciplinary transverse axis on the coviability in a transdisciplinary aims.

Olivier Barrière, as project manager, implements experiments in close cooperation with stakeholders, local elected and national technical institutions to achieve concrete practical results as local law, as environmental convention, socioecological resilience pact, pastoral pact. He also teaches environmental law to future managers of natural lands at the University of Sciences of Montpellier.

Mohamed Behnassi, PhD is specialist in Environment and Human Security Law and Politics. After the obtention of his PhD in 2003 from the Faculty of Law, Economics and Social Sciences, Hassan II University of Casablanca for a thesis titled: *Multilateral Environmental Negotiations: Towards a Global Governance for Environment*, he accessed to the Faculty of Law, Economics and Social Sciences, Ibn Zohr University of Agadir, Morocco, as Assistant Professor (2014). In 2011, he obtained the status of Associate Professor and in 2017 the status of Full Professor. He served as the Head of Public Law Department (2014–2015) and the Director of the Research Laboratory for Territorial Governance, Human Security and Sustainability (LAGOS) (2015- present). In addition, Dr. Behnassi is the Founder and Director of the Center for Environment, Human Security and Governance (CERES) (former North-South Center for Social Sciences (NRCS), 2008–2015). Dr. Behnassi is also Associate Researcher at the UMR ESPACE-DEV, Research Institute for Development (IRD), France. In 2011, he completed a US State Department-sponsored Civic Education and Leadership Fellowship (CELF) at the Maxwell School of Citizenship and Public Affairs, Syracuse University, USA, and in 2014 he obtained a Diploma in Diplomacy and International Environmental Law from the University of Eastern Finland and the United Nations Environment Programme (UNEP), Finland. Dr. Behnassi has pursued several postdoctoral trainings since the completion of his PhD.

His core teaching and expertise areas cover environmental change, human security, sustainability, climate change politics and governance, human rights, CSR, etc. He has published numerous books with international publishers such as *Environmental Change and Human Security in Africa and the Middle East* (Springer 2017); *Vulnerability of Agriculture, Water and Fisheries to Climate Change* (Springer 2014); *Science, Policy and Politics of Modern Agricultural System* (Springer 2014); *Sustainable Food Security in the Era of Local and Global Environmental Change* (Springer 2013), *Global Food Insecurity* (Springer, 2011); *Sustainable Agricultural Development* (Springer, 2011); *Health, Environment and Development* (European University Editions, 2011); and *Climate Change, Energy Crisis and Food Security* (Ottawa University Press, 2011). He has also published numerous research papers and made presentations on these at international conferences. In addition, he has organized many international conferences covering the above research areas in collaboration with national and international organizations and managed many research and expertise projects on behalf of various national and international institutions. Dr. Behnassi is regularly requested to contribute to review and evaluation processes and to provide scientific expertise nationally and internationally. Other professional activities include Social Compliance Auditing and consultancy by monitoring human rights at work and the sustainability of the global supply chain.

Gilbert David is a marine and island geographer by training and research director in IRD: UMR Espace-Dev, head of the research team dealing with integrated approaches of nature and society. During his career, he experimented different types of coviability. From 1984 to 1991, he studied the links between reef fisheries and food security of islanders in Vanuatu (South Pacific).

From 1991 to 1996 he was involved in a project dealing with the spatial coviability of New Caledonia. This big island was divided into three parts: an urban center (Noumea), rural areas with nickel mines, and rural peripheries, whose evolution's trajectories were different. How to cope with these differences and how to change these trajectories to reach a viable future for all people of New Caledonia?

From 1997 to 2000, he was involved in the Regional Environment Programme of the Indian Ocean commission (Comoros, Madagascar, Mauritius, Reunion, Seychelles) working on the coral reef action plan in order to cope with the increasing uses of this ecosystem in the Western Indian Ocean region. From 2003 to 2009, he was based in Reunion Island working of integrated coastal zone management (IZCM) and marine protected areas (MPAs), both considered as tools for coviability. After 2 years in Brest, he is now based in Montpellier. IZCM and MPAs are still his main research topics.

Vincent Douzal graduated from Institut national agronomique Paris-Grignon (INA-PG, now AgroParisTech) and École Nationale du Génie Rural, des Eaux et des Forêts (ENGREF). He completed masters in robotics and a thesis in computer science, on perception: mathematical modeling and data analysis of descriptive sensory analysis experiments, including the underlying theoretical framework on perception. Then for several years he was head of a team in a regional public administration in Grenoble, France, conducting civil engineering works in water supply and sanitation, rural electric infrastructure, irrigation, waste incineration plants, waste sorting centers, and public education buildings. He was responsible for the negotiation process leading to the global regional collective planning of waste treatment. He returned to research in a multidisciplinary team spanning from remote sensing to spatial modeling and social implications of using spatial and temporal data for collective decision-making, with the mandate of defining the bases of information systems for scholarly and general issues on natural hazards. His current work focuses on traceability in computer systems, a problem tightly connected to the representation of time in information systems, with many implications on memory preservation and cultural heritage, long-term digital archiving, hypermedia and collaboratory systems, and of course, at the very basis, reproducibility of scientific results.

Mireille Fargette graduated at both Université Aix-Marseille III (France), where she obtained a Master in Biology and Ecology, and Ecole Nationale Supérieure d'Agronomie (Montpellier SupAgro, France), where she obtained a Master and a PhD in Agronomy. She has been working for IRD (Institut de Recherche pour le Développement) on scientific questions in tropical Southern countries, through national and international scientific programs. She has worked abroad for part of the time. She has published more than 50 publications in indexed scientific journals and participated (posters or papers) in 68 scientific congresses. She first focused on the plant domain (biology, parasitology, ecology) and biodiversity related to agro-systems. More recently her main interests have encompassed the systemic relationships between societies and their environment (including man and societies relationships, on the one hand, and their links with the environment and "natural systems," on the other hand). Her main scientific interests (e.g., dynamics and viability of complex systems, Southern territories, scientific observatories) call for an interdisciplinary approach and rely on ontological approaches and symbolic modeling in order to share knowledge and representations.

Thérèse Libourel is Professor Emeritus in the Department of Computer Science at the Montpellier University. She participated in the creation of UMR ESPACE-DEV (IRD, UM, UAG, UR) and assured co-direction and direction thereof from January 2011 to September 2014 after having been a member of the Montpellier Laboratory of Informatics, Robotics and Microelectronics (LIRMM/CNRS). She received her PhD from Montpellier University in 1992 and she got an HDR (accreditation to supervise research – standing for "Habilitation à Diriger des Recherches" in French) in January 2003. She has served as a program committee president and member for several conferences in information systems and databases (BDA, INFORSID, CASSINI, SAGEO, ICEIS). She is a member of the GDR Magis (Methods and Applications for Geomatics and Spatial Information). Thérèse Libourel has been studying and teaching information systems, databases, and software engineering for nearly 15 years (in various Master University Montpellier 2 and International Master ICT USTH). Her research interests specifically include complex system modeling in various fields (Biology, Geographic Information System, Robotics), particularly in the area of object modeling paradigm where she gained expertise in a number of key issues such as object methodology, data models, data evolution, metadata, information integration or modeling UML and its evolution. As a result of her strong implication in various projects, she is closely collaborating with different research partners from various disciplines (agriculture, environment, biology, geography, etc.) and consequently gained a significant experience in interdisciplinarity.

Maud Loireau is a research engineer at IRD (Institut de Recherche pour le Développement) with a competence in international agro-development (ISTOM engineer) and geography (PhD, University of Paul Valéry – Montpellier 3). She has been working for more than 25 years in arid areas in Saheli and Northern Sahara, and generally in threatened territories (arid zones vs desertification, pioneer frontier vs deforestation and urban extensions, mountains areas changing vs tourism).

Her research consists in constructing, formalizing, and using conceptual frameworks (observatory, landscape, viability, attachment to places) and methods (data collection, co-construction and formalization of shared knowledge, modeling) to characterize the links between the systems involved and their spatiotemporal footprints, between the societies concerned and the place they live in.

This research is conducted on fragile territories and/or in transition under constraints from global changes with management issues, in order to develop ways to observe, analyze, and monitor the dynamics of these territories, and ultimately contribute to enrich and share knowledge, to facilitate discussions and negotiations between actors, accompanying the manager's decisions and actions.

With some 60 scientific publications including book chapters, journal articles, and conference proceedings, her scientific expertise, mainly focused on the issue of desertification, has also contributed to consolidate a network of partners in southern and northern countries; she animates the networks "Scientific Societies-Environments Observatories," and "Relations Between Society-Environment and Ecology: From Knowledge to Action," within the frameworks of MAGIS (http://gdr-magis.imag.fr) and of the SFE (https://www.sfecologie.org), respectively; she encourages communication between scientists, politicians and civil society as a member of the CSFD (http://www.csf-desertification.org).

Laurence Pascal is lecturer at the University of Montpellier, co-initiator and the specialty of Master "Tropical Plant Biodiversity" and responsible for license teaching units and master, "Plant Physiology and operation of plants," "Perfume, aromas, and biomolecules of living," "co-evolution and symbiosis," "biological heritage." She participates in the inter-university exchange, as a member of the scientific and administrative councils of the "Pôle Universitaire de Guyane," by supervising numerous training courses, and the creation of thematic school of tropical biodiversity in French Guyana.

As a researcher in the team "biocultural interactions and coevolution" of CNRS in Montpellier since 1995, she has joined the Joint Research Unit "Microorganisms Environment Interactions Plantes" (EMPI/IRD, CIRAD, Montpellier University) to study the tolerance of plants and ecosystems to environmental constraints in a context of global change. Her research is based on an ecological approach to chemistry relationships that plants have with their biotic and abiotic environments around contemporary issues (maintenance of ecosystems, their biodiversity and species communities where chemical communication is paramount). Her research explores the diversity of species interactions, with particular emphasis on their dynamic and scalable operation in response to current environmental changes.

Catherine Prost is Associate Professor of the Geography Department at the Institute of Geosciences (Federal University of Bahia, Brazil). She completed a postdoctorate at the Institute of Research for Development (IRD-Montpellier, France). Her core expertise areas cover artisanal fishing, territorial and environmental management, and protected areas. She has published papers in accredited journals and communicated several oral presentations in relevant Brazilian conferences.

In addition, Dr. Prost has organized three conferences: the first, second, and third Seminaries about Coastal Spaces, in September 2011, May 2013, and October 2016, respectively, at Salvador, Brazil. Dr. Prost is also a Lead Manager of several research projects such as "Communitarian Management of Natural Resources in the Reserve of Collect Bay of Iguape" and "Environmental Management and Territorial Conflicts in the Baianese Maritime Reserves of Collect" both sponsored by the National Centre of Research.

Voyner Ravena-Cañete is an anthropologist, associate professor at Federal University of Pará – UFPA/Brazil, resident at the Institute of Biological Sciences, working both in the Postgraduation Program in Aquatic Ecology and Fishing and Postgraduation Program in Sociology and Anthropology at UFPA. She graduated in History (1991), has a Master's degree in Anthropology (2000), and PhD in Sustainable Development in the Humid Tropics, at UFPA. She develops research and extension projects regarding especially environmental issues that involve traditional populations at the rural Amazon area. Between 1998 and 2005, she carried out research about Amazon peasantry and their ways of living, focusing kinship as strategy to access and use of natural resources. Between 2006 and 2012, she extended her studies to regional sceneries that involved traditional riverine populations in the Amazon and their specific patterns of pluriactivity. Since 2010, she has been developing studies about fishing populations and their way of living, especially within coastal and estuarine environment.

She presently coordinates the research project "Empowerment, local ethos and natural resources: social cartography as strategy to action planning in RESEX's saliferous marines in Pará," funded by UFPA and aiming at traditional fishing populations within the Amazon north coast microregion. She develops research and extension activities in the Mocapajuba extractive reserve, in São Caetano de Odivelas, studying and nurturing the strengthening of the native population's ethos. She works as a consultant in Anthropology for Amazon traditional populations, especially in the fields of natural resources, memory, reciprocity, sociability, and kinship.

Frédérique Seyler is Director of Research at IRD (Institute of Research for Development) and Acting Director of UMR Espace-Dev, specializing in the study of interrelated dynamics of the environment and societies. She has a recognized experience on soil-water relationships in lateritic landscapes of Central Africa and the Amazon, studied from both remote sensing analysis (visible, near infrared, and radar) and field data collection. Her areas of expertise are mainly the spatial variability of soils, spatial hydrology (using radar altimetry). She has coordinated several partnership programs in Africa and South America (Brazil and Colombia) and organized transfer of the methods developed, mainly in the field of remote sensing of water resources. Frédérique Seyler is vice-president of the International Commission for Remote Sensing of the International Association of Hydrological Sciences (IAHS). She has coedited two IAHS books: *Improving Integrated Surface and Groundwater Management and Vulnerable in a Changing World*, and *GRACE, Remote Sensing and Ground-Based Multi-scale Methods in Hydrology*.

Serge Morand is an evolutionary ecologist, and his main researches focus on parasites and pathogens and their roles in conservation ecology, public and animal health. He is concerned at the role of biodiversity as risk and insurance for zoonotic emerging infectious diseases. He is conducting projects on the impacts of global changes, including climate change and land use land cover change, on the links between biodiversity and health in Southeast Asia, using wildlife-borne diseases as a model. Dr. Morand is a CNRS (French National Center for Scientific Research) research director. He is actually an Adjunct Professor at Kasetsart University, Faculty of Veterinary Technology, and at Mahidol University, Faculty of Tropical Medicine (Thailand), where he teaches disease ecology. He is the author and coauthor of more than 450 articles and chapters, and author and coeditor of 13 books on these domains, with the last ones in 2017:

- by Routledge EarthScan, *Biodiversity Conservation in Southeast Asia: Challenges in a Changing Environment* (Morand, S., Lajaunie, C., Satrawaha, R., Eds.), a book that provides theoretical overviews and challenges for applied research in living resource management, conservation ecology, health ecology, and conservation planning in Southeast Asia.
- by Elsevier, *Biodiversity and Health Linking Life, Ecosystems and Societies* (Morand, S., Lajaunie, C., Authors), a book that fills the gap between the ecology of health and the concepts supported by international organizations, such as EcoHealth and One Health and demonstrates how ecological sciences, environmental sciences, medical sciences, and social sciences may contribute to improve human health through conserving biodiversity and the services it provides to societies.

List of Abbreviations and Acronyms

AECM	Agri-environmental-climate measures
AEM	Agri-environmental measure
APP	Areas of permanent preservation
ARC	Avoid-reduce-compensate
AWB	Artificial water bodies
CAP	Common Agricultural Policy
CCD	Colony collapse disorder
CDB	Convention on Biological Diversity
CDM	Clean development mechanism
CEREMADE	Centre for Research in Mathematical Decision
CERES	Center for Research on Environment, Human Security and Governance
CIRAD	Center for International Cooperation in Agricultural Research for Development
CITES	Convention on International Trade in Endangered Species of Wild Fauna and Flora
CIVAM	Centres d'Initiatives pour Valoriser l'Agriculture et le Milieu Rural
COP	Conference of Parties
CRPMEME	Regional Committee of Marine Fisheries and Aquaculture
CRU	Climatic research unit
CSR	Corporate social responsibility
CUMA	Coopérative d'Utilisation du Matériel Agricole
EC	Council regulation
EFNCP	European Forum on Nature Conservation and Pastoralism
ENCOP	Environment and conflict project
FAO	Food and Agriculture Organization
FFEM	French Fund for the Global Environment
GAEC	Good agricultural and environmental conditions
GDP	Gross domestic product
GEF	Global environment facility
GIZ	Deutsche Gesellschaft für Internationale Zusammenarbeit

HDI	Human development index
HMWB	Highly modified water bodies
IBAMA	Brazilian Institute of the Environment and of Renewable Natural Resources
ICMBio	Chico Mendes Institute for the Biodiversity
IEEP	Institute of European Environmental Policy
IFRECOR	French initiative for coral reefs
INDCs	Intended nationally determined contributions
INTC	Intermediate nitrate-trap crops
IOC	Indian Ocean Commission
IPBES	Intergovernmental Scientific and Political Platform on Biodiversity and Ecosystem Services
IPCC	Intergovernmental Panel on Climate Change
IRD	Research Institute for Development
IUCN	International Union for the Conservation of Nature
IUGS	International Union of Geological Sciences
LAGOS	Research Laboratory for Territorial Governance, Human Security and Sustainability
LEK	Local ecological/environmental knowledge
MAB	Man and biosphere
MAE	Mesure agro-environnementale
MDGs	Millennium development goals
MEA	Millennium ecosystem assessment
NAZCA	Non-state actor zone for climate action
NEPAD	New partnership for Africa's development
NRCS	North-South Center for Social Sciences
NRP	Natural regional park
NWCZ	North West Coast Zone
OREMSE	Ontology, coral reefs, mangroves, environmental services
PA	Protected areas
PACIM	Passeurs de cultures, passeurs d'images
PB	Planetary boundary
PMA	Protected marine areas
PPES	Payment for the preservation of eco-systemic services
PPF	Peace Parks Foundation
PRIO	International Peace Research Institute
SDGs	Sustainable development goals
SEBRAE	Serviço Brasileiro de Apoio às pequenas e Micro-Empresas
SES	Social-ecological systems
SNUC	National System of Conservation Units
TEEB	The Economics of Ecosystems and Biodiversity
TEK	Traditional ecological/environmental knowledge
TFCA	Transfrontier conservation area
TKW	Traditional knowledge and wisdom
UFBA	Universidade Federal da Bahia

UFPA	Universidade Federal do Pará
UML	Unified modeling language
UNCTAD	United Nations Trade and Development Conference
UNEP	United Nations Environment Program
UNFCCC	United Nations Framework Convention for Climate Change
UNTB	Unified Neutral Theory of Biodiversity
USAID	United States Agency for International Development
WWF	World Wild Fund for Nature
YCW	Young Christian Workers
ZAT	Temporary artistic zone
ZNIEFF	Faunistic and floristic interest
ZUP	Priority urban area

List of Figures

List of Tables

The original version of this frontmatter was revised: surname and first name of the authors interchanged in the Reference section of the General Introduction which has now been corrected. The correction to this chapter is available at https://doi.org/10.1007/978-3-319-78497-7_28

Chapter 1
Introductory Chapter: An Interweaving to Be Formalized, the Biosphere Faced with the Relationship Between the Human and the Non-human

Olivier Barrière, Catherine Prost, Voyner Ravena-Cañete, Vincent Douzal, Mireille Fargette, and Jean-Pierre Aubin

Contents

O. Barrière · M. Fargette (✉)
IRD - French National Research Institute for Sustainable Development, UMR ESPACE-DEV, Montpellier, France
e-mail: olivier.barriere@ird.fr; mireille.fargette@ird.fr

C. Prost
The Federal University of Bahia, Salvador, Bahia, Brazil
e-mail: cathprost@yahoo.com

V. Ravena-Cañete
The Federal University of Pará, Belém, Brazil
e-mail: ravenacanete@gmail.com

V. Douzal
IRSTEA, National Research Institute of Science and Technology for Environment and Agriculture, Montpellier, France
e-mail: vincent.douzal@teledetection.fr

J.-P. Aubin
VIMADES, (Viabilité, Marchés, Automatique et Décision); Paris Antenna, 14, rue Domat, 75005, Paris, France
e-mail: aubin.jp@gmail.com
http://vimades.com/aubin/

O. Barrière et al. (eds.), *Coviability of Social and Ecological Systems: Reconnecting Mankind to the Biosphere in an Era of Global Change*,
https://doi.org/10.1007/978-3-319-78497-7_1

This introductory chapter delineates the challenge of formalizing a concept-paradigm and provides specific examples that could be considered as models of socio-ecological viability. We investigate the benefits that the concept of coviability brings to the concept of "sustainable development". Contributions will be given throughout the book and will be summarized in a final chapter, which will define the concept-paradigm through a transdisciplinary approach.

In the Anthropocene era, societies build their own environment.[1] They even go as far as jeopardizing the future well-being, even the fate of making (and that of life itself) on the planet by crossing thresholds and reaching tipping or rupture points[2] (Lenton et al. 2008), forcing the ecological system into a new state and the Earth system into a new era, with no possibility of returning to initial conditions (Steffen et al. 2007; Rockström et al. 2009). The disconnection between the political and economic development of societies on the one hand and ecological processes on the other, results in the societies' failure to see that they are part of the biosphere (Folke et al. 2011). Separating societies from the biosphere is globally explained by the poor integration of nonhuman factors in the reproduction of societies, considering the environment as an externality to the human species, and this mainly through dominating Wester culture. The idea of externality itself, underlining a culture-biosphere dichotomy, is a challenge: it is essentially founded on the ontology of the bio-cultural paradigm specific to this society (Morin 1973). Indeed, more or less significant depending on the socio-cultural representations, the societies-nature dichotomy reflects the paradigm of the relationship between Man and the biosphere (see Part I, below). Formalizing this relationship will enable to go beyond this significant dichotomy by "greatly sharing" in order to participate in the cross-disciplinary scientific bases of socio-ecological viability (see Part II, below).

1.1 Between Man and the Biosphere, a Question of Paradigm

Through globalization, the prevailing economic system locally impacts most societies that adopt this model, either by participating in it or by suffering from it. The Amerandian Wayana people, for example, are torn between the French Republic, which promotes their "development" and simultaneously stresses the need to maintain their lifestyle (described as "traditional" by this model) whilst anchoring them down with a stated need of private land, a tourist development

[1] A geological period, in which human actions become a geophysical force and impact the planet: a geology of mankind (Crutzen 2002). More specifically, the Anthropocene is "an era of addiction, where producing means/goods has become the purpose of existence. It is an era of acceleration, where growth, based on the endless cycle of production and consumption, must continuously produce additional unnecessary objects for those who already have too much. This is the rationale of productivism" (Sinaï 2011).

[2] In ecology, a critical point beyond which the system abruptly shifts from one state to another, or in social sciences the tolerance threshold or socio-dynamic critical point.

scheme, and so on. (See PAG charter 2014,[3] Barrière and Faure 2012). The clash of socio-cultural, economic and religious paradigms can be seen anywhere in the world, whether in urban areas, rural areas, or in large biomes (forests, deserts, the Sahel, oceans, etc.). This generates different levels of acculturation and loss of native cultures with forms of transformation of socio-cognitive profiles and schemes to the detriment of the diversity of being and thinking. The United Nations respond with a declaration on the rights of indigenous peoples adopted in 2007, which recognizes indigenous peoples' right to exist with their differences and cultural specificities.[4] However, in the earlier Johannesburg Earth Summit which took place in 2002, the following question was clearly raised: "Should indigenous peoples be encouraged to integrate the global marketplace without sacrificing their cosmologies, which are fundamentally respectful of the ecosystem?" (UNEP 2002:18). The answer lies in the challenges of a recognition which surpasses legal rhetoric made up of declarations, good constitutional intentions, zones of rights of use and other "legal tinkering" (Barrière 2011, 2017; Stavenhagen 2013; Bélier 2013; Filoche 2011).

The development of a profile of man's position in the biosphere plays an important role in the challenges of conservation politics of natural environments. These policies shift from a notion of "balance" of ecological systems to that of an "a capacity to adapt" (see below) to anthropic effects, based on "co-change" (Blandin 2013). However, it is within a "habitat" system (*oikos*) that the effects of natural and human functions commingle through a recurrent conflictual paradigm between ecology and economy (Dahl Lyon 1996 and his eco theory[5]; Helm 1991). Between adaptation and paradigm, the adaptive cycles of social, ecological or socio-ecological systems are linked together in terms of growth, accumulation, restructuring and renewal, a hierarchical interrelation by cycles defining a "panarchy", according to Holling (2001).[6] The diversity of these cycles underlines an interdependence that opens before us an "era of transformation, in which ecosystem management must build and sustain ecological resilience as well as the social flexibility needed to cope, innovate, and adapt" (ibid. 404). The interrelation of

---

[3]Online: http://www.parc-amazonien-guyane.fr/le-parc-amazonien-de-guyane/la-charte-des-terri - toires/

[4]"Indigenous peoples have the right to observe and revitalize their cultural traditions and customs. They includes the right to preserve, protect and develop past, present and future manifestations of their culture, such as archaeological and historical sites, crafts, designs, ceremonies, technologies, visual and performing arts and literature"; "Indigenous peoples have the right to promote, develop and maintain their institutional structures and their distinctive customs, spirituality, traditions, procedures; practices and, in the case where they exist, juridical systems or customs, in accordance with international human rights standards" (Articles 11-1 and 34 of the Declaration on the Rights of Indigenous Peoples, No. 61/295 adopted by the United Nations GA on 13/09/07).

[5]Defined as "any natural or man-made functional system with internal integrity and distinct feature and behaviour enclosed within clear boundaries."

[6]"Panarchy is the hierarchical structure in which systems of nature and humans as well as combined human-nature systems and social-ecological systems are interlinked in never-ending adaptive cycles of growth, accumulation, restructuring, and renewal." p. 392.

man with the biosphere leads to the concept of "ecological solidarity" recently established as a legal principle by the French legislator.[7]

Several examples can illustrate the paradigm issue of the relationship between humans and the biosphere (Sect. 1.1.1). These lead us to society-nature interrelations (Sect. 1.1.2) and to question the concept of sustainable development (Sect. 1.1.3).

1.1.1 Selected Models of Socio-ecological Viability

For instance, three models of socio-ecological viability illustrate the idea of a new era. The first model deals with the relationship between economics and ecology in pastoral practices. The second model is bio-ecological and ethological; it refers to the socio-physiological field of honeybee colonies, *Apis mellifera*, based on both rich scientific apidological works and a practical research in beekeeping. The third model is that of an emerging circular economy, refusing the linear "extract, produce, consume, discard" system.

1.1.1.1 Pastoralism: Between Production and Preservation

Pastoralism includes all livestock activities that encourage, through extensive grazing, the use of spontaneous fodder resources from natural areas in order to ensure all or part of animal feed. Pastoral areas comprise extensive and seasonal pastures. It is defined as an "in-between" called *saltus* (in opposition to *ager*, cultivated lands and to the *silva*, the forest): "rangelands include the total of non-cultivated farming space" the production of which is dedicated to herbivorous livestock (cattle, sheep, goats, horses, donkeys) (Hubert et al. 2010, 28).

Rangelands include by definition a diversity of very heterogeneous environments, especially when it comes to sheep and goat pastoralism. Pastoral activity is often carried out on the outer rims, which are areas of low (uncultivated) agronomic interest, scarcely accessible or excluding other uses. The pastoral area is defined by the complexity of the herd's daily needs, which require the herd going through: (a) numerous plots of land which are located in multiple rims, that are discontinuous, polygonal, fragmented, odological (itineraries, pathways, roads, routes and so on); (b) many interstitial areas situated between pastoral elements of the local area and present in the landscape such as sheepfolds, rangelands, trails, hedges, ditches, paths, and linking spaces between rangelands and so on) or (c) topocentric focal points such as watering points, stubble fields, huts, pens, and so on.

[7]Law No. 2016-1087 of 8 August 2016 for the recovery of biodiversity, nature and landscapes: "The principle of ecological solidarity which calls for taking into consideration living organisms and natural or managed environments, in any public decision-making having a significant impact on the local environment, the interactions of ecosystems" (Article 2).

In Europe, pastoral land is considered as a valuable form of agriculture for conserving biodiversity through the use of "unimproved" vegetation (neither sown nor fertilized).[8] Thus, by their characteristics, the environments supporting pastoralism are likely to harbor high levels of biodiversity (species and habitats) that render them spaces with "high natural value".[9] This confers on pastoral practices an essential factor of ecological sustainability while the viability of pastoralism depends on "grazing resources" provided by ecological systems: the diversity of resources and factors mobilized to ensure the production provides the extensive livestock farming with the necessary means for adapting to environmental variability both spatially and temporally (Landais and Balent 1993). The result is a symmetrical and interdependent viability between the pastoral system and the ecological system.

However, how is such coviability acquired? By the functioning of the pastoral system itself, in the herding of animals and their access to pasture. Empirically, grazing on pastureland is included in a herd's daily schedule. First, it is organized in terms of the calendar of the utilization of an area, structured as "districts" (Savini et al. 2010, 86), which remains very dependent on spatio-temporal management. Then, it is organized by the spatial behavior of the herd, its "bias" (behavioral homogeneity), or its socio-physiology, that the shepherd must anticipate without thwarting.[10] Several factors are taken into account by the shepherd in the management of the flocks, which concern the availability of pasture (the shepherd/herdsman assesses resources left in "how many days in advance/ahead", ibid.), accessibility, diversity (types of vegetation and undergrowth fruits), its geographical situation

[8]Instruction from the "High Natural Value" Impact Indicator, orientation document 2007–2013, European Commission of Agriculture and Rural Development & European Network "Evaluation of Rural Development", p. 5.

[9]Concept developed in the early 1990s by a group of experts from the London Institute of European Environmental Policy (IEEP), World Wild Fund for Nature (WWF) and the European Forum on Nature Conservation and Pastoralism (EFNCP), which observed that some agricultural systems had a positive impact on biodiversity, especially farms using extensive production techniques without the use of synthetic inputs. Progressively, the HNV concept was enriched with works describing different forms of HNV agriculture in Europe. In the early 2000s, the projects carried out at the request of the European Environment Agency (Andersen 2003) and the European Commission (IEEP 2007) led to defining three types of agricultural zones of high natural value: (1) those with a large proportion of semi-natural areas (very extensive valorization), (2) those based on a mosaic of low-input agricultural areas with a high proportion of landscape elements (e.g. grass strips between plots, hedges, wood and shrubs) (3) those harboring rare species or a large proportion of the European or world population of a species (the development of an indicator on agricultural zones of high natural value, European Environment Agency, Andersen 2003). In 1998, the European biodiversity strategy defined the objective of "conservation and sustainable usage of agro-ecosystems" whose sub-objective is promoting and supporting "low input agricultural systems especially in areas of high natural value". In 2006, the European Community's strategic orientation document for rural development, for the period 2007–2013, shows its commitment to "respect biodiversity" and "preserve ecosystems" and includes measures to "preserve agricultural and forest ecosystems with high natural value". Rural Development Regulation No. 1698/2005, Axis2, provides aid for non-productive investments (art. 36-a-6) of which areas of high natural value benefit (Art. 41-b: investments in exploitation, as long as they enhance the public utility of a Natural zone 2000 or other areas of high natural value to be defined in the program").

[10]"The bias... is the result of the rhythm of the animals, of their habits, of their compulsion, and of the characteristics of the mountain ..." (Savini et al. 2010, 89).

(sunshine, shade, etc.) and access to a watering point. The whole is conditioned by encouraging the of animals' yearning, a level of what is grazed in regards to the resource's capacity to renew itself, in every season and over several years. The challenge of flock/herd management focuses on their appetite (Meuret 2010a, b, 149), which calls for real know-how in managing the economic (production) vs ecological (conservation) relationship. Strategic choices must therefore be made between the of environment's state, i.e. pasture, that must ensure the long-term feedstock of flocks/herds and the development (weight gain) of the animals. The compromises to be made are realized in the management of the herd which focuses, as we have seen, on the access to the diversity of pastures through scheduling, rotation, and a precise duration in space and time. The coviability model of social and ecological systems is lies here.

This compromise seems to be a recurrent challenging factor in order to achieve coviability. Another situation is the association of livestock production through grazing and maintenance or development of biodiversity on the same land, which requires choices reflecting a strategy of compromise. An example is provided by the modeling of the production of an agro-ecosystem (swamp grassland forage) and its ecological performance (maintaining the development of two species of waders) in Europe (Tichit et al. 2007; Sabatier et al. 2008, 2010).

The grazing method determines the sustainability of a community of birds: the model proposes scenarios considering the viability of the wader's breeding systems (defining its risk of extinction) and the grazing strategies guaranteeing satisfactory feeding for livestock (defining the viability of the livestock system). It turns out that grazing maintains the population of birds while the grazing strategy (grazing intensity or mowing period) determines its existence: threshold effects appear showing a temporal variation over the period and grazing intensity. However, excluding livestock in the spring, as advocated by other authors (Hart et al. 2002, cited by Tichit et al. 2007), is not a relevant option for reconciling economic challenges with ecological objectives. Indeed, the compromise associating production and conservation lies not in the exclusion of a system but in the methods of harvesting the ecological resource. It is a question of amounts and the periodicity of interventions in order to allow wader nesting: "It appears.... that the proportion of different uses is the lever of arbitration and therefore of the conciliation between production and conservation" (Sabatier 2010, 29). The author also points out that it is impossible "to simultaneously maximize a farm's productive and ecological performance". The coviability of systems therefore requires making adjustments to the interventions (of environment's "uses") in order to find a form of equilibrium leading to conciliation concessions (preservation of birds at a cost *vis-à-vis* production): "the best ecological performances are reached at intermediate levels of productive performance" (ibid.).

The second example of the socio-ecological sustainability model discussed now is a more bio-ecological one, starting from an animal organism dependent on the human system, the latter being free in choosing its viability.

1.1.1.2 Honeybee (*Apis mellifera*), a Socio-ecological Superorganism

A bee does not exist without the colony it is part of. The colony is constituted of much more than the sum of all its bees. The different individuals which constitute it depend, for their characteristics, on the socio-physiological qualities of the whole colony. The individual is entirely at the service of the colony, which is itself at the service of the species' reproduction. Being a complex biological system *par excellence*, a bee colony builds its collective behavior by the permanent assembly of the elements of each of its components, giving it a capacity to adapt to determining aspects of the environment by their flexibility (in the short term) and their ability to adapt (on the long-term) (Tautz 2008, 248; Tautz 2009).

What particularly characterizes the complex adaptive system (see below) is the interrelationship between the diversity of actors "who simultaneously and continuously act and react to each other's actions" (ibid.). Decision-making is not centralized on the queen's pheromones, as one might think: the coordinated (collective) behavior emanates from a collaboration and competition between all bees. It can be understood as a kind of "spirit of the beehive" (Maeterlinck 1901; Tavoillot and Tavoillot 2015), resulting from social evolution (Page and Hölldobler 2013) and defined by a form of "democracy" (Seeley 2010). The notion of superorganism given/attributed to the colony of bees thus consists in "its capacity to self-organize and evolve" (ibid.). In 1925 Eugène Marais proposed, for the first time, that communities of social insects, such as termites, ants or bees, form "meta-organism" composed of specialized organisms (queen, workers, soldiers, and so on) (Marais 1973). The functional organization that constitutes the bee's superorganism is defined by Robin F. A. Moritz and Edward E. Southwick as "a superorganismic unit with organisms arranged in at least two non-uniform types and differentiated into sterile and reproductive organisms with different functions" (Moritz Robin and Southwick 1992, 4).

The functioning of the whole honeybee organism is in constant "dynamic equilibrium" that depends upon regulatory processes. For example, the brood nest temperature results from social homeostasis perceived in its architectural details (such as the percentage of empty cells and their distribution) that condition the temperature and its capacity to maintain this without large variations. The notion of homeostasis is understood here as auto-regulation against endogenous and exogenous disturbances, a dynamic rather than static stability.[11] The capacity of

[11]Here, we retain Walter Cannon's definition of homeostasis which dates back to 1915: "the set of organic processes which act to maintain the stationary state of the organism, in its morphology and in its internal conditions, despite external disturbances" in Durand 1994, 21. See also Rosnay (de), 1975. And finally "for a return to stability, the adaptive response must correspond as exactly as possible to the disturbance. In a simple system, a specific detector is connected directly to a response. In a complex system, an "image" of the disturbance is constructed, compared to a reference corpus, which leads to a secondary choice of a response. (...) At the base of homeostasis, there is regulation, but a regulation operates only in a very limited

auto-regulation[12] allows the system to maintain itself: the challenge is to maintain the cohesion between the structure, formed by the set of physical components of the organism, and the organization resulting from the interrelations among the components of the system (Penelaud 2010), i.e. the colony. It is therefore autopoietic: "organized as a network of production processes of the components that (a) continuously regenerate, through their transformation and interactions, the network which produced them, and that (b) constitute the system as a physical unit in the space where it exists, specifying the topological domain where it occurs as a network" (Varela et al. 1974, 188).[13] However, the autonomy of the bee colony does not denote a form of "self-sufficiency." The species' imperative to reproduce engages the colony in a regular traumatic process of "reformatting": abandoning the hive allows the old queen (3 years old or more) with part of her members to reestablish another colony. Before the departure, the hiving-off process generates a change (new queen) through the preparation of royal cells which will provide a young queen to the orphaned colony. However, the future of the swarm that leaves its hive rests more or less on the artificial system that surrounds it. Factually, while the superorganism establishes itself, it specifies the environment on which it depends: "they simultaneously co-occur, linked to one another by their structural coupling" (Penelaud 2010) because they belong to each other. A model of coviability emerges from this, which is not always completed (sic).

The process producing the system provides it with an organizational autonomy through auto-regulation and laws that are specific to its functioning and reproduction. For the bees reunited in a colony, we have emphasized that the whole exceeds the sum of the parts (this defines the complexity of the system itself); and conversely the whole determines the behavior of each bee. This reciprocal interrelationship gives rise to an interweaving of feedbacks (in loops[14]) and emphasizes that social homeostasis is defined by attaining a stability or a balance which can only be realized by common actions (the construction of the honeycomb, ventilation,

area. Therefore, multiple regulation operations are necessary, but they must be mutually coupled between themselves"(Tabary 1993).

[12]Homeostasis is intrinsic to the system: "Changes in the environment trigger reactions in the system or affect it directly, resulting in internal system disturbances. Such disturbances are normally kept within narrow limits, because automatic adjustments within the system take effect and, in this way, ample oscillations are avoided, the internal conditions being kept more or less constant" (Aubin 2010, 98 quoting Walter Cannon).

[13]Stability or homeostasis is the result of a retroactive process that is situated in the relationships between the elements that produce and delimit them: "the idea of autopoiesis is based on the idea of homeostasis and develops it in two important directions: by transforming all the references of homeostasis into internal references to the system itself, and conversely by asserting that the identity of the system, which we understand as a physical unity, stems from the interdependence of processes. These systems produce their identity. They distinguish themselves from their environment: that is why we call them autopoietic, the Greek autos (oneself) and poiein (to produce)" (Varela 1989, 45).

[14]As, for instance, in temperature regulation which influences the characteristics of the future bee, or the feedback loops related to nectar harvesting and honey consumption, (cf. Seeley 2010, 2014).

nest hygiene, etc.) of all the members of the colony (Tautz 2009, 249). The specific "laws" of the system (Aubin 2010) resulting from its auto-regulation apply themselves autonomously but not without relationships with the environment.

Open to the world around it, the honeybee is dependent on it particularly for finding food (pollen, nectar, honeydew, propolis) and for its water needs. The factors that are exogenous to the colony have a direct impact on its resistance capacities, its homeostatic rebalancing, its adaptation, and even its resilience. On the one hand, parasitic disturbances affect bees as they are vectors of pathogenic organisms (fungi, unicellular organisms, bacteria, viruses)[15] that participate in the development of infections that manifest themselves as diseases.[16] In the colony collapse disorder (CCD)[17] the Israeli acute paralysis virus (IAPV) has been strongly correlated with this syndrome (Chen et al. 2014; Prisco et al. 2011; Ribière et al. 2008; Cox-Foster et al. 2007). A combined sum of pathogenic, chemical and apicultural factors would be the most probable explanation according to the scientific community, which generally refuses to point at one single origin.[18]

On the other hand, disturbances come from environmental contexts that undergo biodiversity degradation and direct deterioration of biocenoses and biotopes: through the destruction of environments, their artificialization, through an agro-ecosystem developing single crop farming, through (unintentionally) importing exotic species,[19] and through exposing bees to toxic chemical molecules, such as systemic pesticides[20] (developed in 1991). The most common systemic pesticides,

[15]principally: Varroa ectoparasitosis (*varroa destructor*), wax moth (*Achroia grisella* and *Galleria mellonella*), the small hive beetle (SHB *Aethina tumida*), and the parasitic fly (Tachinidae and Phoridae).

[16]These are *Noséma apis Zander*, the CBPV virus (Chronic bee paralysis virus, the black illness), chalk brood (calcified brood or ascophaerosis or mycosis), chilled brood, American foulbrood (*Paenibacillus Larvae*), European foulbrood (several Bacteria: *Streptococcus pluton, Bacillus alvei, Streptococcus faecalis*...), saccade brood, acariosis (*Acarapis woodi Rennie*).

[17]Colony collapse disorder (CCD) results in the unexplainable loss of workers noticed since 1998 in Europe and 2006 in the USA with a loss of 50–90% of hives across the United States (Johnson 2007).

[18]"the worldwide losses of honey bee colonies continue to puzzle researchers and the beekeeping industry" (Ratnieks Francis and Carreck 2010); "what is clear is that researchers must look beyond simple one-factor causes of bee decline and losses. This research is multifactoral and more challenging" (CCD Steering Committee 2012, 7); "(...) a combination of existing stresses that may compromise the immune system of bees and disrupt their social system, making colonies more susceptible to disease and collapse" (Johnson 2010).

[19]Such as the Asiatic Hornet (*Vespa velutina*), introduced "accidentally" in Europe in 2004, from China.

[20]Systemic pesticides are absorbed by the plant by penetrating into its tissues contrarily to other pesticides that remain on the treated surface of the foliage. They are carried by the sap: the molecules of the product release their active substances as the plant grows to protect it from pests; they are also found in the nectar and pollen of the flower causing the death of foragers. These systemic insecticides have become the most widely used group of insecticides worldwide, with a market share now estimated at around 40% of the world market (The task force of Systemics pesticides: http://www.tfsp.info). Their long-term persistence (several years) and low-dose toxicity

neonicotinoids (Imidacloprid, Clothianidin and Thiacloprid) and the Fipronil which include nicotine component, have neurotoxic effects[21]: "Neonicotinoids and fipronil are found in nectar and pollen of treated crops such as maize, oilseed rape and sunflower and also in flowers of wild plants growing on farmland. They have also been detected at much higher concentrations in guttation drops exuded by many crops. In bees, consumption of such contaminated food leads to impaired learning and navigation, raised mortality, increased susceptibility to disease via impaired immune system function and reduced fecundity, and in bumblebees, there is clear evidence for colony-level effects. Studies of other pollinators are lacking. Bees in farmland are simultaneously exposed to some dozens of different agrochemicals, and some act synergistically. The impact of chronic exposure of non-target insects to these chemical cocktails is not addressed by regulatory tests and is very poorly understood" (Lexmond et al. 2015, 3). Large-scale application of neonicotinoids, their persistence in soil and water and their potential for absorption by cultivated and wild plants expose pollinators to moribund concentrations throughout the year. Bee colonies are particularly concerned by the spread of neonicotinoids, which are highly neurotoxic to honeybees and wild pollinators[22] (Van der Sluijs et al. 2013, 299).

Thus, the strong interdependence between human society, its modes of agricultural practices, of urbanization, of management and of the "sanitation" of its environment, and the nonhuman "bee" system, underlines the reality of viability between systems. This can be observed in the way the bee manages its internal stability, which is suffers attacks of various sorts, namely mechanical, chemical and biological, which combine to threaten the homeostasis or even the resilience of the colony. A recent study (Perry et al. 2015) reveals that the plurality of combined stress factors affects the social organization of the colony in its labor distribution.

It is now necessary to highlight some elements on the life cycle of these hymenoptera. Shortly after its "birth", a bee takes on the function of cell cleaner; after 5–6 days, the bee becomes a nurse taking care of the brood; from 5 and 20 days-old, the bee participates in constructing the honeycomb. The average age of the ventilating bees is 18 days-old (but they all participate in it), the guardians are 12–25 days-old, and around the age of 3 weeks the worker becomes a forager, the last

(Bonmatin et al. 2015) expose non-targeted organisms such as invertebrates (Pisa et al. 2015) and have serious environmental consequences: "a growing body of evidence that persistent, low concentrations of these insecticides pose serious risks of undesirable environmental impacts" (Simon-Delso et al. 2015).

[21] "we conclude that non-lethal doses of the three neonicotinoids tested either block the retrieval of exploratory navigation memory or alter this form of navigation memory" (Fischer et al. 2014).

[22] The authors specify: "At realistic field exposure levels, neonicotinoids produce a wide range of adverse moribund effects in honeybee colonies and bumblebee colonies, affecting colony performance through impairment of foraging success, brood and larval development, memory and learning, damage to the central nervous system, susceptibility to diseases, hive hygiene etc. Neonicotinoids synergistically reinforce infectious agents such as *N. Ceranae* and exhibit synergistic toxicity with other agrochemicals" (Van der Sluijs et al. 2013, 300).

function of its life that lasts only for a few days (4–5 days). The life span of a spring and summer bee is 30–45 days, and is much longer in winter when bees stay inside (Frisch 2011, 67; Clément and Le Conte 2011). The change of function occurs as physiological maturation occurs, but an old bee may undertake a previous task in response to the needs of the colony. The colony's needs condition the foraging and nectar gathering in order to feed the colony in nectar, pollen, propolis and bringing water.

However, the stress generated by the colony's environment (pathogens, pesticides, environmental degradations) disturbs this internal physiological order in that the youngest bees are forced to become foragers too early to overcome the insufficiency of adult workers, thereby causing their death.[23]

The importance of the honeybee pollination service for human societies is speaks for itself: 60% of world production comes from crops that do not depend on animal pollination, but 35% of crops depend on pollinators, 80% of which are honeybees, and 5% are not evaluated (Klein et al. 2007). On the one hand, the honeybee depends on the state of the environment and agrosystems; and on the other, the vitality of apicultural populations mirror this state. The health of a bee colony provides a valuable indication of ecosystems' health. The bee is thus considered a true "sentinel" (Clément 2009) of the viability of the socio-ecological system. In fact, it is a bio-indicator of the coviability of social and ecological systems, as the relationship of the colony to human activities is strong.

The third example of a socio-ecological viability model is economic in nature: the linear system "extract-produce-consume-discard" is no longer viable if one considers population increase, access to resources, their rarity, and the ecological status of the planet.

1.1.1.3 The Circular Economy, a Socio-ecological Alternative for the Planet?

"Rethink the future" emanates from a concern of governments and manufacturers about the availability of resources for the future. This justifies a rapid turnaround toward a new industrial model that depends less on energy inputs and raw materials (World Economic Forum 2014). The transition, enacted in different places in the world (Gunther 2014), to a virtuous circle eliminating the very idea of waste is based on the infinite reuse of materials and objects over two cycles (that of biological nutrients and that of technical materials).[24] This socio-ecological

---

[23]"we do not know why colonies transition so rapidly from a state of apparent health to failure. It is well known that individual bees react to nutritional and pathogen stresses by foraging precociously: our study explains how colony failure arises from the social responses of individual bees to stress. We used radio tracking to monitor performance of bees and found that workers who begin foraging prematurely perform very poorly" (Perry et al. 2015).

[24]"The circular economy is a generic term for an industrial economy that is, by design or intention, restorative and in which material flows are of two types, biological nutrients, designed to reenter

model of circularity defines in essence coviability. It severely disrupts production and consumption patterns that impact ecosystems so much, by being based on restitution, or regeneration by eliminating the very idea of waste: "any object that has reached the end of its cycle should be converted into something else in the same way that, in the natural world, the waste produced by one species feeds another species" (Gunther 2014). Evidently, the paradigm of the circular economy is not innocent; it does not call into question the consumerist capitalist system, but it proposes a convincing pattern in the face of the price explosion of raw materials, their extraction costs, and that of food products (Heck and Rogers 2014), the savings to be made (Ellen MacArthur Foundation 2012), and those related to damages due to pollution in the environment and to public health. In the end, both the consumer society and the environment win (Li Gan_shun 2002).

China has recently adopted in its new economic and social planning the implementation of circular economy at the level of companies, industrial parks, cities and regions (Zhijun and Nailing 2007). The European Commission is launching a transition program towards a "zero waste" circular economy by promoting a constant re-use of products (European Commission 2014). In Europe, improvements in resource productivity along value chains[25] could reduce input requirements by 17–24% by 2030 (Meyer et al. 2011). The main stages of a circular economy model, which make waste a resource, offer prospects for sustainable growth, including its potential to create poles of innovation and to create jobs: raw materials, design, production/re-manufacturing, distribution, consumption/use/reuse/repair, collection, recycling/residual waste, raw materials . . .

1.1.2 Questioning the Societies-Nature Interrelationships

Placed end-to-end, the models of socio-ecological viability, pastoralism, bee colonies, the circular economy, demonstrate that the interrelationships between systems go beyond the myth of compartmentalization through their lived experiences, their instability and their capacity to adapt. However, the threshold of resilience (see below), beyond which there is collapse or extinction, is not necessarily far off, as demonstrated by the bee model. The ability for homeostasis and to manage resilience ensures the viability of the system through human actions: "because human actions dominate social-ecological systems, the adaptability of such systems is mainly a function of the individuals and groups managing them. Their actions influence resilience, either intentionally or unintentionally

the biosphere safely, and technical nutrients, which are designed to circulate at high quality without entering the biosphere". Available online: http://www.ellenmacarthurfoundation.org

[25] Value is the amount that customers are willing to pay to obtain the product or service. The value chain allows a company to build its competitive advantage (see Chaptal de Chanteloup C., 2015, "La chaîne de valeur de l'offre", De Boeck).

(Berkes et al. 2003). Their capacity to manage resilience with intent determines whether they can successfully avoid crossing into an undesirable system regime or succeed in crossing into a desirable one" (Walker et al. 2006). The system becomes "undesirable" as soon as one-steps out of the area of stability or the possible spatio-temporal field where resilience is possible, due to insufficiently regulated or totally unregulated behaviors. The management practiced by individuals or groups stems from a regulation, a framing and an orientation of behaviors, decision-making, and operating modes.

The viability of the system may impose radical changes, changing paradigms, such as the case of the circular economy. Profit as an objective is always there but on a rethought basis of a product's perpetual cycle instead of finiteness. Pastoral activity is reinvented in its relation to land whereas the system of land ownership reduces it by disposal rights (Barrière 2015). The era of transformation highlighted by Holling (2001), cited above, refers to the viability of systems that are by definition interdependent, because with globalization, societies encounter the limits of the environments and that of growth based on unlimited resources.[26] Viability becomes an issue that involves the reinvention of a form of existence, growth (or even "decline" for some, Hartus and Virard 2015; Meadows et al. 2002; Passet 1992). If being viable means existing, the nature/society coexistence is in fact only the existence of a joint intertwined viability, one defining socio-ecosystems and the conditions of their viability.

The myth that creates a disconnection between culture and nature is therefore an unsafe paradigm, as the ecological imperatives remind us in national and international public policies (for instance through the Millennium Development Assessment[27] that underlies international conventions or agreements on climate change). This paradigm of separation is disputed in the stated ecological crisis. The reconnection of societies to the biosphere reflects the era of a form of "reconciliation" between man and his own nature, between societies and their ecosystems. The challenge lying here may be to transcend the paradigm of capital, which is centered on *Homo economicus* (Cohen 2012), in order to reason more in terms of interrelationships, and common heritage rather than in terms of a quantified evaluation and the services rendered, effectively disconnecting the biosphere from societies and enhancing the man/nature dichotomy by the artefact of the environment (which we have already emphasized in the general introduction). The disconnection emanates from the broken links between man and his nature. The viability of socio-ecological systems depends on the links that define humans in the biosphere. Would the harmony therefore be reduced to simple problem of management and planning? (Hong et al. 2007, 3), or would it be situated in

---

[26]The idea of an endless, sustainable growth is not sustainable (sic). According to René Passet "a development based on a simple economic logic is self-destructive by destroying the environment that carries it. This situation requires a co-operation of economic development and the biosphere" (Passet 1992, 397).

[27]Goals n°7 for 2015: ensure environmental sustainability (United Nations 2015).

interdependent links to be reconsidered? These links of socio-ecological solidarity (Mathevet et al. 2010; Mathevet and Bousquet 2014) define the question of survival, not that of the planet facing the human species development, but rather the survival of the human species depending on its capacity to maintain a sustainable future from the entire biosphere..

This problematic was clearly raised almost 20 years ago by our Canadian and Swedish colleagues: "when we wish to emphasize the integrated concept of humans-in-nature, we use the terms *social-ecological system* and *social-ecological linkages*" (Berkes and Folke 1998). They were concerned about the manner in which the local social system had developed management practices based on ecological knowledge to deal with the dynamics of ecosystems.

In this book, we intend to investigate these links in terms of connected viabilities, and even more, since they are combined. If management practices and the social and ecological mechanisms used to reinforce resilience are at the heart of the issue of coviability, we intend to examine further this "paradigm" concept, both empirically and theoretically and with much pragmatism. This book is an attempt to do so by presenting a diversity of in-the-field cases and themes.

However, this deeper examination inevitably refers to the concept of sustainable development, which was conceived as a realistic utopia "searching for another model of development" (Jollivet 2001; Ducroux 2003). The emergence of the coviability paradigm revisits the very idea of sustainable development.

1.1.3 Revising the Concept of Sustainable Development Against the Concept of Coviability

The processes of resilience and adaptation participate in scenarios, which are shaped within this new man/nature relationship, in which the concept of sustainable development appears to be central in outlining new categories and approaches in this relationship.

When preparing for the second major global environmental conference (Rio convention of 1992), a commission headed by the Swedish Minister Gro Harlem drafted a report entitled "Our Common Future," which was at the origin of the expression "sustainable development". Further introduction of the expression into the diplomatic sphere led to some polysemy, because of the need to create the widest possible consensus (Porto-Gonçalves 2006). The environmental thematic is rapidly imposing itself in science and the international academic debates are gaining magnitude on the matter, especially from an epistemological perspective (Leff 2010). In this context, the work of Nicolau Georgescu-Roegen (1971), Rachel Carlson's *Silent Spring* (1969) and Garret Hardin's "Tragedy of the Communes" (1968) are important references, as they express the concern by scientists about the environmental and the limitations observed in the models of economic growth.

The foundations for the concept of sustainable development were planted. This is based initially on the classical economy, for which growth is synonymous to development. Despite efforts by economists such as Leff (2009a, b, 2010), Cavalcanti (2004, 2010, 2012) and Martinez-Alier (2007) to provide some theoretical maturing of the concept, the critics of this concept focus on the indiscriminate use of the term. Cavalcanti (2004) insists that the various meanings of the term stem from the concept of development itself that has been extensively debated within the human sciences.

The concept of sustainable development reflects the manner by which the model of Western civilization thinks of the relationship between man and nature. For environmental economics, the prospect of optimizing the use of natural resources at low environmental impact levels could be ensured by technological advances, which would allow the expansion of the productive process over time. However, in this perspective, nature and its processes are not considered as limiting because of the prevailing but arguable perception that uninterrupted growth is sustainable thanks to technological advances.

To overcome these limitations, ecological economics suggests reconsidering the prospect of development as growth, taking into account the limits of nature. It is important to emphasize that, unlike environmental economy, such a point of view refers to thermo-dynamic propositions (Georgescu-Roegen 1971; Leff 2009b, 2010, 2014; Cavalcanti 2004, 2010, 2012; Daly 2002).

Sustainable development is thus artificially maintained and as a result places itself at a dead end. Daly (1990, as quoted by Cavalcanti 2012) distinguishes growth, i.e. a quantitative increase on a physical scale, from development, i.e. a quantitative improvement of potentialities. For Cavalcanti (2012), development means change, evolution and non-growth, understood as increase or expansion. Thus, development may include growth, but it may go beyond that perception.

In capitalist societies, especially in the context of globalization, and despite the theoretical efforts undertaken, development continues to be amalgamated with growth and is understood as the material improvement of living conditions. Furthermore, Santos (2006) accuses globalization of perversity since the production of goods and services is not translated into an extensive/equal increase in social welfare. Conversely, the NGO Oxfam published a study in January 2016 indicating that the share of world heritage held by the richest 1% of the world rose from 44% to 50% between in 2009 and 2016 and would exceed 50% in 2016 (OXFAM 2015). This concentration of wealth achieved at the expense of an ever-increasing number of poor and marginalized people is confirmed by annual reports on wealth by Credit Suisse (Credit Suisse 2015). The confusion between development and growth is also perverse when it comes to sustainability.

Sustainable development has certainly the merit of considering the factor of time by supporting the scheduling of human activities so that they avoid harming future generations. However, excessively focusing on diachronic issues (and the concerns for future generations) may result in insufficient (or even absent) consideration of synchronic problems. It is essential to analyze an action on several geographical and above all ecological scales (Daly 2002) and not only historical. The concept of sustainable development has been based on undefined beliefs on the exploitation of natural resources despite the limitations of ecological processes, thus turning the concept into an empty discourse.

Moreover, the options for action in favor of "economic development" differ a lot depending on the objectives of the social or economic agents or the public authorities. In this perspective, Martinez Alier (2007) distinguishes several environmental movements: (a) the "myth of the wild", (b) sustainable development, and (c) ecologism of the poor. The first movement, the "myth of the wild", intends to create protected areas, integral preservation, which can be justified locally but which still does not contradict the current mode of production (and consumption). In addition, it can coincide with capitalist interests to preserve natural resources as natural capital (Becker 2007), just the time to know how to exploit them and transform them into resources. The second movement is sustainable development which Martinez Alier calls "eco-efficiency" since it is based on a techno-centric perspective to solve problems. From this angle, time appears as a solution. Yet, Rodrigues (1998) states that, on the contrary, with the evolution of the world, time becomes an accumulation of problems due to aggravating environmental impacts. Evolution brings along pressure or even scarcity of resources considered renewable (such as air in the Chinese cities (Rodrigues 1998)). It is therefore recommended to rethink the structures of society, which sustainable development does not question in its discourse. Enrique Leff (2009b, 2010, 2014) criticizes this perception since it is based on mechanical rationality and short-term valorization. He rather defends a social re-appropriation of nature, which converges with the third environmentalist movement quoted by Martinez Alier (2007) as the ecologism of the poor.

The current era thus tends to agree on a consensus in favor of ecology. However, decision-making for large-scale initiatives suffers from conflicting governmental forces whose concern is displaying growth figures to their electorate. Meawhile there are countless examples of measures that may seem to be favorable but which in reality are based on technological strategies that create further problems. How can we explain this paradox? Through, for instance, processes that do not consider the unequal distribution of benefits and costs on the members of society, which generate environmental conflicts (Acselrad 2009). It is also necessary to observe the extension of the commodification of nature instead of promoting integrated policies to create a balance with exploited ecosystems. Thus, the measures promoted by the various Earth Summits, since 1972, turn out to be opportunistic measures for the market. What about the "4 for 1000" initiative (http://4p1000.org), which may be diverted from its initial goals from the moment the search for funds for action is solicited, which, according to the capitalist model, would risk taking the reins of decision-making and reap the benefits from such action (going as far as the revisited risk of land grabbing)?

The Clean Development Mechanism (CDM[28]) is an example of a measure suffering from many contradictions, with only three states respecting their promises.[29] According to Porto-Gonçalves (2006, 340), the "general orientation of the measures and policies hitherto adopted... is to diminish the political responsibilities of a universal nature, which the State has historically been the guardian of, and to resort to voluntary, concentrated and market-based solutions". By creating a global market, the CDM was created to enable companies in the rich countries to invest in the Third World in order to avoid reporting greenhouse gas emissions and to save credits refered under the Kyoto protocol (Porto-Gonçalves 2006). Hence, we could defend the need to replace purely economic rationality with a more eco-technological one based on the principle of negentropic productivity (Leff 2009a, b, c, d; Cavalcanti 2012).

Even if examples of opportunistic systems for environmental compensation are getting more and more numerous, it is though important to think about the theoretical advances on the thematics. After the Second World War, data on economic growth showed that material advances raised human well-being and happiness to levels slightly above basic needs first expected with respect to education, health, food and housing. Once these necessities were satisfied, the curve of human happiness did not follow the trend that the economic growth indicated (Daly 2002). The latter was criticized, especially when economic activity is thought of as a path toward happiness, towards an "enjoyment of life", as the ultimate end of the economic process (Cavalcanti 2012). The notion of prosperity then started to emerge and highlight the forgotten perspectives of human life, perspectives relating to productive processes distinct from the prospect of growth or economic development capable of being measured. Prosperity is understood to be less connected with a perspective of accumulation of valuable wealth and more related to the consideration of what is intangible. This defines the human condition: being well, feeling good, sharing experiences (Jackson 2013). In fact, prosperity is possible only when it is intertwined with the prosperity of the other (ibidem 8–9).

Thus, prosperity, living well and happiness appear as the ultimate/goal/result of the economic process. The issue of human happiness, in fact, occupies a central position with respect to the limits of nature on productive processes. Within this context the experience of Bhutan, in addition to the experiences of the South American peoples, especially Ecuador (Gudynas and Acosta 2011a, b; Vallejo et al. 2011), stand out under the auspices of "living well" (Cavalcanti 2015). In Bhutan, the development of the country and the decisions in relation to national interests rely on the search for human happiness, while respecting the limits of materials and energy available for productive processes (Cavalcanti 2015). In the case of Ecuador, social and indigenous movements refuse the oil production in Yasuní National Park

[28]The clean development mechanism was established in the article 12 of the Kyoto Protocol to help developing countries achieve their greenhouse gas reduction targets in relation to the 1990 level between 2008 and 2012.

[29]Norvège, Suisse et Luxembourg. Cf. Porto-Gonçalves 2006, p. 338.

and have adopted a position against any economic compensation in exchange from developed countries in order to keep the oil underground. This refusal is founded on the argument that nature is a subject of rights (Gudynas and Acosta 2011a). In this context, "living well" does not mean accumulating material goods, but taking what is necessary from nature in order to live while keeping a harmonious relationship with it (Gudynas and Acosta 2011a). This opposes the only rationale of Western life, which strongly imposes itself on a global scale.

In the name of development based on a relationship between science, technology, industry and profits, the neo-liberal system goes all the way to find natural resources that allow its maintenance by increasing production levels and increasingly supporting economic growth. Proposals to reject pressure on natural resources, as is the case in Ecuador, are new ideas. It is noteworthy that this state of mind comes peoples of the Southern hemisphere, with different alternatives than those of neo-liberalism. Those marginalized by history, indigenous peoples and local (native) communities, raise a challenge which proposes "living well" and, with it, an opportunity for human beings to cohabit, in their diversity of cultural values, with the nature. "Living well" could thus be seen as a path to coviability.

1.2 The Axes of a Transdisciplinarity: Viability and Regulation

Two principal axes structure the argument presented in this volume: viability and regulation. These two keywords assist in/take part to defining a paradigm that mathematics have already worked on. In the meanwhile, we approach coviability via an interdisciplinary, even transdisciplinary exploration.

In this editorial work, many disciplines were invited to fuel the concept of coviability, each with its own approach to the issue and without taboo, dirigism or dogmatism.[30] The challenge consists of transcending environmental and social determinism while at the same time developing in the study of socio-ecological systems a transversal approach that is transdisciplinary by overcoming disciplinary boundaries ("interdisciplinary is becoming the buzzword in science", Whitfield 2008). The goal is to avoid the traps of the disciplinary reductionists (Horlick-Jones and Sime 2004)[31] and their challenged, fragmented knowledge on a complex subject that by definition calls for transdisciplinarity (Hirsch et al. 2006). Transdisciplinarity deals with research and organizational problems that belong to complex and hetero-

[30] As Jean-Pierre Aubin points out in this chapter (below: Mathematical viability and coviability; and in the Chap. 3), "no one is the owner of the terms of viability and co-viability, or of their synonyms, which became polysemic. We therefore add the adjective "mathematical" to clarify the meaning given to this term in mathematics.

[31] "(...) much of what is interesting and important about the character of risk tends to be lost by the generalising, decontextualising and reductionist tendencies of discipline-based research".

geneous domains: "beyond complexity and heterogeneity, this mode of knowledge production is also characterized by its hybrid nature, non-linearity and reflexivity, transcending any academic disciplinary structure" (Lawrence and Després 2004).

The guiding thread of the analysis is structured around interactions and interdependencies (1.2.1) and on what maintains any system, its rules, norms and "functioning laws" (1.2.2).

1.2.1 The Challenge of Viability: Interactions and Interdependences

Viability means maintaining the evolution governed by evolutionary system within a range of constraints. For example, in a fishery, a major constraint is the threshold of removing fish from the stock.

Mathematics pioneered the concept of coviability in 1996 at a seminar organized in France under the title "Ecosystems and Viable Development", organized by Marie-Hélène Durand[32] (the Chap. 3 in the first part of this book details this emergence). The mathematical definitions are presented first, followed by those by other authors.

1.2.1.1 Mathematical Viability and Coviability

From Vocabulary to the Problem

The Mathematical Vocabulary of Coviability

Nobody owns the terms viability and coviability or their synonyms. Here they will have to be rigorously defined in the context of mathematics.

The "states" of a "system" form a set, usually a "vectorial space" (once its components have a unit of measurement). A "(vectorial) hyperspace" is formed by subsets (called "environments") of a vectorial space.

An "evolution" is a function, which at any moment in time associates a state of the system, in the case of a vectorial space or an environment, in the case of a hyperspace. The evolution of a vector naturally leads to the joint evolution of its components, but it is not a matter of coevolution in the mathematical sense.

By definition, a "coevolution" is a "joint evolution of a state and an environment". It is "viable" if, at any moment in time, the state belongs to the environment. When the environment is constant, the evolution is considered "viable", to explain that the interactions between state and environmental evolutions as compatible. Coviable evolutions are "sustainable" if the duration of the temporal window is infinite.

[32]Of the French Research Institute for Development, IRD-UMR GRED.

Sustainability never concerns living systems of which all states are born and die within a finite time even if some of them extend themselves through their lineage. We assume therefore that durations are finite and strictly positive, "ephemeral", the windows of zero duration being the "moments". More importantly, the duration of an evolution (its age, as it were) often has to be calculated and is part of the viability problem. The evolutions are "known" for the retrospective temporal windows (of past knowable moments in time) and can only be "foreseen" when the temporal windows are prospective (of unknown future moments in time).

Before 1974, there were no words to mathematically translate the title of Jacques Monod's *Chance and Necessity*. "Uncertainty" and "viability" imposed themselves to reduce polysemy and were chosen arbitrarily in the contemporary everyday language. There were no mathematical techniques capable of ensuring the coviability of a joint development of a state and an environment. The (mathematical) theory of viability refers to the set of mathematical concepts and implications between them. "Theory" is considered thus as a chain of proofs, or *enchainures de preuves* to use the beautiful expression of Savinien Cyrano de Bergerac, and not an explanatory or dogmatic corpus, which denotes the other meaning of theory. Although these techniques were motivated by economic, social, biological and cognitive sciences, it is up to the specialists in these disciplines to validate the "mathematical metaphors" they provide.

How are the evolutions of states and environments produced? We have to define a class of "evolutionary motors" that associate to each initial state and to each initial environment a set of coevolutions that start from them. This set may be empty which is an information in itself, even if it is sad, may contain only one evolution ("deterministic") or a set made of more than one evolutions ("uncertainty"). It seems more serious to translate "motor of evolutions" as "the evolutionary (morphological) system" although the image is less striking.

The probabilistic and stochastic translation of uncertainty calls for "means of viable evolutions" that have no reason to be governed by the motor of evolution, which strips it from any relevance. However, uncertainty can be divided into several types: "contingent," when a property is satisfied by at least one coevolution, "tychastic," when it is valid by all of the possible evolutions, "impulsive," when the motor of evolution is accompanied by an impulsive system which renders a state viable when it is no longer about to be so. The list of examples of uncertainty is far from exhaustive.

The Mathematical Problem of Coviability

- Frustration: There is no reason why an arbitrary evolutionary system should produce coviable evolutions of a state and an environment;
- What can be done about this? Take one's mathematical knife to carve

  1. The largest subset of environments and their states from which begins at least one coviable evolution (called "viability core" (long-lasting or ephemeral);

2. The "regulator" set of all viable feedbacks) that provides the largest subset of "*regulons*" providing the velocities of states and environments that drive coviable evolutions.

The mathematical construction of regulators, deduced from the "theorems of viability", is much more complex than that of viability cores, which describe only the first step for the solution to viability's problems. There is no need to know the hypotheses or the demonstrations to build these regulators. Hypotheses are the price to be paid so that the viability's property is satisfied.

The real task consists of validating these hypotheses in each discipline so that it reaches a consensus, and therefore provide meaning for those that adopt it to find mathematical metaphors. For example, the regulator may provide a mathematical metaphor for "legal rules": they impose constraints on the social behavior of actors, which act retroactively on them by making them evolve in order to ensure the viability of the societies they constitute.

Algorithms exist to approach viability cores and, above all, regulators, but the types of software that use them require only small numbers of variables to survive Bellman's "curse of dimensionality". It is therefore necessary to beware of the dangers of quantitative techniques that sacrifice an infinite number of variables to keep only a few variables: if an assertion is false for a sacrificed system, it is also false for the initial system but the validity of one conclusion for a small number of variables is not a good sign.

It does not really matter because qualitative mathematics provides above all stories to tell (see Aubin 2010, in particular section 12.2, page 802, concerning coevolution and morphogenesis).

Thus, mathematics teaches us that the viability of a system depends on its ability to maintain itself through its self-organizing processes, within varied and changing environments. This implies a capacity to adapt to a variety of exogenous situations: the relationship to the environment cannot be avoided. Since the latter is populated by other autonomous systems, "viability is at the same time a coviability which depends on the interactions with these other systems": which refers to coevolution mechanisms. Thus, "living beings are mutually adequate and viable" (Bourgine 1996).

Viability, which can be defined differently by authors,[33] brings together all the conditions to maintain the system over time. It is therefore to be linked to its sustainability. Moreover, in spatialized time, the constraints define a vital space called "viability core" which is "the subset of the constraints' domain within which it is possible to maintain the system throughout time" (Sabatier 2010, 19). This domain of constraints is "the subset of the states which respect the constraints at a moment in time" (ibid.).[34] Seeking a viability core with viable trajectories is

[33]Cf. For a formulation in terms of mathematical viability, see Sophie Martin 2004.

[34]Examples of viability constraints as quoted above by the author, inthe example of a socio-ecological viability model of/for grazing: food constraint related to productive grazing, food

equivalent means seeking the set of conditions to maintain the system over time (ibid.).

When managing natural resources, state and control constraints are generally defined to represent the conditions of a system's sustainability (Martinet and Doyen 2007). A system's viability may also reflect the capacity to reproduce with sufficient certainty.[35] However, uncertainties participate in the functioning of systems such as pastoralism on semi-arid rangelands, which gives rise to management strategies that depend on subjective factors of socio-cultural order. Viability thus reflects societal values and social standards (in addition to stochastic events such as an earthquake or volcanic eruption).[36]

In 1999, a fishery research program that focused upon modeling approaches in a marine environment laid down the definition of viability through coviability (Le Fur et al. 1999) by proposing new approaches to fishery management. These works succeed in containing the notion of viability within interdependent systems participating in the exploitation system of fisheries: "the viability of fishing activities may be defined as the ability to reproduce, at various time scales, the main elements that constitute them and the functionalities that structure them. These elements are the exploited resources, the ecosystems which host them, the economic elements (production units, structures of marketing and markets), the technical elements (technology, know-how), and the social elements (fishing communities, families, community and professional organizations)." The authors demonstrate that the conditions for viability are both endogenous and exogenous. These conditions reflect resilience through "the ability to exploit organisations and resource stocks in order to cope with changes in their environments (in a broad sense) as well as to produce changes likely to ensure their sustainability" (Le Fur et al. 1999).

1.2.1.2 A Society-Nature Dichotomy Made Illusory by the Co-evolution of Systems

The innate ability to acquire, which is elaborated during hominization, is at the origin of the human cultural mechanism, of human nature. Indeed, the complexity of

---

constraints related to ecological grazing, trampling constraint, constraint of grass height, constraints of usage proportion (ibid.).

[35]For Baumgärtner Stefan, Quaas Martin F., 2009, p. 2 008, the concept of viability has a different meaning: "viability means that the different components and functions of a dynamic, stochastic system at any time remain in a domain where the future existence of these components and functions is guaranteed with sufficiently high probability".

[36]"This emphasizes that viability is not a purely objective property of some system that could be determined on purely scientific grounds, as it appears to be in some ecological applications. Rather, viability is a normative criterion specified for a given ecological-economic system, reflecting societal norms and values" (Baumgärtner, Quaas 2007). Let's also quote the stochastic viability theory (Aubin 1991), which is one case of tychastic viabilityn and better describes uncertainties without any statistical pattens.

the human brain and socio-cultural complexity can only fit into one another (Morin 1973, 99). This integration has made Man "a cultural being by nature because he is a cultural being by culture" (ibid. 100). The nature of the human species is consequently defined in its existential paradigm, which refers to its relation to the nonhuman. We deduce from this the sterility the nature/culture ontological dichotomy; for the ecosystem[37] is not a setting but a "true actor" (Morin 1973, 213) of the coevolution of societies with their environment, in that the latter constitutes a habitat for the very existence of human beings. Interweaving social systems with ecological systems leads to the notion of socio-ecosystem (Berkes and Folke 1998; Walker et al. 2006; Folke et al. 2005; Young et al. 2006, ...)[38] marking the biosphere with an strong anthroposystemic[39] identity.

The man/nature interrelationship, or rather the place man occupies within the biosphere, depends thus on socio-cultural schemes defining a paradigmatic "relation to the world" (Berque 1996). The role of the ontological relationship between the human and the nonhuman is therefore a subjectivization specific to each system of thought, to each culture. Depending on the cultural position, the living system as a whole appears as either dual or non-dual. The man/nature dichotomy consequently results from a myth produced by a psycho-cultural bond uniting the human to the nonhuman through the relation to the invisible. As a consequence, thoughts and beliefs (including "religious foundations) produce diverse representations and "ordering" of the world.

Separating the social system from the ecological system would then only be an artificial anthropocentric construct[40] that extracts man from the biosphere. The socio-ecological system thus defines the biosphere, in which human and nonhuman components and their relationships are integrated.[41] The biosphere itself is embedded in the geophysical environment of the globe. To allow an analysis to be carried out, in a consciously ethnocentric manner, we use the societies/nature distinction in order to investigate the complexity of the combined super-system made of reciprocal interactions (see Liu et al. 2007).

This complexity is reinforced by the complexity of social systems, which, beyond the spatial and temporal dimensions, also integrates the dimension of socio-cognitive schemes and the resulting psycho-cultural representations

[37] Seen locally as an ecological niche and globally as what surrounds man.

[38] and legal socio-ecology: Barrière 2008.

[39] Defined as "an interactive system between two sets constituted by one or more natural and/or artificial ecosystem(s) and one or more sociosystem(s) within a given geographical area and evolving over time; this allows interactions between human societies and natural environments. (Lévêque et al. 2003).

[40] "This perspective sees the nature-culture split as arbitrary, a distinction that masks a continuum of lower to higher order processes, an artefact of the human brain, that is itself natural" (Westley et al. 2002, 104). According to the authors, this is possible because Man is able to construct abstract hierarchy (symbolic constructions), is capable of self-reflexivity and is able of some technological grasp onto the planet.

[41] "Most so-called "natural" systems can be called socio-ecological systems ... " (Décamps 2007).

(see Westley et al. 2002). Even if the biosphere has undergone major changes throughout Earth's history, it also has undergone other major changes during the last centuries and millennia, mainly because of man; the present socio-ecological system is an "anthropized version" of the biosphere. This system, presented as the "biosphere", is defined as an evolutionary and adaptive complex system made of subsystems which are dynamic multiple regimes systems interwoven in a larger system; each of them operates on a particular scale. The whole is characterized by discontinuities between sizes and by changes in scales. As a consequence, the resilience of ecological or social systems is more dependent on the functions of their own elements than on their identity, and also on their distribution within and through scales (Garmestani et al. 2009; Allen et al. 2005).

The evolutionary and adaptive aspects have to be specified. Complex systems enclose a large number of interacting, non-linear, local and global, endogenous and exogenous elements[42] that result from evolutionary and adaptative processes. The main features of a complex adaptive system are situated in its feedbacks (circular causality),[43] be it positive (amplifying) or negative (stabilizing) and in its linear or increasing, or unpredictable changes. The components of the system collectively modify their environment which in turn exerts constraints and modify their states or dynamics. The "richness" of a system results from its inherent capacity for change through its adaptive potential. This adaptive capacity depends on endogenous control and on the internal sensitivity of the system to external disturbances, i.e. its vulnerability to unexpected or unpredictable shocks (Holling 2001, 394).[44]

The biosphere evolves going as far as radically modifies the surrounding physicochemical conditions (such as the composition of the atmosphere), the species involved and their cooperative or competitive interactions: "these evolutions and adaptations are part of a historical process of the biosphere's evolution since they are often induced by changes in the environment and ecosystems, that are themselves caused by the evolution of the species that constitute them. A sort of evolutionary spiral appears here: each evolution calls for others, and the co-adaptations that gradually take place throughout this process correspond to a repeated increase in the complexity of organisms" (Lesne 2008).

Organisms influence each other, leading to a common evolutionary history; the interactions between organisms lead to permanent evolutionary races (Van Valen Lee 1973). By examining the modes of interaction between groups of organisms with a close and evident ecological relationship, such as plants and insects, Paul

[42]For Cumming and Collier (2005), "a complex system is a network of components connected by various dynamical relations that include inputs, outputs, and external constraints".

[43]Retroactive loops also called reflexive interactions: when a component interacts with itself either directly or indirectly through the chain of interactions with other components. This feedback induces a non-linearity of the system's behavior.

[44]According to Holling 2001: "Potential, or wealth, determines the degree to which a system can control its own destiny, as distinct from being caught by the whims of external variability. Resilience, as achieved by adaptive capacity, determines how vulnerable the system is to unexpected disturbances and surprises that can exceed or break that control".

Ehrlich and Peter Raven introduce the term coevolution as the evolution of two or more entities (Ehrlich and Raven 1964). The latter are the result of "evolution in two or more evolutionary entities brought about by reciprocal selective effects between the entities" (Ehrlich and Raven 1964).[45] Therefore, in order to survive, a species depends on the existence of another species (case of predation, parasitism, mutual aid, cooperation, and so on), and any modification of a species affects the species on which it interdependent.[46] Recent works show that coevolution leads to the appearance of complex traits and to promoting evolvability: "because coevolution is ubiquitous in nature, our results support a general model whereby antagonistic interactions and natural selection together favor both increased complexity and evolvability" (Zaman et al. 2014). Humanity's relationship to the biosphere is made up of new processes in the game of coevolution by the suppression of predators, by domestication, alteration of the lifestyles of many species etc., and by acting on ecosystems and the climate: "the consequences of human activity resulting from a transgenerational culture including medical, veterinary and technical knowledge (cloning, genetic engineering, ...) are (...) to increasingly remove people from natural selection processes," emphasizes Claude Combes (1995), who refers to the various factors that intervene in the host-parasite relationship and which humans expect to "control" by working on the various factors involved in the host-parasite relationship.

In this way, man creates a niche for himself, a man-made niche, the artificialization of which he pursues ever further. The adaptation conducted with the help of intelligence is continuing and might be transformed, through the use of some more or less thoroughly thought technologies, become a chimerical race to intentionally emancipate the human condition from its belonging to the biosphere. In the meanwhile, the complexity of societies and their interactions is constantly increasing and the niche gaps and, consequently, the discrepancies between the "things that are possible" (that an individual can individually claim with regard to/from the group, which he can integrate into his personal life project) are constantly expanding.

If coevolution leads to the notion of interrelation and interdependence, the viability of human societies depends on its own cultural diversity (Universal Declaration on Cultural Diversity, Unesco 2002) or ethno-diversity, partly dependent on biodiversity. The interdependence of cultural diversity and biological diversity is a

[45]For a more detailed definition: "The term coevolution is used to describe cases where two (or more) species reciprocally affect each other's evolution. So for example, an evolutionary change in the morphology of a plant, might affect the morphology of an herbivore that eats the plant, which in turn might affect the evolution of the plant, which might affect the evolution of the herbivore...and so on". (http://evolution.berkeley.edu/evolibrary/article/evo_33)

[46]The case of food networks, for example or "above ground communities" of living organisms that communicate and interact with "below ground communities" via the plant unit (and the communication of its airial and underground organs).

principle asserted by the United Nations (UNEP 2002).[47] It has already been given considerable legal force in the international Convention on Biological Diversity in 1992 (article 8j[48]) which insists on the relationship between indigenous or local populations with biodiversity and the need for an appropriate balance between the conservation and the sustainable use of biodiversity: "recognizing that many local communities and indigenous peoples intrinsically and traditionally depend on biological resources. Their traditions are founded on these resources, so it is desirable to ensure the equitable sharing of benefits arising from the use of knowledge, traditional innovations and practices relevant to the conservation of biological diversity and the sustainable use of its components..." (1992 Biodiversity Convention Preamble). The Nagoya Protocol (of 2010), implementing this convention, emphasizes the interdependence between genetic resources and traditional knowledge for the sustainability of these communities' means of subsistence.

As mentioned in the general introduction of this present book, this rhetoric should not entrap us in a "primitivistic" view which would lead to believing that the so-called traditional societies are by definition close and respectful of nature. Indeed, the argument "is based on analogies and cannot therefore claim the rank of an operational concept" (Kohler 2011), at least as it stands. "Native" cosmologies are not fundamentally preservationist because the very notion of nature does not exist in the ecological relationship (Descola 1985, 2005), and also because if a community organizes itself to "manage" a resource, it is not to preserve this or that element or species, but because they are useful in preserving the future of the group. It is not a question of legitimacy perceived as intrinsic to each component of nature but as related to a sense of usefulness that is recognized. The relationship of participation (Lévy-Bruhl 1922; Leenhardt 1947) and of alliance (Kemlin 1917) do not have a conservation goal but they can lead to this. The question of bioculture is to investigate whether "a culturally diverse humanity is better equipped to produce biological diversity than a standardized humanity" (Thomas 2011). At the international level, the principle of the cultural and biological diversity interdependence is likely to be seen as a petition of principle (Kohler 2011). However, this principle is part of a resolutely reflexive approach of Western societies (Thomas 2011). In this, it opens up some prospects of a post-modern paradigm combining cultural diversity with the project of giving more space to the nonhuman.

[47]"Human diversity is inseparable from natural diversity. These are goods that guarantee the prosperity of present and future generations and that of the planet. They constitute the foundations of sustainable development "(...) "It is nowadays widely recognized that a homology exists between biodiversity and cultural diversity".

[48]"Subject to the provisions of its national legislation, (each party) shall respect, preserve and maintain the knowledge, innovations and practices of indigenous and local communities that embody traditional lifestyles which are relevant for the conservation and sustainable use of biodiversity and fostering its wider application, with the agreement and participation of the holders of such knowledge, innovations and practices, and encouraging the equitable sharing of benefits arising from the use of such knowledge, innovations and practices" (Art. 8j of the Convention on Biological Diversity (CDB), 1992).

Interactions and interdependencies lead to the fact that the viability of a system depends on its "operating laws" (Aubin et al. 2011). In this way, regulation is the driving force of coviability between social and ecological systems.

1.2.2 The Challenge of System Regulation, Their Maintenance and Reproduction

Regulation is a key element in the general theory of systems according to which any organized system, consisting in a set of interdependent and interacting elements, is constantly confronted with the factors of imbalance and instability arising from the environment (Bertalanffy 1968). Regulation covers all the processes through which systems maintain their "stationary state" by "canceling" the effect of external disturbances and results in some stability, homeostasis and resilience.

Indeed, identifying and analyzing elements is not enough to understand a "whole" that is based on the concept of interrelation, totality, organization or organism. The different levels of autonomy lead to autopoiesis (self-reproduction) or to a capacity for maintenance, existence by adaptation, homeostasis (maintaining the relative internal stability despite changes in the surrounding environment), or finally to resilience (the ability to resist and "absorb" internal disturbances or external shocks while retaining identity and a capacity to reorganize, see below).

For human societies, Niklas Luhmann proposes a theory of social systems and insists on the difference between the social systems that self-replicate (autopoietic[49]) and their environment (that which is external to it, for example, politics or non-politics). Each system preserves its distinctiveness in with respect to a complex, contingent and constantly changing environment with which it is in contact. Consequently, social systems are self-referring, self-centered (Boisvert 2006) and complex because of the diversity of agents and interactions between agents and their environment. Thus, regulation is in charge with the homeostasis of society through: (a) the cohesion processes that ensure the group's survival despite the diversity of interests that exist within it; (b) the use of more flexible mechanisms of coordination and integration (Chevallier 2001) required in adaptation processes of contemporary societies facing increasingly complex problems. When considering regulatory function, Law appears both as a means to conduct regulation, and as a particularly complex system if refering to its heuristic dimension in terms of juridicity (Le Roy 1999; Barrière 2007, 2008a, b, 2011; Ross 2004; Vanderlinden 1996; Rouland 1991).

---

[49]"[...] autopoietic systems are systems that are defined as unities, as networks of productions of components that recursively, through their interactions, generate and realize the network that produces them and constitute, in the space in which they exist, the boundaries of the network as components that participate in the realization of the network" (Luhmann 1990: 3).

For ecological systems, Holling (1973) defines two distinct properties orchestrated by regulatory elements, resilience and stability. Resilience determines "the persistence of relationships within a system, is a measure of the ability of these systems to absorb changes of state variables, driving variables, and parameters, and still persist. In this definition resilience is the property of the system and persistence or probability of extinction is the result. Stability, on the other hand, is the ability of a system to return to an equilibrium state after a temporary disturbance" (Holling 1973, 17). Resilience studies address the non-linear dynamics of the system, the periods of progressive change, those of rapid change, and how these dynamics interact at different temporal and spatial scales (Folke 2010).

For a formulation in terms of mathematical viability, we can consult the work of Sophie Martin (especially: Martin 2004).[50]

In general, resilience is defined as a system's capacity to absorb disturbance before restructuring its functions and identity.[51] However, the concept is not limited to resistance to disturbance or shock. According to Carl Folke (2006, 259), resilience is defined by reconfiguration of structures and processes caused by perturbation and generates new trajectories of the system. In this sense, resilience provides a capacity to adapt by self-reconfiguring. The important aspect is what happens, i.e. the dynamics that follow the system's disturbance. Indeed, the system relies on its own capacity of self internal organisation. The resilience of the socio-ecosystem therefore integrates the idea of adaptation and adaptability.[52] In this respect, socio-cultural schemes, knowledge acquisition, learning and endogenous organization contribute in the ability to absorb the disturbance. In its broad acceptance, law is situated at this level, that is within the group's capacity to react by transforming or innovating its internal regulation.

However, the attention that discourse gives to resilience (in public policy and in research) is not innocent and may translate a deviation. Indeed, in order to provide an operational, understandable and intuitive definition of resilience (Folke et al. 2002, 2006), *Resilience Alliance* (http://www.resalliance.org) sees as handicap to vulnerability, its negative counterpart: "the desire to highlight the 'positive' aspect of resilience (Klein et al. 2003) is closely related to the search for applicability. Whereas vulnerability denotes a negative connotation, an inability, resilience is presented as a desirable property of a system towards which management should aim" (Djament-Tran et al. 2011). Djament-Tran and Reghezza-Zitt (2012), who are specialists in urban resilience, highlight this non-neutral semantic shift from the concept of vulnerability to the concept of resilience, which is related to political issues. Because the two terms are not interchangeable, such a shift leads to a

[50]and the web site: https://inra-dam-front-pad.brainsonic.com/index.php/player-html5-e86bfea832775953b76407f0877a6112.html

[51]"Ecological resilience is the amount of disturbance that a system can absorb before it changes state" (Gunderson et al. 2002, 530).

[52]"Adaptability is a part of resilience. Adaptability is the capacity of a SES to ajust its responses to changing external drivers and internal processes and thereby allow for development within the current stability domain, along the current trajectory" (Folke et al. 2010).

"new model of responsibility" emphasizing the involvement of individuals in the process that is supposed to lead to resilience: the discourse is centered on actors and no longer on victims (Djament-Tran et al. 2011). Resilience, understood in this way, leads to promoting adaptation as an injunction; it "is promoted as a more flexible strategy than prevention"[53] (ibid.),; with risk management shifting from collective solidarity, towards individual adaptation. The injunction to resilience, an intrinsic property of systems becomes a political leitmotiv, and leads to "postulating a kind of resilience teleology that makes societies inherently resilient (...). By giving a teleological vision of resilience, we end up essentialising vulnerability (...). Resilience then reactivates a naturalizing aim, which can go as far as social Darwinism" (ibid.).

The benefit of questioning coviability consists of stepping back from a concept and to resituate it with respect to viability. Precisely, maintaining the stability of the system through homeostasis is necessarily combined with a dynamic of resilience to ensure viability (otherwise the system disappears). However, adaptation is another key element in understanding the dynamics of socio-ecological systems in the long term. To adapt, etymologically, "apt to attach itself," "to adjust," is to change behavior. There is therefore a malleable state and a process[54,55] (bringing a suppleness and necessary flexibility) as a capacity of the system to react to information received from its environment. This endogenous process allows the system either to handle the novelty or to modify its internal structure to assimilate this novelty. The result is a modified living organism, human organization or "object" so as to remain functional under new conditions. The concept of adaptation is thus "intertwined in a conceptual loop that combines self-organization, eco-organization and evolution" (Morin 1985).

Adapting to environmental changes, including climate change, may be defined not only as a set of adjustments[56] (which is only an occasional response to an occasional event), but as profound social, behavioral[57] and biological[58] changes

---

[53]According to the authors it even becomes central in 2007, through the adoption by 168 countries of the policy framework 2005–2015 "Building the resilience of Nations and communities to disasters" (ONU World Conference on Disaster Reduction (2005, Hyogo-Japan) (Online: http://www.unisdr.org/2005/wcdr/wcdr-index.htm.)

[54]"state-Adaptation" and "process-adaptation," Piaget (1967), quoted by Simonet (2009, 394). "Adaptation refers to a process whereby a subject, when recording a variation in the environment, modifies the parameters of an object refering to a specific model in order to accomplish a specific task" (Vieville 2005).

[55]Defined by Lambert and Rezsöhazy (2004, 319) as the "characteristic of systems that are able to coherently and autonomously distort themselves to respond to internal or external demands" quoted by Simonet (2009, 394).

[56]As the IPCC (Intergouvernemental panel on climate change) report defines it (2001 Report, page 173) "an adjustment of human and natural systems to a new and changing environment".

[57]Behaviors can be neurobiological responses or reactions to environmental stimuli.

[58]One of the key elements of the adaptation process is biological or behavioral variation, which involves/implies chosing and decision-making (Bates 2005).

that limit negative impacts and maximize profitable effects in the long-term.[59] The challenge in order to adapt to these changes and ensure viability, consists in reducing both the exposure and vulnerability[60] of social or ecological systems to environmental hazards. Thus, adaptability is a critical element of the process of adaptation (Adger and Vincent 2005).

There are four factors that are likely to affect the adaptability of societies: (a) knowing how to live with change and uncertainty; (b) developing a "diversity of resilience"; (c) combining different types of knowledge for learning; and (d) creating opportunities for self-organization (Plummer and Armitage 2010). In this register, Jared Diamond (2005), situating societies in their context, identifies five causes that have contributed to the collapse of the societies he studied and did not adapt: (a) environmental damage, (b) climatic change (c) hostile neighbors, (d) dependency ratios with trading partners, and (e) responses brought by societies to these situations, according to their own values, or according to their capacity to adapt, to maintain internal stability, to the degree of vulnerability and resilience capacity.

What characterizes most of these causes is the place occupied by regulation. The survival of societies depends on the law enforced within them, activating their internal and external relationships.

Thus, as a true spine for the viability of systems, the normative field which delineates, guides and evaluates an action (by endogenous and exogenous norms) and the matrix of the generation of practices (i.e. the legal habitus) (Barrière 2012, 2015) contribute to the maintenance and sustainability of the socio-ecological system.

Systems do not have sufficient autonomy to depend only on themselves. Due to their interactions and interdependencies between each other, we have previously seen that their viability is conceivable only in terms of coviability. According to Folke et al. (2002), the interweaving of systems renders ineligible the implicit assumption that ecosystem responses to human actions are linear, predictable and controllable; and it also renders ineligible the assumption that human and natural systems can be treated independently from one another.[61] Their coexistence reflects

[59]Adaptation must be understood as "a long-term process able to reduce the vulnerability of social systems to crises" (Burton et al. 1993), cited by Simonet (2009, 397), adapatation is a determinant of vulnerability (Adger and Vincent 2005).

[60]Vulnerability is a probabilistic concept that tells about the proximity of a subject with a damage (Schröder-Butterfill and Marianti 2006, 11). It reflects the fragility of a system as a whole and concerns the chance/uncertainty (Veyret and Reghezza 2006). Social vulnerability, also known as "organizational vulnerability", expresses a society's ability to anticipate hazard, deal with emergency, adapt its behavior in times of crisis, and rebuild itself. It is directly linked to the resilience and functioning of societies (Wisner et al. 2004), cited by Barroca et al. 2013.

[61]"However, evidence that has been accumulating in diverse regions all over the world suggests that natural and social systems behave in nonlinear ways, exhibit marked thresholds in their dynamics, and that social-ecological systems act as strongly coupled, complex and evolving integrated systems" (Folke et al. 2002).

a socio-ecological coviability, which can only be preserved through management integrated at different scales.[62] We touch here upon a new heuristic paradigm, which encompasses both states and dynamics of the whole Homeostasis, autopoiesis, resilience, adaptability and vulnerability are the structuring concepts of a whole; each taken separately fails in defining the system. All these concepts contribute to the understanding of the viability of socio-ecological systems; the coviability paradigm brings them together, which is based on the assumed fact that the biosphere is defined by they multiple entangled and complex human and nonhuman living worlds.

References

Acselrad H (2009) Conflitos ambientais no Brasil. Relume Dumará, Rio de Janeiro

Adger WN, Vincent K (2005) Uncertainty in adaptative capacity. C.R. Geosci 337:399–410. https://doi.org/10.1016/j.crte.2004.11.004

Allen CR, Gunderson L, Johnson AR (2005) The use of discontinuities and functional groups to assess relative resilience in complex systems. Ecosystems 8:958–966. https://doi.org/10.1007/s10021-005-0147-x

Andersen e. (ed) (2003) Developing a high nature value farming area indicator. Final report. European Environment Agency, Copenhagen, 75 p. http://edepot.wur.nl/3918

Aubin JP (1991) Viability theory. Birkhäuser, Boston

Aubin J-P (2010) La mort du devin, l'émergence du démiurge. Essai sur la contingence, la viabilité et l'inertie des systèmes. Beauchesne/le miroir des savant, Paris

Aubin J-P, Bayen A, Saint-Pierre P (2011) Viability theory, new directions. Springer, New York

Barrière O (2007) Régulation environnementale en Afrique: l'action du droit pour une co-viabilité des systèmes sociaux et écologiques. In: E Le Roy (dir.) Cahiers d'Anthropologie du Droit 2006, Le droit en action. Karthala, Paris, pp 125–174, 307 pages

Barrière O (2008a) Legal aspects of the co-viability of social and ecological systems in African arid zones: an anthropological approach to environmental law. In: Lee C, Schaaf T (eds) The future of drylands. Springer/UNESCO Publishing, Paris, pp 583–597

Barrière O (2008b) L'intégration du droit dans la dialectique sociétés-écosystèmes, le droit d'une socio-écologie. In: Barrière O, Rochegude A (eds) Foncier et environnement en Afrique: des acteurs au(x) droit(s). Karthala, Paris, pp 39–128

Barrière O (2011) Pluralisme juridique et patrimonialisation: entre paradigmes de l' "appropriation" et du "patrimoine commun". In: Serge Mam Lam Fouck, Isabelle Hidair (dir) La question du patrimoine en Guyane française. Diversité culturelle et patrimonialisation. Processus et dynamiques des constructions identitaires. Ibis Rouge Editions, Matoury, pp 43–75

Barrière O (2012) Changement climatique et droit de l'homme: vers un droit de la coviabilité des systèmes sociaux et écologiques? In: Dormoy D, Kuyu C (dir.) Changements climatiques et droits humains. Espérance, Marcq-en-Baroeul, pp 227–250

[62]"ecosystem management is neither local nor central in origin, but requires multi-level management and matching of social and ecological dynamics across scales. The findings indicate that the adaptive co-management approach have the potential to serve this need and build capacity for resilience in social-ecological systems. We conclude that such management may contribute to expanding desirable stability domains of social-ecological systems, making them more robust to change (Olsson 2003).

Barrière O (2015) Repenser le droit de l'environnement dans une conception renouvelée du développement durable: prospective d'un "droit de la coviabilité" des systèmes sociaux et écologiques. In: Blaise S, David C, David V (eds) Le développement durable en Océanie, vers une nouvelle éthique?, Actes du colloque international des 23–26 avril 2013 Koné, (Province Nord de la Nouvelle-Calédonie). Presses Universitaires de Provence & Presses universitaires d'Aix-Marseille, Aix-en-Provence, pp 215–242

Barrière O (2017) Human relationship to the land from a legal perspective as a human and environmental security challenge (Chap. 14). In: Behnassi M, McGlade K (eds) Environmental change and human security in Africa and the Middle East. Springer, Cham, pp 259–304. https://doi.org/10.1007/978-3-319-45648-5_14

Barrière O, Faure J-F (2012) L'enjeu d'un droit négocié pour le Parc Amazonien de Guyane. Nat Sci Soc 20:167–180

Barroca B, Maryline D, Irène M (2013) De la vulnérabilité à la résilience: mutation ou bouleversement? EchoGéo 24, Online: http://echogeo.revues.org/13439

Bates DG (2005) Human adaptive strategies: ecology, culture, and politic. Pearson Education, Toronto

Baumgärtner S, Quaas MF (2007) Ecological-economic viability as a criterion of strong sustainability under uncertainty. University of Lüneburg Working paper series in economics, No. 67. Leuphana-University, Lüneburg

Baumgärtner S, Quaas MF (2009) Ecological-economic viability as a criterion of strong sustainability under uncertainty. Ecol Econ 68:2008

Becker BK (2007) Amazônia: geopolítica na virada do III milênio, 2nd edn. Garamond, Rio de Janeiro, 168 p

Bélier I (2013) La reconnaissance internationale des peuples autochtones. In: Bélier (ed) peuples autochtones dans le monde. L'Harmattan-SOGIP, Paris, pp 13–37

Berkes F, Folke C (1998) Linking social and ecological systems for resilience and sustanability. In: Linking social and ecological systems, management practices ans social mechanisms for building resilience. Cambridge University Press, New York, 4

Berkes F, Colding J, Folke C (eds) (2003) Navigating social-ecological systems: building resilience for complexity and change. Cambridge University Press, Cambridge

Berque A (1996) Etre humains sur la terre. Gallimard, Paris

Bertalanffy (von) L (1968/2012) Théorie générale des systèmes. Dunod, Paris

Blandin P (2013) Towards EcoEvoEthics, in the structural links between ecology, evolution end ethics: the virtuous epistemic circle, Boston studies in the philosophy of science, vol 296. Springer, Dordrecht, pp 83–100

Boisvert D (2006) Niklas Luhmann: la théorie des systèmes sociaux. Asp Soc 13(1):55–82

Bonmatin J-M, Giorio C, Girolami V, Goulson D, Kreutzweiser DP, Krupke C, Liess M, Long E, Marzaro M, Mitchell EAD, Noome DA, Simon-Delso N, Tapparo A (2015) Environmental fate and exposure; neonicotinoids and fipronil. Environ Sci Pollut Res 22:35–67. https://doi.org/10.1007/s11356-014-3332-7

Bourgine P (1996) In: Rialle V, Fisette D (eds) Modèles d'agents autonomes et leurs interactions co-évolutionnistes, Penser l'Esprit: des Sciences de la Cognition à une Philosophie Cognitive. Presses universitaires de Grenoble, Paris, pp 421–443

Burton I, Kates RW, White GF (1993). The Environment as Hazard, Guilford Press, 290 p

Carlson R (1969) Primavera silenciosa, 2nd edn. Editora Melhoramentos, São Paulo

Cavalcanti C (2004) Uma tentativa de caracterização da economia ecológica. Ambiente e Sociedade, Campinas 7(1):149–156

Cavalcanti C (2010) Concepções da economia ecológica: suas relações com a economia dominante e a economia ambiental. Estudos Avançados, São Paulo 24(68):53–67

Cavalcanti C (2012) Sustentabilidade: mantra ou escolha moral? Uma abordagem ecológico-econômica. Estudos Avançados, São Paulo 26(74):35–50

Cavalcanti C (2015) A concepção de desenvolvimento do Butão. Revista Ciência & Ambiente 50:5–19

CCD Steering Committee (2012) Colony collapse disorder, progress report. United States Department of agriculture
Chaptal de Chanteloup C (2015) La chaîne de valeur de l'offre. De Boeck
Chen YP, Pettis JS, Corona M, Chen WP, Li CJ, Spivak M et al (2014) Israeli acute paralysis virus: epidemiology, pathogenesis and implications for honey bee health. PLoS Pathog 10(7):e100426. https://doi.org/10.1371/journal.ppat.1004261
Chevallier J (2001) La régulation juridique en question. Droit et Société 49:827–846
Clément H (2009) L'abeille sentinelle de l'environnement. Alternatives, Paris
Clément, H, Le Conte Y (2011) Traité rustica de l'apiculture. Rustica éditions
Cohen D (2012) Homo economicus, Prophète (égaré) des temps nouveaux. Albin Michel
Combes C (1995) Interactions Durables. Ecologie et Evolution du Parasitisme. Masson, Paris, 518 p
Cox-Foster DL, Conlan S, Holmes EC, Palacios G, Evans JD, Moran NA, Quan P-L, Briese T, Mady, Geiser DM, Vince, Dennis (van) E, Kalkstein Abby L, Andrew, Jeffrey, Junhui, Cui L, Stephen K, Fredrik J, Michael, Jeffery S, Lipkin WI (2007) A metagenomic survey of microbes in honey bee colony collapse disorder. Science 318(5848):283–287. https://doi.org/10.1126/science.1146498
Crédit Suisse (2015) Global wealth report 2015. Crédit Suisse, Zurich, 64 p
Crutzen P (2002) Geology of mankind. Nature 415(6867):23
Cumming GS, Collier J (2005) Change and identity in complex systems. Ecol Soc 10(1):29 Online: http://www.ecologyandsociety.org/vol10/iss1/art29/
Dahl LA (1996) The eco principle. Zed Book Ltd, London & New Jersey
Daly HE (2002) Desenvolvimento sustentável: definições, princípios, políticas. Cadernos de estudos sociais, Recife 18(2):171–184
Décamps H (2007) La vulnérabilité des systèmes socioécologiques aux évènements extrêmes: exposition, sensibilité, résilience. Nat Sci Soc 15:48–52
Descola P (1985) De l'Indien naturalisé à l'Indien naturaliste: sociétés amazoniennes sous le regard de l'Occident. In: Cadoret A (ed) Protection de la nature: histoire et idéologie. De la nature à l'environnement. L'Harmattan, Paris, pp 221–235
Descola P (2005) Par-delà nature et culture. Gallimard, Paris
Diamond J (2005) Collapse, how societies choose to fail or succeed. Viking, New York
Djament-Tran G, Reghezza-Zitt M (dir) (2012) Résiliences urbaines, les villes face aux catastrophes. édition du manuscrit, 360 p
Djament-Tran G, Le Blanc A, Lhomme S, Rufat S, Reghezza-Zitt M (2011). Ce que la résilience n'est pas, ce qu'on veut lui faire dire. HAL Archives ouvertes hal-00679293
Ducroux A-M (2003) Les nouveaux utopistes du développement durable. Autrement, Paris
Durand D (1994) La Systémique. PUF, Paris
Ehrlich PR, Raven PH (1964) Butterflies and plants: a study in coevolution, reviewed work(s). Evolution, 18(4):586–608, Society for the Study of Evolution Online: http://www.jstor.org/stable/2406212
Ellen MacArthur Foundation (2012) Towards the circular economy: Economic and business rationale for an accelerated transition. Report
European Commission (2014) Communication from the commission to the european parliament, the council, the european economic and social committee and the committee of the regions. Towards a circular economy: a zero waste programme for Europe, Brussels, 2.7.2014 COM(2014) 398 final
Filoche G (2011) Les amérindiens de guyane française, de reconnaissances disparates en bricolages juridiques. L'exemple des kali'na d'awala-yalimapo, Journal de la société des Américanistes 97(2):343–367
Fischer J, Müller T, Spatz A-K, Greggers U, Grünewald B et al (2014) Neonicotinoids interfere with specific components of navigation in honeybees. PLoS One 9(3):e91364. https://doi.org/10.1371/journal.pone.0091364
Folke C (2006) Resilience: the emergence of a perspective for social–ecological systems analyses. Glob Environ Chang 16:253–267

Folke C (2010) Resilience: the emergence of a perspective for social–ecological systems analyses. Glob Environ Chang 16:253–267. https://doi.org/10.1016/j.gloenvcha.2006.04.002
Folke C, Carpenter S, Elmquist T, Gunderson L, Holling CS, Walker B (2002) Resilience and sustainable development: building adaptive capacity in a world of transformations. Ambio 31(5):437–440
Folke C, Hahn T, Olsson P, Norberg J (2005) Adaptive governance of social-ecological systems. Annu Rev Environ Resour 30:441–473
Folke C, Carpenter SR, Walker B, Scheffer M, Chapin T, Rockström J (2010) Resilience thinking: integrating resilience, adaptability and transformability. Ecol Soc 15(4):20 [online] URL: http://www.ecologyandsociety.org/vol15/iss4/art20/
Folke C, Jansson A, Rockström J, Per O, Carpenter SR, Chapin SF, Crépin A-S, Daily G, Danell K, Ebbesson J et al (2011) Reconnecting to the biosphere. Ambio 40(7):719–738
Frisch (von) K (2011 (1927)) Vie et mœurs des abeilles. Albin Michel, Paris
Garmestani AS, Allen CR, Gunderson L (2009) Panarchy: discontinuities reveal similarities in the dynamic system structure of ecological and social systems. Ecol Soc 14(1):15. [online] URL: http://www.ecologyandsociety.org/vol14/iss1/art15/
Georgescu-Roegen N (1971) The entropy law and the economic process. Harvard University Press, Cambridge (EUA)
Gudynas E, Acosta A (2011a) El buen vivir o la disolución de la idea del progreso. In: Rojas M (coord.). La medición del progreso y del bienestar. Propuesta desde América Latina. Foro consultivo científico y tecnológico. AC, México, DF (México)
Gudynas, E, Acosta A (2011b) La renovación de la crítica al desarrollo y el buen vivir como alternativa. Utopía y Praxis Latinoamericana/Año 16. N° 53 (Abril-Junio, 2011) p 71–83. Revista Internacional de Filosofía Iberoamericana y Teoría Social/ISSN 1315-5216. CESA – FCES – Universidad del Zulia. Maracaibo-Venezuela
Gunderson LH, Holling CS (eds) (2002) Panarchy, understanding transformations in human and natural systems. Island press, Washington, DC
Gunderson L, Holling CS, Pritchard L, Peterson GD (2002) Resilience. In: Mooney HA, Canadell JG (eds) Encyclopedia of global environmental change. Volume 2, The earth system: biological and ecological dimensions of global environmental change. Scientific Committee on Problems of the Environment (SCOPE), Chichester, pp 530–531
Gunther M (2014) Vers un monde sans déchets. ENSIA, Saint Paul, USA, in Courrier International, n°1258, Economie circulaire, Rien ne se perd, tout se transforme, Un monde sans déchets vu par la presse étrangère, p 29
Hardin G (1968) The tragedy of the commons. Science, New York 13:1243–1248
Hartus P, Virard M-P (2015) Croissance zéro, comment éviter le chaos? Fayard
Heck S, Rogers M (2014) Ressources revolution: how to capture the biggest business opportunity in a century. New Harvest, Boston/New York
Helm D (ed) (1991) Economic policy towards the environment. Blackwell Publisher, Oxford/Cambridge, 326 p
Hirsch HG, Bradley D, Pohl C, Rist S, Wiesmann U (2006) Implications of transdisciplinarity for sustainability research. Ecol Econ 60:119–128. https://doi.org/10.1016/j.ecolecon.2005.12.002
Holling CS (1973) Resilience and stability of ecological systems. Annu Rev Ecol Syst 4:1–23. https://doi.org/10.1146/annurev.es.04.110173.000245
Holling CS (2001) Understanding the complexity of economic, ecological and social systems. Ecosystems 4:390–405
Hong S-K et al (2007) Landscape ecological applications in man-influenced areas. Linking man and nature systems. Springer, Dordrecht, 535 p
Horlick-Jones T, Sime J (2004) Living on the border: knowledge, risk and transdisciplinarity. Futures 36:441–456. https://doi.org/10.1016/j.futures.2003.10.006
Hubert B, Deverre C, Meuret M (2010) Deux siècles de changements radicaux pour les parcours du Sud de la France. In: Michel Meuret (dir) Un savoir-faire de bergers, éditions Quae, pp 27–41
Institute for European Environmental Policy (IEEP) (2007) HNV Indicators for Evaluation, Rapport final pour la DG Agriculture, Avis de marché 2006- G4-04

IPCC (2001) Bilan des changements climatiques, rapport de synthèse, Troisième rapport d'évaluation. Groupe intergouvernemental d'experts sur l'évolution du climat, Genève, www.ipcc.ch

Jackson T (2013) Prosperidade sem crescimento: vida boa em um planeta finito. Planeta Sustentável, Abril Editora, São Paulo

Johnson R (2007) Recent honey bee colony declines, CRS report for Congress. Congressional Research Service

Johnson R (2010) Honey bee colony collapse disorder, CRS report for Congress. Congressional Research Service

Jollivet M (2001) Le développement durable, de l'utopie au concept. Elsevier, Paris

Kemlin JE (1917) L'alliance chez les Reungao. Bulletin de l'École française d'Extrême- Orient 17:1–119. https://doi.org/10.3406/befeo.1917.5321

Klein RJ, Nicholls RJ, Thomalla F (2003) Resilience to natural hazards: how useful is the concept? Environ Hazards 5(1–2):35–45

Klein A-M, Vaissière BE, Cane JH, Steffan-Dewenter I, Cunningham SA, Kremen C, Tscharntke T (2007) Importance of pollinators in changing landscapes for world crops. Proc R Soc B Biol Sci 274(1608):303–313. https://doi.org/10.1098/rspb.2006.3721

Kohler F (2011) Diversité culturelle et diversité biologique: une approche critique fondée sur l'exemple brésilien. Nat Sci Soc 19:113–124

Landais E, Balent G (1993) Introduction à l'étude des pratiques d'élevage extensif. Etudes et Recherches sur les Systèmes Agraires et le Développement 27:13–34

Lawrence RJ, Després C (2004) Introduction. Futures of transdisciplinarity. Futures 36:397–405. https://doi.org/10.1016/j.futures.2003.10.005

Leenhardt M (1947) Do kamo: la personne et le mythe dans le monde mélanésien. Gallimard, Paris

Le Fur, Curry, Laloe, Durand, Chaboud (1999) Co-viabilité des systèmes halieutiques. Nat Sci Soc 7(2):19–32

Leff E (2009a) A reapropriação social da natureza. In: Leff E (ed) Saber Ambiental: sustentabilidade, racionalidade, complexidade, poder, 7th edn. Vozes, Petrópolis, pp 65–82

Leff E (2009b) Democracia ambiental e desenvolvimento sustentável. In: Leff E (ed) Saber Ambiental: sustentabilidade, racionalidade, complexidade, poder, 7th edn. Vozes, Petrópolis, pp 56–64

Leff E (2009c) Educação ambiental e desenvolvimento sustentável. In: Leff E (ed) Saber Ambiental: sustentabilidade, racionalidade, complexidade, poder, 7th edn. Vozes, Petrópolis, pp 236–252

Leff E (2009d) A cultura como mediação de processos econômicos e ecológicos. In: Leff E (ed) Ecologia, capital e cultura. Vozes, Petrópolis, pp 98–142

Leff E (2010) Epistemologia ambiental, 5th edn. Cortez, São Paulo

Leff E (2014) La apuesta por la vida: imaginación sociológica e imaginários sociales en los territorios ambientales del sur. Siglo XXI Editores, México

Lenton TM, Held H, Kriegler E, Hall JW, Lucht W, Rahms -torf S, Schellnhuber HJ (2008) Tipping elements in the Earth's climate system. PNAS 105(6):1786–1793

Le Roy E (1999) Le jeu des lois, une anthropologie "dynamique" du Droit. LGDJ, Paris

Lesne A (2008) Complexité du vivant, sélection naturelle et évolution. Nat Sci Soc 16:150–153; et 17, 55-56 (2009)

Lévêque C, Muxart T, Abbadie L, Weil A, van der Leeuw S (2003) L'anthroposystème: entité structurelle et fonctionnelle des interactions sociétés – milieux. In: Lévêque C, van der Leeuw S (eds) Quelles natures voulons-nous? Elsevier, Paris, pp 110–129

Lévy-Bruhl L (1922) La Mentalité primitive. Alcan, Paris

Lexmond MB (v), Bonmatin J-M, Goulson D, Noome DA (2015) Worldwide integrated assessment on systemic pesticides global collapse of the entomofauna: exploring the role of systemic insecticides. Environ Sci Pollut Res 22:1–4. https://doi.org/10.1007/s11356-014-3220-1

Li Gan_shun (2002) To achieve "double-win" of economy and environment by developing circular economy. J Hebei Univ (Philos Soc Sci Inst Jpn Stud), 03, Baoding, China

Liu J et al (2007) Complexity of coupled human and natural systems. Science 317:1513. https://doi.org/10.1126/science.1144004

Luhmann N (1990) Essays on self-reference. Columbia University Press, New York

Maeterlinck M (1901) La vie des abeilles. Eugène Fasqeulle éditeur, Paris

Marais E (1973/1925) The soul of the white ant. Penguin Books

Martin S (2004) The cost of restoration as a way of defining resilience: a viability approach applied to a model of lake eutrophication. Ecol Soc 9(2):8. [online] URL: http://www.ecologyandsociety.org/vol9/iss2/art8/, Published here under license by The Resilience Alliance

Martinet V, Doyen L (2007) Sustainability of an economy with an exhaustible resource: a viable control approach, Working Papers 22177, Institut National de la Recherche Agronomique, Francer

Martinez Alier J (2007) O ecologismo dos pobres. Contexto, São Paulo, 380 p

Mathevet R, Bousquet F (2014) Résilience environnement, Panser les changements socio-écologiques. Buchet-Chatel, Paris

Mathevet R, Thompson J, Delanoë O, Cheylan M, Gil-Fourrier C, Bonnin M (2010) La solidarité écologique: un nouveau concept pour une gestion intégrée des parcs nationaux et des territoires. Nat Sci Soc 18(4):424–433

Meadows D, Randers J, Meadows D (2002) Limits to growth. The 30-year update. Chelsea Green Publishing, Post Mills

Meuret M (2010a) Un savoir-faire de bergers. éditions Quae, Paris, 333p

Meuret M (2010b) Stimuler l'appétit lors des circuits de garde: échange d'expériences entre un berger et un chevrier. In: Meuret M (ed) Un savoir-faire de bergers. éditions Quae, Paris, pp 149–165

Meyer B, et al (2011) Macroeconomic modelling of sustainable development and the links between the economy and the environment, gws Research Report 2012/1 (ENV.F.1/ETU/2010/0033)

Morin E (1973) Le paradigme perdu: la nature humaine, Points Essai, Paris, 249p

Morin E (1985) La Vie de la vie, La Méthode, t. 2. Le Seuil, Paris

Moritz Robin FA, Southwick EE (1992) Bees as superorganisms: an evolutionary reality. Springer/Heidelberg GmbH & Co, Berlin

Olsson P (2003) Building capacity for resilience in social-ecological systems. Doctoral dissertation, Department of Systems Ecology Stockholm University, Sweden, 37p

OXFAM (2015) Insatiable richesse: toujours plus pour ceux qui ont déjà tout. Oxfam International, Oxford, 2015, 14 p. Online: https://oxfam.qc.ca/communiques/2015-01-19_les-1-les-plus-riches-poss-deront-plus-que-le-reste-de-la-population-mondiale

Page RE, Hölldobler B (2013) The spirit of the hive, the mechanisms of social evolution. Harvard University Press, Cambridge, MA

Passet R (1992) Le copilotage du développement économique et de la biosphère. Tiers-Monde 33(130):393–416

Penelaud O (2010) Le paradigme de l'énaction aujourd'hui. Apports et limites d'une théorie cognitive "révolutionnaire". PLASTIR 1:18

Perry CJ,Søvika E, Myerscough MR, Andrew B (2015) Rapid behavioral maturation accelerates failure of stressed honey bee colonies. PNAS (Proceedings of the National Academy of Sciences of the United States of America), 112(6):1643–1644, Edited by Gene E. Robinson, University of Illinois at Urbana–Champaign, Urbana, IL https://doi.org/10.1073/pnas.1422089112

Pisa LW, Amaral-Rogers V, Belzunces LP, Bonmatin JM, Downs CA, Goulson D, Kreutzweiser DP, Krupke C, Liess M, Mc Field M, Morrissey CA, Noome DA, Settele J, Simon-Delso N, Stark JD, Van der Sluijs JP, Van Dyck H, Wiemers M (2015) Effects of neonicotinoids and fipronil on non-target invertebrates. Environ Sci Pollut Res 22:68–102. https://doi.org/10.1007/s11356-014-3471-x

Plummer R, Armitage D (2010) Integrating perspectives on adaptive capacity and environmental governance. In: Derek A, Ryan P (eds) Adaptative capacity and environmental gouvernance. Springer, Berlin, pp 1–19
Porto-Gonçalves CW (2006) A globalização da natureza e a natureza da globalização. Civilização brasileira, Rio de Janeiro. 462 p
Prisco et al (2011) Varroa destructor is an effective vector of Israeli acute paralysis virus in the honeybee, Apis mellifera. J Gen Virol 92(Pt 1):151–155. https://doi.org/10.1099/vir.0.023853-0
Ratnieks Francis LW, Carreck NL (2010) Clarity on honey bee collapse? Science 327:152. https://doi.org/10.1126/science.1185563
Ribière M, Olivier V, Blanchard P, Schurr F, Celle O, Drajnudel P, Faucon J-P, Thiery R, Chauzat M-P (2008) Les effondrements de colonies d'abeilles: le cas du CCD ("colony collapse disorder") et du virus IAPV (Israeli acute paralysis virus), Virologie (John Libbey eurotext), vol 12, issue 5 (Afssa, Unité pathologie de l'abeille, 06902 Sophia-Antipolis)
Rockström J, Steffen W, Noone K, Persson A, Chapin SF III, Lambin EF, Lenton TM, Scheffer M et al (2009) A safe operating space for humanity. Nature 461:472–475
Rodrigues AM (1998) Produção e consumo do e no espaço. Problemática ambiental urbana. Disponível em http://creativecommons.org/licenses/by-nc-sa/2.5/deed.pt. Também publicado em Rio de Janeiro: Hucitec. 193 p
Rosnay (De) J (1975) Le macroscope, vers une vision globale. Seuil, Paris
Ross A (2004) Introduction à l'empirisme juridique, LGDJ-Bruylant, 231p
Rouland N (1991) In: Jacob O (ed) Aux confins du droit. Les Presses universitaires d'Aix-Marseille, Aix-en-Provence, 318p
Sabatier R (2010) Arbitrages multi-échelles entre production agricole et biodiversité dans un agroécosystème prairial, Thèse de doctorat, Institut des Sciences et Industries du Vivant et de l'Environnement de AgroParisTech, directeurs Muriel Tichit et Luc Doyen, 226p
Sabatier R, Doyen L, Tichit M (2008) Evaluer l'effet des seuils de chargement sur les performances productives et écologiques de troupeaux valorisant des prairies à forts enjeux de biodiversité. Rencontre Recherche Ruminants (3R)/INRA 15:213–216
Sabatier R, Doyen L, Tichit M (2010) Modelling trade-offs between livestock grazing and wader conservation in a grassland agroecosystem. Ecol Model 221:1292–1300
Santos M (2006) A natureza do espaço. EDUSP, São Paulo, 494 p
Savini I, Landais E, Thinon P, Deffontaines J-P (2010) Des pratiques d'un berger expérimenté à la construction d'outils d'aide à la gestion d'alpages. In: Meuret M (dir) Un savoir-faire de bergers. éditions Quae, pp 81–100
Schröder-Butterfill E, Marianti R (2006) A framework for understanding old-age vulnerabilities. Ageing Soc 26(1):9–35
Seeley TD (2010) Honeybee democracy. Princeton University Press, Woodstock
Seeley TD (2014) Honeybee ecology. A study in the adaptation in social life. Princeton University Press, Woodstock
Simonet G (2009) Le concept d'adaptation: polysémie interdisciplinaire et implication pour les changements climatiques. Nat Sci Soc 17(4):392–401
Simon-Delso et al (2015) Systemic insecticides (neonicotinoids and fipronil): trends, uses, mode of action and metabolites. Environ Sci Pollut Res 22:5–34
Sinaï A (2011) Fukushima ou la fin de l'anthropocène. Sortir d'urgence de l'inanité de notre mode de croissance, Le Monde, 18.03.2011, on line : http://www.lemonde.fr/idees/article/2011/03/18/fukushima-ou-la-fin-de-l-anthropocene_1495211_3232.html
Stavenhagen R (2013) La protection des droits autochtones: le défi de la mise en œuvre. In: Bélier (dir) peuples autochtones dans le monde, Les enjeux de la reconnaissance. L'Harmattan-SOGIP, pp 7–12
Steffen W, Crutzen PJ, McNeill JR (2007) The anthropocene: are humans now overwhelming the great forces of nature? Ambio 36(8): 614–621. URL: http://www.jstor.org/stable/25547826
Tabary J-C (1993) Théorie de la Connaissance et Autonomie Biologique. thèse, Available online at: http://cerveau.pensee.free.fr
Tautz J (2008) The buzz about bees : biology of a superorganism, Springer

Tautz J (2009) L'étonnante abeille, ed. De Boeck & Larcier
Tavoillot F, Tavoillot PH (2015) L'Abeille (et le) philosophe: Étonnant voyage dans la ruche des sages, Odile Jacob
Thomas F (2011) Cosmologies, diversité bioculturelle et préservation de l'environnement. Nat Sci Soc 19(2):129–132 Online: www.cairn.info/revue-natures-sciences-societes-2011-2-page-129.htm
Tichit et al (2007) A co-viability model of grazing and bird community management in farmland. Ecol Model 206:277–293
UNEP (2002) Diversité culturelle et biodiversité pour un développement durable, Table ronde organisée conjointement par l'UNESCO et le PNUE le 3 septembre 2002 à Johannesburg (Afrique du Sud) à l'occasion du Sommet mondial pour le développement durable, On line: http://unesdoc.unesco.org/images/0013/001322/132262f.pdf
Vallejo MC et al (2011) La iniciativa yasuni-itt: desde una perspectiva multicriterial. Quito: Programa de las Naciones Unidas para el Desarrollo, PNUD; Organización de las Naciones Unidas para la Alimentación y la Agricultura, FAO
Vanderlinden J (1996) Anthropologie juridique, Dalloz
Van Valen Lee (1973) A new evolutionary law, Evolutionary law, Vol 1, 1–30, Chicago, USA
Varela FJ, Maturana HR, Uribe R (1974) Autopoiesis : The organization of living systems, its characterization and a model. Biosystems 5(4):187–196
Varela FJ (1989) Autonomie et connaissance. Essai sur le vivant. Seuil, Paris
Veyret Y, Reghezza M (2006) Vulnérabilité et risques. L'approche récente de la vulnérabilité, Responsabilité et environnement, n°43, 9–14. On line : http://www.annales.org/test-site/test-site/re/2006/re43/Veyret.pdf
Vieville T (2005) Quelques idées sur le concept d'adaptation, On line : http://www-sop.inria.fr/members/Thierry.Vieville/cours/adaptation.html
Walker BH, Gunderson LH, Kinzig AP, Folke C, Carpenter SR, Schultz L (2006) A handful of heuristics and some propositions for understanding resilience in social-ecological systems. Ecol Soc 11(1):13 [online] URL:http://www.ecologyandsociety.org/vol11/iss1/art13/
Westley F, Carpenter SR, Brock WA, Holling CS, Gunderson LH (2002) Why systems of people and nature are not just social and ecological systems. In: Gunderson Lance H, Holling CS (eds) Panarchy, understanding transformations in human and natural systems. Island Press, Washington
Whitfield J (2008) An indifference to boundaries. Nature 451:872–873. https://doi.org/10.1038/451872a
World economic forum (2014) Towards the Circular Economy: Accelerating the scale-up across global supply chains. Report, prepared in collaboration with the Ellen MacArthur Foundation and McKinsey & Company, Geneva
Young OR, Berkhout F, Gallopin GC, Janssen MA, Ostrom E, van der LS (2006) The globalization of socio-ecological systems: An agenda for scientific research. Glob Environ Chang 16(2006):304–316. https://doi.org/10.1016/j.gloenvcha.2006.03.004
Zaman L, Meyer JR, Devangam S, Bryson DM, Lenski RE et al (2014) Coevolution Drives the Emergence of Complex Traits and Promotes Evolvability. PLoS Biol 12(12):e1002023. https://doi.org/10.1371/journal.pbio.1002023
Zhijun F, Nailing Y (2007) Putting a circular economy into practice in China. Sustain Sci 2:95–101. https://doi.org/10.1007/s11625-006-0018-1

Part I
Towards the Theoretical Foundation of Coviability

Preview 1: A Multidisciplinary Reading

Discussing coviability induces a diversity of approaches that these first eleven chapters highlight. Each of these works, whether monodisciplinary or interdisciplinary, contributes to understanding what the "co" prefix is linking and what kind of link(s) it defines. Thereby, an explanation of the concepts used is undertaken. The challenge of this part is to present this plurality of disciplinary prisms on the concept of coviability, and thus contribute to nourish an interdisciplinary or even transdisciplinary perspective around this new paradigm.

The genesis of coviability begins here with the concept of viability approached by mathematics. It is defined as a set of constraints in which the system is maintained. Beyond these boundaries, sustainability is no longer assured. Evolutions and dynamics are "regulated" within a given environment.

The first fundamental step in the definition of coviability is thus carried out in the field of mathematics. It was during a seminar in the South of France in 1996 that the idea of "eco-viability" or "coviability" was defined as the joint evolution of a vector and the subset to which this vector belongs.

It should also be noted that that whatever the discipline involved in this first part of the volume, the idea of control systematically re-emerges to demonstrate that coviability consists in "controlling the trajectory". The issue of regulation is clearly emphasized and considered as a keyword throughout the chapters, expressed in one way or another.

The question of bringing the concept of coviability closer to that of the link between social and ecological systems challenges the fantasy of man's domination over nature and makes it possible to take the completely opposite approach. The very idea of "nature" leads to that of cultural diversity. The world opens up to plurality with the notion of coviability: we will find as many socio-cognitive representations of both nature and societies. "Each one has his own nature": this affirmation provides the type and modalities of relationships, whether among humans and non-humans or between humans. To the power of "profit" and "service" that produces the predominant globalized system of relationship to the biosphere, the paradigm of coviability, discussed in this part, emerges to perceive the Western scheme (the naturalistic ontology) as non-universal.

This non-universality stems from the fact that the challenge discussed here is related to the 'co' (which etymologically means 'together'). From now on, the "coviability of social and ecological systems" raises these kinds of questions: do we start from a dichotomy that can quickly slide into an opposition? This is the ambiguity that should be alleviated. Should we rely on this social and ecological dichotomy to get out of the deadlock? However, this ambiguity seems ineluctable even with the attempt to specify that the social would be the human and the natural (the non-human). So, from the beginning, what do we mean?

It is conceivable that it is not a question of being tied to this "sharing" but rather to making it a marker: according to the designated ontology of society-nature relationships (as we see, it is difficult to be disconnected from this combination) the division is consumed (naturalism), or not at all (animism, totemism, analogism). The "co" is understood as non-rigid, mixing the human in the non-human and, according to the paradigms, generating more or less distance between the two entities. Yet, if humanity is claimed to be consubstantial with nature, there are two entities that make it one, two entities that belong to a whole: a planet Earth, with a Biosphere and a Humanity.

Each chapter of this rather theoretical part, while often referring to concrete situations and empirical works, revolves around a form of socio-ecological integration. Indeed, coviability evokes the man/nature antagonism only to overcome it by emphasizing the structure conformed to "with" that is inscribed in the etymology of the "co". The readers will make the judgment whether some of the authors of the chapters fall into this trap, rightly or wrongly.

Coviability cannot be dictated. The need for criteria or indicators is needed to demonstrate whether or not we are in the sphere of coviability. But what makes "coviability"? Science requires knowledge, on the basis of questioning, analysis, and thinking; it's our starting point. This first part does not claim to give a definite, exclusive or final answer. It aims rather to present the conceptual frameworks and to link the disciplines in order to stimulate the interdisciplinary emergence of the concept-paradigm of socio-ecological coviability. The understanding will be clarified over the works presented here. There is no prejudiced assumption about the results of this research exercise, which in itself formalizes an interdisciplinary approach. So let us dare to attempt this promising adventure . . .

In this part, eleven chapters are attempting to lay the first foundations of socio-ecological coviability on a disciplinary plurality bringing together in this first enterprise ecology, mathematics, systems engineering, law, anthropology, biology, geography, and philosophy:

List of des chapters in Part I:

Chapter 4 A Mathematical Approach to Coviability: Concept, Modeling and Control
Chapter 5 The Relationships Between Man and His Environment: A Systemic Approach to the Viability of System Earth
Chapter 6 Socio-ecological Viability and Legal Regulation: Pluralism and Endogeneity
Chapter 7 The Paradigm of Coviability Defined by the Adequacy Between Social Usefulness and the Ecological Function: The Legal Challenge of the Socio-ecological Connection
Chapter 8 Local ecological knowledge and the viability of relations with the environment
Chapter 9 A Biological Approach to Coviability: Biotics Interactions and Dynamics of Biodiversity
Chapter 10 A Geographical Approach of the Socio-ecosystem Coviability
Chapter 11 A Rupture Between Human Beings and Earth: A Philosophical Critical Approach to Coviability
Chapter 12 When Coviability Meets Ecosystem Services: The Case of Reunion Island's Coral reefs

An ontology of coviability and a synthesis of this transdisciplinary exercise will be the subject of the last two chapters of the volume, just before the concluding chapter.

The intention here is to build an ontology of coviability. This ambition requires a work of construction beyond disciplines, an attentive listening open to collaboration and a continuous maturation in the requirement of thinking. Such an enterprise is time-consuming and the present research is capturing this challenge on the medium term. The present volume is just the primary milestone. As an early outcome of this exercise, a first trans-disciplinary summary based on the contents of this volume will be attempted in the last two chapters. However, it will be necessary to consider them as the instigation of more ambitious work to be pursued.

From now on, it should be noted that a first experiment is attempted, which founds the basis of a summary by relying on an analytical reading of each text, followed by the construction of heuristic maps. A heuristic map, often referred to as a cognitive or mental map, is a diagram that allows the associative path of thought to be visually represented. To that end, for each chapter, the few keywords or "reminder words", that are particularly relevant and necessary for understanding and memorizing the text, are retained. These words serve as a basis for each map, which is both a snapshot of the reader's feelings, with all the ambiguity of an external or non-specialist interpretation, and a vision of the essential terms of the subject, with the links connecting them (refinement or correlation).

Three ways are therefore possible to achieve this:

- Each author is responsible for feeding, from his writing, the undertaken ontological work. Backed by a specialist of the method, he/she participates in the ontological construction. However, the author can rightly consider that the elicitation work that he/she is producing, is already the needed research;

- A single reader seizes the whole and thus tackles a set of fields of which he/she is not necessarily specialist. This course of action presents at the same time the advantages and disadvantages induced by the uniqueness of the reader, in particular the induction of different biases according to the theme and disciplines;
- A plurality of readers makes it possible to blur the variable biases of understanding while reading the meaning given by the author, because of the meanings understood according to the individual and his disciplinary affiliation.

We chose here the second way, assuming the "risk of bias" presented above.

Using the Freemind software, we have drawn the maps presented below for the chapters in Part I (and in the introductions of Parts II and III). The Chap. 26 summarizes this "Transversal ontology analysis".

The Chap. 2 summarizes, through a very general vision, the major issues around coviability and relies on the concepts of system, viability, interaction and coevolution of natural and human systems (Fig. Preview 1.1).

Chaps. 3 and 4 from a mathematical perspecive, deal with the theory of viability and that of systems and their control. The theory of viability is based on the notion of evolution (deterministic or multiple) implemented by an evolutionary engine based on regulation laws (Fig. Preview 1.2).

The theory of systems and their control expresses coviability as a harmonious evolution of connected systems, which evolve together towards one or more equilibrium states (Fig. Preview 1.3).

Chapter 5 proposes the systemic perspective of the Earth system. The latter is a complex system in which various interactions exist within the environment in which it is immersed. The viability of this system depends on the coviability of these interactions (Fig. Preview 1.4).

Chapter 6 describes the legal and anthropological aspects declined in terms of legal regulation system (Fig. Preview 1.5).

Chapter 7 discusses the paradigm of coviability in terms of the adequacy between the social utility (relationship to natural elements and the satisfaction it provides in

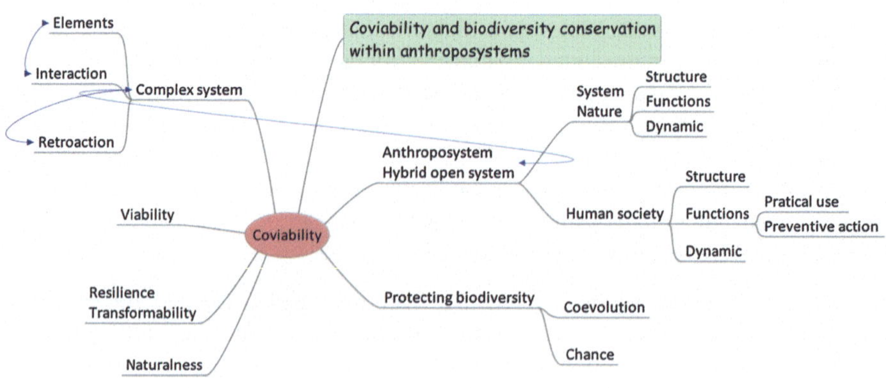

Fig. Preview 1.1 Mind map of Chap. 2 (®Thérèse Libourel)

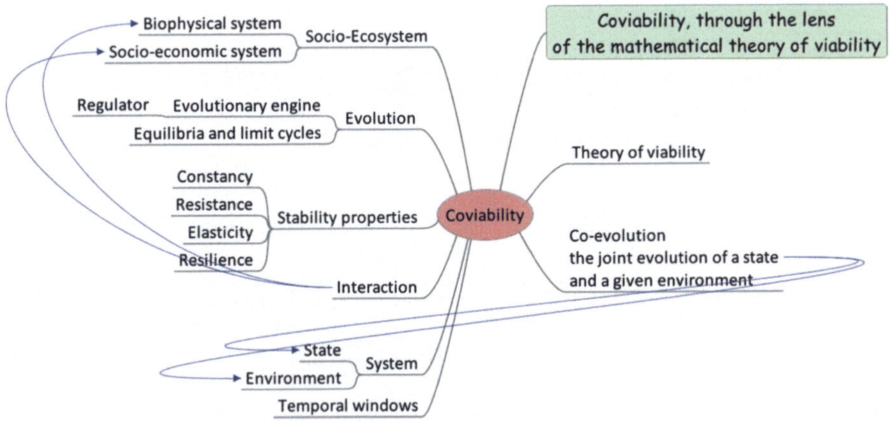

Fig. Preview 1.2 Mind map of Chap. 3 (®Thérèse Libourel)

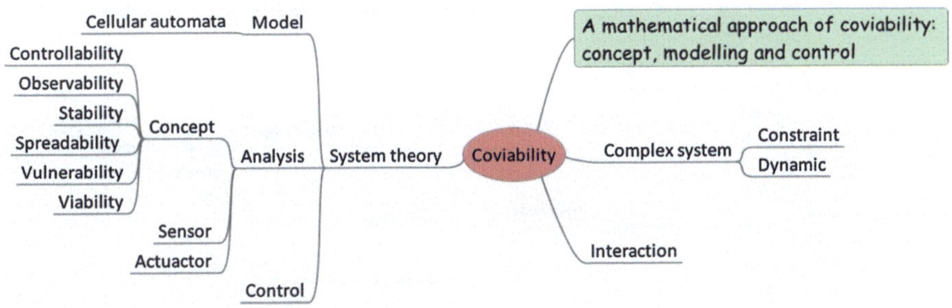

Fig. Preview 1.3 Mind map of Chap. 4 (®Thérèse Libourel)

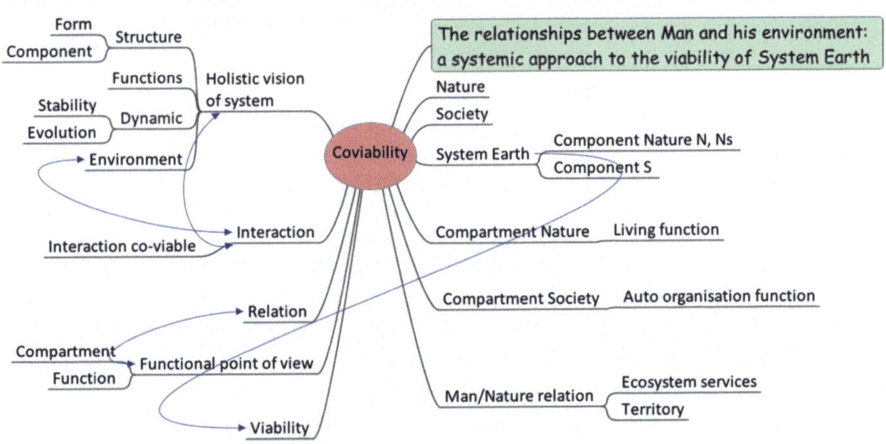

Fig. Preview 1.4 Mind map of Chap. 5 (®Thérèse Libourel)

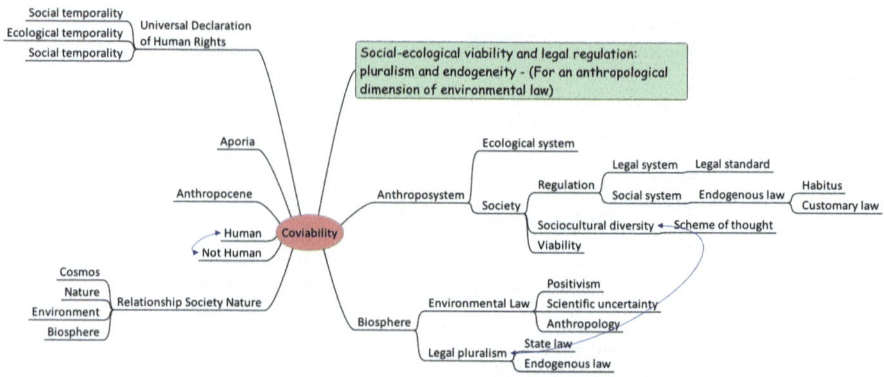

Fig. Preview 1.5 Mind map of Chap. 6 (®Thérèse Libourel)

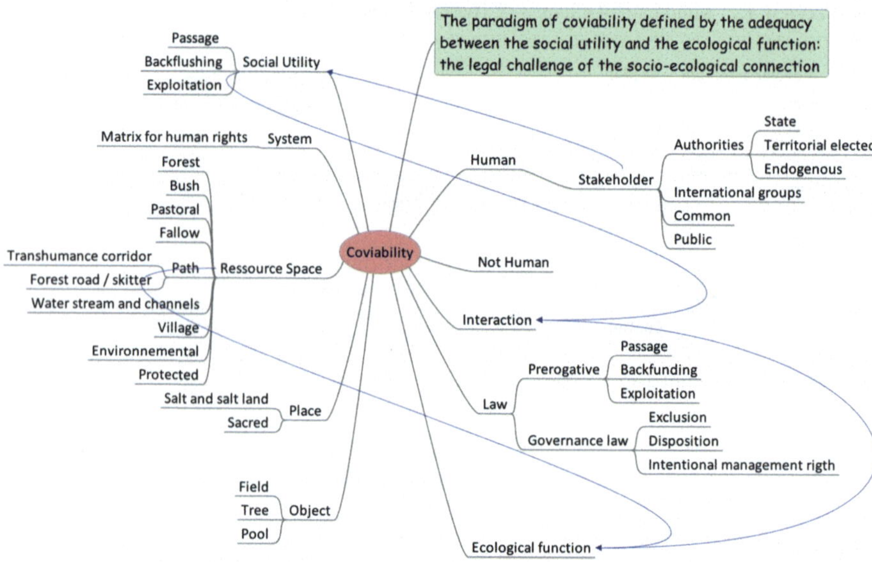

Fig. Preview 1.6 Mind map of Chap. 7 (®Thérèse Libourel)

meeting societal needs) and the ecological function. The reading focuses on a matrix of the law of utilities (Fig. Preview 1.6).

Chapter 8, dedicated to the ethno-ecological vision, reviews the fundamental systems and their dynamics, with coviability at the heart of negotiations based on practices, uses and knowledge (Fig. Preview 1.7).

Chapter 9 on the biological approach, deals with interactions between biotic and abiotic elements and highlights the various dynamics of biodiversity (Fig. Preview 1.8).

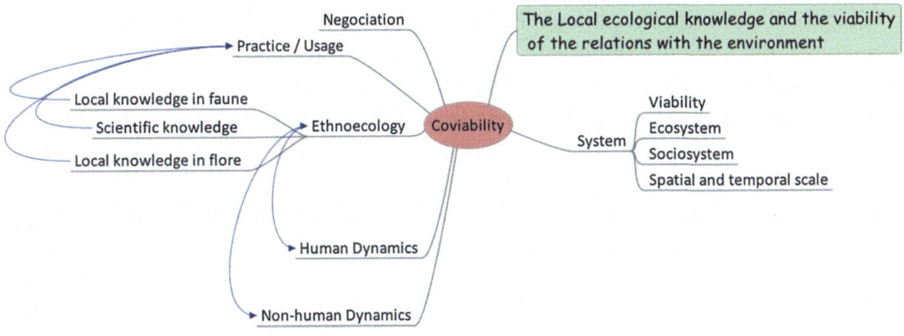

Fig. Preview 1.7 Mind map of Chap. 8 (®Thérèse Libourel)

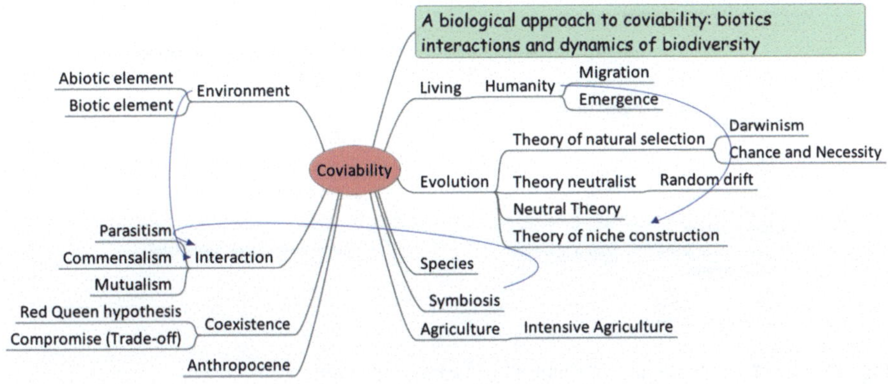

Fig. Preview 1.8 Mind map of Chap. 9 (®Thérèse Libourel)

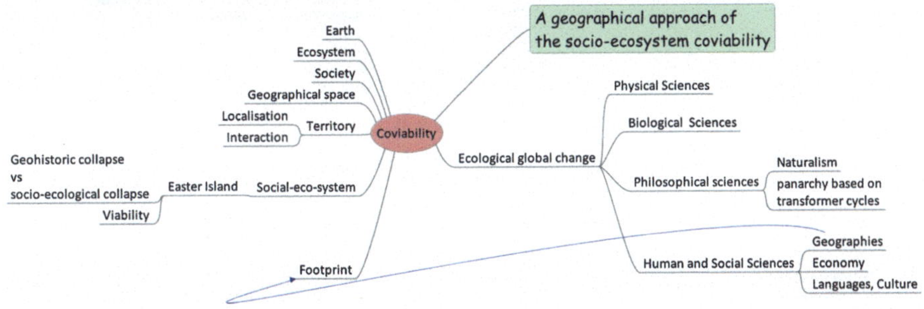

Fig. Preview 1.9 Mind map of Chap. 10 (®Thérèse Libourel)

The geographical perspective of Chap. 10 analyzes the relationships of societies with the Earth via the concept of geographical space defined as the "social product and the system of relationships between sites/places". The reasoning is supported by the representative case of Easter Island revisited using this approach (Fig. Preview 1.9).

Chapter 11 provides a philosophical perspective flowing from critical theory.[1] It analyses the various ontologies echoing the perceptions between Nature and Society[2] and the approach proposed by Marx which highlights the rupture of the interaction between society and nature triggered by capitalism. The modern way of coviability should be based on new intellectual, legal and political referentials (Fig. Preview 1.10).

Chapter 12 focuses on the importance of the feedback loop between ecosystem and socio-systemic services in the human-nature coviability (Fig. Preview 1.11).

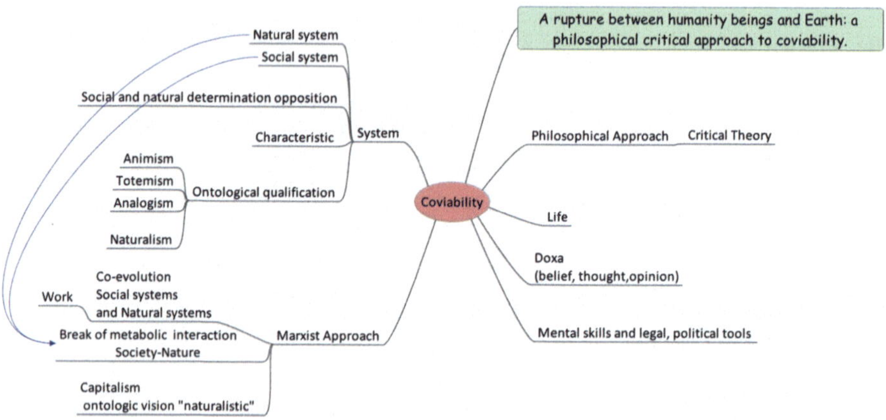

Fig. Preview 1.10 Mind map of Chap. 11 (®Thérèse Libourel)

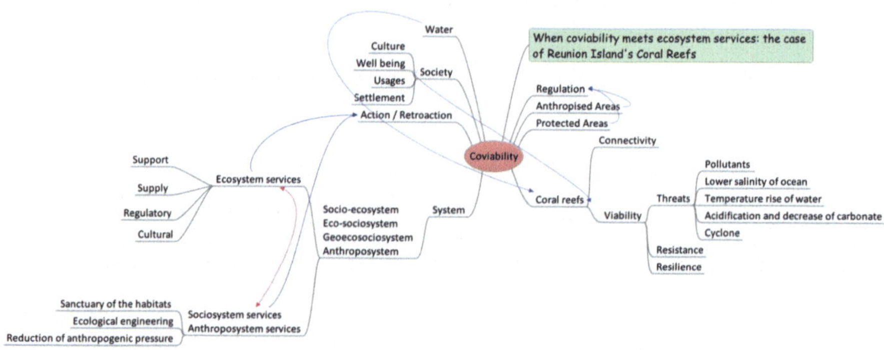

Fig. Preview 1.11 Mind map of Chap. 12 (®Thérèse Libourel)

[1]The term "critical" here is understood in the Kantian sense of a thinking on the limits of the claims of certain types of knowledge, be they metaphysical, philosophical or scientific, according to Alvarenga, Raphael & Carre, Louis (2008), "Théorie critique", in V. Bourdeau & R. Merrill (dir.), *DicoPo, Dictionnaire de théorie politique*.

[2]Descola Philippe (2005), Par delà nature et culture, Gallimard.

Chapter 2
Coviability and Biodiversity Conservation Within Anthroposystems

Christian Lévêque

Contents

2.1 Introduction

One of the great concerns of the beginning of the twenty-first century is the environment and the protection of the biodiversity impacted by human activities. If the destruction of emblematic environments such as the Amazon rainforest or the forest of Borneo challenges us because, in the eyes of Westerners, they are still «wild» environments, what does the protection of nature and biodiversity mean in societies who have co-built their environment over the long-term, following the European

C. Lévêque (✉)
French Research Institute for Sustainable Development (IRD), Marseille Cedex 02, France

O. Barrière et al. (eds.), *Coviability of Social and Ecological Systems: Reconnecting Mankind to the Biosphere in an Era of Global Change*,
https://doi.org/10.1007/978-3-319-78497-7_2

example? Is there any sense, or legitimacy, in considering biodiversity conservation from a coviability perspective? And in this case, what are the objectives that could be fixed in the long-term, and what difficulties are we coming up against?

2.2 What's Nature to Some People Isn't Nature to Others: Each to His Paradigm

Over the centuries, the relationship of man to nature has given rise to many exegeses on the part of philosophers. Their reflections, sometimes impregnated with religious assumptions, have namely focused on the man/nature or societies/nature divide. Behind these debates lies the question of the nature of man and of his position in the animal world (Descola 2005, 2011).

There may be a fairly general consensus on wanting to live in an idealized environment: no pollution, a good management of natural resources, a luxuriant and welcoming nature, etc... Who has never been tempted by this myth of a harmonious humanity, maintaining good relations with a paradisiacal nature, in which all of the troublesome species and all of the problems of everyday life would be obscured? But going back to reality is more prosaic . . . .

For many people, nature is a refuge value, synonymous with health, beauty and the sublime. It is this more or less elaborate dream that some militant ecological and/or conservationist movements with environmental ethics propose to us. Depending on their sensitivities, they refer to "virgin" or "pristine[1]" nature, of a life "in harmony with nature", of "wild nature", of "naturalness". For scientists, there was the mirage of the climax, the ultimate and ideal stage of a balanced nature if man did not impede its dynamic (Lévêque 2011). For managers and some scientists, there are now these notions of "good ecological condition", "biotic integrity" and "ecosystems health".

The concept of ecological integrity was outlined by Leopold (1949) on an ethical basis in order to preserve the "integrity, stability and beauty of biological communities". Strictly speaking, preserving the integrity of an ecological system equates to maintaining the components of the system (its structure) and their interactions (its functions). The loss of a component or an interaction supposes the loss of system integrity because something has changed. The concept of "ecosystem health" is quite similar to that of biotic integrity. It corresponds to a typical organicist approach, popular at the beginning of the twentieth century in ecology, based on the metaphor according to which ecosystem functioning is comparable to that of living organisms (Rapport 1995; Rapport et al. 1998; Boulton 1999). The good environmental condition of the European Water Framework Directive is based on physical, chemical and biological parameters. It gives a preponderant place to biology, and biodiversity is erected as justice of the peace. The good condition refers to a standard condition of a system which is not or which is little impacted by human activities, which is the focus of debate among scientists (SDAGE Rhône-

Méditerranée 2011; Bouleau 2013). All these concepts are based more or less on the idea of a system in equilibrium, untouched by any human activity, which would be the reference of "true" nature. Pristine: is another expression for a nature, which is untouched by any human intervention.

But man lives nature differently in his daily life. Nature is of course beautiful. It provides resources, and everything we need. But it is also a nature fraught with danger, with its predatory animals, diseases among man and cattle, crop pests, excesses ... And human societies have devoted a significant proportion of their means to protect themselves from the perils of nature and biodiversity. It is this ambivalent relationship of interest and fear of nature (Terrasson 1991) that has structured the relationships between societies and their environment. Conservationist discourses distil their vision of an idyllic nature, which is not that of the citizen.

There are therefore very different projects at the level of societies, between a nature imagined by theorists, and a nature lived and desired by citizens. At this moment in time, it is the first group that imposes its views by surfing on a stereotypical speech propagated by major nature conservation NGOs, based on dramatization, and widely adopted by the media. The result is concepts and regulations that are out of line with the realities in the field and citizens expectations, such as "good ecological condition" or "identical" compensation measures for destroyed ecological systems.

Let us remember that whilst there are ecological systems from which man is absent, there are no (at least, not yet . . .) human societies on earth which live independently of an ecological system from which they exploit the resources . . . . It is clear that man is dependent on the nature from which he is born, and which he needs to survive, while protecting himself. It is in this ambivalent relationship that our relationship with nature is positioned, and not in an idyllic or rural vision of a nature in which man would be absent.

2.3 The Anthroposystem: A Conceptual Framework to Study the Interactions Between Social and Ecological Dynamics

For a long time, European societies have modified and shaped environments and landscapes by using land to develop their agricultural or industrial economy. They have also perfected techniques and developed the appropriate facilities to exploit natural resources and cope with their environment constraints (what are sometimes called techno-systems). They have somehow forged their own cultures and social organizations in different bioclimatic contexts, through various uses of nature and its local resources.

This finding that nature and biological diversity, at least in Europe, are the result of a long process of human/environmental co-evolution, has led to the development of the concept of anthroposystems, which distinguishes itself from both the purely natural (ecosystem) and purely social (socio-system or socioeconomic-system) or geographic (geo-system) approaches (Bertrand and Beruchachvili 1978).

The concept of anthroposystem is consequently defined as an interactive system between two sets constituted by one or several socio-systems and natural ecosystems and/or artificial ecosystems in a given geographical space and evolving over time. The anthroposystem is therefore a hybrid system. It is also an "open" system, interacting with other anthroposystems, whose functioning is subject to the forcings of external factors (Lévêque and Muxart 2004; Lévêque and van der Leeuw 2003).

The Notion of "System"

The concept of "system" is an intellectual construction designed to understand and explain the interactions within a complex studied object. Three broad general principles underlying the properties of a system can be identified:

- The system consists of a set of elements that interact, in such a way that the functioning of each of them is conditioned by the others. These elements are therefore not isolable. In the societies/nature system, if man acts to modify his environment, the environment in turn influences social, cultural and technological behaviors.
- If each element of the system contributes towards its operation, the system acts in return on its components. The principle of feedback concerns the feedback from the whole system to each of the parties. There are positive feedback loops that accelerate action in the same direction and negative loops that act in the opposite direction.
- The interacting elements of the system constitute an entity from which new properties "emerge" in relation to the properties of each of the elements. It is the principle according to which "the whole is more than the sum of the parts". In a cell, life is an emerging property that is not only the sum of the properties of the organs. However, the emergence principle is difficult to identify in ecological systems (Blandin 2009).

In summary, the complex systems theory leads to the recognition that the system is an organized entity, often hierarchized, able to self-regulate, with an adaptation or even an optimization of the structures in order to respond to external perturbations.

2.4 From Co-evolution to Coviability

According to various authors (Aubin 1996; Aubin et al. 2011), the viability of a system depends on its autonomy, that is, on its ability to maintain itself through its self-organizing processes, in varied and changing environments. This implies good adaptability to a variety of situations. Some speak about resilience, which is defined as the reconfiguration of structures and processes following a disturbance, consequently generating new system trajectories (Folke 2006). Viability is also defined in the coupling between natural and social components (Catanzano and Rey 1997). As a first approximation, we can think that this definition applies to the global functioning of anthroposystems which, by essence, pose as a scientific object, the interactions between the dynamics of natural or artificial systems and those of human societies.

But what about the application of the coviability concept to biodiversity protection that is only one element of anthroposystems, just like fisheries for example? To speak of coviability with respect to interactions between systems, refers to the mechanisms of coevolution. For a biologist co-evolution means that there are constant adjustments to adapt to changes in the environment, without there being a definite objective, because biological evolution does not follow a pre-established plan. Evolution leaves much to chance and contingency (see below). The protection of biodiversity is not situated within this context however. On the contrary, we propose to take action to modify the system trajectory in a desired direction. Some authors (Weber 1995) have stressed the importance of setting long-term objectives within a coviability perspective! In other words, coviability implies that the anthroposystem has to be steered in order to meet the objectives that society has set itself, so that each of the subsystems finds its interest there, with of course some concessions, because each element cannot just pursue its own dynamic. We could therefore say that co-viability is steered co-evolution...

If applied to biodiversity protection, this coviability/co-evolution is part of a dynamic approach that presupposes permanent modes of adaptation in relation to the objective that is being set. This objective can itself evolve in space and time. In this context we can no longer speak of autonomous self-organized systems. There is a need for steering in order to practice an integrated and adaptive management at different scales of time and space. If we take the example of fisheries, fish populations have their own dynamics, the viability of which can no longer be demonstrated when man does not intervene. It is not certain, however, that the viability of human societies is ensured without fishing, but this is open to discussion. Whatever the case, the coviability of a fishing system requires that a certain number of ecological and social rules be respected in order to maintain a reproductive stock over the long term. Such rules can change if the environmental or economic context also changes.

Protected areas are one of the responses provided by conservationist movements in order to protect biodiversity (Aubertin and Rodary 2008). We will not expand on this topic here, except to say that they can be part of a use perspective (which is not the case at present) if they are given a "recreational use" role, as opposed to spaces forbidden to the public...

More generally, current biodiversity protection in the context of coviability must therefore be part of a reflection on the uses of our ecological systems, with a view to multi-functionality. It is by maintaining or further developing these uses according to needs and circumstances that we will maintain or not the diversity of environments and species that are dear to us, if it is our societal project to preserve them.

However, the question of the perpetuation of such an approach within the context of the global changes currently underway must be posed. This once again raises the definition of long-term objectives. We can repeatedly speak of biodiversity protection, but which protection and in what way for systems that are continuously evolving on different spatial and temporal scales?

2.5 What Biodiversity Are We Looking to Protect?

Within a coviability approach, it is essential to clearly define the objectives. In this context, the debate about nature representations is fundamental (Liarsou 2013; Casetta and Delord, 2014; Gunnell 2009). In terms of biological diversity protection in Europe, there are two main types of paradigms:

Either it is considered that biodiversity in Europe is a co-production of societies/nature, and that its protection should be considered within the framework of heritage management, in which case it is legitimate to intervene on the dynamics of species and ecological systems, through practices and uses. Or we fix ourselves the objective of a return to a non-anthropic nature and to a pristine biological diversity, in which case we have to erase the facilities developed and let "nature regain its rights" according to the accepted expression ... This should result in drastic changes on the economic level. However, many nature protection or ecological system restoration projects advocate the return to a previous state, ideally little impacted by human activities ... In broad terms, in one case we speak of a "cultural" nature, in the other of a "natural" nature.

2.5.1 A Hybrid or Cultural Nature in Europe

Without going back to the many exegeses to which the term "nature" has given rise, let us simply recall that nature, in the strict sense, is that which has not been produced or modified by man. "Virgin" nature symbolizes, especially nowadays, the true, the authentic, the wild environment... However, what we refer to as "nature" in France are actually landscapes which have been shaped and constructed by man for thousands of years ... our rural landscapes. (Mathieu and Jollivet 1989; Micoud 2005). "The environment is nature, and nature is the countryside," Mathieu and Jollivet (1989) used to say when speaking about the notion of the environment to city dwellers. The forests that previously invaded the continent have been replaced by a mosaic of diversified territories: fields, meadows, vineyards, planted forests, scrublands etc. A diversity of environments creating a heterogeneity which is eminently favorable for biological diversity to flourish ....

Over the centuries, territorial development corresponded to needs related to agricultural or industrial activities: the use of biological diversity and ecosystems products; the use of services (driving forces of water (mill) or air, the self-treatment capacity of water, etc.). But also the desire to protect oneself against natural risks (floods, fire, ...) or diseases (swamp drying). Without counting the many plant and animal species that man has introduced and which have become naturalized on our soil.

As a result, biodiversity in European regions is no longer "wild". It can even be said that most of our so-called "natural" environments, such as the Camargue, the Sologne, the Landes and Mediterranean forests, the Alpine tundra and, of course, our marvelous hedged farmlands, are actually shaped and managed by man in response to particular uses. Their temporal dynamics is closely dependent on these uses. Our nature is obviously not "virgin": it is a hybrid nature that owes as much to the action of man as to spontaneous ecological processes.

It is within this context that Perrein (1993) proposed the notion of "bio-heritage", which consists in techniques and practices, through the use of resources or ecological systems, which serve to maintain these environments as they are. This concept of bio-heritage is goes hand in hand with the notion of "bio-history". *"Bio-history is the daughter of ecology by recognizing man as one of the determining factors in ecosystem structure and functioning in the same way as soil or climate" (Perrein* 1993*).*

2.5.2 The "Naturalness" of Militant Movements

The representation of "wild" nature, the "wilderness" of Americans, is at the heart of environmentalist movements. This nature, conceived as part of a territory free of human influence, this "virgin nature", is clearly a myth in the majority of European countries whose territories has been subjected to the action of man for thousands of years. But this representation is still going strong and has numerous followers.

What is sometimes called the return to naturalness has its supporters (Schnitzler et al. 2008) who also speak of "decolonized nature". This naturalness is associated with the state of spontaneous nature, independent of human activities, although it integrates anthropic heritage (Lecomte 1999). This naturalness *"attributes ... a strong intrinsic value to the spontaneity of every process, even if there is a "loss of biodiversity "or at least a modification of biodiversity* (Schnitzler et al. 2008).

This naturalness option is also presented as the will to protect the nature of the future: we have to accept new trajectories, proposed by nature alone, without human intervention other than observation. Naturalness is therefore a bet on the future insofar as the environments have dynamic trajectories with assemblies of species that it is difficult to predict and much less to control.

This naturalness option is to some extent related to the public attraction to the wilderness, nevertheless with its limits. Indeed, if wild nature makes our citizens dream, they don't want species that disturb them, the first which spring to mind being symbolic species, such as the wolf, but also predators in general, along with all species that create nuisances or are responsible for diseases among humans, domestic animals or cultivated species.

Compared with this principle of spontaneous naturalness, certain people have stated that laissez-faire in this matter leads to the closure of environments and therefore to changes in biodiversity. We are at the heart of the debate: maintaining a patrimonial diversity (often emblematic) or accepting the risks of an uncontrolled evolution, but in which one finds intrinsic processes corresponding to "true" nature. It should be noted, however, that the intrinsic value attributed to these processes is an ethical value and not a scientific one.

We may observe a certain ambiguity when we speak of a spontaneous nature: what happens when naturalness opens the door to a certain number of undesirable species? Conservationist environments are constantly shunning the issue of species once qualified as "harmful", now called "pests" to present an idealized image of biodiversity. However, considerable sums of money are spent every year to combat a part of biological diversity in order to preserve human, domestic animal and cultivated plant health. Why are these two major biodiversity-related areas not communicating?

2.6 Some Methodological Difficulties Which Oppose the Concept of Coviability

The notion of coviability is confronted with a number of methodological difficulties, including uncertainties as to the temporal evolution of systems under the influence of exogenous hazards, but also owing to their own dynamics. The role of chance is clearly underestimated in most reflections, and anticipation is difficult because of the many factors involved. This raises problems of decision-making in a situation of uncertainty, within the framework of a policy based on objectives that have been set. If the biodiversity concept is intellectually appealing, is it operational?

2.6.1 Chance, Conjuncture, Contingency

One difficulty facing the coviability concept is the role of chance. We generally have a negative view of chance, because it evokes phenomena that we cannot predict or explain. Yet, the paleontologist J. Gould (1991) has largely talked about the positive role of chance and contingency in evolution.

The ecology of origins, impregnated with the mechanistic and deterministic vision of physical systems, had set itself the objective of identifying general laws allowing, as in physics, to simply describe the functioning of ecological systems. In reality, the exercise proved difficult because there are no laws in ecology that are comparable to those of physics, as John Lawton (1999) pointed out. The history of ecological systems is largely random (Pavé 2007). Their nature and the species they host are indeed the result of a history that was far from smooth ... If we refer to the history of biological diversity in Europe, chance and conjuncture played a major role in the recolonization of land released from the ice by current flora and fauna (Lévêque 2011). We must realize that hazards of all natures are part of the functioning of ecological systems as we have been shown by retrospective ecology.

"The role played by chance must challenge us on the scientific perception of nature. The term "ecosystem" is indicative of a deterministic, structured vision, where chance is seen only as background noise which blurs our perception. However, it plays a central role: ecology must recognize it and then carry out its "Copernican revolution", change its paradigm".[1]

In short, ecological system dynamics do not only depend on deterministic laws, and chance is omnipresent, which is very inconvenient for managers because they do not know how to manage it ... Recognize the role of chance and instability is a little like opening Pandora's box for managers who find the principle of the balance of nature far more comfortable. Under these conditions, it is very difficult to predict the long-term evolution of ecological systems.

Contingency is also a major feature of ecosystem ecology. In effect, a great source of confusion resides in the globalization and the amalgamation of situations from which hasty generalizations are drawn, which no longer correspond to the realities in the field. Virtual ecology has promoted these globalizing processes that have nothing to do with the wide variety of local situations. Indeed, each region, even every territory, has its climatic and evolutionary history. The situation in Europe is nothing like that of the Amazon or Australia! And each region has its history of human relations with nature, according to population density and the value systems of the societies concerned.

Clearly, the same discourse cannot apply to European biodiversity and biodiversity in other continents. The status of biological diversity cannot be caricatured by extrapolating local and cyclical observations to the whole world. However, this is what conservationists' movements are constantly doing.

[1] http://www.alain-pave.fr/une-aventure-scientifique-et-humaine/

2.6.2 Uncertainties Related to Temporal Trajectories

A difficulty with coviability lies in the paradigms that are used to address the functioning of ecological systems. For a long time we have based ourselves on the concept of the balance of nature. And although it is known that equilibrium (or stationarity) is a decoy in ecology, we continue to reason, in science as well as in the domain of management or conservation, as if a "balance of nature" state, and therefore an immutable biodiversity, which must be preserved as it is (Lévêque 2013), really existed. Field ecologists are well aware that natural environments are not frozen, and that they change over time. The term "resilience" refers to the ability of an ecosystem to maintain its essential functions in spite of the disruptions it is subjected to. In fact, this concept of resilience, which recognizes the variability of ecological systems, has long remained anchored in the notion of stability, since it speaks of returning to the initial state! More recently, the concept evolved towards the question of "transformability", which is the ability to create a new system, different from the previous one, when economic and social ecological conditions evolve (Folke 2006; Mathevet and Bousquet 2014).

Temporal dimension is a highly structuring element: in fact, the anthroposystem is the product of a history and its future lies within the continuity of previous dynamics, but also in the consideration of of changes and new or random factors which in turn, will interact with the previous ones. This evolution is the result of factors which, most often, are exogenous and which are of a higher hierarchical level than those acting within anthroposystems. In this context, the coviability of each subsystem would also be adapted to these global changes and not just to their own interactions.

We are therefore clearly in the co-evolution perspective, which implies both compromise and adaptation, but also choices that can lead to abandoning certain paths in favor of others.

The climate has continued to change over the millennia. It has been one of the main factors in biodiversity evolution. While anthropogenic forcing seems to be the main driver of the changes currently underway, the solutions to this problem are still very difficult to implement! Inevitably, according to the IPCC, we must therefore expect significant climate changes over relatively short time scales compared to the human scale. For example, the Explore 2070[3] program (MEDDE 2012) envisages a possible increase in mean air temperatures of between +1.4 °C and +3 °C, a decrease in summer rainfall over the whole of France, on average to the order of between −16 and –23%, as well as a drop of between 10 and 40% in average annual river flows according to simulations.

The magnitude of these changes must lead us to question the future of our anthropized natural systems, and the relevance of conservation/restoration measures that are being put in place, sometimes at great expense, without anticipating the future. In France, the Camargue is currently subject to severe erosion, which is partly the result of a reduction in sediment inputs from the Rhône resulting from the construction of dams on the river. For decades, it has been protecting itself behind dikes, but their effectiveness is questionable should sea level rise as predicted. If sandy belts were to be destroyed, the retreat of the coastline in the Saintes Maries-de-la-Mer region would be approximately 13 km, and the low submersible lands extending to the Vaccarès Pond would disappear under the waters of the Mediterranean (Duvat 2012).

In summary, long-term goals can always be set, but it is unclear whether or not they will still be relevant in a few decades time, when the overall context has changed. The particularity of adaptive systems is that we cannot extrapolate trends because their trajectory is not predictable.

2.7 How to Respond to Changes in Use and Practices?

Biodiversity is primarily based on habitat diversity. European landscapes and ecological systems are the result of developments for utilitarian or security purposes. By dividing up and opening landscapes, man has created heterogeneity, which is favorable for the existence of a rich biological diversity.

The way in which land use planning is envisaged and the manner in which the agricultural matrix will be managed in the twenty-first century constitute a major challenge for the preservation of biological diversity.

2.7.1 Maintaining Practices?

Many of our countryside's emblematic species are dependent on agricultural practices and on areas managed by agriculture. A rather frequent situation is the abandonment of certain uses of land, as practices have considerably evolved over time. If these environments are no longer managed to provide the expected services, we must question their future ... Should we maintain the type of management to which they were subjected in order to preserve them? Or, insofar as they are artificial systems, do we have the right to make them evolve into new types of uses and new types of ecological systems? And which ones? Bearing in mind that changing the use can also mean a change in biological diversity!

While it cannot be totally generalized, it appears that the abandonment of management practices often leads to a reduction in biodiversity itself, in particular by the scrubbing and closure of environments for terrestrial systems. As pointed out in a report on the Mediterranean forest, "*working to maintain ecological biodiversity requires working mainly to maintain the diversity of practices and practitioners*".[2]

[2]http://www.developpement-durable.gouv.fr/IMG/pdf/Explore2070_4pages_Hydrologie_surface.pdf

So is it the decrease in the diversity of practices and practitioners that is causing the loss of biodiversity itself[3] . . . ?

Various European agri-environmental measures have been put in place to perpetuate the ecological systems resulting from economic activities, which have now been abandoned. The limestone hillsides of the Seine estuary (France) were used extensively during the nineteenth century for agricultural and livestock activities (Dutoit and Alard 1995). The dry grasslands are particularly rich in plant species and protected species and have been classified as a Natural Area of Special Ecological, Faunistic and Floristic Interest (ZNIEFF in French law).

Abandoned by agriculture since the 1950s, these grasslands are becoming wasteland and are consequently losing some of their remarkable species. To try to remedy this situation, the Haute Normandie Conservatory of Natural Sites has carried out restoration operations. The maintenance phase is most often carried out by extensive itinerant grazing with rustic animals, ie sheep, horses, cows. The aim of these actions is to maintain open environments, consequently allowing the expression of numerous plant and animal species, which are a testimony to site biodiversity. These measures are undoubtedly interesting, but have a cost ... and their scope is contingent.

We also find ourselves confronted with slightly Kafkaesque situations, as in the case of compensatory measures. In the case of Notre Dame des Landes (France), compensatory measures concerning areas that would eventually be destroyed by the construction of the airport pose an interesting problem. Following the Loire-Bretagne Master Plan for Water Development and Management (France), these compensations should result in the "recreation or restoration of equivalent wetlands from the point of view of functionality and quality of biodiversity». It should be noted that in view of the implementation of this project, for the past 50 years farmers have maintained extensive farming practices, while the various types of habitat in the surrounding watersheds have been modified owing to agricultural intensification.

In summary, the habitats and species in the areas to be compensated correspond to what existed at the time that the land was frozen in anticipation of the construction of the airport. Therefore, systems we are intending to compensate are obviously not virgin natural systems, but rather agricultural systems, which are consequently anthropized. If we pursue this train of thought, it would be necessary to recreate the agricultural conditions of the 1950s in order to compensate the areas to be transformed ... Whether we are for or against the construction of this airport, which is not the subject here, the application of the law poses almost insurmountable problems! (de Billy et al. 2015).

[3]https://www.google.fr/webhp?source=search_app#q=foret+m%C3%A9diterran%C3%A9enne+usages

2.7.2 *Main Criticisms of the Compensatory Measures Proposed for the Construction of Notre Dame des Landes Airport (France) (de Billy et al. 2015)*

The finalization of the compensatory measures raises a number of concrete problems. The compensation of the wetlands due to be destroyed during the construction of the Notre Dame des Landes airport is a case study for the application of the ARC (Avoid-Reduce-Compensate) Directive. Firstly, it was necessary to ensure that the first two terms (avoid-reduce) had been respected, which led to some points of the initial plan being revised. However, discussions namely focused on the definition and application of compensatory methods, and the proposals of the consulting firm responsible for the compensation case gave rise to severe criticism by the Committee of Experts responsible for evaluating them:

- Lack of information and anticipation on the availability of land near to the proposed development. The farming profession is opposed to the compensation project owing to its "double penalty" (loss of agricultural land for the construction of the airport and changes in agricultural practices for compensatory measures).
- Lack of a systemic approach, notably with regard to the landscape organization, and the initial state of the site is is known to be insufficient.
- Fixit vision of ecological systems. The initial state and the assessment of the need and the compensation response are based on the principle that ecological systems are stationary. They do not take into account the temporal dynamics of ecosystems.
- Lack of prospective and anticipatory vision regarding the possible future of wetlands in the coming decades in connection with climate change, which raises the question of the relevance of certain actions, proposed in a context of climate change and the possible reduction in rainfall.
- The regulatory texts recommend "identical compensation" which is often utopian. In an attempt to answer these questions, methodologies based on ecological functionalities have been proposed. However, such texts are incomprehensible to the general public.
- The risk of the failure of ecological engineering works is not taken into account, whereas feedback indicates that objectives are rarely met. The evaluation of the effectiveness of compensatory measures requires the simultaneous implementation of test and compensation site monitoring to check compliance with the initial objectives. This has not been provided for.

For aquatic systems the European Water Framework Directive refers to "good ecological condition". Is it a legitimate concept on the scientific level or is it simply a utopia? Scientists remain dubious. The difficulty lies in the definition of the state of reference that is set as an objective with respect to restoration. How can we precisely define it to enable managers to refer to it and ensure it is socially acceptable? These are all governance issues, which are relatively little debated insofar as the single mindset of conservationists, widely relayed by the media, tends to impose itself in politics and evade disputes (Projet BEEST 2011).

The RM & C Water Agency provides an operational interpretation of good condition. It is not a condition, which would correspond to situations where anthropogenic constraints are absent. The good condition of waters corresponds to "the conditions allowing the proper functioning of the ecological processes, in particular the presence and the maintaining of aquatic, floristic and faunistic communities ... Consequently, the good condition assumes a certain level of human activity and guarantees a certain balance between activities and uses".[4] This definition is close to the notion of good ecological potential also referred to by the European Directive for highly artificial water bodies.[5] But it must be recognized that these definitions are quite vague in comparison with chemical standards that give precise (and sometimes questionable) figures on acceptable doses in waters.

2.7.3 *Back to Naturalness?*

The developments and ecological systems consequently created can of course be called into question in the context of an approach that advocates a return to an earlier state supposedly little impacted by human activities. This is currently being considered with the concept of the ecological continuity of watercourses. But we must clearly specify what we expect from these restorations. This is the highly debated issue of baseline in ecological restoration programs. Is it a nature that excludes man? Or is it a nature developed by and for man? Do we want to make French territory a protected area, a preserve for militant lobbies? Do we want to look for new uses of natural environments? For wanting to keep ecological systems as they are is unrealistic in the long-term.

The plantation in France of the Landes forest in the nineteenth century had the objective of fixing dunes that threatened certain villages as well as promoting marsh areas for agro-pastoral use by planting pines whose resin was sought after by industrialists. However, as usage evolves over time, resin extraction is no more than a sideline activity. We can discuss the choices that were made in the nineteenth century, but the Landes forest has become a legacy and we can wonder about the reception that a deforestation program would receive were we to return to the landscapes of yesteryear.

[4]SDAGE Rhône Méditerranée. 2011. Qu'est-ce que le bon état des eaux? Note du Secrétariat Technique du SDAGE. http://www.eaurmc.fr/?eID=dam_frontend_push&docID=1823&xtor=RSS-4

[5]Artificial water bodies (AWB) or Highly modified water bodies (HMWB).

2.7.4 Changes in Allocation and Use?

The uses of anthropized systems have of course evolved over time. In some cases, there is no abandonment but a change in use (Papy et al. 2012). Consequently the Sologne and Brenne Ponds might have lost their importance for fish production, but they are extremely sought after for hunting, their rental being very lucrative. Without this transfer of use, it is a safe bet that many of these ponds would have been drained to grow maize.

The steam engine has supplanted the water mills, but electric power production has taken over in the case of dams on watercourses. While some ecologists and militant movements are calling for the restoration of ecological continuity, many citizens are opposed to the leveling of dams that have become a part of local heritage and an element of economic development. Another paradox is that some dams have also become sites of biodiversity. An emblematic example is the Der-Chantecoq reservoir dam in the Marne (France) which was built at the expense of hedged farmlands, but which has become a major site of naturalness and a European hub for migratory birds.

Another example is that of the numerous gravel quarries, the product of industrial activity, which are dotted throughout the country. Purists refuse to consider them as wetlands, whereas when they have been developed accordingly they can accommodate numerous aquatic species.

In Europe, whilst the contribution of ecological systems to economic activity has lost some of its importance, it remains very strong. At the same time, citizens are increasingly concerned with their living environment. Increasing urbanization is accompanied by a growing demand for green areas and relaxation spots. This explains the flourishing of peri-urban restoration programs, notably through the reinvestment of riverbanks.

2.8 Biodiversity Protection and Anthroposystem Coviability?

It is clear that the agro-pastoral system, which in the eyes of us Europeans is a reference with respect to nature, is endangered by intensive agriculture and land-use practices for urbanization. The mosaic of habitats that had been created over the centuries is being standardized, namely with the removal of hedges. And the disappearance of certain uses and associated practices also contributes towards the creation of undifferentiated systems. We can feel a sense of nostalgia for our past landscapes, but the real question to be asked in a world where the climatic context as well as the uses of ecological systems are quickly changing is this: what natures do we want? A compromise between our dreams and the reality in the field?

The nature and biodiversity referred to by certain militant movements and ecologists constitute virtual, idealized ecological systems that they call natural, built around an ideology: the Garden of Eden. Such an ideology would require erasing any developments or facilities to return to naturalness, which presupposes a drastic change on an economic level. It is therefore very difficult to talk about coviability. This scenario, which excludes man as a player in his environment, is in reality highly improbable.

On the contrary, what the citizen dreams of is a gardened, orderly nature in which all the nuisances caused by bothersome species would be absent; a nature that man uses, but which also meets his aesthetic and recreational expectations (Cottet-Tronchère 2010; Sirost et al. 2012). If we accept that European nature is linked to the uses of ecological systems, its maintaining necessarily passes through these uses. It is in relation to this expectation that we can speak of coviability. In reality, this is currently the case in certain conservation operations, where agricultural practices consistent with the existence of a flora and fauna having witnessed previous activities, are perpetuated as part of a desire to preserve them. This is also what some conservationist movements would like to put into perspective: the conservation of what already exists. But one question remains: if the uses change, what happens? Are we going to let nature express itself freely? Or will we look for other uses?

Nature conservation requires the demonstration that it is useful for men, as opposed to gestures of exclusion. Of course this question of utility does not exclude ethical or aesthetic considerations. In particular, the European citizen is sensitive to heritage aspects. But it is fundamental for citizens to know why money is being spent in this area and not in others, and what society has to gain from it, in order to enable populations to adhere to these choices. Unfortunately, more often than not, an esoteric language is used to speak to citizens about a virtual nature they cannot understand. In Europe, the citizen appreciates what is tangible: picking berries or mushrooms, fishing, discovering beautiful landscapes and green spaces in which he can relax, etc. He lives and breathes nature, he doesn't intellectualize it. Biodiversity is an abstract notion for him.

As regards biodiversity protection, the roads of coviability are imprecise and must be permanently adapted between ecological and social systems. Each bifurcation implies compromises between economic and social realities and more or less utopian value systems. But in reality it is not a matter of imposing its law on nature, or of giving it free rein. Coviability is trying to steer the trajectory, with moderation, but without fear or favor.

References

Aubertin C, Rodary E (eds scientifiques) (2008) Aires protégées, espaces durables? IRD Editions

Aubin JP (1996) Une métaphore mathématique du principe de précaution. Natures Sciences Sociétés 4(2):146–154

Aubin J-P, Bayen A, Saint-Pierre P (2011) Viability theory: new directions. Springer, Heidelberg/New York

Bertrand G Beroutchachvili N (1978) Le géosystème ou système territorial naturel. Rev. Géogr. Pyrénées et du S. O. t. 49(fasc. 2):17–180

Blandin P (2009) De la protection de la nature au pilotage de la biodiversité. Editions QUAE

Bouleau G (2013) Le bon état des eaux: sociologie politique de l'écologie fluviale française et européenne (1960–2010). Congrès AFSP Paris 2013. www.congres-afsp.fr/st/st20/st20bouleau.pdf
Boulton AJ (1999) An overview of river health assessment: philosophies, practice, problems and prognosis. Freshw Biol 41:469–479
Casetta E, Delord J (eds) (2014) La biodiversité en question. Enjeux philosophiques, éthiques et scientifiques. Éditions Matériologiques, Paris, 272 pp
Catanzano J, Rey H (1997) La recherche halieutique entre science et action: réflexions sur fonds de crise. Nature Sciences Sociétés 5:19–30
Cottet-Tronchere M (2010) La perception des bras morts fluviaux. Thèse Université Jean Moulin Lyon 3
de Billy V, Tournebize J, Barnaud G, Benoît M, Birgand F, Garnier J, Lesaffre B, Lévêque C, de Marsily G, Muller S, Musy A, Zimmer D (2015) Compenser la destruction de zones humides. Retours d'expérience sur les méthodes et réflexions inspirées par le projet d'aéroport de Notre-Dame-des-Landes (France). Natures Sciences Sociétés 23:27–41
Descola P (2005) Par-delà nature et culture. Editions Gallimard, Paris, 624 p
Descola P (2011) L'écologie des autres. L'anthropologie et la question de la nature. Quae
Dutoit T, Alard D (1995) Mesures agri-environnementales et conservation des pelouses sèches: premier bilan en Seine-Maritime. Courrier de l'Environnement de l'INRA 25:63–70
Duvat V (2012) Ces îles qui pourraient disparaître: éditions Le Pommier
Folke C (2006) Resilience: the emergence of a perspective for social–ecological systems analyses. Glob Environ Chang 16:253–267
Gould SJ (1991) La vie est belle. Les surprises de l'évolution. Seuil, coll. "Points sciences", 484 p
Gunnell Y (2009) Ecologie et sociétés, éd. A. Colin, 480 p
Lawton J (1999) Are there general laws in ecology? Oikos 84:177–192
Lecomte J (1999) Réflexions sur la naturalité. Le Courrier de l'environnement de l'INRA 37:5–10
Leopold A (1949) Almanach d'un comté des sables. Paris Aubier, 290 p
Lévêque C (2011) La nature en débat. Cavalier bleu, collection idées reçues
Lévêque C (2013) L'écologie est-elle encore scientifique? Ed Quae
Lévêque C, Muxart T (2004) Anthroposystème in hypergéo.eu
Lévêque C, Van der Leeuw S (éditeurs scientifiques) (2003) Quelles natures voulons-nous? Pour une approche socio-écologique du champ de l'environnement. Elsevier, Paris, 324 pp
Liarsou A (2013) Biodiversité. Entre nature et culture, préface d'André Langaney. Sang de la Terre, Paris
Mathevet R, Bousquet F (2014) Résilience, Environnement. Penser les changements socio-écologiques, Buchet-Chastel
Mathieu N, Jollivet M (sous la direction de) (1989) Du rural à l'environnement. La question de la nature aujourd'hui. ARF, Editions L'Harmattan, Paris
MEDDE (2012) Vulnérabilité des milieux aquatiques et de leurs écosystèmes. Etude des zones humides. Explore 2070 Eau et changement climatique, 144 p. + annexes
Micoud A (2005) La biodiversité est-elle encore naturelle. Ecologie et Politique 30:17–25
Papy F, Mathieu N, Férault C (2012) Nouveaux rapports à la nature dans les campagnes. éd. Quae, 192 p
Pavé A (2007) La nécessité du hasard. Vers une théorie synthétique de la biodiversité. EDP Sciences, Les Ulis
Perrein C (1993) Histoire de l'entomologie et biohistoire. La lettre de l'Atlas entomologique régional (Nantes), 15–16. 2, Octobre 1993. Nantes (ISSN 1260-0520)
Projet BEEST (2011) Synthèse – vers une approche multicritère du bon état écologique des grands estuaires. GIP Seine-aval. http://www.seine-aval.fr/wp-content/uploads/2017/01/L%C3%A9v%C3%AAque-2011-Synth%C3%A8se-BEEST.pdf
Rapport DJ (1995) Preventive ecosystem heath care: the time is now. Ecosyst Health 1:127–128
Rapport DJ, Costanza R, McMichael AJ (1998) Assessing ecosystem health. TREE 13:397–402
Schnitzler A, Génot JC, Wintz M (2008) Espaces protégés: de la gestion conservatoire vers la non-intervention. Courrier de l'environnement de l'INRA 56:29–43

SDAGE Rhône Méditerranée (2011) Qu'est-ce que le bon état des eaux? Note du Secrétariat Technique du SDAGE. http://www.eaurmc.fr/?eID=dam_frontend_push&docID=1823&xtor=RSS-4
Sirost O, Femenias D, Birot L, Bouilln D (2012) Sequana. Les constructions sociales de l'estuaire de la Seine. Rapport Seine aval 4, GIP seine Aval
Terrasson F (1991) La peur de la Nature. Sang de la Terre, Paris
Weber J (1995) Gestion des ressources renouvelables: fondements théoriques d'un programme de recherche. GREEN

Chapter 3
Coviability, Through the Lens of the Mathematical Theory of Viability

Jean-Pierre Aubin and Marie-Hélène Durand

Contents

3.1 Scientific Metaphors of Life Sciences

The (mathematical) theory of viability was born in the mid-1970s at the CEREMADE[1] at the University of Paris-Dauphine, out of a frustration caused by life sciences' exclusive use of equilibrium concepts, optima, probabilistic translation of the hazard concept, etc.

[1]The Centre for Research in Mathematical Decision (CEREMADE) is a mixed research unit (UMR No. 7534, CNRS and Université Paris-Dauphine) was devoted at the time to studying the application of mathematics in scientific disciplines as diverse as economics, management, finance, cognitive science, epidemiology, biology in the framework of evolutionary and control systems, as well as data analysis and the theory of classification, image and signal processing, etc. The main purpose was the mathematical formulation of these problems, their mathematical analysis, the design of numerical computation algorithms and their practical implementation in the context of interactions with business and industry.

J.-P. Aubin (✉)
VIMADES, (Viabilité, Marchés, Automatique et Décision); Paris Antenna, 14, rue Domat, 75005, Paris, France
http://vimades.com/aubin/

M.-H. Durand
French Institute of Research for Sustainable Development (IRD), UMR GRED, Montpellier, France
e-mail: marie-helene.durand@ird.fr

O. Barrière et al. (eds.), *Coviability of Social and Ecological Systems: Reconnecting Mankind to the Biosphere in an Era of Global Change*,
https://doi.org/10.1007/978-3-319-78497-7_3

- Equilibriums, states that do not evolve, do not exist in the living world in which organisms born and die in finite time, even if they expand in their lineages. No more than the asymptotic behavior of evolutions is relevant to the living world;
- Cognitive sciences have cast doubt on the translation of rationality with the use of static and inter-temporal optimization criteria that require a teleological vision presuming, a priori, knowledge of desired goals, a vision which does not form a consensus;
- Probabilities provide measurement tools (expectancy, variance and higher order moments) of the errors caused by deterministic dynamic systems which are 'randomly' disrupted by an invisible hand rolling dices. Probabilities take into account averages which are rarely achieved, especially in micro-economy where time and duration are neglected, and when this is not the case, differential equations are supposedly known. Whereas, outside of physics, they cannot be known. As for awareness of spatial variations, this is hardly touched upon. Neither do we know these evolutionary equations, utility functions or other criteria to be optimized. Building evolutionary systems *inductively* from observations is a task for the future, thanks to the advance of numerical processes of more and more massive data that digital computers offer nowadays. Inventing mathematical evolutionary systems to *deduce* the evolutions of living organisms they govern, is too hazardous a task.

These criticisms are not new. Hayek (and more recently Taleb for a denunciation of probabilities' abuse) has already expressed them, due to lack of adequate mathematical tools (Brahic and Terreaux 2009; Griffon and Griffon 2010, 2011; Hayek 1978, 1993; Pavé 2007; Taleb 2008). However, only those forged through the prism of physical sciences remained ongoing, obstructing the possibility for other analysis (Le Moigne 1983). However, there is no reason why mathematics motivated by lifeless sciences should be applicable to life sciences. Physicist Eugene Wigner has already explained in 1960 in a famous article that "the unreasonable effectiveness of mathematics in the natural sciences" was a miracle. In this article, "natural" means "physical nature", so that this miracle is even more miraculous in life sciences, which has led to what is commonly known as the (mathematical) theory of viability (Aubin 1985, 1991, 2010; Aubin et al. 2011; Barbault and Weber 2010; Griffon and Griffon 2010, 2011; Rojey 2013; Sueur 2012).

So, it was necessary to define the space of all elements called "the states of a system"; their subsets, interpreted as "environments" defined by "viability constraints "and consider to which extent a state belongs to it or not. Viability constraints defining environments were already taken into account to study equilibria and optima in a static framework (Aubin and Frankowska 1990).

Since time plays a crucial role, it became inescapable to study the joint "evolution" of both states and environments, and guarantee that at every instant ranging over a temporal window, an evolution remains in the environment. There were no words to describe this property that had received little attention by mathematicians up to that time, so were forged the terms "coviability" to describe such a joint evolution, or just "viability" when the environment does not evolve (Aubin and Lesne 2006).

Jacques Monod's book entitled 'Chance and Necessity' had a decisive influence when it was released (Monod 1971). Viability could mathematically translate the concept of necessity studied in his book. The question, however, was how to translate chance?

This term's polysemy required a more restrictive definition. The mathematical translation of chance by probabilities was not appropriate due to the irreversible and unique nature of the evolutions of living organisms, which radically distinguishes itself from the evolutions of inert matter as described in physics. There is no probability for the living because the evolutions of living beings are irreversible, as are those of their engines which frequently disappear and, since experiments are impossible to conduct, it is hard to measure the frequency of errors which do not exist due to the lack of an evolution of reference.

The term "uncertainty" provided an alternative to the restrictive concept of probability. Accordingly, we could modify the book's title to: "Uncertainty and Viability".

How can we mathematically translate uncertainty? By building an evolutionary system which associates a set of evolutions with every instant of a temporal window and with every state. It can be an empty set (sad news, but important information), contain only one evolution (determinism), or many. Hence, uncertainty can be measured by the size of this set of evolutions.

Uncertainty can also be defined in various ways: "contingent", when a property is satisfied by at least one co-evolution, "tychastic",[2] when applicable for all possible evolutions, "impulsive", when the evolutionary engine is accompanied by an impulse system that instantly makes a state viable when it is about to no longer exist (Aubin 2010, 2013a, b; Aubin and Dordan 2016; Durand et al. 2013). The list of uncertainty examples is far from being exhaustive.

3.2 A Tychastic Encounter

With the arsenal having been forged and a few PhD students having been trained, the CEREMADE's "Viabilité-Jeux-Contrôle team at Université Paris-Dauphine was formed around this new research theme, bringing in its wake the development of various new mathematical tools. These are "mathematics motivated" by economics and biology and may be prone to becoming applied mathematics if the users of these "mathematical metaphors" validate them according to their relevance in their respective disciplines (Aubin 1985). Metaphors rather than peremptory

[2]Tychastic uncertainty, or tyche means 'chance' in Greek personified by the goddess Tychy whose goal was to disrupt the course of events, in a good or bad way. This denomination which describes fortuitous events was suggested by Charles Peirce in 1893 in his article "Evolutionary love".

"models" assuming that the modeler knows functions to optimize, differential equations governing the evolutions or *a priori* probability measures. A new research direction involves detecting *a posteriori,* from observed evolutions, the dynamics that reproduce them more or less faithfully (relational analysis).

This is the reason why these researchers have been open to specialists from other fields such as economics, biology, cognitive sciences, control, and environmental sciences, so as to forge mathematical metaphors.

Hence, after a random or a "tychastic" encounter, to use Charles Peirce's terminology (Peirce 1893), among the authors of this article at the beginning of the 1990s, scientific contacts multiplied and soon turned friendly with Philippe Cury, Christian Mullon, Jean-Philippe Terreaux, Jacques Weber, and many others. This is how the seminar "Viable Development" was born. It was organized by Luc Doyen, Hélène Clément-Pitiot, Daniel Gabay and Gérard Weisbush from the Ecole Normale Supérieure in "la rue d'Ulm" (Paris), which extended to the Waikiki bistrot, fueled by punch. This heterodox seminar hosted every week a diversified audience: historians, demographers, anthropologists, economists, biologists, ecologists, cognitive sciences' specialists, etc. Yet, this seminar cannot be associated with the kind of pluridisciplinarity constantly summoned nowadays. These meetings were organized solely on the basis of mutual curiosity and good will, even enthusiasm, they were free and open, which is rare to find today. The arguments were lively but listened to with mutual tolerance. All this led to fascinating encounters and sometimes to partnerships or joint projects when affinities and motivation managed to convince people that it was worth spending precious time on this...

The second author of this article joined a scientific community in Montpellier devoted to finalized works in life sciences and, more generally, development studies. She decided to decentralize this Parisian seminar under a residential format. The seminar "ecosystems and viable development"[3] was held in June, 1996 in Mèze. It was not yet another conference on the environment and sustainable development. The main objective was not to provide immediate answers to practical questions, but to present the theory of viability or at least the metaphors that this (mathematical) theory provides to researchers, at a time when traditional approaches, mainly focused on the search for equilibria, were being challenged. It was also about discovering and debating the meaning of common problematic issues, those of evolution phenomena and interactions between dynamics of different dimensions, located at different temporal or spatial scales, and generally on the way of approaching their formalization and the relation between theory and research or empirical observations through diversified contexts, methods and experiences. It was a success and the lively debates resulted in a shared decision to create a new research field called "coviability" or "eco-viability". The Mèze seminar was

[3]In attendance were: Peter Allen, Martine Antona, Jean-Pierre Aubin, Christophe Béné, François Bousquet, Jean Cartelier, Christian Chaboud, Philippe Cury, Serge Diebolt, Luc Doyen, Marie-Hélène Durand, Daniel Gabay, Ghislain Géniaux, Michel Griffon, Francis Laloé, Jean Lefur, Stéphane Luchini, Lydia Mellul, Jean-François Noel, Hélène Clément-Pitiot, Patrick Saint-Pierre, Juliette Rouchier, Jacques Weber, Gérard Weisbuch.

followed by a summer school held in Orléans 1998, organized by Christian Mullon, Luc Doyen and Marie-Hélène Durand, and dedicated to training on the concepts and mathematical tools of viability, for the most motivated scientists. Luc Doyen with Christophe Béné in halieutics, Alain Rapaport and Jean-Philippe Terreaux on forest sciences, then Michel de Lara and other colleagues from Inra, Irstea, Cired, and the Museum National d'Histoire Naturelle took over and many PhD students were trained. (Alvarez et al. 2013; Béné et al. 2001; Bernard and Martin, 2013; Cury et al. 2005; DeLara and Doyen 2008; DeLara and Martinet 2009; Domenech et al. 2011; Doyen et al. 2007, 2013; Durand et al. 2017; Hardy et al. 2013; Martinet et al. 2007; Mullon et al. 2004; Mullon 2013; Pereau et al. 2012; Rapaport et al. 2006; Regnier and De Lara 2015; Terreaux 2018; Wei et al. 2013)

"Coviability" is now known, largely cited and often invited in the field of bio-economics but it is not that simple. Because of lack of other global representation tools, coviability is sometimes perceived as a means to analyze the dynamics of "Socio-Ecosystems", complex objects undergoing a pluridisciplinary investigation, encompassing dynamic interactions of a biophysical system with socio-economic systems. Up to now, research works on coviability issues have revamped "bioeconomic models", mainly to deal with management problems, and suggested other "evaluation criteria", taking into account ecological and economical dimensions, which are often more relevant than those provided with classical deterministic or stochastic approaches (Gourguet et al. 2015, Mouysset et al. 2014, Griffon and Griffon 2010, 2011; Sabatier et al 2010, 2015; Tichit et al. 2007). It took almost 20 years to popularize this particular acceptance of coviability and much more time is still needed to first, explore and develop other aspects and computation tools which are still missing, then, and above all, to train new generations in these non-standard mathematics before being able to satisfy the diverse disciplines worried about attracting these mathematicians to their fields. The notion of coviability is very attractive, but we should pay attention to illusions generated by the metaphorical strength, and amplified by the polysemy of words.

3.3 Towards a Mathematical Version of Coviability

A vector's evolution naturally involves a joint evolution of its components, which are typically governed by a differential system. Of course, we can give to the vector's diverse components climatic, biological and economical interpretations, variables that evolve together. However, this is not the meaning that we give in mathematics to the concept of coviability. Nowadays, coviability designates a joint evolution of *a vector and an environment*, a subset to which this vector belongs. Nobody owns the terms "viability" and "coviability", neither their synonyms, which have become polysemous. Therefore, we add the adjective "mathematical" to specify here the meaning given to this term in mathematics. Indeed, when a mathematician engages in a field into which he is invited, he starts by impoverishing the polysemy hidden under the language so as to select a meaning which can be associated with a definition and mathematical processing.

These mathematical metaphors can be created in two ways:

- Either search in the mathematical corpus concepts and methods available at a given moment, which constitute the field of pure mathematics (avoiding every contact with nature other than physical) and to "apply" them;
- Or when not appropriate, create new concepts, identify assertions that connect them and forge new tools, a domain called motivated mathematics (they are not specific to mathematics, Antoine Danchin, for example, took over this terminology in biology). The absence of the expression "motivated mathematics" in the usual vocabulary indicates that they are not familiar.

Physics deals with concepts endowed with measurement units, so they make use of numbers and evolve in vector spaces. Yet, elements perceived by the human brain and described with words are rarely quantifiable because they are the result of an abstraction to which we can't link measurement units. They conveniently fall under a formalization through the set theory. We introduce the hyperspace of a vector space, defined as the family of all its environments, defined qualitatively, which are only "quantifiable" through arbitrary and not "consensual measures".

The "kilotheorem unit", for example, by which we would measure the activity of mathematicians does not exist, no more than bibliometrics (yet invented nevertheless by physicists) for evaluating scientific work. Weigel 1669 "pantometric curse", a contemporary of Leibniz, which considers only scientific that which can be measured, intensified by the power of computers (digital and no longer analogical), leads to the distortion of an observation by measuring it with digital and meaningless indicators. The "Big" of *big data* will never be big enough to describe concepts lacking measurement units and to lock us in vector spaces of "big" dimension.

This is the first pitfall under which our attempts to "mathematize" the world falter. The concept of the real number for example, is the least "real" of the numbers that human minds have invented throughout their history; which is a problem when it comes to using them to describe the "real" world. Nevertheless, the first attempt was to measure anything and everything, and this is how probabilities, statistics and other "models", etc., have undergone the *pantometric* curse from which we have to extract ourselves (Hayek 1978).

Having no faith and disobeying became necessary to remind us that if the mathematical activity that all human beings practice with varying degrees of intensity (in the same way as language, faith, obedience, among the wide variety of capacities with which they are endowed) has roots in numbers. Their abusive use to create metaphors may be dangerous, while a "qualitative" mathematics, free from numbers, which mainly became set theory, needs to be further developed. Moving from numbers measuring the size of sets to the sets themselves is a painful cognitive, but necessary, rupture, (Dordan 1995).

3.4 Mathematical Coviability

First, we have to specify more precisely what we mean by the polysemous concept of time (See Aubin to appear; Longo and Montévil 2013), so as to be able to speak about evolution in mathematical terms. We have to admit that this word has two meanings at least, that of date or chronological time, and that of duration, of 'time' passing by. Both are necessary to define a "temporal window" and to measure it by its duration, a positive number, and by a date, a real number, denoting a temporal window whose duration is equal to zero. Temporal windows are thus described by two real numbers, a date and a duration. The relationship of the order of real numbers enables to define a date posterior to another by the duration which separates them. The duration is measured in seconds (whose definition evolves with time, which is no longer linked to a perception of ephemerides). The number 0 is a legitimate candidate to be at the origin of durations of time, even if for the time being, it has not been measured by physicists who have been unable to exceed the yoctosecond (nor to measure distances below the attometer). Mathematicians, however, accept to believe that both instants and points exist at least in their mind.

By contrast, for dates to be represented by real numbers, a "time origin" must be defined. Contrary to durations, this is nothing more than consensual, such as Christ's birth, for instance, or that of the "big bang". Therefore, the notion of time, twinned in duration and date, requires an act of faith. We admit this.

For each date, we distinguish retrospective temporal windows, ending in this date, prospective temporal windows, starting at this date. Evolutions may be "known" on retrospective temporal windows (of known past instants) when the perception of events is memorized and remembered, but they can only be "predicted" when temporal windows are prospective (of future unknown instants). A date such as that of the beginning of a temporal window can be chosen, but this requires the ability to predict the future, what human minds love to do under the influence of a "pythic drug". It is less tiring and compelling to foretell the future than to remember the past, and learn lessons from history.

Evolutions on temporal windows of infinite duration can be qualified as "perennial" and their viability properties too. Those on temporal windows of finite duration are "ephemeral". What is more, the duration of an evolution (to some extent, its age) is often calculated, and is an integral part of the viability problem. Mathematically speaking, it is much easier to work on temporal windows of infinite duration, and it is within this framework that the first viability theorems were formulated and proved. It was only in the year 2000 that a convenient hypothesis was found enabling their adaptation to temporal windows with finite duration.

Whether anchored in religious (creationist) beliefs or having its roots in celestial mechanics created by scholars of the seventeenth and eighteenth century, i.e. the "best of the worlds" of Leibniz, faith in determinism led to the almost exclusive use of the prospective temporal window with infinite duration, starting at a conventional origin equal to zero by convention, and thus with infinite horizon. It is in this context that the evolutionary systems were studied.

As far as the evolutionary systems governing living organisms are concerned, we feel it is more judicious to study evolutions on retrospective temporal windows with a finite duration, prescribed or calculated. Let us, thereby, make a clean sweep of the future.

We now come to the representation of the space of states of the system. We can separate them in two categories, "vector spaces" and "hyper-spaces".

Vector spaces are the most familiar. Vectors are sequences of numbers, each representing a unit number of a quantifiable entity (the number of mosquitoes, lions, etc., which depends on biologists' taxonomy). We may subject them to familiar operations, add and subtract them, multiply them by numbers, take their minimum or maximum, but their multiplication is less natural to define, because the "tensor product" of vectors is a "tensor", object which is no longer a number (as the scalar product), therefore less used. These operations are preserved by linear operators, even multilinear, so that this more familiar framework has been developed for two centuries by a multitude of mathematicians who created the linear world, a kind of mathematical heaven out of which we are chased to address a non-linear world.

Non-linear doesn't mean very much, because this term comprises boundless possible structures. We add another one, a structure known under diverse synonyms, such as a "hyper-space": that of subsets of a given space, among which space itself is the biggest, and the empty set is the smallest. Although initiated by Boole, their study has been relatively neglected and we know much less about their properties than the linear ones. We can take as a starting space, for instance, all the conceivable legal articles and for a hyper-space those of legal codes defined as sets of legal articles. Neither of them is measurable, except for measuring a legal code by the number of its articles, which hardly makes any more sense than evaluating a researcher by bibliometric indices. They benefit, though, from Boolean operations (complementary, intersection), i.e., logical properties, which, by now, can be "fuzzy field" in a variety of ways.

We come to a mathematical definition of coviability, which requires speaking of viability first, then involving the joint evolution of vectors of a vector space and subsets in their hyper-spaces, demanding that at every instant, the vector belongs to a set (an element of hyperspace).

Speaking mathematically of evolution implies talking about "engines" that produce them. These entail notions of speed, acceleration, jerk (a name given to the third derivatives by physicists). We will retain here only the notion of speed, the first derivative in relation to time, which appeared for the first time in 1637 under the ingenious quill of Pierre de Fermat. He defined it with the help of the notion of tangent to the graph of function, neglected since Gottfried Leibniz characterized it as the limit of the average speed on temporal windows when their duration tends towards 0; limits which only exist in the human mind (that are still mathematicians), since a zero duration has not yet been measured. Despite all the progress made by the classical differential calculation, it was necessary to go back to the origins, and follow Fermat by considering mainly the notion of tangent (to any subset of a vector space). It appeared that the central concept was the regulators or (regulation maps) governing viable evolutions, which could be characterized with the viability theorems and computed with the viability algorithms. Here again, progress consisted in unlearning, returning to important branching of scientific itineraries, including, in our case, the one opened up by Fermat which had to be cleared and deciphered all at the same time.

In the beginning, evolutionary engines were described by differential equations, which associate with every initial vector the speed of one or various evolutions starting from it, or with every terminal vector an evolution which leads to it, or, even better, an evolution that links these initial and terminal states. It is then sufficient to introduce one or two vectors to obtain evolutions. We may find it curious that the first accomplished tasks to study the evolution were to deny it, either by searching for the equilibria of an evolutionary system, which are vectors that do not evolve (stationary states), or by examining the asymptotic behavior of evolutions when time converges to an unattainable infinity. In this latter case, the state converges towards a (viable) "limit cycle", hopefully reduced to a providential equilibrium, and otherwise, to a "real cycle" to delight supporters of an eternal return, or rather, a "helicoïdal time" conciliating this cyclic aspect with the arrow of time. Cyclic evolutions have been the object of predilection since celestial mechanics has been an important source of motivation. In fact, all these developments took form because they correspond to situations that are easily calculated, and not because they necessarily correspond to what interest us most (d'Holbach 1770). Then came the time to introduce constant parameters which some physicists or facetious aliens changed, as if they were rheostats. When evolutions depend on a given parameter, the question of stability (a dangerous polysemous vernacular word used for defining many different concepts) arises. The concern was then to know how the characteristics of evolutions were or were not preserved when this officially constant parameter was modified by an invisible shaking hand. Among these evolutions, equilibria and limit cycles also depend on such parameters, they are born, are or are not unique, may divide or disappear. It was the mathematical reign of bifurcations and catastrophes.

Control specialists made visible this hand which chooses this parameter at each instant to reach an objective, which might encompass realities as different as quenching one's thirst on planet Mars, for instance, or more prosaically, driving one's car: correcting by a steering wheel's rotation the distance between the car wheels and the road side. Evolving, these parameters became a control, in an open-loop if its evolution is calculated before, or a closed-loop, if a regulator (like the centrifugal governor of James Watt), also referred to as feedback, is built (or most of the times guessed or imposed, linearly in addition).

However, does this hand exist in the field of life sciences? Does it have a final (teleological) goal? Is it optimal? We may doubt it, and if we do, we can restrict our ambitions so that it would be viable only, as it concerns living systems, and if a pilot existed, even an intelligent "designer", his "hand" wouldn't deal with those parameters considered as commands, but would be happy to just regulate them in order to make some evolutions viable. It is no longer a question of activating them to search optimal decisions, but to search for "decisions taken on time" (kairos). Such parameters are parameters for regulating and not piloting to the extent that it was tempting to replace "controls" by "regulons". The concern is mainly "adaptation to constraints", of homeostasis in biology, of eudemonism (in ancient Greek *eudaimonismos,* "good genius", stipulating that the search of happiness is the natural rationality driving the evolution of humanity, in order not to confuse these constraints with the static concepts of utility or Paretian ophelimity) or scarcity in economy, etc., of viability, to summarize them all (Aubin et al. 2011; Aubin and Catté 2002).

Initially, the evolution of environments was neglected and the question of viability searched for "regulons" which allowed evolutions to remain in a given environment. Simply put, this is about building mathematically and algorithmically the (set valued) regulators associating a set of controls or "regulons" with every state, depending on the nature of the problem, so that the evolution would be viable at every instant (the notion of regulator or feedback introduced by Watts was used by Ludwig Von Bertalanffy for general systems, Von Bertalanffy 1968).

However, we observe that the environments do not remain the same over time: they also evolve. Their evolution hadn't been studied due to lack of an operational notion of the velocity of the evolution of sets. This has only been carried out 20 years ago. An environment's evolution velocity is no longer a vector, but a "vector field", which transforms the environment into an infinitely close neighborhood of the environment, just as velocity does for vectors.

We can define morphological evolutionary systems governing the evolution of environments allowing us to define coviability. In particular, we can construct mathematically (for the time) regulators associating, with every environment and with every vector belonging to it, the whole set of "regulons" which govern the joint evolution of environments and of the vectors required to belong to it (Aubin 2000). This requires an important mathematical investment (Lorenz 2010).

3.5 Are Mathematics Useful?

Since the viability theorems could be adapted to this new mathematical framework, this definition of coviability could naturally be accepted. But much remains to be done. Yet, what do we mean by a mathematical solution? It can be "qualitative", pointing out assertions that a concept implies another, the only ones we can prove for systems with a large number of variables, or "quantitative", when measurements can be taken, and calculations made. This can be achieved through analytical formulas, or numerical algorithms better suited to increasingly powerful digital computers making these analytical formulas outdated. To achieve this, the "big data" universe must be discernible by the human mind, agreed upon, and make sense accordingly.

As for viability problems, these algorithms compute at each moment not only vectors, but sets as well. These operations consume (exponentially) more computer memory and computational time. They are even more subject to the "dimensionality curse" that Bellman deplored. So, we have to pay attention to the dangers of quantitative techniques which sacrifice an infinite number of variables in order to retain only some of them: if an assertion is wrong for a sacrificed system, it is also true for the initial system, but the validity of a conclusion for a small number of variables says nothing worthy of note.

Could these results be "applied"? It will be firstly up to a discipline's potential users to validate these mathematical metaphors, even though they have been motivated by the disciplines in question. Being unable to take into account and analyze a large number of quantitative observations, the results that might be obtained will therefore be above all qualitative, and will have to be corroborated by other observations and analyses, also qualitative. It does not matter, because qualitative mathematics provides primarily stories to tell. Fairy tales? It is up to the reader whether to believe these or not.

To conclude, existing mathematics can do little, even though this "little" is already "much" and which must be handled with a circumspection dictated by a cognitive precautionary principle.

Acknowledgements The authors dedicate this work in memory of Jacques Weber (1946–2014), a faithful and enthusiastic participant of the seminar "viable development" and a witness of the historical development of coviability as a concept. He was a great source of inspiration and support.

This chapter was written during the course of the program ANR GAIA-TROP on the governance and viability of tropical agro-systems. Valérie Angeon, Samuel Bates, Anya Désilles, Jean-Louis Diman, Audrey Fanchone, Harry Ozier-Lafontaine, Sophie Martin and Patrick Saint-Pierre, are also among those involved in this reflexion on coviability.

References

Alvarez I, Martin S, Dordan O, Litrico X, Saint-Pierre P (2013) Indicateurs de sécurité et de restauration dynamiques. In: Ancey V, Avelange I, Dedieu B (eds) Agir en situation d'incertitude en agriculture, regards pluridisciplinaires au Nord et au Sud. Peter Lang, Bruxelles, pp 309–326

Aubin J-P (1985) Motivated mathematics. SIAM News 18:1, 2 & 3

Aubin J-P (1991) Viability theory. Birkhäuser, Boston/Berlin

Aubin J-P (2000) Mutational and morphological analysis: tools for shape regulation and morphogenesis. Birkhäuser, Boston

Aubin J-P (2010) La mort du devin, l'émergence du démiurge. Essai sur la contingence, la viabilité et l'inertie des systèmes. Éditions Beauchesne, Paris

Aubin J-P, Bayen A, Saint-Pierre P (2011) Viability theory: new directions. Springer, Heidelberg/New York

Aubin J-P (2013a) Conjurer l'angoisse d'un futur inconnu. In: Bouamrane M, Antona M, Barbault R, Cormier M-C (eds) Rendre possible, Jacques Weber, itinéraire d'un économiste passe-frontières. Éditions Quae, Versailles, pp 157–166

Aubin J-P (2013b) Time and money, how long and how much money is needed to regulate a viable economy. Lecture Notes in Economics and Mathematical Systems, 670, Springer

Aubin J-P, Catté F (2002) Bilateral fixed-point and algebraic properties of viability kernels and capture basins of sets. Set Valued Anal 10:379–416

Aubin J-P, Dordan O (2016) A survey on Galois stratifications and measures of viability risk. J Convex Anal 23:181–225

Aubin J-P, Frankowska H (1990) Set-valued analysis. Birkhäuser, Boston

Aubin J-P (to appear) La valeur n'existe pas. À moins que [...]. Essai sur le temps, l'argent et le hasard

Aubin J-P, Lesne A (2006) Analyse morphologique et mutationnelle: des outils pour la morphogenèse. In Bourgine P, Lesne A (eds) Morphogenèse. Belin, pp 162–177
Aubin J-P, Bayen A, Saint-Pierre P (2011) Viability theory, new directions. Springer, Heidelberg and New York, Springer-Verlag
Barbault R, Weber J (2010) La Vie, quelle entreprise! Pour une révolution écologique de l'économie. Edition Le Seuil, Paris
Béné C, Doyen L, Gabay D (2001) A viability analysis for a bio-economic model. Ecol Econ 36:385–396
Bernard C, Martin S (2013) Comparing the sustainability of different action policy possibilities: application to the issue of both household survival and forest preservation in the corridor of Fianarantsoa. Math Biosci 245(2):322–330
Brahic E, Terreaux J-P (2009) Évaluation économique de la biodiversité. Edition Quae, Paris
Cury P, Mullon C, Garcia S, Shannon L (2005) Viability theory for an ecosystem approach to fisheries. ICES J Mar Sci 62(3):577–584
d'Holbach (1770) Système de la nature ou des lois du monde physique et du monde moral, Corpus des œuvres de philosophie. Fayard (réédition 1990), Paris
De Lara M, Doyen L (2008) Sustainable management of natural resources, mathematical models and methods, Environmental science and engineering. Springer, Berlin/Heidelberg
De Lara M, Martinet V (2009) Multi-criteria dynamic decision under uncertainty: a stochastic viability analysis and an application to sustainable fishery management. Math Biosci 218(2):118–124
Domenech PA, Saint-Pierre P, Zaccour Z (2011) Forest conservation and CO2 emissions: A viable approach. Environ Model Assess 16:519–539
Dordan O (1995) Analyse qualitative. Masson, Milan
Doyen L, De Lara M, Ferraris J, Pelletier D (2007) Sustainability of exploited marine ecosystems through protected areas: a viability model and a coral reef case study. Ecol Model 208:353–366
Doyen L, Cissé A, Gourguet S, Mouysset L, Hardy P-Y, Béné C, Blanchard F, Jiguet F, Pereau J-C, Thébaud O (2013) Ecological-economic modelling for the sustainable management of biodiversité. Comput Manag Sci 10:353–364
Durand M-H, Desilles A, Fronville A (2013) Incertitude contingente, adversité tychastique. In: Ancey V, Avelange I, Dedieu B (eds) Agir en situation d'incertitude en agriculture, regards pluridisciplinaires au Nord et au Sud. Peter Lang, Bruxelles, pp 297–308
Durand M-H, Desilles A, Saint-Pierre P, Angeon V, Ozier-Lafontaine H (2017) A viability approach for agro-ecological transition, the example of soil preservation in French West Indies. Nat Resour Model 30(3):e12134
Gourguet S, Thébaud O, Jennings S, Little LR, Dichmont CM, Pascoe S, Deng RA, Doyen L (2015) The cost of co-viability in the Australian Northern Prawn Fishery. Environ Model Assess 21:1–19
Griffon M, Griffon L (2010) L'homme viable. Edition Odile Jacob, Paris
Griffon M, Griffon L (2011) Pour un monde viable: Changement global et viabilité planétaire. Edition Odile Jacob, Paris
Hardy P-Y, Béné C, Doyen L, Schwarz A-M (2013) Food security versus environment conservation: a case study of Solomon Islands' small-scale fisheries. Environ Develop 8:38–56
Hayek F (1978) Coping with ignorance, the Ludwig von Mises memorial lecture, Imprimis 7. Hillsdale College, Hillsdale
Hayek FA (1993) La présomption fatale. Presses Universitaires de France
Le Moigne J-L (1983) La Théorie du Système Général, Théorie de la Modélisation. PUF, Paris
Longo G, Montévil M (2013) Biological time, symmetries and singularities. Springer, Berlin
Lorenz T (2010) Mutational analysis, a joint framework for Cauchy problems in and beyond vector spaces, Lecture notes in mathematics, vol 1996. Springer, Berlin
Martinet V, Thebaud O, Doyen L (2007) Defining viable recovery paths toward sustainable fisheries. Ecol Econ 64:411–422
Monod J (1971) Le hasard et la nécessité. Édition du Seuil, Paris

Mouysset L, Doyen L, Jiguet F (2014) From population viability analysis to co-viability of farmland biodiversity and agriculture. Conserv Biol 28:187–201
Mullon C (2013) Network economics of marine ecosystems and their exploitation. CRC Press, Boca Raton
Mullon C, Cury P, Shannon L (2004) A viability model of trophic interactions in marine ecosystems. Nat Resour Model 17(1):71–102
Pavé A (2007) La nécessité du hasard: Vers une théorie synthétique de la biodiversité. EDP Sciences, Les Ulis
Peirce C-S (1893) Evolutionary love, The Monist, 3, 176–200. Repris dans Hartshorne and Weiss (eds) (1958) Collected papers of Charles Sanders Peirce, 6. Harvard University Press, Cambridge, MA
Pereau JC, Doyen L, Little R, Thebaud O (2012) The triple bottom line: meeting ecological, economic and social goals with individual transferable quotas. J Environ Econ Manag 63(3):419–434
Rapaport A, Terreaux J-P, Doyen L (2006) Sustainable management of renewable resource: a viability approach. Math Comput Model 43:466–483
Regnier E, De Lara M (2015) Robust viable analysis of a harvested ecosystem model. Environ Model Assess 20(6):687–698
Rojey P (2013) Entre utopie et principe de réalité. L'Harmattan, Paris
Sabatier R, Doyen L, Tichit M (2010) Modelling trade-offs between livestock grazing and water conservation in a grassland agroecosystem. Ecol Model 221:1292–1300
Sabatier R, Oates LG, Jackson RD (2015) Management flexibility of a grassland agroecosystem: a modeling approach based on viability theory. Agric Syst 139:76–81
Sueur C (2012) Viability of decision-making systems in human and animal groups. J Theor Biol 306:93–103
Taleb N (2008) Le Cygne noir: la puissance de l'imprévisible. Belles Lettres, Paris
Terreaux J-P (2018) N′oublions pas le futur. Valeurs, justice et taux d′actualisation. Ethics and Econ 15(1):66–80
Tichit M, Doyen L, Lemel JY, Renault O (2007) A co-viability model of grazing and bird community management in farmland. Ecol Model 206(3–4):277–293
Von Bertalanffy L (1968) Théorie générale des systèmes. Dunod, Paris, 2012
Wei W, Alvarez I, Martin S (2013) Sustainability analysis: viability concepts to consider transient and asymptotical dynamics. Ecol Model 251:103–113
Weigel E (Erhardi VVeigelii) (1669) Idea matheseos universæ cum speciminibus inventionum mathematicarum

Chapter 4
Mathematical Approach of Coviability: Concept, Modelling and Control

Abdelhaq El Jai, Samira El Yacoubi, Marie Claude Simon El Jai, Morgan Mangeas, Vincent Douzal, and Abdel Samed Bernoussi

Contents

4.1 Introduction

This volume explores how the concept of coviability can be a fruitful approach for us to apprehend and develop representations of a family of phenomena that surround us. To be more specific about this family, among the phenomena in our world, we may distinguish two broad categories: Those lasting systems, or objects,

A. El Jai (✉) · S. El Yacoubi · M. C. Simon El Jai
IMAGES, UMR ESPACE-DEV, UPVD, Université de Perpignan Via Domitia, Perpignan, France
e-mail: eljai@univ-perp.fr; yacoubi@univ-perp.fr; mceljai@univ-perp.fr

M. Mangeas
IRD - French National Research Institute for Sustainable Development, UMR ESPACE-DEV, Montpellier, France
e-mail: morgan.mangeas@ird.fr

V. Douzal
IRSTEA, National Research Institute of Science and Technology for Environment and Agriculture, Montpellier, France
e-mail: vincent.douzal@teledetection.fr

A. S. Bernoussi
GAT, FST Tanger, Tanger, Morocco

O. Barrière et al. (eds.), *Coviability of Social and Ecological Systems: Reconnecting Mankind to the Biosphere in an Era of Global Change*,
https://doi.org/10.1007/978-3-319-78497-7_4

that we simply describe as being stable, without any specific condition (e.g., *stable* nuclides). And those, that are seen to live on by keeping on maintaining dynamical conditions, e.g., stars, living beings, ecosystems, social phenomena; in the biosphere they appear at all scales.

The latter are those which require a theory on our part, if we want to understand, explain their enduring, their behaviour, and possibly influence them efficiently and in a direction at our convenience. We know that these systems can diverge, collapse, disappear; and if our models cannot escape at least a core of deterministic behaviour, in most cases the enduring of phenomena is not seen as completely deterministic. Deterministic means static: there is only one evolution path. Whenever the path splits, what is called a bifurcation in models of dynamical systems. What may tilt the behaviour on one side or the other is seen either as a perturbation (noise), or as the effect of the free will of an experimenter to act upon the system.

Noise enters our models in the form of 'random effects': we add a term that is supposed to sum up all the effects from the ambient context that are not explicitly accounted for in the deterministic components of the model. This noise is 'undirected', it is not better characterised than by its statistical distribution, and in practice is produced on our computers by a pseudo-random generator, our best aproximation to account for undetermined phenomena. The overall effect on the model is that different runs will not always take the same path, whenever a bifurcation is possible. It was a widening realisation in science from about the 1980s that many very simple differential equations, or their time-discrete correspondent, iterated maps, can display a high sensitivity to initial conditions: that extremely close initial values will not take the same evolution branch, and end up at completely different positions in the possible space.

The action of a human operator enters our models in the form of the addition of a *control* term (usually denoted u in control theory) that can be fixed according to a given objective or a constraint set.

As explained all along the book, we expect the coviability to be the right approach to study the long term connections between human societies and elements of the biosphere and to assess the impacts of environmental, ecological and social changes. To make the coviability concept useful in modelling and control of alive systems and their coexistence and durability, we consider an approach using systems theory. More precisely by considering a convenient spatio-temporal model, its analysis and control. This is the purpose of this chapter. We try to make it easy for those who are not familiar with the mathematics jargon. Few mathematics for complex behaviour and real systems.

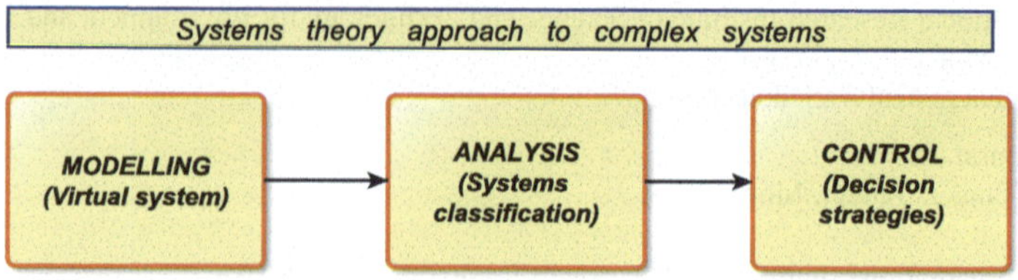

- For modelling we firstly consider a lumped system described by a set of ordinary differential equations, which may be continuous or discrete, deterministic or stochastic. Then we move to distributed parameter systems, i.e. systems evolving in time and space. The usual mathematical tool is based on partial differential equations (Curtain and Zwart 1995). Due to the complexity of the system we will consider cellular automata (CA), which may govern very complex systems. Cellular automata are models easily implementable and could illustrate the evolution of such systems. We can say that CA consists in very simple mathematics that generates very complex dynamics.
- In system analysis, we give an overview on different concepts associated with specific cases in systems evolution. These concepts are motivated both by human dynamics and biogeographical behaviour in various systems like populations dynamics, vegetation evolution, disease expansion, fire and desertification, etc.
- Finally, the control problem consists to know how it could be possible making coviable a system which is not. The problem will be stated in a general approach.

4.2 Spatial Concepts in Distributed Parameter Systems: From Controllability to Coviability

The study of complex real systems often leads to consider new approaches and tools that allow a better knowledge of the system evolution. Furthermore, it can help in the prediction of its evolution or in the choice of controls. Systems theory remains the best context for this exploration. For this purpose, in the above scheme *Modelling* → *Analysis* → *Control*, we need to focus on the analysis step. It consists in a set of various concepts, which have been explored from the 1950s, first for lumped systems and later for distributed parameter systems.

Among these concepts are the usual controllability, observability, stability and so on. The controllability is a basic and important concept which means that a system can be steered to a desired given profile, while the observability is related to the possibility of state reconstruction based on the knowledge of the system dynamics together with output measurements. The stability is very important and explores the asymptotic equilibrium behaviour of the system.

These concepts are different when the systems are distributed in space and time. Furthermore the space variable brings various new ideas, which do not exist in the case of lumped systems. Additionally in real distributed systems, actuators and sensors (El Jai and Berrahmoune 1984; El Jai and Pritchard 1988; El Jai 1991; El Jai et al. 1993; Afifi and El Jai 1994) have a space existence defined by their location and their space distribution, i.e. how the actuators act really in space. A wide literature is devoted to these considerations; see the work of El Jai and Pritchard (1988), El Jai et al. (1984, 1993, 1994, 1995) and the references therein.

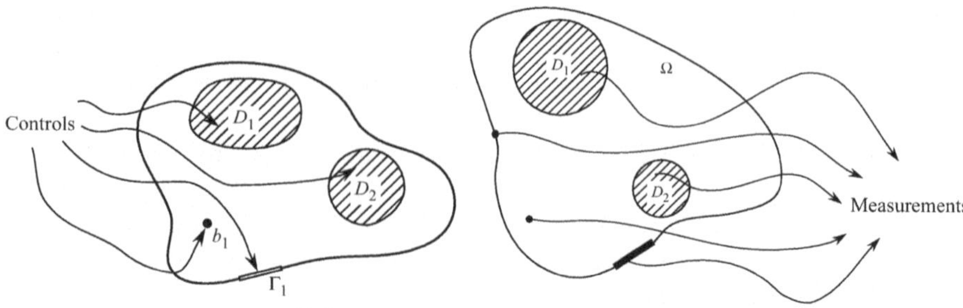

Various types of actuators and sensors.

Now if we consider various applications in ecology, environment, and more generally in populations dynamics, the study of such concepts is not adapted to the whole space. That is why the regional approach in distributed systems analysis has been developed (Afifi et al. 2011, 2012, 2013a,b,c; Zerrik and El Jai 2014; Afifi and El Jai 2015). The regional aspect means that we focus our objectives on what happens in a given desired area. Various works have been developed and enhance the meaning of such notion, see the work summarized in the book by Afifi et al. (2012) and the wide references therein (El Jai et al. 1995; Afifi and El Jai 1995; Zerrik et al. 1999, 2000; Afifi et al. 2000, 2001, 2002).

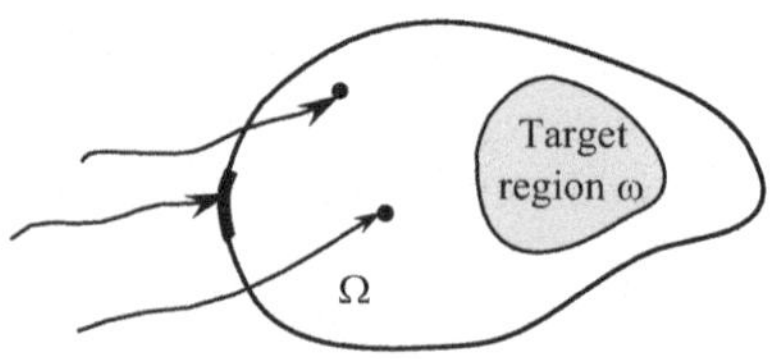

The system evolves in the whole space Ω while the target objective is the region ω.

Motivating applications have been considered. The stabilizability is considered when one has to find a control that makes stable a system. This also may have a regional sense. By duality the detectability has also be considered and widely explored. Various researchers have also studied detection of unknown sources. It consists to reconstruct a disturbance, using measurements given by an output function. The sensors, which allow such reconstruction has been called, spy sensors. The spy sensors allow the knowledge of a disturbing source using the system dynamics together with the output function. There are various papers with academic applications in Afifi et al. (1994, 1995, 2000, 2001, etc.)

All these notions together with biogeographical phenomena have naturally led to what is called spreadability. It consists to study space-time systems in which a given property increase (or decrease) in space, that is to say the property is satisfied at time t in an area ω_t which expands and occupy a larger area. As an example, consider the space where the vegetation biomass density is greater than a given value. This applies for desertification, pollution, disease, and any population expanding in space. Considering real systems, many types of spreadability have been considered and explored in the literature. This is the case of the usual spreadability (the (ω_t) are included in each other) the measure-spreadability (the areas of the (ω_t) are increasing), and other more sophisticated spreadability approaches; see El Jai and Kassara (1994), Bernoussi (2007, 2010), and Bernoussi et al. (2015) and the figure hereafter.

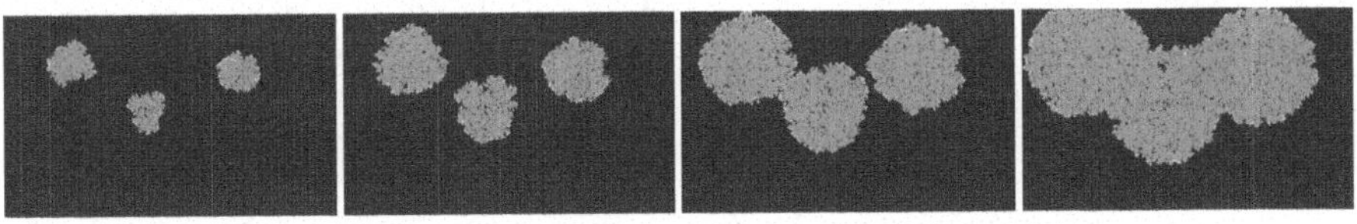

Example of a spreadable phenomenon (from left to right).

The system is said to be resorbable when the given property is decreasing. Now from systems theory approach one may ask if it is possible to protect a given region from the spreadability (of an undesired phenomena). The way the region is covered by the spreadable phenomena leads to define the vulnerability index. When this is not possible we say that this given region is vulnerable. Some authors have tried to find controls that protect from spreading phenomena. An illustrative situation is that of a spreading disturbance which combines both spreadability and space compensation, in the sense that one have to find controls which reduces the spreading disturbance, see Bernoussi et al. (2015).

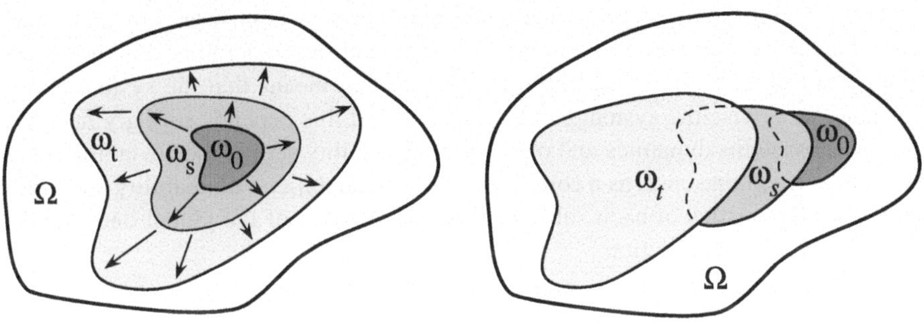

Usual spreadability (left) and measure-spreadability (right).

The exploration of such concepts in system analysis gives more information on the system and how it evolves. There are two consequences of that additional knowledge. On one hand, it can help for a better representation by an adequate model (this is the case of spreadable systems). On an other hand, an objective is to be reached. This objective is often defined by a cost function to be minimized, and which may define a regulation or tracking problem, a minimum energy or maximum precision one, or simply desired states to be reached, etc. The knowledge of these systems behaviour (controllable, spreadable, stable, etc.) makes easier the resolution of the minimum cost problem. As an example, for linear systems, a tracking problem (finite time) or a regulation one (asymptotic case) has a unique solution when the system is controllable.

An other concept in systems analysis is that of viability (Aubin 1991, 2001; Aubin et al. 2011). While the stability concerns systems evolving in infinite time

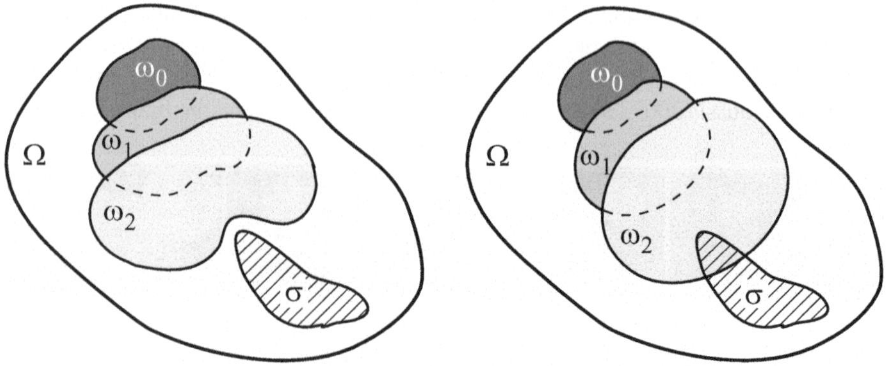

The region σ is not vulnerable (left) and vulnerable (right).

(asymptotic concept), the viability is a particular case for finite time evolution. The viability of a dynamical system means that the state of the system remains in a feasibility prescribed set. The coviability may be considered as an extension to systems with more components and constraints. The coviability can be described by a set of connected subsystems; each subsystem has its own dynamics and state constraints. The coviability means that the relations between all the subsystems remain viable. That is to say that the connected subsystems evolve together towards one (or more) equilibrium state.

The coviability is more complex when the dynamics is governed by equations evolving in time and space (distributed system), and the constraints also may have space conditions. If the feasibility set is empty, that means that the system is not coviable, otherwise the system can be coviable and this depends strongly on what are the subsystems dynamics and on what the feasibility set is. The coviability may be seen, to a certain sense, as a combination of spreadability, vulnerability and other proper characteristics of each subsystem which is a part of the global one. This is developed in the next section.

4.3 Coviability: Modelling and Control

The viability of a system means that the state of the system satisfies a certain property. Equivalently it may also mean that the state remains in a prescribed assigned constraint set. We can say that the coviability is an extended viability in the following sense. The prefix *co* denotes *with*, that is to say the simultaneous viability with something. There is a parallelism in the coviability approach. More precisely, if we consider many systems (or subsystems of a global system), each one evolving with its own dynamics and constraints. These systems are assumed to be interconnected and interacting with each other. From mathematical point of view, the coviability of these systems describes the viability of a relation between these systems, each one satisfying its own dynamics and constraints. The coviability means that there is a harmonious evolution in the parallel simultaneous evolutions of the systems. Now we consider the mathematical formulation.

4.3.1 Problem Statement

Consider a system (S) which state evolves in a state space Z. The state of the system is denoted z, $z \in Z$. The choice of the state space Z depends on the problem and the needed regularity of the state. Consider now a subset K of Z, which elements satisfy a given constraint.

The viability of the subset K means that the states which are in K, while obeying to the system dynamics, evolve remaining in K. When they remain in K for infinite time we say that K is globally viable.

More precisely denote the state of the system at time t : $z(t) = (z_1(t), z_2(t), \cdots, z_n(t))$ or simply $z = (z_1, z_2, \cdots, z_n)$ and consider the system (S) given by the state equation

$$(S) \qquad \begin{cases} \dot{z} & = \varphi(t, z_1, z_2, \cdots, z_n, u) \\ z(t_0) = z_0 = (z_0^1, z_0^2, \cdots, z_0^n) & \leftarrow \text{(initial state)} \end{cases} \tag{4.1}$$

augmented with the output function

$$y = \psi(t, z_1, z_2, \cdots, z_n) \tag{4.2}$$

and a coviability relation which can be stated by a function f

$$f(z_1, z_2, \cdots, z_n) = 0 \tag{4.3}$$

In the above statement :

- $\varphi = (\varphi_1, \varphi_2, \cdots, \varphi_n)$ is given and describes the system dynamics. Equivalently, the system (S) may be written as follows

$$(S) \qquad \begin{cases} \dot{z_1} & = \varphi_1(t, z_1, z_2, \cdots, z_n, u) \\ \dot{z_2} & = \varphi_2(t, z_1, z_2, \cdots, z_n, u) \\ \cdots & \quad \cdots \quad \cdots \quad \cdots \\ \dot{z_n} & = \varphi_n(t, z_1, z_2, \cdots, z_n, u) \\ z(t_0) = z_0 = (z_0^1, z_0^2, \cdots, z_0^n) & \leftarrow \text{(initial state)} \end{cases}$$

each of the φ_i corresponds to the dynamics of z_i; all the dynamics may be interconnected or not, depending on the considered problem.

- $y = (y_1, y_2, \cdots, y_q)$ is an output function which gives measurements via q sensors. The output may be written as follows

$$\begin{cases} y_1 &= \psi_1(t, z_1, z_2, \cdots, z_n) \\ y_2 &= \psi_2(t, z_1, z_2, \cdots, z_n) \\ \cdots & \cdots \ \cdots \ \cdots \\ y_n &= \psi_n(t, z_1, z_2, \cdots, z_n) \end{cases}$$

where y_i is associated to the measurement given by the ith sensor. The measurements may also depend on the control exciting the system.
- The term u is a control exciting the system. It will be made precise later.

Remarque 3.1

(i) The coviability function f may be a real function or a vector function $f = (f_1, f_2, \cdots, f_\ell)$, depending of how the coviability is considered for each real application. In the examples of the next section, f is a real function.
(ii) A simple choice of f would be

$$f(z_1, z_2, \cdots, z_n) = \alpha_1 z_1 + \alpha_2 z_2 + \cdots + \alpha_n z_n \tag{4.4}$$

which corresponds to a linear combination of the states. Depending on the considered application one has to choose the coefficients (α_i). From mathematical point of view, the above combination assumes that the states are in the same space. In the examples we consider the coviability defined with (4.3).
(iii) The coviability condition may be more complex than (4.3). Indeed, it may be defined by inequalities. This is the case, in certain applications, where the state must not exceed (or be less than) a given value. Additionally f may be a vector function and defines constraints for each state of the system, i.e. $f = (f_1, f_2, \cdots, f_\ell)$, with

$$\alpha_i \leq f_i(z_1, z_2, \cdots, z_n) \leq \beta_i \ , \quad \text{for } i = 1, \cdots, \ell \tag{4.5}$$

the parameters α_i and β_i depending on the considered problem.
(iv) From practical point of view, the coviability can be explored by considering the mathematical relation between the states (z_i), given in (4.3), which must be satisfied for all times $t \geq t_0$.

The system (S) may be seen as a set of n subsystems which states z_i have their own dynamics. The dynamics are interconnected by means of the function φ.

Consider the set K, a subset of the state space Z, $K \subset Z$. Thus we have the following intuitive definition

Definition The system (S) is said to be coviable if, from any initial state $z^0 = \left(z^0{}_1, z_2{}^0, \cdots, z_n{}^0\right)$ **in** K, then the future states $z(t) = (z_1(t), z_2(t), \cdots, z_n(t))$ remain **in** K, while satisfying the coviability relation (4.3), for all $t \geq t_0$.

We can notice, from the above definition, that

(1) *The coviability implies the viability if we consider* $f = 0$,
(2) *The coviability can be defined differently by introducing the relation* (4.3) *in the constraint set* K.
(3) *If a system* (S) *is viable, then for its coviability, one has to check if the relation* (4.3) *is satisfied,*
(4) *The output function* (4.2) *is useful to have information on the state of the system and its evolution. Additionally it can be used for a feedback control problem,*
(5) *The above definition holds for lumped and distributed systems.*

We use to note :

- $Viab(K)$ the set of states which are viable (or K-viable),
- $Viab_u(K)$ the set of states which are viable (or K-viable) under the effect of the control u,
- $NViab_u(K)$, the viability kernel, the set of states for which there exist a control u which leave them viable. This shows clearly that $Viab(K) \subset NViab(K)$ and $Viab_u(K) \subset NViab(K)$.

As explained in the scheme *Modelling* $\rightarrow$ *Analysis* $\rightarrow$ *Control*, what would be the control problem (i.e. the decision strategy) in the coviability context ? In fact taking into account the right approach for the long-term connections between human societies and the biosphere, we can simply say that

$$\textbf{Control problem} \equiv \textbf{Coviabilisation}$$

In the case where the system is not coviable, the control problem to be considered consists to find a control which makes the system coviable. Consider the controlled system (S) which state is $z = (z_1, z_2, \cdots, z_n)$

$$(S) \qquad \begin{cases} \dot{z} &= \varphi(t, z_1, z_2, \cdots, z_n, u) \\ \\ z_0 &= (z_0^1, z_0^2, \cdots, z_0^n) \end{cases}$$

where the control term u is to be find. The control problem can be stated as follows.

$$(P) \qquad \begin{cases} \text{Find } u \in \text{ admissible control set } \mathcal{U} \\ \\ \text{which makes the system } (S) \text{ coviable} \end{cases}$$

For the solution of the above control problem, one should precise the systems data, the coviability constraints, together with the set of admissible controls $\mathcal{U}$. The above problem can be considered for both lumped and distributed systems. In the distributed case, all the parameters and states depend on the space variable. Under convenient conditions the solution to the above problem may exist and, in some cases, be unique or not. This needs a huge amount of mathematics.

The problem (P) may be defined using the output function (4.2), and then find a feedback control which depends on the output.

In connection with the previous concepts, the set of viability K may be defined by the region where the states are not overrun (invaded) by a spreadable phenomena. This leads to a combination of spreadability, vulnerability and viability.

In the next section, we show the techniques used for the viability and coviability. This is illustrated through a lumped system and a distributed one.

4.3.2 *Illustrative Examples*

We give hereafter three examples of dynamical systems and explore their viability and coviability. The first system is lumped and the second one is distributed. A third example concerns the case where we are concerned by regional (in space) coviability. All the examples are concerned by two dynamical systems; but higher order cases can be easily considered. These systems may be considered as illustrative cases of competition between species.

4.3.2.1 A Two-Dimensional Lumped System

Consider the system given by the state equation

$$\begin{cases} \dot{z}(t) = Az(t) \quad t \in]0, T] \\ z(0) = z_0 \end{cases} \tag{4.6}$$

where

$$z(t) = \begin{pmatrix} z_1(t) \\ z_2(t) \end{pmatrix} \quad \text{and} \quad A = \begin{pmatrix} 1 & 4 \\ 2 & -1 \end{pmatrix} \tag{4.7}$$

The system (4.6) describes the dynamics of two connected states z_1 and z_2. It can be easily shown that the solution is given, for $t \in]0, T]$, by

$$\begin{cases} z_1(t) = \frac{1}{3}(e^{-3t} + 2e^{3t})z_0^1 + \frac{1}{3}(2e^{3t} - 2e^{-3t})z_0^2 \\ z_2(t) = \frac{1}{3}(e^{3t} - e^{-3t})z_0^1 + \frac{1}{3}(2e^{-3t} + e^{3t})z_0^2 \end{cases} \tag{4.8}$$

For the viability. Consider the constraint set

$$K = [0, 60] \times [0, 30] \quad \text{and} \quad T = 1$$

Then the state (z_0^1, z_0^2) is K-viable if

$$\begin{cases} 0 \leq z_1(t) \leq 60 & t \in]0, 1] \\ 0 \leq z_2(t) \leq 30 & t \in]0, 1] \end{cases} \tag{4.9}$$

As the states $z_1(t)$ and $z_2(t)$ satisfy the above inequalities, the set $Viab(k)$ needs to be computed. However we illustrate the result by considering

$$K_1 = [0, 2] \times [0, 2] \subset Viab(K)$$

which is satisfied because

$$\begin{cases} 0 \leq z_0^1 \leq 2 \\ 0 \leq z_0^2 \leq 2 \end{cases} \Rightarrow \begin{cases} 0 \leq z_1(t) \leq 60 & t \in]0, 1] \\ 0 \leq z_2(t) \leq 30 & t \in]0, 1] \end{cases} \tag{4.10}$$

For the coviability. Assume that the coviability function (4.3) is given, in this case, by

$$f(z_1, z_2) = z_1 - 2z_2 = 0$$

The above relation means that the coviability needs the state z_1 to be twice the state z_2.

Assume that the initial state is given by

$$z_0^1 = 2 \quad \text{and} \quad z_0^2 = 1 \tag{4.11}$$

The solution of the system is

$$\begin{cases} z_1(t, z_0^1) = 2e^{3t} & t \in]0, T] \\ z_2(t, z_0^2) = e^{3t} & t \in]0, T] \end{cases} \tag{4.12}$$

and thus the initial state $(z_0^1 = 2, z_0^2 = 1)$ is coviable because

$$\begin{cases} (i) \;\; (z_0^1, z_0^2) \in K \\ (ii) \;\; (z_1(t, z_0^1), z_2(t, z_0^2)) \in K & t \in]0, 1] \\ (iii) \; z_1(t, z_0^1) - 2z_2(t, z_0^2) = 0 & t \in]0, 1] \end{cases} \tag{4.13}$$

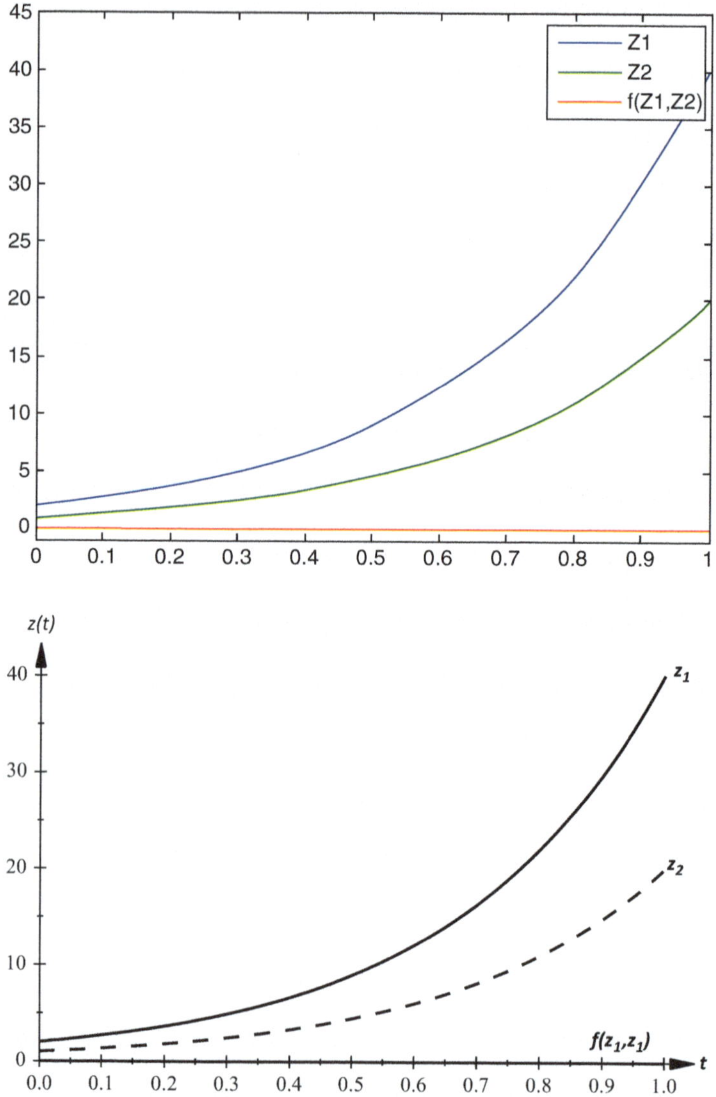

Fig. 4.1 The states z_1 (continuous) and z_2 (dashed) satisfy the coviability relation f

The states are shown in the Fig. 4.1. They evolve in time, because the system is lumped. In the Fig. 4.1, we can see that, at any time t, the state z_1 is equal to twice the state z_2, which means that $f(z_1, z_2) = z_1 - 2z_2 = 0$. Consequently the coviability of the states z_1 and z_2 is satisfied.

4.3.2.2 A Spatio-Temporal Example

Consider the system with two components z_1 and z_2 described by the equation

$$\begin{cases} \dfrac{\partial z_1(x,t)}{\partial t} = xz_1(x,t) + u(x,t) & x \in \Omega, \quad t \in I \\ \\ \dfrac{\partial z_2(x,t)}{\partial t} = 2xz_2(x,t) + u(x,t) & x \in \Omega, \quad t \in I \end{cases} \tag{4.14}$$

where the domain is given by $\Omega =]1, 2]$ on the time interval $I =]0, 1]$.

Autonomous case.

Here we assume that the control $u = 0$. The solution of (4.14) is

$$\begin{cases} z_1(x,t) = e^{xt} z_0^1(x) & x \in [1,2] \ , t \in]0,1] \\ \\ z_2(x,t) = e^{2xt} z_0^2(x) & x \in [1,2] \ , t \in]0,1] \end{cases} \tag{4.15}$$

For the viability. Let the constraints set be given by $K = [-5, 15] \times [-10, 70]$. Thus we have

$$\begin{cases} -5 \leq z_1(x,t) \leq 15 & x \in \Omega \ , t \in I \\ \\ -10 \leq z_2(x,t) \leq 70 & x \in \Omega \ , t \in I \end{cases} \Leftrightarrow \begin{cases} -5 \leq z_0^1(x) \leq 15e^{-2} & x \in \Omega \\ \\ -10 \leq z_0^2(x) \leq 70e^{-4} & x \in \Omega \end{cases} \tag{4.16}$$

Therefore the viability domain of K is

$$Viab(K) = [-5, 15e^{-2}] \times [-10, 70e^{-4}]$$

For the coviability. Consider the function (4.3) which defines the coviability relation between the states. Assume it is given by

$$f(z_1, z_2) = z_1 - \frac{3}{2} z_2 = 0 \tag{4.17}$$

This relation means that the coviability is ensured if the state z_1 is equal to $3z_2/2$.

Assume that the initial state ($z_0^1(.)$, $z_0^2(.)$) is given by

$$\begin{cases} z_0^1(x) = -x & x \in [1,2] \\ \\ z_0^2(x) = -\frac{2}{3}x & x \in [1,2] \end{cases} \tag{4.18}$$

Thus the initial state $(z_0^1(.), z_0^2(.))$ is K-viable, but is not coviable because it does not satisfy the relation (4.17). Indeed, we have

$$f(z_1(x,t,z_0^1(.)), z_2(x,t,z_0^2(.))) \neq 0 \ ; \quad x \in [1,2] \ , \ t \in]0,1]$$

Controlled case: coviabilisation.

We consider the control u given by

$$u(x,t) = 2x^2e^{-xt} \quad , \ x \ \in [1,2] \ , t \ \in]0,1] \tag{4.19}$$

The solution of the system is then given by

$$\begin{cases} z_1(x,t) = e^{xt}z_0^1(x) + v_1(x,t) & x \ \in [1,2] \ ; t \ \in]0,1] \\ \\ z_2(x,t) = e^{2xt}z_0^2(x) + v_2(x,t) & x \ \in [1,2] \ ; t \ \in]0,1] \end{cases} \tag{4.20}$$

with

$$\begin{cases} v_1(x,t) = xe^{xt} - xe^{-xt} \\ \\ v_2(x,t) = \frac{2}{3}x[e^{2xt} - e^{-xt}] \end{cases} \tag{4.21}$$

For the viability. Consider the constraints set $K = [-5, 15] \times [-10, 70]$, we have

$$\begin{cases} -5 \leq z_1(x,t) \leq 15 & x \ \in \Omega \ , t \ \in I \\ \\ -10 \leq z_2(x,t) \leq 70 & x \ \in \Omega \ , t \ \in I \end{cases} \Leftrightarrow \begin{cases} -5 \leq z_0^1(x) \leq \frac{1}{2}e^{-2} & x \ \in \Omega \\ \\ -10 \leq z_0^2(x) \leq -3e^{-4} & x \ \in \Omega \end{cases} \tag{4.22}$$

and then

$$Viab_u(K) = [-5, \frac{1}{2}e^{-2}] \times [-10, -3e^{-4}]$$

For the coviability. Consider again the coviability constraint function

$$f(z_1, z_2) = z_1 - \frac{3}{2}z_2 = 0$$

and the initial state ($z_0^1(.)$, $z_0^2(.)$) given by

$$\begin{cases} z_0^1(x) = -x & x \ \in [1,2] \\ \\ z_2(x) = -\frac{2}{3}x & x \ \in [1,2] \end{cases} \tag{4.23}$$

We have

$$f(z_1(x,t), z_2(x,t)) = z_1(x,t) - \frac{3}{2}z_2(x,t) = 0$$

This leads easily to

$$xe^{xt} - xe^{2xt} = \frac{3}{2}e^{2xt}z_0^2(x) - e^{xt}z_0^1(x) \tag{4.24}$$

Consequently the state $(z_0^1(.), z_0^2(.))$ is K-viable and furthermore it satisfies the relation (4.24). This shows that the state becomes coviable under the effect of the control $u(x, t)$.

The states are shown in Fig. 4.2. They evolve in space and time, because the system is distributed. In Fig. 4.2, we can see that, for any time t and space x, the

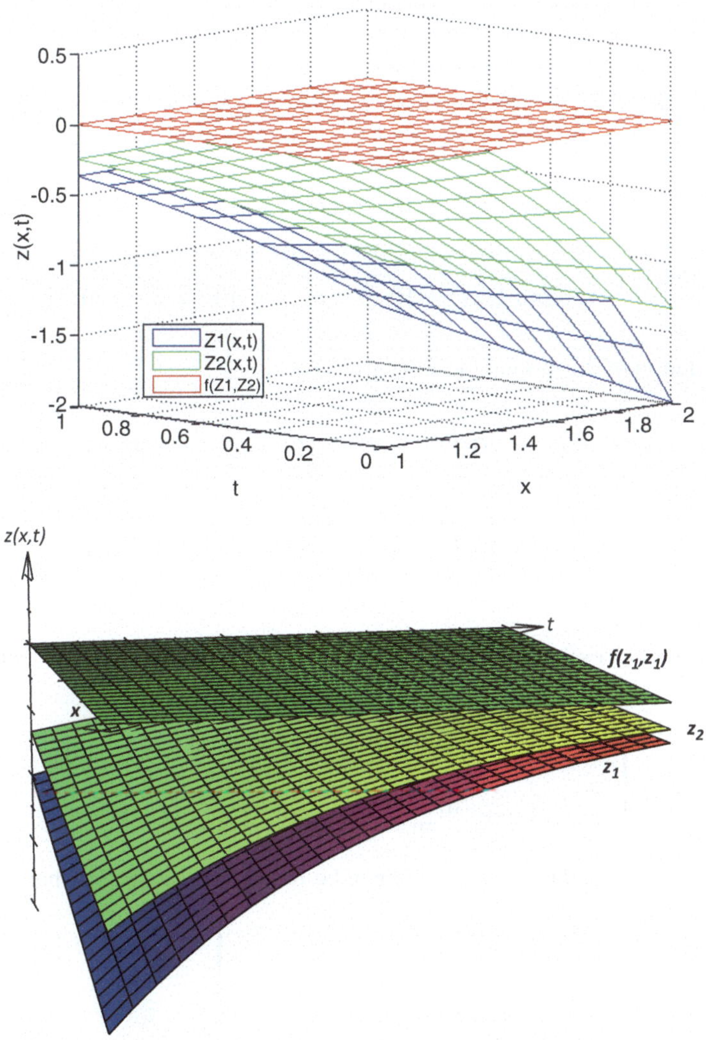

Fig. 4.2 The states z_1 (blue) and z_2 (green) satisfy the coviability condition

state z_1 is equal to 3/2 of the state z_2, which means that $f(z_1, z_2) = z_1(x,t) - 3/2 z_2(x,t) = 0$. Consequently the coviability of the states z_1 and z_2 is satisfied over the whole space Ω and for any time t.

4.3.2.3 An Example of Regional Coviability

We have explained previously what is the regional approach in systems analysis. For the coviability, it means that we consider a spatio-temporal system and we are interested by the coviability in a given region. More precisely we assume that the system is coviable everywhere except in that given region and we focus on a control strategy which makes it coviable even in that region. We give hereafter an example of a distributed system defined in a domain Ω, which will be made coviable, with a convenient control, in a given subregion ω, with $\omega \subset \Omega$.

Let Ω be given by $\Omega =]0,4[$, $\omega =]3,4]$, and . The system i governed by the state equation

$$\begin{cases} \dfrac{\partial z_1(x,t)}{\partial t} = -2xz_1(x,t) + 10u_1(x,t) \quad x \in [0,4] \ , \ t \in]0,1] \\ \\ \dfrac{\partial z_2(x,t)}{\partial t} = -3xz_2(x,t) + 10u_2(x,t) \quad x \in [0,4] \ , \ t \in]0,1] \end{cases} \tag{4.25}$$

The solution of the system (4.25) is given by

$$\begin{cases} z_1(x,t) = e^{-2xt} z_0^1(x) + 10 \displaystyle\int_0^t e^{-2x(t-s)} u_1(x,s) ds \quad t \in]0,T] \ , \ x \in [0,4] \\ \\ z_2(x,t) = e^{-3xt} z_0^2(x) + 10 \displaystyle\int_0^t e^{-3x(t-s)} u_2(x,s) ds \quad t \in]0,T] \ , \ x \in [0,4] \end{cases} \tag{4.26}$$

Autonomous case.

Consider the case where the control $u = 0$. The solution of system (4.25) becomes

$$\begin{cases} z_1(x,t) = e^{-2xt} z_0^1(x) \quad t \in]0,T] \ , \ x \in [0,4] \\ \\ z_2(x,t) = e^{-3xt} z_0^2(x) \quad t \in]0,T] \ , \ x \in [0,4] \end{cases} \tag{4.27}$$

For the viability. Let the constraint domain be $K = [0,15] \times [0,7]$. Thus we have

$$\begin{cases} 0 \le z_1(x,t) \le 15 \quad t \in]0,T] \ , \ x \in \sigma \\ \\ 0 \le z_2(x,t) \le 7 \quad t \in]0,T] \ , \ x \in \sigma \end{cases} \Leftrightarrow \begin{cases} 0 \le z_0^1(x) \le 15 \quad x \in \sigma \\ \\ 0 \le z_0^2(x) \le 7 \quad x \in \sigma \end{cases} \tag{4.28}$$

Therefore the viability domain of K is

$$Viab(K) = [0, 15] \times [0, 7]$$

For the coviability. Let the coviability function be given by

$$f(z_1, z_2) = z_1 - 2z_2 = 0$$

For the initial state $(z_0^1(.), z_0^2(.))$ given by

$$z_0^1(x) = \begin{cases} xe^{-\frac{1}{x-3}} & \text{if } x \in]3, 4[\\ e^{\frac{1}{x-3}} & \text{if } x \in]0, 3[\end{cases} \quad and \quad z_0^2(x) = \begin{cases} \frac{x}{2}e^{-\frac{1}{x-3}} & \text{if } x \in]3, 4[\\ e^{\frac{2}{x-3}} & \text{if } x \in]0, 3[\end{cases} \tag{4.29}$$

which shows that the initial state $(z_0^1(.) \quad , z_0^2(.) \ )$ is $K - viable$, but is not regionally coviable because the relation $f(z_1(t, z_0^1(.)), z_2(t, z_0^2(.))) = z_1(t, z_0^1(.)) - 2z_2(t, z_0^2(.)) = 0$ is not satisfied in ω.

Controlled case.

We consider the control given by

$$\begin{cases} u_1(x,t) = x^2 e^{-3xt - \frac{1}{x-3}} \chi_{]3,4[}(x) & t \in]0, T] \\ u_2(x,t) = x^2 e^{-2xt - \frac{1}{x-3}} \chi_{]3,4[}(x) & t \in]0, T] \end{cases} \tag{4.30}$$

where χ_A denotes the characteristic function of A. The solution of the controlled system (4.25) is

$$\begin{cases} z_1(x,t) = e^{-2xt} z_0^1(x) + v(x,t) \ \chi_{]3,4[}(x) & t \in]0, T] \\ z_2(x,t) = e^{-3xt} z_0^2(x) + v(x,t) \ \chi_{]3,4[}(x) & t \in]0, T] \end{cases} \tag{4.31}$$

with

$$v(x,t) = x(e^{-2xt - \frac{1}{x-3}} - e^{-3xt - \frac{1}{x-3}}) \ , \quad t \in]0, T]$$

For the viability. The considered constraint domain is given by $K = [0, 15] \times [0, 7]$. We have

$$\begin{cases} 0 \le z_1(x,t) \le 15 & x \in \sigma \ , t \in]0, 1] \\ 0 \le z_2(x,t) \le 7 & x \in \sigma \ , t \in]0, 1] \end{cases} \Leftrightarrow \begin{cases} 0 \le z_0^1(x) \le 15 - 4e^{-1} & x \in \sigma \\ 0 \le z_0^2(x) \le 7 - 4e^{-1} & x \in \sigma \end{cases} \tag{4.32}$$

We obtain

$$Viab_u(K) = [0, 15 - 4e^{-1}] \times [0, 7 - 4e^{-1}]$$

For the coviability. Consider the coviability relation $f(z_1, z_2) = z_1 - 2z_2 = 0$ and the initial state $(z_0^1(.), z_0^2(.))$ defined by (Fig. 4.3)

$$z_0^1(x) = \begin{cases} xe^{-\frac{1}{x-3}} & \text{if } x \in]3, 4[\\ e^{\frac{1}{x-3}} & \text{if } x \in]0, 3[\end{cases} \quad and \quad z_0^2(x) = \begin{cases} \frac{x}{2}e^{-\frac{1}{x-3}} & \text{if } x \in]3, 4[\\ e^{\frac{2}{x-3}} & \text{if } x \in]0, 3[\end{cases} \tag{4.33}$$

Then we have

$$\begin{aligned} f(z_1(x, t, z_0^1(x)), z_2(x, t, z_0^2(x))) &= z_1(x, t, z_0^1(x)) - 2z_2(x, t, z_0^2(x)) \\ &= \begin{cases} 0 & \text{if } x \in]3, 4[\\ \neq 0 & \text{if } x \in]0, 3[\end{cases} \end{aligned} \tag{4.34}$$

Consequently the state ($z_0^1(.)$, $z_0^2(.)$) is K-viable, and moreover it satisfies the coviability condition on the region ω, therefore it is $\omega - coviable$ for the relation f under the effect of the control $u(x, t)$.

4.3.3 Remark on Modelling Approach of Coviability

The above examples have been considered to show how the concept of coviability works. In these examples the solution of the state equation was analytic and makes the study easier. For more complex real spatio-temporal systems, one has to consider numerical models. For that purpose, the most useful approach can be based on cellular automata. Cellular automata models have been used to simulate other concepts in distributed systems (as spreadability).

Cellular automata can model complex systems because of their implementation simplicity. They are based on discrete approach (time, space, states) and can generate very complex behaviour.

The evolution of CA can lead to homogeneous, periodic, chaotic or to more complex steady states.

In the case of coviability, it is recommended to consider CA. On one hand, it may govern very complex dynamics, as nonlinear or delay system. On an other hand the implementation and simulation of cellular automata models may be done on any home computer.

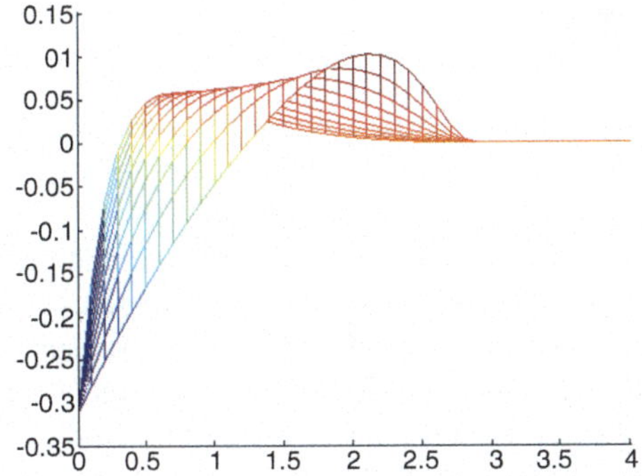

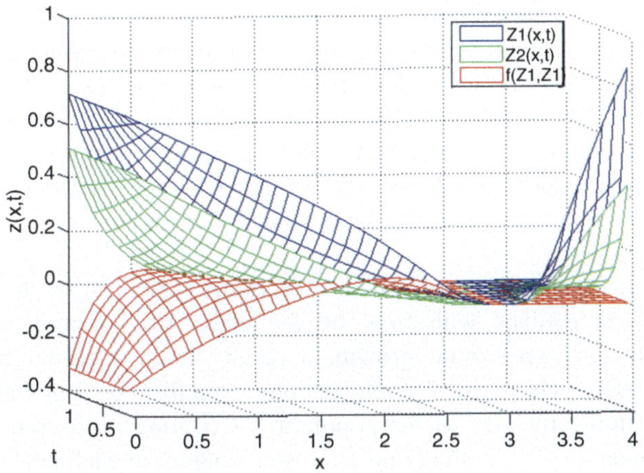

The coviability function which vanishes in $\omega = [3{,}4]$.

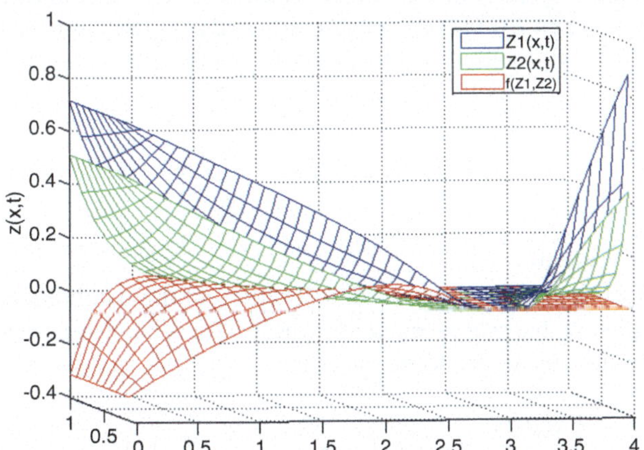

Fig. 4.3 The states satisfy the coviavility condition only in the region $\omega = [3, 4]$

4.4 A Real System Case

Numerous real systems can illustrate the coviability concept and approach. And the more the system is global the more its components number grows. Hereafter we give the example of fishery management which integrate, in a simplified approach, two components which are fish resource and socio-economic management.

Dynamic modelling of fishery management that integrates resource biology and ecology as well as socio-economic aspects offers an appropriate application layout to illustrate the coviability concept. We will refer to the fleet dynamics model based on Smith's model (Smith 1969), derived from the Gordon-Shaefer population dynamic model (Gordon 1953, 1954; Schaefer 1954) with the added assumption that long-run fishing levels are proportional to profit and then to the exploited biomass. In other words, fish biomass depends on a growth rate, the carrying capacity of the environment and a fishing effort, which itself is proportional to available biomass. Fishermen and fish resource are then involved in an intricate process for which when the resource collapses, the fisheries also collapses. From this point of view, in this application example, we can talk about coviability dynamic and attempt to express the coviability conditions using the framework introduced within the previous sections.

The following equation was obtained by Verhulst (1847a,b) by this reasoning: a population $B(t)$ evolves with time by the effect of a reproduction rate r, yielding initially an exponential growth, a result obtained earlier by Malthus: $dB(t)/dt = rB(t)$. The term B could denote a number, or a global mass, any quantity representing the size of the population. Counting in discrete time steps, with two descendants, $r = 2$, the population size doubles at each generation. Thus quickly the population would grow up to exhaust the resources the environment can provide, called the carrying capacity κ. A corrective term is introduced in the equation such that, when $B(t)$ is very small compared to κ, the growth is exponential at the intrinsic rate r, since $B(t)\kappa$ is close to zero, but the growth levels to a null rate as $B(t)$ approaches κ, since $(1 - B(t)\kappa)$ tends to zero.

$$\frac{dB(t)}{dt} = rB(t)\left(1 - \frac{B(t)}{\kappa}\right) \tag{4.35}$$

This is a lumped system.

Now consider this equation, where B will represent the fish biomass in a given ecosystem, coupled with a second equation where $f(t)$ is the fishing effort of a fleet of vessels determined by an economic model:

$$\begin{cases} \dfrac{dB(t)}{dt} = rB(t)\left(1 - \frac{B(t)}{\kappa}\right) - qB(t)f(t) \\ \dfrac{df(t)}{dt} = \eta\,(pqB(t) - c)\,f(t) \end{cases} \tag{4.36}$$

Where η, the fleet dynamic parameter η expresses how strongly the fleet reacts to the revenue level $(pqB(t) - c)$ offered by the fish stock. This unit revenue level growth with p, the price of a unit of fish, and q its catchability coefficient, and is depleted by c, the unit cost of fishing effort.

The profit $\pi(t)$ derived from fishing is a function of total sustainable revenues and total costs and is defined by:

$$\pi(t) = (pqB(t) - c)\, f(t) \tag{4.37}$$

The above statement (4.36) needs the knowledge of the initial states $B(0)$ and $f(0)$, and can be seen as a particular case of the general model (4.1) of the previous section. For the equivalence one has to consider the variable change $z_1 = B$ and $z_2 = f$, thus with

$$\begin{aligned}\varphi_1 &= rz_1\left(1 - \tfrac{z_1(t)}{\kappa}\right) - qz_1(t)z_2(t)\\ \varphi_2 &= \eta\,(pqz_1(t) - c)\, z_2(t)\end{aligned}$$

we retrieve the general statement given in (4.1)

$$\dot{z} = \varphi(t, z_1, z_2) = \begin{pmatrix}\varphi_1(t, z_1, z_2)\\ \varphi_2(t, z_1, z_2)\end{pmatrix} = \begin{pmatrix} rz_1\left(1 - \frac{z_1(t)}{\kappa}\right) - qz_1(t)z_2(t)\\ \eta\,(pqz_1(t) - c)\, z_2(t)\end{pmatrix}$$

Using the set of parameters defined in Table 4.1, the process behaves as shown in Fig. 4.4. From an initial state $B(0)$ corresponding to a fish biomass without human pressure, $B(t)$ converges towards a value around 1,250,000 tonnes. Similarly, after a few oscillations, the fishing effort $f(t)$ and the profit $\pi(t)$ converges toward the values 578 and 0 respectively.

In accordance with the previous definition of the viability, the system is said K-viable (the system may evolves in) using the following constraint set:

Table 4.1 Parameters for the dynamic bioeconomic model (Gordon-Schaefer & Smith)

Parameter	Variable	Value
Intrinsic growth rate	r	0.36
Catchability coefficient	q	0.0004
Carrying capacity of the system	κ	3,500,000 tonnes
Price of the target species	p	60 US$/tonne
Unit cost of fishing effort	c	30,000US$/yr
Initial population biomass	$B(0)$	3,500,000 tonnes
Initial fishing effort (vessels)	$f(0)$	1
Fleet dynamics parameter	η	0.000005

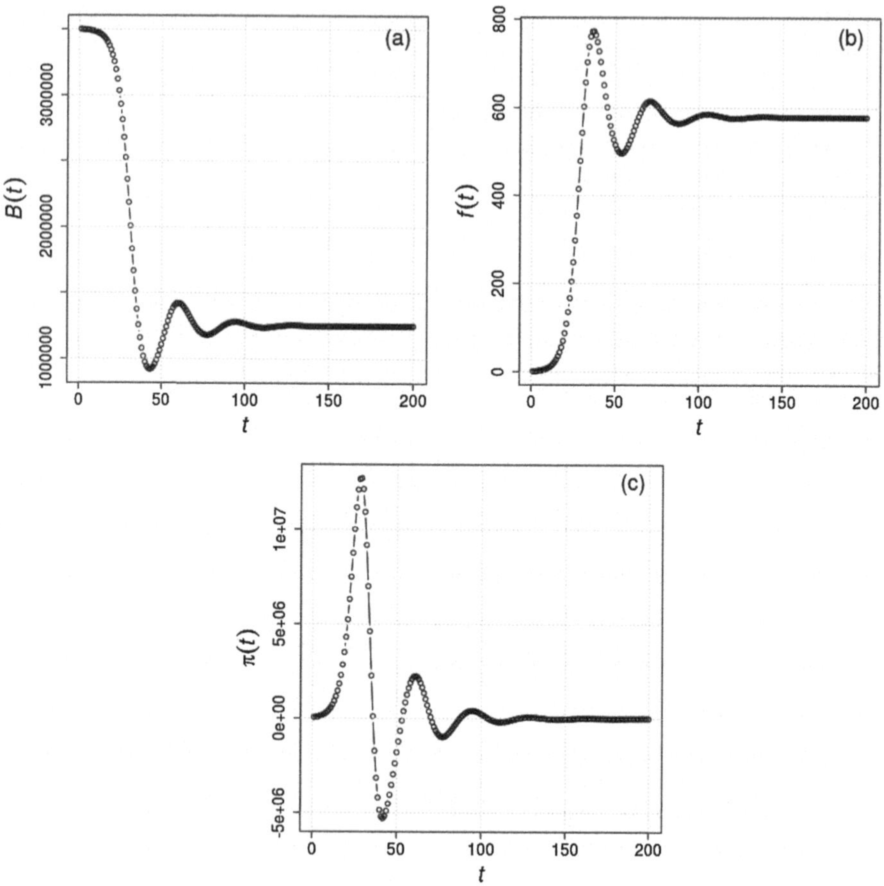

Fig. 4.4 Evolution of the biomass (**a**) the fishing effort (**b**) and the profit (**c**)

$$K = \begin{cases} 9{,}19{,}465 \leqslant B \leqslant 3{,}500{,}000 \\ 0 \leqslant f(t) \leqslant 800 \end{cases} \tag{4.38}$$

A coviability statement could be defined as a system of inequalities that ensures a minimal profit α without exerting unbearable pressure on the resource (biomass should stay above β).

$$\begin{cases} (pqB(t) - c)\, f(t) \geq \alpha \\ B(t) \qquad\qquad\quad \geq \beta \end{cases} \tag{4.39}$$

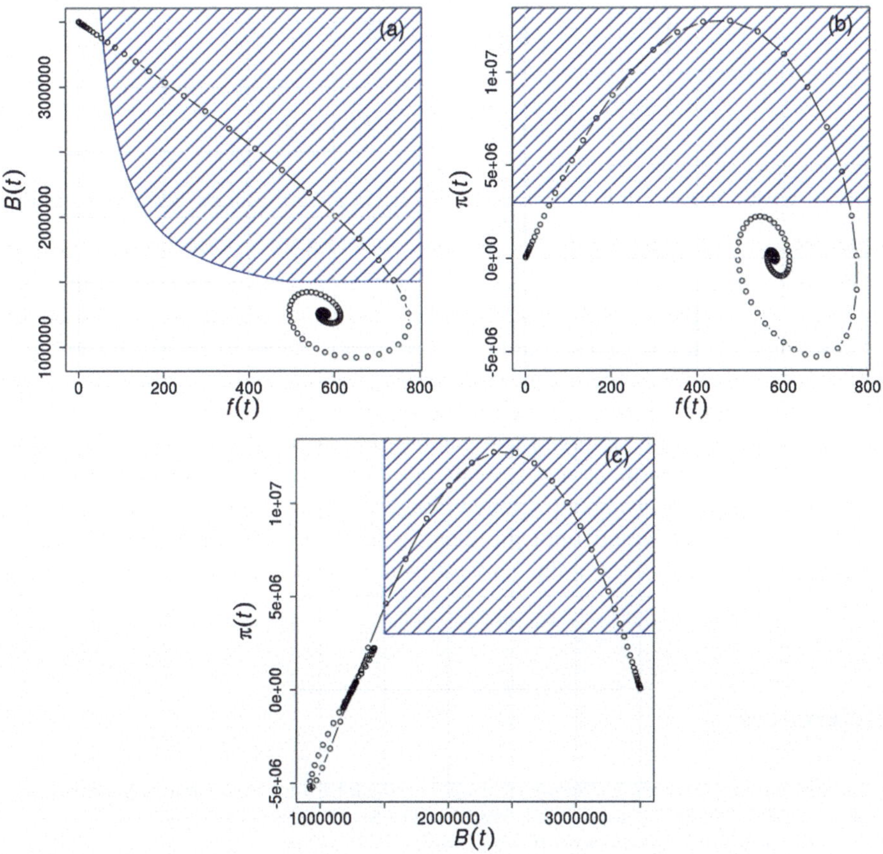

Fig. 4.5 Evolution of the biomass versus the fishing effort (**a**) the profit versus the fishing effort (**b**) and the profit versus the biomass. The blue hatched areas correspond to values that respect the coviability constraints

The terms α and β could be defined by expertise or estimated from observations. Here, we chose the following settings for illustration purposes: $\alpha = 3{,}000{,}000$ and $\beta = 1{,}500{,}000$.

The above constraints conditions (4.39) can also be seen as a particular case of (4.5) with $\ell = 2$.

The dynamic of the fish biomass ($B(t)$) versus the fishing effort ($f(t)$) can be analysed Fig. 4.5, so the behaviors of the profit versus the fishing effort and the profit versus the biomass. It should be noted that the system may evolve outside of the areas corresponding to the coviability statement (4.39) (the blue hatched areas).

At this stage, a control u can be added in order to limit the fishing effort and to leave the whole system coviable in accordance with the equation (4.39).

$$\begin{cases} \dfrac{dB(t)}{dt} = rB(t)\left(1 - \frac{B(t)}{K}\right) - qB(t)f(t) \\ \\ \dfrac{df(t)}{dt} = \eta\,(pqB(t) - c)\,f(t) + u(t) \end{cases} \tag{4.40}$$

It is worth interpreting this term u. It has been introduced additively in the equation, thus effecting directly $\dfrac{df(t)}{dt}$, as if it had a direct effect on the growth or depletion of the fishing effort. Since it results from human decision, it must be informed by visible effects. In practice, fishers and policy makers become aware of an alarming collapse of stocks. From then on, management rules may be decreed. Hence $u(t)$ is a null function in time until some conscious action becomes necessary. Management rules can be expressed based on some even rough estimate of $B(t)$, perhaps resulting from observed fish catches, provided all the effect of variable fishing effort can be eliminated. A realistic rule can be summed up in the catchability coefficient, e.g., authorising only limited time slots, constraining the size of catches. Another component would limit the fleet, in number, size, catch power, overall resulting in a control of $f(t)$, the simplest form being to saturate f at a maximum value.

References

Afifi L, El Jai A (1994) Strategic sensors and spy sensors. Int J Appl Math Comput Sci 4:553–573

Afifi L, El Jai A (1995) Spy sensors and detection. Int J Syst Sci 26(3):1447–1463

Afifi L, El Jai A (2015) Systèmes distribués perturbés. PUP, 342 pp

Afifi L, El Jai A, Merry M (2000) Detection and sources reconstruction in a tube. Int J Syst Sci 31(2):149–159. PUP, Université de Perpignan, 346 pp

Afifi L, Chafiai A, El Jai A (2001) Spatial compensation of boundary disturbances by boundary actuators. Int J Appl Math Comput Sci 11:9–20

Afifi L, Chafiai A, El Jai A (2002) Regionally efficient and strategic actuators. Int J Syst Sci 33(1):1–12

Afifi L, Amimi N, Magri EM, El Jai A (2011) Regional domination in distributed systems. Int J Appl Math Sci 6(19):913–924

Afifi L, El Jai A, Zerrik E (2012) Systems theory: regional analysis of infinite-dimensional linear systems. PUP, Perpignan, 425 pp

Afifi L, Joundi M, Magri EM, El Jai A (2013a) Domination in controlled and observed distributed parameter systems. Intell Control Autom 4(2):217–226

Afifi L, Joundi M, Magri EM, El Jai A (2013b) An extended regional classification of distributed systems. Int J Math Anal 7(28):1363–1377

Afifi L, Joundi M, Amimi N, Bahadi M (2013c) Asymptotic classification of distributed linear systems. Appl Math Sci 7(51):2537–2554

Aubin JP (1991) Viability theory. Birkhauser, Boston

Aubin JP (2001) Viability kernels and capture basins of sets under differential inclusions. SIAM J Control Optim 40(3):853–881

Aubin J-P, Bayen AM, Saint-Pierre P (2011) Viability theory: new directions. Springer, Berlin. ISBN:978-3-642-16683-9

Bernoussi A (2007) Spreadability and vulnerability of distributed parameter systems. Int J Syst Sci 38(4):305–317
Bernoussi A (2010) Spreadability, vulnerability and protector control. Math Model Nat Phenom 5(07):145–150
Bernoussi A, Amharref M, Belfekih A (2015) Permanent protector control and viability for a class of distributed parameter systems. Accepted, to appear
Curtain RF, Zwart HJ (1995) An introduction to infinite-dimensional linear systems theory. Texts in applied mathematics. Springer, New York
El Jai A (1991) Distributed systems analysis via sensors and actuators. Int J Sensors Actuators 29(1):1–11
El Jai A, Berrahmoune L (1984) Localisation d'actionneurs frontières pour la contrôlabilité de systèmes paraboliques, vol 298, Série I, N 8. Comptes rendus de l'Acadèmie des Sciences, Paris, pp 177–180
El Jai A, Kassara K (1994) Spreadable distributed systems. Math Comput Model 20(1):47–64
El Jai A, Pritchard AJ (1988) Sensors and actuators in distributed systems analysis. Ellis Horwood series in applied mathematics, vol 47. Wiley, New York
El Jai A, Simon MC, Zerrik E (1993) Regional observability and sensors structures. Int J Sensors Actuators 39(3):95–102
El Jai A, Pritchard AJ, Simon MC, Zerrik E (1995) Regional controllability of distributed systems. Int J Control 62:1351–1365
Gordon HS (1953) An economic approach to the optimum utilization of fishery resources. J Fish Res Bd Can 10:442447
Gordon HS (1954) The economics of a common property resource: the fishery. J Polit Econ 62:124142
Schaefer MB (1954) Some aspects of the dynamics of populations important to the management of commercial marine fisheries. Bull Inter-Amer Trop Tunna Comm 1:2756
Smith VL (1969) On models of commercial fishing. J Polit Econ 77:181198
Verhulst PF (1847) Notice sur la loi que la population suit dans son accroissement. Correspondance mathématique et physique de l'Observatoire de Bruxelles 10:113–121. https://play.google.com/store/books/details
Verhulst PF (1847) Deuxième mémoire sur la loi d'accroissement de la population. Mémoires de l'Académie Royale des Sciences, des Lettres et des Beaux-Arts de Belgique 20:1–32
Zerrik E, El Jai A (2014) Stabilité des systèmes dynamiques. PUP, 302 pp
Zerrik E, Badraoui L, El Jai A (1999) Sensors and regional boundary observability of parabolic systems. Int J Sensors Actuators 75:102–117
Zerrik E, Boutoulout A, El Jai A (2000) Actuators and regional boundary controllability for parabolic systems. Int J Syst Sci 31(1):73–82

Chapter 5
The Relationship Between Man and His Environment: A Systemic Approach of the Viability of "System Earth"

Mireille Fargette, Maud Loireau, and Thérèse Libourel

Contents

5.1 Introduction

A category-based dual vision has been adopted by some, in order to quote and differentiate Nature and Man, often setting them against one another whether by reasoning or by action (the thinking Man, objectivizing what surrounds him, manipulating "things" and making them "his", for his own ends, Descartes 1637; d'Alembert 1751). Nature and Man have also been presented by the archetypal visions of the great founding texts on origins (see Genesis) or in the founding myths of societies, which address ancestors and divinities, without necessarily having a clear break between these terms, and with, perhaps, a less fractured vision of the

M. Fargette (✉) · M. Loireau
IRD - French National Research Institute for Sustainable Development, UMR ESPACE-DEV, Montpellier, France
e-mail: mireille.fargette@ird.fr; maud.loireau@ird.fr

T. Libourel
Université Montpellier, UMR ESPACE-DEV, Montpellier, France
e-mail: therese.libourel@umontpellier.fr

O. Barrière et al. (eds.), *Coviability of Social and Ecological Systems: Reconnecting Mankind to the Biosphere in an Era of Global Change*,
https://doi.org/10.1007/978-3-319-78497-7_5

world (Descola 2010, 2011). Beyond this category-based vision, we have recently used terms (for instance, biosphere, biodiversity, humanity, society, and so on). Despite their formulation in the singular, they cover a plural, diverse or collective meaning. As for the vision proposed by the scientific world, and more particularly the globalizing prevailing Western vision of the second half of the twentieth century, this has preferred to bestow the technical notions of biological systems, ecosystems, and even sociosystems and anthroposystems.

The representation of Man's relationship to nature or to his environment, through the vague semantic that accompanies it, belongs to a "vast nebula", which, depending on tendencies, does or does not produce a number of terms: viability, sustainability, sustainable management, biodiversity and biosphere, resilience, adaptation, coviability, ecosystem service, monetarization, ecological engineering, etc.

Current findings show that all elements of nature are affected by Man. This effect touches, for instance, cultivated plants or domestic animals whose genetic composition has been transformed over millennia. Landscapes (and the underlying systemic functioning) are marked by man's "signature" over all continents, including the 7th continent.[1] Given the finite nature of the globe and resources, today's complex interactions (notably climate change, soil degradation, land and water pollution) question the sustainability of the whole.

Actors are numerous and diverse (citizens, NGOs, local communities, scientists, politicians, investors, etc.). They issue opinions, decisions, proposals, and so on, and each of them calls upon competencies in fields as diverse as ethics, scientific knowledge, the exercise of responsibility, the exercise of know-how and so on. In this context, if the issue is about "Managing" the questions are: What must be done? How must it to be done? What steps to be taken? On what scale do we reason them?

Above all, what semantic coherence should be attributed to the exchanges, to the recognized responsibilities, to the decisions taken and the actions that follow, when the semantic content of the terms used is poorly understood or even polysemous?

In a context which belongs to us all insofar as our common future is inscribed and engaged within it (IGBP program 2006, 2010; Méda 2013; Morin 2015), how is it possible to hear, to be heard and to dialogue without a shared and precise vocabulary?

We aim to explain first what we call "Man's relationship to his environment.[2]" The arguments put forward help unravel the ambiguity of the underlying visions and elicit[3] the notions of coviability, viability and perenniality of this relationship. Our

[1]Large area of the Pacific ocean, emerged ex nihilo because of man, where non-(or slowly) degradable waste (e.g. plastic) accumulates.

[2]The environment is initially taken in its basic meaning, i.e. as being what surrounds man.

[3]The action of helping experts to formalize their knowledge to allow saving and/or sharing them.

approach is based on a theoretical systemic vision, especially in terms of viability, in order to formalize interactions generated by this relationship. We also aim to abstract several visions from different angles (semantic relativism) that man can have on the relationship. In this construction, we seek to identify what contributes to system viability and to position and clarify the concept of coviability, which is the purpose of this work.

5.2 Method

5.2.1 Systemic Approach

Cambien (2008) writes that "the history of systemics, from its birth to its advent as a school of thought, took place throughout the second half of the twentieth century. The study of systems is not new. The emergence of systemics is based on an inter fertilization of ideas from different disciplines and on the determining role of a certain number of American and French scientists. The latter are indeed aware of the need for a theoretical synthesis of the set of laws that are at the basis of systems science. We find among them, H. Simon, H. von Foerster, J. Forrester, E. Morin, I. Prigogine, H. Atlan, J.L. Le Moigne.

Systemics contains two main lines of thought. The first is based on an analytical and mechanistic approach with a direct filiation to cybernetics when it comes to describe processes. It intends to predict future trajectories using tested and verified deterministic models. Various works, especially in mathematics and engineering, belong to this line of thought. The second addresses questioning through a holistic approach that aims at a global understanding of phenomena coming from a world recognized as being complex. The present and the past inhabit the approach, with a possible "dive into the future" which does, however, recognize the limitations imposed by complex worlds. All these approaches have experienced their ups and downs because of their respective advantages and obstacles... Ecological and societal questions can be addressed by each of these approaches; this depends in particular on the discipline or scale considered (de Rosnay 1975; Lovelock 1979; Levin 1998; Gunderson and Holling 2001; Odum and Barret 2006; Hall and Clark 2009; Kleidon 2009; IGBP programme 2006, 2010; Aubin and Durand 2019). Our approach is based on a holistic vision that emphasizes the importance of the overall intelligibility of the system, in order to enrich, disentangle, and discuss the question asked here. In this respect, we retain the main principles of the systemic triangulation (Donnadieu et al. 2003), i.e. structure, functioning, dynamics (Fig. 5.1a).

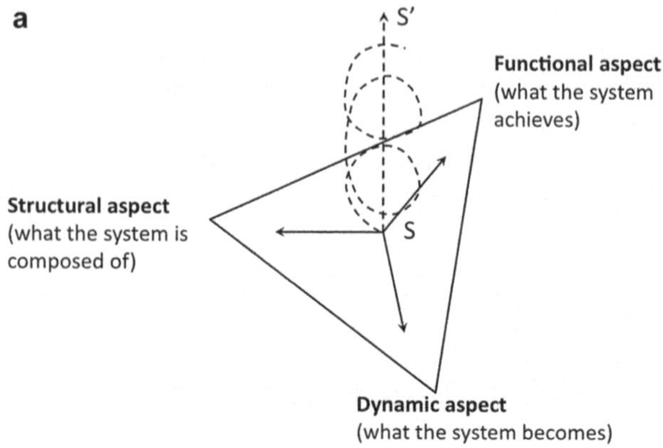

Fig. 5.1a The systemic triangulation. (Adapted from Donnadieu et al. 2003)

5.2.2 *Approach by Compartmentalization*

The methodological difficulty associated with complex systems and universes leads us to proposing the compartmentalization approach in order to study some of their structural entities or processes. By using the scientific knowledge available, the method allows a complex set of entities to be extracted, not necessarily directly assimilated to an s.s. system (i.e. in its theoretical exemplary character). We define a *compartment* as a compromise that reflects an angle[4] on the system, associated to a function. The approach we have developed is described in Sect. 5.3.1.6.

5.2.3 *Object Modeling Approach*

Following Minsky's vision (1995): "*To an observer B, an object A* is a model of an object A to the extent that B can use A* to answer questions that interest him about A.*" and in order to rigorously establish various analyses, we produce models of representation with the help of UML (Unified Modeling Language) formalism.

UML is a standard of Object Management Group. It is the result of merging three formalisms initially dedicated to software engineering (object-oriented approach) and developed between 1980 and 1995 by Booch et al. (2005). UML usage has become widespread. It can adapt, because of its versatility, to a wide range of applications while remaining easily translatable into code for the purpose of further computational developments.

In its most recent version, UML offers a set of graphic notations (Booch et al. 2005).

[4]Here the term angle does not mean "making an opinion", possibly subjective. Rather, it means "having an angle of view" which, once accepted, logically extracts the parts of the system relevant to it.

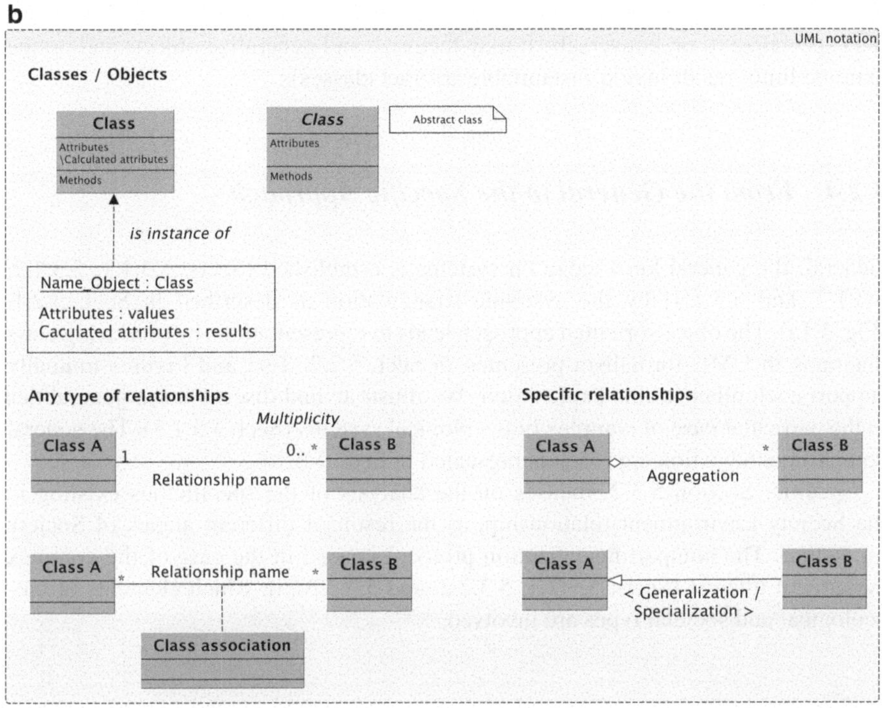

Fig. 5.1b UML Chart/scheme legend

For our purposes, (Fig. 5.1b) only the key needed to understand the descriptive models elaborated is detailed here. These models are based on the representation of classes, objects and relationships.

A class allows the description of a group of similar objects. The latter are instances of the class. The intentional definition of the class describes a list of properties[5] and methods (in terms of operations/calculations) that must be shared by the instance objects of the class. It is also possible to give the extensive definition of the class by describing the whole set of its instances (i.e. the objects generated by this mold through instantiation).

The classes may be connected to each other by arbitrary relations to which a name is assigned and for which the multiplicities are defined (0 .. *, *, 1), specifying how many instance objects of each class can be connected via links instantiating the relationship in question. If additional information about the relationship needs to be considered, it is necessary to use association classes (which reify the relation). Specific relationships are also proposed: the Aggregation relationship that connects one aggregate class (or everything) to its constituents or parts (the instances of the aggregate class are objects composed of constituent objects) and finally the Generalization/Specialization relationship that allows classifying the objects of the

[5]Which may be derived from a calculation.

observed context (the classificatory reasoning may start from a general category then specialize it; or, conversely, it may factorize and generalize objects and, to the extreme limit, result in non-instantiable abstract classes).

5.2.4 From the General to the Specific Approach

General: the general knowledge on systems is established (Sects. 5.3.1.1, 5.3.1.2, 5.3.1.3, and 5.3.1.4) by the systemic triangulation as described in Sect. 5.2.1. (Fig. 5.1a). The object-oriented approach leads to representations by producing class diagrams in UML formalism presented in Sect. 5.2.2. Text and Figures mutually support each other in this explanation. We illustrate and discuss these generalities in the particular case of complex bio-ecological systems (Sect. 5.3.1.5). The general compartmentalization approach is presented in Sect. 5.3.1.6.

Specific: Section 5.3.2. focuses on the analysis of the specificities existing in the Society-Environment relationship, as the result of different angles of Society on Nature. The compartmentalization protocol is used in the case of the complex systems of interest (Sects. 5.3.2.1, 5.3.2.2, and 5.3.2.3), in which elements of bio-ecological and societal types are involved.

5.2.5 Principle of Assertions

Throughout the analysis, we develop 37 "assertions",[6] statements and comments, deduced step by step from the preceding developments. They participate in the construction of the concepts of Viability and Coviability. So that they can be found quickly, they have been framed in the text.

5.3 Proposition – Analysis

5.3.1 System

5.3.1.1 Holistic Presentation

Systemic triangulation (Fig. 5.1a) uses three concepts to define a **system** (Fig. 5.2a) in a holistic view, i.e. the system considered as a "whole": structure, functioning and dynamics.

[6]In linguistics, as much as in philosophy, an assertion represents a statement presented as true within the framework of a specified theory.

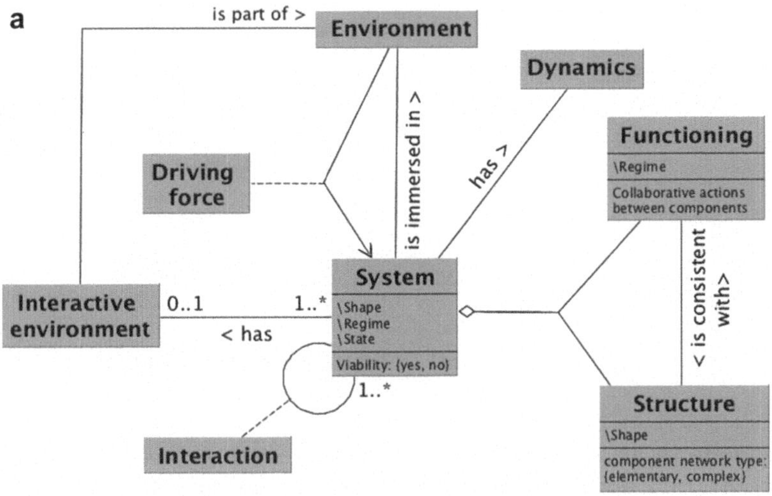

Fig. 5.2a Holistic model of concept "System" (UML formalism)

A **structure** whose **shape** is consistent with a functioning **regime)**, constitutes a system. Conversely, it is a **functioning regime**, in accordance with a shape (structure,) that makes a system. Whether focus should be made on structure, or on the functioning in order to discuss, study or describe a system, is only a matter of choice. These two concepts are in fact inseparable. They converge to form a system. "Being present" (and having a shape) or "being at work" (and functioning with a particular regime) are two facets of existence... and of the "life" of the system. Throughout Sect. 5.3, it will be demonstrated, how structure calls upon the functioning and, conversely, how it is almost impossible to discuss functioning without evoking structure.

Any system S (shortened to S in the remainder of the paragraph) is immersed in an **environment** (term taken in its systemic meaning). The latter either provides S with a stable context or subjects it to possible **driving forces** when it changes for reasons independent to S. The reciprocality is invalid: a system neither confers stability to its environment nor does it apply force to it. The system can act on other systems (1 to several) and, conversely, can undergo the action from these systems (1 to several). These are **interactions** (between S and other systems). We call such interacting systems as functionally neighboring systems.[7] We use the term "neighboring systems" only to indicate that they interact. The neighboring systems of S belong to its environment; they also possess a special status since S is able to act on its neighbors (Fig. 5.2b). These neighboring systems constitute what we call in Fig. 5.2a, the **interactive environment** of S. Fig. 5.2b shows instances of Fig. 5.2a.

[7]Mentioning neighborhood does not prejudice the proximity of systems in terms of typology, or from a spatial angle.

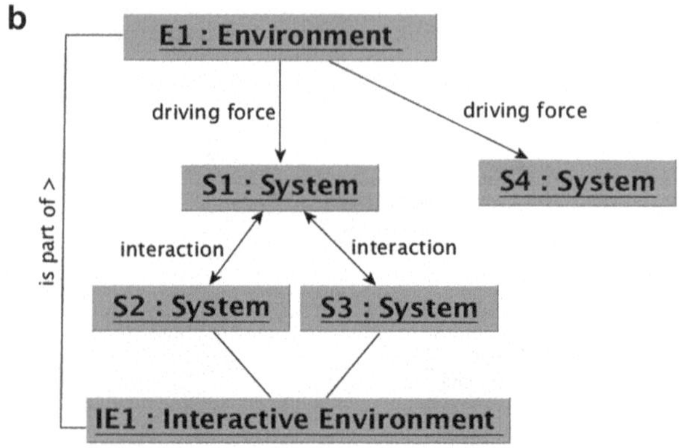

Fig. 5.2b S1, instance of system, its environment, its interactive environment (UML formalism)

The **state** of the system is the consequence of the good (or bad) adequacy (consistency) of a structure's shape with a functioning regime. The state is the result; in a way the "footprint" found on the structure or the "impact" of the functioning mode on the structure.

Assertion 5.1
A supposedly viable system requires coherence between structure and functioning. A "bad" state[8] jeopardizes the viability of the system.

Due to pressures (by environmental driving forces or interactions with neighboring systems), the system may change. Changes in shape/regime/state of a system describe its **dynamics** (what the system becomes, its history).

The focus "adopted" provides a holistic model of the considered system S: a "whole" characterized by its (structural) shape, its (functional) regime, its changes in state (structure/functioning; history), its interactive environment[9] and its environment. By doing so, the systemic level to which observations or reasoning are operated is defined.

5.3.1.2 Structure

The structure describes the arranging of the various components of a system. The meta-model (Fig. 5.3a) represents a **system** abstraction which gives us the possibility to zoom in and out during the discourse concerned in order to highlight

[8]An unusual state, whose shape and regime are apparently inadequate for each other (not adapted or not yet adapted: confrontation with reality, together with resilience or adaptation, will "judge" this state: towards existence or collapse).

[9]Term emerging from our reasoning to distinguish it from the term Environment.

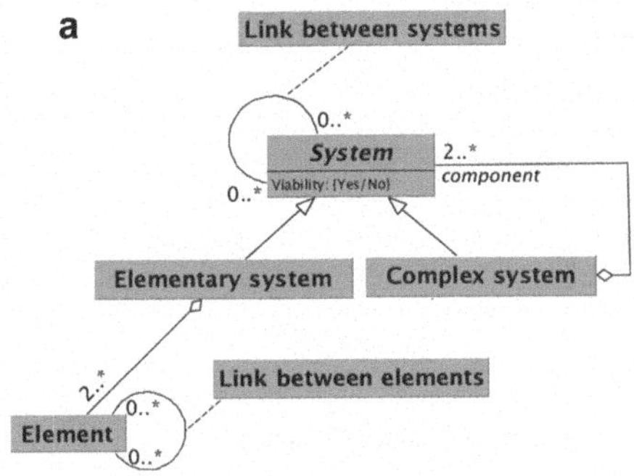

Fig. 5.3a Structural meta-model of concept "System" (UML formalism)

the system's nature (complex, elementary), its **components** (i.e. the systems constituting it), and the potential complexity of its structure. At the systemic level, terminology considers the elementary system as the component of minimal granularity (below which it is not possible to zoom).

Two or more **elements** can aggregate to form an **elementary system**. By iterative aggregations, a **complex system** corresponds to a higher level assemblage of components that are more or less complex (elementary or complex systems). **Links between elements or between systems** intervene, whether they are within the system or between systems (from 0 to n). These structural links participate in more or less complex constructions and lead to functional interactions (Fig. 5.2a) of different natures (as we will see in Figs. 5.4 and 5.5). The structural type (Fig. 5.2a) of the system is defined by studying its components (Fig. 5.3a), their (intra-systemic) links, and the shape of the whole (holistic view).

Assertion 5.2
The presence of more or less complex links between the components of a system (internal links) induces the concept of Coviability between its components, provided that they are themselves systems. From a holistic view, to say that a complex system S is viable means that Sa and Sb are coviable within this so-called integrated system S (if we assume that this system is constituted only of two subsystems, Sa and Sb). In this case, the distinction between the concepts of viability and coviability is only a matter of systemic level (Fig. 5.3b).

Assertion 5.3
The existence of external links, between non-integrated systems, induces a new version of the concept of Coviability. The difficulty lies in identifying the completeness of the systems and links concerned.

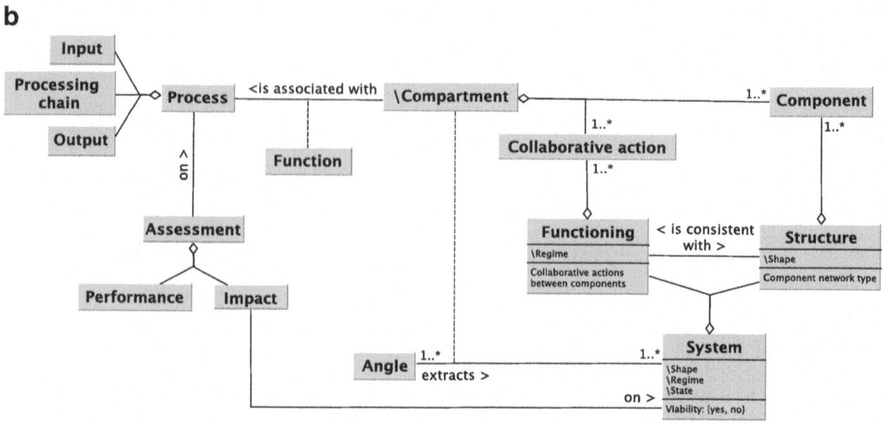

Fig. 5.3b Coviability, and viability (UML formalism)

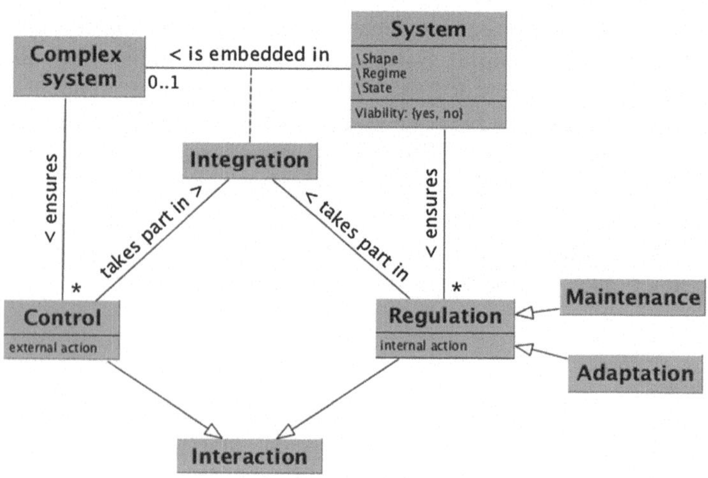

Fig. 5.4 Functional model of concept "System" (UML formalism)

5.3.1.3 Functioning

The presentation explores the basics of a system's functioning in terms of its viability, and possible coviability with other systems.

In Fig. 5.4, the observation point is located either inside or outside (when zooming in or out depending upon the discourse) of the boundary of the system being considered. As a result, the interactions located inside or outside this boundary are observed.

In a holistic view (the system being a "whole"), the system "ensures" the execution of all collaborative operations between its components[10] (Fig. 5.2a). This execution results in a system, and the system both functions and regulates itself. Inside the boundary of the system, this is a set of **interactions** between components: the **regulation** of functioning together with **maintaining** the structure in working order.[11] Regulation is part of the execution.

Outside the boundary of an n ranked system, the latter can take its place (functional niche) in an n + 1 ranked complex system through **integration**.[12] Considering the n ranked-system, the n + 1 ranked system exerts **control**-type interactions. By upscaling (iteration on the rank), that which is called external control will be called internal regulation.

The action of one system on another (interaction) is called regulation between components when the systems in question belong to the same integrated system. In a holistic view, self-regulation refers to the set of interactions when this results in the integration of the interacting components and in maintaining the whole in a viable state. Integration is a harmonious blend of controls and regulations. By iteration, ever more complex systems can be built, endowed with this principle of self-regulation.

Control and regulation are views from two angles of the same reality. The distinction lies on the systemic rank adopted and on the passage (at this given systemic level) through the line delineating between "inside" and "outside" (the system's boundary). It is by integrating (elementary and/or more or less complex) systems, i.e. by successful iteration of these games of interaction (via adjustment, adaptation – See Sect. 5.3.1.4), that the emergence of a system of higher systemic rank takes place. The increase in systemic complexity occurs, going hand in hand with the process of integration and emergence.

Assertion 5.4

Control and regulation are key concepts in terms of the capacity to emerge and then the viability of a system or the functional coviability of its components. A structure functioning in an integrated manner guarantees the viability of its components despite the "pressures" coming from the environment.

[10]The term "component" refers to an elementary system or a complex system (aggregate level) located inside the (more or less complex) system studied.

[11]Similar to autopoiesis in biological systems. To the idea of repair is added that of construction, of synthesis of the structure, either by the reproduction of the whole or the synthesis of its parts.

[12]Another way of saying this is to say that a rank n-system S has its place in an n + 1 ranked-complex system S′ is equivalent to saying that S′ and S interact in such a way that there is integration of systems.

Assertion 5.5
Auto-regulation is therefore one of the aspects of viability.[13] This relates to the principle of homeostasis (see Claude Bernard's work in physiology in the nineteenth century; see in particular Bernard 1865, 1877) of the system which ensures keeping the system "as it is".

To speak of system robustness calls for the notion of completeness. Ideally, it corresponds to completion of integration; and factually, it corresponds to a more or less advanced degree of completion.

Assertion 5.6
The viability of a system is based in part on the quality (richness and "complexity") of the bonds (i.e. interactions between components) that construct it in a coherent manner. These bonds give it the robustness which will provide it with a degree of resistance (i.e. its capacity to remain unchanged when facing extrinsic hazards).

5.3.1.4 Dynamics

Through dynamics (Fig. 5.5), we are interested in the transformation of a system over time.

Dynamics account for change and it is through dynamics that the notion of time "becomes tangible": with the evidence of change in shape, regime, state, or even type of system. In a temporal referential, there is a "before" (e.g. "preceding state", at moment "t") and an "after" (e.g. "next state," at moment "t + 1"). Change is observed, is calculated (differential of shape, of regime, and of state). If a system is **stable** (no difference), the absence of change artificially portrays what is commonly referred to as "immobile time" (time that does not seem to move on).

The level of integration of a complex system, when it is not completed,[14] gives way to pressures, whether coming from the environment (driving forces) or from neighboring systems.

Pressure is thus experienced by a system, which exists outside the particular integrating relational frame. There are a variety of consequences for putting pressures on a system.

[13] We shall see in Sects. 5.3.1.4 and 5.3.1.5 that the other side is related to adaptation/adaptability.

[14] More or less complete, for it is rarely this way in the case of complex systems, see Sect. 5.3.1.5.

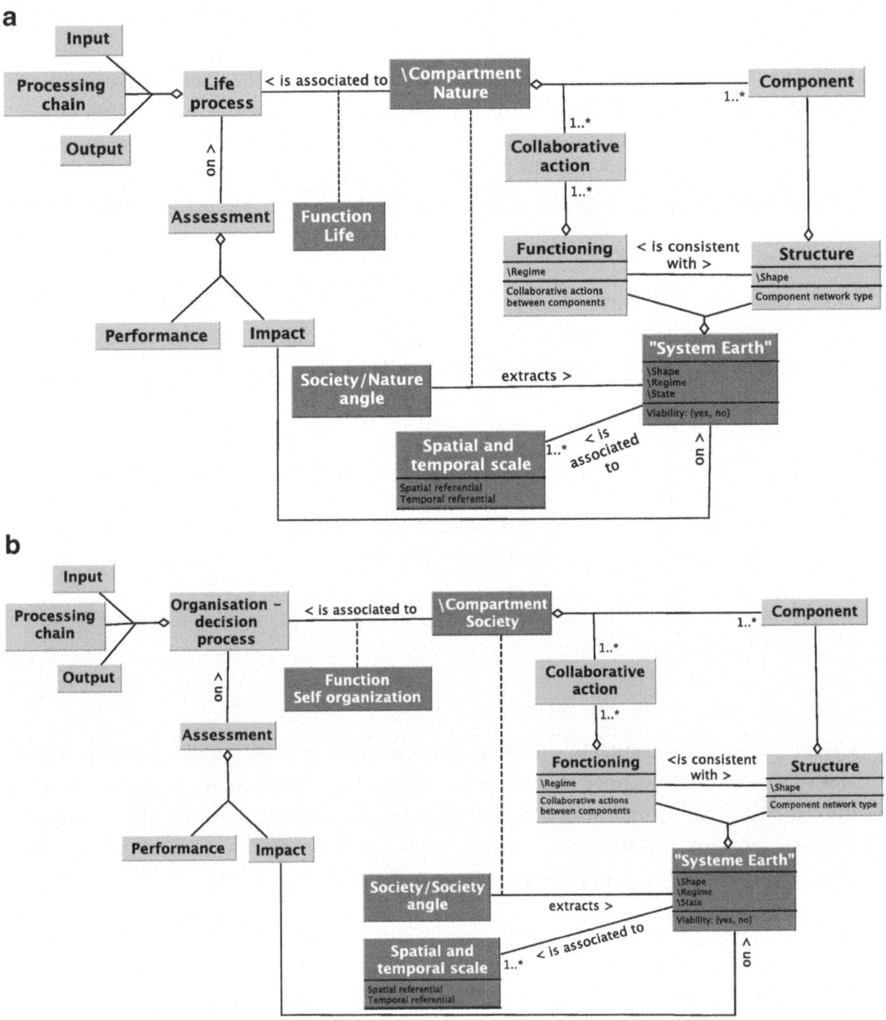

Fig. 5.5 Dynamic model of concept "System" (UML formalism)

- So called stabilizing pressures contribute in maintaining the system "*as it is*". These intervene on System S (of systemic rank n), that is independent or partially integrated, and whose boundary may be partially unclear. Although regarded as "external" to S, some systems contribute in stabilizing it, to its shape, to its regime... Such pressures are similar in nature to regulations (internal to the system) and are seen as external just because the systemic rank n + 1 is not integrated. Consequently, System S largely depends, for its stability, on an external contribution (e.g. synergic or competitive).[15]

[15]For example, the interactions linking species within a trophic network.

Assertion 5.7
Stability is therefore the consequence of both (internal – see above) self-regulation and stabilizing (external) pressures. If stability is not completely guaranteed, the resilience capacity of a system corresponds to its capacity to recover more or less quickly its initial properties. It testifies to the proper functioning of the regulation/stabilization mechanisms (homeostasis[16]).

- Other pressures act substantially the same on the system but gradually result in a change in the structure and/or functioning of the system. There is an **evolution**, with change being achieved through **adaptation**.

Assertion 5.8
Adaptation preserves the system, but simultaneously, the latter is gradually led toward other shapes and regimes, so that the shape is again adapted to the regime and vice versa. Adaptation ensures the good state of the system and therefore its viability. Adaptability represents the second aspect[17] of viability.

Assertion 5.9
Under constant pressure in an interval of time (i.e. stable environment), the stabilizing pressures only preserve the adaptations previously acquired. We thus see the "juggling" between stability through self-regulation and gradual change through adaptation... which help keeping the system alive while responding to the pressures from the environment.

- However, evolution not only happens gradually. It can adopt a more abrupt formula and proceed by quasi-rupture or even **rupture**. There is then a change

[16]Homeostasis is the result of effective resistance or resilience. Resistance or resilience are like springs (which, after extension return to their initial state). However, the spring for resilience would be more flexible (the return would take longer, the initial state would not be totally recovered) than that of resistance (in absolute terms, resistance would be a wall). This is a very mechanistic image

[17]An image for these two sides of viability: under pressure, on the ridge of a mountain, descent would be made either on one slope or the other; a return to the point of departure (resistance, resilience) or, on the contrary, exploration/invention of novelty by experiencing the other side. Should we talk about resilience (a "second type")? About emergence? Biologists are constantly confronted with this problem: when does one species become another (i.e. transition from one system to another)? In any case, if there is change, there is adaptation ... or emergence.

of system, change in its type and place (Figs. 5.2a, 5.3a and 5.4) it occupies in relation to "neighboring" systems or in a system (more or less integrated) of higher rank.

In all cases, evolution is at the origin of change, be it oriented and gradual or more or less abrupt, especially corresponding for the latter, to a "new order", possibly spectacular and whose origin is not known in advance. These changes concretize new combinations (shape – regime) for the system or even the emergence of new systems.

5.3.1.5 By Way of a Transition: The Case of Complex Bio-Ecological Systems

On the bases of scientific knowledge in ecology (gathered and organized e.g. in Begon et al. 2007; Dajoz 2003; Ramade 2003; partly Holling 2001), biological and ecological systems, examples of complex systems, elicit comments on the concepts related to the issue of this article:

- About structure, functioning, dynamics ... and history:

In a world of partially integrated complex systems, very open from a functional perspective, the difficulties in describing systems particularly lie in the definition of the boundary as mentioned above. Difficulties also arise when including dynamics: (a) during gradual changes,[18] at which point is a change in system recognized rather than just a change in "shape" (typical question in evolution of species); (b) is a dip into history (in geological or historical terms) necessary to understand the system, its past trajectory and future tendencies?

- About viability:

Assertion 5.10
Being viable means having a sufficiently stable foundation while being adaptable: this seems a dilemma (change/no change). Thus, a very (or even "excessively") integrated system is certainly "very" viable as it could be very resistant but also "hardly" viable because its adaptability would be reduced and costly (calling into question deconstruction, i.e. renunciation of certain, previously acquired structures). It is therefore "very fragile" from this angle.

[18] By adaptation for a better fit of the shape/regime combination.

Based on this dilemma, life (a viable system) takes on a fragile character (vulnerability) and the turnover becomes the ransom of "too much success".[19] On a higher level, this dilemma accounts for the characteristic of such systems, and their dynamic equilibrium (equilibrium incessantly challenged and in perpetual movement).

Assertion 5.11
Vulnerability and adaptability are complementary concepts to viability concept.

Assertion 5.12
To be endowed with capacity for life after a crisis, then to be viable does not only mean self-maintenance (and reproduction) but also creating and inventing again. This capacity is based on:

(a) wealth of resources: the wealth in components (or even their partial apparent duplication, i.e. substitutable structures) and their complexity in organization,
(b) plasticity in their links: the richness of exchanges and the diversity of flow paths (flexible functioning).

Assertion 5.13
The capacity for life, within the general dynamics, is resistance[20] (status quo), adaptability (adaptation inducing a gradual change) or resilience (permitting/providing a response via a new shape, a new regime of the system or a post-crisis return to the initial shape/regime). The emergence (death and life by renewal) of new shape/regime formulas, new types of systems, is the product of either a gradual evolution or a rupture-revolution.

[19] Perfection is relative: an evolutionary cul-de-sac.

[20] Resistance works for the homeostasis of the system as it is. Dynamics question this state to find a new one, for which it will then be necessary to conserve (homeostasis again, but for the sake of a new state).

Assertion 5.14
Our definition of co-viability.

In a systemic approach, coviability requires at least two systems that are both viable. By putting them together and in interaction, they remain viable. They improve their synergies and/or minimize their competition: this may lead to the emergence of an integrated system. Inference: if they do not "know" each other (i.e. no interaction), the question of coviability does not arise; they only coexist.

5.3.1.6 The Compartment: An Assumed Compromise Between Simplification and Imperfection

In the systemic approach,[21] one of the difficulties[22] lies in the definition of a system's boundary, which corresponds to its "structural and functional perimeter." Another difficulty consists of grasping its organization, in its completeness and complexity. One way of overcoming this difficulty is to introduce the concept "Compartment" (Fig. 5.6).

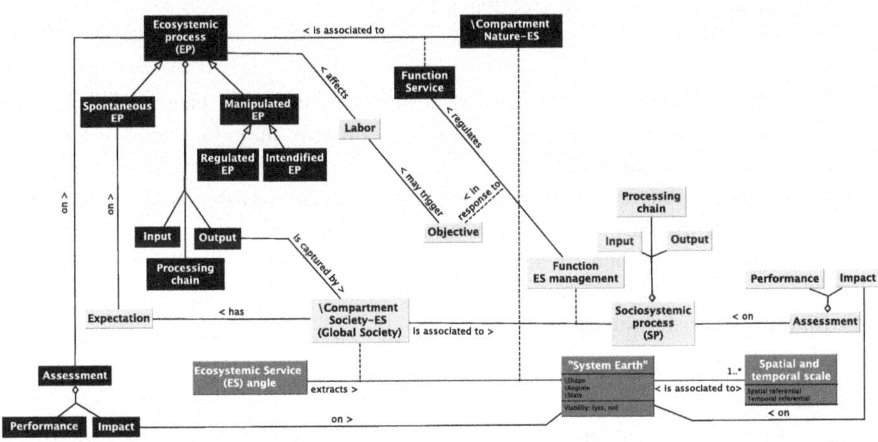

Fig. 5.6 Compartment model (UML formalism)

[21] When we simplify, we take the risk of making a mistake, of caricaturing, or being biased... We must be aware of this, accept it, not forget it when the time comes. The compartmentalization approach (a pragmatic approach to delimit a system) is a way of doing things without forgetting the risks and uncertainties involved in this action.

[22] Except for some man-made systems.

A compartment is associated with an angle[23] of the system and with an identified **function**, which is related to a process. For any angle a compartment is extracted by calculation or through a filter placed on the system. The components which belong to the filter's complement (in the sense given by the set theory) are not necessarily seen as a compartment, except in certain special cases (see Sect. 5.3.2.2). The process (related to the function identified) calls upon only some of the **components** and **collaborative actions** that are based in the system. This is this part which is named Compartment. It consists of a **processing chain** which allows **input** (from the system) and **output** (towards the system). An **assessment** can be estimated related to the process. It refers both to the **performance** of this process with respect to the function identified and to the **impact** that this functioning (at the given regime) has either on the structure of the compartment involved or more broadly on the system in question or on other "neighboring" systems (neighbor as defined in Sect. 5.3.1.1).

Assertion 5.15
The impact resulting from the assessment of a "compartment's" process may affect the state and potentially the viability of the system from which the compartment is "extracted" (see Sects. 5.3.2.2 and 5.3.2.3). It may also trigger a particular dynamic: impact leading to a cascade of other consequences, "virtuous or vicious circles" (feedback loops).

Due to the complexity involved and sometimes/often the lack of knowledge that accompanies it, the compartment representation may be "incomplete" or "imperfect". In some cases, it may even be considered "arbitrary." In all cases, the compartment is "relative" to the angle that delimits its outline. It is an admitted partial vision, for the purpose of reasoning or experimentation.

In a single system, one or several compartments can be individualized and, consequently, one or several functions. Compartmentalization allows access to the multifunctionality of systems.[24]

Assertion 5.16
It is only possible to address the viability of a compartment if it is a system. By definition, a compartment is only a part in a whole (the system from which it is extracted), and nothing proves that it is a subsystem of this system.

[23]Here the term angle does not mean "making an opinion", possibly subjective. Rather, it means "having an angle of view" which, once accepted, logically extracts the parts of the system relevant to it.

[24]The system may be analyzed from several angles. Each angle makes it possible to extract a compartment and a function.

5.3.2 *Society – Environment Relationships*

In this section, we first consider Man's organization into Societies, and the fact that he depends on an "environment" in its most basic sense, i.e. what surrounds Man. It should also be noted that as soon as a Society views a system, it gives it **spatial**[25] **and temporal**[26] **scales**. The system deploys itself not only in a temporal frame of reference (the dynamics of the system leads to it) but also in a spatial referential.

We are going to follow the compartmenting protocol (according to the meaning given to compartment as a concept in Sect. 5.3.1.6.) in order (1) to take an interest in Society's views on Nature, and (2) because the protocol also helps us describe some of the more general characteristics of the Society – Environment relationships and clarify the semantic content of the term Environment and the consequences that ensue in the context of this work.

We develop, in Sects. 5.3.2.2 and 5.3.2.3, compartmentalization according to two angles called "Ecosystemic Service" and "Territorial" angles. Both prevail in the twenty-first century's Western culture, and are becoming widespread in the present era of globalization. This does not restrict the importance of other positions such as Descola's (2010, 2011) in anthropology.

5.3.2.1 A Return to the Systemic Framework

The Fig. 5.7 is in accordance[27] with the Fig. 5.3a. It models an integrated complex system[28] that we call "**System Earth**", an instance of a complex system with spatial and temporal scales, and is based on interactions between Society-type and Nature-type components (i.e. systems). We observe that if the existence of **Society-type components** (hereinafter referred to as **S**) is undeniable, there are no Nature-type components in the strict sense, **but only more or less anthropized Nature-type components** (hereinafter referred to as **NS**). The latter have all been exposed to Man's impact, for more or less a long time and more or less intensively. Those responsible for such impact acted from close or from far. For example, Chansigaud (2013) or Diamond (2005) re-transcribe part of this eventful, even dramatic, history.

---

[25] While in pure systemics, a spatial dimension is not always necessary.

[26] "We never bathe twice in the same river," (Eraclite d'Ephèse).

[27] The model in Fig. 5.7 conforms to the meta-model in Fig. 5.3a.

[28] With the changes currently underway, this integration may be questioned.

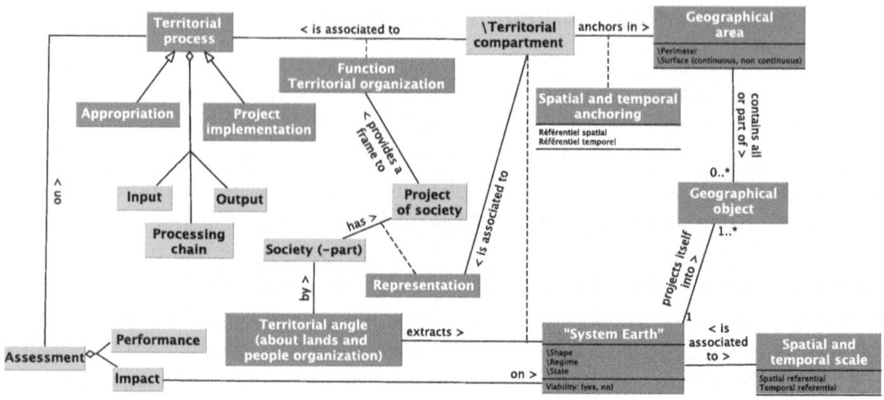

Fig. 5.7 "System Earth" model (UML formalism)

Assertion 5.17
The viability of the Society's relationship to Nature stems from interactions between systemic Society-type components (S) and more or less anthropized Nature-type components (NS), and this at multiple systemic levels.

There are, in summary, several comments to be made on this model:

- In a strict sense, there is no Nature-type component (i.e. *system*). Indeed, any ecosystem is the base of ecosystemic relations, but it is also the base for human-caused impact relations.
  → **Consequently**, there is no question about Nature which, especially concerning nature's dynamics or future, is not also a question about Society.[29]
- Man has claimed Nature as his own and impacts every component of it. The system which accounts for the Society-Environment relationship is based on what we have previously referred to as "System Earth".[30] This is a complex system in which Society's relationship to Nature exists at all levels of systemic integration.
- When "Environment" is understood in a systemic way (i.e. one way action: driving force(s) on a system by its environment, and no reciprocal action), then the environment of societal components (i.e. of social systems) lies beyond the stratosphere. It should be noted that in environmental sciences[31] the term environment corresponds at most to the interactive environment (the systemic

[29]It is assumed, as before, that there exists "social systems" in the strict sense of the term.

[30]We will subsequently adopt the terminology "System Earth" to refer to this level.

[31]The environment: all the natural components of Planet Earth and all the phenomena and interactions that occur i.e. everything that surrounds man and his activities (Wikipedia).

term defined in Sect. 5.3.1.1) for societal components (since they impact it) and not to their environment (in the systemic sense).

- The term Environment takes on additional complexity. If we take society as a whole (hereinafter referred to as "Global Society"), the Environment corresponds to the complement of Compartment Global Society (in the sense given in the set theory) with respect to System Earth. In this case, it is *de facto* equivalent to the concept of the environment as defined by environmental sciences (Gouyon and Leriche 2010). However, if the society considered corresponds to only a part of Global Society, then the environment of this partial society, which still corresponds to the complement of the partial Society compartment (one part of Global Society), now includes parts of S and NS type components of "System Earth".

  → **Consequently**, when a part of global society (this part of society will subsequently be called territorial society, and noted below as a society-part, see Sect. 5.3.2.3) raises questions on Nature and its relationship to Nature, the NS components and the S components of its environment become automatically involved, all inextricably linked to it accordingly to the structure presented in Fig. 5.7.

Assertion 5.18
As for the viability of Society's relationship with Nature, it questions the viability of "System Earth", whether the question is posed by Global Society or a part of it. This is why we prefer to use the term "NS Environment" in the first case and "S and NS Environment" in the second case. This cannot be assimilated to the coviability of Man and Nature, since these latter correspond to categories as seen by man and not to systems, components of "System Earth".

- Because Nature does not exist in the strict sense and the boundaries of the S and NS components are difficult to define at different systemic levels, the use of compartmentalization (as a proxy) may allow an evaluation of performance-impact assessments.

The Fig. 5.8a presents the **Compartment Nature**, which was extracted from a generalist's angle of Society on "System Earth" and is associated to **Function Life**. Compartment Nature[32] is actually the result of a "calculation", which filters what the angle extracts from the overall system. It is made up of NS-type components (or parts of components).

[32]In the twenty-first century, the Compartment "Nature" is not a system. There is no global ecosystem.

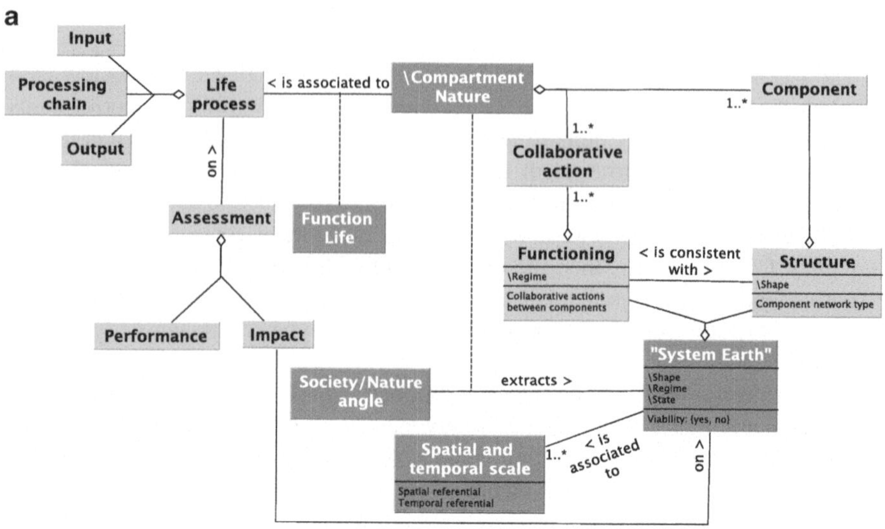

Fig. 5.8a Nature compartmentalization as seen by Man (UML formalism)

Man commonly associates "Function Life" with the fact that Nature provides him with resources with which to live. In ecology,[33] it comes across as the notion of habitat together with the whole set of its functions, regardless of the spatial-temporal referential.

The Fig. 5.8b presents **Compartment Society**, extracted from another generalist angle of Society on "System Earth". It highlights the reflexive aspect of the process (man being interested in himself) and corresponds to **Function Self-organization**. Compartment Society is, as above, the result of a calculation carried out on the overall "System Earth" according to the angle. It corresponds to Global Society. Within it, all sorts of exchange and decision processes take place, following political, cultural, legal and economic, considerations.

This compartment is made up of Society S type components or parts of components. Compartment Global Society is not a system. It is through self-regulation (see Sect. 5.3.1.3) that Global Society might be seen as a sociosystem. To become so, the self-organizing function should succeed in performing as self-regulation (systemic term/vocabulary). In any case, whether it is the Society's self-organizing function or the regulation function of a sociosystem, a humanistic view on Man and Society invites to confer both rationality and generosity of this function, hence a strong utopian component/part,[34] when it is concerned with/comes to equity between men and between societies, to their happiness and well-being (Sen 1987; Gasper 2008; Stiglitz et al. 2008).

[33]In Greek "oikos": house, habitat

[34]Utopia: "something" that does not exist but may be attained.

b

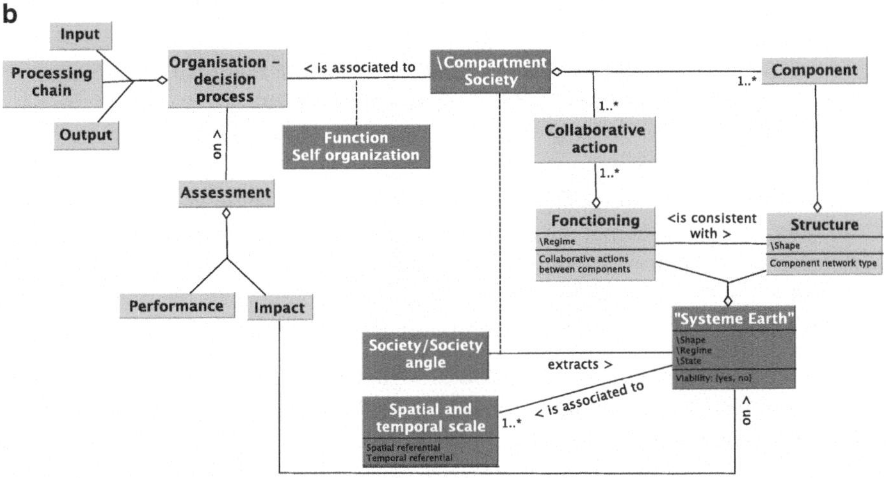

Fig. 5.8b Society compartmentalization as seen by Man (UML formalism)

Assertion 5.19
Compartment Nature is associated with Function Life, whose quality derives from the performance-impact assessment calculated on a global level. Compartment Society is associated with Function Self-organization whose quality derives from the performance-impact assessment calculated on a global level. The stability of "System Earth" results from the adjustment of these two functions.

- In terms of dynamics, to say that "System Earth" changes (e.g. climate change, soil degradation (Nahon 2008), globalization of trade) implies that pressures (exerted by its systemic environment) change (i.e. new pressures), and/or that certain deregulation (internal to "System Earth") is taking place. Within integrated (in a Holocene configuration) "System Earth", the dynamic response of any component (which normally exerts regulations and controls on other system components) now exerts pressures and impacts on these other components.[35] Step by step, each component will see its shape/regime formula change (adaptation in response) and induce pressures on others (see Sect. 5.3.1.4).

[35] We have seen in Sect. 5.3.1.3 that the viability of an integrated system includes regulations and controls, having as origins the systems of lower and higher systemic ranks, respectively.

Assertion 5.20
It is difficult, if not impossible, to approach the viability issue of components belonging to "System Earth" other than on a global level.

- According to a Nature-Society compartmentalization angle and by taking into account, in the reasoning, the combined impacts on "System Earth", what future is there for the latter? Rupture? Or Adaptation? Will the current Holocene-type Nature – Society relationship, and its biophysical components, be preserved? Or are we entering a new era, the Anthropocene[36] (Bonneuil and Fressoz 2016)?

With what actions will Man contribute? Will self-regulation by "System Earth" be possible? And, to use a striking formula, does this concern the adaptation of Man to himself (Toussaint et al. 2013)?

Assertion 5.21
Through focusing on the two angles expressed in Fig. 5.8a and Fig. 5.8b, two compartments are extracted from "System Earth." To ensure the perenniality of the coexistence of Nature and Society compartments, "System Earth" must be viable. This perenniality seems to involve the adaptation of man[37] to himself (Toussaint et al. 2013), the adjustment of society to itself.

Indeed, if Man is at the origin of the changes observed at the level of "System Earth", the viability issue of this system poses the question of the adaptation and adaptability of Man, organized in Society, to Man as an agent of pressure. The self-regulation of "System Earth" seems to depend on the plasticity of Society in the face of events for which man is responsible. This is a matter of a new operation for the self-construction/self-organization of Society (and societies), making sense when it comes to contributing to System Earth's regulation.

The remainder of our paper presents two angles by Society on "System Earth" that highlight the Society-Environment relationship according to a global and a local reading.

[36]The IGBP participated to the emergence of the term Anthropocene which transcribes significant atmospheric changes and a modified relation of man to nature. There is a debate as to whether it corresponds to a geological era (the definition of geological eras is based on stratigraphy works (Elmi and Babin 2006).

[37]Considered as a system.

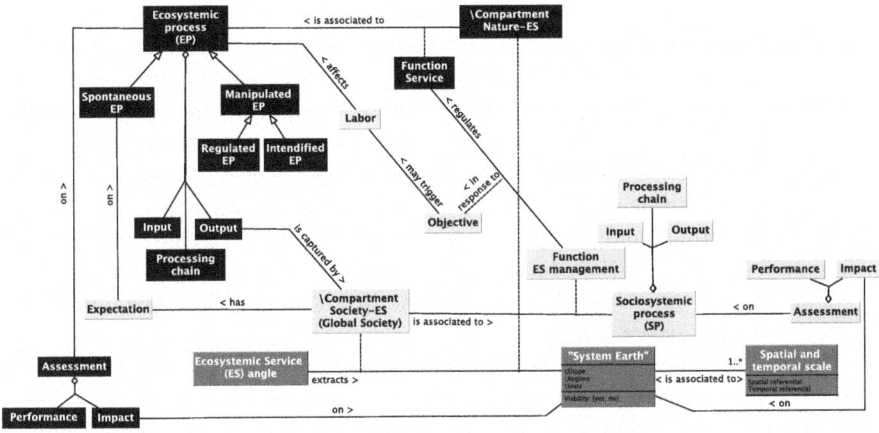

Fig. 5.9 "Ecosystemic Service" angle (UML formalism)

5.3.2.2 The "Ecosystemic Service" Angle for a Reading of the Society-Environment Relationship

The emphasis here is on the global dimension of the Society-Environment relationship.

Based on the compartmentalization protocol (Sect. 5.3.1.6 and Fig. 5.3a) and on the reasoning developed in Sect. 5.3.2.1, Fig. 5.9 proposes the "**Ecosystemic Service**"[38] **angle** (referred to as ES in the remainder of the text), a view of Society on "System Earth".

This angle is "complex" since it encompasses a view of Society on Nature (Fig. 5.8a) and on itself (Fig. 5.8b). As a result, two **compartments "Nature-ES"** and "**Society-ES (Global Society)**" are extracted and two functions identified. **Function Service** is one specialization of Function Life and directly reflects a relationship between Society and Nature, while Function "ES management" is a specialization of the self-organizing function and deals with the regulation of the relationship between Society and Nature. It is hoped that such management be both rational and respectful of humanist values. The related functional framework, presented here through compartmentalization, corresponds to "System Earth" (see Sect. 5.3.2.1), and the two compartments extracted in this way complement each other (set theory) to form System Earth.

More precisely:

1. With each Function "Service" that Society identifies (for example, Service Supply) is associated a compartment of "Earth System" that ensures the targeted

[38] We will only take into consideration systemic (structural, functional) aspects in this presentation. We have deliberately set aside the monetarization aspect of ES which represents a (good/bad) way of approaching the negotiation aspect; an aspect whose sole principle is discussed here.

Ecosystemic Process (EP). As seen above, the processing chain includes an input and a **useful output** captured by the global Society.

The relationship of Global Society to Nature (Nature taken in the sense of the NS seen above) is based partly on the Society's **expectations** on spontaneous Ecosystemic Processes ("**Spontaneous EP**"), for example ecosystemic services related to "System Earth" regulation and self-maintenance: climate regulation, water and air purification, nutrients and mineral element cycles, soil maintenance and fertility, and primary production (according to scientific community terminology Weber 2012). In this case, it goes hand in hand with a notion of gratuitousness (gifts from Nature) and the diversity of representations of the Society in this respect is to be noted.

The Global Society's relationship with Nature (NS) are also based on actions to meet the **objectives** of this Society.

The **Labor** (e.g. agricultural labor) provided by members of Society, a more or less specialized type of work, more or less organized and at various systemic levels, also intervenes on the process (Manipulated Ecosystemic Process: **Manipulated EP**). Work has an objective, an intention on the product of the EP, i.e. on the useful output. It is often **intensification** (increasing the output quantity, see intensified **EP**), e.g. Western intensive agriculture, especially that of the last two generations. The know-how deployed intervenes directly to manipulate the regime: intensify it in order to increase **performance**.

This interaction may be regarded as engineering, where the process (whether spontaneous or requiring extra work) and Compartment Nature-ES are assimilated to a "machine" (its functioning and structure, respectively).

Compartment Nature-ES, which is already anthropized, sees its anthropization level further increased by this action-work. In doing so, there is the risk of indirectly intervening on the system, disrupting it, deregulating it (**impact**).

2. The relationship between global Society and Nature (NS) reach an even greater degree of complexity/integration when they relate, not directly to the product of the EP (see above), but to the processing chain itself, with the intention of "preserving" it, "maintaining" it, or even "repairing" it. The ecosystemic process is then managed (**Regulated EP**).

This is the case for elaborate and highly integrated agrosystems, for instance, traditional agrosystems (Mollard and Walter 2008; Fargette et al. 2019, chapter 29, volume 2 of the 'book', on the oasis agrosystem) which, throughout the history of societies, became complex and adapted to the constraints of the environment while preserving its production potential, even when fragile.

This is also the case today when agro-ecology or ecological engineering attempt:

- *a posteriori* to remedy (repair action) to an observed deterioration (due to the impact of another action),
- in anticipation of a foreseeable potential impact, to prevent this risk (preventive action).

It is therefore a matter of managing (**Function ES management**) the actual or potential impact: the regime is not the only element prone to being manipulated; it is also possible to manipulate the controls and regulations themselves of the shape/regime couple (see Figs. 5.2a and 5.4). The Function ES management is here nothing but a specialization of Function Self-organization in Fig. 5.8b. The ecological engineering approach uses theories and protocols of conservation, protection, remediation, restoration, rehabilitation... of the "environment" (environmental sciences). This management targets, with varying degrees of success, the stability/sustainability or reconstitution of the process (and/or structure) in its systemic context. The task is complex because it has to deal with the intricacy of reality; from a "simplifying" representation by compartment, one returns to a systemic and therefore holistic approach, with all the difficulties due to the multiplicity of interactions, and for which most are often misunderstood.[39]

Assertion 5.22
It is difficult (for humans) to "invent" a sustainable, lasting ecosystem. Indeed, the actions of societies on an ever more anthropized (human-transformed) Nature (NS), have effect of simplifying ecosystems (into number of species, into number of links). This leads to their weakness or increase of vulnerability. Dajoz (2003, p. 470 and subsequently in chapter 21), Kéfi et al. (2016), Lurgi et al. (2016) and Gravel et al. (2016) discuss the stability of ecosystems according to their biodiversity. Systemically and thermodynamically speaking (Kéfi 2012), the objective of reconstitution is ambitious and, unfortunately, often unrealistic. The objective of prevention seems more accessible.

Each function "ES Management" (remediation, protection...) that Society identifies is associated with a Compartment Society-ES of "System Earth" that is in charge with reaching the targeted socio-systemic process (e.g. implementation of the work). We will next see in what way and how this Society-ES compartment enlists the whole of Society (Global Society).

3. Formulating an objective, which involves work whether on the level of intensifying the Service or managing it, affects both the Nature-ES and Society-ES compartments and their associated ecosystemic and sociosystemic processes (i.e. organizing the work for production or management, organizing workload division and work implementation...). In addition, particularly with regard to management objectives, the need for organization (or even reorganization) may strongly solicit Society and may affect living conditions and even the Society's choices.

[39]Possible existence of chaotic modes often observed in complex systems. They are by definition difficult to control or are even unmanageable.

Assertion 5.23
Managing Society's relationship to Nature is not an "external" or a "technical" issue only (i.e. a subject working on an object). It is rather a question of Society in its own right, an "interne/internalized" question, which may have a structural impact on Compartment Society and on the socio-systemic processes it hosts.

The representation of Society's relationship to Nature according to the ES angle is above all very functional (which allows, on a global level, the systemic terminology to be used).

Assertion 5.24
Adaptation (of the system, through Society's self-organizing function) plays an important role, especially by using actions related to Function Management. Some modes of functioning, even those apparently well established, may be challenged. The re-foundation, if it proves necessary, would even require structural changes! As a result, Compartment Society and Compartment Nature would be modified.

4. The positioning of the whole problematic is "global," i.e. "System Earth". As evidenced already by: the complexity of ecosystemic interactions ("butterfly effect", the effects of which are widespread); and the interweaving of Society's relationship to Nature (and that of S and NS type components at all systemic levels, Fig. 5.7), especially Society's expectations and objectives on "Compartment Nature". It is difficult to apprehend a compartment without too much bias[40] other than on a global level.

The notion of assessment calls for the notion of time, the interval over which the assessment is calculated. It is applied at the "global" level[41] since the impacts affect the "System" (see Fig. 5.6), and they may thus go beyond the only compartment considered.

[40]We will therefore be aware, in Sect. 5.3.2.3, dealing with the territorial compartment, of how significant the risk of bias is.

[41]The transposition from the systemic global level to the geographical global level is intuitive.

Assertion 5.25
Performance and impacts of the ecosystemic processes linked to Function Service, and those of the socio-systemic processes linked to Function ES-Management, are assessed at global level.[42]

As a result, Society's challenges and decision-making processes are reasoned and negotiated on a global level. Similarly, actors must belong to and represent Society in its entirety (and diversity). Similarly again, adaptation (from the capacity to accept questioning and challenging to the ability to provide solutions) concerns, and must involve, the whole of Society. The organization (self-organization) equally implies both an adjustment of each society to its environment (see Sect. 5.3.2.1), and an adjustment between societies, for a global adaptation. The negotiation between societies that must take place belongs to the process of self-organization and therefore to the mechanism of adaptation. The unfolding of protocols of consultation, negotiation, consensus, compromise, and ultimately the choice of "adapted" objectives and decisions-actions are theoretically derived from the self-organizing process itself. International conventions on the environment, including those on climate change,[43] biodiversity,[44] and desertification,[45] fall under this mechanism. The declination of negotiations in delimited areas and separate associated conventions should not make us forget that the issue is unique and global ("System Earth") and is thus addressed to the whole of Society.

Again, it is clear that any question of Nature is a matter of Society, including that of ecosystemic services, whose management needs to be addressed and negotiated. We could imagine and also invent societies (perhaps more complex) that, on the basis of equitable relationships between compartments of Society, would be both effective in managing Society-Nature Relations and thus participating in their perennial existence and coexistence. This would be adaptation as challenged by utopia.

Concerning actions, in particular those with an ES management objective (to ultimately improve Society's relationship with Nature), the technical and elaborated proposals (choice of know-how, engineering development), and the negotiated social proposals (contribution, compensation, equity – taking into account the quality of working conditions and income, and even access to work and work distribution among/between populations), typically fall within the scope of the adjustment (by the organizing function) of Society's compartments, for purposes of global adaptation. Similarly, the development of lifestyles (lesser pressure on NS components, e.g. by decreasing resource capture or decreasing discharges and

[42]We shall see later that this assertion does not exclude the recognition of particularity/diversity/specificity of environments or societies.

[43]http://newsroom.unfccc.int

[44]https://www.cbd.int/convention/

[45]www.unccd.int

waste) are adaptations of Society's components that allow an improvement in Society's relationship to Nature (see Assertion 5.19).

Assertion 5.26
Function Management of Society's relations to Nature, as a means of adaptation by adjustment, is indeed a Society self-organizing function. It involves Society as a whole and cannot dismiss any of its Society components either for ethical or simply systemic reasons.

Finally, the challenges of adaptation (from negotiation to action, including the definition of objectives) are of various kinds: (a) involve Society in its entirety and in all its diversities (cultural *s.l.*, economic, and so on); (b) take into consideration its geographical fragmentation as well as its diversity; (c) integrate the need for knowledge and mutual recognition; (d) work not only on assessments (of performance and impacts) and on the choice of actions (decision), but also on implementation and support mechanisms. The challenge today for Global Society is to be sufficiently "inventive" and "reactive" to allow the adaptation of "System Earth" and/or propose a new order in the course of its dynamics (see Fig. 5.5) by perhaps calling for changes in both frameworks of reflection and action.[46] Fields of thought, such as the ethics of development (Gasper 2014), the relationship of Man to his (natural and social) environment and coviability, could help outline criteria in order to better appreciate the actions carried out and the interactions involved.

A set of sociosystemic processes is involved. They are diverse in their quality and in their scale of implementation: those linked to ES intensification or ES management, those related to the definition of objectives and to the organization of work when implemented, those related to negotiation and decision-making. They all belong to the self-organization of partial societies and, more ambitiously, to that of the so-called Global Society.

5.3.2.3 The "Territorial" Angle for a Reading of the Society-Environment Relationship

Elements of Definition

In this section, we deal with the Society-Environment relationship in a non-global way according to the "territorial" angle, expressed not by Global Society, but by

[46]For example, the 17 Sustainable Development Goals (SDGs), for which it is stated that the inclusive nature of the Sustainable Development Program is an ethical imperative: http://www.un.org/sustainabledevelopment/fr/2016/01/16/le-caractere-inclusif-du-programme-de-developpement-durable-est-un-imperatif-ethique-selon-lonu/

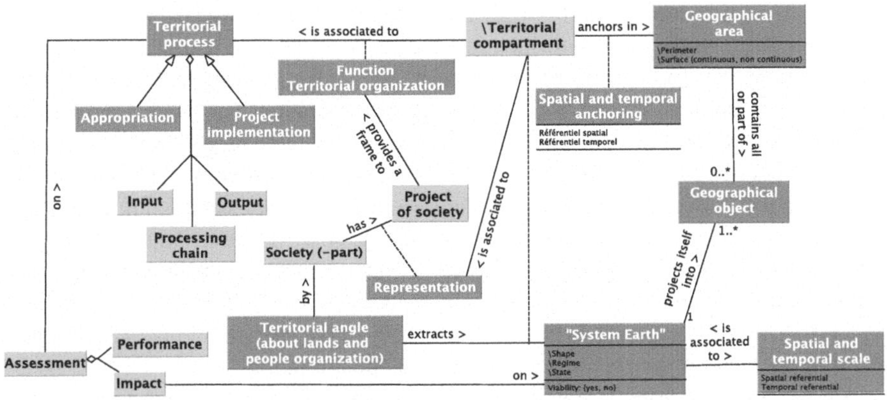

Fig. 5.10 Territorial angle (UML formalism)

a part of Society, hereinafter referred to as **society(−part)**, (Fig. 5.10). This angle takes into account the geographical dimension. It concerns an area and it adopts a **spatial scale**. For each territorial angle, the *ad hoc* geographical focus may range from the smallest (land, commune, etc.) to a most general one (nation, region, etc.) or even a global one,[47] e.g. the "global village" (Marshall McLuhan[48]).

The **territorial angle** by the society(−part) extracts from "System Earth" the **territorial compartment**, which assumes/carries out the **Function Territorial Organization**. Function Territorial Organization emanates from the view Society has on itself as explained in Fig. 5.8b. It is nothing else than a specialization of Function Self-organization shown in Fig. 5.8b. It organizes the society(−part) by organizing the **territorial processes** attached to this compartment. As we will see below, it has the specificity (especially in its "elaborate" form, i.e. the one that extends towards a systemic framework) to apply, not only within the territorial compartment, but also to be connected with other territorial compartments[49] of "System Earth", those which emanate from other territorial angles from other parts of Society.

A "territorial compartment", whatever the society(−part) it is related to, is a complex and composite set with the specificity, at the same time as being extracted, of being attributed a **geographical area**. The area may be continuous or discontinuous, is delimited by a **perimeter** and **anchored** in a **spatial and temporal** referential. Therefore, the compartmentalization[50] approach is accompanied here by a second procedure, the "Spatial Projection" of the system into geographical objects,

[47]This geographic global level (the globe) meets the global systemic level ("Earth System").

[48]https://fr.wikipedia.org/wiki/Village_planétaire

[49]Which originate from other territorial angles? by other parts of Society (i.e. other society(−part).

[50]Including multi-compartmentalization to account for the multi-functionality offered by System Earth with regard to the expectations/objectives of society(−part).

and the further "confrontation" of all these objects as being inside vs outside the perimeter. Ultimately, the territorial angle creates a geographical area, the territorial compartment, and retains as its "members" the only geographical objects[51] (or parts of geographical objects) that fall within its **perimeter**. This is an aggregate of geographical objects, possibly spatially superimposed. Objects are derived from the spatial projection of some "S" and/or "NS" components (or portions of components) belonging to "System Earth". They co-deploy, co-exist and possibly communicate[52] not only within the territorial perimeter but also beyond. The exchanges that occur, internally as well as beyond the perimeter, result from systemic interactions. The territorial compartment is therefore porous by definition. It will be so, more or less, and this is what will make it more or less individualized, autonomous, and even showing an "intention to reinforce" separation and autonomy.

The territorial compartment is associated with a geographical area showing a more or less composite structure. Intra and extra-compartmental exchanges are involved. These exchanges implicate interactions[53] (between systems), coherent at their own systemic levels.

Assertion 5.27
The coherence of systematic interactions and the viability of their systems and their interactions operate without reference to the spatial reality as retained by the territorial compartment.[54] The coherence of systematic interactions and the viability do not coincide with (or seldomly) the territorial compartment perimeter.

The complement[55] of the territorial compartment with respect to "System Earth" may also be called the Environment (with the meaning as explained in Sect. 5.3.2.1.). It consists of parts of S and NS type components. The complement may also be split, projected into other territorial compartments *a priori* not spatially overlapping.[56] Thus, exchanges can take place between territorial compartments.

[51]Themselves resulting from classical compartmentalization, see Sect. 5.3.1.6.

[52]Beware, this is not systemic in the strict sense; therefore, there is no interaction in the systemic sense. The interaction takes place on an ad hoc systemic level and not at the level of the compartment, which is by definition "arbitrary," hence the use of the verb "communicate." They are relationships, not systemic interactions.

[53]Systemic vocabulary.

[54]The territorial compartment has its share of arbitrariness and even irrationality. It is not the case with systemic systems.

[55]As in the set theory.

[56]Since: as many societies(−part) adopting a territorial angle as territorial compartments extracted from "System Earth" and distributed on the geographical globe which corresponds to "System Earth". The question becomes more complicated when more than one society(−part) projects its project on the same geographical area.

Furthermore, the society(−part) which emits the angle is itself complex. It includes social categories based on various criteria (professional, cultural, etc.). Each one of them can have its own system of representations according to which it founds its life project and embrace a territorial compartment.[57] This compartment may thus be associated with several representations. However, territorial identity of society(−part) (and of its various social categories) relies on sufficient sharing among the different member representations, or at least on the fact that they do not contradict each other. This follows the principle that a minimum of social cohesion seems necessary for a society as it is.[58] The project of society is (at least in part) induced by society(−part)'s representations.

The attractiveness of a place and, conversely, the attachment of a society(−part) to the place depends in particular on the suitability of the place for the project of the society; they also depend on the (biophysical or immaterial) specificities of the place (Lacoste and Salanon 1999; Pitte 2010). Geographical S and NS type objects (resources in raw materials, water, soil, biodiversity, topography, establishments and historical sites, "places of worship", sites rich in representations, linked to Nature, History, Culture and Civilization) in the territorial compartment can be either quite common and ubiquitous (they can be found anywhere; they are not specific to this compartment), or, on the contrary be unique or special and rare. Thus, the territorial compartments may be very diverse and different from each other.

Finally, in regards to action, **territorial processes** (social processes) for implementing projects combine symbolic actions (**Appropriation**) and tangible actions (Delimitation, **project Implementation**). They use the territorial geographical area and the geographical objects it contains (derived from S or NS type components, see above) in order to meet objectives, formulated in projects, particularly [but not exclusively, see below] those falling within the scope of the **Society's**[59] **Project**.

The Function **Territorial Organization** related to the territorial compartment maintains the geographical objects present in the compartment and the processing chains they constitute and organizes them into services. In so doing, there is part actions of creation/conception, part actions of regulation/management. Services, as specializations of Function Life, "must" have as objectives the creation/improvement of living conditions,[60] and should comply (more or less, but as close as possible to) to the society's[61] project. This therefore concerns services[62] in general, which must be managed and if necessary invented. Territorial processes

[57]Related here to the territory by Di Meo (1998), "the territory bears witness to an economic, ideological and political appropriation by groups which give themselves/propose/tell a particular representation of themselves, of their history, and of their singularity."

[58]If not, the risk is inconsistency an degradation, or no existence at all.

[59]i.e. society(−part).

[60]it should be possible to "evaluate" the quality of life, see below.

[61]Of society(−part).

[62]Services relate traditionally to the tertiary sector; they include those organizing the primary and secondary sectors.

are organized by the territorial organizing function at two levels of intervention: (1) organizing by creating, by inventing; and (2) organizing the management as precisely and efficiently as possible), one having a higher level of abstraction than the other.

The Society-Environment Relationship Within a Territorial Compartment

In a territorial compartment, the relationship between society(−part) and the Environment (see Sect. 5.3.2.1) and its NS type components is achieved through a double repertoire. On the one hand, through the attractiveness/attachment to a territorial compartment; in reality, the attachment to a "place", and on the other hand, through the Ecosystemic Services repertoire (see Sect. 5.3.2.2). Each one of them has an abstract, symbolic, rational or irrational part, in addition to some very material contingencies. The first is distinguished from the second by the historical dimension (past, project for the future); whereas the second is much more immediate, contemporary and applied to the present. They are all the more complementary to each other.

The view presented in Sect. 5.3.2.2, which brings the "functions related to Compartment NS Nature" into the abstract "Service" framework, formalizes their use (use of ecosystemic Service, Function Life). It also authorizes their management (ES Management, Organizing Function). Society, from a territorial angle may identify such ES services in its society project. It thus gives them a place in the primary sector (e.g. agriculture, mining, and so on): usage by product capture and/or ecosystemic process manipulation, ES management (see Sect. 5.3.2.2).

Also, the particularity, the specificity and the scarcity of the NS geographical objects singularize the ES services and identify them as more or less relevant (service and disservice) in society's[63] project. Hence, it is possible to find an ES interpretation to the attractiveness/attachment of society(−part) to (intra-perimeter NS) environment.

Assertion 5.28
There are often inconsistencies between the areas involved in systemics and the territorial geographical area. Viability calls for systemic coherence that generally needs to be sought at other geographical levels (up to the global level if necessary). This is especially true for all the Society-Environment relationships of "ES" and "ES Management" types, as we saw in Sect.

(continued)

[63]The situation is more complex when two societies(−part) have relations to the Environment in the same geographical area. The concerned Environment is then S + NS!! ... Immediately, therefore, it is no longer a matter of managing "only" ESs, but, relations between societies directly.

5.3.2.2. We have also shown that the organization of ES management solicits Society, through its organizing function, which, at one stage, will imply societal questioning/remodeling and contributes in this way to the adaptation of "System Earth". The global systemic framework is required in both cases (see Sect. 5.3.2.2).

Assertion 5.29
The "co-viability" of Compartments society(−part) and Nature attached to a territorial compartment, even if the relevant areas sometimes happen to match, cannot be considered as a rule and will, in most cases, be an abuse of language that disconnects this from systemics. Indeed, even if it is possible in some cases to envisage that the territorial compartment contains all the systemic components [S][64] as well as [NS][65] in interaction, this situation is not general. The question of their perennial coexistence within the territorial compartment is part of systemic reasoning, that is generally beyond the local level and which can only be addressed at a more general or even global level.

We have seen in Sect. 5.3.2.1, Assertion 5.19 that, at the global level, the coexistence of the two compartments calls for the adjustment of their functions: Function Life (attributed to Nature's compartment) and Society's organizing function (for better management of life functions). How can the issue at the territorial level be heard? And how can the link between global and territorial scales be organized?

The Territorial Assessment

The notion of territorial assessment offers a "toolbox" which allows a number of calculations. Particularly, in the case of Society-Environment relationships, the following operations are possible:

- Census of impacts caused: The territorial assessment highlights/takes into account the impact on "Earth System", caused by the territorial processes related to the Organizational Function of the territorial compartment: impact of the compartment on itself (by reflexive loop) but also on its complement[66] in "Earth System".

---

[64]This would require that a society(−part) be hardly permeabile to external influences (see the work on the oases (Fargette et al., part 3, this volume who observes and discuss these external influences).

[65]This would be if the entire ecosystems fitted into the territory.

[66]Set theory.

Assertion 5.30
Considering the impact in the assessment, i.e. taking into account consequences that happen beyond the function performance and beyond the perimeter, is a step towards approaching the overall coherence (i.e. moving towards the systemic approach).

- Census of incurred impacts: we have also seen that the complement of the territorial compartment is itself populated by territorial compartments, created by the views of other societies. The territorial compartments therefore communicate with each other, in particular by impacting each other.

Assertion 5.31
The impacts of extra-territorial origin, occurring in the compartment, have consequences on the living conditions offered in the compartment, whatever the performance of the Organizing function. The impacts are also to be taken into account in the part of the assessment dealing with the quality of life (see below).

- Performance: The territorial assessment takes into account performance, i.e. the result obtained by the territorial Organizing function. The ("calculated") result takes into account the achievements related to the objectives set by the project. It also takes into account the "internal perception" by the society(−part).

When it comes to organizing territorial processes, performance affects the quality of life (at least society(−part)'s) and the result is more or less in line with the social project.[67] Moreover, as seen in point 2, incurred impacts, impacts originating from other territorial compartments, may affect the quality of life. It is important to also take them into account when assessing the quality of life, which may become either degraded or improved.

Assertion 5.32
Quality of Life and Criteria: For a society, the notions of harmony, well-being, happiness (Stiglitz et al. 2008) participate in/reflect the "Quality of Life," revealed in a tangible manner (lived, measurable) or in an intangible manner (as thought, uttered feeling) to society(−part). In particular, the Quality of

(continued)

[67] Society(−part).

Life cannot hide under the bushel the equity of distribution resulting from the organizing function. Everyone"[68] has "a right" (Universal Declaration of Human Rights[69]) to challenge if this is not the case.

- Responsibility: Similarly, one cannot ignore the impact (whether positive or negative) that the territorial organizing function, operating in a given geographical area, may have on the Quality of Life of other societies in other geographical areas. Therefore, the contribution (positive or negative) of each compartment (and of society/responsible actors) must be recognized (see incurred impacts).

Assertion 5.33
The concepts of "Quality of Life" and of "mutual Responsibility" complement and enrich the notion of "System Earth" viability. Taking into account these two notions contributes in enhancing man's dignity, from the individual to every component of Society. Events that threaten the Quality of Life or ignore mutual responsibility put at risk the viability of "System Earth" in its entirety and in a form that would ensure Society's perenniality.

Assertion 5.34
The territorial assessment, as we have defined it, is a tool that records, calculates, and then provides information. It contributes thereby to the Organizing and Managing Function for the viability of "System Earth".

The Territorial Compartment: A Global Society Organizing Function
Transmission Framework, from Decision-Making to Implementation

1. Identification of the strengths and weaknesses of the territorial compartment

The assets of territorial compartments, formulated in "functions" (e.g. the different "Ecosystemic Services Functions") and a wealth in resources[70] can be taken into account through the unit area that corresponds to the territorial compartment. For instance, these assets, while anchored at a local level via the territorial compartment, are important in terms of systemic strategy with respect to "System Earth"'s adaptability (its potential to adapt). These assets are attributed to territorial areas and societies(−part). They have the potential to weigh on negotiations and decisions when everyone's contribution to the adaptation process

[68]In one society(−part), in the global Society.

[69]http://www.un.org/fr/documents/udhr/

[70]Natural, human, economic, financial.

of "System Earth" is recognized (e.g. concerning compensations recognizing contributions to carbon sequestration).

- Territorial compartments represent the node where "natural" risks are experienced and affect the quality of life. Some of them result from ongoing changes in "System Earth" (e.g. risks linked to climate Change and its consequences). There are also risks directly generated by society (e.g. political instability) and their consequences: an increased fragility of Nature compartments, precariousness of Society compartments, migration starting points for environmental or political reasons . . .).
- By fixing a granularity to the reasoning, the territorial compartment makes it possible to find out whether the causes of poor quality of life are local (due to internal processes to the territorial compartment, in which case action must be taken locally) or exogenous (and this must be said . . . and heard by others).
- Similarly, the notion of equity (and especially that of inequity) may, in part, be assessed by comparing geographical compartments. Undertaken in this way, it may highlight the disparities (cases of opulence and extreme poverty) that the global scale could erase.

2. Transfer of Scale

The territorial compartment is relevant for performing transfers of scale from local to global levels and *vice versa.* As a grid for carrying out assessments and gearing action, it makes it possible to implement decisions (top down). It also gives society(−part) the ability to testify and transfer information towards the global level (to the whole) about the experience and feelings at the local level (bottom up). Therefore, it belongs to the transmission/listening apparatus, before decisions are made for the Organization and the Management of Function Life and, at the highest level, before the harmonization of societies and contributes to Function Organization's adjustments. *In fine*, the territorial compartment contributes to working the Function "Reasonable Organization" that global society equips itself with.

Assertion 5.35

The territorial compartment gives body to the effort of management, to its area and its actors, to how they are organized within society(−part) and their involvement through the work[71] implemented in accordance with projected actions. The territorial compartments are a transmission belt for action, be it organization, reorganization or the invention of the management of the environment.

[71] The right to work is part of fundamental rights.

The territorial compartment:

- provides information on its "internal" quality of life;
- reports on actors' responsibility regarding the caused impacts, be it "internal" or "external" to it;
- communicates: from "the inside to the outside;" it is a way of "expressing" and telling others about one's own wealth and vulnerabilities; it participates in (global) negotiations, leading to (global) decision-making (which precedes (local) action);
- receives, transmits information "from the outside to the inside", internalizes it, embraces it, and translates it by organizing the local implementation of the action;
- implements, puts into effect locally.
- In summary, the territorial compartment (1) has a tool for "measuring/identifying (including surveys)", (2) plays an interfacing role between the local and the global levels, and vice versa, (3) is a link for implementing action.

3. The Contribution of the Territorial Compartment to the Management of the Society – Environment Relationships

Regarding the Society – Environment relationship and its organization/regulation, the systemic reasoning (to which we are nearing through the territorial assessment as detailed above), and the fact that its coherent framework is the "System Earth", cause the following:

- Governance, coordination, integration (and the accompanying negotiations), decision, in order to be coherent, should be reasoned on a global level. The decision does not belong to the territorial level. However, governance for a global purpose is not a blank check in order to impose the implementation of any sort of territorial action or impose any sort of impact on a territorial compartment. On the contrary, territorial adjustments should be possible (and/or it should be possible to comment on and discuss possible amendments to the decisions made in this overall/global framework), especially with an objective of equity to remedy the inequalities observed.
- In terms of management, it is not necessary to have a balanced assessment of the Function Life at the level of the territorial compartment; an unbalanced territorial assessment is not necessarily synonymous with bad management. The Organizational Function expects "reasonable" results, but the systemic approach has shown that the interactions and impacts go beyond the geographical area in which the action appears to be located.
- Exchanges have always existed between territorial compartments. Globalization is increasing this tendency. We can observe how much the territorial compartment is open to/ subjected to "somewhere else other than the one territorial compartment" and "to others" and how much each compartment intervenes on others, in a large process of globalization.[72]

[72]The initiation of this movement is very old.

- When the global governance (United Nations, Conference of the Parties[73] ...) makes a decision or draws up frameworks for agreement and action, it is essential to recognize the importance of the symbolic part of adhesion of Society(−part) to the new organizational project, which is supposed to lead to better management. Its incorporation, assimilation, into society's project (that of society (−part)) is paramount to ensure its double coherence, i.e. in a territorial rationale while still responding (this is essential, as it is a question of adaptation) to a global rationale.

In order not to take the risk of being only a transmission belt or an executing element in this vast apparatus, the territorial compartment must be able to voice its "say" to and be heard by (about its wealth as much as its vulnerabilities) the democratic bodies. It is also necessary that such bodies exist ... and be recognized ....

5.4 Discussion

From a dual categorical vision Society-Nature or Society-Environment, we have seen how their coexistence may be repositioned in terms of the viability of "System Earth". First, we have seen that the compartmentalization protocol[74] applied to "System Earth" provides a means for approaching a systemic framework. Second, we have seen that the territorial assessment emphasizes two essential notions, impact and quality of life. And third, we have seen that the territorial compartment[75] is at the core of a change in scale, from the global systemic level to the geographic levels regarding decision-making or actions implemented. It must be equipped with bodies for transmitting the flow of information and decisions, from top down as well as bottom up.

When it comes to improving the Society-Environment relationship, the expected results are firstly achieved by direct management of this relationship. However, as we have seen in the functional (and adaptive) context of "System Earth", it is also logical and systemic to call upon possible changes in the organization of the Society itself.

The choice of society's objectives, in terms of the organization of Society, directly depends on the "intelligence, heart and freedom" blend/alchemical art,[76] the privilege of men and societies. It is by acknowledging such qualities that we have

[73]COP: Conference of the Parties.

[74]Reminder: by definition, the compartment is not systemic. However, in practice, it can sometimes be this.

[75]Without prejudging its size.

[76]A mysterious blend, for are we capable of analyzing the parts of the heart vs egoism, freedom vs. confinement, intelligence vs blindness, attention to self-interest vs. being subject to fear, that participate in making choices and taking action?

named one function as "Reasonable Organization", by postulating that such utopia is also reasonable! While respecting the "intelligence, heart and liberty" triptych,[77] how may systemic reasoning contribute to and clarify the definition of such choices in terms of relations between Society and the Environment?

The consideration of the Society-Environment relationship, a relationship now positioned in a systemic context, confronts in a contradictory manner two temporalities to which the components of "System Earth" refer to: the geological scale (geological eras, life history, evolution of species and habitats) and the historical scale (history of humanity, construction of societies, on the happening of civilization).

- The plasticity of societies is potentially high (rapid dynamics, historical scale). Although it cannot be said that societies are in some constant and performing improvement process, we can notice that they are undeniably able to rapidly increase their complexity (e.g. technologies relating to structures and functioning, e.g. knowledge concerning representations).
- The constantly evolving plasticity of the living does not lack quality and inventiveness. The dynamics however refer to a geological time scale. They are not very compatible with the historical scale, except in cases of simplification of structure[78] where phenomena often accelerate.

Assertion 5.36
In terms of adaptability and reactivity on an historical scale,[79] it is better to rely, in all rationality, on the adaptation of societies rather than on the adaptation of Nature's components (whose anthropization is even more forced by certain methods of engineering).

New phenomena take place, the best as much as the worst, (e.g. land grabbing, citizen movements, positions taken by society's leading figures: Méda 2013; FRB 2015; Pape François 2015; Hulot 2015; Rocard 2015). New mechanisms are being built (e.g. ecological compensation, carbon market); meetings and negotiations are taking place (e.g. COP 12,[80] COP 21[81] in 2015). The inventiveness and plasticity of the human and social phenomena are at work and, as we have demonstrated, the Society-Environment relationship is a fact of Society.

---

[77]Without interference or totalitarianism.

[78]Often going hand in hand with increased vulnerability, or even actual degradation.

[79]And all the more, given the urgency of the current issues.

[80]COP12: Conference of the Parties, Twelfth session, of the United Nations Convention to Combat Desertification, Ankara (Turkey), 12–23 October 2015.

[81]COP21: Conference of the Parties, 21st session, of the United Nations Convention on Climate Change, Paris (France), 30 Nov–11 Dec. 2015.

In addition, solutions to improve "Function Management" are to be sought upstream of the relationship between Society and Environment, i.e. within societies, in the Society Organizing function (and societies, at the level of territorial compartments).[82]

Assertion 5.37
If the challenge is to rebuild the Society-Environment relationship, this indeed concerns, in terms of adaptation, a re-founding of society's relationships. A question arises however: will the resources of inventiveness "intelligence, heart and freedom" be mobilized for respecting and improving everyone's quality of life and for establishing equity between all?

Choices regarding re-foundation depend strongly on anthropological (diversity of cultures, Descola 2010, 2011), philosophical and "existential" positioning. Each position delineates a moral, ethical, social and legal framework ... in order to think about Man, his history, his condition and his dignity and to make a decision about his rights, duties (relationship of Man to his environment, relationship of Man to Nature, relationship between men). The point here is not to develop issues of other specialities but to show their anchoring point in our own reasoning. The disciplines are permeable to each other and the specialities communicate with, challenge and respond to each other. They can and must work together; Braudel (1969) already insisted on this point. There is room for all disciplines in this reasoning, including the overview of Man's history and the knowledge acquired throughout Man's history. It also includes the acquisition of knowledge, in particular the learning of Nature, encompasses engineering, and of course the reflexive view when man observes himself... and learns from himself and about himself (politics and action, democracy and power, ethics, solidarity and justice, aesthetics, ideal...).

Co-viability and coexistence, in their turn, concern coexistence between groups in Society. In this way, the question concerning Society-Environment relationships is inevitably addressed. However, raising the question of Society-Environment relationships without taking into account the relationships between compartments of Society does not adequately pose the question. This will only bring false answers (at best partial and more than likely, inadequate answers) in terms of perenniality and therefore of viability.

Many see that humanity and "System Earth" are reaching a crossroads. The choice of new values, founding the Society-Environment relationship and the Organization of Society, are to be invented, some of which have been discussed in this chapter. In particular, the notion of "Service" l.s. could regain its primary

[82]Without denying the technical solutions downstream of these choices, and of which they are the consequences.

characteristic of gratuity, rather than remain limited to its economic meaning prevailing today. Similarly, the notion of "wealth" could take on a different aspect, not only through an ethic of sharing in accordance with a humanist vision, but also by taking into account the non-economic wealth contained in the diversity of men, their projects and their capacity; the wealth resulting from their work. The economically focused vision, which may be narrow and incomplete and even misleading, would be revisited. "Marginalized persons" and forgotten lands would then play their full role in such a re-foundation.

In summary, this chapter uses a systemic logic to investigate the term coviability, and the link between Man and Nature as questioned within such framework. It has been observed that this questioning includes analyzing the perenniality of the relationship between Society and the Environment, and analyzing "System Earth's" viability. It has also been observed that indeed a re-founding of this relationship needs to be carried out and that it implies, in terms of adaptability, all of the resources in societies, especially the humanist part contained in each one.

A subsequent article (see Fargette et al., 2019, chapter 29, volume 2 of the 'book') takes up the theoretical and methodological elements set out here to investigate and clarify a general question of society posed today: the future of oases in North Africa.

References

Aubin J-P, Durand M-H (2019) Chapter 3: Coviability, through the lens of the mathematical theory of viability. In Barrière O et al. (eds) Coviability of social and ecological systems: reconnecting mankind to the biosphere in an era of global change. Springer, Cham, pp xx–xx

Begon M, Townsend CR, Harper JL (2007) Ecology: from individuals to ecosystems. ref à 4ième édition. July 2005, ©2005, Wiley-Blackwell, 752p

Bernard C (1865) Introduction à l'étude de la médecine expérimentale, 1865, Rééd. Champs, Flammarion, Paris

Bernard C (1877) Leçons sur le diabète et la glycogenèse animale JB Baillière, 1877

Bonneuil C, Fressoz JB (2016) L'évènement Anthropocène: la terre, l'histoire et nous. Seuil

Booch G, Rumbaugh J, Jacobson I (2005) Unified modeling language user guide, 2nd ed. Addison-Wesley Professional. Object Technology Series

Braudel F (1969) Ecrits sur l'histoire. Avant-propos. Champs, Flammarion, Paris

Cambien A (2008) Une introduction à l'approche systémique; appréhender la complexité. Coll. Rapport d'Etudes. Ed CERTU www.certu.fr, 83pp

Chansigaud V (2013) L'homme et la nature; une histoire mouvementée. Delachaux et Nieslé, 272p

D'Alembert J (1751) Discours préliminaire de l'encyclopédie

Dajoz R (2003) Précis d'écologie; 7ème édition, Dunod, 615 p

De Rosnay J (1975) Le macroscope. Seuil, Paris, 314pp

Descartes R (1637) Le discours de la méthode

Descola P (2010) Diversité des natures, diversité des cultures. Bayard Ed, 85p

Descola P (2011) L'écologie des autres; l'anthropologie et la question de nature. Paris, Ed Quae, coll.Sciences en questions, 110p

Di Meo G (1998) Geographie sociale et territoires. Collection Fac Geographie. Nathan Universite, Paris, 317 p

Diamond J (2005) Collapse. How Societies Chose to Fail or Succeed – Effondrement. Comment les sociétés décident de leur disparition ou de leur survie. Paris, Gallimard (nrf essais), 648 p
Donnadieu P, Durand D, Neel D, Nunez E, Saint-Paul L (2003) L'approche systémique: de quoi s'agit-il? Synthèse des travaux du groupe AFSCET "Diffusion de la pensée systémique"
Elmi S, Babin C (2006) Histoire de le terre
Fargette M, Loireau M, Libourel T (2019) Chapter 29: The future of oases in Maghreb as through the prism of a systemic approach: which viability and coviability? In Barrière O et al (eds) Coviability of social and ecological systems: reconnecting mankind to the biosphere in an era of global change. Srpinger, Cham, pp 13–53
François P (2015) Loué sois-tu; sur la sauvegarde de la maison commune. Encyclique, Bayard Editions, 215p
FRB (2015) Prospective scientifique pour la recherche française sur la Biodiversité. Rapport, 58pp. http://www.fondationbiodiversite.fr/images/documents/Prospective/Prospective_2015_web3.pdfhttp://www.fondationbiodiversite.fr/images/documents/Prospective/Prospective_2015_web3.pdf!
Gasper D (2008) Denis Goulet and the project of development Ethics: choices in methodology, focus and organization. J Human Dev Capabilities 9(3):453–474
Gasper D (2014) Ethics and development. In: Potter RB, Desai V (eds) The companion to development studies, 3rd edn. Routledge, Abington, pp 47–50
Gouyon PH, Leriche H (2010) Aux origines de l'environnement, 497p
Gravel D, Massol F, Leibold M (2016) Stability and complexity in model metaecosystems. Nat Commun 7:12457
Gunderson L, Holling CS (2001) Panarchy: understanding transformations in human and natural systems. Island Press, Washington, DC, 507pp
Hall A, Clark N (2009) What do complex adaptative systems look like and what ae the implications for innovation policy? United Nations University, Working Papers Series, #2009–046
Holling CS (2001) Understanding the complexity of economic, ecological and social systems. Ecosystems 4:390–405
Hulot (2015) Osons. Eds "Les liens qui libèrent". Fondation Nicolas Hulot pour la nature et l'homme
International Geosphere-Biosphere Programme (2006) IGBP Science plan and implementation strategy. http://www.igbp.net/download/18.1b8ae20512db692f2a680006388/1376383126825/report_55-IGBP.pdf
International Geosphere-Biosphere Programme (2010) IGBP Strategic Vision. http://www.igbp.net/download/18.2709bddb12c08a79de780001017/1376383018329/IGBPDraftvision27September.pdf
Kéfi S (2012) Ecosystèmes et transitions catastrophiques. Regards Croisés n° 37, Société française d'Ecologie. http://www.sfecologie.org/regards/2012/10/19/r37-hysteresis-sonia-kefi/
Kéfi S, Miele V, Wieters EA, Navarrete SA, Berlow EL (2016) How structured is the entangled Bank? The surprisingly simple organization of multiplex ecological networks leads to increased persistence and resilience PLOS Biology. https://doi.org/10.1371/journal.pbio.1002527
Kleidon A (2009) Non-Equilibrium thermodynamics and maximum entropy production in the Earth System: Applications and implications. Naturwissenschaften 96:653–677
Lacoste A, Salanon R (1999) Eléments de biogéographie et d'écologie. Nathan, collection fac, 318pp
Levin SA (1998) Ecosystems and the biosphere as complex adaptative systems. Ecosystems 1:431–436
Lovelock JE (1979) Gaïa, a new look at life on earth. Oxford University Press, Oxford
Lurgi M, Montoya D, Montoya JM (2016) The effects of space and diversity of interaction types on the stability of complex ecological networks. Theor Ecol 9:3. https://doi.org/10.1007/s12080-015-0264-x
Méda D (2013) La mystique de la croissance; comment s'en libérer. Editions Flammarion, 265pp
Minsky M (1995) Matter, mind and models. Rev. Version of the essay in "Semantic Information Processing" ed Marvin Minsky. MIT Press, Cambrdge, MA

Mollard E, Walter A (2008) Agricultures singulières. Nouvelle édition [en ligne]. Montpellier: IRD Éditions, 2008 (généré le 20 octobre 2015). Disponible sur Internet: http://books.openedition.org/irdeditions/2834. ISBN: 9782709917834

Morin E (2015) Penser global; l'humain et son univers. Editions de la Maison des Sciences de l'Homme, Paris, 128pp

Nahon D (2008) L'épuisement de la terre; l'enjeu du XXIe siècle. Editions Odile Jacob , 235p

Odum EP, Barret GW (2006) Fundamentals in ecology, 5th ed. Thomson International, 598pp

Pitte JR (2010) Le génie des lieux. CNRS Editions, collection Débats, 60pp

Ramade F (2003) Elements d'écologie; Ecologie fondamentale, 3ème édition, Dunod, 690p

Rocard M (2015) Suicide de l'Occident, suicide de l'humanité? Flammarion – Documents Flammarion, 18 février 2015, 432p

Sen A (1987) Ethique et économie. PUF, Paris

Stiglitz JE, Sen A, Fitoussi J-P (2008) Rapport de la Commission sur la mesure des performances économiques et du progrès social. www.stiglitz-sen-fitoussi.fr

Toussaint J-F, Swynghedauw B, Boeuf G (2013) L'homme peut-il s'adapter à lui-même? Marges d'adaptation de l'espèce humaine face aux changements environnementaux. Editions Quae, 187p

Weber J (2012) Biodiversité et production de richesse, une érosion inéluctable? In: Toussaint JF, Swyghedauw B, Bœuf G (eds) L'homme peut-il 'adapter à lui-même? Quae, pp 150–155

Chapter 6
Socio-ecological Viability and Legal Regulation: Pluralism and Endogeneity – For an Anthropological Dimension of Environmental Law

Olivier Barrière and Mohamed Behnassi

> *Earth, humanity's home, constitutes a whole, marked by interdependence (;) the existence and the future of humanity are inseparable from the natural environment.*
> *Humanity has a right to the preservation of common goods, especially air, water and soil, and a right to universal and effective access to vital resources. Future generations have a right to the transmission of these resources (Article 8).*
> *Humanity has a right to free choice to determine its destiny. This right is exercised by taking into account the long term, and in particular the cycles inherent in humanity and nature in collective choices (Article 10).*
> *The present generations are responsible for resources, for ecological equilibrium, for common heritage and the natural, cultural, tangible and intangible heritage. They have a duty to ensure that this bequest is preserved and that it is used with prudence, and equity (Article 12).*
>
> (Draft of the Universal Declaration of the Rights of Humankind, President of the French Republic at the United Nations, December 2015)

O. Barrière (✉)
IRD - French National Research Institute for Sustainable Development, UMR ESPACE-DEV, Montpellier, France
e-mail: olivier.barriere@ird.fr

M. Behnassi
Center for Research on Environment, Human Security and Governance (CERES), Ibn Zohr University of Agadir, Agadir, Morocco

O. Barrière et al. (eds.), *Coviability of Social and Ecological Systems: Reconnecting Mankind to the Biosphere in an Era of Global Change*,
https://doi.org/10.1007/978-3-319-78497-7_6

Contents

6.1 Introduction: The Legal System in the Interrelationship of Social and Ecological Systems: A Legal Regulatory Challenge for the Inseparableness Between Man and Planet Earth

The Universal Declaration of the Rights of Humankind introduces the principle of the continuity of humankind. The very fact of formalizing this biological object, the reproduction of the human species, does not evoke indifference. In the Declaration, the sustainability of the human species is based on responsibility and solidarity between generations for the concomitant safeguard of both Planet and Man. Future generations, constituted of beings who are not yet conceived but only predicted, expected or planned, thus acquire a legal existence made concrete through rights of future generations' (Gaillaud 2011). Humanity recognizes its duty to avoid disappearing with the obligation of maintaining its "homeland," the Earth. The foundation of principles, rights and duties set forth in the Declaration on the future of humanity anchors itself in relationships of interdependence and inseparability between the human and the nonhuman. If the timeframe of the relevant rights shifts from the existence of the individual (lifetime) to the survival of the species (indefinite duration), it will be obviously at the expense of an agreed upon, regulated relationship: human beings confronted with duties *vis-à-vis* their future. In other words, it concerns the manner by which the future invests in the present through a regulation imposing the right of existence and a duty to sustain the human species. Social temporality is thus combined with ecological temporality. The intimate relationship between man and the biosphere does not arise from a dichotomy but from a biological integration of the inseparableness of beings to their environment.

Social and ecological systems collaborate with each other and ultimately merge, but without assimilation between the human and the nonhuman. Thus nature is not humanity's object but it involves itself fully in it. This is why nature does not constitute that which "surrounds" man, but rather its intrinsic functioning process called "anthroposystem" (Lévêque et al. 2003). Hominids live and die in the nature they create, produce and construct. They remain only one component of this "anthropo-ecological" system, from which flows a relationship between the human and the nonhuman.

The interweaving of functionalities formalizes itself in a set of subsystems, human societies and ecosystems. The links formed between them revolve around constraints of dependencies, connectivity, correlations, feedbacks and of causal relationships. Everything is connected; and that which connects also dominates. This link depends on the types of regulation that we apprehend here. Within societies, regulation is marked by a legal component. It is often related to "law," but not exclusively as regulating has an economic component and derives from social conventions, morals, religions, and so on; but where does this law come from and where does it originate? (Sect. 6.1.1.1). This fundamental question leads us to its structure as a system. The legal system within a society constitutes a social subsystem (Sect. 6.1.1.2) which merges with the ecological system to form one entity, taking into account man's interdependence with the biosphere (Sect. 6.1.1.3).

6.1.1 Legal Regulation and Society: Positivism and Legal Anthropology

The objective of regulating is defined in the viability of social systems (their maintenance, development, sustainability, reproducibility). Marked anthropocentrism deals implicitly with ecological sustainability while dependence on ecosystems imposes itself by an evident recognition (Earth Charter June 29, 2000,[1] Constitution Bolivia, Mother Earth Rights[2]). We thus reach an aporia: norms intervene in a more or less coercive manner but by investigating society outside an ecological context: hence the need to reconnect human beings to the biosphere of which they are part of and in which they participate. The particularly strong hold humanity exerts on

[1] "The future at once holds great peril and great promise. To move forward we must recognize that in the midst of a magnificent diversity of cultures and life forms we are one human family and one Earth community with a common destiny. We must join together to bring forth a sustainable global society founded on respect for nature, universal human rights, economic justice, and a culture of peace. Towards this end, it is imperative that we, the peoples of Earth, declare our responsibility to one another, to the greater community of life, and to future generations". "Recognizing that all beings are interdependent and every form of life has value regardless of its worth to human beings". (Principle 1.1a).

[2] Law N 071 of 21 December 2010: "To promote the harmony of Mother Earth in all fields of its relationships with human communities and ecosystems "(art.9).

the planet has generated a new geological era, the Anthropocene, leading us to discuss the anthroposystem. Rather than the absence of a man/nature dichotomy, the simultaneous existence of humans and nonhumans, results in the paradigm of socio-ecological coviability. This expression of joint viability, of an inter-viability, translates itself into a single system, a super-organism which is the biosphere. The challenge of environmental law is thus raised by the socio-ecological coviability, for example, the coexistence of human and ecological rationales which imposes interdependence between the human and the nonhuman.

However, what is a legal regulating system[3]? It is a set of constraining provisions supported by values. Values are preferences and appreciations. Law ensures values through norms and *legal habitus* (defined below). The question of division between the social and the legal is then addressed. Orthodoxy requires that sociology targets "that which is" (Sein in German) and law be based on "that which must be" (Sollen in German). By investigating the distinction between Sein and Sollen, Hans Kelsen (1953, 1960, 1979) guarantees an autonomous field of study for law, and *a contrario* for sociology. Thus, the science of law deals with what ideally "must be"; whereas, sociology concerns the being, concrete facts and conduct determined by the legal order. This distinction between "that which is" and "that which must be" is certainly not as clear as the works of many authors suggest.

Maurice Hauriou (1856–1929), one of the first theoreticians of the social dimension of law, questions the position of sociology in the faculties of law (1893). Georges Gurvitch (1894–1965) brings to light the idea of a social law participating in legal pluralism (1935, 1940). Henri Lévy-Bruhl (1884–1964), a Romanist and historian of law, notes that law determines social relations (1961). Max Weber (1864–1920) examines the rationality of law (1986), followed by André-Jean Arnaud who looked at law and its relationship with society (Arnaud et Dulce 1998, Law betrayed by sociology; 1981, Where Is sociology of law going?; 1998). Pierre Bourdieu explores principles that generate practices (1986). The frontier from law to social issues does not create unanimity; it generates schools of thought with active participation from Jean Carbonnier (1978), Jacques Commailles (1994), Chazel et Commaille (1991), Renato Treves (1995), and so on.

At the same time, the anthropology of law is actively developing on the basis of a so-called "primitive" law analysis (Karl N. Llewellyn and E. Adamson Hoebel 1941), and of "legal customaries" of French West Africa (the Historical and Scientific Studies Committee of French West Africa (1939); see Elias T. Olawale on the nature of African customary law (1961), Kouassigan Guy Adjété (1982)). It is in this way that Michel Alliot situates law as a mirror to anthropology (2003), and he involves in his reasoning Norbert Rouland (1988, 1991), Jacques Vanderlinden (1983, 1996), Etienne Le Roy (1999, 2011), Louis Assier-Andrieu (1996), Rodolfo Sacco (2008, 2009), Régis Lafargue (2003, 2010), Gilda Nicolau et al. (2007). The path to legal analysis has thus been mapped by choosing legal science over

[3]Clam and Martin 1998; Chevallier 1983, 1998, 2001; Romano 1919; Ost 2000; Barrière 2007; Chazel et Commaille 1991.

anthropological and sociological sciences. This path differs from an American practice which adopts a reading of the law solely through anthropology (Sally Falk Moore (1978, 2005), Laura Nader (1997, 2002), Alain Pottage and Martha Mundy (2004), June Starr et Goodale (2002), followed by French authors such as Jean-Pierre Jacob (2007). However, they all question the primacy of state law no matter how they approach it.

The very notion of a legal order is discussed in its social context. The challenge consists of transcending fiction to behold the real world, as Bernard Edelman's title (2007) correctly suggests, "Quand les juristes inventent le réel" ("When Jurists Invent the Real"). In his general theory of law, Santi Romano integrates the idea of a plurality of legal orders (1919). Jacques Chevallier reflects on postmodern law (1998). In philosophy of law, Alain Papaux invests on "law in situation" (2006). Criticism on legal rationality, as a mytho-logical drift, dates back to 1980 (Lenoble and Ost). In addition to Michel Van de Kerchov, these thinkers locate the legal system between order and disorder (1988). The reflection questioning the monolithic character of law is fruitful and goes so far as re-conceiving the thought of law (Simone Goyard-fabre 2007). If starting from 1894 Gabriel Tarde worked on the transformations of the law (1994), the dynamic never ceased to progress and continued with Jean Clam and Gilles Martin (1998) and their emulators.

Legal anthropology is fully integrated in the genesis of a dynamic trend, driven by the need to revisit and redefine the Kelsian approach to law which limits law to an order of constraint in the form of a normative hierarchy (Kelsen 1979). Etienne Le Roy investigates this need in his two founding works. One of these works explores a dynamic anthropology of law (1999) while the other examines the regimes of land appropriation (2011). We cannot claim being able to refer to all waves and authors, but it is still necessary to emphasize that the reading of the law cannot be based solely on positivism. We have entered a legal pluralism, or as others put it, a "plurijuridism" (Bergel 2005) by surpassing the state monolith. The law situates itself within a plurality, or even a diversity, of law-generating sources. André-Jean Arnaud (1998) spotlights norm production by speaking of "poly-centricity." Within an extreme approach, law becomes "solubilized" in the social body (Belley 1996).

The empirical approach adopted by Perrin (1997) and Ross (2004), for whom a norm is valid only if it is socially essential, is the norm that we favored in order to develop a capacity to "break away from the codes".

The question of the position of legal regulation within society examines the interrelationship of the legal and social systems. It leads to defining the amplitude of the legal system against global changes, in particular climatic change, ecological urgency, globalization, and social and economic challenges present on all levels. There is no longer any legitimacy in containing the law exclusively within laws as suggested by the increasing importance of regulation emerging from the need for viability, interweaving social aspects in all their cultural diversity with ecological aspects.

6.1.2 Law as a Social Subsystem

The coexistence of a state's legal paradigm and that of social groups defines a legal duality. To this complexity, one may add a complicated plurality of legal systems in each order. As the legal order of any society reflects its cultural scope, that which society is, this scope becomes a system and translates the legal order into norms, becoming a system; it translates into norms and a habitus the frameworks of thinking and of being of a society, i.e. its socio-cultural paradigm. In any society, what becomes law are the social facts elevated to a legal rank to meet the needs of a group's viability: its maintenance, its sustainability, its growth. The law, as a subsystem of a social system, exists in the objective of sustainability. It is a subsystem which cannot claim uniqueness because it is specific to each society with its own bio-geographical, social, historical, political and religious environment. Thus, each society has its own legal regulation, but this does not mean that components are not the same between systems.

The intrinsic law of the social system depends on the latter and therefore cannot claim self-determination (Ost 1991, 142) bypassing its environmental context that includes both the socio-cultural and the ecological components. Within the hypothesis of a normative auto-production affirmed by certain authors,[4] the legal system regenerates itself totally independently. However, it is necessary to verify if the auto-regeneration of a legal system exists in reality, that is, does not underline fiction or even a myth added to that of the disconnection between law and society.

The myth of an isolated or almost isolated legal system leads to redefining the law. Schools of thought diverge a lot on this position: on the one hand, the law of an orthodoxy that refutes the social element (such as Kelsen 1960) and ensures the formal reproduction of the group with its norms (legislated or not, recommended or declaratory rules, and so on) (Thibierge 2003, 2008, 2009) and, on the other hand, the legal habitus (Bourdieu 1986, see below).

Why not remain with juridical positivism, with the dogma of the hierarchical order of norms established by the State, with a so-called legal objectivity?[5] Why should we wish to go beyond the exiting framework of legality? The answer is the legitimacy and the very definition of the legal system. Everything is a matter of paradigm: the one that restricts the world to what it must be, the "must be" (Sollen),

[4]Law, seen as an autopoietic subsystem by the positivist paradigm (state law) restricting the law to laws, locks itself within the context of a self-replicating legal system, making it quasi-autonomous. In this representation, the basis of the validity of a norm comes from its creation which is established in conformity with a higher norm, which defines a system of normative hierarchy: auto-reproduction leads to an auto-creation of the law: "we observe that these (legal) norms are themselves created by legal acts, and that in turn these acts receive their legal significance from other norms "(Kelsen 1960); "The law regulates its own creation" (ibid). Thus, the law as a system leads to an order (the legal order) which reproduces itself because of its internalized process of evolution (cf Teubner 1984, 1993).

[5]"The inadequacy of the Kelsenian positivist theory to the mutations of positive public law prompts us to reconstitute the current theory of law to deal with such mutations. As for the supposedly scientific nature of pure theory, it has never been tested as much as it is today" (Sabete 1999).

leading to the question of the validity of a norm; and the paradigm which transcends this framework in order to "be" (Sein), that which is, leading to the question of the effectiveness of a regulation made of norms and provisions. The preferred hypothesis advances that the legal arena cannot be confined to clear-cut outlines and restricted acts of will. Social complexity introduces the legal field to a judicial arena emerging from an often excessive simplicity, or even from a caricature of human behavior. The correlation of Sein with Sollen makes it possible to disengage from the dogma of the law and to take a step back through a holistic view of the law.

6.1.3 *Law in the Socio-ecosystem*

In the contemporary context of modernity and postmodernity (Latour 1991), of globalization, ecological urgency, migration, and climate change, reading the law can hardly be left in the hands of positivism alone.

The legal system is part of the regulating system which integrates social and ecological systems. Its dimension emerges from state convention for an exhaustive dimension defining law within a socio-ecosystem. We are able thus to integrate the human factor in bio-ecological science and vice versa: what is a social system without an ecological system? It is impossible to dichotomize, because the human and the nonhuman coalesce within the existence of the biosphere. The scientific dimension of the environment limits the manner by which the environment is understood, since, by definition, the environment is that which surrounds man. The human environment relates to a legal regulation which cannot extract itself from the human, and hence the social, context. Consequently, the social (including politics and economics) is at the heart of the environmental problem.

The legal system is defined as a system of regulation that every society initiates within itself and whose purpose is the viability of the social system from which it emanates. However, the viability of the social system cannot be dissociated from the viability of the living system of the nonhuman existing in the all-encompassing system: the biosphere. Therein, man is consubstantial to nature. Environmental law, which is said to intervene in regulating society's relationship with its environment, cannot be exempt from this consubstantiality. Its *raison d'être* (purpose) is thus based, by definition, on a principle of coviability of social and ecological systems, which determines the coexistence of the worlds within the socio-ecosystem.

The aim of this chapter is to formalize the legal content of coviability's paradigm. To do this, we will first address the two main dimensions of environmental regulation relating to ecological viability while being anthropocentric. The first dimension characterizes the technical branch of law which is part of ecological science; the second, still not included or hardly, determines the anthropological dimension of human sciences (Sect. 6.2). We will investigate how the combination of these two dimensions can characterize the socio-ecological link and formalize it legally. The challenge of environmental law is to rethink itself, on the basis of viability, through a law of coviability. The "encompassing by law" (judicial proceedings) of the socio-ecological connection results in releasing environmental law from its anthropocentrism by re-identifying the relationship between society and nature, within anthropocenian reality, through ecocentric and biocentric borrowings (Sect. 6.3).

6.2 From the Consubstantiality of Ecological and Human Sciences to Environmental Law

What is environmental law made of and what should it be made of? The answer to this question is handled differently depending on the country and continent it relates to. Environmental regulation focuses on the effects of human activities on the living environments of societies. In this context, human beings participate in an integrality, a living whole, that is the biosphere. Added to this whole, societies operate according to their cosmogony, their place within the cosmos, or according to their mode of ontological relations to that which is not human. The scientific rationality of René Descartes, "men must become masters and possessors of nature" (Discourse on the Method 1637) aims at getting rid of speculation to ensure certainties, especially through technical methods. Environmental law initiated by the international community and States was greatly inspired by it. Indeed, the descriptive natural order defines that "which is" and the *prescriptive* legal defines that "which must be". Empirical reality cannot produce a norm or a value; the must-be (Sollen) can only come from outside of being (Sein). In the (pure) theory of law, reality and law belong to two distinct domains (Kelsen 1953, 1960, 1979). Within this rationale, law is defined by the set of rules of which the State is the author and the depositary, what is called positive law. This legal positivism characterizes the law of the environment, which thus takes a strongly technical and scientific connotation.

The absence of anthropological considerations in the positive law of the environment makes it, however, profoundly anthropocentric on an ecological basis. The explanation is simple: the dichotomy established between the social system and the ecological system. It is only by overcoming this duality of culture and nature that it becomes possible to reach the grail of a consubstantiality of ecological and human sciences with environmental law.

International law emerges from this naturalism by solemnly affirming in 1982 that "humanity is part of nature and life depends on the uninterrupted functioning of natural systems which are the source of energy and nutrients. Civilization has its roots in nature, which has shaped human culture and influenced all artistic and scientific works, and it is by living in harmony with nature that man gains the best opportunities to develop his creativity, relax and spend leisure time" (World Charter for Nature, adopted and proclaimed by the General Assembly of the United Nations on 28 October 1982).

The transition from a law with an ecological dimension only to (1) a law integrating an anthropological dimension requires (2) opening legal thought.

6.2.1 *The Ecological Dimension of Environmental Law: Technical and Scientific Properties, a "Legal Staging of Nature"*[6]

The aim of national and international (positive) environmental law is protecting the environment. This protection results directly from a scientific observation, biodiversity degradation and climate change, resulting in pollution and nuisance laws and a law of nature. This latter contributes to a reading of the real by science. Man and nature are dichotomized in environmental law which owes its existence to scientific rationality and thus to an objectivity characterizing it.

6.2.1.1 An Order and a Rationality for an Automaton World

During the last few centuries, modern science announced the existence of general, unalterable and eternal laws of nature. Order and rationality are supposed to reign, the world viewed as an automaton (Prigogine and Stengers 1986). Since humans are the only beings endowed with reason, they are also the only species capable of discovering and dominating the rationality of nature. Modern science in this way validates the Cartesian credo which suggests that man must behave as master and owner of nature. It also bases itself on an absolute separation between the subject and the object (nature is perceived as a passive "object" while man is perceived as an active "subject"). Modern science can thus claim thus a strictly neutral position in relation to the things it claims being capable of revealing as they are. This perception denotes the *adequatio rei et intellectus* principle (adequacy of the thing and the spirit). The scientist is perceived as a neutral and direct representative of an object, so in this scenario s/he position themselves outside society. This allows the modern scientific enterprise to be convinced that it does not take part in the political and legal mediation that characterizes intersubjective relationships. According to Latour (1991), two major divisions are therefore at the center of modern scientific design: the one created between subject and object (or between culture and nature) and the one between science and society. The resulting science is enveloped with double power knowing no modesty: an unlimited power over the object or nature which also a power in relation to the rest of humanity. In other words, it evokes the power to speak and be free from doubt and discussion; the power to be above all suspicion (Gutwirth 2001).

6.2.1.2 A Law Heavily Impregnated with Science

Environmental law is heavily impregnated with sciences due to the purpose of this law which concerns nature and pollution (Prieur 2001). Therefore, the rules of the

---

[6]Gutwirth et Ost 1999.

human activities that may affect the environment are subject to scientific knowledge and will depend on its reliability and its progress. This led to environmental law being, first of all, a "technical" law, elaborated by engineers and examined by traditional legal experts on non-law. This law remains the product of political will and reflects society's needs (ibid). Its scientific interpretation and its technical aspect are merely the reflection of political or economic constraints or timidity. Environmental law has developed well as a legal creation with its institutions and its general principles (Prieur 2000), and with increasing codification in several countries. Moreover, this law has increasingly become independent of science which, translating uncertainty or ignorance, is no longer the compulsory reference. In this context, it is environmental law itself which initiated the precautionary legal principle, a principle dissociating the law, administrative decisions, and scientific truth and seizing health and consumer laws over the last few decades. Referring to Aristotle's writings, Prieur (2001) asserts that law – thinking precisely of the law – is not an expression of science but of "prudence", for the ultimate goal of the law is not searching for scientific truth but reaching the well-being of society and human beings.

6.2.1.3 Law as a "The Notary of Scientific Norm"

Regulating the science-law relationship is an essential issue for environmental law (Gutwirth 1996). Modern sciences and law share a paradigm: with an absolute split, a radical dualism between the subject and the object. Through scientific practice, the scientist assumes an unrestrained mastery over nature, which, in law, extends the quasi-sovereign status of ownership. With the arrival of modern times, science imposed itself in the science-law relationship because it self-proclaimed itself a possessor of an absolute, objective, and timeless truth, a field free from any socio-political arbitrariness and challenge. Contemporary environmental law clearly reflects this double enslavement; formalized by the dependence of objects under the ownership system and technocratic management law thus loses its role of mediator to be reduced to that of the notary of scientific norms (Ost and Gutwirth 1996, 8).

Environmental legal provisions are defined in a specific way in the sense that they are based on the "scientific analyses" of a situation, serving thus as justification for the rules of action (Barrue-Pastor 1986). It is a verified approach, whether it be for the inclusion of a plant or animal species on a list of legally protected species, delimitating a space or an ecologically sensitive environment, or defining a threshold of critical pollution loads determining the intervention of law. Henceforth, the problem of scientific expertise and, *in extenso*, of the relationships of knowledge and power, has started to wrap itself around the legal corpus of environmental law, structured in a consubstantial manner. Science, thus a tutelary Fig. 6.1 and a sprawling power over all living beings, exerts a decisive influence on environmental law, up to the point of confining its normative process into an eco-power (Naim Gesbert 1996 and 1999).

Yet, a reaction may come from science itself as it becomes now more aware of the fact that reality is not necessarily rational, thus creating a problem in its

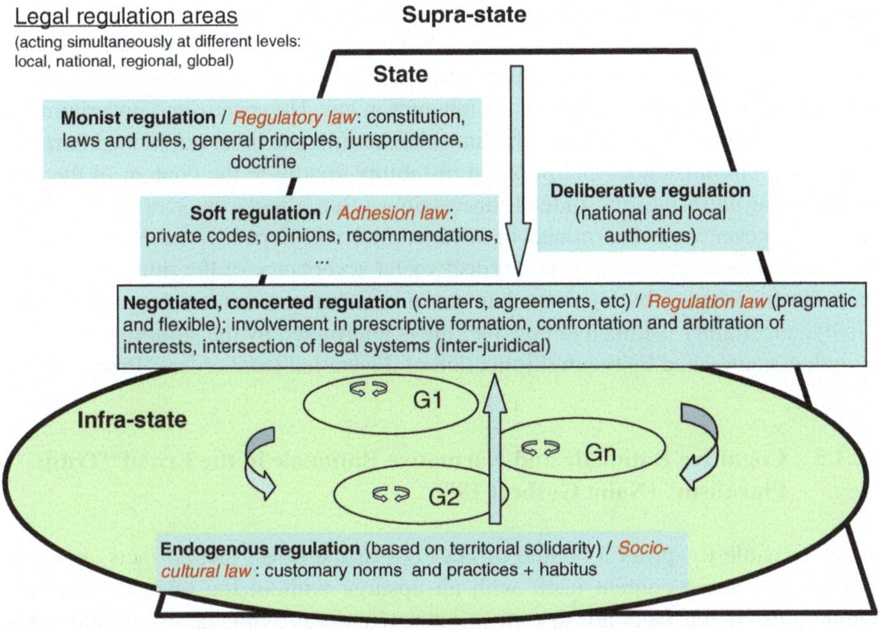

Fig. 6.1 The different scales of the legal landscape emphasizing the plurality of legal systems

claim to objectivity and this, with a new "pluralism of truths" between which arbitration is necessary. In fact, the production of truths responds to socio-political interests and values as well as to a pure logic of knowledge in the sense that the productions of modern science may even appear as "hybrids" (a mixture of nature and culture), whose effects must be controlled. It is only by succeeding in "mastering our mastery" of nature that the law will ensure its function as mediator which must not result in aligning itself with fact but in instituting meaning, nor in strengthening dominance – here, the power of techno-science – but in creating counter-powers everywhere (Gutwirth 1996).

6.2.1.4 The Role of Uncertainty

With planetary ecological problems, we increasingly understand that the more knowledge grows, the greater the gap is between what we know and what we would like to know in order to manage and make decisions. Uncertainty has thus gained a place in debates that it did not have before. We are no longer evaluating the margins of error in the analysis of known phenomena but rather, asking ourselves what could happen, and it does indeed occur that we are confronted with diverging, yet equally plausible, scenarios. Uncertainty is primarily of all a scientific nature. Decision-makers increasingly believe that scientists no longer know, and so they turn to experts, commissioned to give an answer, an opinion. It is therefore important to take stock of this scientific uncertainty and to know its outcomes (state of technology, etc.). When addressing this subject matter, Prieur (2001) believes that environmental law must take into account scientific uncertainty, by admitting that the latter is not in itself an anomaly since it may even be inherent in law. However, such uncertainty is still not be exempt from affecting the future of the rule, the latter not being alterable. Changes in circumstances and political instability condition the content of the rule and increase its instability. Indeed, uncertainty often affects the scope of the rule since it is never well determined in advance and will depend on the practices of the actors, which will more or less reflect social acceptance of the rules. The scope may also be affected by legal mechanism of reservations inside international law. Finally, uncertainty is linked to how the rule is interpreted by the administration and the judge, aggravated by the multiplication of dovetailed legal orders (Prieur 2001).

6.2.1.5 Cognitive Rationale and Normative Rationale in the Era of "Truth Pluralism" (Naim Gesbert 1996)

If it is possible to sprout a bundle of laws out of the driest factual facts, environmental law cannot content itself with an upsurge without the relevant scientific foundations. If the facts are dry, raw, *a priori* neutral, with the era of pluralism of truths that contributes in making them multiple and confusing, the rule of environmental law cannot deny the ecological interpretations of reality, i.e. this new vision which constantly rediscovers *Terrae Incognitae*[7] (Naim Gesbert and Gutwirth 1995). Globality, complexity, and processes are not abstract intellectual pipe dreams. Nature, too, possesses its laws. The pluralism of truths, a modern Fig. 6.1 of the science in question, requires that reason responds to its ontological questions on reality by referring to "possible worlds," but without providing copies of a world that has always existed; science is then science *stricto sensu* (in the strictest sense of the term) only if it is constituted of an apodictic certainty (Sandkuhler 1994). Knowledge of empirical certainty could be only one of many forms of knowledge. Thus, does ecological reality penetrate with difficulty in a global and reliable model because, according to these authors, chaos turns out to be the universal rule and derogatory order to the real, an exception? Henceforth, we must master the collision between natural history and human history (Chesnaux 1992). According to Edgar Morin (1980), this collision simultaneously poses the problem of thought as (complex) thought, the problem of nature as nature, and the problem of humanity as humanity.

In the context where endemic uncertainty reigns, the paradigm of a multiple, plural and equivocal truth emerges, whose conceptual implications in law must

[7]A *terra incognitae* (from Latin meaning "unknown land") is a territory that has not yet been explored by Man, or by European explorers, travelers and merchants. This expression is linked to discovery and to large areas and is therefore frequently used in the fields of knowledge and research.

be taken into account. Science, today anchored in society since it is no longer omniscient, is partially powerless to explain the cosmic order and to impose truth. Consequently, it offers a latitude for law to develop which is then conceived as a space for reflection and choice, a potential for law to encompass the man-nature relationship in terms of responsibility and precaution (Naim Gesbert 1996).

Environmental law thus orchestrates man's relationship with "his environment" in the form of rules, norms, orientations, planning that are more or less coercive in nature. However, since the 1992 Rio de Janeiro Earth Summit, there has been a more social or anthropological approach to environmental law that recognizes non-Western cultures, "traditional ways of life," and even the rights of native peoples.

Article 8J of the International Convention on Biological Diversity (CBD in 1992)[8] recognizes non-Western lifestyles (native communities and local communities). The Convention's guidelines for the flagship concept of sustainable use of biological diversity (Addis Ababa, in 2004) recognizes "local customs and traditions" and the strengthening of local rights (practical principle 1 & 2). The seventh meeting of the CBD Working Group on Article 8J on customary use (Montreal, October 31–November 4, 2011) promoted a new action plan on sustainable customary use, whose objective is to protect, respect and safeguard traditional knowledge, innovations and practices of native peoples and of local communities when it comes to sustainable use and biodiversity conservation.[9] Other conventions have also preconfigured the recognition of local populations such as the 1973 Polar Bear Convention,[10] the International Labor Organization Convention No. 169 (1989), the Convention for the Safeguarding of the Intangible Cultural Heritage (Unesco, Paris 17 October 2003),[11] the Convention on the Protection

[8]"Subject to national legislation, respect, preserve and maintain knowledge, innovations and practices of indigenous and local communities embodying traditional lifestyles relevant for the conservation and sustainable use of biological diversity and promote their wider application with the approval and involvement of the holders of such knowledge, innovations and practices and encourage the equitable sharing of the benefits arising from the utilization of such knowledge innovations and practices. (art. 8j, CBD in 1992).

[9]cf. Report of the 11th conference of CDB parties, Hyderabad, 2012, p.173 & s.

[10]Agreement on conservation of polar bears, Oslo, 15.11.1973 (between: Canada, Denmark/Greenland, Norway, the Soviet Union/Russia, and the United States): art.3: "Subject to the provisions of Articles II and IV any Contracting Party may allow the taking of polar bears when such taking is carried out: (...) d) by local people using traditional methods in the exercise of their traditional rights and in accordance with the laws of that Party; (...)".

[11]"**Recognizing** that communities, in particular indigenous communities, groups and, in some cases, individuals, play an important role in the production, safeguarding, maintenance and re-creation of the intangible cultural heritage, thus helping to enrich cultural diversity and human creativity, (...)"," we mean by "intangible cultural heritage" the practices, representations, expressions, knowledge, skills – as well as the instruments, objects, artefacts and cultural spaces associated therewith – that communities, groups and, in some cases, individuals recognize as part of their cultural heritage. This intangible cultural heritage, transmitted from generation to generation, is constantly recreated by communities and groups in response to their environment, their interaction with nature and their history, and provides them with a sense of identity and continuity, thus promoting respect for cultural diversity and human creativity (art.1).

and Promotion of the Diversity of Cultural Expressions (Unesco, Paris 20 October 2005),[12] and finally the United Nations Declaration on the Rights of Indigenous Peoples adopted on 12 September 2007,[13] which fully recognizes the rights of the people in their territory. Genetic resources and local knowledge are subject to obligations to enable access and equitable sharing of benefits and consent of the concerned populations (Nagoya Protocol 29 October 2010).

The real challenge consists of implementing these conventions (see Bellier 2013). For example, in the case of the Amazonian Park of Guyana, people's lifestyles are well recognized in the Park Charter,[14] which defines collective rights[15] and provides numerous exemptions to the multiple regulatory bans.[16] However, even if the leitmotif of taking into account "local realities" and "co-construction" is repeatedly stated in the Park's charter,[17] the objective remains a top-down sustainable development (of state origin), ethnocentric (Western-centered), based on the settling of populations of nomadic origins (Barrière and Faure 2012).

Thus, on the scientific level – as well as philosophical, ethical, legal and economic levels – the instrumentalist design of nature is deeply rooted in Western culture. Pairs of dualist concepts, such as spirit/matter, nature/culture, subject of law/object of law, and rationality/irrationality are vectors. Property law, "laws of the market" and the epistemological principle of control and possession, *inter alia*, that were established in several regions (Western countries, ex-colonies) on the scientific, legal, cultural and sociological levels since the advent of capitalism, precede the ecological awareness that began in the 1960s. Such observation is

[12]For purposes of the Convention, "Cultural diversity" refers to the manifold ways in which the cultures of groups and societies find expression. These expressions are passed on within and among groups and societies. Cultural diversity is made manifest not only through the varied ways in which the cultural heritage of humanity is expressed, augmented and transmitted through the variety of cultural expressions, but also through diverse modes of artistic creation, production, dissemination, distribution and enjoyment, whatever the means and technologies used (art.4). The objectives of the Convention are in particular"(a) to protect and promote the diversity of cultural expressions; (. . .) (d) to foster interculturality in order to develop cultural interaction in the spirit of building bridges among peoples; (e) to promote respect for the diversity of cultural expressions and raise awareness of its value at the local, national and international levels; (. . .) (art 1).

[13]Essential text responding to natives' demands . . . on paper: "Respecting native knowledge, cultures and traditional practices contributes to a sustainable and equitable development of the environment and its proper management" (considering that) "natives, people and individuals, have the right not to be subjected to forced assimilation or destruction of their culture" (art. 8)," natives have the right to maintain and strengthen their special spiritual ties with lands, territories, waters and coastal areas and other resources they traditionally own or occupy and use, and to assume their responsibilities of the matter toward future generations" (art. 25), "to recognize the rights of native peoples in what concerns their lands, territories and resources" (art. 26 & 27).

[14]Decree of creation of the Park n ° 2007–266 of 27 February 2007.

[15]Article R170–56 of State Domain Code recognizes the rights of collective use to the populations. These rights of use relate to so-called "traditional" subsistence activities, including hunting and fishing (Decree No. 87–267, amended by Decree No. 92–46 of 16 January 1992).

[16]art. L331–15–3 of the French Environment Code.

[17]Charter of 30 October 2013, approved by Decree No 2013–968 of 28 October 2013.

clearly evident in the weaknesses of environmental law which was included in the game only after the cards were distributed (Gutwirth 2001).

Environmental law has, until recently, privileged nature itself at the expense of it links with man and society. It is therefore useful to examine if environmental law enables the paradigm of coviability (see Aline Treillard and Jessica Malkoviak, *infra* in this book) which, by definition, is not limited to a remarkable nature that environmental law has seemed to favor until recently.[18]

The analysis of the scientific origin of environmental law has led us to a concrete example in the Amazon that tackles the question of the anthropological dimension of the law in the "legal staging of nature".

6.2.2 The Anthropological Dimension of Law: Legal Pluralism and the Definition of Law Through a Juridicity

The anthropological dimension of law offers a perspective transcending fiction for a legal empiricism that includes the law as it is, not as it should be. We cross thus the threshold from the provision to the description. The essence of legal empiricism is concerned with that which exists, that which "gives" experiences. The law, therefore, is not defined solely by what is presented by an institution, a deliberate law influenced more by idealism than realism: "practiced law maintains a shady, often antinomic, almost always dialectical relationship with legal dogmatism" (Perrin 1997, 100). In a process connected with the effectiveness of law's existence in all its assertions, the diversity of the dimensions of the juridical arena translated into a juridicity is taken into account. This juridicity falls within the written, oral, gestural spheres, of attitudes, practices, representations, and judgments. Realism opposes any metaphysical conception of law by the study of existing, actual, effective and not fantasized law.[19] Therefore, it develops in terms of legal pluralism whose distinctiveness consists of rejecting dogmatic thinking of State monopoly in the production of law.

The norm, whether compulsory or not, may emanate either from a social ("bottom-up") or an institutional ("top-down") source[20] through either repeated practices followed by everyone or through models derived from paradigmatic

[18]Cf. ongoing thesis by Aline Treillard: "La construction d'un régime juridique de la nature ordinaire" (viva in 2019), University of Limoges/CRIDEAU under the direction of Professor Jessica Malkoviak.

[19]On legal empiricism, cf. Perrin 1997; Millard 2002; Serverin 2002; Ross 2004.

[20]Normative creation does not depend solely on an institutional procedure: "the group possesses the capacity of creating norms that come to legal life as a spontaneous social phenomenon" (Perrin 1997, 44). For Perrin (and us), it is therefore unwise to call these spontaneous legal phenomena "infra-laws".

identities.[21] Two structuring elements stand alongside each other: understandable norms that define by themselves the law for legal positivism, and habitus that are difficult to grasp, unrecognized and unacceptable for legal orthodoxy. The norms, called customary (we prefer "endogenous", see below), reflect both an identity matrix and consensual "ways of doing things," all deriving from the social body and differ according to residential groups or lineages. This juridical arena of social origin also comprises a non-normative part which fully participates in regulation, constituted of all these individual provisions for practicing and for behavior, laid down in models, in legal habitus.

The leap from positive law to social law (Gurvitch 1931, 1935, 1940)[22] characterizes legal anthropology with a search for a homeostasis, or even a legal resilience: adapt or change, without renouncing identity. Legislative normativity attempts to meet the objectives of sustainable development (see the National Constitutions) in order to ensure the survival of society: its adaptation and anticipatory behavior to global changes, including environmental and climatic changes. However, does the legislator alone have the necessary capacity and means for this? Should we not go beyond the assumption of legislative rationality to add that of the (anthropological and scientific) empirical rationality of socio-ecosystems?

The contribution of legal anthropology opens three windows in social complexity: (Sect. 6.2.2.1) understanding the plurality of orders and legal systems, (Sect. 6.2.2.2) understanding legal endogeneity, (Sect. 6.2.2.3) defining "post-modern" legal regulation, the law going beyond the legislative framework.

6.2.2.1 Superposing a Plurality of Orders and Legal Systems: Legal Pluralism

Legal pluralism is based on the idea that legal mechanisms of different orders or systems may be applied to the same situation (Vanderlinden 1993; 583). This phenomenon is widely recognized and commented upon by a doctrine of jurists, sociologists and anthropologists. The (already cited) authors of this doctrine reject the very idea of a universal law system and of an exclusive state order. Forming norms and legal frameworks "from within" social bodies continues despite the colonial attempts to "purge" customs.[23] We may cite here the so-called procedure

[21] As an identity matrix of the group. The paradigm is a type of representation of the world, a worldview model which rests a base defined as a cultural production or an intellectual current.

[22] For the author, "(. . .) the fate of the idea of social law is linked (. . .) to the struggle against "fetishism of the law", a vestige of legal individualism glorifying the authoritarian will of the State that dominates all the others. However, since the beginning of the twentieth century, a whole doctrinal wave . . . has increasingly dispossessed state law, not only of its supposed monopoly and its supremacy, but also of its creative force . . . " (Gurvitch 1931, 214).

[23] "Legal sociologists understand the diversity of legal orders, ranging from positive to infra-law. Common rights fail to be stifled by state rules. The customary laws of colonized countries have not disappeared by imposing foreign standards (. . .)" (Ribordy 2010).

of "customary land laws" supposed to purge these laws in order to achieve a (Western) civil-law type land ownership.[24] We thus witness an artificial creation of land ownership by a procedure invented specifically to generate land ownership that did not exist before. It is certainly not the best way to preserve social order by ignoring the very nature of social groups. Since social life is based on a plurality of regulations, "the law must therefore value the complementarity of differences. Every law is relative: there is a plurality of legal sources – inter-legality (inter-normativity, Perrin 1997) – a crossroad of different legal orders" (Ribordy 2010).

Indeed, legal pluralism is a direct consequence of socio-cultural diversity. The available diversity of law systems offers the outline of legal pluralism. That which is plural does not refer to the multiplicity of legal sources, but to the scheduling of that which "makes a law" within a social unit. The normative formation of the system is carried out according to the group, with interpretations specific to the socio-political system that gives rise to it. Creation modalities of legal provisions generate legal orders, which define super-systems. The colonial state order opposes thus the so-called customary legal orders. However, the state system has a diversity of legal systems within it (Ost and Kerchove 2002), which States often tend not to recognize under the pretext of monopolizing the creation of law. In this case, the law is its set of rules and nothing else. Everything emanates from connections to power: "the tension between the monistic tendency of the dominant group and the pluralist societal and legal reality is more or less important depending on the case, which leads to the analysis of legal pluralism in various degrees. In contemporary Western societies, the tendency of the State to monopolize the law incites it to disseminate an ideology that denies the reality of legal pluralism..." (Rouland 1993, 450). Thus, most legal experts consider the State the sole source of law. This consubstantiality between the State and the law means that there are societies without law, which is unacceptable at least from sociological, anthropological and legal anthropology viewpoints (see Alland and Rials 2003, 1158). Whereas, the very purpose of legal anthropology is to be found in the study of non-state legal systems (see above).

This positivistic stance raises questions on the regulation of social behaviors that cannot be conceived without a legislator. Yet, socio-cultural representation defines a way of "seeing the world": to be, to think, to believe, to act, to react, to decide ... This mental structure consists of cognitive "frameworks". These terminologies are related to cognitive and social psychology which promotes the notion of "paradigm". The paradigm may be defined as a framework of thought[25] or a model of thought (Canguilhem 1978[26]) and perhaps more: a socio-cultural and/or institutional matrix. There are several ways of seeing the world, not one

[24]Decree No. 55–530 of 20 May 1955 on the reorganization of land and property in French West Africa and French Equatorial Africa, see Decree No. 56–704 of 10 July 1956 laying down the conditions for the application of the Decree of 20 May 1955.

[25]Thomas Kuhn, 1972, cited by Rumelhard Guy (2005).

[26]According to the author, "analogies or models are not presented by themselves. They are chosen, and the inspiration for the choice sensitizes the latent presence of paradigmatic values, of collective apperception frameworks, characteristics of a determined cultural space-time".

single way. Indeed, there is a plurality of worlds that societies generate by their social representations[27] (a set of cognitive elements) derived from grids of reading reality, that is, mental frameworks. We emphasize that mental frameworks constitute a veritable individual theory of reality. From the individual's viewpoint, "by shaping his perception of opportunities, this theory influences his types of behavior. In other words, individuals do not act according to a set of objective information, but according to the knowledge they have of it. This knowledge depends on available information interpreted through the prism of mental schema" (Wirtz 1999, 26). According to Vergnaud (2007), a framework organizes action, conduct, and more generally activity. It is "not a conduct, but a constituent of representation whose function is to engender activity and conduct in a situation." He thus defines "an invariant form of organizing activity and conduct for a given category of situations". In this way, the framework expresses "a cognitive entity which generates the activity of the subject: both behavioral activity (gestures, looks, verbalizations, and so on) and underlying cognitive activity (perception and information gathering, reasoning, adaptation, meta-cognitive control, and so on). The function of the framework is thus to generate the two fundamental registers of the activity: thought and the conduct it stimulates.

The relationship between a framework and legal regulation is a causal relationship. The conceptual elements that allow grasping all juridical aspects outside the law are legal pluralism (diversity of systems and legal orders), juridicity (the set relating to law) (a), and endogeneity ("that which occurs inside") (b). They link the framework to law by being based on the assumption that anything coming under the law constitutes social fact.[28]

The social foundation of law and its dependence on a social body limit its autonomy making it overcome a minimalist definition. Let us start with the social facts that derive from socio-cultural paradigms, the "ways of life" based on such paradigms. The latter generate a collective process giving common sense to the legal system (juridical order), which in its turn depends intrinsically on a (socio-ecological) environment. This dependence also limits the autonomy of the legal system (see note 3, Teubner 1984, 1993). In fact, normative validity cannot be restricted to formal criteria of hierarchy between norms because of its empirical dimension and its axiological legitimacy. Based on values, this normative validity lies in the origin of norm's formation itself: plural and circular; that is to say, it adheres not only to a top-down relationship, but also to an bottom-up one. The model of the American realists, for whom effectiveness validates the norm, i.e. "the

[27]"A social representation is an organization of socially constructed opinions, relative to a given object, resulting from a set of social communications, making it possible to control the environment and to appropriate it according to symbolic elements specific to its group memberships" (Roussiau et Bonardi, cited by Moliner et al. 2002, 13).

[28]Among others: Romano 1919; Arnaud 1981; Rouland 1988, 1991; Luhmann 1994; Le Roy 1999; Barrière and Barrière 2002; Sacco 2008.

bottom validates the top", strengthens these remarks.[29] Indeed, "(...) it is necessary to take into account the actor, his imaginary investments, his legitimacy judgments and his practical behaviors (...) the autonomy of the law appears as a "dependent autonomy" (Ost 1991, 161).

It must however be emphasized that introducing the subjective into the legal objective (sanctioned command norms for Hans Kelsen 1960) is inconceivable in the positivist logic which refutes any legal axiology: the value of the law situates itself in the matching norm for a penalty. At the risk of bringing "impurity" into law, the latter still cannot disregard society for the simple reason that it emanates from it. Thus, axiology[30] does indeed exist in environmental law, if only through its founding principles, the value given to the notion of nature, biodiversity, landscape and so on. The very notion of the environment becomes a societal issue, in France, in the Constitutional Environment Charter (2005).

Legal pluralism confronts monist thought for which the law only exists through the State, rejecting any form of juridicity (any other expression of law through laws). Yet, the juridical landscape does not fit within this restriction. Taking into account the diversity of socio-cultural frameworks by the law falls within an institutional reality summarized in the following diagram. Between legality and local legitimacy, here, this is about opening a horizon to a holistic legal paradigm.

The interfacing levels between local and global contexts generate a relationship between the different paradigms that coexist between subsystems of State and endogenous orders. The territorialized social groups (G1, G2, G3) simultaneously constitute a social order and are part of a State order. The challenge resides more in the linkups to be made, the creation of bridges between orders and systems rather than in exclusion.

The anthropological dimension of law leads inexorably to penetrating societies and investigating them further from the inside than from the limited examinations of their surface.

6.2.2.2 Endogenous Law

Endogenous law defines itself through the ways of doing things, customs, and the ways of being, the habitus. Customary norms and legal habitus constitute an endogenous law, a law originating from the social body and constituting a system.

Customs are understood to be a set of oral rules which gain a binding force through constant and repeated use. More specifically, customs are a "set of legal

[29]Realism is a wave of the general theory of law, which is defined as "an attitude or approach, wishing to describe the law as it "really is (. . .). It claims to design law as an empirical object and not as a set of ideal entities with a mandatory value" (Troper 2007.) The realistic term is "taken in an extremely banal sense: to shift discourse on the law to discourse on what exists, on what is real. This obliges us to define empirically (and precisely) the facts of reality that are designated as law» (Millard 2014).

[30]Science or theory of intrinsic values.

practices that have gained a binding force in a given socio-political group through the repetition of peaceful public acts over a relatively long period of time" (Gilissen 1982, 20).[31] Thus, the notion of custom reflects both the essence of law and the essence of culture which render it universal (Andrieu 2001, 68).

The question that arises in social immersion concerns the functioning of a group, the behavior of individuals, and particularly the motor of regulation. In the end, the question is about social viability. Therefore, the definition of the law questions that which "makes the law," that which shapes a behavior. The question exposes a challenge of deciphering "that which is not seen." If understanding the legal system becomes accessible with legal documents, interpretation becomes particularly opaque, difficult without physical support or beyond the written word. Even possessing written works does not make it possible to avoid looking further on. The pedestal of orality includes observing the facts, behavior, the analysis of discourse, the words and the representations which support them. Even in societies of very oral origin, which nowadays increasingly integrate written works, any transcription of orality can only distort the endogenous juridical reality.[32] Certainly, if the rules can be transcripted from the practices, the next task is to understand how these practices are conditioned by the social mold.

As a result, beyond legal and customary juridical norms, we take into account the existence of legal habitus.[33] Here, we shall attempt to clarify this notion. The individual has a "psychic envelope"[34] that will condition his thought activity, his representations and his affects through a cognitive entity that will generate his behavioral activity (ways of being: gestures, views, verbalizations, and so on), underpinned by cognitive activity (perception, reasoning, adaptation, learning, and so on). This is how the cognitive process (intervening in all conduct as information gathering) allows a regulation of action (Paquay et al. 2001). On the individual's personal level, this regulation is translated into frameworks as a cognitive entity which generates the activity of the subject whose function is to generate the two fundamental registers of this activity: thought and the conduct it creates (Vergnaud

[31] In France, the notion of custom has existed since the early Middle Ages. It denotes "an unwritten law, introduced by the practices and acts of continual repetition of men or practitioners, which have been used publicly without opposition from the majority of the people, the time necessary to prescribe it" (definition dated XVI by Philippe Wielant, quoted by Gilissen 1982, 21).

[32] Cf. Example of customs (Committee for Historical and Scientific Studies of French West Africa 1939).

[33] See the works of Etienne Le Roy, who presents the law as a tripod comprising legal habitus, alongside rules and customs, 1999, p. 189 & s.

[34] The psychic envelope is a "container for the activity of thought and affects" (Anzieu 1987 in Ciprut 2007, 35). We are supported by the psychic, familial, cultural, social and environmental envelopes: "If a human being comes into the world after being carried for nine months in the maternal uterine envelope, the human being continues to be carried by different envelopes, internal and external, throughout his/her existence" (Liechti M (2007), 35, "Le cadre culturel et les enveloppes psychiques").

2007).[35] Frameworks of perception and action, which enable the individual to produce a set of new practices adapted to the social world in which he finds himself, define provisions that Pierre Bourdieu (1980, 1986) conceptualized in the notion of "habitus".[36]

The habitus is determined by a system of practical social provisions, a matrix of representations, judgments, and affects shared by the actors who belong to the same cultural framework.[37] Some of these provisions are of legal nature as they deal with provisions relating to social reproduction, to a group's vital aspects found in social facts and involving its survival and its vital perspectives of the future (biological, ecological, economic, cultural) such as a balance in the group's internal relations (its harmony), identity transmission, relationship to the environment, to resources, and so on.

The reading of the legal regulation resulting resulting from the group itself, through customs and habitus, brings to light an "endo-regulation." Constructed and shaped by territorial actors,[38] endo-regulation constitutes a real living law[39] which establishes a legal output "from the bottom."

Thus, legal endogeneity suggests that the law takes its meaning from the social arenas that it aims to regulate (Edelman 2002, 2007). Accordingly, the author develops a theory of legal endogeneity articulating a process by which the daily practices of an organization, routines and structures subtly influence legal thinking, legal categories, and legal logic.[40] For Gérard Timsit, legal endogeneity refers to one of the most striking phenomena in the recent transformation of normativity all over the world.[41] In international law, this endogeneity is shown through recognizing the

[35]For the author, "the framework is an invariant form of organization of activity and conduct for a class of predetermined situations". The function of the framework is "to generate activity and conduct in situations".

[36]The habitus defines dispositions which are regular behaviors founded objectively from the logic of practice (Bourdieu 1986: 40); "... habitus produces practices (. . .) it ensures the active presence of past experiences which, deposited in each organism in the form of a perception framework, thought and action, tend to ensure conformity of practices and their consistency over time "(1980, 91) more surely than all formal rules and all explicit norms.

[37]"Habitus is a system of dispositions and of competences specific to a particular field of cultural production, the matrix of perceptions, judgments and actions shared by all actors who belong to this field and deployed by them in their struggle to monopolize resources or forms of "capital"" (Howes 1996, 28).

[38]"(. . .) African communities have generated, through their history, their own land systems whose coherence and logic can only be understood by looking at them for what they are and not for what they lack" (Ouédraogo 2011, 79).

[39]Cf. Michaïlidis-Nouaros 1982.

[40]Edelman 2002.

[41]The author gives the following definition, one which is closely related to our approach: "the appearance of an auto or endo-normativity. Instead of being organized around the pole and the notion of public power as it was the case until now, this normativity is now organized around the opposite pole: that of civil society and the associations that are supposed to represent it or to translate its intentions, worries and concerns. It is supposed to be a normativity based on development and respect shown by the very actors who formulated rules (in the form, for

rights of indigenous peoples (Convention of the International Labor Organization No. 169, 1989, the United Nations Declaration on the Rights of Indigenous Peoples adopted on 12 September 2007,[42] which fully recognizes the rights of populations on their territory) and by national legislators in taking into account the cultural specificities of local populations (but does this include recognizing endogenous legal systems?). For example, the Burkina Faso legislator explicitly recognizes local property conventions as "inspired by customs, local real estate usage and practice, developed at the local level and seeking to… take into account the diversity of ecological, economic, social and cultural contexts in a rural environment."[43] In Benin, the law arising from recognized local practices and costumes may serve as a guarantee for loan granting.[44] The Malian Land Code of 22 March 2000 recognizes, by "confirmation", the "customary laws exercised collectively or individually on unregistered lands" (art.43). However, taking people's rights into account through the notion of customary rights is not synonymous with legal pluralism: this is more a question of integrating endogenous regulation elements into positive law.

France recognizes that customs in New Caledonia are not subject to the civil code allowing the emergence of a true status of "customary law". This is an example of legal pluralism. In Australia, the judgment of Mabo v. Queensland No.2 1992 (Cth) reversed the doctrine of "terra nullius" ("no-one's land"): the judgment recognizes traditional property rights of the Meriam populations to their eastern Torres Strait islands; the judgment also states that native title existed for all natives in Australia prior to Cook's Instructions and to the establishment of the British Colony of New South Wales in 1788. This decision altered the basis of land tenure in Australia.[45] Following the Mabo judgment, the Australian Commonwealth Parliament adopted in 1993 the Native Title Act which recognizes and protects native land rights,

example, of codes of conduct or good practices) and which they themselves apply. We find here the decentralization, the non-hierarchization that regulators demand, making it possible to base and anchor the legitimacy of the rules of operation through various techniques. These techniques aim in particular at guaranteeing (but in a different way than in traditional normativity) transparency, predictability, impartiality of the actors' action - which enables them to obtain – that on which legitimacy is analyzed – a recognition by their peers (other regulators) and by actors of the system" (Timsit 2005, 85).

[42]Essential text responding to natives' demands … on paper: "Respecting native knowledge, cultures and traditional practices contributes to a sustainable and equitable development of the environment and its proper management" (considering that) "natives, people and individuals, have the right not to be subjected to forced assimilation or destruction of their culture" (art. 8)," natives have the right to maintain and strengthen their distinctive spiritual relationship with their traditionally owned or otherwise occupied and used lands, territories, waters and coastal seas and other resources and to uphold their responsibilities to future generations in this regard" (art. 25), "to recognize the rights of indigenous peoples concerning their lands, territories and resources" (art. 26 & 27).

[43]Art. 6 defining local land charters of Act No. 034-2009/AN of 16 June 2009 on rural land tenure (Burkina Faso).

[44]Art.9 Act n°2007-03 of 16 October 2007 on rural land tenure.

[45]cf. Merle 1998; Lafargue 1994, 1996, 1999.

including the creation of a "right to bargain" (art. 43). We may have reached here an interval: between the absorption by the predominant system and pluralism and the recognition of a legal system other than the State's (coexistence between a Western system and an endogenous system).

Recognizing a legal order that does not adhere to positivism (state law) allocates, through native laws and property laws of the original inhabitants, a legal endogeneity. The law is not monotone so it is defined in terms of "juridicity".

6.2.2.3 A Holistic Definition of the Law, Juridicity

The previous analyses results in a law system being conceived as a model of behavior. The complexity of juridicity leads us to two slopes of law ontology: the rational part with legal norms, and the empirical part with customary norms and provisions of practice (habitus). This initiatory course, carried out in the den of legal pluralism, stands on three foundations: (a) a realism which encourages going further than the fiction of positive law by going in the right direction, by starting from society because "if society re-makes the law every day, the law has never reshaped society,"[46] (b) the complexity of the law, through adopting a holistic reading of a law and avoiding a narrow reading that reduces law to a State norm; (c) the legal paradigm being transformed into postmodern law, even into a post-Western law (transcending the paradigmatic framework of the West). The legal paradigm, of syncretic nature, includes new attributes: material, as a mode of expression (pluralist, relativistic, pragmatic), and organic, as a mode of creation (concerted, negotiated, proximity). It promotes thus a law of "regulation" (Chevallier 1998), constituting "juridicity" (cf. Le Roy 1999). Analyses of real situations allow us to move in this direction. The latter are presented in the following section, synthesized by a matrix.

This diagram (Fig. 6.2) draws a behavior model which includes three law sources, giving rise to dovetailed systems: (a) that of norms, legal or endogenous, more or less interconnected; (b) that of habitus, derived from the social body in its framework practice of socio-cultural origins (filiation and identity); (c) that of an in-between system, syncretic, an endogenous, social order and a order for which legislation has been passed. The former occupies a more or less integrated place in the State order. The in-between system defines a law of regulation, the product of a collusion from that which is rational to empiricism.

The complexity of the presented system of legal regulation deviates from the exclusivity of a legal output logic which is limited to the state. The challenge of legal pluralism is expressed in the relationships between orders and legal systems, in the bridges to be forged between them. This is about coexistence by transcending "the observation of pluralism to consider its construction" (Chassot 2014, 147). The example of New Caledonia clarifies these remarks by endowing the island with a customary code, with a margin of legislative autonomy (Gindre 2008; Lafargue

[46]Jean Cruet, 1908, cited by Arnaud André-Jean 1981, 137.

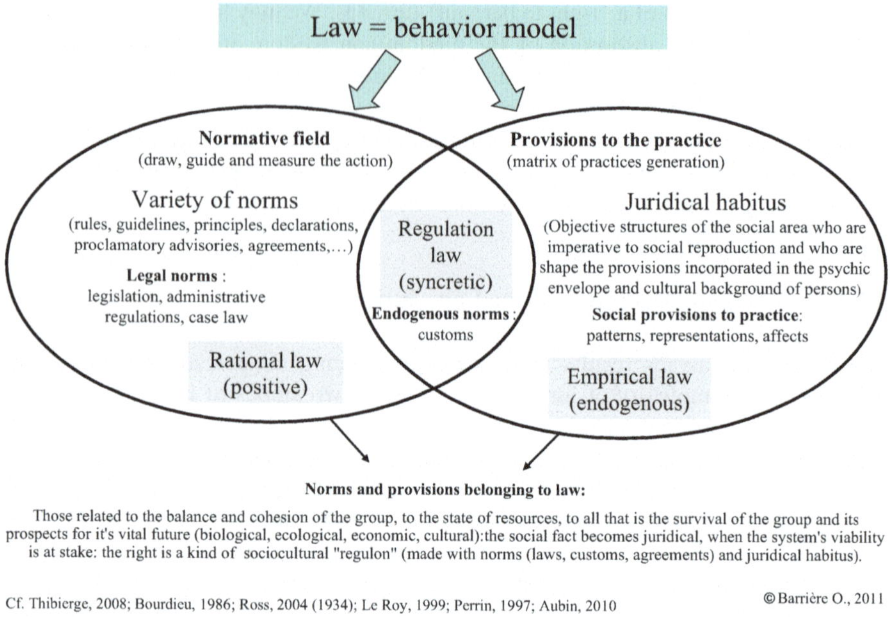

Fig. 6.2 The legal system in its complexity, as behavior model

2003). Finally, isn't the objective to preserve cultural diversity by allowing everyone to remain what they are? A truly pluralistic society "must seek a *modus vivendi* where the State preserves social order while providing citizens with access to their beliefs and their atavistic way of life" (Sheleff Leon 2000, 6).

In his diversity, Man is involved in the viability of the socio-ecological system through a social bond that binds him to sociability and through the ecological bond that sustains him in the biosphere. These two bonds unite in order to achieve the viability of social groups within their environment. If then, social sustainability depends on that which binds the human to the non-human, what makes it possible to achieve this socio-ecological viability? Revisiting the very notion of "environment" leads us to another dimension, that of the biosphere, the part of the planet where life has developed. Ariadne's thread indeed lies within.

6.3 From an Anthropocentric Paradigm, the "Environment", to the Paradigm of Socio-ecological Viability, the "Biosphere"

The extent of man's impact on the planet is such that humanity has become a force of telluric amplitude. A new human period began since the industrial revolution in the 1800s (Crutzen et Stoermer 2000; Crutzen 2002; Steffen et al. 2007) or even earlier, around 1600 (according to Lewis and Maslin 2015). Entering the Anthropocene, deeply linked to the capitalist model, characterizes a veritable "geological revolution" of anthropic origins. This global event goes far beyond an "environmental crisis" (thinking of the Anthropocene, "means abandoning the hope of an exit from the crisis"): we are facing an exploitation of the Earth system by a species that has seized it to respond to its unique growth needs (Bonneuil and Fressoz 2013). Changes in the biosphere are major. With climate change and the damage in biodiversity, human influence is affecting the bio-geochemical cycles of water, nitrogen, phosphate and carbon. Human population growth, from 900 million people in 1800 to more than 7.4 billion to date (2016),[47] results in an annual consumption of 1.5 times what the planet can provide[48]: "in just eight months humanity consumes the entire annual ecological budget of the planet".[49] The anthropization of the planet affects 83% of the unfrozen surface area of the globe (Ellis 2011), and the average atmospheric carbon dioxide concentration has exceeded 400 parts-per million (ppm) throughout the Northern Hemisphere in April 2014.[50] In the Southern Hemisphere, concentrations reach 393–396 ppm due to lower population density and economic activity. On the eve of the Industrial Revolution, the average worldwide level of the atmospheric carbon dioxide concentration was 278 ppm. The anthropization of the planet also exposes ecosystems to dispersed molecules of synthetic organic chemistry, hydrocarbons, plastics, endocrine disrupters, organochlorine pesticides and now neonicotinoids, radionuclides, and more.

Man's position in the nonhuman becomes therefore ambiguous: an isolated or a contributing entity? (Sect. 6.3.1). The answer lies in the ontological analysis of society-nature division (Sect. 6.3.2).

6.3.1 The Earth, "Environment" or "Habitat" of Humanity?

The Earth forms a system called Gaia, Greek goddess of the Earth in the Hesiod cosmology, a mythical Fig. 6.3 theorized by science (Lovelock 1979/2016, 2010; Latour 2015). The Earth system is that of the living in which man finds himself. We are still speaking of the environment; that which surrounds man: the object of resources. However, it also happens to be an object of development and shaping in the image of a humanity that superimposes itself in a sphere of life where it is not the only living being. The harm (degradations, nuisances) that humanity causes to the other nonhuman entities are "externalities" to it. That which is not human is external to it. This nonhuman, called environment, remains separate from the human. By generating the era of the Anthropocene, can humans still detach themselves from the Earth system and the biosphere that man depends on so much to remain viable,

---

[47] http://www.worldometers.info.

[48] Living Planet Report 2014, WWF; Bonneuil and Fressoz 2013.

[49] See: http://www.footprintnetwork.org/en/index.php/GFN/page/earth_overshoot_day/.

[50] According to the World Meteorological Organization, press release 991, 26 May 2014.

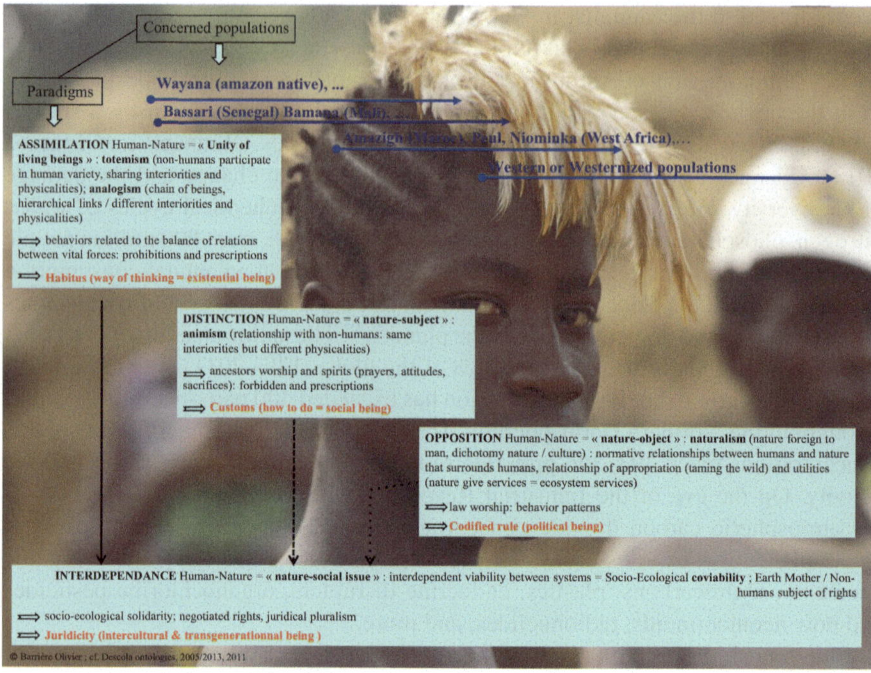

Fig. 6.3 Socio-ecological paradigms, sources of legal paradigms

i.e. to exist and to last? The Anthropocene is "an event, a point of no return" towards the Holocene (Bonneuil and Fressoz 2013: 36).

By entering the Anthropocene more than 200 years ago, humanity discards its relationship with that which surrounds it for a relationship with what he is, i.e. part of the biosphere. We observe here the unfolding issue, that of man's reconnection to the earth system, which is the biosphere, a living space endowed with dynamic processes. Defined by Vernadsky Vladimir Ivanovich in 1926 (Soviet Union), the notion of biosphere is understood from bio-geological and ecological perspectives, based on the assumption that life is a geological force that transforms the Earth (Vernadsky 1929). Vernadsky was the first to scientifically consider the impact of human activity on climate, anticipating the idea of Anthropocene (Bertrand 2014), but at the time of his study it was believed that nature had inexhaustible regeneration capacities.

International law and national laws very often obey the West. The financial capitalist logic imposes itself all the more by globalization in the legal order of Nations. The very notion of "development" translates this state. It is based on two imposed truths: an endless growth, supposedly "sustainable" ("durable" in French), and a Western universal model, one that eradicates any difference and therefore all diversity.

With its development, the Anthropocene incites us to behold a world excluded from infinity, a planet whose resources are limited and an anthropization harming an already shortening future. The recent Paris Agreement, based on the COP21 on climate, corroborates these remarks by acknowledging that "climate change represents an immediate and potentially irreversible threat to human societies and to the planet (...)" and considering that "climate change is a matter of concern for all mankind".[51]

The model of sustainable development is based on a global economy that imposes its own patterns of thought with a form of so-called universal regulation. Even if the international community explicitly recognizes cultural diversity (*op.cit. supra*), intangible cultural heritage,[52] non-Western lifestyles (Article 8J of the International Convention on Biological Diversity of 1992, *op.cit. supra*), the referential model remains Western. Modernity is defined in the consumerist society (based on the values of possession) which often opposes non-capitalist societies, called "traditional" or "customary" societies (based on values of being and connections). Since their legal systems are not State-specific, they are rarely identified in legal scope. The fact that the legal orthodoxy of the West rejects the endogenous legal order, specific to social groups, it underlines a rejection of legal pluralism.

Yet, being aware of legal diversity is necessary to foresee possibilities of reconnecting man to the Earth system: if the Western world has immersed the planet into the Anthropocene era, other human communities have remained connected to the biosphere through their own connection to nature. The socio-cognitive diversity of relationships to the nonhuman generates legal diversity of societies-nature ontologies (Sect. 6.2). Indeed, the relationship of two-thirds of humans with the land follows patterns of living that are different from the Western model of land ownership system, often considered unique and which colonial history has not succeeded in generalizing.[53]

6.3.2 A Diversity of Societies/Nature Ontology: From that Which "Surrounds" Humans to Man's Position Within the "Biosphere"

The foundation of the social system, or the social construct, emanates largely from the man/nature relationship.[54] These relationships are shaped collectively

---

[51] Adoption of an agreement having legal power developed under the convention and applicable to all parties on 12 December 2015, Conference of the Parties to the Framework Convention on Climate Change (1992), 21st session, Paris, 30 November-11 December.

[52] Convention on the Protection and Promotion of the Diversity of Cultural Expressions (Unesco, Paris, 20 October 2005); Convention for the Safeguarding of the Intangible Cultural Heritage (Unesco, Paris, 17 October 2003).

[53] "Two-thirds of humanity is not yet fully determined by the rules of a globalized market and continue, more or less systematically, to prohibit or ignore the sale of land" (Le Roy 2011, 348).

[54] "The principles of the construction of social reality must be sought first in the relationships between humans and their natural environment" (Descola, 1992), cited by Charbonnier 2015, 286.

against the external world of societies by two essential dimensions, subsistence and knowledge (Charbonnier[55] 2015: 281). By traveling, transforming and ordering the world, by territorializing it and thus socializing it, nature has become "a social fact of the highest importance" (ibid). To say that nature derives from culture produces a paradox. Two perspectives justify it. The first is elaborated in the mental construction of the human/nature relation, in which nature becomes a social element and *vice versa*, that is, the social becomes accessible to nature. The second perspective concerns the social which is no longer defined solely by human interactions but by hybrid interactions (ibid). Ever since mankind left the Holocene geological era, its impact on the planet has deeply anthropized the Earth, to the extent of making it an anthroposystem.[56]

Within the biosphere, a diversity of ontologies on the human/nonhuman relationships depicts an humanity far from being homogenous. In the Western and Westernized worlds, the dichotomy is still deeply rooted in cognitive representations, but the ecological problematic changes things. Nature as an "object" regains a "subjective" status. It becomes a subjective actor of humanity, which conditions its future, and a victim of often irreversible degradation. Indeed, man is suffering the full effects of the world economic order[57] on the biosphere (climate change, with violent climatic events). The "end of the great sharing" (of the societies/nature dichotomy) becomes inevitable (Charbonnier 2015).

Our fieldwork (explained in detail in the next chapter of this book) draw us closer to the ontological molds[58] developed by Philippe Descola (1996, 2005/2013, 2006, 2011, 2012). His conceptualization of continuities and discontinuities between humans and nonhumans offers us sets of physical and moral qualities shared by humans and nonhumans. The diversity of social groups, studied in relation to nature, leads us to identifying the degree of integration, proximity or distance between societies and the ecosystems within which they live. As for us, we were more concerned with deciphering the aspects of legal regulation.

The following figures present the societies-nature/law relationships, with a concern for objectivity and without favoring a "mode of identification" or a "distinctive pattern of world composition" (Descola 2005).

By looking at the existing modes of regulation within social groups, we proceeded by a transect of socio-cultural frameworks or paradigms underlying the

[55]In his thesis, the author takes up the works of Durkheim (The Elementary Forms of Religious Life 1960), Lévis-Strauss (La pensée Sauvage 1962), Descola (Beyond Nature and Culture 2005).

[56]Lévêque et al. 2003; See Christian Lévêque's contribution in this book.

[57]An economic order called into question recently by a Report, dated 9 August 2010, of the Secretary-General of the United Nations: "An overview of the major economic and general policy issues that must be resolved internationally to ensure supported economic growth and sustainable development, which should be equitable and inclusive, in addition to the role of the United Nations in this regard with a view to establishing a new international economic order ".

[58]"An ontology, for me, is simply the established result of a mode of identification" (Descola 2014, 237).

representations. These are the ontologies of Philippe Descola[59] These socio-cultural paradigms create legal paradigms (presented below). The identification methods, frameworks integrating practices (Descola 2014, 222), result from socialization: they "govern most of our skills, and thus proceed from a 'know-how', which is effective because it is unconscious, rather than proceeding from 'knowing that'" (Ibid, 221). There are four of them in total (totemism, analogism, animism, naturalism) (Descola 2005/2013), which we have positioned in the following table and in the following diagrams, plus another that we added, "coviabilism". This last, derived from the coviability of social and ecological systems, represents a fifth inclusive framework, which already exists or emerges in other framework. However, with globalization confronted with the effects of the Anthropocene era, source of ecological urgency, the ecological imperative imposes itself as a new paradigm in perceptions, representations, political acts, international and national legal norms, global agreements the last of which is the Paris agreement (Cop21 on the climate, November 2015), and so on. "Coviabilism" defines an ontology through nature/social interdependence that participates in the paradigm of living-being unity and which approaches totemism through the continuity of interiorities and physicalities. The resemblance (or continuity) stems from the socio-ecological solidarity relating the human to the nonhuman.

The ontologies of human/nonhuman relationships support socio-ecological paradigms (Assimilation, Distinction, Opposition, Interdependence) giving rise to sources of law (habitus, customs, codified norms, juridicity) that constitute three legal paradigms: endogenous law (Social), monist law (state), pluralistic law (negotiated). The last combines endogeneity and the State. Finally, four regulation modes emerge: behaviors rooted in the balance of relationships between vital forces, giving rise to prohibitions and provisions, the cult of ancestors and spirits, giving rise also to prohibitions and prescriptions, the cult of the law with models of behavior, and socio-ecological solidarity based on a negotiated state-related and pluralistic law (defined below) (Table 6.1).

Now that this discussion has been completed, constructing a situation to provide substance and express a certain empiricism of the analysis is necessary. Indeed, as we have already explained, this results from fieldwork in legal anthropology. In

[59]The ontological formulas developed by Descola proceed from a combination of interiority and physicality. "Interiority" should be understood as "the range of properties commonly associated with mind, soul or consciousness - intentionality, subjectivity, reflexivity, affects, ability to signify or to dream. We can also include the immaterial principles supposed to cause the effect, such as breath or vital energy, as well as more abstract notions such as the idea that I share with others the same essence, the same principle of action or the same origin" (Descola 2005, 168). "Physicality" concerns "the external form, the substance, the perceptual and sensorial-motor physiological process, or even the temperament or the way of acting in the world insofar as they manifest the influence exerted on conduct or habitus by body moods, diets, anatomical features or a particular mode of reproduction" (Descola 2005, 169). For the author, physicality is not the materiality of bodies, but "it is the set of visible and tangible expressions that dispositions belonging to an entity take whenever they are deemed to result from the morphological and physiological characteristics Intrinsic to this entity" (Descola 2005, 169).

Table 6.1 From the relationship of socio-ecological paradigms to regulation paradigms

Society-nature relations	Cosmos	Nature	Environment	Biosphere
Socio-ecological paradigms	Assimilation: Unity of living beings	Distinction: nature-subject	Opposition: nature-object	Interdependence: nature- social issues
Integrating framework (ontologies): Interiority and physicality	Totemism (interiority continuity and physicality), analogism (discontinuity interiority and physicality)	Animism (interiority continuity and physicality discontinuity)	Naturalism (interiority discontinuity and physicality continuity)	Coviabilism (interiority continuity and physicality continuity)
Regulation modes	Equilibrium of relations between vital forces: Prohibitions and prescriptions	Cult of ancestors and spirits: Prohibitions and prescriptions	Cult of law: Behavior models	Socio-ecological solidarity: Negotiated regulation, legal pluralism
Sources of law	Habitus	Customs	Codified norms	Juridicity
Legal paradigms	Endogenous law	Endogenous law	Monist law	Pluralistic law (endogenous and monistic)
Institutional frameworks	Existential being	Social being	Political being	Intercultural and transgenerational

Caption: continuity = resemblance/discontinuity = difference

addition, the "concerned" populations are examples of research carried out over 15 years (see the table further down).

6.3.2.1 A Variety of Socio-ecological Paradigms

This diagram shows combined socio-ecological paradigms associated with social contexts, and the legal paradigms resulting from them. It is impossible to assess the proportion of populations associated with these socio-ecological paradigms; as it is the case for populations that are very often undergoing major transformations, or even mutations (of course, by this we do not mean "evolution") under globalization. For the Bassari in Senegal, for example, "the unity of living beings" is gradually adhering to animism, and for those who flee the constraints of "customs" (migration outside the Bassari territory) or those who no longer believe in it adhere to naturalism. The previous chief of the Ethiolo village, Tchandenine Bendia (known as Tchan Tchan for those close to him) who died a few years ago to join his ancestors, admitted to me that he increasingly wondered about this, to the point of only refering to the tribe's ancestors (Barrière et al. 2005).

The paradigms situated outside the cult of the law concern a large part of the world population due to the fact that naturalism does not spread over the whole planet. The estimated number of native peoples[60] around the world is around 400 million,[61] but what about the number of "local communities"?[62]

6.3.2.2 Interdependence Paradigm

The (above) diagram places the paradigm of interdependence in relation to the other paradigms in various intensity. The different ontological molds join this logic of interdependence because no matter what place man has in the cosmos, *vis-à-vis* his environment or in relation to the biosphere, object or subject, nature becomes a social issue. For example, the French legislature integrates the natural social issue by adopting the concept of ecological solidarity through the law of 14 April 2006 on National Parks. The international community in the framework of Cop21 on Climate Change (the Paris Agreement of 12 December 2015) raises climate effects to the legal level, a major concern for the future human beings.

6.3.2.3 Negotiated Law

Negotiated law defines, within a regulatory arena, syncretic values and behavior patterns of "being" and "must-be." They are concerted between local, regional

[60]The Universal Declaration of the Rights of Indigenous Peoples (UN 2007) does not define the meaning of indigenous peoples. For the World Bank, indigenous peoples are the peoples who "are culturally distinct societies and communities. The land on which they live and the natural resources on which they depend are inextricably linked to their identities, cultures, livelihoods, physical, and spiritual well-being" (http://www.worldbank.org). A definition in 1982 is given by a working group in the United Nations (on Indigenous populations): "Indigenous populations are composed of the existing descendants of peoples who inhabited the present territory of a country wholly or partially at the time when persons of a different culture or ethnic origin arrived there from other parts of the world, overcame them, and by conquest, settlement or other means, reduced them to a non-dominant or colonial situation; who today live more in conformity with their particular social, economic and cultural customs and traditions than with the institutions of the country of which they now form a part, under a State structure which incorporates mainly the national, social and cultural characteristics of other segments of the population which are predominant. Although they have not suffered conquest or colonization, isolated or marginal groups existing in the country should also be regarded as covered by the notion of "indigenous populations" for the following reasons: they are descendants of groups which were in the territory at the time when other groups of different cultures or ethnic origins arrived there; precisely because of their isolation from other segments of the country's population they have preserved almost intact the customs and traditions of their ancestors which are similar to those characteristics as indigenous, and they are, even if only formally, placed under a State structure which incorporates national, social and cultural characteristics alien to theirs". (U.N. Economic and Social Council Commission on Human Rights, Preliminary Report of the Problem of Discrimination Against Indigenous Populations. U.N. Document E/CN.4/Sub.2/L.566, Chapter 11).

[61]Review conducted in 2011 and revised in 2013 by Patrick Kulesza (GITPA – www.gitpa.org) and by Irène Bellier (CNRS/SOGIP). Total estimated by the UN according to Bellier 2013, 21.

[62]Local community can be defined as a population, a group living on a geographic territory associated with a sense of belonging, sharing an environment (see o "Sense of Community, McMillan and Chavis 1986).

and national actors for a law situated in an intercultural perspective, one which is simultaneously legitimate in the eyes of the (local) community and the Nation (Barrière and Faure 2012). Negotiating the law consists of reversing the normative creative source: starting from the bottom to reach the top, a bottom-up relationship rather than a top-down one. In this case, the actors mobilize to develop their own territorial stewardship. This law, initiated on the basis of local values, associates for its development all the institutions intervening on the territory.

The diversity of ontologies, or integrating frameworks combined with the diversity of regulation modes (orders and legal systems) depicts the landscape of a plurality of worlds. However, beyond this heterogeneity of the social-ecological, planet Earth is unique. A single planet for at least four paradigms defining the nature/society relationships within cosmogonies or a capitalist logic: Cosmos, Nature, Environment, Biosphere. The cosmos represents the universe as an organized system, an ordered whole. It is an immense space (entity, being) whose every part is connected to the other; all its parts are subject to the same laws, and function in a similar manner (Barbault 1961, 20). Nature, in this context, refers to that which does not belong to humans (that which the paradigm of coviability calls into question). The environment defines what surrounds man; and the biosphere defines that which integrates humanity, the space of the living world. A final diagram summarizes this perception (Fig. 6.4).

The disappearance of the "great sharing" between Nature and Society has not yet been realized. Nevertheless, the Western world is revisiting its representations as it faces ecological urgency. If one chooses to somewhat depart from the Western shroud or to open it to other perspectives "beyond nature and culture", the grail of socio-ecological coviability may not be inaccessible. We demonstrate this in the following chapter of the book: "The paradigm of coviability defined by the adequacy

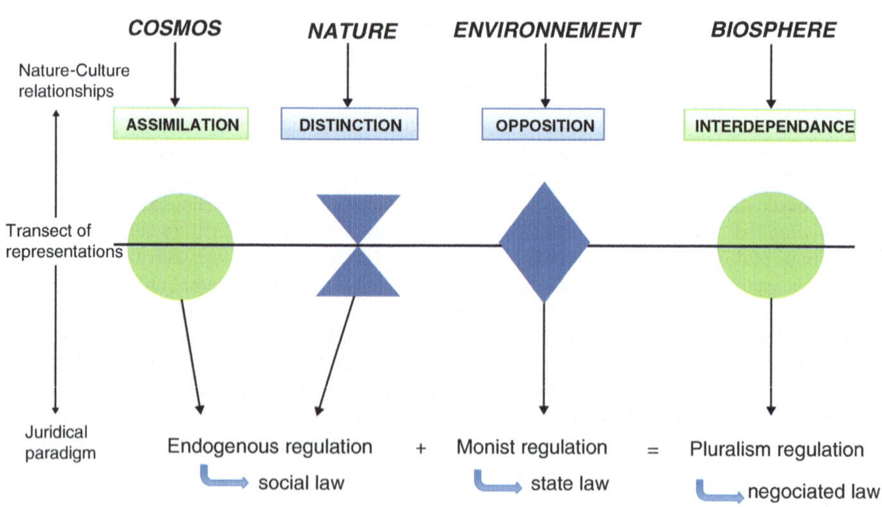

Fig. 6.4 From a transect of representations (Socio-cognitive frameworks) to legal paradigms

between social usefulness and the ecological function: the legal challenge of the socio-ecological connection" (O. Barrière and T. Libourel).

6.4 Conclusion

From environmental law to the law of coviability: the challenge of the socio-ecological connection

> Biological norms and social orders are intertwined; the rights of nature, the laws of nature, and human rights cannot differ from each other. They are active in the same world. (Ribordy 2010)

We live on the same planet, of course, but each lives in their own world ... Whatever the case, the challenge of the socio-ecological connection is that of a coviability law. Right from the start, our position is such that it takes into account the humans in their diversity. Legal pluralism and the holistic approach to law encompass the complexity of the socio-ecological connection because of the multitude of facts and the ways of being that bind the human to the nonhuman. Practices and representations form the foundation of the legal systems that truly reflect societies. As an essential element of any social group, the law (seen in terms of legal regulation) articulates this connection which creates solidarity in view of the interaction between social usefulness (translating human need) and the ecological function translating the state of the environment and therefore of the availability of resources. Furthermore, coviability is defined, from a legal anthropological standpoint, as the regulation that law brings to the relationship between the social group (social system) and the environment, considered by humans as a land-resource (Barrière and Barrière 2002). This regulation supports the paradigm of coviability itself by translating the socio-ecological connection, made of a whole social and ecological order.

Connecting the local to the global depends on a bridge between legal paradigms, that of the State and those of social groups. Co-management experiments (Brown 2010), more particularly of negotiated law, show that this bridge can be achieved through negotiation processes in the normative formation that permits rallying social legitimacy to national law (Barrière and Faure 2012). Revisiting the legal relationship to the biosphere requires this relationship.

References

Alland D et Rials S (2003) Dictionnaire de la culture juridique, Presses Universitaires de France, Paris

Alliot M (2003) Le droit et le service public au miroir de l'anthropologie. Karthala, Paris

Arnaud A-J (1981) Critique de la raison juridique, où va la sociologie du droit? LGDJ, Paris

Arnaud A-J (1998) Le droit trahi par la sociologie. LGDJ, Paris

Arnaud A-J, Dulce JF (1998) Introduction à l'analyse sociologique des systèmes juridiques. Bruylant, Bruxelles
Assier-Andrieu L (1996) Le droit dans les sociétés humaines. Nathan, Paris
Assier-Andrieu L (2001) Penser le temps culturel du droit. Le destin anthropologique du concept de coutume. L'Homme 160:67–90
Aubin J-P (2010) La mort du devin, l'émergence du démiurge. Essai sur la contingence, la viabilité et l'inertie des systèmes. éd. Beauchesne/le miroir des savant, Paris
Barbault A (1961) De la psychanalyse à l'astrologie. Seuil, Paris
Barrière O (2007) Régulation environnementale en Afrique: l'action du droit pour une co-viabilité des systèmes sociaux et écologiques. In: Le Roy E (dir) Le droit en action, Cahiers d'Anthropologie du Droit. Karthala, Paris, pp 125–174
Barrière O et Barrière C (2002) Un droit à inventer. Foncier et environnement dans le delta intérieur du Niger, éditions IRD, collection "À travers champs", 476 p
Barrière O, Faure JF (2012) L'enjeu d'un droit négocié pour le Parc Amazonien de Guyane. Natures Sciences Sociétés 20:167–180
Barrière C, Barrière O, Soto P, Rousseau E (2005) Bassari, de l'ocre à la lumière. IRD/Romain Pages, Paris
Barrue-Pastor M (1986) Alibi paysager et législations montagnardes d'aménagement rural (XIXe et XXe siècle). In: La nature et le rural, Protéger et gérer des coûts et des règles. Colloque national, Strasbourg, p 3
Belley J-G (dir) (1996) Le droit soluble. Contributions québécoises à l'étude de l'internormativité. LGDJ, Paris
Bellier I (2013) La reconnaissance internationale des peuples autochtones, In: Bellier I (dir) Peuples autochtones dans le monde, Les enjeux de la reconnaissance. L'Harmattan, Paris, pp 13–37
Bergel JL (dir) (2005) Le plurijuridisme. Presses universitaires d'Aix-Marseille, Aix-en-Provence
Bertrand G (2014) Vernadsky's philosophical legacy: a perspective from the Anthropocene. The Anthropocene Rev 1(2):137–146. https://doi.org/10.1177/2053019614530874
Bonneuil C et Fressoz J-B (2013) L'événement anthropocène, La terre, l'histoire et nous. Seuil, Paris
Bourdieu P (1980) Le sens pratique. Les Editions de minuit, Paris
Bourdieu P (1986) Habitus, code et codification. Actes de la recherche en sciences sociales 64:40–44
Brown R (2010) Co-management schemes: legalities and fish. In: Ribordy F-X (dir) La nature et la loi. Le pluralisme juridique dans la gestion de la nature/Nature and Law, Legal puralism in environmental stewardship, Presses de l'université Laurentienne, Sudbury, pp 81–112
Canguilhem G (1978) Le second règne de la nature. In: Delaporte F (ed) Préface. Flammarion, Paris
Carbonnier J (1978/1994) Sociologie juridique. Presses Universitaires de France, Paris
Charbonnier P (2015) La fin d'un grand partage. Nature et Société, de Durkheim à Descola. CNRS éditions, Paris
Chassot L (2014) Essai sur le pluralisme juridique, l'exemple du Vanuatu. Presses Universitaires d'Aix-Marseille, Aix-en-Provence
Chazel F et Commaille J (dir) (1991) Normes juridiques et régulation sociale. LGDJ, Paris
Chesnaux J (1992) Maîtriser la collision entre l'histoire naturelle et l'histoire humaine. In: Écologie Politique n°2, p 127
Chevallier J (1983) L'ordre juridique. In: Chevallier J et Loschak D (dir) Le droit en procès. Presses Universitaires de France, Paris, pp 7–49
Chevallier J (1998) Vers un droit postmoderne? In: Jean Clam et Gilles Martin (dir) Les transformations de la régulation juridique, LGDJ, Paris, pp 21–46
Chevallier J (2001) La régulation juridique en question. Droit et Société 49:827–846. En ligne: https://www.cairn.info/revue-droit-et-societe-2001-3-page-827.htm; http://www.reds.msh-paris.fr/publications/revue/html/ds049/ds049-07.htm

Ciprut M-A (dir) (2007) Migration, blessure psychique et somatisation. éd. Médecine et Hygiène, Association Pluriels, Chêne-Bourg
Clam J et Martin G (dir) (1998) Les transformations de la régulation juridique. LGDJ, MSH/Réseau européen Droit et Société, recherches et travaux n°5
Comité d'études historiques et scientifiques de l'Afrique Occidentale Française (1939) Coutumiers juridiques de l'AOF, t1 Sénégal, sérieA n°8. Larose, Paris
Commaille J (1994) L'esprit sociologique des lois. Presses Universitaires de France, Paris
Crutzen PJ (2002) Geology of mankind. Nature 415:23
Crutzen PJ, Stoermer EF (2000) The Anthropocene. IGBP Global Change Newsl 41:17–18
Descola P (1996) Societies of nature and the nature of society. In: Kuper A (dir) Conceptualizing Society, vol 109. Routledge, London
Descola P (2002) L'anthropologie de la nature. In: Annales. Histoire, Sciences Sociales. 57e année (1):9–25. https://doi.org/10.3406/ahess.2002.280024. http://www.persee.fr/doc/ahess_0395-2649_2002_num_57_1_280024
Descola P (2005) Par delà nature et culture. Gallimard, Paris
Descola P (2006) Beyond Nature and Culture, Proc Br Acad 139:137–155. Online: http://old.eu.spb.ru/en/news/files2007/descola.pdf
Descola P (2011) L'écologie des autres. L'anthropologie et la question de la nature. Editions Quæ, Versailles Cedex
Descola P (2012) Beyond nature and culture. Forms of attachment, Journal of Ethnographic Theory 2(1):447–471. https://doi.org/10.14318/hau2.1.020
Descola P (2013) Beyond nature and culture. University of Chicago Press, Chicago. Online: https://itp.nyu.edu/classes/interspecies/texts/philippe-descola-beyond-nature-and-culture_ch1.pdf
Descola P (2014) La composition des mondes. Flammarion, Paris
Edelman LB (2002) Legality and the endogeneity of law. In: Selznick P, Krygier M, Kagan R, Winston K (dir) Legality and community: on the intellectual legacy. Rowman & Littlefield, Lanham
Edelman B (2007) Quand les juristes inventent le réel. Hermann, Paris
Ellis EC (2011) Anthropogenic transformation of the terrestrial biosphere. Philos Trans R Soc 369(n°1938):1010–1035
Gaillaud E (2011) Vers un droit des générations futures. LGDJ, Paris
Gilissen J (1982) La coutume, Typologie des sources du moyen âge occidental. Brepols, Turnhout
Gindre David C (2008) Essai sur la loi du pays calédonienne. La dualité de la source législative dans l'état unitaire français. L'Harmattan, Paris
Goyard-fabre S (2007) Repenser la pensée du droit. Librairie philosophique J. Vrin, Paris
Gurvitch G (1931) Le temps présent et l'idée du droit social. Librairie philosophique J. Vrin, Paris
Gurvitch G (1935) L'expérience juridique et la philosophie pluraliste du droit. Pédone, Paris
Gurvitch G (1940) Éléments de sociologie juridique. Aubier éditions Montaigne, Paris
Gutwirth S (1996) Science et droit de l'environnement: quel dialogue? In: Ost F et Gutwirth S (eds) Quel avenir pour le droit de l'environnement. Publication des Facultés universitaires Saint-Louis, Bruxelles
Gutwirth S (2001) Trente ans de théorie du droit de l'environnement, Environnement et Société/26, Normes et environnement, pp 5–17
Gutwirth S et Ost F (1999) Avant-propos. In: Eric N-G (ed) Les dimensions scientifiques du droit de l'environnement. Contribution à l'étude des rapports de la science et du droit. Bruylant, Brussels, pp 13–16
Hauriou M (1893) Revue général du droit, édition Thorin et fils
Howes D (1996) De l'oralité et de la lettre de la loi. Droit et société 32:27–49
Jacob JP (2007) Terres privées, terres communes. Gouvernement de la nature et des hommes en pays winye. éd. IRD, Paris
Kelsen H (1953/1992) Qu'est-ce que la théorie pure du droit? Droit et Société 22:551–568. Online: http://www.reds.msh-paris.fr/publications/revue/pdf/ds22/ds022-15.pdf
Kelsen H (1960/1999) Théorie pure du droit. LGDJ, Paris
Kelsen H (1979/1996) Théorie générale des normes. Presses Universitaires de France, Paris

Kerchove (van de) M, Ost F (1988) Le système juridique entre ordre et désordre. Presses Universitaires de France, Paris
Kouassigan GA (1982) Les droits fonciers coutumiers. In: l'Encyclopédie juridique de l'Afrique. ed. Les nouvelles éditions africaines, Abidjan
Lafargue R (1994) La Révolution Mabo ou les fondements constitutionnels du nouveau statut des Aborigènes d'Australie. Revue du Droit Public et de la Science Politique 5:1329–1356
Lafargue R (1996) La Fédération australienne à l'épreuve du Titre Indigène: le Native Title Act 1993. Droit et Cultures 32:85–106
Lafargue R (1999) La révolution Mabo et l'Australie face à la tentation d'un nouvel apartheid. Journal of legal pluralism and unofficial Law 31(43):89–134
Lafargue R (2003) La coutume judiciaire en Nouvelle-Calédonie. Aux sources d'un droit commun coutumier. Presses universitaires d'Aix-Marseille, Aix-en-Provence
Lafargue R (2010) La coutume face à son destin. Réflexion sur la coutume judiciaire en Nouvelle-Calédonie et la résilience des ordres juridiques infra-étatiques. LGDJ, Paris
Latour B (1991) Nous n'avons jamais été modernes. Essai d'anthropologie symétrique. La Découverte, Paris
Latour B (2015) Face à Gaïa, huit conférences sur le nouveau régime climatique. La découverte, Paris
Le Roy E (1999) Le jeu des lois, une anthropologie "dynamique" du Droit. LGDJ, Paris
Le Roy E (2011) La terre de l'autre. Une anthropologie des régimes d'appropriation foncière. LGDJ, Paris
Lenoble J et Ost F (1980) Droit, mythe et raison, essai sur la dérive mytho-logique de la rationalité juridique. Publication universitaire des Facultés de St Louis, Bruxelles
Lévêque C, Muxart T, Abbadie L, Weil A et Van der Leeuw S (2003) L'anthroposystème: entité structurelle et fonctionnelle des interactions sociétés-milieux. In: Lévêque C, Van der Leeuw (dir) Quelles natures voulons-nous? pour une approche socio-écologique du champ de l'environnement. Elsevier, Paris, pp 110–129
Lévy-Bruhl H (1961/1990) Sociologie du droit. PUF – Que sais-je 951, Paris
Lewis SL, Maslin MA (2015) Defining the anthropocene. Nature 519:171–179. https://doi.org/10.1038/nature14258
Liechti M (2007) Le cadre culturel et les enveloppes psychiques, In: Ciprut Marie-Andrée (dir). Migration, blessure psychique et somatisation, éd. Médecine et Hygiène, Association Pluriels, Chêne-Bourg, pp 35–40
Llewellyn KN et Hoebel EA (1941/1999) La voie Cheyenne, conflit et jurisprudence dans la science primitive du droit. LGDJ, Paris
Lovelock J (1979/2016) Gaïa: a new look at life on earth. Oxford University Press, Oxford
Lovelock J (2010) La terre est un être vivant: l'hypothèse Gaïa. Flammarion, Paris
Luhmann N (1994) Le droit comme système social. Droit et société 11(12):53–67
McMillan DW, Chavis DM (1986) Sense of community: a definition and theroy. J Community Psychol 14:6–23
Merle I (1998) Le Mabo case. L'Australie face à son passé colonial. Annales Histoire, Sciences Sociales 53e année(2):209–229
Michaïlidis-Nouaros G (1982) Droit vivant et droit naturel. Introduction à des problèmes cardinaux de la sociologie juridique et de la philosophie du droit. Éditions Fres Sakkoulas, Athènes
Millard E (2002) Une théorie empirique du droit est-elle possible? Droit et société 50:13–18
Millard E (2014) Réalisme scandinave, réalisme américain. Revus (J Const Theory Philos Law), 24, Le réalisme scandinave dans tous ses états, 81–97, Online: http://revus.revues.org/2915. https://doi.org/10.4000/revus.2915
Moliner P, Rateau P, Cohen-Scali V (2002) Les représentations sociales, Pratique des études de terrain. Presses Universitaires de Rennes, Rennes
Moore SF (1978) Law and anthropology, a reader. Blackwell, Malden
Moore SF (2005) Law as process, an anthropological approach. Routledge & Kegan Paul, London
Morin E (1980) La méthode II. La vie de la vie. Seuil, Paris

Nader L (1997) Law in culture and society. University of California Press, Berkeley
Nader L (2002) The life of the law, anthropological projects. University of California Press, Berkeley
Naim GE (1996) Expertise scientifique et droit de l'environnement, In François Ost et Serge Gutwirth, Quel avenir pour le droit de l'environnement. Publication des Facultés universitaires Saint-Louis, Bruxelles, p 43
Naim Gesbert E et Gutwirth S (1995) Science et droit de l'environnement: réflexions pour le cadre conceptuel du pluralisme de vérités. Revue Interdisciplinaire d'Études Juridiques (RIEJ) n°34:33–98
Naim-Gesbert E (1999) Les dimensions scientifiques du droit de l'environnement. Contribution à l'étude des rapports de la science et du droit. Naim-Gesbert Eric, Bruylant
Nicolau G, Pignarre G, Lafargue R (2007) Ethnologie juridique. Autour de trois exercices. Dalloz, Paris
Olawale ET (1961) La nature du droit coutumier africain. Présence africaine, Paris
Ost F (1991) Le droit comme pur système. In: Bouretz P (dir) La force du droit – Panorama des débats contemporains. Le Seuil, Paris, p 139 et s
Ost F (2000) Le droit comme recours? La régulation juridique de l'environnement et les dichotomies de la rationalité occidentale. In: Renard D, Caillosse J, Béchillon D de (dir) L'analyse des politiques publiques aux prises avec le droit, série politique n°30. LGDJ – Droit et société, Paris
Ost F et Gutwirth S (1996) Quel avenir pour le droit de l'environnement. Publication des Facultés universitaires Saint-Louis, Bruxelles
Ost F et van de Kerchov M (2002) De la pyramide au réseau? pour une théorie dialectique du droit. Publication facultés universitaires St Louis, Bruxelles
Ouedraogo HMG (2011) De la connaissance à la reconnaissance des droits fonciers endogènes, in Etudes rurales. EHESS 187:79–93
Papaux A (2006) Introduction à la philosophie du "droit en situation". De la codification légaliste au droit prudentiel. LGDJ, Paris
Paquay L et al (2001) Former des enseignants professionnels. De Boeck Université, Bruxelles
Perrin J-F (1997) Sociologie empirique du droit. Helbing & Lichtenhahn, Bâle
Pottage A et Mundy M (dir) (2004) Law, anthropology and the constitution of the social. Making persons and things. Cambridge University Press, Cambridge
Prieur M (2000) Le code dans l'environnement. AJDA n°12:1030
Prieur M (2001) Incertitude juridique, incertitude scientifique et protection de l'environnement. In: Institut Fédératif Environnement et eau, Incertitude juridique, incertitude scientifique: Actes du séminaire de l'Institut Fédératif "Environnement et eau", Limoges, 5 avril 2000, Presses Universitaires de Limoges, Limoges, pp 9–16
Prigogine I et Stengers I (1986) La nouvelle alliance. Métamorphose de la science 26. Gallimard/Essais, Paris
Ribordy F-X (2010) Introduction. In: Ribordy F-X (dir) La nature et la loi. Le pluralisme juridique dans la gestion de la nature/Nature and Law, Legal puralism in environmental stewardship. Presses de l'université Laurentienne, Sudbury
Romano S (1919/1975) L'ordre juridique. Dalloz, Paris (réédition 2002)
Ross A (2004) Introduction à l'empirisme juridique. LGDJ, Paris
Rouland N (1988) Anthropologie Juridique. Presses Universitaires de France, Paris
Rouland N (1991) Aux confins du droit. éd. Odile Jacob, Paris
Rouland N (1993) Pluralisme juridique. In: André-Jean A (dir) Dictionnaire encyclopédique de théorie et de sociologie du droit. LGDJ, Paris, pp 459–450
Rumelhard G (2005) Problématisation et concept de paradigme. Approche épistémologique, psychologique, sociologique. ASTER-Institut français de l'éducation, Recherches en didactique des sciences expérimentales, n°40, Problème et problématisation, pp 205–223, Online: http://ife.ens-lyon.fr/publications/edition-electronique/aster/RA040-09.pdf

Sabete W (1999) La théorie du droit et le problème de la scientificité. Quelques réflexions sur le mythe de l'Objectivité de la théorie positiviste. Archives de philosophie du droit et de sociologie juridique 43:302–326. Online: http://www.philosophie-droit.asso.fr/APDpourweb/216.pdf

Sacco R (2008) Anthropologie juridique. Apport à une macro-histoire du droit. Dalloz, Paris

Sacco R (2009) Le droit africain, Anthropologie et droit positif. Dalloz, Paris

Sandkuhler HJ (1994) Le livre de la nature dans l'esprit de la culture, Cassirer est le nouvel esprit scientifique. In: Bulletin la Société Française de Philosophie, n° 1, janvier-mars, p 2

Serverin E (2002) Quels faits sociaux pour une science empirique du droit? Droit et société 50:59–69

Sheleff L (2000) The future of tradition: customary law, common law andlegal pluralism. Franck Cass Publishers, London

Starr J et Goodale M (dir) (2002) Practicing ethnography in law. New dialogues, enduring methods. Palgrave MacMillan, New York

Steffen W, Crutzen PJ, McNeill JR (2007) The Anthropocene: are humans now overwhelming the great forces of nature? Ambio 36(8):614–621

Tarde G (1894/1994) Les transformations du droit, étude sociologique. Félix Alcan/Berg International, Paris

Teubner G (1984) Autopoiesis in law and society: a rejoinder to Blankenburg. Law Soc Rev 18(2):291–301

Teubner G (1993) Law as an autopoietic system. Blackwell Publishers, Oxford/Cambridge

Thibierge C (2003) Le droit souple. Réflexion sur les textures du droit. Revue trimestrielle de droit civil 4:599–628

Thibierge C (2008) Au cœur de la norme: le tracé et la mesure. Pour une distinction entre normes et règles de droit. Archives de philosophie du droit t51:341–371

Thibierge C (2009) Rapport de synthèse. In: Association Henry Capitant (ed) Le droit souple. Dalloz, Paris, pp 141–161

Timsit G (2005) La cumulativité des connaissances dans le domaine de la science juridique. Revue européenne des sciences sociales XWLIII-131(2005):81–86

Treves R (1995) Sociologie du droit. Presses universitaires de France, Paris

Troper M (2007) Le réalisme et le juge constitutionnel. Cahier du Conseil constitutionnel 22

Vanderlinden J (1983) Les systèmes juridiques africains. Presses Universitaires de France, Paris

Vanderlinden J (1993) Vers une nouvelle conception du pluralisme juridique. Revue de la recherche juridique 2:573–583

Vanderlinden J (1996) Anthropologie juridique. Dalloz, Paris

Vergnaud G (2007) Définitions du concept de schème. Recherches en Education n°4:7–22

Vernadsky VI (1929/2002) La Biosphère, Paris, Librairie Félix Alcan, 323p. – Rééd.: Paris, Seuil, coll. "Points/Science"

Vernadsky VI (1938/1997) The transition from the biosphere to the Noösphere, Scientific Thought As a Planetary Phenomenon,/Translated from Russian by B.A.Starostin. – Moscow, Nongovernmental Ecological V.I.Vernadsky Foundation, Online: https://www.21stcenturysciencetech.com/Articles_2012/Spring-Summer_2012/04_Biospere_Noosphere.pdf

Weber M (1986) Sociologie du droit. Presses Universitaires de France, Paris

Wirtz P (1999) Evolution institutionnelle, schémas mentaux et gouvernement des entreprises: le cas Krupp – Thyssen. Finance Contrôle Stratégie, Association FCS 2(1):117. Online: https://halshs.archives-ouvertes.fr/halshs-00748139

Chapter 7
Legal Challenge of the Socio-ecological Connection: The Paradigm of Coviability Defined by the Adequacy Between Social Usefulness and the Ecological Function

Olivier Barrière and Thérèse Libourel

Contents

7.1 Introduction

The question of joint social and ecological viability has led us to think of law in terms of the adequacy of existing regulations relating to human activities vis-a-vis the environment. The anthropo-legal approach, conducted on a diversity of

O. Barrière (✉)
IRD - French National Research Institute for Sustainable Development, UMR ESPACE-DEV, Montpellier, France
e-mail: olivier.barriere@ird.fr

T. Libourel
Université Montpellier, UMR ESPACE-DEV, Montpellier, France
e-mail: therese.libourel@umontpellier.fr

O. Barrière et al. (eds.), *Coviability of Social and Ecological Systems: Reconnecting Mankind to the Biosphere in an Era of Global Change*,
https://doi.org/10.1007/978-3-319-78497-7_7

situations, has allowed a relationship between social usefulness and the ecological function to be deciphered.

The notion of usefulness leads us to the relationship between a thing and the satisfaction that it provides when meeting needs. The usefulness of anything is the measurement of the overall satisfaction that an individual draws from it. The maximum level of usefulness may depend on the quantity or the quality of the thing. Usefulness is often confused with the interest, the benefit, the concept of the service of a thing. In fact, the quality of what is useful is confused with the ability to ensure a certain function. The latter is at the root of a desire or a need (physiological or social). Far from being a physical property of a thing, usefulness is a reflection of the importance that an individual attaches to this thing and upon which he assumes that his subjective well-being depends. In economics, usefulness implies thus an appreciation (Romeuf 1958), the concept is therefore subjective by nature. Consequently, we define usefulness as the satisfaction provided by a thing, expressed by an advantage. This advantage shapes a "service" which brings something tangible or intangible. In the transect of the ontological representations of the society-nature relationship (introduced in the previous chapter of this book, Barrière and Behnassi), the level of usefulness is the fundamental basis influencing the relationship to natural and land resources; and consequently, the system of regulation which is its expression However beware, because this analysis using usefulness seems very anthropo-centered, as it is a usefulness defined in terms of human needs.[1] We are very much aware of this, but we are, however, starting from an anthropological and biological perspective that gives the notion of usefulness a dimension far exceeding that of materiality. The angle here is holistic: if usefulness is defined by a response to needs, it integrates the need of viability. In addition, the non-human is so called "useful when it provides viability to humans thereby allowing humans to exist and to continue. This does not mean, however, that the value given to non-human "resources" should confine itself to an instrumental quality, because we are in the context of hybrid ontologies of society-nature relationships positioned between assimilation and opposition (see the previous chapter in this book, Barrière and Behnassi).

The analysis of the society-nature relationship in Africa and French Guiana based on five field situations (excluding the West) will be used as a starting point. These examples have allowed a table to be built, described here.

The contexts of these examples offer another representation showing that not all things can be owned in themselves. The relationship to the world is based on other values for which a thing (land, forest, pasture...), the object of usefulness (growing cutting down, picking, grazing, fishing...), cannot be an asset but rather a generational and intergenerational link; a link of social cohesion and socio-cultural continuation of the group on which the usefulness of a thing are based. The separation

[1]"Humans are considered the only beings to be equipped with a value and all the rest of the living and non-living creatures only have an instrumental value, measured by their usefulness to humans" (Maris 2011).

between usefulness and a thing characterizes this paradigm: an "asset" thing (land base, *fundus*) cannot be owned and thus takes on the status of a "common asset", due to the underlying objective of passing it on, while the useful things that are the object of intangible property rights (prerogatives), are likely to enter the legal arena.

The table allows the viability of local social groups to be interpreted (1). This will result in the definition of socio-ecological coviability (2). The challenge here is to define the type of man's legal relationship, as a social group and an individual, to spatial resources.

7.2 Analytical Process and Method of Real, Contextualized Situations: Deciphering and Table

Interpreting endogenous systems[2] requires a table listing ranked rights in which legal regulation is taken into account in its broadest sense. In so doing, we go beyond the legislative framework which is occasionally, very rarely or never implemented (excluding the West).

Therefore, in-depth deciphering is required to analyze the law practiced. Its formalization is mostly the spoken word, in discourse and in practices. The understanding of a non-written, non-legislative law, leads to reasoning by means of an interpretation table that has to be developed *in situ*.

Our reasoning here has been based on real rights, and administration and superintendence rights (which correspond to rights of custody in a rationale of transmission to future generations). These rights of a collective or individual type come in a variety of forms within local regulations. The latter are based on the value of usefulness and not on the land value that supports them. Consequently, actors are holders of function rights, i.e. "the right to" hunt, cultivate, graze, fish, collect, cut down, pass on etc. which constitute a cluster of prerogatives. The purpose of the ownership of land focuses on these prerogatives, in contrast to the Western rationale which consists in the ownership of land regardless of the how it is used, in order to exercise exclusivity and ownership. Beyond an ethnocentric viewpoint, territorialization is thus based on exploitation methods which express the modalities of spatio-temporal action that we can identify as intangible assets subject to ownership: I have the right to act on an area such as hunting or farming, that I can sell, exchange, alienate ... but the land itself cannot be owned.

The deciphering of five situations in the field, in Africa and in French Guiana, gives rise to a table that we introduce here in a simplified way (Table 7.2). The analysis in legal anthropology led to an interest in the legal regimes of different socio-cultural groups not falling under property laws or property rights. We have moved away from the Shlager Edella and Ostrom Elinor model, dating from 1992, which in economy rolls out an" ownership rights" model. According to us, this model cannot be used here, on the one hand for the tropical zone, without risking

[2]System "coming from the inside" of the social group, of the society.

Table 7.1 Rights of access to natural resources: how to read the internal legal relationship of communities to land and renewable natural resources

Regulation	Types
Individual prerogatives (subjective law)	
Crossing / passage	**Operational**
(access to, passage through & parking = the right to enter and sometime park)	(temporary action of presence in a place)
Withdrawal	**Operational**
(collect natural resources elements = extraction = free product (spontaneous, not the product of a human action) = fishing, hand picking, collection of forest resources, ...)	(subsistence action)
Exploitation	**Operational**
(production: work on natural resources = farm, lumbering, ...)	(production action)
Governance norms (objective law)	
Exclusion	**Stewardship**
(determine who has a prerogative and how this prerogative may be transferred = control on resources)	(distribution of powers, allocation of land and resources)
Alienation	**Stewardship**
(sell or lease either or both of the above rights)	(transfer of land and resources, transfer of rights in utilities)
Intentional	**Stewardship**
(guide the behavior of actors: negotiations, incentives, forms of contracts or arrangements, gift, ... regulations)	(regulatory practices: responsibility for the defense of the general interest)

ethnocentric transposition and on the other due to the confusion made by the authors with no legal expertise, between possession and property (Shlager and Ostrom 1992). The idea of a cluster of rights remains.

The adopted interpretation table (Table 7.1) was closely developed with the reality in the field using a legal anthropology approach and not a political science or institutional economy one. This is the result of an empirical reading of how law has been applied locally in the areas studied: the initial assumption was to consider the civil code as unsuitable for analyzing the endogenous legal systems of the groups studied (non-Western).

A simplified general table is established (Table 7.2 below) from the five local tables on the rights to resources[3] each corresponding to fieldwork carried by ourselves (from 1993 to 2009)[4] by following a nearly similar protocol on each site.

[3]Barrière and Barrière 1997; Barrière 2003, 2012; Barrière and Faure 2012.

[4]More specifically: the interior delta of the Niger River in Mali (in 1995), the gum zone of Chad (1998), the Bassari country in Eastern Senegal located at the edge of the Niokolo Koba National Park, in 1999, the territory of the Aït Zekri tribe in the Moroccan High Atlas Mountains, Morocco (2007) and the Wayana Territory in French Guiana, Maripasoula municipality (in 2009).

Table 7.2 Table of usefulness law allowing an interpretation of economic functions (land-resources) and social usefulness (actors)

Actors (subject of rights & decision makers) / **Land-ressources (rights issues)**	**Endogenous authorities** chieftaincy (family, lineage, household, farm, village) & Masters (land, water, pasture)	**Common** community members (lineage, tribe, residence) + Private in Common	**Public** (stranger, follower, married woman)	**Internal groups** (association, committee, farmer)	**Territorial elected authorities** (regional council, rural council)	**Authorities State**
field object	●▲Ϟ👽★	■●▲Ϟ	■●▲	▲Ϟ	Ϟ👽	▲Ϟ
tree object	●▲Ϟ👽	■●▲Ϟ		👽	Ϟ👽	
pond object	●▲Ϟ👽★		■●▲	▲Ϟ		▲Ϟ
fallow land	●▲Ϟ👽★		■●			
bush land	●▲Ϟ👽		■●▲			
forest land	●▲Ϟ👽★	■●▲Ϟ	■●		Ϟ👽	▲Ϟ
pastoral land	●▲Ϟ👽★	●▲	■●▲	👽Ϟ	Ϟ👽	
path land (corridor & transhumance track, forest path)	●▲Ϟ👽	■●▲Ϟ	■●▲			
stream and channel area (water ways)	●▲Ϟ👽★	■●▲Ϟ	■●▲	▲Ϟ👽		▲Ϟ
village area	Ϟ★	■●▲Ϟ				
« environnemental » land					Ϟ👽	
« protected » land						▲Ϟ
salt place & salty earth	Ϟ👽	●				
sacred place	Ϟ👽					

Legend: right to cross (passage) = ■ right to take (extraction) = ● right to exploit (production) = ▲ right to exclude = Ϟ
right to alienate = ★ right to incentive (intention) = 👽

Explanation: ● **or/and** ▲ **or/and** Ϟ **or/and** 👽 **...**
No highlight = Prérogatives = ; Grey highlight = Gouvernance

Overview bringing together the realities of an endogenous legal nature of five territorialized local communities (interior delta of the Niger River in Mali, Bassari of Senegal, Chadian gum zone, the Tamazirte of the Moroccan High Atlas Mountains, the Wayana of French Guiana) (1993–2009)

The areas concerned are located in West Africa (dry mountains of Eastern Senegal in "Bassari country"; the interior delta of the Niger River, in the Sahel, Arabic gum area in Chad), Morocco (the Aït Zekri tribe on the southern slope of the High Atlas Mountains), and in French Guiana (Wayana Amerindians of High Maroni, in the Amazon, humid tropical forest). The overview performed on these areas of study has resulted in the overview given here in the form of a table,[5] expressing a system named by us as "usefulness law"

It is thus from real situations that a conceptual analysis has been carried out, that takes the form of a table system. Therefore, we have used examples whose data are not results in themselves, but elements participating in the analysis, in the demonstration of what socio-ecological coviability may be. The prototype given here is not an end in itself, but a step in opening up new research opportunities. It is by considering situations excluding the West that the table below has been drawn up

This table shows the control of the territory (governance[6] in red) and the access to resources (prerogatives in blue) in relation to the "land-resource"[7] considered.

[5] See the complete table published in Barrière 2017.

[6] Here governance is understood as a "decision-making process, regulation of practices, in terms of actions and interventions in a territory and implementation of public policy" (Barrière 2005).

[7] Based on the "land-resource" concept, developed by us on the interior delta field of the Niger River (Mali) in 1996: "Even the substance of the renewable resource is the main element, which is the reason why its qualification prevails and appears in the land-resource concept." The universality

Understood as a territorialized area, i.e. socialized by a specific group, the land is a space defined by its functionalities. In this case, it is attached to the resources giving rise to the concept of "land-resource", justifying the terminology of "useful land". The territory is organized around land-resources allowing the extension of law and legal process to include the land-resource relationship which "gives right to". The legal relationship emerges from the distribution and the delimitation (allocation and assigning) of resources defining the supporting land. The prerogatives focus on the resources and the land based on rights crystallizing the two-fold social link between the land and the resources. This man-land-resource relationship is reflected by a cluster of interests expressed through a series of legal relationships clearly distinguishing the law from its object.

The law on the usefulness of the thing is inherently based on a relationship of "a link" and not on a relationship of "a possession" in that the portion of land is the materialization of the link between the generations. Therefore, all the useful uses is what constitutes the territory, which is thus not necessarily a polygon drawn on the land. In this context, ownership focuses on the usefulness and not on the thing itself, due to the fact that the territorialized thing participates in the group's existence; the individual is only part of the common. The legal forms encountered in a diversity of situations in the Southern hemisphere illustrates a relational mode to land which is different from the model defined by the capitalist system, of naturalist ontology (discussed in the previous chapter).

The table shown establishes relationship of three components: the player, the land-resource, and Law. Its objective is to translate a reciprocal dependency by connecting components structuring the system.

7.2.1 The Social Usefulness Defines the Socio-ecological Link Relayed by "Translators", Who Are the "Actors"

Therefore, in the table shown, the usefulness defines the link to nature by the satisfaction of human needs, which are as much physical, spiritual, psychological, intellectual, moral etc., as material. The usefulness, determining thus social viability, exceeding the usage and exchange value, builds the relationship between the social and ecological systems: it expresses the relationship of societies to eco-systems.

---

of land-resource concept is not related to an eco-system as a whole, but to one of these elements: grass, fish, topsoil, hunted animal species (the game), trees, their fruits and products. In fact, the renewable resource itself represents only a support accessory, but it forms a whole with the latter. However, we cannot consider the resource without its land and that is why it is difficult to prefer the land over the resource. Moreover, since the land supports multiple resources, it is likely to be the subject of a plurality of "land-resource", which reflects the multi-functionality of land. The land-resource constitutes a legal qualification of universality which cannot dissociate the land element from the resource element. Each land-resource thus constitutes a volume that is not completely independent, because the different resources occupy the physical spaces that more or less intersect and overlap. In contrast, these land-resources imply a distinction of the holders of rights to access, exploit and manage resources" (Barrière 1996, 119; Barrière and Barrière 1996, 162, 1997, 22).

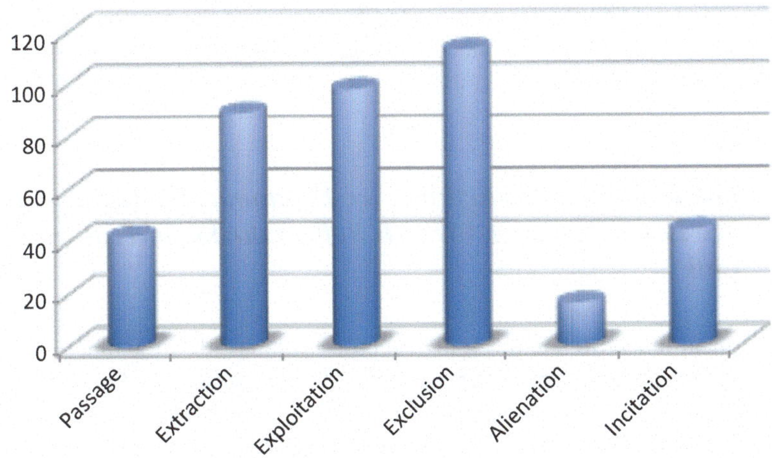

Fig. 7.1 Occurrences of different rights of uses. (From the raw data of the entire table)

The useful relationship to the environment may be a response to an objective need,[8] which is reflected in a form of management (practices) and modes of governance (decisions).

This socio-ecological link passes through "translators", who are the actors of the social system: in this case community members, individually, statutorily or in groups and the decision-makers (authorities). Here, the social system actors are those who are involved in the environment: directly (so-called operational actor), or through a governance mode (the regulation). These two types of actors translate the social usefulness (Fig. 7.1).

Actors categories are here of three types: public, common and internal groupings. The public includes the foreigners to the territory, who do not fall within the scope of the territorial community, but we have associated the married women as their independence is restricted (sic). The private actors are included in the common category because they are dependent on it. The specificity of internal groups deserves to be maintained. The definition of the common deserves a particular attention: what is shared by the community members is common. The territorial relationship is strong links of lineage or residence. Here, we are not in a situation of "common assets" as in the West, nor of common things as defined by the article 714 of French Civil Code,[9] but on a status of "common heritage". This status does not give rise to any ownership (thus without alienation or possible right of disposal), and

[8]Unlike Pareto's (1896) approach who prefers the term "ophelimity" (from the Greek term "ophellimos" meaning "useful") for the subjective satisfaction of needs which is an economic usefulness of an asset or a service experienced by a given agent in a given moment of time: a relationship of caused convenience in order for a need or desire, legitimate or not, to be satisfied. It is distinguished from social usefulness, which concerns all types of satisfaction.

[9]"There are things that belong to no-one and whose usage is common to all".

maintains the land and the resource under the control of the territorialized group. Meanwhile, access rights are granted (passage, removal and exploitation) and the exercised right of exclusion.

7.2.2 The Ecological Functions Are Economically Translated by "Land-Resource" Which Reflect the Socio-ecological Link

The ecological functions are the biological processes that allow the functioning, the evolution and the maintenance of ecosystems. These processes are linked to the life of the ecological system and participate in the definition of the land-resource, which is only its social translation. The renewable natural resource, biotic component of the ecosystem understood by man, is located in a geographic or odologic area (as pathways science). Dependent on such an area, the resource cannot be created without a spatial relationship, a territorial anchor and a bio-ecological dimension. In our approach, this last dimension integrates the opportunity of the existence of the resource whose vocation is to be already to define land-resource ("espace-ressource" in French).

In the table shown, from socio-cognitive practices and representations, ecosystems do not provide "services" to man, but host biological "processes" which offer resources needed by men. This distinction is not inconsequential, and even less so, useless. In fact, human activities are economically defined by removal, exploitation, production (usefulness) with distribution, exchange and consumption. The land-resource concept is thus the result of spatialization and the economic identification of ecological functions:

Ecosystem ⇒ Ecological function ⇒ Social usefulness ⇒ Land-resource

7.2.3 The Regulation Is Legally Formalized by "Rights"

Rights maintain the socio-ecological link around two relationships: action (translated by prerogatives), and choice (translated by modes of governance). In the relationship to land, we are not here in a "relationship over" a land entity (of ownership), but in a "relationship of" (of doing). It is difficult to get rid of western thought and to understand and accept another rationale. Therefore, the relationship to land is not a relationship to a property, but to thing with controlled access, limited by a power connexion on the land, made up of six types: right of passage, right to removal, right of exploitation, right of execution, right of disposition and right of intentional management.

The socio-ecological link revolves around four structuring elements: Land, Object, Player and Rights. Figure 7.2 shows, in UML formalism,[10] the relationships between these four structuring elements forming the socio-ecological system. This relational system is a representation model of the structures within the territories. The examples of these lands offer items and connections which, of course, are specific to the territories in question. We thus provide here a framework model capable of integrating different realities.

The question is on how this model can be generalized. The assumed generics of the latter come in a variety of forms by establishing objective connection between Land-Actor-Rights in which the object is the resource in question:

- The "Land-resource" element: the land-resource combines the ecological function with the socialization of the land concerned defining a territory; it is divided here into seven categories: forest, pastoral, bush, fallow land, village, environment and protected. The heterogeneity that results from this categorization is justified as it comes from both local socio-cognitive representations and from a national classification. The empiricism of the approach has forced us to use endogenous set-ups, but also legislative ones, with the risk of keeping to ethnocentric fiction (Colleyn 2005). Therefore, we understand that this categorization will be different depending on the real cases. The land-resource is crossed by networks, includes specific locations and contains objects.
- The "Object" element: it differentiates itself from the land itself to cover only, here in the encountered situations, tree, field and pond. The object is a representation of the supporting resource: the tree = fruits and wood; the field = soil and crops (harvest product); pond = water, fish and plant (materials and pasture). The resource is either in the land, in the object, or in the endogenous right.
- The "Actor" element in our case is made up of four items: that which is common to all, the public, groupings and the authorities (endogenous, elected, State). Private status exists, but is located in the common status, which would not be the case in the West. This "Actor" element is key in the system because it positions the statutes and related rights. This represents the focal point of the regulation.
- The "right" element rolls out the regulation on two fronts: (a) the prerogatives (passage, removal, exploitation) available to the directly "active" or "operational" actors on the ecological functions (translated here in terms of land-resource); (b) right to governance (exclusion, provision, – management right – intentional) that stewardship actors have over operational actors, and indirectly over the ecological functions. Let's take a quick glance on the right to transfer: this right to transfer exists, but is submitted only to endogenous authorities which allows for market transmissions often limited in time, rarely definitive, and restricting the acquisition of a right of access to a specific statute or a formalized relationship.

[10]Unified Modeling Language.

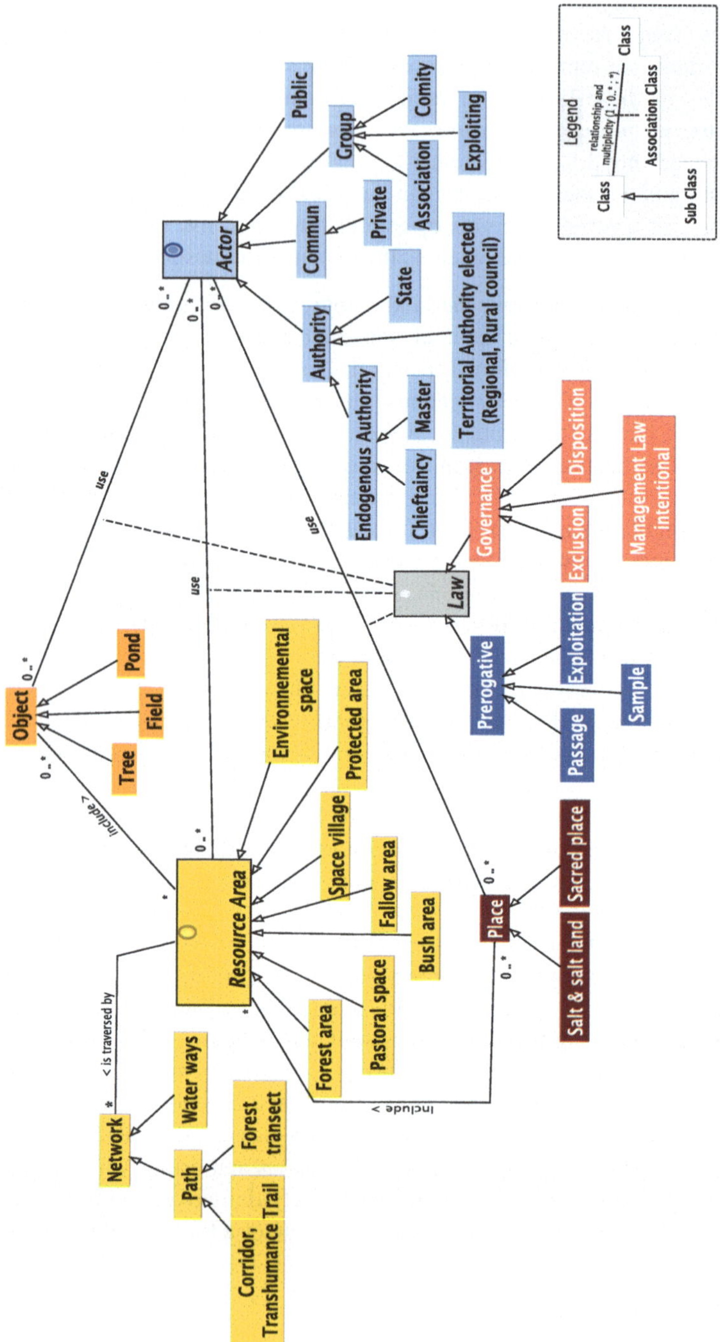

Fig. 7.2 Actors-Land Resource-Right relational system

The prerogatives (three in number) associated with the rights of governance (three in number) represent the carving up of the law of uses: Passage, Removal, Exploitation, Exclusion Transfer, Intentional. A legitimate question that arises is that of their occurrence in the table. Figure 7.1 features this occurrence, which highlights the position of the regulation by the exclusion that will determine the direct pressure on the environment due to exploitation and removal. The competition on the resource grows when it is less available: the pressure is regulated through exclusion, mode of central governance. In the table, this right of exclusion is not in the hands of just one person, but shared by all field actors who will all intervene at their level. There is no delegation of rights, but rather a form of distribution according to the resource and the player's status.

The distribution of rights (on land and objects) is more structural (generator of group coherence) than functional (fulfill a function), it can be conventional among groups. It will essentially depend on the usefulness relationship (needs, necessities, subsistence, development, etc.) to the ecological function (the state of the resource, the degree of pressure on the resource, the responses of the environment, etc.). The player, with his/her status (member of the community, authority, foreign, member of a grouping, etc.), is instrumental in this inter-relationship by expressing the needs, the response to which will depend on the right enjoyed, and consequently, the decisions made and practices implemented.

Below, is what the table shows, which can be read in the relationships between the governance rights (in red) and the prerogative right (blue):

- The endogenous authorities' governance is focused on removal and exploitation through the right to exclude, to transfer and to intentionally manage;
- Governance by community members is carried out on passage, removal and exploitation by the right to exclude prerogatives;
- The public does not have power of governance, but enjoys prerogatives (passage, removal, exploitation);
- The internal groupings (modern institutions) intervene only upon exploitation by the right to exclude and the right of intentional management;
- Governance by elected authorities focuses on exclusion and intentional management, without having the prerogative right;
- Governance by a state authority focuses on exclusion by the right of exploitation.

This reading informs us on several points of the legal regulation carried out:

- Governance power is concentrated around the endogenous authorities;
- The operational actors implement direct control through exercising the right to exclude; the right of intentional management is specifically exercised by the internal groupings (who are operational actors);
- The control of the access to resources is carried out by the community: the access to the resource is open but controlled and strongly regulated; therefore, there is no free access but a pressure of local institutions instead;

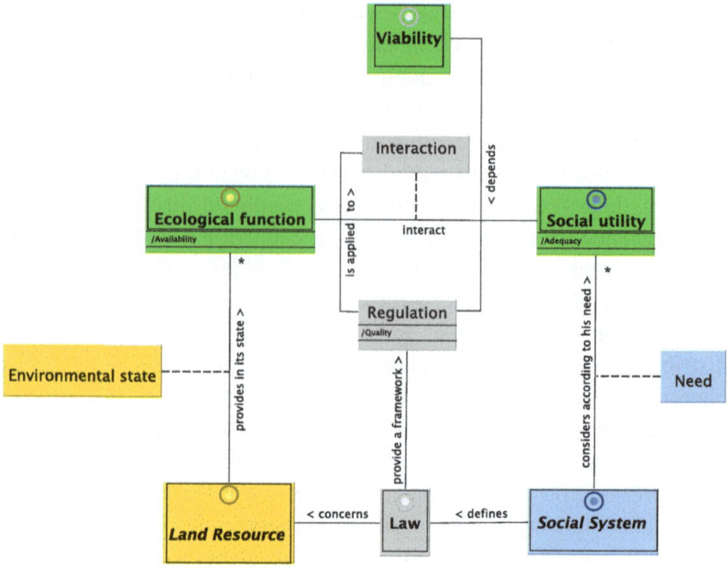

Fig. 7.3 The viability resulting from of the continuum between what is social and what is ecological: the importance of the regulator Law, heart of the socio-ecological system

- The democratic institutions (elected) are national scale relays deriving their legitimacy from law; their mission focuses on the intentional directed management of environmental practices for the implementation of national principles.

Overall, the decision-makers thus intervening on the territories are not confined to the governance and can exercise power over prerogatives of access to resources, according to the decision-maker. The governance is diffuse (more or less shared), but non-natives have only restricted prerogatives. What particularly emerges is that communities control their territory including the right of passage managed by them. However, this conclusion has of course exceptions (specificities which cannot emerge here) as the case of national parks or the area so-called "environmental".

The challenge of the regulation for viability focuses on the nature-society continuum. This continuum is expressed through the social usefulness interaction-ecological function (Fig. 7.3). Regardless of the concerned area of the globe, for a given society-nature system, the –land-resource provides a plurality of ecological functions, according to the state of the territory in question. According to its needs (tangible, intangible, psychological, cultural, etc.) the social system judges the usefulness of these. As a result, ecological functions and social utility are both in interaction. However, this interaction can prove to be dangerous if, for instance, the needs keep growing and exhaust the resource, and prejudicially affect the function. Therefore, the social system defines the rights which deal with the land-resource, and hence, initiates a regulation of the social ecological-usefulness function interactions in order to ensure the viability of society. A viability provided by a concordance between the ecological functions and the social usefulness. The adequacy sought is coherence between the pressure exerted by societies on resources and their ecological functions. The adequacy reflects an adaptation supported by the values that the social group grants to the ecological functions (translated in terms of land-resource) that generate the legal regulation. The coviability of social and ecological systems will depend on such adequacy.

7.3 The Legal Paradigm of Coviability: The Socio-ecological Link for Which the Social Usefulness Adapts to the Ecological Function

Coviability defines a viability link, which attaches and holds the social systems together with the ecological systems. Although different, humans are related to the non-human (Sect. 7.3.1) in the relationship to the living within the biosphere. However, the nature of socio-ecological coviability fundamentally accomplished through regulation. In fact, by associating the social usefulness to the ecological function, the challenge of coviability falls under the capacity to adapt (Sect. 7.3.2). Finally, we will be confronting coviability with land ownership system in which the usage is opposed to ownership (Sect. 7.3.3).

7.3.1 The Socio-ecological Link

The socio-ecological link represents the Ariadne's thread, a guideline which defines the actions to be taken to achieve the goals. This link is both a means and a goal which coalesce. The objective is viability. Moreover, the viability of one is the viability of the other and this is why we talk about coviability, as previously highlighted. This socio-ecological link, even the quintessence of coviability, is situated at the heart of the coviability concept-paradigm, because, in essence, coviability defines the actions to be taken to remain viable.

On the transect of man-nature ontologies, introduced in the previous chapter (Barrière and Behnassi), the connection of societies to the biosphere depends on the existing link between the human and non-human. The latter is either "continuous" or "discontinuous" according to the relationship of assimilation, distinction or conflict in the relationship between man and nature (cf. *infra*). Consequently, this link translates a degree of interdependence experienced and recognized by the societies towards the living world surrounding them. This interdependence falls firstly under a social dependency to its environment, because we could retort that the latter may not need humans (cf. deep ecology-biocentrism, Naess 1973). This is objectively true, but in the Anthropocene era in which planet Earth is positioned, the ecological system is for many the result of human activity. So much so, that this results in us talking about a "socio-ecosystem", or more explicitly of an "anthroposystem". Even though the human system depends mainly on the ecological system (the habitat), the latter is somewhat its product, because man is consubstantial with the nature (cf. the introductory chapter of this book).

One cannot be considered without the other. Such interdependence generates a form of solidarity between social and ecological systems due to the link of existence and viability, which unites them. Even the idea of socio-ecological solidarity is derived from coviability as well. We deduce a mutual dependence leading to "an imposed limit to freedom of action in the battle for existence", source of "cooperation modes" (Leopold 1948/2000).

7.3.2 Regulate to Remain Viable

Regulate to remain viable is to adapt the usefulness (rights to the environment) to the ecological functions (viability of environments and thus of resources they include).

The ability of a living creature to live, to develop or the capacity to survive defines the viability of its environment. Nevertheless, the latter has a temporal dimension and leads to evolution in time, of a living creature or a system. Existing means to stay alive, hence to last, to reproduce and to grow. Sustainability or long-term viability requires the ability to change and to adjust to situations in order to survive.

Viability is defined by mathematics as an arena of acceptable constraints for the maintenance of the system, called the environment, beyond which viability is no longer guaranteed. Adaptation to the constraints of viability results in having to find "regulons" which allow changes to remain in an environment" (Aubin and Durand, in this book). Viability regulation will be translated by a concept of socio-cultural "regulon" (Aubin 2010) which is a variable of standards (laws, customs, agreements) and habitus (Barrière 2011).

The regulation factor, or the regulon, serves for the adaptation or the homeostasis that allows the living being, or the system, to persist. The usefulness introduced in the table (Table 7.2) corresponds to the natural resources necessary for the human beings. The social stake is in the organization of the pressure exerted on this resource, which is carried out by the legal regulation. Therefore, usefulness translates a regulation on the resource. This complicity between the usefulness and the right is the connector that inter-relates the usefulness factor with the right to the land-resource. In addition, we have previously seen that the land-resource is the social translation of the ecological function. Hence, the adequacy of the social usefulness to the ecological function defines the socio-ecological coviability.

In ecology and socio-anthropology, the viability of a system (or a living creature) dependent on the usefulness (the physiological, sociological needs, etc.), the regulation (or the right) is the result of the usefulness because the needs are constrained by the environment.

The non-Western systems step back while distancing themselves from Western references. The regulation modes operated accomplish legitimacy through their effective implementation and operational effectiveness and efficiency, maintaining the viability of a system still in force. However, the global changes (environmental,

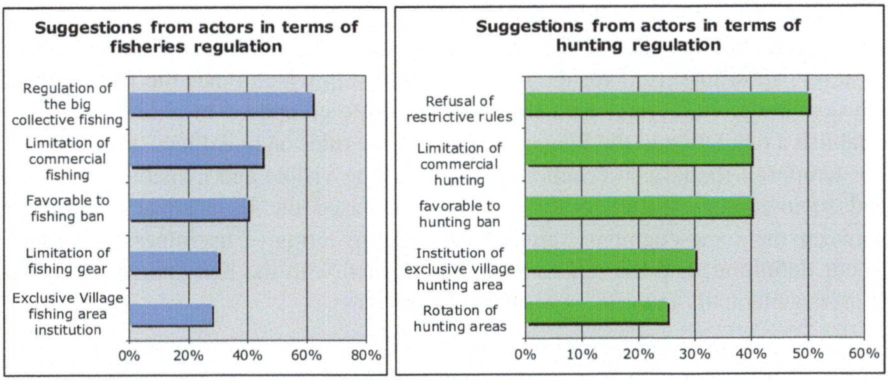

Fig. 7.4 Regulations suggested by the Wayana Amerindians (French Guiana) on fishing and hunting: an illustration of the adequacy of social usefulness to ecological function

climate, social and economic due to globalization) profoundly affect the viability of these socio-ecological systems.

Let us take the example in the Amazon of the Wayana Amerindians, in French Guiana, who were provided with a national park in 2007.[11] The conducted surveys[12] (Barrière and Faure 2012) are built on an interview guide intended to prompt spontaneous and flexible responses by directing them as little as possible. These responses expressed on the specific needs of regulation are summarized in two charts (Fig. 7.4) concerning hunting and fishing, introduced below. The suggestions originate from those interviewed, they were not "suggested" to them.

The suggestions include standards on the adaptation to pressures exerted on the resources. Nothing in these responses exists in their current mode of operation.

Looking at theses field investigation results, the Wayana teach us that it would perhaps be possible to construct a regulation less based on the general and impersonal standard than on an intrinsic logic to the group, and especially to the individual. The link to the environment is not necessarily communicated by a collective dynamic only (the general interest), but can also be built on another reference, moving from the direct relationship of the individual with the sphere of

[11]Decree No. 2007-266 of 27 February 2007 establishing the National Park called "Parc Amazonien de Guyane".

[12]Survey conducted in 2009 among the 39 decision-makers distributed over the four Wayana villages, with approximately 900 residents: Antecum Pata (10), Twenke- Talwen (13), Kayode (10), Elahe (6). The Captains and the Gran man, with the support of local officers of the 'Parc Amazonien de Guyane' (PAG), helped us to determine the list of decision-makers per village (traditional leaders, head of the association, local reference persons). We have made the choice to add the representatives of the PAG inhabitants' councils, well aware that it is an artificial institution provided by the park.

living, i.e., from the vital force (*akwuli*) to the cosmos. The general interest, the public good, falling within a very Western approach, may not from the outset (hardly or never) substitute itself for the special relationship which binds the Indian, their physical body, vital force, with their sphere of life composed of plural spirits. To establish a regulation in the Wayana country game rules on tree dimensions must be set: temporal, spatial and cosmic, or intermingling visible and invisible, combine; and socio-ecological coviability can only be based on the adequacy (balance) between the socio-cognitive values (expressed in terms of usefulness according to our definition) and the relationships maintained with the non-human (what we express generically in terms of ecological functions).

7.3.3 In the Regime of the Law of Properties, the (Collective) Use Is Confronted with the Right of Ownership (Individual)

In Roman-German civil law,[13] the asset is a thing, which is the subject of ownership: "assets are the things whose usefulness justifies the ownership" (Zenati-Castain and Revet 2008, 18) or "the things that are used to serve man" (Terré and Simler 2014, 11). A possibility of ownership is thus needed to make an asset from a thing (Carbonnier 1995, 93). The ownership relationship thus defines the asset[14]; consequently, the asset is a thing object of ownership. In addition, the right to ownership itself, "absorbing all the usefulness of the thing", is confused with the latter (ibidem).

Here, we are in a legal system which reflects a Western relationship to the world according to its own cultural and economic reference. If this system seems to increasingly dominate the planet through globalization, and a form of commercial imperialism, a significant share of the world population, however, does not practice this paradigm: the socio-cultural contexts limit or prohibit the ownership of various things. We have taken in the introduced table (Table 7.2) examples that do not fall under the system of the law of ownership, which is rather rare. This choice has enabled to shift from a very ethnocentric view, and to move away from a Western rationale anchored in a capitalist relationship to the world.

However, let us take an example in the Cevennes (France, Department of the Gard), an area of inter-communality including 16 towns and villages (bringing together nearly 6000 inhabitants for a surface area of 475 km^2). The legal paradigm of the coviability (social usefulness + ecological function) has been the subject of

---

[13]As opposed to the English Common Law.

[14]"Assets are physical objects for which a demand exists, over which ownership rights can be established and whose ownership can be transferred from one institutional unit to another by other means of transactions on markets" (INSEE, online: http://www.insee.fr/fr/methodes/default.asp?page=definitions/biens.htm.)

field trials, which resulted in the adoption of a neighborhood regulation, an "inter-municipal pastoral pact", on 13th of May 2015. The challenge is between the respect of individual freedom, through private property, and the general interest of an area to be managed in order to benefit all. The local project is to maintain and develop a pastoral activity in order to preserve its living agro-pastoral landscapes that have recently been included as a World Heritage site by UNESCO. Private use cannot prevail over the general interest of maintaining the open environment, and thus an ecological function to be preserved, closely dependent on pastoral practices.

In practice, the pastoral use (the pasture) of the area is directly confronted with the system of the property law as an easement of the property of others. The enjoyment refers to the benefits and various advantages attached to the possession of a property or an asset: this concerns the right to enjoy the advantages of an asset (the *jus fruendi*), and more widely the right to use the owned thing (the *usus*). Legally, the enjoyment is thus defined in the exercise of a right on the enjoyment of the asset: we are holders of a personal right to crop growing, harvesting, pasture, etc. on a property that is not ours (privately or collectively). The enjoyment (the owner's) is therefore conflicts with the right of use (of the non-owners). This right to use concerns the property of others. This distinction is essential in law. Consequently, the owner of land has a right to grazing, which is an expression of the enjoyment of his property. Excluding him, this right is translated into a right of use.

Based on the problematic of access to land for grazing purposes, the question of the land ownership system is raised in regards to the use of the land. In fact, the ownership of the land requires finding the owners, free to do what they like with their property, to enter into a negotiation and contractual relationship (for either acquisition or use) which will always remain precarious or costly. The property landscape puzzle enters quickly into the complexity of land ownership, whose land boundaries are not always clear, and whose owners are often difficult to identify. In parallel with the reality of local and micro-local situations, the public policies, whose first stake is the public's interest, are confronted with the expression of private interests. It is at the local level that the perspective can be viewed; the local stake exceeds the addition of individual interests, and the collective interest is not the sum. From this rationale the idea of a local project, which is based on values shared collectively, emerged. WithUNESCO having just registered as a World Heritage site the agro-pastoral cultural landscape of Causses and Cevennes,[15] livestock breeders of the inter-communal "Causses, Aigoual, Cévennes Terres solidaires" group contacted their local councilors to suggest them to work on the objective of maintaining and developing the vertebral column of the territory, the pastoral activity.

Working groups, public meetings and putting into contact with one another local actors working on inter-communal aspects, contributed to the co-creation of specifications (which makes compulsory). The co-creation has progressively validated the versions of a pact, which have followed. The pact consists of four

[15]2011, see: http://whc.unesco.org/fr/list/1153; and: http://www.causses-et-cevennes.fr.

articles and a long preamble, which lays the foundation for a local area stewardship. If the vocation of the pact is not regulatory, the fact still remains that it provides specifications through the formalization in law of local values through planning and management, and standards on the behaviors and practices impacting the agro-pastoral landscapes. The area is being given the status of common asset to those who are live in it,[16] building it and giving it life (Preamble). Pastoralism has been proclaimed a collective interest and the territory has been granted a "for pastoral use" status, thereby underpinning directed public policies. Legal innovations are emerging to preserve the ecological functions of different environments of the area: pastoral easement, and in the urban planning instruments, a zoning specifically for grazing has been established (AP – agricultural pastoral and NP – Pastoral Natural).[17]

In the land property system, the matching of social usefulness to ecological function is made possible by a local regulation (a negotiated right) for the collective interest without prejudice to individual freedom. The given example of the inter-municipal pastoral pact Causse Aigoual Cevennes, is proof.

7.4 Conclusion

From the socio-ecological link to the definition of socio-ecological coviability

The definition given in this chapter to "coviability" (the adequacy of social usefulness to ecological function) is closely found, in French law "for the recovery of biodiversity, nature and the landscapes" of 8 August 2016). In fact, the relationship between social usefulness and ecological function is inserted in the revision of article L110-1 of the French Environment Code by the association of eco-systemic services for social uses, as one of the aims of sustainable development: "(. . .) the protection of services they provide and uses related to it". The link between the services and the uses are established by two new principles: (a) the sustainable use "according to which the practice of uses is an instrument at the service of the conservation of biodiversity" (I-7°); b) the complementarity "between the environment, agriculture and forestry, according to which the agricultural and forest areas carry specific and varied biodiversity. As to agricultural and forestry activities, they can be a vector of eco-systemic interactions. The latter guarantee, on the one hand, the preservation of ecological continuities and, on the other hand, environmental services that use the ecological functions of an ecosystem to restore, maintain or create biodiversity" (I-8°). The French legislator brings the concepts of "complementarity" and "interaction" which take part in the coviability paradigm. However, if the latter does not employ the interdependence concept, it includes in

[16]Through the pastoral and the global pact of the UNESCO, see previous footnote.

[17]See the text of the pastoral inter-municipal pact Causses Aigoual Cévennes, online: http://www.caussesaigoualcevennes.fr/connaitre-communes/

the environment code a third additional principle, that of ecological solidarity.[18] In line with the concept introduced in 2006,[19] not defined at the time, the legislator concentrated on "(. . .) the interactions of ecosystems, living creatures and natural or developed areas" (I-6°). Nonetheless, we remain in a rationale of solidarity but only among ecological systems. Therefore, there is only little man-environment interdependence. The idea of interdependence refers to reciprocal dependency. Recently, the French Act of 8 August 2016 (n° 2016–1087) "for the recovery of the biodiversity, nature and landscapes" establishes the principle of the ecological solidarity without even defining it. Clearly indefinable but representing a higher interest, the principle of ecological solidarity "calls for the taking into account of, (. . .), the interactions of ecosystems, living creatures and natural or developed environments" (Article 2). We are still in the rationale of 2006, unless humans are among the "living creatures".

In the ontologies of society-nature relationships, in the previous chapter (Barrière and Behnassi), the interweaving of man with nature is such that it leads to thinking about ecological systems as more or less anthropized during the anthropocene era. The weight of human beings on the planet determines the state of ecological systems to the point of talking about interdependence. For instance, the preservation of wetlands or open environments depends on human intervention in order to avoid them from filling-in or being invaded by forest. The translation of this interdependence is manifested in the socio-ecological link founding the coviability paradigm. Therefore, we can imagine the perspectives of a socio-ecological solidarity, since the dependence here is mutual.

The mathematicians (Aubin 2010; Bourgine 1996; and the different mathematical articles in this book) provide an objective definition of coviability. From the legal anthropology point of view, with surely an aspect of subjectivity, regulation by law comes from societies. However, this subjectivity remains relative and "made objective" since human destiny is shared with that of ecological systems in which it evolves and participates.

Let us emphasize a value which is not directly from our field work. It focuses on the intrinsic character of eco-system elements. International law recognizes it in the Convention on Biological Diversity (1992). This is an ethical form but one which is fits in with the concept of social usefulness (which is, as we mentioned, not a synonym of utilitarianism) through society-nature ontological relationships.

All ambiguity of the relationship between regulation scales is now clear: the international level is not the national level, which is not the local level. Despite the tautology, but this obvious fact fully participates in the problem of environmental law, its level of shaping affecting its degree of effectiveness and method of implementation. International level seek to integrate national ones, which itself must

[18]In the draft biodiversity law of 2006 (op.cit.) revising act. L110-1 of the French Environment code.

[19]Act No. 2006-436 of 14 April 2006 related to national parks, marine natural parks and the regional natural parks.

integrate local aspects, through methods which cannot circumvent the rationale of local legitimacy; which makes sense for local actors, for the local areas in question, and not necessarily for legislators who establish law and within the international institutions where multilateral conventions are drafted.

References

Aubin J-P (2010) La mort du devin, l'émergence du démiurge. Essai sur la contingence, la viabilité et l'inertie des systèmes, éd. Beauchesne/le miroir des savant

Barrière O (1996) Gestion des ressources naturelles renouvelables et conservation des écosystèmes au Sahel: le foncier-environnement, Thèse Droit Université Paris1

Barrière O (2003) Droit et gestion de l'environnement en pays Bassari (Sénégal oriental); réflexions sur un droit de l'environnement au Sénégal. Revue canadienne Droit et Société/Can J Law Soc 18(1):73–101

Barrière O (2005) Une gouvernance environnementale dans une perspective patrimoniale: approche d'une écologie foncière. In: Eberhard C (dir) Droit, gouvernance et développement durable, Cahiers d'Anthropologie du Droit. LAJP- Karthala, Paris, pp 73–98

Barrière O (2011) Pluralisme juridique et patrimonialisation: entre paradigmes de l'"appropriation" et du "patrimoine commun. In: Serge Mam Lam Fouck et Isabelle Hidair (dir) La question du patrimoine en Guyane française. Diversité culturelle et patrimonialisation. Processus et dynamiques des constructions identitaires. Ibis Rouge Editions, Matoury, Guyane, pp 43–75

Barrière O (2012) Nature juridique de l'agdal, de la propriété collective au patrimoine commun. In: Auclair L, Alifriqui M (dir) Agdal Patrimoine socio-écologique de l'Atlas marocain. éd. Institut Royal de la culture Amazighe (IRCAM) et IRD, pp 207–242

Barrière O (2017) Chapter 14: Human relationship to the land from a legal perspective as a human and environmental security challenge. In: Behnassi M, McGlade K (eds) Environmental change and human security in Africa and the Middle East. Springer, Cham, pp 259–304. https://doi.org/10.1007/978-3-319-45648-5_14

Barrière O, Barrière C (1996) Approches environnementales: systèmes fonciers dans le delta intérieur du Niger: de l'implosion du droit traditionnel à la recherche d'un droit propice à la sécurisation foncière. In: Le Roy E, Karsenty A, Bertrand A (eds) La sécurisation foncière en Afrique: pour une gestion viable des ressources renouvelables. Karthala, Paris, pp 127–175. http://www.documentation.ird.fr/hor/fdi:010065117

Barrière O et Barrière C (1997) Le foncier-environnement, fondements juridico-institutionnels pour une gestion viable des ressources naturelles renouvelables au Sahel, Études Législatives n 60. éd. FAO, col., Rome

Barrière O, Faure JF (2012) L'enjeu d'un droit négocié pour le Parc Amazonien de Guyane. Natures Sciences Sociétés 20:167–180

Bourgine P (1996) Modèles d'agents autonomes et leurs interactions co-évolutionnistes, Penser l'esprit: des Sciences de la Cognition à une Philosophie Cognitive. Rialle V, Fisette D (dir), Presses universitaires de Grenoble

Carbonnier J (1995) Droit civil, Les biens (monnaie, immeubles, meubles), T3, PUF

Colleyn J-P (2005) Fiction et fictions en anthropologie, L'Homme, Vérités de la fiction, 3, 175/176, pp 147–163, Online: https://www.cairn.info/revue-l-homme-2005-3-page-147.htm

Leopold A (1948/2000) Almanach d'un comté des sables. Flammarion, Paris

Maris V (2011) De la nature aux services écosystémiques. Une comodification de la biodiversité, Ecorev n 38, Quelle(s) valeur(s) pour la biodiversité? Revue critique d'écologie politique, Online: ecorev.org

Næss A (1973) The shallow and the deep, long-range ecology movement. Inquiry 16:95–100. Online: http://www.alamut.com/subj/ideologies/pessimism/Naess_deepEcology.html
Pareto V (1896) Cours d'économie politique, F. Rouge Editeur, Lausanne, 2 tomes; Online: http://www.institutcoppet.org/2012/05/08/cours-deconomie-politique-1896-de-vilfredo-pareto
Romeuf J (1958) Dictionnaire des sciences économiques, Puf
Shlager E, Ostrom E (1992) Property-rights regimes and natural resources: a conceptual analysis. Land Econ 68(3):249–262
Terré F et Simler P (2014) Droit civil, Les biens. Dalloz
Zenati-Castain F et Revet T (2008) Les biens. PUF, Paris

Chapter 8
Local Ecological Knowledge and the Viability of the Relationships with the Environment

Catherine Sabinot and Nicolas Lescureux

Contents

> Because of the sheer complexity and specificity of ecosystemic interrelationships and fluctuations, it is not unreasonable to expect that optimal strategies for sustainable resource management are generally best defined by local practitioners with close and long-term experiences of these specificities, and with special takes in the outcome. (Hornborg 1996: 54)

> Faced with an ecological crisis whose roots lie in this disengagement, in the separation of human agency and social responsibility from the sphere of our direct involvement with the non-human environment, it surely behoves us to reverse this order of priority. (Ingold 2000: 76)

If viability characterizes that which is able to live and develop, coviability refers to a system whose elements are able to live and develop together, to contribute to life and to the development of the system and its different components. Therefore, reflecting on the notion of coviability requires observing and analyzing interactions and

C. Sabinot (✉)
IRD - French National Research Institute for Sustainable Development, UMR ESPACE-DEV, Nouméa, Nouvelle-Calédonie, France
e-mail: catherine.sabinot@ird.fr

N. Lescureux
National Centre for Scientific Research (CNRS), UMR5175 Centre d'Ecologie Fonctionnelle et Evolutive (CNRS-UM-UM3-EPHE), Montpellier Cedex 5, France
e-mail: nicolas.lescureux@cefe.cnrs.fr

O. Barrière et al. (eds.), *Coviability of Social and Ecological Systems: Reconnecting Mankind to the Biosphere in an Era of Global Change*,
https://doi.org/10.1007/978-3-319-78497-7_8

interdependencies between human dynamics and non-human ones. The connection between human dynamics and ecological dynamics attracts the attention of ethnoecology. One of the main interests of this disciplinary field is to investigate the interaction of societies with their environment, *a fortiori* in the interdisciplinary context that this book initiates.

Therefore, why is it important to take a look at the concept of coviability in terms of ethnoecology? How would this disciplinary field contribute to the definition and analysis of coviability? Lastly, how can the field contribute to discussions on the indicators of coviability? The aim of this chapter is to provide some answers to these three questions. By considering the knowledge, norms, and values that are related to the environment as the result of numerous interactions of human societies with non-human elements of their environment, we aim to demonstrate, firstly, that the study of such knowledge as well as that of the ecological and socio-cultural frameworks of their practical application, allow us to understand the relationship between humans and non-humans. We then describe how the various modes of relationships between societies and environment are adjusted, adapted and negotiated following the pace of socio-environmental transformations. Finally, we investigate the appropriateness of using local ecological knowledge as a possible indicator of coviability.

8.1 Ethnoecology: Understanding the Relationships Between Humans and Non-humans via the Study of Ecological Knowledge

Ethnoecology developed on the basis of ethnobiology during the 70s and 80s.[1] The principles of ethnobiology developed by Berlin et al. (1973) still provide the analytical vocabulary of comparative analysis. The descriptive report of nomenclatures and local classifications forms the foundation from which wider problematics are developed. Notably, ethnoecology investigates how local knowledge on flora and fauna is translated into daily life through customs and practices. The anthropologist and ecologist Victor Toledo thereby promoted the concept of ethnoecology, whose main concern is the relationship between knowledge and action, especially in relation to the management of territory and resources (Toledo 1992). The concept of TEK (Traditional Ecological/Environmental Knowledge) as well as other concepts such as IK (Indigenous Knowledge), TKW (Traditional Knowledge and Wisdom) and LEK (Local Ecological/Environmental Knowledge) include this notion. They do not limit themselves to providing an inventory of plants and animals, along with their uses; rather, they accumulate knowledge on soils, climates, plant and animal communities, succession stages, and so on.

Ethnoecology therefore is not only the study of ecological knowledge of a human group. Rather, it involves understanding the manner by which this knowledge

[1] For a history of ethnoecology, see (Hunn 2007)

influences the way of being in the world of the individuals who form the group. This knowledge is related to particular perceptions of the environment, to practices, to actual world views, and to dynamic and diverse ways of thinking and organizing the world. These ways are impregnated with multiple modalities of relations to others, both humans and non-humans. On the one hand, ethnoecology describes the ways of categorizing, classifying, and organizing information into a system of knowledge; and on the other, it builds a comparative and symmetrical analysis between various societies, including ours (Fossier and Gardella 2006). Insofar as knowledges are part of specific socio-cultural contexts and reveal them, ethnoecology in this context allows to simultaneously highlight: (1) elements of the functioning of societies, (2) environmental elements of these societies and (3) elements concerning relations between societies and their environments.

The analysis of relations between societies and their environment led a growing number of authors to suggest that it is impossible to separate nature from culture in many societies. Moreover, a society is not homogeneous and does not come under a single ontology (Descola and Pálsson 1996; Ellen and Katsuyoshi 1996; Ingold 1996; Friedberg 1997; Descola 2013; Weber et al. 2013). As Hviding suggests, "*assumptions about a universal nature-culture dualism cannot constitute a meta-language for the comparative analysis of relations between people and the environment*" (Hviding 1996: 179). It appears that the non-humans[2] are often viewed by local actors not only as objects of study but also as interactive subjects, subjects of interaction with others, that is actors in relations. Not only can they be the initiators of production and adjustments of ecological knowledge, but they also participate in establishing education of attention (Gibson 1986; Ingold 2004) which implies that knowledge is accessible only to those who already have a certain intimacy with their environment, an intimacy which implies *dwelling* in a common environment with non-humans (Ingold 2005). The ecological knowledges can then be defined as an *engagement*: not the result of a construction but a consequence of dwelling, not a viewpoint built *on* the world but a viewpoint adopted *in* the world (Ingold 1996).

Scientific knowledges does not oppose itself to local knowledges, especially in a world whose massive communication capacities relatively break the boundaries between different knowledge systems (Dove et al. 2007). Although they do not always succeed in practice (Latour 1993, 2001), scientific knowledges base themselves mainly on observation and objectification, and therefore on a non-engagement that Ingold calls "*disengagement*" with the object of their study related to a "*Western metaphysics of the alienation of humanity from nature*" (Ingold 2000: 76). The inability or unwillingness of researchers to recognize their engagement, that is to say not only the impact they may have on the subject of their research but also the impact that this object can have on researchers and on the results of their research, is related to the historical path of Western sciences. This path led them

[2]The non-human category includes tangible and intangible elements capable of being in relation with humans (animals, plants, objects, geological elements, artifact, and so on).

to create a dichotomy between Nature and Culture (Latour 1993; Descola 2013) and to consider non-humans as passive objects obeying natural laws while humans, although being also subject to the laws of Nature, are emancipated from a number of these laws thanks to social life, technics, and so on.

It follows that building ecological knowledges in Western science remains largely based on external observations of broad phenomena, whose objective and explanatory general laws are investigated. The non-humans are considered passive objects that are subjected to the laws of Nature from which they cannot escape. Their particular behavior disappears beneath these general laws; and from all the possible explanations of their behavior, only those involving the lesser complexity are retained, see for instance the famous Morgan's Canon (Morgan 1903). Humans are investigated differently by the social sciences, as these latter tend to describe the complexity of human societies and adopt qualitative methods to better capture at a small scale the details of the functioning of societies or communities. On the one hand, generalization stifles variability and, on the other hand, objectification does, when it decontextualizes the emergence of knowledge. Objectification does "*alter the relationships between person and world by subordinating or eclipsing the non-objectifiable, local specificities which render meanings everywhere so implicit and inextricable*" (Hornborg 1996: 53). Differently put, despite their undeniable analytical capacities, objectification and generalization fail to access a number of factors inherent in the relationship between humans and their environment in addition to factors explaining them. Ultimately, they neither explain the interactivity of these relations nor the engagement required for their implementation. The aim here is not to reject scientific knowledges and their peculiarities (such as objectification and generalization), but rather to value the set of knowledges built through interacting with the world. Caillon and Degeorges (2007) discuss this need to reconcile the biological and cultural dimensions of biodiversity (the taro in this case), especially for purposes related to conservation and deontology.

We believe that the valorization of these knowledges is inescapable regardless of the applied objectives of research. The willingness to understand the relationships between humans and non-humans, one which is the closest to the experienced and perceived realities, requires this valorization. Indeed, it appears that despite their asserted position as observant entities, scientists in fact interact with the world. They participate not only in understanding the territories, but also in constituting and transforming them.

The concepts of engagement and dwelling as we have defined them give a specific vision of territory and landscape, which should not be reduced to their human or ecological components. By incorporating the knowledges and practices of human communities who occupy them, ethnoecology alongside cultural geography, law, and anthropology, takes into consideration the existence of a territory which integrates and transcends the only ecosystemic component. Indeed, the concept of territory includes not only the diversity of the non-human, but also the diversity of the relationships between humans as well as the relationships between humans and non-humans. In addition to other elements, a territory is revealed by knowledge, perceptions, and present or past practices (Bonnemaison et al. 1999; West 2005;

Sabinot 2008; Sabinot et al. in press). Thus, this concept allows us to consider the dynamic and patrimonial dimensions of territories shared by humans and non-humans.

8.2 Constantly Adjusted and Negotiated Knowledge, Markers of Dynamic Relationships Between Man and His Environment

Far from being reduced to socio-centric cultural constructions, ecological knowledges appear as co-constructions arising from negotiations with non-human elements in the environment (Bird 1987; Walsh 2002; West 2005). These elements are not passive objects waiting to be categorized, classified, and used; they rather attract, provoke, act, react to, and interact with humans, especially when these elements are living entities. Indeed, "*whereas inanimate objects afford actions (to obtain the use values or to avoid the dangers), animate objects afford interactions, and socialized objects afford proper (as against improper) action and interaction*" (Reed 1988: 112).

One of the particular characteristics of local ecological knowledges or – more precisely – ecological knowledges related to practices, to an *engagement* with environmental elements (Ingold 1996), is that they are constantly *tried out* and *renewed* through regular relations, often daily interactions. These knowledges are experienced by individuals who build them, who enliven and transform them; these knowledges are also adjusted to the rhythms of social, economic, political, symbolic, and ecological changes. In doing so, the behaviors (*s.l.*) of the interacting elements and humans are modified by the relation, giving thus a mutual, dynamic, and historical dimension to this relation. However, as Ingold points out in one of his founding papers: "*while both humans and animals have histories of their mutual relations, only humans narrate such histories. But to construct a narrative, one must already dwell in the world and, in the dwelling, enter into relationships with its constituents, both human and non-human*" (Ingold 2000). In other words, the narrative of the mutual relationships between humans and non-humans can result only from those who experience them in a common environment.

Local ecological knowledges reveal the mode of being in the world of a society, and more specifically of a human group or groups sharing a specific living place, belonging to the same generation, or which is characterized by an occupation or a community of practice – CoP (Greenfield and Lave 1982). Much of these knowledges are not "simply" transmitted; and when specific learning contexts that involve several individuals are created, these contexts become fields where knowledges are constituted and renewed at the crossroads of human–human and human–non-humans interactions (Lave and Wenger 1991; Ingold 1993; Pálsson 1994; Sabinot 2007). In fact, these constantly questioned and renewed knowledges participate in the process of education of attention.

Local knowledges exist everywhere and are characterized by their ability and propensity to change and adapt. They are particularly constructed, processed, and confronted to others, both humans and non-humans, notably by practitioners of the environment, those who frequent the environment on a daily basis. They are farmers, livestock breeders, hunters, fishers, managers, and so on; and they sometimes combine several of these practices. The richness, frequency, and regularity of these practitioners' interactions with their environment ensure the richness of their knowledges as well as their replenishment adjusted to the modifications of the environment, the modification of interactions among non-humans and between humans and non-humans. Numerous studies on the knowledge and expertise of fishers all over the world show, for example, that an excellent understanding of elements, with which fishers interact, allows the fishers to construct and adjust their ecological knowledge on a daily basis (Acheson 1981; Pálsson 1998; García-Quijano 2007; Sabinot 2007; Lauer and Aswani 2009 entre autres). Learning how to fish is not acquired through listening but through practicing, through directly experiencing the interactions with the elements, the transformations that affect each element as well as the interactions between all the elements; that is by dwelling in the world and not by observing it. Similarly, the knowledges of Kirghiz breeders and hunters on animals like the wolves are constantly renewed because wolves' behaviors keep changing (Lescureux 2006; Lescureux and Linnell 2010).

These knowledges may be marked either by mild or brutal adjustments. The emergence of new materials may lead to major transformations of a tool, which leads to using it in another ecological context; and consequently, the emergence of new environmental knowledge is provoked. If knowledge may emerge from practices, it is also validated by them (Lescureux 2006; Sabinot 2008). Major political or ecological upheavals may also lead to major changes in the relational context, which may require a renegotiation of the relationships between human groups and particular elements of their environment. Thus, the major political and socio-economic changes in the countries of the former Eastern Bloc, after the fall of the Berlin Wall, affected the relationships between some societies and their environment down to the smallest details. Particularly, the fall of the Soviet Union and the collapse of Yugoslavia led to changes in farming and hunting practices in the countries concerned; these changes affected the relationship between breeders, hunters, and wolves in ways that depend on ecological, historical, social, and cultural contexts (Lescureux 2006; Lescureux and Linnell 2013). On the other side of the Atlantic, in Cuba, this same revolution which appears as a social, economic, and political crisis led to major changes in the knowledges and practices of fishing, which changed the relationship between humans, and between fishers and their everyday environment (Doyon 2007). Everywhere on the planet, major upheavals of the environmental context also lead to renegotiating the terms of the relationships between human groups and their environment. Anthropologists are increasingly describing these renegotiations against the transformations induced by climate change through mobilizing adaptation and resilience concepts (Lazrus 2005; Crate and Nuttall 2009; Hastrup and Rubow 2014).

Finally, ecological knowledges are tested through the production of scientific, legal, and administrative knowledge. Sometimes, these last questions, disturb and reconfigure the locally produced knowledge through regular interactions with the environment; and other times, they highlight and strengthen local knowledge. In New Caledonia, the redefinition of knowledge on the green sea turtle, *Chelonia mydas,* in a context of formulating its protection policies, illustrates the dynamic dimension of the knowledge itself and its attributed value (Bernard et al. 2014; Sabinot and Bernard 2016). Having a strong symbolic, cultural, and customary status in the Kanak context, the turtle is classified as a vulnerable species on the red list of endangered species of the IUCN. Emblematic in various ways, the green turtle, whose international status has been translated into various local environmental policies since 1977, is currently at a crossroad of administrative and customary pathways. Since 2008, the Environmental Code of New Caledonia's South Province considered the turtle as a fully protected species for which exceptions are possible for customary and scientific purposes. Exceptional "ecological," scientific, and ritual dimensions are recognized but the third dimension is barely defined in the text itself, which creates difficulties within tribes and within the provincial department which has to determine the customary or non-customary character of the received application for exemption. Legal and scientific knowledge are unable to solve the dilemma; provincial officials call upon ethnoecological and anthropological knowledge that based themselves on ecological knowledge produced locally, translating not only the relationship between humans and non-humans but also between humans. The ecological knowledges related to the green turtle, *do ngûû*, translated literally as "true turtle" in Numèè language, are shaken by new knowledges produced on the scientific and legal levels. The study of the transformation of these knowledges allows us to understand the changes that affect the hierarchical relationships between the clans, in a context of rapid transformation of knowledges and practices in food-production and traditions. In response to the "turmoil" produced by exogenous knowledge and standards, the knowledges produced locally are reconsidered, reclassified and even reinforced by the "holders" and "producers" of such knowledges. Today, they are at the heart of new discussions within the provincial institution, as well as in the Environmental Advisory Council bringing together customary authorities of New Caledonia's Southern area or their representatives.

The confrontation between various types of knowledges reaffirms the quality, the legitimacy, and the locally-adapted character of local knowledges. Beyond the knowledges, the provincial and tribal, collective and individual legitimacies confront and redefine each other. Renegotiating norms regarding animal management is associated to redefining knowledges and values along with their expression. Constructing pluralistic or integrative norms is then discussed, thus underlining the debate between two opposite conceptions of law. As noticed by other authors (Escobar 1998; Doyon and Sabinot 2014; Rosillon 2014), when implementing environmental standards that are dictated or initiated by the outside, such reconfigurations often emerge, that lead to consider local knowledges, norms and values in a different way, and to the possibility of their intermingling with standards and knowledge belonging to often normative global categories.

8.3 Local Ecological Knowledge As an Indicator of Coviability?

By focusing on the diachronic dimensions of the relations between humans and non-humans on a shared territory, ethnoecology alongside with other disciplines (such as ecology, archeology, history, geography) investigates the viability of these relationships and therefore the coviability of these associations on a larger or smaller time scale.

Defining coviability without any spatial or temporal scale remains a difficult task. Even if our knowledge on societies and on ecology is constantly growing, many parameters still escape us. The analytical validity of collected data does not necessarily provide them with a predictive characteristic. Can we speak of viability in a perpetually changing world? Furthermore, there is major difficulty in defining the coviability of social systems and ecosystems, which are not usually conceived on the same spatial and temporal scales. Even with shifting boundaries, an ecosystem is spatially defined by interactions among the species that constitute it. However, it is rare for a society, a socio-system, to coincide spatially with the said ecosystem. Either the socio-system inhabits a part of the ecosystem, or it spreads over several ecosystems. The viabilities of relationships therefore need to be investigated from multiple perspectives, especially as the transformations that affect them both, happen at different paces. The viability of the relationship between humans and non-humans, if understood through ecological knowledges, aims at considering the socio-system as the scale of analysis. Indeed, in the present state of knowledge, only humans – with whom we share a common language – are capable of reporting in details and in a conscious way their relationship with non-humans.

In any case, ecological knowledges and their dynamic function appear as privileged indicators of environmental changes and may allow societies to adjust their behavior within the ecological and socio-cultural frameworks, which are more or less constraining and more or less dynamic. If they are analyzed as coviability indicators, ecological knowledges must be understood as adjustments of societies with their environment. They are temporary and incessantly renewed. Nevertheless, it appears that some "arrangements" (some "bricolages" in french) between human groups and their environment are maintained and show a certain adaptability, without necessarily disturbing the major characteristics of the system in place. They reveal a way of relating to the environment which could serve as a model of coviability. As long as they do not jeopardize the sustainability of the relationship between a society and its environment, these adjustments allow the society to carry on in a territory which may itself be redefined through interactions with other societies.

The adjustments in local knowledges falls under a process of negotiation subjected to the need to maintain the viability of several elements within the system. Additionally, it is no longer possible to consider systems as being closed. Thereby, the viability of a system is not only related to the coviability of its constitutive components but also to its own coviability with external elements. Can we speak thus of a coviability of different cultures,[3] of different relations to the world on reducing spaces or intertwining territories? How can one's relationhips to the world, one's relationhips to the environment be adjusted to the new standards which are built on scales that significantly exceed those of one's territory?

The conducted studies in multi-cultural contexts produced by migrations for economic, political or climatic reasons may fuel these questions. Reflections on the appropriation of territories and the exchange/production of knowledges that may arise from the encounter of migrants and natives, show that arrangements and adjustments are numerous and varied. They may go through conflictual stages over larger or smaller time scales. In this context, the study of the management of bamboo fishing territories (to catch small catfish) in southern Gabon, more precisely, the Banio Lagoon is an illuminating example (Sabinot 2007, 2008). These places are usually inhabited and managed by the *Vili* women in the area. Since they happen to marry Senegalese migrants, these latter acquired new ecological knowledge, while building a certain legitimacy on the territory, enabling the creation of new types of breeding ponds which require access to a shallow lagoon area neighboring the village. Downstream of the brackish lagoon, the "oyster diving," which was previously practiced only by Gabonese, Congo and Benin women, has, for the last couple of decades, been practiced by some men. Symbolic boundaries related to the force of the water spirits associated with oysters, aquatic depths, and women, as well as social and technical boundaries (separation of genres and organization of transmission between generations or between peers and within groups), were overcome. This led to a new spread of ecological knowledges, leading to a transformation of the modalities of their transmission and finally modifying the relationship to places and beings inhabiting them, humans or non-humans.

8.4 Conclusion

As stated in the introduction, the concept of coviability was until now not used in our disciplinary field. In this chapter, we chose to discuss the ecological knowledges and their construction, transmission and negotiation processes. Indeed, studying and analyzing ecological knowledges appear essential for understanding the socio-environmental transformations from the viewpoint of society and apprehending how social systems respond to the modifications of ecological systems and of global or nearby social systems. We do not claim, in this short chapter, to question the relevance of the concept of coviability for our disciplinary approaches or, conversely, to question the relevance of ethnoecology in contributing to the understanding of the concept of coviability. Only if it is further borrowed and discussed in the field of ethnoecology, as well as questioned on the field by many of us, may we be able to carry on with these early reflections.

[3]See Barrière Catherine's chapter in this book.

Nevertheless, this chapter has shown that ethnoecology, which puts the study of local ecological knowledge at the heart of its topics, allows the viabilities of larger or smaller time scale relationships to be understood as well as observing and analyzing them, while also considering the engagement of the researcher in the world that s/he tries to understand. Ethnoecology seeks also to describe and analyze knowledge by adopting synchronic and diachronic approaches. Therefore, understanding the dynamics of knowledges and their learning and transmitting processes render it possible to appreciate the dynamics of relations' viability. We suggest that local ecological knowledges, revealed by ethnoecology, may be considered as coviability indicators, or at least indicators of the viability of relationships between humans and non-humans. It is interesting to note that this knowledge tends nowadays to be integrated (in a more or less a relevant and fortunate way) in the reflections on ecological crises at the planetary level, particularly within the Intergovernmental Science-Policy Platform on Biodiversity and Ecosystem Services, IPBES.

Finally, it is obvious that one disciplinary field – whichever one it is – cannot afford to understand the complex concept of coviability alone. Collaboration is necessary not only between different disciplines, but also between different knowledge systems, which implies respecting other knowledges and recognizing the subjectivity of knowledges in general and of their various meanings. Thus, shouldn't we follow Hornborg's thoughts when he suggests that "*Once we recognise that human subjectivity, along with the subjectivity of all the other species, is an aspect of the very constitution of ecosystems, we have a solid foundation for the conclusion that the destruction of meaning and the destruction of ecosystems are two aspects of the same process*" (Hornborg 1996: 53)?

References

Acheson JM (1981) Anthropology of fishing. Annu Rev Anthropol 10:275–316

Berlin B, Breedlove DE, Raven PH (1973) General principles of classification and nomenclature in folk biology. Am Anthropol 75:214–242

Bernard S, Lacombe S, Lancelot L, Herrenschmidt J-B, Sabinot C (2014) Dynamiques des habitudes, des pratiques et des savoirs relatifs à l'usage et à la gestion du littoral et de la mer dans un contexte de pression industrielle sur le milieu et de changements sociaux. Programme LIVE (Littoral vivrier et environnement). IRD-Océanide-CCCE, Nouméa

Bird EAR (1987) The social construction of nature: theoretical approaches to the history of environmental problems. Environ Rev 11:255–264

Bonnemaison J, Cambrézy L, Quinty-Bourgeois L (1999) Université de Paris IV: Paris-Sorbonne, and O.R.S.T.O.M. (Agency: France). Le territoire, lien ou frontière? Harmattan, Paris

Caillon S, Degeorges P (2007) Biodiversity: negotiating the border between nature and culture. Biodivers Conserv 16:2919–2931

Crate SA, Nuttall M (eds) (2009) Anthropology and climate change. From encounters to actions. Left Coast Press, Walnut Creek

Descola P (2013) Beyond nature and culture. University of Chicago Press, Chicago

Descola P, Pálsson G (1996) Nature and society. Anthropological perspectives. Routledge, London

Dove MR, Smith DS, Campos MT, Mathews AS, Rademacher A, Rhee S, Yoder LM (2007) Globalisation and the construction of Western and Non-Western knowledge. In: Sillitoe P (ed) Local science vs global science. Approaches to indigenous knowledge in international development. Berghahn Books, New York, pp 129–154
Doyon S (2007) Fishing for the revolution. Transformations and adaptations in cuban fisheries. Marit Stud 6:83–108
Doyon S, Sabinot C (2014) A new 'conservation space'? protected areas, environmental economic activities and discourses in two Yucatán biosphere reserves in Mexico. Conserv Soc 12:133–146
Ellen R, Katsuyoshi F (1996) Redefining nature: ecology, culture and domestication. Berg, Oxford
Escobar A (1998) Whose knowledge, whose nature? biodiversity, conservation, and the political ecology of social movements. J Political Ecol 5:53–82
Fossier A, Gardella É (2006) Entretien avec Bruno Latour Tracés. Revue de Sciences humaines 10
Friedberg C (1997) Diversité, ordre et unité du vivant dans les savoirs populaires. Nat Sci Sociétés 5:5–17
García-Quijano CG (2007) Fishers' knowledge of marine species assemblages: bridging between scientific and local ecological knowledge in Southeastern Puerto Rico. Am Anthropol 109(3):529–536
Gibson JJ (1986) The ecological approach to visual perception. Lawrence Erlbaum Associates, Hillsdale/New Jersey/London
Greenfield P, Lave J (1982) Cognitive aspects of informal education. In: Wagner D, Stevenson H (eds) Cultural perspectives on child development. Freeman, San Francisco, pp 181–207
Hastrup K, Rubow C (eds) (2014) Living with environmental changes: waterworld. Routledge, Abingdon/New York
Hornborg A (1996) Ecology as semiotics. In: Descola P, Pálsson G (eds) Nature and society: anthropological perspectives. Routledge, London/New York, pp 45–62
Hunn E (2007) Ethnobiology in four phases. J Ethnobiol 27:1–10
Hviding E (1996) Nature, culture, magic, science. In: Descola P, Pálsson G (eds) Nature and society: anthropological perspectives. Routledge, London/New York, pp 165–184
Ingold T (1993) The temporality of landscape. World Archeol 25:152–174
Ingold T (1996) Hunting and gathering as ways of perceiving the environment. In: Ellen R, Fukui K (eds) Redefining nature: ecology, culture and domestication. Berg, Oxford, pp 117–154
Ingold T (2000) From trust to domination. An alternative history of human-animal relations. In: Ingold T (ed) The perception of environment. Essays on livelihood, dwelling and skill. Routledge, London/New York, pp 61–76
Ingold T (2004) Beyond biology and culture. The meaning of evolution in a relational world. Soc Anthropol 12:209–221
Ingold T (2005) Epilogue: towards a politics of dwelling. Conserv Soc 3:501–508
Latour B (1993) We have never been modern. Harvard University Press, Cambridge
Latour B (2001) Le métier de chercheur, regard d'un anthropologue. INRA, Paris, p 103
Lauer M, Aswani S (2009) Indigenous ecological knowledge as situated practices: understanding fishers' knowledge in the Western Solomon Islands. Am Anthropol 111:317–329
Lave JW, Wenger E (1991) Situated learning: legitimate peripheral participation. Cambridge University Press, Cambridge
Lazrus H (2005) The role of knowledge in global climate change governance: modes of legitimation in Tuvalu. University of California, International and Area Studies, Berkeley
Lescureux N (2006) Towards the necessity of a new interactive approach integrating ethnology, ecology and ethology in the study of the relationship between Kirghiz stockbreeders and wolves. Soc Sci Inf 45:463–478
Lescureux N, Linnell JDC (2010) Knowledge and perceptions of macedonian hunters and herders: the influence of species specific ecology of Bears, Wolves, and Lynx. Hum Ecol 38:389–399
Lescureux N, Linnell JDC (2013) The effect of rapid social changes during post-communist transition on perceptions of the human – wolf relationships in Macedonia and Kyrgyzstan. Pastoralism Res Policy Pract 3:4

Morgan CL (1903) An introduction to comparative psychology. Walter Scott Publishing, London
Pálsson G (1994) Enskilment at sea. Man 29:901–927
Pálsson G (1998) Learning by fishing: practical engagement and environmental concerns. In: Berkes F, Folke C (eds) Linking social and ecological systems. Management practices and social mechanisms for building resilience. Cambridge University Press, Cambridge, pp 48–66
Reed ES (1988) The affordances of the animate environment: social science from the ecological point of view. In: Ingold T (ed) What is an animal. Routledge, London, pp 110–126
Rosillon C (2014) La gouvernance environnementale au Sud: regard sur un projet de conservation des tortues au Sri Lanka. ethnographiques.org
Sabinot C (2007) Des hommes, des cultures, des savoirs et des savoir-faire en mouvement sur le littoral gabonais. Dynamique des savoirs et savoir-faire. Mobilités, immobilismes. L'emprunt et son refus, pp 69–82
Sabinot C (2008) Dynamique des savoirs et des savoir-faire dans un contexte pluriculturel. Étude comparative des activités littorales au Gabon. Muséum National d'Histoire Naturelle, Paris
Sabinot C, Bernard S (2016) An emblematic marine species at a crossroads in New Caledonia: Green Turtle. Resources, boundaries and governance: what future for fisheries in the pacific? Fache E. et Pauwels S
Sabinot C, Chlous F, David G, Dégremont M (in press) Knowing and truly appreciating the diversity of values according to marine environment. In: Germain L, Mongruel R, Ridoux V, Thiebaut E, Sabinot C, Walter C, Watremez P (eds) Multi-goal Marine Portected Areas: a new way to manage the sea? Springer
Toledo VM (1992) What is ethnoecology? origins, scope and implications of a rising discipline. Etnoecologica 1:5–21
Walsh A (2002) Responsibility, taboos and 'the freedom to do otherwise' in Ankarana, Northern Madagascar. J R Anthropol Inst 8:451–468
Weber J, Betsch J-M, Cury P (2013) À l'interface hommes-nature: les ressources renouvelables. In: Bouamrane M, Antona M, Barbault R, Cormier-Salem M-C (eds) Rendre possible. Jacques Weber, itinéraire d'un économiste passe-frontières. Quae Editions, pp 81–93
West P (2005) Translation, value, and space: theorizing an ethnographic and engaged environmental anthropology. Am Anthropol 107:632–642

Chapter 9
Biotic Interactions, Coviability and Dynamic of Biodiversity

Laurence Pascal, Catherine Moulia, and Laurent Gavotte

Contents

9.1 Introduction

Every living organism on our planet is in direct and often complex interaction with the different elements that make up its environment, whether they are abiotic (physical character) or biotic (all of the other species living in this environment).

L. Pascal (✉)
University of Montpellier, Montpellier, France
e-mail: laurence.pascal@umontpellier.fr

C. Moulia · L. Gavotte
Evolutionary Sciences Institute of Montpellier (ISEM 5554), University of Montpellier, Montpellier, France
e-mail: catherine.moulia@univ-montp2.fr; laurent.gavotte@umontpellier.fr

O. Barrière et al. (eds.), *Coviability of Social and Ecological Systems: Reconnecting Mankind to the Biosphere in an Era of Global Change*,
https://doi.org/10.1007/978-3-319-78497-7_9

The ability of species to deal with these different elements and the biological balances resulting from them are the fruit of different evolutionary histories that will lead to the emergence of new trait. Therefore, the most integrated inter-specific associations which exist in the living world, symbioses, are particularly conditioned by the evolution of each partner species and their interactions. Symbiosis constitutes a lifestyle, which, for two organisms of different species, consists in living in a condition of sustainable interaction in space and time. This interaction may be neutral (commensalism), advantageous for one of the two partners and costly for the other (parasitism) or advantageous for the two protagonists (mutualism). Thus, one of the partners (in the case of parasitism) or both of them (in the case of mutualism), gain a definite advantage from the interaction, to the extent that often, they can no longer survive alone! This close and total dependence in symbioses represents a perfect example of long-term **coviability** in the living world as we are going to demonstrate in this chapter.

We will start by introducing the theoretical concept according to which they operate, that of the appearance and maintaining of biodiversity. We will consequently try to understand the general mechanisms of the Evolution of living organisms to highlight how, owing to their central role in biodiversity, symbioses wholly correspond to a biological definition of coviability. We will conclude by showing how man destroys this "coviability" of natural systems in many examples and try to consider the lessons to be learned to ensure the future of humanity.

9.2 Biodiversity and Evolution

9.2.1 Evolutionary Theories

To understand the existence and the functioning of symbioses, it is necessary to properly understand the mechanisms of the evolution of species. Several often complementary theories exist to describe and explain these complex mechanisms.

The topic of evolution first came to light during the nineteenth century and is the result of many works, the most famous and complete of which are attributed to Darwin (1859) and Wallace (1870). The **theory of natural selection,** continued to develop alongside discoveries during the twentieth and twenty-first centuries, which namely included the most up-to-date knowledge in genetics. Consequently, the variable nature of the genetic information support, DNA,[1] allows the appearance of mutations, a part of which will cause the appearance of new traits, which, depending on the environmental conditions in which the organism finds itself, will favor, or not, the individuals expressing them. . It is solely the aspect of success of the offspring of these individuals, modified or not by new traits and defined according to the notion of **fitness** (which represents the capacity of survival and

[1]Desoxyribonucleic Acid: genetic information carrier molecule.

reproduction of an individual), which will determine their evolutionary future. As specifically described by Jacques Monod (1973), two mechanisms, firstly the **chance** for the appearance of new traits owing to DNA mutations and secondly, their **necessity** in the environment, result in biological diversity. The best-known example highlighting natural selection was described by Charles Darwin himself after his expedition around the world. In 1835, Charles Darwin amassed several finches on the various archipelago islands. A study of these specimens showed that they belonged to 13 species which morphologically, were both very different yet very similar. The main features differentiating these birds were the size and shape of the beak. Darwin proved that these differences corresponded to the plant resources consumed by the different species (Darwin 1859). Ecological and modern molecular biological works have since proven that Darwin's hypotheses were based on this group of species (Fig. 9.1) (Rands et al. 2013).

Similarly, many very similar plant species have specific adaptations allowing them to exploit different soil types. Consequently, the four species of Passiflora belonging to the *Passiflora vitifolia* complex are differentiated according to their adaptation to four types of soil (Gentry 1981). In some species of the *Protieae genus*

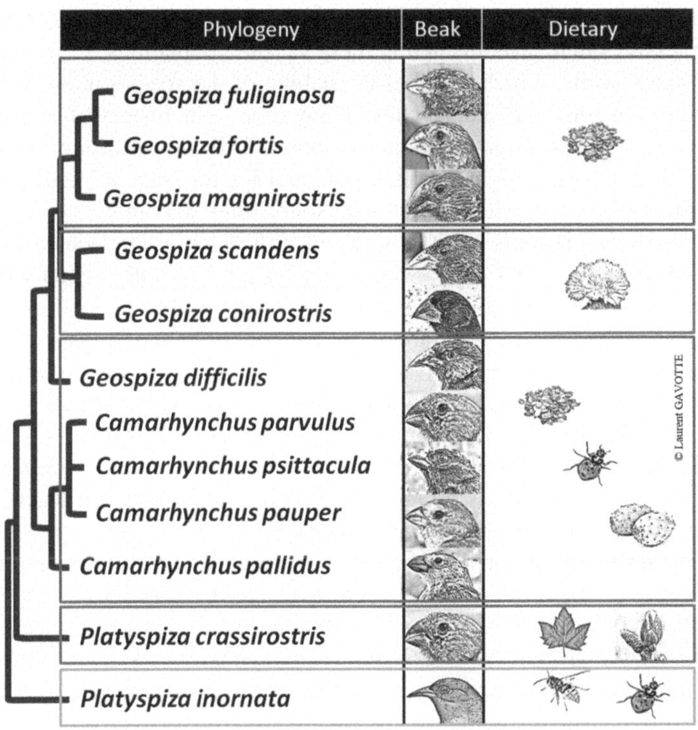

Fig. 9.1 Phylogeny of Darwin's finches in the Galapagos Islands associated with the shape of their beaks and diets. (Adapted from RAND et al. 2013)

(Fine et al. 2004), phylogenetic studies indicate that the heterogeneity of soils has been an important factor in the evolution and speciation of this group. Finally, both the biotic and abiotic environment in which an organism lives determine over the selection pressures to which it will be submitted during its evolution; the resulting natural selection therefore seems an essential driver in the adaptation and evolution of species according to Darwinian Theory.

One of the interesting concepts in Evolution and Ecology is that each species possesses an **ecological niche** which is defined by a set of environmental conditions, allowing it to live and multiply (Vandermeer 1972; Hanski 2001; VanderMeer 2008). The structures of niches therefore determine the species's colonization potential, but also determine the various **selection pressures** that guide species's evolution. For example, the different species of penguins which colonized extremely varied niches in the southern hemisphere (from Antarctica to the Equator), present with extremely diverse traits in line with their environment. Consequently, Cape penguins (*Spheniscus demersus*) have few feathers, some parts of the head have none at all, as they spend a large part of the day in water to protect themselves from the Equatorial heat (Frost et al. 1976). In contrast, Adelie penguins (*Pygoscelis adeliae*), which have many feathers, gather in dense groups during the cold winter and have a complex respiratory system that permits a part of thermal energy to be saved when they exhale (Chappell and Souza 1988). Each species has been subjected to different selection pressures that have resulted in these various adaptations.

Furthermore, whilst a niche defines the habitat of the species at a given time *t*, it can evolve over time and space. Thus, at any time in its history, this species can find itself outside of its assumed habitat at time *t*. This passage from one niche to another (with, a priori, a higher growth potential for this species) will permit the colonization of the space, within a so-called source-sink dynamic.

The niche theory permits habitats to be identified but is inadequate to explain the spatial distribution of species, because it gives a fairly static vision which puts forward the idea that communities live in equilibrium with their environment. We can consider the environment as a landscape, a set of "Habitat patches" (constituting the niche) for a given species, surrounded by an "off-niche" space (outside of the habitat, not conducive to the development of the species). Described in this way this landscape defines a meta-population, a set of species populations living in patches. The extinction of a species depends on its density, on the surface area of the patches forming the meta-population and on migration between patches which will be governed by the distance between them.

The **neutral theory**, initiated by the Japanese geneticist Motoo Kimura in the late 1970s (Kimura 1968), constitutes a break with Darwinian concepts by highlighting the central role of **chance** in evolution to the detriment of natural selection. This theory explains the diversity of organisms via a phenomenon known as the genetic **drift** that describes the maintaining, the fixing or the disappearance of a mutation by (random) stochastic phenomena. The smaller and more isolated the population, the stronger the drift. It is this mechanism which best explains the great diversity observed in island systems because each island, for species that cannot cross the sea, shelters a population isolated from others, which will *drift* and therefore present with unique characteristics (Blondel et al. 1999; Eibl et al. 2001). The neutralist theory, even if it breaks with Darwinian Theory, is not in opposition with the theory of natural selection. It has a different mechanism to explain all or a part of the evolution of organisms (on islands, the natural selection of the characteristics which have appeared by mutation will also continue). Currently, even if the debate is not closed, the two visions are no longer exclusive and it is more about determining the respective roles of the drift and of natural selection in the appearance of the biodiversity. The **neutral theory of biodiversity,** commonly called **UNTB,** "Unified Neutral Theory of Biodiversity", developed by Hubbell (2001), reconciles the two neutralist and niche approaches by bringing an ecological vision, which considers species as being equivalent for a same trophic level within a niche, and explains the distribution of species by demographic stochasticity (random variation in the survival rates of individuals in a given population) and by the capacity of species to immigrate. The UNTB applies to studies on relationship systems between species and to the mechanisms that shape the networks of species (Bascompte et al. 2003). Krishna and her collaborators (2008), point to the fact that the interlinking of species within networks can be easily explained when taking into account specific traits due to interaction (such as the importance of particular fruits in the diet of a frugivorous), the importance of the relative abundance of species, which are all characteristic factors of a niche, consequently proposing the addition of the niche to the neutralist theory, "a neutral niche theory", concerning the interweaving of species within networks.

A mosaic of niches with defined parameters would therefore condition the evolution of species. However, an organism is not neutral for its environment: owing to its simple metabolism, it draws molecules from this environment and modifies them for its own use. A plant or an arthropod will modify the texture of the soil by extending its roots or by digging galleries; a beaver will create a dam and change the dynamics of a river etc. The **theory of niche construction**, which is more complete than the niche theory, assumes that an organism will be influenced by its environment and conversely, that it will modify this environment, and consequently more or less create its own evolutionary niche (Day et al. 2003).

Humanity is the most obvious and important modern examples of niche construction. Man has built his niche by modifying the environment according to a cultural process deeply affecting his evolution. Cultural practices, selection factors according to an interaction between gene and culture, have structured the human genome throughout its history (Laland et al. 2010). For example, West African forest was partially modified by Kwa populations who cut into local vegetation in order to free up the territory needed to plant root crops. These plants are grown according to a cultural practice that structures the space into a network of ridges between which water stagnates during the rainy season, favoring the installation and development of malaria-carrying mosquitoes in these regions. A strong selection of the sickle (S) hemoglobin allele (*HbS*), conferring resistance to malaria, is observed among Kwa populations according to a much higher frequency than among neighboring populations who do not practice this type of agriculture (Fix 2003). Other works analyzing such genetic variations among human populations support these results, and reveal that 100 or so genes would be concerned by recent positive selections in response to human activity (Boyd and Richerson 1995; Beja-Pereira et al. 2003).

9.2.2 Dynamics of Biotic Interactions

Let's go back to the niche construction theory. In effect, it brings us to the interesting concept that this theory changes over time, and to a still more challenging idea that this dynamic is intimately linked to species interaction dynamics.

This is the case in the trophic interactions of the predation or herbivore type. In the *Eucalyptus* genus for example, the quality of the leaves varies significantly depending on the environment, and according to the life cycle of different species, consequently offering a variable quality of food to phytophagous species, whose diversity in the niche evolves over time and space (Steinbauer et al. 2015). The balance between primary and secondary foliar metabolites, which reflects previous ecological conditions, shows the action of species on the construction of the niche, participating in resource dynamics by acting on the fitness of individuals and their evolution.

We can also identify excellent illustrations of niche construction thanks to interactions between species in closer, more intimate relationships, for example in the case of plant-ant associations, which are often referred to as mutualistic. Let's take the case of the *Passiflora*- ant model, where a guild of ants attracted to the foliar nectar protects leaves from herbivores thanks to repeated patrols (Labeyrie et al. 2001). Ants live according to specific preferential associations with *Passiflora* growing in sympatry. These associations are structured by the formic acid content of the foliar nectar of *Passiflora* and according to whether they belong to the group of *Formicineae* or non-*Formicineae* ants (any other group of ants). Ants belonging to the Formicineae subfamily, which are the most aggressive towards herbivores, produce formic acid themselves and are the only species able to consume nectar containing this acid. *Passiflora* producing this type of nectar such as *Passiflora glandulosa*, consequently provide a new food niche reserved for ants belonging to the *Formicineae* subfamily, and ensure an effective protection against herbivores whilst producing less nectar (Fig. 9.2). The ants from this subfamily detect volatile formic acid at great distance and invade the plants that produce it. Following an attack by herbivores [or simulated attacks by herbivores, as in the experiment presented in Fig. 9.2a, where leaves are damaged (at 21:30 orange arrow), as would be caused by a herbivore], the foliar nectars of *Passiflora glandulosa* produce a very large amount of volatile formic acid. Does this acid play the role of chemical mediator in attracting *Formicinae* ants and/or in repelling non-*Formicinae* ants and herbivores? Its presence in the nectar repels ants with the exception of *Formicinae ants*, which consequently benefit from a preferential food source. A very demonstrative example of niche construction!

By entering the world of symbioses in the strictest sense of the term and as defined in the introduction to this chapter, we can note that symbionts are also able to

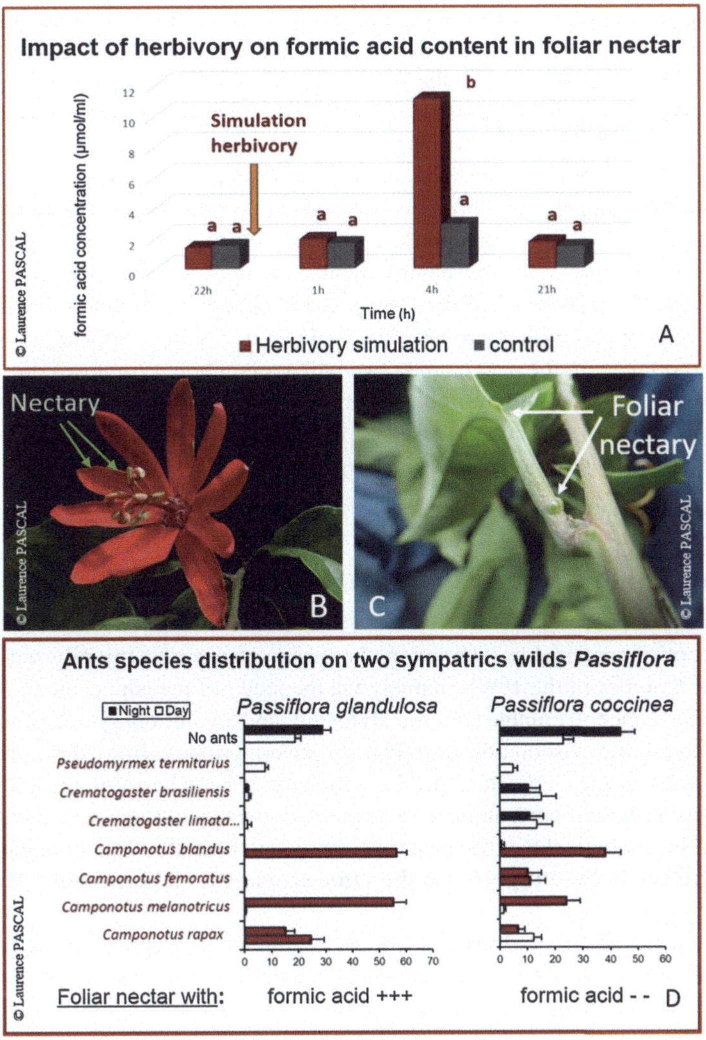

Fig. 9.2 *Passiflora* and its protective ants
Leaf and budflower nectars attract protective ants against herbivores (**b**, **c**). Foliar nectar produces formic acid including following an attack by herbivores. Nectar composition is detected in 20 stems following a simulated herbivore attack over a 24 h cycle (**a**). Formic acid is a volatile compound which attracts *Formicineae ants* over a long distance *Passiflora glandulosa* nectar contains 10 times more formic acid than that of *P. coccinea* and is preferentially consumed by the most aggressive *Formicinae* ants (in red histograms, **d**), (data and photos from Pascal L.)

modify (at least locally) the biological characteristics of their hosts (which are none other than their niches), in order to adapt them to suit their physiological needs. Parasites are particularly effective in these changes. An example of this are the, larvae of parasites belonging to the *Trichinella* genus, which include species capable of infesting several species of mammals, including humans, and form cysts in the muscles of their host. In addition to changes to the muscle cell to their advantage, around their feeder cell these larvae provoke the production of new blood vessels by producing molecules mimicking those produced naturally by mammals resulting in angiogenesis (Kang et al. 2011). This provides them with a greater amount of nutrients in their host's blood. Similar mechanisms can be found in the interactions between *Agrobacterium* genus parasitic bacteria and their host plant (Zhu et al. 2000).

The theory of **niche construction** permits an analysis of the evolutionary processes in keeping with the environmental changes induced by all biological species. Via this approach, we discover a world living in a dynamic equilibrium, based on inter-specific interactions whose role, if not central, appears to be of the utmost importance. In the next chapter, we will explain how symbioses, interactions which are in place on a long-term basis in the evolutionary sense (a long duration in the history of life), constitute typical examples of biologic associations within which "we co-live", in other words, perfect examples of **coviability** in biological systems.

9.3 Sustainable Interactions and Coviability

The concept of sustainable interactions between species was initially put forward by Claude Combes in the 1990s, namely via the study of parasite ecology (Combes 1997). This concept emphasizes the dynamic and evolutionary relationships of complex symbiont association systems, on several scales, from the gene to the community of species. Parasitic species, which a priori are harmful, may, by a set of interactions with their environment, control the dynamics of the populations present and install an ecological balance ensuring biological diversity is maintained, making the **coexistence between species in the same ecological niche possible** (Hatcher et al. 2006).

Whilst mutualistic symbioses allow an organism to acquire new capabilities that its genome does not contain, parasites are no less important in the history of evolution and in the implementation of biodiversity. Therefore, all organisms devote a large part of their resources to defending themselves against other organisms whether they are predators, competitors or parasites. The better we understand the diversity of life and the structure of genomes, the more it seems that the major part of evolution can be attributed to the selection pressures exerted on each species by other species belonging to the ecosystem in order to appropriate, divert or steal resources. Many major events in the diversification of life can consequently be traced thanks to the appearance of new species in interactions. The consequences of these events are extremely far-reaching, and in some cases, the genomes of species in interaction are so closely related, it is easy to forget how central these relationships were for maintaining life during evolution. We can mention mitochondria at the origin of eukaryotic cells and their energy synthesis, chloroplasts and the origin of plants and the production of organic matter (autotrophic), lichens, Mycorrhizae and *Rhizobium* which intervene in the process cycles of mineral elements; intestinal symbioses which ensure the digestion of complex molecules in animals (heterotrophy) or the existence of the immune system to combat the parasitic "intruders". It is now common knowledge that all of the organisms we know of live in symbiosis (whatever the modality) with one or more symbionts, and that these relationships from one organism to another ensure the survival of all of the species present on our planet, as we know them. This **integrative dimension of species** highlights not only an unsuspected hidden aspect of the biodiversity of organisms, but also the fact that living organisms can never live alone in a given environment.

For example, without knowing or desiring it, man is responsible for maintaining hundreds of different species. Whether at his expense with a huge number of parasites for which our species is an indispensable resource (thousands of virus species, bacteria, fungi, unicellular eukaryotes or invertebrates) or to his advantage with hundreds of species sharing a mutualistic symbiotic life with us, at the forefront of which we find all of the organisms hosted in our digestive tract that help us to assimilate food resources but which also synthesize molecules which are essential for our metabolism.

The concept of coviability is therefore a reality which is intrinsically linked to the ecology of symbiotic species given that all organisms live in close conjunction with a number of other organisms and these interactions are essential for the survival, reproduction and evolution of these different organisms. Ecological and evolutionary mechanisms tend towards dynamic balances (which are therefore unstable over time) which contribute towards maintaining different species and therefore towards biodiversity. This concept of coviability is pushed to its extreme within the context of symbioses, and applies to both mutualistic and parasitic interactions. In both of these cases, the existence of a strict dependency between two organisms results in the implementation of mechanisms enabling them to live (with benefits or disadvantages) in the same ecosystem. This coviability is conditioned by co-evolutionary processes which have resulted in the current situation and which tend toward an optimization of the characteristics of both species within the context of symbiosis.

The level of coevolution and the mechanisms implemented depend on several factors: (i) the interaction mode; (ii) the density of this interaction (all or a part of the populations of each partner involved in symbiosis); (iii) the duration of the symbiosis on the evolutionary scale (indexed to the number of generations of each organism); (iv) the evolutionary plasticity of each of the partners and (v) the other necessities of each of the organisms in their interactions with their biotic and abiotic environments.

Mutualism allows two organisms to achieve phenotypes that they would be unable to achieve alone. It **therefore implies an interdependence between the two organisms**. Against intuition (because this dependence is only positive in one sense), this is also the case for parasitic relationships. Indeed, the parasite is entirely dependent on the existence of its host (s). At the level of evolution, a parasite that provokes the disappearance of its host species, provokes its own demise. It will therefore put in place a balanced virulence (of the cost) of parasitism that will vary according to numerous factors. *Ophiocordyceps* fungi manipulate the brain of host ants that irreversibly bite a support at a height and remain stuck to it, which facilitates the dispersal of the fungus. The specificity of these fungi was laboratory tested on many species of ants which had never been infected and on ants already affected by the host-parasite relationship in a natural environment. All of the infected ants die following contact with the parasite, whilst only the specifically associated host-ant survives and is merely manipulated by the fungus. The fungus is capable of *recognizing* the brain of its specific host-ant. It reacts heterogeneously according to the brain of the different ant species present producing toxic molecules that are lethal for all species, with the exception of the species it is manipulating. In the case of the latter, it produces a chemical cocktail to control its host, consequently reducing its virulence (de Bekker et al. 2014).

The mechanisms to maintain the coviability of two interacting organisms have been the subject of many studies whose interpretations are summarized according to two concepts that we will explain and illustrate, the **Red Queen theory** and the **trade-off theory**. We will conclude this section on the evolutionary mechanisms of durable interactions with the reminder that man, like every species in our biosphere, is involved in symbioses that have influenced and continue to influence his biological history.

9.3.1 The Red Queen Theory

Regardless of the interaction modalities between symbionts and their hosts, there is a necessary implementation of control mechanisms for symbiosis. Indeed, even in the context of mutualism, it is necessary for the host, but also for the symbiont, to regulate its population and adjust the molecular dialogues and metabolic exchanges that exist between the two partners. In the case of parasitism, the host seeks to eliminate the parasite via its immune system whilst the parasite seeks to avoid this immune system, but also to manipulate its host in order to obtain the necessary elements for the realization of its life cycle. These relations imply specific selection pressures for each of the partners leading to their parallel evolution in response. However, each adaptation of a partner will change its relationship towards the other but will also change the related selection pressures. So step by step, an evolutionary race begins and a more or less important coevolution between the partners of a symbiosis is put in place. We can therefore speak of an "arms race"; where each party seeks to maintain its edge within the context of symbiosis. However, within these integrated systems, nobody can really prevail at the risk of destroying the association. We therefore observe a sort of **evolutionary standstill** which corresponds to the theory developed by Van Valen (1973). The latter, under the hypothetical name of the Red Queen, adopts the literary image of Lewis Carroll (1865) in "Alice's Adventures in Wonderland", where the *Queen of Hearts* organizes a race whose goal it is to run on the spot. The image perfectly illustrates these specific evolutionary races that aim to **maintain an organism in its niche,** a niche which is constantly changing and therefore never permitting a perfect adaptation.

This "arms race" in symbiotic systems results in the implementation of extremely complex mechanisms, such as the immune system of vertebrates for example which involves hundreds of genes, organs and specifically dedicated cell lines and consumes a large share of the metabolism of individuals. Mutualistic symbioses are therefore not left on the sidelines to optimize interaction with its symbionts. Dedicated structures can be found among many species, such as the foliar nectars which constitute a food source for the symbiotic partner (see Fig. 9.2, *Passiflora glandulosa*), hollowed out thorns that grow on the acacias of the *Acacia cornigera* species in which mutualistic ants settle to protect the plant, or root nodules of leguminous plants that host nitrifying bacteria.

9.3.2 Trade-Offs

In symbiotic systems, coviability implies the existence of a powerful mechanism governing the possible adaptations of each of the organisms: this constitutes the concept of ***trade-offs***. The term is borrowed from economic vocabulary and conceptualizes the "choices" made by an individual in the allocation of the resources he possesses (Tilman 2011). Within the context of ecology and evolution, unlike the economy, these *compromises* are never voluntary and result from the evolution of each species and the limits of its phenotypic plasticity. In effect, resources, regardless of their type, remain limited for organisms. For example, the (metabolic) energy resources that an organism can mobilize must be partly affected to basic cell function. If a surplus exists, it can be assigned to growth (making new cells) or to the reproduction of the organism (fabrication of a new organism). When there is symbiosis, dialogue and maintenance or the fight against the symbiotic organism, require a mobilization of resources that must consequently be taken from other functions. For example, the metabolic cost of immune reactions to combat parasitic organisms is extremely important. In some extreme cases, the host organism reduces the metabolic flow allocated to its reproduction or its growth to orient it towards the immune system.

These *trade-offs* also apply to parasites which have to *juggle* between reproduction and the mechanisms that allow them to avoid the immune response of the host. As for mutualists, they have to limit their own reproduction in order not to become too costly for their host and therefore maintain the symbiotic balance. Conversely, a mutualist, which would limit the reproduction of its host to maintain symbiosis, would risk orienting the association towards a parasitic relationship. For example, in the protection mutualisms of plants against herbivores, myrmecophytes plants (literally ant plants) are engaged in a *sustainable* relationship with mutualistic ants, to whom they provide shelter (domatia: accommodation structure in the host plant) and food (foliar nectar). These extremely widespread plant-ant symbioses involve 100 kinds of plants and 40 ant species and have evolved independently several times (McKey et al. 2005; Rico-Gray and Oliveira 2007).

When the plant enters its reproductive phase, its energy is redirected towards flowering, to the detriment of the foliar nectar. This metabolic compromise consequently leads to a "parasite" response by the mutualistic ant: it castrates the plant by sectioning its flower buds, with the objective of reorienting resources to its advantage (Gaume et al. 2005).

Examples of these constraints are extremely numerous throughout the parasite-mutualistic continuum and even beyond the maintaining of organisms within their niche, as shown by trials to transplant 20 tree species into clay and sandy soils (Fine et al. 2004). The results revealed that the specialization for their habitat not only the result of a physiological response, but also of an interaction between pressure from the herbivore and the soil type, according to a *trade-off* between faster growth (*advantageous* in clay soil) and a better defense against the herbivore (advantageous in sandy soil).

9.3.3 Are Humans a Symbiotic Partner Like any Other?

As highlighted by several of the above-mentioned examples, the human species does not present with any specificity in terms of symbiosis. Each of us lives according to sustainable interactions with several thousand species and more than one hundred thousand billion symbiotic organisms, ten times more than the number of our own cells. From our time of birth, our mutualistic symbionts namely colonize our skin (Schommer and Gallo 2013) or our digestive systems (Walter and Ley 2011). The diversity and the abundance of symbiotic species (mainly eubacteria and archaebacteria) is specific to each individual and does not depend on his or her genome but rather on individual history. These mutualistic communities are essential and provide us with an effective protection against other potentially parasitic organisms by occupying all of the niches available in our bodies. However, these organisms also provide us with an **extended phenotype**,[2] particularly as regards immune and metabolic capacities. These mutualistic communities vary depending on the organisms with which we are in contact, and it appears that our lifestyles particularly affect them. Recent studies show that our modern lifestyles (sedentary, over-hygienic or over-use of antibiotics . . .) have reduced the diversity of intestinal symbionts in many populations, inevitably leading to sometimes pathological imbalances (Moeller et al. 2014). Today, it is proven or suspected that these imbalances are at the origin of many *modern* pathologies, whether digestive (Crohn's disease) (Verdú et al. 2015) or otherwise (asthma, immunity . . .) (Fujimura and Lynch 2015), or even pathologies affecting cerebral development (Foster et al. 2015). These implications are of such importance that a new path

[2]The extended phenotype is an integral concept in the phenotype of an organism (from its form to its finest physiology) the set of the manifestations of its own genome and of those of its symbionts in a given environment.

of promising treatment is emerging in light of this knowledge: the transplantation of digestive microbiota between healthy and sick individuals (Borody et al. 2015). This procedure aims to treat both metabolic and auto-immune diseases as well as (Bromberg et al. 2015) fighting recurrent bacterial infections which resist antibiotics (Brandt et al. 2011).

Obviously not all of our symbionts are mutualistic and we have to fight a large number of parasites which set up home in our organisms, causing more or less serious illnesses. Whether arthropods, nematodes, fungi, unicellular eukaryotes, bacteria or viruses, these parasites have an important influence on history, ecology and human evolution as well as on all species. The history of civilization has been marked by widespread epidemics which have sometimes played a major role in the history of humanity. In ancient times, the plague (which probably included the plague, smallpox, typhus and other diseases) sometimes played the role of justice of the peace among Mediterranean civilizations. For example, the Athenian domination after Age *of Pericles,* began with the *Plague of Athens* (-430 to-426); or the demise of the Byzantine Empire with the ravages of the *Justinian Plague* (541–767). And more recently, the Napoleonic- *Russian Berezina was* largely due to the ravages of typhus and other diseases transmitted by lice (Peterson 1995). These parasites have also structured human populations for many years, limiting the dispersion or the development of human activities in some ecosystems. For centuries, inter-tropical Africa was totally inhospitable to European explorers and sparsely populated by local ethnic groups mainly due to malaria and other tropical fevers. We can note that the parasites from domestic animals have also indirectly influenced the development of human societies. Consequently, for centuries, throughout Western and Central Africa, the presence of animal trypanosomiasis has limited the development of livestock farming and protein resources among a few rare breeds with little developed parasitic disease regulation mechanisms (trypanotolerance). More generally, our parasites have influenced our evolution and continue to do so via the evolutionary "*Red Queen* type" race that we are running against them. The effectiveness of our immune systems owes them a good deal.

Over time, the ecological and social history of our species has led us to more or less successfully settle in all of the ecosystems available on our planet. To do this, major migratory movements have marked our history and continue to do so today. In the Paleolithic period, our species and its close relatives migrated slowly across Africa, Asia and Europe and then took advantage of the Ice Age and the fall in the level of ocean waters to reach America or Australia (Finlayson 2005). Since this bygone age, colonization and/or migration towards new unchartered (or inhabited) territories have never stopped. This human cosmopolism and this more or less large-scale migration dynamic inevitably placed populations in contact with a multitude of organisms, some of which had evolved in the absence of human beings. Man and these organisms have therefore evolved independently outside of the *Red Queen* mechanism. The human species is therefore continually exposed to a multitude of new, potentially parasitic organisms which are badly-adapted and which can provoke new epidemics of parasitic disease, a phenomenon identified within the concept of **emergence** that characterizes the appearance and the adaptation of new parasites for a given species. We have therefore opened up the "meeting filter" with all of these organisms, as defined in Combes (1997), and our species is more or less learning to co-evolve alongside them.

We do not only encounter these parasites, we also transport them with us among new human populations during events such as former migration or colonization or during other movements of multiple origin. In this case, emergence is not in the human species, but in populations which until this time had remained unscathed and were therefore "naïve" from a co-evolutionary point of view. One of the saddest examples in our history is the penetration of Europeans into South America with their parasites that ravaged Amerindian populations. Our current movements (for example in 2013, 232 million people changed countries, spanning various distances, equating to 3% of the world population (United-Nations 2013) can also promote the displacement of parasites among populations. The most recent visible examples of such displacements of human parasites are those of viruses transmitted by mosquitoes of the *Aedes,* Dengue and Chikungunya genera, which after extending beyond the borders of Asia, were scattered throughout tropical areas and are starting to appear in continental Europe (Devaux 2012). The door has always been open to parasites and it would seem that the opening is only getting wider....

"Coviable" associations whose origins we have already explained and the mechanisms which maintain them over time therefore appear as powerful drivers of the evolution of species. However, beyond evolution and the adaptations of specific partners of symbiosis, in this chapter we will see how they are also drivers of innovation for living organisms and taxonomic and ecological biodiversity.

9.4 Coviability and Biodiversity Dynamics

9.4.1 Symbioses and Major Innovations Concerning Living Organisms

The more knowledge we acquire on symbioses, the more we require examples of the direct involvement of symbioses in generating biodiversity. Due to the extended phenotype conferred by certain symbionts, species are able to colonize or create new niches that will not be accessible to individuals from the species which do not realize the adequate symbiosis. Given that the drift acts on these new populations, they can be at the origin of new species.

Symbiosis is at the origin of one of the most fundamental stages in the evolution of living organisms and consequently, of biodiversity. Our ability to use molecular oxygen as a fuel for our metabolism is an extended phenotype that is given to us by a symbiont which is so deeply integrated into our cells it has long been regarded as being a part of them: **mitochondria**. In reality, all multi-cellular organisms (fungi, plants and animals) possess mitochondria. They were acquired by the ancestors of eukaryotic organisms (possessing a nucleus containing chromosomes), by a symbiosis which was almost certainly mutualistic from the outset, between bacteria involved in breathing – an oxidative phosphorylation of carbonaceous molecules – and prokaryotic photosynthetic or chemosynthetic organisms which remove these carbonaceous molecules via saprophytism or predation (Sagan 1967). This ability to use a molecule which is accessible in large quantities everywhere in the atmosphere and the oceans has enabled both the colonization of new niches (no terrestrial organisms existed at this time), but also the exploitation of a much more effective energy source than those used until then by living organisms (photosynthesis, chimiosynthesis or fermentation). These new resources have probably enabled the emergence of multi-cellularity and the exponential increase in the size of organisms. Indeed, if organisms which recover their energy via chimiotrophism, phototrophism or fermentation still exist, they have all remained unicellular, with the exception of algae and plants. The latter carry out photosynthesis to make their carbonaceous matter but they breathe oxygen converted into chemical energy by their mitochondria. Furthermore, algae and plants appeared following another very integrated symbiosis with photosynthetic bacteria which produced another eukaryotic organelle: **chloroplasts**. This symbiosis allows them, by extended phenotype, to synthesize their own carbon molecules from carbon dioxide, water and photons, and consequently to free themselves of the need to take these molecules from other sources (organisms). The diversification of such organisms has created new niches offering new opportunities for other species that, as time passes, will become modern herbivores or frugivorous.

The entire biodiversity of multicellular organisms therefore has its origin in a succession of symbioses with bacteria that have given them new ecological capacities, providing them with a multitude of new niches.

In some cases, parasitism itself is responsible for sometimes spectacular adaptive radiations. Therefore, the effects of parasitic intracellular bacteria of arthropods, such as the *Wolbachia* genus, are responsible for the appearance in the environment of populations of a same species that cannot reproduce with each other owing to their infection status by the bacterium. This effect, known as the **cytoplasmic incompatibility effect,** renders embryos sterile following fertilization between the sperm of an infected male and the oocytes of an uninfected female. This phenotype consequently creates two reproductively isolated populations even if they share the same niche. The drift acting in the same way as for geographically isolated populations, *Wolbachia* may therefore be at the origin of new species (Rokas 2000; Vavre et al. 2003). Bearing in mind that a same species of arthropods can be infected by multiple strains of *Wolbachia,* extremely complex systems of incompatible subpopulations (Vavre et al. 2002), may be responsible for several new species at the same time. For example, the *Culex pipiens* mosquito present in the South of France, is infected by one dozen different bacterial strains, and presents with very complex reproductive patterns among its various subpopulations (Atyame et al. 2014). When we know that between 20 and 70% of arthropods species are potentially infected by bacteria of the *Wolbachia* genus, we are entitled to question the relationship between the extreme diversity of this taxonomic group and the potential implications of these symbionts in biodiversity. Many researchers are now working on speciation mechanisms induced by these symbionts and believe that a sizeable share of the ten million species of insects has been generated by Wolbachia bacterium (Engelstadter and Telschow 2009).

This evolutionary dynamic of symbioses is found in the original history of green lineage, through mycorrhizas, an endosymbiotic association between a fungus and the roots of a plant, present in all plant groups. Their story seems to be more recent than that of plants, suggesting a shared diversity between terrestrial plants and their symbiotic partners (Simon et al. 1993). Their extremely ancient origin would suggest that they have played a key role during the colonization of land by plants, even though the first land plants had no roots. These plants were unable to adapt to the terrestrial environment and its constraints, namely to the water constraint. Mycorrhizal symbiosis could have been a water absorption optimization factor (Jany et al. 2003; Richard et al. 2011). Their presence is reported in symbiotic associations of *Rhynia* fossils, primitive vascular plants from the Devonian period (Boullard and Lemoine 1971). Glomerales fungi would have provided water and minerals to feed plants devoid of roots, and would have certainly participated in the synthesis of photo-protective compounds such as as phenolic compounds of the flavonoid type (Taylor et al. 1995). The evolution of pseudo-roots into veritable roots would have been stimulated in order to facilitate the installation of the fungus. Currently, symbioses between plants and Glomerales are most common, and descend directly from the primitive symbioses which first conquered the land environment (Morton 1990; Schüßler et al. 2001). Owing to their symbiotic association, *mycorrhizaes* would have permitted higher, more evolved plants to take a *leap* forward in evolution by combining their genomes with those of plants and consequently multiplying their functions.

9.4.2 Symbioses and Ecological Diversity

The adaptation of plants to contrasting environments (light intensity, soil) is partly made possible thanks to their biochemical innovative capacity and the synthesis of increasingly complex compounds such as polyphenols, which would namely have allowed plants to adapt to extremely dry land conditions during their terrestrial conquest (Macheix et al. 2005). This capacity can explain the ability of plants to constantly colonize new areas, such as Philodendrons, hemi-epiphytic plants from the tropical Amazonian rainforest, which, through the production of more than 60 terpene volatile chemicals compounds, can resist a radical change in their environment during a forest opening, as a result of a natural phenomenon known as "chablis" (open space created by the felling of one or more trees), or an anthropogenic event. Consequently, along with many other "pioneer" plants, they participate in 'forest healing', a driving force phenomenon for natural regeneration (Fig. 9.3g). These epiphyte plants have petioles and aerial roots, which produce a very heavy, heady scent. These organs are traversed by a network of channels secreting essential oil which is responsible for their scent and which is produced

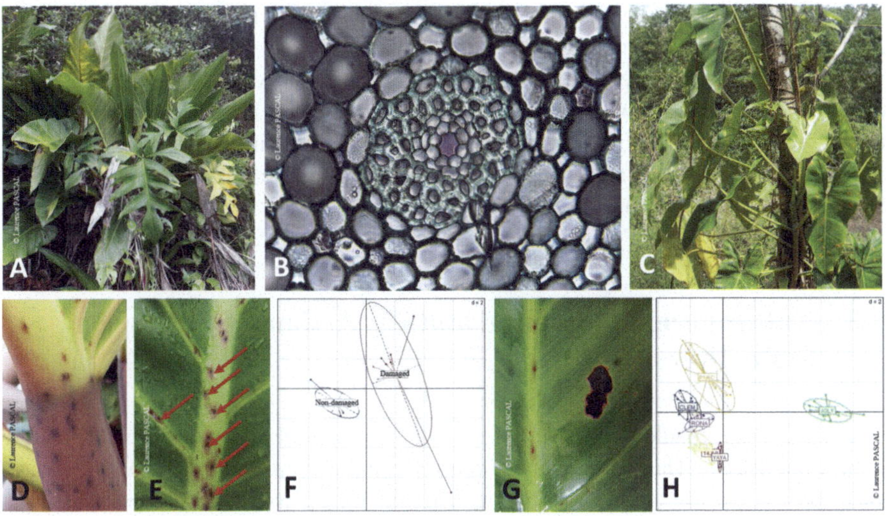

Fig. 9.3 Scent of *Philodendron*, ants and herbivores
Philodendrons, hemi-epiphytic plants from tropical forests, live in communities of species and play an important role in the regeneration of open pioneering fronts (**a**). *P. solimoesense* (**c**) shows a great adaptive plasticity, and grows in many contrasting environments (more or less dry, warm, good light). The scent emitted by their aerial roots and their petioles is composed of more than 60 volatile terpenes and differs according to the environment. An ACP shows the distribution of individuals from different populations (**h**), depending on the composition of their scents and namely, the importance of the ratio of sesquiterpenes (anti-herbivore compounds) versus monoterpenes (compounds known for facilitating resistance to water stress). Axis 1 separates "Joly" population individuals (in green) living in "arid" areas on the coast and producing scents rich in monoterpenes, from other populations. The petioles (d) and leaf veins (e) of *P. melinonii* are covered with nectar glands, which attract anti-herbivore ants (*Camponotus* SP., red arrows, photo **e**). Moreover, petioles and aerial roots have a set of canals which secrete essential oil (cross-section of aerial root of *P. melinonii*, toluidine blue color, picture **b**), and which plays a role in repelling herbivores. The ACP clearly shows the separation of individuals having been attacked by herbivores on leaves (**g**) and non-herbivores, according to the sesquiterpenes content of their scent which is stronger in herbivore individuals (**f**) (Photos Pascal L)

following an injury to the plant (Fig. 9.3b). The composition of the scent produced varies when the plant is under attack by herbivores (Fig. 9.3f). The plant's capacity to produce chemical compounds is stimulated by the pressure from a set of biotic (pathogens, herbivores, or symbionts, pollinators) and abiotic agents (light intensity, high temperatures, water stress, etc.). This ability allows them to create relationships with the most varied environments, but also to exchange signals with living organisms close by. A set of traits that provide plants with many advantages to enable them to adapt to their environment, and sometimes explains their precise confinement corresponding to the habitat of their pollinator or disseminator.

The chemical diversity produced by plants is the result of interactions with their environment. This diversity will be even greater when plants interact with a multitude of associated generalist species, as opposed to in response to an attack by a small number of specialists (Speed et al. 2015).

In the text above we have consequently demonstrated the synonymy of the concept of symbiosis with that of coviability for the living systems composing our biosphere. However, the impacts of man on his natural environment seem to deeply question the regulations and the dynamic balances of these "coviable" systems and lead us to ponder the future of biodiversity and our species.

9.5 Anthropocene: Mankind and His Actions with Respect to Biodiversity

The actions of mankind and their impacts on natural environments are more than established. They are also difficult to quantify as their magnitude is so vast and occurs on various scales, at different times, from the individual to the ecosystem. Examples of modifications to eco-systems owing to man's actions, whether voluntary or otherwise, are numerous.

From the time that domestication appears, morphological, physiological and behavioral modifications (when it comes to animals) irreversibly change species, some of which will completely disappear from the wild. The majority will colonize new distribution areas following human migrations, to the potential detriment of endemic wild species (Mazoyer and Roudart 2002).

9.5.1 Introduction of Species: For Better or for Worse

As a consequence, one of the oldest, most significant and visible impacts is the result of the movement **of species** from one ecosystem to another, whether for the purpose of cultivation, breeding, pleasure or unintentionally. This phenomenon has recently intensified owing to the *modern* lifestyle that notably facilitates quick travel over short or long distances. Through the example of the yam, introduced during man's travels throughout history, we will see that impacts can sometimes be positive, such as increasing the biological diversity of local crops (Pascal and Motte-Florac 2008) – or negative, because the ecological balance which exists in the associations between native species is displaced. This is the case of the Argentine ant, introduced into 15 countries all over the world during the lilac-rose trade in the XIX and XX centuries (Sunamura et al. 2009), which is ranked among the 100 worst invaders on our planet. In the natural environment, this ant replaces native ant colonies and disrupts or even destroys, owing to its abundant presence, associations between local species, such as the *Ferocactus viridescens, cactus* which is then no longer pollinated by its pollinator, which is ousted from flowers by this aggressive ant (LeVan et al. 2014).

Other cases of species transfer directly affect our agriculture and can have very significant consequences. This is the case, for example, of cowdriosis, caused by the *Ehrlichia ruminantium* bacterium, which infects numerous mammals (particularly sheep, cattle and goats). This disease, which is transmitted by ticks namely belonging to the *Amblyomma genus*, is endemic to sub-Saharan Africa, where it can cause widespread mortality to livestock (up to 90% in certain conditions), although it seems to be native to South Africa (Allsopp 2010). Livestock movements, which have become increasingly important with the advent of globalized trade, have served to move hosts, bacteria, and vectors into different geographical zones and namely to the Caribbean islands, where they have caused major damage (Kelly et al. 2011) leading to economic restrictions put in place by the USA, which is particularly concerned about importing the disease to its continent (Burridge et al. 2002).

Man has always sought to improve the diversity of the plants he cultivates by introducing new plant species from distant continents, during conquests or commercial shipments, or even during the terrible journeys made by slave traders. All periods of human history reflect population movements and cultural exchanges, which begin with exchanges of food products, a particularly salient domain for the identity of a population and a cultivated man-plant relationship.

In French Guiana, a mosaic of people from various origins and cultures has been shaped by centuries of successive arrivals. The different lifestyles of all of these cultural groups have been largely preserved owing to the difficulty of communications. However, the opening up of certain regions has changed things and the effects are noticeable on markets, places of exchange between the different groups (Pascal and Motte-Florac 2008). The biological diversity of plants sold on markets in French Guiana, illustrates the blending of species (adoption of species, the names given to them, how they are used, grown) or resistance at work. Thanks to exchanges, the diversity of food resources has been vastly increased, as in the case of the yam introduced from Africa (wrongly named *Dioscorea cayennensis*, because it was described for the first time in Cayenne) and from Asia (*Dioscorea alata*) which has *acclimatized* to the new "American environment". The cultivated diversity introduced added to the already wide variety of native crops (*Dioscorea trifida*), shaped for centuries by Amerindians from the Amazon. Their ancestral agricultural practices consist in collecting many spontaneous tubers of yams in the savannahs-rocks and forest undergrowth, and planting them in their fields. Hybrids consequently form spontaneously between the wild forms introduced in the field and the cultivated plants. Through such cultural practices, farmers develop new varieties by sexual multiplication and vegetative propagation, and select the best genotypes consequently preserving the potential for future adjustments (Fig. 9.4) (Pujol et al. 2005).

This vegetative propagation is all the more important to fix specific morphological traits such as the color of the tuber, namely due to the presence in the flesh of pigments from the chemical family of phenolic compounds (anthocyanin, flavonols . . .). Colors are recognized and appreciated by consumers as they are associated with a particular taste (relative sugar content) and good nutritional quality. The color and its intensity in the tuber flesh depends on the genotype. It is also produced in response to the cultural environment and can therefore

Fig. 9.4 The American wild yam, *Dioscorea trifida*
Grows in the understory of primary forest bordering the savannahs-rocks (**a**), and liana stems climbing up trees (**b**). Traditional Creole or Amerindians farming practices using stick-support (**c**), contrast with the productivist agrosystems of Hmong population where yam grown directly on floor (**d**). Wild and cultivated American yam forms differ from each other and from the D. rotundata, *D. cayennensis*, and *d. alata* species introduced (**e**), owing to the content of different derivatives of Quercitine in their tubers (flavonoids with a strong antioxidant power) (Photos L. Pascal)

be modulated by cultural practices (phenotypic plasticity). Tuber color choice is variable depending on the ethnicity of the farmer and the consumer, which creates a dynamic between biological and cultural diversity and offers the guarantee of a diversified food resource (Fig. 9.5).

Species with a great phenotypic plasticity have the ability to adapt to different environments, which facilitates their growing (for cultivated species) and harvesting (for wild species) because they are more widely available. They are consequently more widely used. Several multi-disciplinary studies combining field investigations on the traditional use of medicinal plants and their biochemical quality were able to show the impact of the environment and namely lighting (Zhang and Bjorn 2009) and edaphic factors (Radix et al. 1998), on the production of active ingredients by these plants. The diversity of terroirs obliges plants to innovate and play the "sorcerer's apprentices", in order to survive and remain in their environment. This serves to further increase the specificity of local knowledge and traditional uses. The definition of the "terroir" (local territory), which encompasses the processing practice concerning the product extracted from local crops, consequently makes perfect sense.

Within this context, we can understand how important **it is for populations to maintain an agro-biological wealth in their fields**, as it is the only way of avoiding the health issues encountered in modern crops owing to their excessive homogeneity (Blanco et al. 2013).

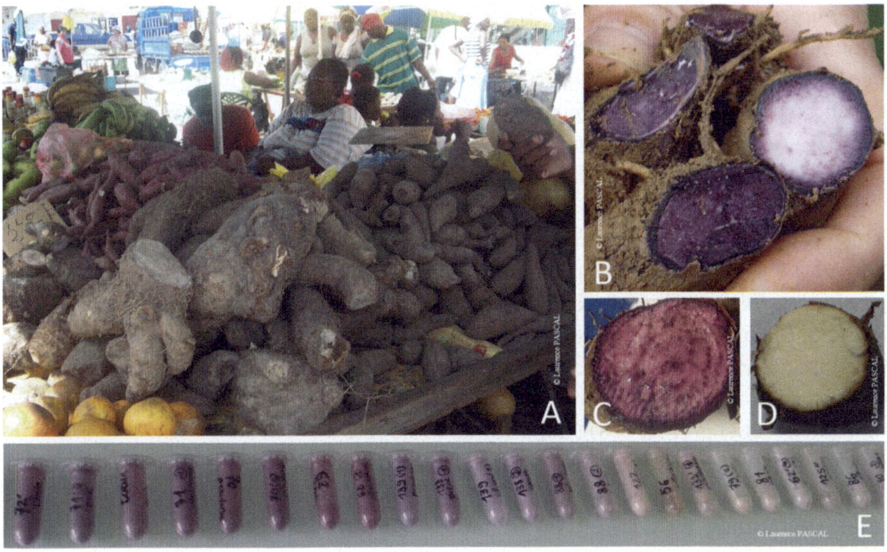

Fig. 9.5 The American yam, *Dioscorea trifida,* is sold on markets in French Guiana mixed with the tubers of yams introduced from Africa, *D. rotundata* and *D. cayenensis,* and from Asia, *D. alata* (**a**). It is called the purple yam by farmers owing to its accumulation of anthocyanic pigments in the tuber (**b**, **c**). The anthocyanin content of the tuber is modulated by environmental and cultural parameters. Only cultivated yams are purple; wild forms all have a white tuber (**d**) before being planted. Every ethnic group has specific farming practices to modulate the production of anthocyanin in tubers (**e**) (Photos Pascal)

9.5.2 Modern Agriculture and Loss of Biodiversity

A good example of this is that of **arabica coffee** (*Coffea arabica*), which was spread around the world from a genetic material of very reduced origin, and consequently presents with a poor genetic base which is also attributed to its autogamous[3] reproduction (Teressa et al. 2010). Its low genetic diversity jeopardizes the evolution of C. *Arabica* populations in response to environmental changes, consequently reducing their chances of persisting in the long term and posing a problem as to their resistance to numerous pests such as nematodes, rust agents, and coffee berry borers (Frankham et al. 2002). New biotechnologies have permitted *improved* plants to be developed in order to create cultivars which are better adapted to specific environments or farming practices.

In effect, the two most widely used species of coffee for commercial purposes, are *Coffea arabica*, covering more than 68% of global productivity, owing to the quality of its grain and despite the above-mentioned concerns, and *Coffea*

[3]Allogamy: fertilization of an ovum from one individual with the spermatozoa of another. Autogamy: the fusion of two gametes from the same individual for fertilization.

canephora, better known under the name of Robusta. The latter has lower ecological requirements and a greater genetic diversity. It was with the objective of mixing these two species within the same plant, that led to the idea of creating the interspecific hybrid *Coffea* x *arabusta* (Capot 1976). The significant genetic diversity of the hybrid obtained allows it to be cultivated in varied environmental conditions, and also offers the advantage of a high resistance to attackers. It was introduced into French Guiana in order to create a niche outlet based on C. *arabusta*. This outlet includes *C. arabica, C. canephora and* thirty four C. *arabusta* hybrids with variable capacity of adaptation to environment. Ten of these were selected and planted *in situ*, on volunteer farms, spanning the length of the Guyanese coastline, in very contrasting environmental conditions. The different *C. arabusta* hybrids have developed a vast panel of chemical compounds (phenolic and terpene compounds), which varies according to the hybrids and the adaptive response to each of these new environments (Fig. 9.6) (Lepelley et al. 2012).

In older plots of land, where soil is highly degraded due to years of intensive farming (Fig. 9.6f), attacks by coffee berry borers are significant in number. This mode of farming with long crop rotations does not ensure soil renewal, meaning

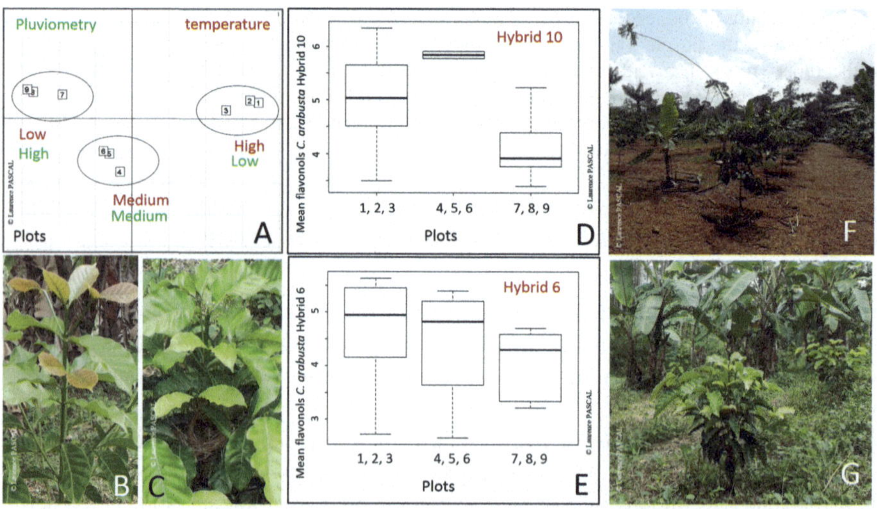

Fig. 9.6 The Coffea *arabusta* hybrid (*C. arabica x C. canephora*) is cultivated along the length of the French Guiana coast, according to a gradient of temperature and rainfall (**a**). In their young leaves, the different C. *arabusta* hybrids produce phenolic compounds of the red anthocyanin (**b**) or yellow flavonols type (**c**). These are produced in response to the strong light and high temperature, in a variable manner depending on the hybrids (**d, e**), and for certain hybrids, show significant production variations (**e**). Field measurements are obtained thanks to Dualex® sensors. Beyond the variable environmental conditions between the East and the West of the Guyanese coast, field (**f**) or agroforestry (**g**) cultivation systems condition the production of secondary metabolites, the development and the production of coffee tree grains, and are unique to the farmers who consequently control the production and the quality of the coffee (Photos and graphs Pascal L)

that soil is of low nutritional quality, but worse still, the repeated domestic uses of land affect the arbuscular fungi involved in root symbioses (Smith and Read 2008). These symbiotic partners play an important role in the tolerance of coffee trees to dehydration (dry season), in their mineral nutrition and contribute towards their good health. The coffee plants growing in these degraded plots are confronted with a double biotic (coffee berry borer attacks and microbiome poverty) and abiotic stress (drought and poor soil quality).

Furthermore, it was able to be shown that the environment (rainfall, altitude) has an influence on the fertility of C. *arabusta* hybrids. More interestingly, their reproductive mode may be a limiting factor in the case of introduced plants. *C. arabusta* hybrids inherited a different percentage of allogamy from their parents (cross-fertilization from different gametes of individuals) according to the share inherited from the autogamous *C. arabica* (fertilization between gametes produced by the same individual), or the allogamous *C. canephora*. Similarly, their grain production is highly variable, certainly due to their reproductive mode which for some of them, requires the intervention of a regular pollinator (bee), absent in their new growing environment. Scents produced by flowers vary significantly among hybrids and attract a diverse fauna of insects (Pascal, unpublished data). The small size of plots and their proximity to the surrounding wild forest, offers coffee plants access to local natural entomofauna. The effectiveness of pollination by these new local "pollinators", is not yet known and would benefit from being tested to better understand variations in grain production among hybrids. The management of these agro (eco) systems for an ecologically *stable* production can only occur on a landscape scale.

Therefore, in this example, man is trying to recreate the genetic diversity and the biotic and abiotic conditions of the natural adaptation underway in living organisms. Studies to understand the adaptability of these plants, which are still in progress, must therefore include the multiple responses to biotic and abiotic stresses and all of the interactions that allow this adaptability.

9.5.3 Intensive Agriculture: Is Coviability Broken?

In California, the monoculture of the introduced Prunus dulcis species (Almond) is a flagship example of the havoc that intensive agriculture can cause to the environment. Those 267,000 ha of pale pink orchards in bloom, which cross some 650 km from the north to the south of this immense American state, are simply not *sustainable*, because the coviability between the plant and its pollinator insect was broken when the almond tree was introduced alone, without its pollinator. Furthermore, local insects cannot pollinate these trees since monoculture and significant environmental degradation have decimated local entomofauna. The population of potential wild pollinators in the great Central Valley of California is 5 times lower than expected (Garibaldi et al. 2011). To compensate, 50–60 billion bees have been transported from all over America, equating to 1.5 million hives (nearly two-thirds of the country's hives) (Tardieu 2009). Pollinators are fed with an energy-rich preparation mainly composed of sucrose, beer yeast and pollen imported from China (!). This diet, which is fed to the bees all year round outside of the pollination period, and without taking account of their biological cycle, has resulted in a dramatic death rate affecting up to 30% of hives (Tardieu 2009). Transported to California from the month of September, the lack of local vegetation enabling bees to feed themselves obliges the breeder to provide bees with food supplements (corn syrup and protein). This intensive agriculture, conducted for the sole purpose of achieving maximum production and with no biological hindsight, is stretching the limits of its own logic. Californian almond plantations could certainly function without domestic bee colonies, or at least partly so, if the agricultural environment was better managed. It is currently impossible to imagine being able to function without *Apis mellifera (bee)* breeding, whilst the ultimate goal is to produce increasing quantities of fruit. Yet, the decline of the bee-keeping sector, mainly owing to pathogen attacks such as the *Varroa destructor* and poisoning by pesticides, raises the question of its "replacement" by wild local pollinators. And this assuming that we can ensure that wild bees are not concerned by the decline of breeding colonies. Furthermore, these wild insects cannot be controlled by breeders and farmers, and the proliferation of colonies is extremely modest with respect to the number of individuals compared to domestic bee hives. It is clear that in larger areas, to meet the objectives of increased cereal and agro-fuel production, we have created spaces which are totally inappropriate to wild bees, where no pollen or nectar is available throughout the year (Naug 2009). To resolve the current crisis, it would be fairer to maintain both populations of bees in our environment. These wild and domestic bees can play complementary (namely by pollinating different plants), and additional roles (pollinating the same plants which improves their fertilization).

This example clearly shows once again that a sustainable or a co-viable management of agro-systems of such magnitude can only be conceived on an integrative scale as regards human activities and the degraded ecosystem component which is not related to these activities (wildlife, non-cultivated flora, free and symbiotic microorganisms, soils, water . . .) and their multiple interactions. Such an ecosystem must be seen as a patchwork of niches allowing the regeneration of a biodiversity dynamic involving interactions between all of the partners from obligatory symbioses to more diffuse interactions but which are still essential.

Unfortunately, it seems that humanity is not committed to following such a path and that the future of the Anthropocene excludes any concept of coviability.

9.6 Conclusion

9.6.1 The Future of the Anthropocene

Public or private initiatives designed to preserve biodiversity are certainly commendable and have to be encouraged. Unfortunately, they often constitute naive responses to the scientific reality of ecology and the evolution of species.

Biodiversity conservation programs are being implemented, namely as regards the conservation of seeds from cultivated plants. These programs are occurring in an era where the vast majority of the planet's seeds are possessed by a very small number of companies, the consequence of the patentability of living organisms. Huge projects are being undertaken such as the manufacturing of a reserve of frozen seeds in an underground container in Svalbard in Norway (Charles 2006). This idea of conserving biodiversity in DNA form and outside of its ecological context, removing the necessary dynamics for its evolution and maintenance, is, as we have seen above, a biological nonsense given that we are unable to ensure their coviability with other environmental players which will almost certainly have changed since their withdrawal from nature. Cultivated plants function as wild plants that are diversifying through adaptive processes and selection. Ensuring sufficient and diversified food resources for future generations involves taking account of the essential maintaining of an evolutionary dynamic for species in a fast-changing world. This is the only mechanism that will allow them to survive in tomorrow's environment, characterized by climate change to which they will have to adapt, and changes in biotic interactions with predators/pathogens which will also have had to adapt.

If the necessary adaptation of our plants and animals to our abiotic and biotic environment to come and to our social needs seems to be an economic and concrete stake, what is the situation concerning the conservation of wildlife and wild flora? For example, wildlife conservation programs are facing multiple challenges: the co-existence of this fauna with man and his activities is a source of conflict that often results in culls. In France, recent examples include wolves *threatening* herds of sheep and ibexes infected with brucellosis which is potentially transmissible. The *wild* territory is reduced through deforestation that is constantly progressing given that land is necessary for crops or for ever-expanding urbanization. Animals are forced into limited and fragmented geographical areas which are often inadequate to provide them with food resources and meet their basic behavioral needs. As a result, they are being forced into our cultures, our homes and our cities. The most important aspect remains the constraints on their evolutionary dynamics however: their numbers and therefore their essential genetic diversity for adaptation are significantly reduced. Some of these animals are *maintained* in parks outside of their natural environment and away from the pressures of natural biotic and abiotic selection, others are facing changes which are too fast for them to adapt *via* evolutionary mechanisms. These species are exposed to a high risk of extinction where the competition for resources with other animals (wild and domestic species) and infectious risks (parasites), become too strong to keep them *running with the Red Queen*... Humanity seems to be choosing, even unconsciously, a world in which biological diversity, natural ecosystems and biotic interactions can be forsaken by *controlling* ecosystems and reproducing endangered species without land or by the long-term storage of the fixed genotypes of particular species. This choice, made a reality by public and private actions to manage ecosystems and biodiversity, serves to gradually reinforce the risks of extinction and consequently fails to provide a solution to the numerous problems that emerge in this unstable

Anthropocene, having eliminated co-existences which are essential sources of dynamic, evolution and consequently of sustainability for our biosphere, weakening the viability of the human species itself.

Indeed, can such a human-centered world, *which excludes nature,* in other words, a world which is outside of all biological interactions, of all natural evolutionary and innovative processes driven by these interactions, really continue to ensure the survival and the adaptation of our species? The issue of re-placing the concept of coviability in all of its biological and socio-economic dimensions, in our activities and in our own diversity of practices, in the place given to what is not of any direct use to us, has become central for our future.

References

Allsopp BA (2010) Natural history of Ehrlichia ruminantium. Vet Parasitol 167:123–135

Atyame CM, Labbe P, Dumas E, Milesi P, Charlat S, Fort P, Weill M (2014) Wolbachia divergence and the evolution of cytoplasmic incompatibility in Culex pipiens. PLoS One 9:e87336

Bascompte J, Jordano P, Melian CJ, Olesen JM (2003) The nested assembly of plant-animal mutualistic networks. Proc Natl Acad Sci U S A 100:9383–9387

Beja-Pereira A, Luikart G, England PR, Bradley DG, Jann OC, Bertorelle G, Chamberlain AT, Nunes TP, Metodiev S, Ferrand N, Erhardt G (2003) Gene-culture coevolution between cattle milk protein genes and human lactase genes. Nat Genet 35:311–313

Blanco J, Pascal L, Ramon L, Vandenbroucke H, Carrière SM (2013) Agrobiodiversity performance in contrasting island environments: the case of shifting cultivation in Vanuatu, Pacific. Agric Ecosyst Environ 174:28–39

Blondel J, Dias PC, Perret P, Maistre M, Lambrechts MM (1999) Selection-based biodiversity at a small spatial scale in a low-dispersing insular bird. Science 285:1399–1402

Borody T, Fischer M, Mitchell S, Campbell J (2015) Fecal microbiota transplantation in gastrointestinal disease: 2015 update and the road ahead. Expert Rev Gastroenterol Hepatol:1–13

Boullard B, Lemoine Y (1971) Les Champignons endophytes du Rhynia Gwynne-Vaughanii: étude morphologique et déductions sur leur biologie. Le Botaniste 54:49–89

Boyd R, Richerson P (1995) Why does culture increase human adaptability? Ethol Sociobiol 16:125–143

Brandt LJ, Borody TJ, Campbell J (2011) Endoscopic fecal microbiota transplantation: "first-line" treatment for severe clostridium difficile infection? J Clin Gastroenterol 45:655–657

Bromberg JS, Fricke WF, Brinkman CC, Simon T, Mongodin EF (2015) Microbiota-implications for immunity and transplantation. Nat Rev Nephrol 11:342–353

Burridge MJ, Simmons LA, Peter TF, Mahan SM (2002) Increasing risks of introduction of heartwater onto the American mainland associated with animal movements. Ann N Y Acad Sci 969:269–274

Capot J (1976) Obtention et perspectives d'un nouvel hybride de caféier en Côte d'Ivoire: l'Arabusta. In 7e Colloque International sur lakhimie des Cafés (Paris: ASIC), pp 449–457

Carroll L (1865) Alice's adventures in wonderland. Macmillan, London

Chappell MA, Souza SL (1988) Thermoregulation, gas exchange, and ventilation in Adelie penguins (Pygoscelis adeliae). J Comp Physiol B 157:783–790

Charles D (2006) A 'forever' seed bank takes root in the Arctic. Science 312:1730–1731

Combes C (1997) Interactions durables. Elsevier Masson, Paris

Darwin C (1859) On the origin of species by means of natural selection, or the preservation of favoured races in the struggle for life. John Murray, London

Day RL, Laland KN, Odling-Smee FJ (2003) Rethinking adaptation: the niche-construction perspective. Perspect Biol Med 46:80–95

de Bekker C, Quevillon LE, Smith PB, Fleming KR, Ghosh D, Patterson AD, Hughes DP (2014) Species-specific ant brain manipulation by a specialized fungal parasite. BMC Evol Biol 14:166

Devaux CA (2012) Emerging and re-emerging viruses: a global challenge illustrated by Chikungunya virus outbreaks. World J Virol 1:11–22

Eibl JM, Plunkett GM, Lowry PP II (2001) Evolution of Polyscias sect. Tieghemopanax (Araliaceae) based on nuclear and chloroplast DNA sequence data. Adansonia Sér 23:23–48

Engelstadter J, Telschow A (2009) Cytoplasmic incompatibility and host population structure. Heredity 103:196–207

Fine PV, Mesones I, Coley PD (2004) Herbivores promote habitat specialization by trees in Amazonian forests. Science 305:663–665

Finlayson C (2005) Biogeography and evolution of the genus Homo. Trends Ecol Evol 20:457–463

Fix AG (2003) Simulating hemoglobin history. Hum Biol 75:607–618

Foster JA, Lyte M, Meyer E, Cryan JF (2015) Gut microbiota and brain function: an evolving field in neuroscience. Int J Neuropsychopharmacol

Frankham R, Briscoe DA, Ballou JD (2002) Introduction to conservation genetics. Cambridge University Press, New York

Frost GH, Siegfried WR, Burger AE (1976) Behavioural adaptations of the Jackass Penguin, Spheniscus demersus to a hot, arid environment. J Zool 179:165–187

Fujimura KE, Lynch SV (2015) Microbiota in allergy and asthma and the emerging relationship with the gut microbiome. Cell Host Microbe 17:592–602

Garibaldi LA, Steffan-Dewenter I, Kremen C, Morales JM, Bommarco R, Cunningham SA, Carvalheiro LG, Chacoff NP, Dudenhoffer JH, Greenleaf SS, Holzschuh A, Isaacs R, Krewenka K, Mandelik Y, Mayfield MM, Morandin LA, Potts SG, Ricketts TH, Szentgyorgyi H, Viana BF, Westphal C, Winfree R, Klein AM (2011) Stability of pollination services decreases with isolation from natural areas despite honey bee visits. Ecol Lett 14:1062–1072

Gaume L, Zacharias M, Grosbois V, Borges RM (2005) The fitness consequences of bearing domatia and having the right ant partner: experiments with protective and non-protective ants in a semi-myrmecophyte. Oecologia 145:76–86

Gentry AH (1981) Distributional patterns and an additional species of the Passiflora vitifolia cornplex: Amazonian species diversity due to edaphically differentiated communities. Plant Syst Evol 137:95–105

Hanski I (2001) Spatially realistic theory of metapopulation ecology. Die Naturwissenschaften 88:372–381

Hatcher MJ, Dick JT, Dunn AM (2006) How parasites affect interactions between competitors and predators. Ecol Lett 9:1253–1271

Hubbell SP (2001) The unified neutral theory of biodiversity and biogeography. Princeton University Press, Princeton

Jany JL, Martin F, Garbaye J (2003) Respiration activity of ectomycorrhizas soil water potential in five beech forests. Plant Soil 255:487–494

Kang YJ, Jo JO, Cho MK, Yu HS, Ock MS, Cha HJ (2011) Trichinella spiralis infection induces angiogenic factor thymosin beta4 expression. Vet Parasitol 181:222–228

Kelly PJ, Lucas H, Yowell C, Beati L, Dame J, Urdaz-Rodriguez J, Mahan S (2011) Ehrlichia ruminantium in Amblyomma variegatum and domestic ruminants in the Caribbean. J Med Entomol 48:485–488

Kimura M (1968) Evolutionary rate at the molecular level. Nature 217:624–626

Krishna A, Paulo R, Guimarães PJ Jr, Bascompte J (2008) A neutral-niche theory of nestedness in mutualistic networks. Oikos 117:1609–1618

Labeyrie E, Pascal L, Delabie J, Orivel J, Dejean A, Hossaert-Mckey M (2001) Protection of Passiflora glandulosa (Passifloraceae) against herbivory: impact of ants exploiting extrafloral nectaries. Sociobiology 38:317–321

Laland KN, Odling-Smee J, Myles S (2010) How culture shaped the human genome: bringing genetics and the human sciences together. Nat Rev Genet 11:137–148

Lepelley M, Mahesh V, McCarthy J, Rigoreau M, Crouzillat D, Chabrillange N, de Kochko A, Campa C (2012) Characterization, high-resolution mapping and differential expression of three homologous PAL genes in Coffea canephora Pierre (Rubiaceae). Planta 236:313–326

LeVan KE, Hung KL, McCann KR, Ludka JT, Holway DA (2014) Floral visitation by the Argentine ant reduces pollinator visitation and seed set in the coast barrel cactus, Ferocactus viridescens. Oecologia 174:163–171

Macheix JJ, Fleuriet A, Jay-Allemand C (2005) Les composés phénoliques des végétaux: Un exemple de métabolites secondaires d'importance. PPUR, Lausanne

Mazoyer M, Roudart L (2002) Histoire des agricultures du monde. Du néolithique à la crise contemporaine. Edition du Seuil, Paris

McKey D, Gaume L, Brouat C, Di Giusto B, Pascal L (2005) The trophic structure of tropical ant-plant-herbivore interactions: community consequences and coevolutionary dynamics. Cambridge University Press, Cambridge

Moeller AH, Li Y, Mpoudi Ngole E, Ahuka-Mundeke S, Lonsdorf EV, Pusey AE, Peeters M, Hahn BH, Ochman H (2014) Rapid changes in the gut microbiome during human evolution. Proc Natl Acad Sci U S A 111:16431–16435

Monod J (1973) Le Hasard et la Nécessité. Editions du Seuil, Paris

Morton JB (1990) Revised classification of the arbuscular mycorrhizal fungi (Zycomycetes): a new order, Glomales, two new families, Acaulosporaceae and Gigasporaceae, with an emendation of Glomaceae. Mycotaxon 37:473

Naug D (2009) Nutritional stress due to habitat loss can explain recent honeybee colony collapses. Biol Conserv 142:2369–2372

Pascal L, Motte-Florac E (2008) Plantes, transferts et résistances; reflet d'un métissage à l'œuvre sur les marchés de la Guyane française. In 133° Congrès National des Sociétés Historiques et Scientifiques (Quebec, Canada)

Peterson RKD (1995) Insects, disease, and military history. Am Entomol 41:147–161

Pujol B, Muhlen G, Garwood N, Horoszowski Y, Douzery EJ, McKey D (2005) Evolution under domestication: contrasting functional morphology of seedlings in domesticated cassava and its closest wild relatives. New Phytol 166:305–318

Radix P, Bastien C, Jay-Allemand C, Charlot G, Seigle-Murandi F (1998) The influence of soil nature on polyphenols in walnut tissues. A possible explanation of differences in the expression of walnut blight. Agronomie 18:627–637

Rands CM, Darling A, Fujita M, Kong L, Webster MT, Clabaut C, Emes RD, Heger A, Meader S, Hawkins MB, Eisen MB, Teiling C, Affourtit J, Boese B, Grant PR, Grant BR, Eisen JA, Abzhanov A, Ponting CP (2013) Insights into the evolution of Darwin's finches from comparative analysis of the Geospiza magnirostris genome sequence. BMC Genomics 14:95

Richard F, Roy M, Shahin O, Sthultz C, Duchemin M, Joffre R, Selosse MA (2011) Ectomycorrhizal communities in a Mediterranean forest ecosystem dominated by Quercus ilex: seasonal dynamics and response to drought in the surface organic horizon. Ann For Sci 68:57–68

Rico-Gray V, Oliveira PS (2007) The ecology and evolution of ant-plant interactions. The University of Chicago Press, Chicago

Rokas II (2000) Wolbachia as a speciation agent. Trends Ecol Evol 15:44–45

Sagan L (1967) On the origin of mitosing cells. J Theor Biol 14:255–274

Schommer NN, Gallo RL (2013) Structure and function of the human skin microbiome. Trends Microbiol 21:660–668

Schüßler A, Schwarzott D, Walker C (2001) A new fungal phylum, the glomeromycota: phylogeny and evolution. Mycol Res 105:1413–1421

Simon L, Bousquet J, Levesque R, Lalonde M (1993) Origin and diversification of endomycorrhizal fungi and coincidence with vascular plants. Nature 363:67–69

Smith SE, Read DJ (2008) Mycorrhizal symbiosis. Academic Press, London

Speed MP, Fenton A, Jones MG, Ruxton GD, Brockhurst MA (2015) Coevolution can explain defensive secondary metabolite diversity in plants. New Phytol 208:1251–1263

Steinbauer M, Wallis I, Davies N, Watson S (2015) Foliar quality of co-occurring mallee eucalypts: balance of primary and secondary metabolites reflects past growing conditions. Chemoecology 25:179–191
Sunamura E, Espadaler X, Sakamoto H, Suzuki S, Terayama M, Tatsuki S (2009) Intercontinental union of Argentine ants: behavioral relationships among introduced populations in Europe, North America, and Asia. Insect Soc 56:143–147
Tardieu V (2009) L'étrange silence des abeilles. Enquête sur un déclin inquiétant. Belin, Paris
Taylor T, Remy W, Hass H, Kerp H (1995) Fossil arbuscular mycorrhizae from the early Devonian. Mycologia 87:560–573
Teressa A, Crouzillat D, Petiard V, Brouhan P (2010) Genetic diversity of Arabica coffee (Coffea arabica L.) collections. EJAST 1:63–79
Tilman D (2011) Diversification, biotic interchange, and the universal trade-off hypothesis. Am Nat 178:355–371
United-Nations (2013) UN Migration Report (United Nations)
Van Valen L (1973) A new evolutionnary law. Evol Theory 1:1–30
Vandermeer JH (1972) Niche theory. Ann Rev Ecol Syst 3:107–132
Vandermeer JH (2008) The niche construction paradigm in ecological time. Ecol Model 214: 385–390
Vavre F, Fleury F, Varaldi J, Fouillet P, Bouletreau M (2002) Infection polymorphism and cytoplasmic incompatibility in Hymenoptera-Wolbachia associations. Heredity 88:361–365
Vavre F, Fouillet P, Fleury F (2003) Between- and within-host species selection on cytoplasmic incompatibility-inducing Wolbachia in haplodiploids. Evolution 57:421–427
Verdu EF, Galipeau HJ, Jabri B (2015) Novel players in coeliac disease pathogenesis: role of the gut microbiota. Nat Rev Gastroenterol Hepatol 12:497–506
Wallace A (1870) Contributions to the theory of natural selection. Macmillan and Co, London
Walter J, Ley R (2011) The human gut microbiome: ecology and recent evolutionary changes. Annu Rev Microbiol 65:411–429
Zhang WJ, Bjorn LO (2009) The effect of ultraviolet radiation on the accumulation of medicinal compounds in plants. Fitoterapia 80:207–218
Zhu J, Oger PM, Schrammeijer B, Hooykaas PJ, Farrand SK, Winans SC (2000) The bases of crown gall tumorigenesis. J Bacteriol 182:3885–3895

Chapter 10
A Geographical Approach to Socio-ecological Coviability

Christophe Grenier

Contents

10.1 Introduction

The negative ecological consequences of planet-wide human activities began in the nineteenth century when industrial capitalism developed while national states and colonial empires were extending: at that time geographers (Marsh 1864; Reclus 1905) were among the first scientists to be aware of these. During the twentieth century, it was mostly naturalists (field biologists, ecologists), physicists, and chemists who investigated the manner by which humans are changing the biosphere, and who launched a series of increasingly urgent warnings regarding the global ecological degradation (Deléage 1991). Conversely, social science researchers were

C. Grenier (✉)
Nantes University, UMR CNRS 6554 LETG-Géolittomer, Nantes, France
e-mail: christophe.grenier@univ-nantes.fr

O. Barrière et al. (eds.), *Coviability of Social and Ecological Systems: Reconnecting Mankind to the Biosphere in an Era of Global Change*,
https://doi.org/10.1007/978-3-319-78497-7_10

less interested by the ecological issue, and when they were, they looked at the matter from an economic perspective, so "*as natural scientists have become more concerned about the detrimental effects of the economic system on the environment, and correspondingly radicalized, asking more and more root questions, social scientists have increasingly turned to the existing economic system as the answer*" (Foster Bellamy et al. 2010, 20). Indeed, many nature scientists gained merit for producing various concordant studies on global change and by warning both political powers and the public opinion about its consequences. However, this represents a double challenge to the social sciences. First, the scientific debate on global ecological change, which is essentially anthropic, is dominated by concepts belonging to biological and physical sciences; so when these concepts are applied to the social sphere, they distort analyses. Second, a growing number of social scientists naturalize the society/environment relations without investigating the concepts originating from biological and physical fields.

As these approaches seem to me scientifically distorted by a naturalizing ideology, I suggest a geographical approach to socio-ecological coviability, which requires a brief preliminary description of some concepts of the discipline. A geographical analysis relying on other social sciences can reveal this naturalizing ideology in three ideas connected to the *social-ecological systems* (SES), a concept forming the basis of socio-ecological coviability. Easter Island is a valid example of an ecological and social "collapse,' so investigating it may help us test the socio-ecological coviability *vis-à-vis* geo-historical facts. This leads us to ask questions on the geographical conditions of a socio-ecological coviability in a globalized world.

10.2 Defining the Necessary Geographical Concepts for Analyzing Coviability

In the vast multidisciplinary scientific field of society/environment relationships, geography may be defined etymologically as a science that studies the "*geo graphies*", or footprints, that societies leave on Earth. Being a social science, geography has an anthropocentric perspective in which the biosphere is considered as an "ecumene": this implies that geography focuses on "ecological, technical, and symbolic" relationships that humans develop with Earth as their home (Berque 1996, 2010).

"Geographical space", the main concept allowing the analysis of the relationships between societies and Earth, may be defined as "a social product and a system of relationships between places" (Brunet et al. 1992), and for geographers, territories are mainly spaces appropriated by a population or a society (Le Berre 1992; Brunet et al. 1992; Lévy and Lussault 2003; Lacoste 2003). "Environment" is a hybrid concept and is both a natural given – an ecosystem, for example – that geographers take into consideration but still do not study (this task belongs to natural sciences); and a human production that develops from the relatively profound transformation of an ecosystem for social ends. Modifying environments, and organizing terrestrial contexts in the form of geographical spaces expanding from a particular place to the world, is a process of planet anthropization. The geographical milieu represents the relationship between a society and an environment (Berque 1990). The milieu is relative because society/environment relationships vary across cultures and throughout history. Environments and spaces – be it urban, rural, touristic, protected, and so on – are social footprints on Earth. They are organized geographies, exposed to mapping at different scales, and investigated through the analysis of territories, environments, and landscapes.[1]

A geographical region is both a spatial entity of variable surface which still intermediates between a particular place and national/global spaces, and a singular part of Earth identified as such by various elements, starting by its inhabitants. Regions have a position specified by the latitude and longitude in an absolute geophysical entity: the globe. Regions have also a situation defined by their connections to other places and spaces within a relative geo-historical entity: the world. Therefore, the location of a region implies taking into account both its position, which determines its natural conditions (such as relief, climate, and biogeography . . .), and its situation, which partly explains its geography. Geographies change over time which is reflected in geo-histories. Indeed, even if regions – or parts of the Earth – always have the same position, regional spaces, "milieux", and environments appear or disappear in the course of history. This occurs according to the geo-historical trajectory of these regions in global spaces as well as their changing situations *vis-à-vis* other places and regions.

These details help in examining the concept of SES, which is the basis of socio-ecological coviability, according to a geographical perspective.

10.3 Biased Coviability: Three Common Ideas About Social-ecological Systems (SES)

The concept of a *Social-Ecological System* was developed to analyze the interactions between societies and ecosystems in order to promote their mutual and sustainable management (Holling 2001). This concept occupies a central position in the theoretical corpus of the *Resilience Alliance*, a think tank first formed around ecologists and environmental economists. Many scientists belonging to *Resilience Alliance* adopt three ideas on SES that embody the naturalizing approach of SES in the social sphere, which distorts the analysis of coviability; therefore, we need to explore these ideas.

[1] A landscape is both the image and the representation of an environment (Brunet et al. 1992).

10.3.1 Do Social-ecological Systems Function as Ecosystems?

The first idea denotes an equivalence laid between social systems and ecosystems present within SES (www.resalliance.org). It is based more precisely on the beliefs of certain founders of the *Resilience Alliance* who suggest that "*human society is part of the biosphere and societies are embedded in ecological systems*" (Folke et al. 2002), or that "*the delineation between social and natural systems is artificial and arbitrary*" (Berkes et al. 2003). Paradoxically, this postulate is the result of an "anthropological naturalism", defined by Ph. Descola (2005) as a "naturalistic ontology", that of "man/nature" dualism of classical modernity. This ontology is based on "the similarities of physical properties and the differences of interiorities" between humankind and nature. In the name of common "physical properties" and at the expense of human "interiorities" – constructed by culture and history – the naturalistic approach of SES affirms that the "*social systems can be thought of as interdependent systems of organisms*" (Anderies et al. 2004); and therefore may be understood both in and as ecosystems.

Prior to the development of the concept of socio-ecosystem, the historian of ecology, P. Acot, in the face of certain naturalists claiming the integration of human societies in ecosystems, stressed that "we are in the presence of a thought that does not establish a separation between nature and society, i.e., a thought connecting Man and his environment without defining Man (. . .); the fundamental consequence of this view is the denial of social sciences "(1988, 211). Social sciences are indeed established on the idea that there is no equivalence between societies and ecosystems because human populations, although part of nature, are not reduced to it. These populations form societies that have largely emancipated the *Homo sapiens sapiens* from evolution in order to integrate our species in history. According to anthropologist M. Godelier (1984), "Man has a history because Man transforms nature"; in other words, Man has a history because he has geography. Through transforming ecosystems to environments and terrestrial width to spaces, societies produce geographies that partially determine the history of human populations. Nowadays, the state of the biosphere and its habitability for humans and other beings, i.e., the social-ecological coviability at various scales, depend on these geographies.

The belief that societies function as ecosystems is the ecological version of an old "ideological naturalism", whose other dimensions are philosophical and economic (Rist 1996; Caillé et al. 2001). This is summarized in an attempt to explain human societies by natural laws. All modern sciences are involved in an "ontological naturalism," but ideological naturalism is only the product of a few scientific movements with political overtones. Ideological naturalism has blossomed in the nineteenth and twentieth centuries among German conservative intellectuals, some ideas of which fed Nazi theories. Some of these ideas belong to intellectuals such as the geographer F. Ratzel (1897) who established the concept of "*Lebensraum*" and who suggested that the state is "an organism grounded in a land." Another example is the philosopher O. Spengler (1921) whose cyclical conception of history considers cultures as organisms. Ideological naturalism is also essential in multiple contemporary avatars of social Darwinism, such as the "sociobiology" founded by the naturalist E.-O. Wilson (Sahlins 1976). This ideology is still present in geography, even if this discipline was built a century ago against naturalistic determinism (Livingstone 1992). Hence, the arguments of the naturalist J. Diamond (1997, 2005), who holds a geography Chair, are considered emblematic of an "environmental neo-determinism" (Judkins et al. 2008).

Nevertheless, the most antique and persistent success of ideological naturalism is present in political economy, where it touches liberal and neoliberal thoughts from T. Malthus to F. Hayek (Passet 2010). The latter shares many ideas with the ecologist CS Holling, one of the founders of the *Resilience Alliance*, among them the notion of unpredictability and, beyond this, the idea of non-historicity of complex systems, which apparently "evolve" cyclically (Walker and Cooper 2011).

10.3.2 Do SES Have a Cyclical History?

A second common idea about the SES results from the equivalence stated between social and ecological systems, which is as ideological marked as scientifically erroneous: their common history would be cyclical in nature. L. Gunderson and C. S. Holling coined the concept of "panarchy" to describe the evolution of complex meta-systems which may simultaneously explain cyclical history of ecosystems and of societies from all temporal and spatial scales. "Panarchy" designates indeed all "*these transformational cycles (which) take place in nested sets at scales ranging from a leaf to the biosphere over periods from days to geologic epochs, and from the scales of a family to a socio-political region over periods from years to centuries*" (Holling 2001, 392). Panarchy denotes therefore a "scientific panacea" based on "transformational cycles" that could be applied both to objects of the natural sciences and to those of social sciences. Nevertheless, this concept seems wrong to me at least with regard to the history of human societies.

For we must not confuse the conception of the history of a society, which may be cyclical, with *the* history of this society, which is not cyclical. For example, economists have described several cycles of activities in the history of capitalism belonging to different times – such as Kondratieff cycles, secular *trends*, etc. However, if this cyclical conception of economic history assists in understanding the internal dynamics of capitalism, it is not adequate for analyzing the global geo-historical process driven by its expansion: the history of globalization is linear and has irreversible consequences. One may give as example the invasion of the Canary Islands by the Spaniards in the fifteenth century: this is considered by the historian of environment A. Crosby (1986) as the starting point of modern European imperialism, for it caused the demise of Guanche people that no panarchic cycle will ever be able to revive. As diverse human societies integrate "global history" and connect to "World space", they partly develop a common history and geography, which forms a geo-historical process: globalization. This latter is linear for anything that appears at some point, be it "new" lands, technical inventions, "modern" ideas and so on, only appears once. Whereas, anything that disappears, be it people, languages, cultures, species and so on, never revives again. Therefore, there is no "eternal return." A cyclical conception of history is not relevant in analyzing human societies and specifically the SES.

Because this linear history is also cumulative in all societies (Lévi-Strauss 1952). It is the case of environmental history, especially when seen from an anthropocene perspective (Crutzen 2002). This geological epoch is a historical period that begins in the early nineteenth century with industrialization and which is characterized first by the accumulation of greenhouse gases in the atmosphere, and then by the vastly anthropic ecological consequences present across the planet. This historical accumulation is accelerating exponentially since the middle of last century (McNeil 2000; Steffen et al. 2007) to the point that some global biophysical boundaries are crossed irremediably or are about to be crossed (Rockström et al. 2009). Hence, the anthropocene may be considered as a "geo-historical event" related to the development of capitalism (Bonneuil and Fressoz 2013), as a consequence of globalization: what then would be the cycles of this planetary environmental history?

10.3.3 Are the Location and Spatial Limits of SES Negligible?

The third idea driven by a naturalizing approach to SES suggests that their location and spatial limits are not important enough to be defined. Indeed, while the investigated social-ecological systems – of real cases rather than models – are generally local and occasionally regional, their borderline is not determined by the spatial extent of the interactions between the social and ecological systems functioning within them. Moreover, the few texts that specify the locations of the *social-ecological systems* do not analyze them (see for example Olson et al. 2004; Liu et al. 2007a).

However, some *Resilience Alliance* authors suggest that because of globalization, local and regional *social-ecological systems* are included in larger SES (Folke 2006), or are connected to ones that are sometimes distant (Liu et al. 2007b). This set of evidence is still rarely considered because, in the naturalizing approach of SES, the latter seem to be utopian, as they function outside real space, the actual world. Consequently, theses SES become irrelevant models for investigating the effects of the Anthropocene and globalization at the regional or local levels. Today, the location of SES in the world is indeed fundamental in the understanding of the socio-ecological interactions function within them, and consequently their coviability. It is important to explain how globalization affects the SES (Young et al. 2006); for this reason, the global environmental history must intersect with that of the world system (Hornborg et al. 2007).

Humanity started to live within "mini-systems" (Wallerstein 2004) that we can consider as isolated SES ensuring the coviability of social and ecological systems as such over very long durations.[2] These multiple SES, or "mini-systems," resulted in a geographical diversification of the Earth. This diversity of geographies or anthropic footprints, which I call "geodiversity" (Grenier 2003, 2008, 2014), results

[2]Those of Australian Aborigines have the durability record of 60,000 years!

historically from different modes of adaptation to a wide variety of ecosystems belonging to these isolated populations and societies. Geodiversity is accompanied by cultural diversity, languages, lifestyles, and so on, and it maintains biological diversity: diversity as "bio-cultural," which is investigated at the regional level (Maffi 2001).

Throughout history, some of these *social-ecological systems* have developed and expanded, and thereby turned into various "world-systems," world-empires and world-economies, present almost everywhere on Earth (Wallerstein 2004). As they integrate world-systems, the SES of mini-systems change greatly; their geographical boundaries fade away and many of them disappear. World-systems are not SES because, from a supra-regional perspective, they bring together various types of societies, actors, spaces and environments, sometimes ones that are very distant. Since "Modern Times," the European world-economy began to form the World system (Wallerstein 1974) activating with it a geohistorical process of globalization which, during periods largely determined by the development of capitalism (Braudel 1979; Wallerstein 1983; Murray 2006), led to a universal "World space" (Dollfus 1994).

By integrating the World system and therefore being located in the world space, regions are increasingly marked by external geographical actors that produce footprints determined by their own interests or goals, and that operate at various scales. By "geographical opening," I mean the deep ecological and social transformations of a region that are caused by its connection to the World system; and by "geographical globalization," I refer to the distribution process of the World system on Earth, of its spaces, milieux and environments. In this way, openings and geographic globalization explain geodiversity erosion and consequently the erosion of cultural and ecological diversity at the global scale: these processes destroy SES and different types of socio-ecological coviability everywhere on Earth.

It is therefore only through a contextualization in historical time and geographical space that the coviability of the SES can be analyzed: this is what the Easter Island case shows.

10.4 From the Coviability of Rapanui's[3] Social-ecological System to the Incorporation of Easter Island in the World System

Some environmental historians consider Easter Island as a model of ecological and social tragedy (Ponting 1991; Bahn and Flenley 1992; Diamond 2005, 2007; Hughes 2009; Welzer 2009). This thesis, still dominant today but being disputed, explains

[3]The Rapa Nui (no "s" in the plural form) are the Polynesian inhabitants of Rapa Nui, which is the current Polynesian name of Easter Island. I use it here to deal with the history of the island before European arrival.

"the collapse of Easter Island" by a deforestation completed in the sixteenth century; therefore, its Polynesian inhabitants are considered to be the only destroyers of the island since external actors arrived only two centuries later. However, the SES of Rapa Nui is an outstanding example of socio-ecological coviability in the long term, and its collapse is the result of the first geographical openings of Easter Island that took place from the eighteenth century onwards.

10.4.1 The Socio-ecological Coviability of the Rapa Nui's SES

Rapa Nui is a small island (166 km^2) located in a very isolated position (see map) and therefore with poor terrestrial biodiversity. It was colonized approximately a 1000 years ago by Polynesian people coming from Marquesas Islands via the Gambier Islands (Kirch 2000, 2010). At that time the island was covered mostly with palm trees forest (Orliac 1998). It had fresh water, a subtropical climate, and fertile volcanic soil; and it was a refuge for vast populations of seabirds. All these natural conditions facilitated the Polynesian establishment (Cristino and Vargas 2006). Archaeologists distinguish three periods in the Rapa Nui history prior to European contact: the first, a period of colonization taking place over approximately two centuries; the second, a cultural peak known as the "moai" period ("Moai" refers to the great statues that represent the ancestors who protected lineage (Metraux 1940)) and lasting three or four hundred years and finally, the period of decline, which began in the sixteenth century and corresponds to a both ecological and social systemic crisis (Bahn and Flenley 1992).

Rapanui society and Rapa Nui Island constitute a well-defined *social-ecological system*. Its terrestrial ecosystem seems closed because of extreme isolation; but it is not as it depends on Earth climate and partly on oceanic action. However, its social system remains closed, which is rare, because the island is completely isolated within the Polynesian area[4]: the Rapanui depend only on the island's ecosystems and on the adjacent sea. Polynesian colonization led to the disappearance of terrestrial avifauna and to the fleeing of seabirds. The heyday of the rapanui society ended with the total deforestation of the island, and its period of decline was accompanied by extreme exploitation and partial degradation of coastal ecosystems. Even if the coviability between social and ecological systems was not faultless in Rapanui's SES, it still allowed the population to grow[5] and create an endemic culture within the Polynesian cultural area. Rapanui society implemented a unique geography at various levels, such as the environment, territorial organization, milieu,

[4]At the beginning of colonization, the Rapa Nui maintained episodic relations with the island communities of Henderson-Pitcairn-Mangareva group, 2000 km to the west, before becoming completely cut off from the world until the arrival of Europeans in the eighteenth century.

[5]The demographic maximum of the island in the early sixteenth century was approximately 10,000 (Kirch 2010), compared to 4000 today.

and so on, in which moai are central (Grenier 2005; Hunt and Lipo 2011), which allowed it to adapt over the long term in extreme insularity.

However, the Little Ice Age caused the SES of Rapanui to enter a systemic crisis. This global climate change, which was reflected in the island by more frequent and pronounced droughts, incited the Rapanui to intensify the construction of protective moai. The construction required further use of palm trees and monopolized population's attention and energy, which led to a growing number of rats consuming seeds and proliferating. These two factors are responsible for the complete deforestation of the island in the sixteenth century (Hunt 2006). The crisis then affected the entire SES of the Rapanui and caused famines, clan wars, and a sharp decrease of the population. Nonetheless Rapanui society proved to be strong, as it adapted to a degrading environment (for example, they fished from the shores because of the lack of wood for canoes). When the Dutch approached Easter Island in 1722, they describe it as the "Garden of Eden" sheltering healthy, vigorous people. The SES of Rapanui proves that there may be coviability between a society without great technical resources and a very depleted ecosystem; they work together as an isolated and resilient "mini-system".

10.4.2 The First Geographical Openings of Easter Island and Its Integration into the World System, or the End of Rapanui's SES

The integration of Rapanui's SES in the World system marks the beginning of its end. The map (see Fig. 10.1) demonstrates that Easter Island, which is considered to be "the most isolated inhabited place in the world," has actually been subject to several geographical openings for three centuries which were fatal to Rapanui's SES.

In the eighteenth century, short stopovers of European navigators – the Dutchman Roggeveen in 1722, Spaniard Gonzalez in 1770, the Englishman Cook in 1774, and the French La Perouse in 1786 – provoked microbial shocks which decimated the island's population, causing a state of endemic war and the definitive abandonment of the worship of moai. Nevertheless, the Rapanui society once again showed resilience by adopting a new geographic adaptation in a devastated and isolated environment, the worship of the Birdman. The episodic visits of American whalers to Easter Island during the first half of the nineteenth century, allowed the surviving Rapanui population to be partly immunized, halting the demographic decline, and to learn monetary economics (Fischer 2005).

However, the ultimate disaster occurred in 1862 as labor smugglers captured 1500 Rapanui– half the population at that time – to make them work in Peru. Fifteen Rapanui return the following year, contaminating the remaining population of the island. These events provoked the destruction of the upper castes and with them most of the Rapanui culture (McCall 1994). The year 1864 marks the beginning of French missionaries assigned by Picpus's Sacred Hearts Congregation, charged

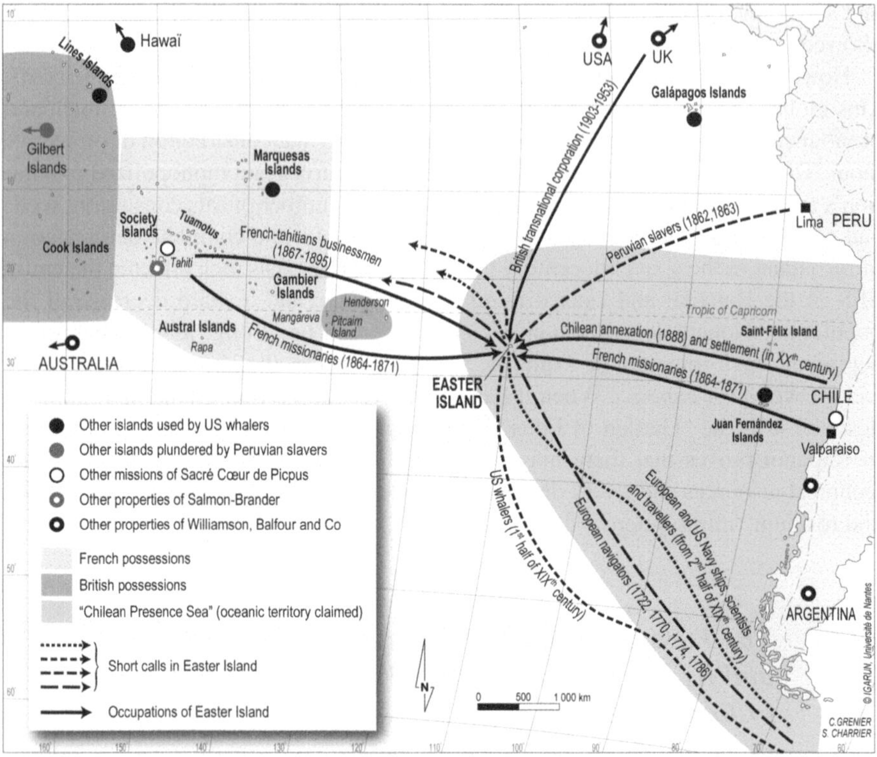

Fig. 10.1 First geographic openings of Easter Island and its integration to the world system

by the Papacy to evangelize eastern Polynesia starting from Valparaiso and Tahiti. The mission provoked the end of an already dying Rapanui culture. Rapa Nui's population itself was hardly in better state, and the missionaries hurried to baptize the Rapanui before they all died. By 1868 only 600 were left . . . (Cools 1973). The island then came under the control of French-Tahitian entrepreneurs, who drove out the missionaries in order to use the land to raise sheep and export wool to Tahiti and beyond. Only over a hundred Rapanui remained by the late 1870s. Even if they were rare, stopovers of Western ships contributed in publicizing the "mysteries" of Easter Island in the world through the stories of travelers (Loti 1899) and through the work of the first anthropologists (Routledge 1919). They launched the making of "curios" for visitors (Porteous 1981), which since then is one of the main economic activities of some Rapanui people.

Chile annexed Easter Island in 1888, as it was in a period of full territorial expansion; whereas the Eastern Polynesia had already been colonized by France and the United Kingdom. However, this insular possession is too distant, small, poor, and isolated to attract Chilean settlers. They only started arriving in in numbers from the 1960s onwards, thanks to a regular airline connection and to the extension of international tourism (Grenier 2002a, b). In 1903, the Chilean government rented Easter Island to a British transnational company, *Williamson, Balfour and Co*, which used it for sheep farming. Its subsidiary, the Exploitation Corporation of Easter Island, detains all the power: the Rapanui were confined to a small space surrounded by barbed wire, which they left only to serve the company (Fuentes 2013). In 1934, the anthropologist A. Metraux (Métraux 1940) reported that Easter Island had 40,000 sheep and 300 Rapanui inhabitants.

10.4.3 The Causes of the Collapse of Easter Island Are Geohistorical, Not Socio-ecological

The concept of geographical opening is well illustrated on Easter Island because the extent of damage it causes in a hitherto isolated Rapa Nui SES is not always related to the duration of the intrusions of external actors. Indeed, the brief stops of the European navigators of the eighteenth century, and of the Peruvian slave-traders in the nineteenth century, are the main causes of the collapse of the Rapanui society. These devastating episodes are followed by the settlement of other foreign actors on the island who complete the destruction of the people, culture, and the ecosystem and, more generally, of a rapanui-specific geography. In their turn, missionaries, French-Tahitian entrepreneurs, and the British transnational company, are responsible for the "geographical globalization" of Easter Island, where they implemented a new geography, one imported from other areas of their respective networks (see map). The traditional organization of Rapa Nui space – its division into clan territories with lands extending from shores to summits – has been replaced by a unique habitat, around the mission, and by sheep farming, which covers the rest of the island. The rapanui environment, especially its system for coping with insularity, disappeared with the spread of Christianity and the settlement of entrepreneurs. The island's ecosystem, already impoverished, has been ravaged by sheep and other species introduced in vast numbers.

Some researchers have questioned the thesis of the collapse of Easter Island before the arrival of Europeans (Peiser 2005; Hunt 2007; Hunt and Lipo 2007), but they only attribute it to "contact with the West" which they do not analyze. Since the collapse did not happen before but after the arrival of Europeans, the causes are not socio-ecological but geo-historical. The social and ecological systems of Rapa Nui Island have indeed been profoundly transformed, the first disappeared while the second has been highly degraded by external actors who, from 1722, introduced and integrated Easter Island to the World system. Since then, rapanui's SES no longer exists.

Thus, we cannot analyze the collapse of Easter Island in the light of the concept of coviability since the history of its social system is not combined with that of its ecosystem. Indeed, the latter has remained static since the sixteenth century until the settlement of livestock companies during the last third of the nineteenth century. The

Rapanui were already disorganized by then: destroyed culture, dying population, and distorted geography because of successive geographical openings. There is no SES without a connection between the social system and the ecological system, and consequently no coviability. The example of Easter Island highlights the difficulty of considering, from the nineteenth century in this case, a coviability of social and ecological systems on a geographical basis.

10.5 Conclusion

10.5.1 A Geodiversity Promising Socio-ecological Coviability

How would it be possible today to affirm the coviability of *Social-ecological systems* while their social and ecological systems are radically altered, even produced by foreign actors, those of globalization and by external agents, those of the Anthropocene? Unless one considers them only as abstract patterns, social and ecological systems have a spatial dimension and are located on Earth. However, since the "planetary globalization," that is since entering a historical period of globalization that has drastic effects on the entire globe, it is no longer possible to avoid analyzing SES and coviability interactions across the regions in which these systems are located. Nowadays, many causes of these regional interactions are to be sought at higher levels: the World and the globe. Consequently, as today "*the planetary social-ecological system is at the top of a hierarchy*" (Gotts 2007), this implies that regional SES no longer exists except for some residual mini-systems that are threatened by overall globalization.

However, regions remain relevant spatial entities for understanding biological or geographical terrestrial diversity. Geographers have known this for a long time (Vidal de la Blache 1899), and ecologists discovered it recently, which led them to develop the "eco-region" concept (Olson et al. 2001). It is on this regional basis that a socio-ecological coviability can be built.

This coviability still has to be defined in the light of the above. The "viable" is that which can continue to live, that which is sustainable. However, anything viable is not necessarily always desirable: Easter Island's history after the systemic crisis of the sixteenth century is a noteworthy example of viability, but is it desirable to be imprisoned on an environmentally ravaged island? Therefore, everything depends on what one means by "viability": a slum or a transgenic soybean plantation may be "viable," but for how long, for whom, and at what social and environmental costs? Likewise, coviability assumes that social and ecological viabilities are related. The example of Easter Island shows that a society can continue to be viable in a degraded environment. Nowadays, many spaces, environments, and landscapes all over the Earth demonstrate that coviability exists between a substantial part of humanity and highly simplified, damaged and/or artificial ecosystems, although this do not seem to be a choice made by their human inhabitants, nor desirable for all living beings.

Therefore, the main objective is to develop adequate socio-ecological coviability for terrestrial diversity of ecosystems, societies, and their geographies. In this geographical approach of coviability, the viability of various societies bases itself on the diversity of terrestrial ecosystems. The viability of these ecosystems in its turn depends on the variety of environmental and spatial human footprints on Earth. From a geographical perspective, ensuring socio-ecological coviability implies conserving the diversity of a terrestrial habitat. This coviability bases itself on geographies that maintain and produce humanly and ecologically viable habitats; they have meaning for their human inhabitants in their cultural diversity while preserving the biological diversity of other living entities. In other words, socio-ecological coviability depends on the preservation and the development of geodiversity. In a globalized world, this requires a certain territorial autonomy of SES to be preserved or developed through managing flows that connect them to the rest of the World, and through controlling the interventions of external actors[6]. Only after fulfilling these conditions can these social-ecological systems, guaranteeing socio-ecological coviability, exist.

References

Acot P (1988) Histoire de l'écologie. Presses Universitaires de France, Paris

Anderies JM, Janssen MA, Ostrom E (2004) A framework to analyze the robustness of social-ecological systems from an institutional perspective. Ecol Soc 9(1):18

Bahn P, Flenley J (1992) Easter Island, Earth Island. Thames & Hudson, London

Berkes F, Colding J, Folke C (2003) Introduction. In: Berkes F, Colding J, Folke C (eds) Navigating socio-ecological systems. Building resilience for complexity and change. Cambridge University Press, Cambridge, pp 1–31

Berque A (1990) Médiance. De milieux en paysages. RECLUS, Montpellier

Berque A (1996) Être humains sur la terre. Gallimard-Le Débat, Paris

Berque A (2010) Des fondements ontologiques de la crise, et de l'être qui pourrait la dépasser. VertigO 10(1.) http://vertigo.revues.org/9384

Bonneuil C, Fressoz J-B (2013) L'évènement anthropocène. Le Seuil, Paris

Braudel F (1979) Civilisation matérielle, économie et capitalisme, XVe-XVIIIe siècles, Le temps du Monde. Armand Colin, Paris

Brunet R, Ferras R, Théry H (dir.) (1992) Les mots de la géographie. Dictionnaire critique. GIP RECLUS/La Documentation Française, Montpellier/Paris

Caillé A, Chanial Ph, Vandenberghe F (2001) Présentation, in Chassez le naturel… Ecologisme, naturalisme et constructivisme, Revue du MAUSS n° 17. La Découverte, Paris, pp 5–21

Cools A (1973) L'île de Pâques et la Congrégation des Sacrés Cœurs. Documentation, Rome, M.S. 271 788–95 (972)-5

Crosby A (1986) Ecological imperialism. The biological expansion of Europe, 900–1900. Cambridge University Press, New York

Crutzen P (2002) Geology of mankind. Nature 415:23

[6]As the historian of environment J. Radkau suggests, "wherever the local population does not have control over its resources and is unable to keep outsiders away, the environment degrades" (2008, 3).

Deléage J-P (1991) Histoire de l'écologie. Une science de l'homme et de la nature. La Découverte, Paris
Descola P (2005) Par-delà nature et culture. Gallimard, Paris
Diamond J (1997) Guns, germs, and steel. The fates of human societies, (De l'inégalité parmi les sociétés. Essai sur l'homme et l'environnement dans l'histoire, Paris, Gallimard, 2000)
Diamond J (2005) Collapse. how societies choose to fail or survive. Penguin Books, New York
Diamond J (2007) Easter Island revisited. Science 317:1692–1694
Dollfus O (1994) L'espace Monde. Economica, Paris
Fischer S (2005) Island at the end of the world. Reaktion Books, London
Folke C (2006) Resilience: the emergence of a perspective for social–ecological systems analyses. Glob Environ Chang 16:253–267
Folke C et al (2002) Resilience and sustainable development: building adaptive capacity in a world of transformations, ICSU Series on science for sustainable development, vol 3. International Council for Science, Paris
Foster BJ, Clark B, York R (2010) The ecological drift. Capitalism's war on the earth. Monthly Review Press, New York
Fuentes M (ed) (2013) Rapa Nui y la compañía explotadora. Rapa Nui Press, Santiago
Godelier M (1984) L'idéel et le matériel. Fayard, Paris
Gotts N (2007) Resilience, panarchy, and world-systems analysis. Ecol Soc 12(1):24. http://www.ecologyandsociety.org/vol12/iss1/art24/
Grenier C (2002a) Océaniques ou Américaines? Analyse comparative du rattachement aux Amériques des Galapagos et de l'île de Pâques. Mappemonde 66:38–44
Grenier C (2002b) How tourism reduces geodiversity and how it could be different: the Galapagos archipelago and Easter Island cases. In: di Castri F, Balaji V (eds) Tourism, biodiversity and information. Backhuys, Leiden, pp 233–255
Grenier C (2003) Gardeer l'espace: la notion de géodiversité et la conservation du patrimoine littoral, Océanis, fascicule 28–1/2: Environnement, politiques publiques et dynamique des activités littorales, pp 233–251
Grenier C (2005) La patrimonialisation comme mode d'adaptation géographique. Galápagos et île de Pâques. In: Cormier-Salem M-C, Juhé-Beaulaton D, Boutrais J, Roussel B (eds) (dir) Patrimoines naturels aux suds, Territoires, identités et stratégies locales. IRD Editions, Paris, pp 475–513
Grenier C (2008) La gestion de parcs nationaux mondialisés dans des régions à forte géodiversité. Corcovado (Costa Rica), Galapagos (Equateur), Rapa Nui (Chili). In: Héritier S. et Laslaz L., dir., Les parcs nationaux dans le monde. Ellipses, Paris, pp 123–142
Grenier C (2014) Géodiversité et mondialisation. Les fondements géographiques de la diversité terrestre et de son érosion, volume scientifique d'Habilitation à Diriger des Recherches, Université de La Rochelle
Holling CS (2001) Understanding the complexity of economic, ecological, and social systems. Ecosystems 4:390–405
Hornborg A, McNeill J, Martinez-Alier J (eds) (2007) Rethinking environmental history: world–system history and global environmental change. AltaMira Press, Lanham
Hughes JD (2009) An environmental history of the world. Routledge, New York
Hunt T (2006) Rethinking the fall of: new evidence points to an alternative explanation for a civilization's collapse. Am Sci 94:412–419
Hunt T (2007) Rethinking Easter Island's ecological catastrophe. J Archaeol Sci 34:485–502
Hunt T, Lipo C (2007) Chronology, deforestation and "collapse": evidence vs. faith in Rapa Nui prehistory. Rapa Nui J 21(2):85–97
Hunt T, Lipo C (2011) The statues that walked. Unraveling the mystery of Easter Island. Free Press, New York
Judkins G, Smith M, Keys E (2008) Determinism within human–environment research and the rediscovery of environmental causation. Geogr J 174(1):17–29
Kirch P (2000) On the road of the winds. An archeological history of the Pacific Islands before European contact. University of California Press, Berkeley

Kirch P (2010) Peopling of the Pacific: a holistic anthropological perspective. Annu Rev Anthropol 39:131–148
Lacoste Y (2003) De la géopolitique aux paysages. Dictionnaire de la géographie. Armand Colin, Paris
Le Berre M (1992) Territoires. In: Bailly A, Ferras R, Pumain D, dir., 1992, Encyclopédie de Géographie. Economica, Paris, pp 617–638
Lévi-Strauss C (1952) Race et histoire. Albin Michel et UNESCO (2001), Paris
Lévy J, Lussault M dir., (2003) Dictionnaire de la géographie et de l'espace des sociétés. Belin, Paris
Liu J et al (2007a) Complexity of coupled human and natural systems. Science 317:1513–1516
Liu J et al (2007b) Coupled human and natural systems. Ambio 36(8):639–649
Livingstone D (1992) The geographical tradition. Blackwell, Oxford
Loti P (1899) L'île de Pâques. Journal d'un aspirant de La Flore, St Cyr sur Loire, Pirot Editeur (2006)
Maffi L (2001) On the interdependance of biological and cultural diversity. In: Maffi L (ed) On biocultural diversity. Smithsonian Institution Press, Washington, DC, pp 1–50
Marsh G (1864) Man and nature. University of Washington Press, Seattle
McCall G (1994) Rapanui. Tradition and survival on Easter Island. University of Hawaii Press, Honolulu
McNeil JR (2000) Something new under the sun. An environmental history of the twenthieth-century world. Norton & Company, New York
Métraux A (1940) Ethnology of Easter Island, vol 160. Bernice P. Bishop Museum Bulletin, Honolulu
Murray WE (2006) Geographies of globalization. Routledge, London
Olson D et al (2001) Terrestrial ecoregions of the world: a new map of life on earth. Bioscience 51(11):933–938
Olson P, Folke C, Berkes F (2004) Adaptive comanagement for building resilience in social–ecological systems. Environ Manag 34(1):75–90
Orliac C (1998) Données nouvelles sur la composition de la flore de l'île de Pâques. Journal de la Société des océanistes 107(2):135–143
Passet R (2010) Les grandes représentations du monde et de l'économie. Les Liens qui Libèrent, Paris
Peiser B (2005) From Genocide to Ecocide: the Rape of Rapa Nui. Energy Environ 16(3–4):513–539
Ponting C (1991) A green history of the world (Le viol de la Terre, Paris, NiL, 2000)
Porteous D (1981) The modernization of Easter Island, Western geographical series vol. 19. University of Victoria, Victoria
Radkau J (2008) Nature and power. Cambridge University Press, Cambridge
Ratzel F (1897) Géographie politique. Fayard, Paris
Reclus E (1905) L'Homme et la Terre, Introduction et choix de textes par B. Giblin. La Découverte, Paris
Rist G (1996) Le développement. Histoire d'une croyance occidentale. Presses de Sciences Po, Paris
Rockström J et al (2009) Planetary boundaries: exploring the safe operating space for humanity. Ecol Soc 14(2):32. www.ecologyandsociety.org/vol14/iss2
Routledge K (1919) The mystery of Easter Island. Adventures Unlimited Press, Kempton, (1998)
Sahlins M (1976) The use and abuse of biology. An anthropological critique of sociobiology (Critique de la sociobiologie, Paris, Gallimard, 1976)
Spengler O (1921) Le déclin de l'Occident. Gallimard, Paris
Steffen W, Crutzen P, McNeill J (2007) The anthropocene: are humans now overwhelming the great forces of nature? Ambio 36(8):614–621
Vargas P, Cristino C (2006) 1000 años en Rapa Nui. Arquelogogía del asentamiento. Editorial Universitaria, Santiago de Chile

Vidal de la Blache P (1899) Leçon d'ouverture du cours de géographie. Annales de Géographie 8(38):98–109

Walker J, Cooper M (2011) Genealogies of resilience: from systems ecology to the political economy of crisis adaptation. Security Dialogue 42(2):143–160

Wallerstein I (1974) The modern world-system 1, capitalist agriculture and the origins of the European world-economy in the sixteenth century. University of California Press, Berkeley, (2011)

Wallerstein I (1983) Historical capitalism (Le capitalisme historique. La Découverte, Paris, 2002)

Wallerstein I (2004) World-systems analysis. an introduction (Comprendre le monde. Introduction à l'analyse des systèmes-monde. La Découverte, Paris, 2006)

Welzer H (2009) Les guerres du climat. Gallimard, Paris

Young O, Berkhout F, Gallopin G, Janssen M, Ostrom E, van der Leeuw S (2006) The globalization of socio-ecological systems: an agenda for scientific research. Glob Environ Chang 16:304–316

Chapter 11
A Rupture Between Human Beings and Earth: A Philosophical Critical Approach to Coviability

Aliénor Bertrand

Contents

11.1 Introduction

Is the 'coviability' of natural and social systems a paradigm offering an adequate response to the ecological crisis? Does it promise to replace the notion of sustainable development the media success of which has been inversely proportional to its ability to counteract the environmental crises?

On the methodical level, the paradigm of "coviability" has three major characteristics:

- a systematic description of natural and social sets, the ontological qualification of these systems remaining open
- an opposition of natural determinations to the social ones, this opposition operating as a structuring principle

A. Bertrand (✉)
CNRS, Faculty of Law and Political Science, UMR 5815, Dynamiques du droit, University of Montpellier, Montpellier, France
e-mail: bertrand.alienor@orange.fr

O. Barrière et al. (eds.), *Coviability of Social and Ecological Systems: Reconnecting Mankind to the Biosphere in an Era of Global Change*,
https://doi.org/10.1007/978-3-319-78497-7_11

- an attempt to come over this opposition by the notion of 'life' contained in the idea of "coviability", this one may be understood either in analogical and rigorous manner, or in a very wide way as a synonym of an aptitude to endure while modifying as a model.

This chapter is a philosophical side step to approach the paradigm of coviability with the tools of critical theory. It offers to correct the effects of too strong an emphasis on systematic modeling that overshadow the fertile legacy of reflexive political history, thus leading to a regrettable intellectual bias. This bias is usually reinforced by the common place which would see environmental awareness being born late in the day, at the end of the twentieth century. However, regarding the relationships between "nature" and "politics", critical thinking deserves to be reconsidered. The main characteristics of the coviability concept, at least, as they have been recalled, enter in particular consonance with the founding texts of Karl Marx: the tradition that is anchored here denounces, indeed, the contradictions jeopardizing what we now call the "sustainability" of the social system. Unlike the prejudice, long led by a perspective of "orthodox" reading (Foster 2011, 24–27), these contradictions do not highlight the non-sustainability of capitalism as a principle of social organization only, but 'simply' its non-sustainability. The 'coviability' of the social 'system' with natural 'systems' was indeed an early concern of Marx, whose assessment was severe on the historical turn taken by the West in the modern age. Capitalism would have broken, in an "irreparable" way, the 'metabolic interaction' between human beings and Earth, i.e. the eternal condition of production.[1] However, Marx believed that capitalism would quickly collapse, giving to a new society the task of restoring this lost 'coviability'. Nevertheless, not only has capitalism not collapsed, but it has also shown an extraordinary capacity to overcome its crises without ever restoring any of the balances broken by modernity: it has imposed its unifying hold everywhere while universally destroying the previous diversity of social and natural systems.

With regard to the foundations of critical thought on "coviability", Marx's thoughts deserve to be put into perspective with the recent research on the anthropology of Nature. The notion of coviability conveys, thus, a subtle paradox: it appears that the very methodical principle of an opposition between "social systems" and "natural systems" reactivates the framework on which capitalism

[1]"So far therefore as labour is a creator of use value, is useful labour, it is a necessary condition, independent of all forms of society, for the existence of the human race; it is an eternal nature-imposed necessity, without which there can be no material exchanges between man and Nature, and therefore no life ». (Marx 1867). Right from his first written piece of work at a young age, Marx dealt with the idea that "the human lives from nature": « Nature is man's inorganic body– nature, that is, insofar as it is not itself human body. Man lives on nature – means that nature is his body, with which he must remain in continuous interchange if he is not to die. That man's physical and spiritual life is linked to nature means simply that nature is linked to itself, for man is a part of nature » (Marx 1844). Let us even cite the manuscripts of 61–63: "actual labour is the appropriation of nature for the satisfaction of human needs, the activity through which the metabolism between man and nature is mediated" (Marx 1861–1863).

rests; Philippe Descola has shown that the opposition of nature and society only made sense within the anthropological regime described as "naturalistic ontology" (Descola 2005) and of no other. Relying on this opposition, the paradigm driven by the notion of 'coviability' may then risk renewing the obstacles which it seeks to overcome ... The critical philosophical approach we outline here invites us to identify the pitfalls and difficulties that must be avoided so that this paradigm has a chance for an effective beginning.

11.2 A Return to Marx

The need to refer to Marx to think the coviability of natural and social systems cannot be regarded as evident. Certainly, Marx made the "social system" *transformation* the horizon of his reflection, opposing such will to transform to the Hegelian thought of the world. Nevertheless, the resulting "absolutization of politics"[2] was often considered as a convenient reduction of the heterogeneity of human life determinations to social power relations, that is, to class struggle alone. It has also been frequently interpreted as an over-determination of the social life by its "economic" aspect. Seduced by this criticism, some ecologists said that Marx really ignored the exteriority of nature by conceiving it as a condition of production only, the factor of "productive forces".[3]

There would be very little to gain today in the reading of Marx if the 'coviability' of natural and social systems only means for him a single adjustment of productive forces to production reports as Serge Moscovici still thought (Moscovici 1969). On the contrary, highlighting the contradictions of capitalism – as a 'social system'-, with nature – as a 'system' and as a cycle-, is one of the fundamental elements of the criticism on which the political transformation is based and for which he wants to work. Marx developed a major analysis of the nature/society dialectic which allows the ongoing environmental crisis caused by capitalism to be understood. More precisely, and unlike the biased economist reductionism of his thinking claimed to impose, Marx had set up the framework of "historical-environmental materialism" taking into account the *co-evolution* of nature and human society.[4] This principle

[2]Miguel Abensour suggests that this is a "political absolute" which Marx shares with the Young Hegelians, and which aims to transform the philosophy to politics, substituting a "phenomenology of the will to the phenomenology of spirit" (Abensour 2012, 104).

[3]For highlighting this type of interpretation, see "the European ecology and the oversight of the history (of capitalism)", (*Europeana n ° 6*, 2015).

[4]It was J.B. Foster who demonstrated it, by refuting the four commonly accepted theories on the reports of Marx and e nature. Here are the four theories: (1) Marx's thought is anti-ecological and the history of the URSS shows this; (2) Marx produced insightful analyses on the reports but succumbed to prometheism, believing that environmental problems would disappear with the abundance; (3) Marx produced an analysis of the ecological degradation in the agricultural domain, but this analysis is separated from the rest of the work; (4) Marx developed a systematic approach to

of this co-evolution is indeed the basis of one of the most important concepts designed by Marx, that of labour. Indeed, in an amazing fidelity to his early texts and vocabulary, Marx defines 'labour' in *Capital* as:

> « Labour is, in the first place, a process in which both man and Nature participate, and in which man of his own accord starts, regulates, and controls the material re-actions between himself and Nature. He opposes himself to Nature as one of her own forces, setting in motion arms and legs, head and hands, the natural forces of his body, in order to appropriate Nature's productions in a form adapted to his own wants. By thus acting on the external world and changing it, he at the same time changes his own nature. (Marx, 1867, 127)
>
> So far therefore as labour is a creator of use value, is useful labour, it is a necessary condition, independent of all forms of society, for the existence of the human race; it is an e ternal nature-imposed necessity, without which there can be no material exchanges between man and Nature, and therefore no life. (Marx, 1867, 31)

The famous accents of *the Manuscripts of 44*, according to which nature is the inorganic human body, can be found here (Marx 1844, 32). Marx stated that, by its production, man-made nature his work and thus became a species being:

> It is just in his work upon the objective world, therefore, that man really proves himself to be a species-being. This production is his active species-life. Through this production, nature appears as his work and his reality (Marx 1844, 32).

In '*Capital*', the concept of labour has replaced the notion of world shaping: the human/nature dialectic is put to service of one of the most fundamental concepts of social analysis. To think of the "coviability" of social and natural systems, it is, thus, necessary to enter into the analysis of structures of work's social process. Defining labour as a metabolic process may seem quite speculative. It is yet an accurate proposal and highly controversial:, Marx's fight against the classical liberal economists who consider natural systems as sources of wealth without 'value' to be taken into account in political economy.. Translated somewhat abruptly into the language of today, this means that Marx rejects any strict (or 'economist') conception of natural resources, as "stocks".

However, we can argue too that he also contested in his principle the idea that natural systems can be defined adequately by the eco-systemic "services" they provide to human societies. Marx developed in a very precise and detailed way a position of this kind when he studied one of the most serious environmental problems of his time, the decline in soil fertility. We need to focus on this matter.

Instead of the classical assertions by economists that deal with the soil fertility as a "natural gift",[5] i.e. –something which is like a (free) economic 'service'-, Marx analyses it as the concrete intersection of two heterogeneous systematic loops, that

---

nature and environmental degradation and raised the issue of environmental sustainability (Foster and écologiste 2011).

[5]« Enfin, les physiocrates, si supérieurs à tant d'égards, n'ont-ils pas imaginé que la rente foncière n'est pas un tribut arraché aux hommes, mais un présent fait par la nature même aux propriétaires? » (Marx 1867).The English version is different "How long is it since economy discarded the physiocratic illusion, that rents grow out of the soil and not out of society?", https://www.marxists.org/archive/marx/works/download/pdf/Capital-Volume-I.pdf, p. 52.

of the natural cycle of nutrients, and that of social organization. These two systems do not maintain symmetrical relationships between each other. In his principle, the natural cycle of nutrients is not necessarily determined by the social system, while social organization is always determined in reverse by the local richness of soils. However, social organization can interfere with the cycle of nutrients and damage it – or improve it – in a major way. Thus, to consider soil fertility as a natural "service" in the way of economists is to conceal the structure that systematically links the nutrient cycle and social organization: it is to ignore the fact that soil impoverishment is the result of the imbalance caused by the antagonism of cities and the countryside, born itself from capitalist agriculture.

As this is often the case, Marx's demonstration is the opposite of what some "economist", vulgate in his thinking, propagated.[6] Social determinations, and even less so, natural determinations, are not reduced to economic determinations; on the contrary, Marx shows in what the claims of so-called "economic science" – those of political economy – hide the social antagonisms and relations of power. In this case, the theories of liberal economists had not only disguised the unfairness of the distribution of profits from capitalist farming for ideological purposes, but also the effects of social antagonisms on natural systems. Certainly, we can give these economists the indisputable fact that the capitalistic "improvement" of soil (temporarily) increased its productivity. But even the definition of this productivity is ambiguous and should not be confused with the amount of food produced by the soil. Productivity is an inseparable economic concept of the capacity of agricultural work to provide added value. However, the capitalistic of land led to the disappearance of many traditional farming and feeding uses, the results of hard toil and non-market activities immersed in the natural cycles. The examination of the destruction of these uses under an ecological perspective shows the commodification of land in the modern era, and this alone, initiated a continuous downward spiral in soil fertility and is therefore, nothing ? "the theft of nutrients".

11.3 Capitalism and the Decline in Natural Soil Fertility: A Metabolic Rupture

Marx develops this fundamental idea by showing what happened when the decline of soil fertility became economically threatening for the capitalist owners: 'theft' of nutrients continued to be practiced ? under a new form, and on a larger scale. Despite an exorbitant price, it was indeed 'cheaper' for the richest owners to obtain fertilizer from the other side of the world than questioning the social relationships of the dispossession 'of lands and workers',[7] which were the basis of their profits. Assisted

[6]Cf. (Harribey 2008, 2009, 123–124).

[7]Cf. *Capital*, Volume I, Book, 1, chapter xv: "It [capitalist production] therefore violates the conditions necessary to lasting fertility of the soil. By this action it destroys at the same time

by an imperialist policy, the artificial merchant remediation to the disappearance of the 'ecosystem service' of land fertility has been an economic godsend opening new markets ranging from the collection of bones from the battlefield to the plundering of guano. The substitution of the purchase of fertilizers to the equilibrium of the nutrient cycle has mechanically worsened the existing social inequalities, has created new ones, and caused unprecedented capitalistic concentrations; it has speeded up both the ruin of small farmers and the degradation of soils, driving away as much the prospect of a social transformation making agricultural production and the natural system of nutrient cycle 'co-viable'. This spiral has led, a century later, to the death of most of the European soils and the virtual disappearance of peasantry. Marx describes how agricultural capitalism first accommodated the destruction of soils, and how it used the sudden appearance of a 'price' allowing the 'ecosystem service' of natural soil fertility to be 'as' if rendered by the importation of guano and the spraying of skeletons. Now if, from the economic point of view, this service may have seemed 'like' it was being 'rendered', it has failed to ensure the viability of the natural systems of soils, which is ecologically irreducible. Socially, it is accompanied by increased exploitation, and/or violent dispossession mechanisms. So, in an archetypal way, Marx, denounced in advance the conceptual mistake according to which an 'ecological service' would be 'replaceable', whether it is free or paid for.

Nevertheless, how was the double theft made to the land and to the workers hidden by the works of liberal economists? The dissimulation of theft made to the workers is the main and *well-known* purpose of the *Capital,* and we will not deal with it here. As to the occultation of land theft, this is covered by a liberal apology of the so-called capitalist improvement of soil. Marx gives at least two different explanations: the first is based on the analysis of the fetishism of commodities specific to the capitalist social relations. The second, less apparent, but may be connected to the first, is based on the special modes of development and appropriation of knowledge in capitalist society.

The treatment of the example of land price assessment in the chapter of *Capital* on '*The Fetishism* of Commodities and its *Secret'* is quite significant (Marx 1867, note 26). The main paralogism of economists comes from the artificial correlation between the population and the land's ability to produce food. This is explained by the fetishization of land as a commodity. Indeed, by arguing that the fall of agricultural productivity was due to the fact that demographic pressure pushed human beings to exploit less and less fertile land, Malthus has, for a long time,

the health of the town labourer and the intellectual life of the rural labourer. (...) Moreover, all progress in capitalistic agriculture is a progress in the art, not only of robbing the labourer, but of robbing the soil; all progress in increasing the fertility of the soil for a given time, is a progress towards ruining the lasting sources of that fertility. The more a country starts its development on the foundation of modern industry, like the United States, for example, the more rapid is this process of destruction. Capitalist production, therefore, develops technology, and the combining together of various processes into a social whole, only by sapping the original sources of all wealth-the soil and the labourer. » (Marx 1867, 330).

prevented the understanding of the deep social, political, and natural dynamics of agricultural production. According to him, Ricardo also justifies the existence of a proportional relationship, quasi-geometrical, between the increase in the population and the fertile lands cultivated, and this until the impossible improvement of unfertile lands. However, not only is this correlation arbitrary, but it is also based on a fetishization of land value; land fertility is, thus, viewed as being independent of the work and the uses by human societies, and only viewed on its rarity in the market, a rarity that is reinterpreted in terms of an arbitrary correlation with population growth... This fetishism attached to commodities is actually a reification of social relations, i.e. a transformation of living relations of domination as they appear in the form of relationships between things:

> There, the existence of the things *quâ* commodities, and the value relation between the products of labour which stamps them as commodities, have absolutely no connection with their physical properties and with the material relations arising therefrom. There it is a definite social relation between men, that assumes, in their eyes, the fantastic form of a relation between things. In order, therefore, to find an analogy, we must have recourse to the mist-enveloped regions of the religious world. In that world the productions of the human brain appear as independent beings endowed with life, and entering into relation both with one another and the human race. So it is in the world of commodities with the products of men's hands. This I call the Fetishism which attaches itself to the products of labour, so soon as they are produced as commodities, and which is therefore inseparable from the production of commodities. (Marx 1867, 48)

The power of the fetishization of land-commodities thereby prevented liberal economists from understanding why the capitalist practices of work had caused a rupture of the fundamental balance of nature. Marx then points to the denial effects of the ignorance of the economists the land uses by peasant uses, and their confusion between exchange value and value of usage. This is something which we wish to consider later, but, in the *Grundrisse*, the main elements of the criticism of Malthus' theories are in place:

> « ...He [Malthus] regards *overpopulation* as being *of the same kind* in all the different historic phases of economic development; does not understand their specific difference, and hence stupidly reduces these very complicated and varying relations to a single relation, two equations, in which the natural reproduction of humanity appears on the one side, and the natural reproduction of edible plants (or means of subsistence) on the other, as two natural series, the former geometric and the latter arithmetic in progression. In this way he transforms the historically distinct relations into an abstract numerical relation, which he has fished purely out of thin air, and which rests neither on natural nor on historical laws. (...)He transforms the immanent, historically changing limits of the human reproduction process into *outer barriers;* and the *outer barriers* to natural reproduction into *immanent limits* or *natural laws* of reproduction. (..)The invention of surplus labourers, i.e. of propertyless people who work, belongs to the period of capital. The beggars who fastened themselves to the monasteries and helped them eat up their surplus product are in the same class as the feudal retainers, and this shows that the surplus produce could not be eaten up by the small number of its owners. (..). Never a relation to a *non-existent* absolute mass of means of subsistence, but rather relation to the conditions of reproduction, of the production of these means, including likewise the *conditions of reproduction of human beings,* of the total population, of relative surplus population. This surplus purely relative:

> in no way related to the *means of subsistence as* such, but rather to the mode of producing them. Hence also only a *surplus* at this state of development. (Marx 1857, Marx and Engels 1978, 92).

The conclusion of Marx's analysis is the opposite to that of Malthus. He shows how the overpopulation phenomenon, which cannot be reduced to simple population growth, is a consequence of capitalism, and concludes that 'surplus workers', i.e. human beings deprived of property and capable of work, are an invention proper to the capital era'(Marx and Engels, 1978, 90). Therefore, with Malthus' criticism, the notion, that could be considered nowadays as an abstract conception of coviability, collapses because it connects parameters allegedly salient but arbitrarily correlated. To believe that 'the' human society weighs on natural ecosystems by its number is to hide the fact that capitalism breaks down the natural balances because of singular social relationships that it imposes and perpetuates. It also means that the representation of these social relations are hidden and which denounce the ideological function of the commodification/fetishization of the land, namely that of a war machine in the service of 'the destruction of the proletariat'.

However, does Marx really give the proletariat the (excessive) power to create its own conditions of existence independently of nature's limits? It would, indeed, be common sense to grant Malthus that there is *ultimately* a weight correlation between a maximum population size and the free eco-systemic services provided by soil fertility. Even if Malthus' social analysis is not accurate and completely omits the structural effects of the class relations, the fact remains that the ability of soil to feed the men will be over. Marx's environmentalist critics hastily conclude that he had not been aware of the finiteness of natural resources. This is not correct. Marx confirms that capitalism 'invents' surplus workers, but he does not argue that the capacity of the land to feed people is infinite. In a demiurgic sense, humanity has no more power to 'go beyond' the limits of nature than a social system has to create' nature; the metabolism between humans and land must instead be established 'on a regulating law of social production', and this may lead to the disappearance of fetishism attached to 'land' as a commodity.[8]

The erroneous objections of some Marx's environmentalist critics ignore that 'production' is not a simple manufacturing process physically open-ended to infinity and describable in terms of exchange transformation of flow materials or even services. Production is a social relationship based on work, which is quite different. In his criticism of Malthus, Marx is also careful in using the notion of 'workers', forsaking the categories that fall under abstract statistical analyses of the population. However, it is precisely because the analysis of 'production' makes the concrete social process of work the heart of the metabolism between human beings and nature, that Marx is able to think about, *at the same time,* the 'viability' of natural and human systems. It is for this reason too that a deeper knowledge

[8]"The religious reflex of the real world can, in any case, only then finally vanish, when the practical relations of every-day life offer to man none but perfectly intelligible and reasonable relations with regard to his fellowmen and to Nature. » (Marx 1867, 52).

of natural cycles teaches us even more about "coviability", than the economic quantification of the fetishized value of fertility (exchange) ever can, including in its contemporary metamorphosis of 'ecosystemic service',. Therefore, Marx contrasts Liebig's *organic chemistry*[9] with the erroneous calculations of the economists. Thanks to the works of Liebig, he is able to rigorously argue that soil fertility, or its infertility, is largely caused by the production relationships which structure the social system, i.e. the relationships of exploitation at work. This statement implies no form of Prometheism, on the contrary, the analysis of the ecological problem of soils fertility loss allows him to show how the social system imposed by capitalism is 'naturally' viable.

While denouncing the fetishism associated to "land as a commodity" and thinking in terms of uses, Marx explains how production relationships of European countryside interfere in the *natural* cycles of fertility. Not only is Malthus' modeling wrong due to the oversimplification of its parameters, but it is entirely based on a faulty economic paradigm. Unlike its biased beginnings in its reasoning, that of land use value which gives it its *real* wealth, fertility, is not correlated in proportion to its market value. The market value of the land is linked only to its potential to generate incomes, not to its natural fertility. On the contrary, the land use value, which is not comparable to a gift or to an 'eco-systemic service', is the result of a complex dynamic combining social uses and natural cycles. Actually it is theoretically and technically possible to generate a huge income from land that has become naturally barren, and this possibility has been extensively carried out by the industrial agriculture dominating Europe a century and a half after Marx's analyses. Not only does capitalist agriculture create surplus workers, but it also benefits from the sterilization of soil that it has caused. This again illustrates the difficulty of a relevant model of the coviability of natural systems and social systems. Even if it is quite easy to significantly refine the mathematization of defined social or natural settings; such sophistication may radically be false, for lack of a proper analysis of existing dynamics, inextricably historical and natural.

11.4 Scholarly Domination of Man Over Nature and the Destruction of Knowledge and Farming Practices

This first argument denouncing the economists' paralogisms hides a second one, more subtle, which implements the foundations of social criticism of scientific practices. Indeed, Marx concedes that economists could not have known the developments of chemistry upon which his own analyses are based. This gap is obviously linked to the contingency of science progress. However, here, there is

[9] John Bellamy Foster quotes the edition of 1862 of the Liebig book proposed by Erland (Marold 2002, 74), (Liebig 1999).

much more than the trivial observation that economists belonging to their time were limited by the science of their era. The erroneous statements of economists show the type of relationship they maintained with science, that of unlimited belief in the 'scholarly domination of man over nature'(Marx 1867, 564), a characteristic of the modern time linked, as we shall see, to the concentration of the means of production. However, such a belief not only has the power to eradicate any clear awareness from the necessarily fragmentary and limited character of scientific knowledge itself, but also any attention to knowledge linked to ancient uses, reduced to folklore or superstition. This type of belief in the scientific 'truth' is radically foreign to the '"proletariat" practices, whose social knowledge, and knowledge in general, are immerged and produced by an intimate knowledge of the changing conditions that allow the survival and the resistance to domination. The blind belief of economists in the limited state of scientific knowledge of their time, associated in defiance of knowledge of the ancient peasant practices of nature, contributed effectively to impose and perpetuate the imbalances caused by the capitalist mode of production. It is remarkable how, in the eighteenth century, the denunciation of 'superstitions' as dogmas is often the reversed mirror of the belief in the 'truth' of emerging science. By assimilating popular knowledge to institutionalized religious beliefs or by considering it as intellectual misery of superstition, the philosophers of the Enlightenment did not only defend the power of reason: they worked also to eliminate the medieval practices of nature which structure the rural world, and this rupture was irreversible. Yet few voices rose against the pyres where tens of thousands of women were murdered (Federici 2014, 293–294). Neither the new uses of the urban proletariat nor the new sciences have allowed what had has been lost from that period (Boumediene 2016) to be restored or reinvented.

However, Marx explains that it is up to the proletariat to allow the change from social fantasy of the 'domination of nature' to new practices of knowledge contained in its limits. Indeed the 'proletariat is bearer of knowledge about a nature different from that of 'the bourgeoisie', whose abstract systematization is linked to class domination. This mental mechanism is based on the identification of conclusions whose truth is necessarily always limited, partial and related to a certain state of knowledge, to one or more universal proposals widely extending beyond the experimental framework or that of actual observation. This rise in generality, itself, occurs through a decontextualisation of material and social conditions epistemologically allowing the validity of observations and experimental protocols. In addition to neglecting these conditions and the unpredictable effects of reducing the parameters on which the experiment is based, such a generalization can also eliminate items of knowledge that escape, or even contradict, the paradigm on which they are based. The fantasy of the domination of nature by man is therefore inseparable from the subjective position produced by social domination: it produces a factory of allegedly 'objectifying' systems but used in practice to suffocate any opposed knowledge rooted in foreign social relationships on capitalist exploitation. The contempt for peasant knowledge is also class contempt, and not only in terms of 'distinction': it is contempt precisely rooted in representations economically useful for relationships of exploitation. The history of industrialization is woven with episodes that prove this: recently, Jean-Baptiste Fressoz described it well by tracing the conflicts between the fishermen and the soda industry. The nascent botany was dogmatically built by the State as reliable knowledge come to industry's rescue: the partial results of a fledgling science were contrasted against assertions triggered by age-old observations of fishermen on the sea-grass ecosystems. Accordingly, the total uprooting of lichens necessary for the production of soda was authorized, depriving the fishermen of their vital resource. These results were proved to be so wrong and their consequences so harmful that the royal power had to back down and again regulate the uprooting of lichens (Fressoz 2012).

When Marx contrasts the use value with market value, it is not for a *purely* theoretical principle, nor only for revolutionary teleology. Nor does consideration to usages have anything to do with reactionary nostalgia for a disappearing world. The legitimization of the respect for knowledge from practices that he outlines holds the vital reality of class relationships. It is because 'the proletariat' is engaged in a practical thought and continuous actions to ensure the shifting conditions of its survival, that its relationship to uses bears a society potentially free of the fiction of knowledgeable domination of men and nature. The dissipation of man's fantasy in dominating nature cannot therefore be 'managed' from above, by virtue of an abstract and systematic scientific vision of the relationships of human societies with natural cycles; it will only emerge from the transformation of social relationships of domination. This does not mean that a potential reactive reversal, term by term, of class social relations, allowing a return to the patterns of previous relationships, or making the dominated of yesterday the dominant of today, would be sufficient to solve the problem. This is not the way the dialectic overthrow of social relationships should be understood. Marx states this very clearly:

> The capitalist mode of appropriation, the result of the capitalist mode of production, produces capitalist private property. This is the first negation of individual private property, as founded on the labour of the proprietor. But capitalist production begets, with the inexorability of a law of Nature, its own negation. It is the negation of negation. This does not re-establish private property for the producer, but gives him individual property based on the acquisition of the capitalist era: i.e., on co-operation and the possession in common of the land and of the means of production (Marx 1867, 542).

Nevertheless, this is not the place to discuss the notions of 'class' and 'proletariat". It is, though, rather important to go back to the error of Marx's historical appraisal, which led him to believe that the contradictions of capitalism such as the action of the working class would allow a quick collapse of the system of exploitation. This belief in a quick working class victory certainly led him to play down the profound effect of the relegation of, and then the disappearance of old small rural farming practices (Federici 2014), as if the destruction of the medieval world had been too equivocal, interspersing the victory of the small peasantry in its struggle against feudal domination and its contemporary defeat. 'Contrary to what we sometimes think, Marx has not underestimated the ability of peasants to defend their interests: he instead describes the free peasants as a class, the '*Yeomanry'*, having succeeded in emancipating themselves from the feudal oppression. How can their historical failure therefore be explained? If the small free peasants, as a class, were finally crushed by the *enclosures*, it is because of the heterogeneous financial and economic imbalances caused by conquests and colonization as well as the power newly acquired by the *bankocracy* that destroyed medieval social relationships. The historical, the historical destruction of free peasantry thus shows that capitalism holds power not only from the foundation of relationships of exploitation, but also from the part which it gains from colonial conquests and its ability to play societies against one another and to completely reverse the local results of class relationships. Today, as during the modern age, accumulation mechanisms related to new forms of colonization transform not only 'workers', but also 'nature' into targets of appropriation.

With international land grabbing facilitated by the States, the extractivist destruction of forests, the patentability of the living organisms, government through public debt, and the financialization of nature are all creating as many new *enclosures* characteristic of the contemporary globalization (Harribey 2013, 83–84; Chouquer 2012). This aspect of the history of capitalism is today the object of multiple works which reexamine the structuring role of imperialism in primitive accumulation: 'accumulation' is revealed to be less of a prerequisite to capitalism, the bygone condition of the destruction of the medieval world, than an intrinsic phenomenon accompanying its entire history since the modern era. By the action of this 'accumulation by dispossession' (Harvey 2010), and despite the fierce resistance that it arouses, so the grip of the anthropological 'production' category always spans even further over the spheres of life and of human societies. A lesson should be learnt from this for the intelligence of the notion of 'coviability' of natural and social systems. Even the notion of 'social system' is indeed invalidated this expansive territorial action which tears apart human groups into discrete 'societies' (Godelier 2007, 26). As this aspect of globalization is, however, relatively well known, we will end by showing how meditation on colonial history and Western domination of the world since modern times also forces us to put the notion of the "natural system" into perspective.

Marx opened the way for a critical analysis of the natural system notion by linking the scholarly domination of nature to the emergence of capitalism. But this critical analysis especially nourishes the anthropology of nature by Philippe Descola (Descola 2005). This invites the opposition of 'nature' and 'society' to be relativized, as a unique structure of the ontological regime of the modern West as opposed to all the other ways of inhabiting the world. However, the discovery of this anthropological relativity in the opposition of 'natural systems' and 'social systems', is undoubtedly the major internal obstacle of the coviability paradigm. An example introduced by Philippe Descola in 2013 in his *Lectures* at the College de France on *Landscape Forms* can illustrate this (Descola 2013).[10] It is based on the interpretation of nature pratices of the peoples of pre-Amazonia that "garden the forest".

[10]Sessions on 20th and 27th of March and 4th, 10th and 27th of April 2013. The video recordings are available on the website of the France College: http://www.college-de-france.fr/site/philippe-descola/course-2012-2013

11.5 Critical Analysis of the Natural System Concept: The Example of Amazon Forest Gardening

Practiced in the Amazon for 8000 years, this pioneering horticulture is indeed both an anthropization of the forest and a positive contribution to the sustainability of regional biodiversity. This irrefutable fact has led the ethnobiologist Darrell Posey to assert that some people, in this case the Kayapo that he has studied, 'manage' nature to such a point that their society could be taken as a coviability model (Posey 1985). The skills of the forest peoples in creating and maintaining the forest gardens would a combination of concern for the future and genuine ecological knowledge dressed in vernacular beliefs about supernatural beings. These plantations allow them to ensure sustainability of both their nomadic way of life and that of the forest. According to Darrell Posey, the Kayapo deliberately plant the forest by choosing the species that they will need in the future, thereby creating wooded patches in the savannah, and planting very slow-growing species in their cleared gardens, or scattering gallery forests along the tracks they use. Philippe Descola disputes this theory. After recalling several ecological arguments against it, he shows that even if these Amazonian peoples know that they are changing the forest, they can in no way be considered as seeking to ensure the "coviability" of their society with the forest ecosystems. Strictly speaking, the Kayapo neither 'plant' nor 'manage' the forest: the choice of species arranged by the order of the gardens or the 'gardening' of the forest can only be explained by the concrete interrelationships that each maintains with all the others as well as with the spirits and all non-humans. It is these relationships that structure both the social organization of the forest peoples and their horticultural practices Therefore, in a same Amazonian cultural area, each collective represents its relationships with gardens and with the forest in a particular way. In regards to the Achuars:

> The garden, an area cultivated by humans thanks to plants of the Nunkui spirit, is an image of forest, area cultivated by the Shakaim spirit, who himself sees the garden of humans as being a forest encroaching on its plantations. The metamorphosis here is, thus, a game of perspectives: in the eyes of the Achuar, the garden which returns to being a forest, is a forest which returns to being a garden in the eyes of the spirits. (Descola 2013, 16)

On the contrary, for the Miraña, the garden is the icon of the demiurge-body, whereas it is the icon of a home-body for the Yukuna and the Makuna. However, what is especially notable in relation to our purpose here, is that the cosmologies of the Amazonian peoples are *all* devoid of representations or concepts of nature and society.

This brief anthropological detour is therefore essential to clarify the paradigm of the 'coviability'. It defuses any attempt of specular views scanning the image of our own naturalistic categories in each ancient society in order to extract abstract normative principles. Anthropology, in contrast, vividly shows the existing confusion between the idea of nature spreading in different life sciences, especially in scientific ecology, and the practical concept of nature which organizes Western social relationships in the modern age, constituting a unique anthropological regime, "naturalist" ontology (Descola 2005). It highlights the singularity of human relationships with their surroundings, relationships that define the ontological regime upon which capitalism is based.

11.6 The Equivocalness of Nature: A Major Epistemological Challenge

The major epistemological challenge of the concept of coviability is therefore not to confuse the idea of nature that conditioned the scientific ecology with the West's practical scheme of nature, which is anthropologically foreign to all other collectives. Even if these two categories probably have common historical roots, they are also fundamentally opposed. In the constructions of scientific ecology upon which the paradigm of 'coviability' rests, the idea of nature is the basis of many demonstrations that highlight the dependence of human beings on the biosphere. In contrast, the practical scheme of nature that shapes all of the social relations of naturalistic collectives contrasts human beings to everything which they are not. It fixes a position of exteriority between human being endowed with subjectivity, and all of what faces them, which is identified as a physical continuum systematically organized by laws, but deprived of interiority. This ontological structure determines the supremacy of the relational scheme of production in the organization of relationships to non-humans [or to "nature"] which regulates the capitalist regime.

The anthropology of nature, forces the meaning of the historic rupture of the modern era to be reevaluated: the study of geographical and synchronic deployment of the diversity of relationships that human societies establish with 'nature' corrects the evolutionary bias mistakenly inherited by a certain history of Marxism in favor of the inevitability of the evolution of social relationships towards the capitalist[11] production mode. If the naturalistic anthropological regime is becoming more widespread today, it is not because of its attractiveness. Chronologically and ontologically, even before the appearance of modern metaphysical dualism or the division of natural sciences and social sciences, the naturalistic anthropological regime has been slowly determined by the growing hold of capitalism. The war of forests has been a decisive element in this. It broke the complex personal relationships of the feudal society and promoted the exclusive extension of the production scheme in relationships with non-humans. As demonstrated especially by Marx, Karl Polany (Polanyi 1983) and Edward P. Thompson (Thompson 2014), it precipitated the 'disembedding' of the market. However, contemporary globalization is also characterized by the metamorphic spread of new mechanisms of primitive accumulation. Moreover, if the major epistemological challenge of the coviability paradigm is to avoid any confusion between the "scientific" concept

[11] Obviously, the history of Marxism is not the only one to give up to this evolutionary aspect.

of nature and the practical scheme that supports it, its political challenge will be to clearly identify the disruptive anthropological mechanisms introduced by this violent movement of *enclosures* for those who are subject to it. With regard to contemporary issues of collective survival, it seems obvious today that the *doxa* that prevailed during the second half of the twentieth century to ensure the 'transition' to capitalism of evil known as the 'Third world', *doxa* still today in the heart of the 'sustainable development' concept, is obsolete. In the West even, and in the new major capitalist powers, it seems very vital to preserve the shreds of practices of 'nature' which are still partially outside the scheme of production and the commodification of the world and to support the emerging practices that resist it.

The possibility of a coviable insertion of human societies into the ecosystems in which they participate and depend upon, is therefore not essentially dependent on a deep systematic understanding of *strictly* social or ecological mechanisms, and even less to the computing sophistication of their models. It has much to do with the way in which the collectives, governed by the Western naturalist regime, will be able to modify the scheme which determines them and dooms them to engage almost only in relationships of profit with non-human beings and humans as well, increasing to infinity the destructive effects of the irreversible metabolic break denounced by Marx. The 'coviability' of social systems and natural systems will depend on mental, political and legal tools enabling practices and uses to be deployed that are resistant to the grip of this naturalism or mixes it. And if we follow the direction indicated by Marx, it cannot, therefore, be treated as an abstract "civilizational" problem. It requires instead an intellectual practice committed to those who really try to escape from their forced assignment to productive functions or from pure and simple dispossession. In contrast to the posture which tends to treat "coviability" as an abstract problem of *systematic* theory, critical foresight requires a scientific *and* political intelligence of practices by nature from 'below'.

References

Abensour M (2012) La démocratie contre l'Etat, Marx et le moment machiavélien. Éditions du Félin, Paris

Boumediene S (2016) La colonisation du savoir, Une histoire des plantes médicinales du « Nouveau Monde » (1492–1750). Les éditions des mondes à faire, Paris

Chouquer G (2012) Terres porteuses, entre faim de terres et appétit d'espace. Actes Sud, Arles

Descola P (2005) Par delà-nature, culture. Gallimard, Paris

Descola P (2013) Les formes du paysage, Cours au Collège de France, année 2012–2013; résumé; séances des 20 et 27 mars et des 4, 10 et 27 avril. http://www.college-de-france.fr/site/philippe-descola/course-2012-2013

Europeana (2015). n°6, (dir. Jean-philippe Pierron) Les philosophies environnementales européennes. L'Harmattan, Paris

Federici S (2014) Caliban et la Sorcière, Femmes, corps et accumulation primitive. Entremonde et Senonevero, Paris

Foster JB (2011) Marx écologiste. Éditions Amsterdam, Paris

Fressoz J-B (2012) L'apocalypse joyeuse. Une histoire du risque technologique. Seuil, Paris
Godelier M (2007) Au fondement des sociétés humaines. Albin Michel, Seuil
Harribey J-M (2008) Valeur-travail, transformations du capitalisme et primat de l'économie : controverses, malentendus et contresens », in Ivain Sainsaulieu (dir.), Par delà l'économisme, La querelle du primat en sciences sociales. L'Harmattan, Paris
Harribey J-M (2009) La lutte des classes hors-sol. À propos du prétendu économisme de Marx. Contretemps, Nouvelle série, n° 1, 1er trimestre, pp 123–133
Harribey J-M (2013) La richesse, la valeur et l'inestimable. Les Liens qui Libèrent, Paris
Harvey D (2010) Le Nouvel Impérialisme. Les prairies ordinaires, Paris
Mårold dans E (2002) Everything circulates: agricultural chemistry and recycling theories in the second half of the nineteenth century. Environ Hist:8
Marx K (1844) Manuscripts, https://www.marxists.org/archive/marx/works/download/pdf/Economic-Philosophic-Manuscripts-1844.pdf
Marx K (1857) Grundrisse, https://www.marxists.org/archive/marx/works/1857/grundrisse/ch12.htm
Marx K (1861–1863), Economic Manuscripts of 1861–63, Volume 30, MECW, p. 33–42, on line: https://marxists.catbull.com/archive/marx/works/1861/economic/ch13.htm
Marx K (1867) Capital, Volume I, https://www.marxists.org/archive/marx/works/download/pdf/Capital-Volume-I.pdf
Marx and Engels (1978) In: Dangeville R (ed) Critique de Malthus. Maspero, Paris
Moscovici S (1969) Le marxisme et la question naturelle. L'Homme et la société, n° 13, Sociologie et philosophie
Polanyi K (1983) La grande transformation. Gallimard, Paris
Posey D (1985) Indigenous management of tropical forest ecosystems: the case of the Kayapó Indians of the Brazilian Amazon. Agr Syst 3(2):139–158
Thompson EP (2014) La guerre des forêts. La Découverte, futurs antérieurs, Paris
Von Liebig J (1999) La chimie agricole de Justus Liebig. CTHS, Paris

Chapter 12
When Coviability Meets Ecosystem Services: The Case of Reunion Island's Coral Reefs

Espérance Cillaurren and Gilbert David

Contents

12.1 Introduction

At the end of the 1960s and the early 1970s, the awareness of people about the impacts anthropogenic activities on the state of the natural environment has grown (Willson and Matthews 1970). In 1972, The United Nations Conference on the Human Environment, known as the *Stockholm Conference,* has highlighted this change. Its final declaration was clear. "*Man is both creature and moulder of his environment*"[1] ... "*The protection and improvement of the human environment is a major issue which affects the well-being of peoples and economic development throughout the world*". On this basis, new paradigms[2] have been developed around the term "ecosystem". Following a multidisciplinary approach, they aim to offer a better understanding of the nature-society relationship. Six close concepts have thus

[1] Report of the United Nations Conference on the Human Environment, Stockholm, June 1972.
[2] We will define the paradigm as a way of understanding the world.

E. Cillaurren (✉) · G. David
IRD - French National Research Institute for Sustainable Development, UMR ESPACE-DEV, Montpellier, France
e-mail: esperance.cillauren@ird.fr; gilbert.david@ird.fr

O. Barrière et al. (eds.), *Coviability of Social and Ecological Systems: Reconnecting Mankind to the Biosphere in an Era of Global Change*,
https://doi.org/10.1007/978-3-319-78497-7_12

emerged: the socioecosystem, the social-ecological system, the eco-socio-system, the anthroposystem, the coupled human and natural system (Liu et al. 2007) and even the geo-eco-sociosystem (Mirault and David 2009) in reference to the geo-system, introduced as a geographic "substitute" for ecosystem concept (Bertrand 1968).

Integrated in space, the geo-system includes both: (a) ecological support (geological substrate, relief, climate, etc); (b) plant and animal communities; and (c) human beings who use and affect these two subsystems. C and G. Bertrand (1992, 2002) highlights that these various abiotic, biotic and anthropogenic elements react to each other and drive the evolution of the geosystem as a holistic entity. Corlay (1993, 1995) has applied the concept of a geo-system to the coast shown as "*an entity combining ecological and socio-economic system and the induced spatial structure*". The term eco-socio-system is only used by a small number of French speaking researchers. Having emerged from urban ecology and environmental education (Sauvé 1994), the term was then used by geographers (Corla 1995, 1998) then by ecologists (Hénocque et al. 1997). Some researchers understand it as a simple association of an ecosystem and a sociosystem (Mirault 2006). Others stress on the interactions between ecology, economy and society and highlight the homogeneity of the space in which these interactions take place.[3] Unlike the socio-eco-system, the anthroposystem and social-ecological system involve a co-evolution of the ecosystem and the sociosystem. Lévêque and Muxard (2004) and Lévêque et al. (2003) define the anthroposystem as… "*a structural and functional entity that takes into account society-environment interactions, and integrating in the same space one or numerous natural subsystems and one or numerous social subsystems, all of them are co-evolving in the long term*".[4] The concept of *social-ecological system* (also called socio-ecological system) is widespread. Its diffusion is especially boosted by the Journal *Ecology and Society* and the *Resilience Alliance*, a network of researchers which aims to promote resilience as a key concept of ecological and socio-economic dynamics. These dynamics are perceived in a non-linear interactive form, in accordance with the pioneering work of Holling, popularized in the book "Panarchy" (Gunderson and Holling 2002). The concept of social-ecological system is closely linked to the concept of ecosystem service (Fig. 12.1).

The concept of ecosystem services emerged in the 1970s and 1980s (Ehrly and Mooney 1983; Méral 2012). At the end of the 1990s and the beginning of the 2000s, their first worldwide monetary valuation published in Nature (Costanza et al. 1997) and the Millennium Ecosystem Assessment led by United Nations (MEA 2005) introduced ecosystem services to the forefront of the international stage. Bridging the scientific community and the decision-makers community, they are now a dominant paradigm in the management and conservation of the environment. They give a new interpretation of the nature/society relationships and are a way in which to internalize biodiversity into the market economy. In an interview with the

[3]http://www.eco-socio-systems.fr/eco-socio-system.html

[4]http://www.hypergeo.eu/spip.php?article270

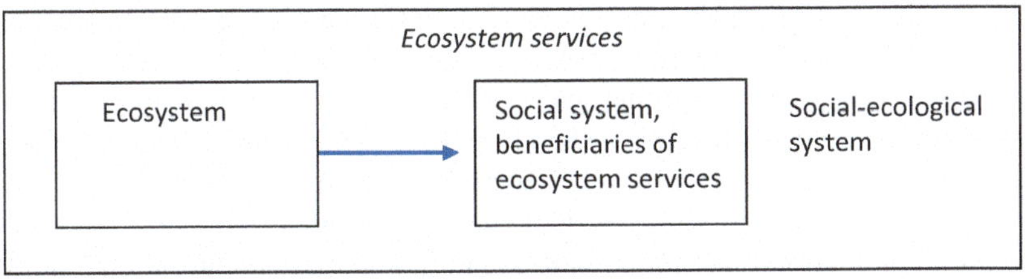

Fig. 12.1 Ecosystem services in a social-ecological system

daily newspaper, La Croix, October 12, 2010, C. Jouanno, a former Secretary of State for Ecology in France pointed out, "we need economic evaluations in order to be seen as serious people" in biodiversity management and conservation. A request of this kind is quite new. Until then, ecosystems were only perceived with a single focus: the protection of biodiversity. Protected areas are the main tool to achieve this goal. The strategic plan of the Convention on Biological Biodiversity has target that by 2020 17% of land areas and 10% of marine areas become protected areas on a world scale.

However, reducing the protection of biodiversity to create protected areas leads to a major risk: a division of planetary space into two entities: protected areas and highly 'anthropized' areas in which the protection of biodiversity would not take place. This spatial division dedicates a total absence of coviability between men and nature. In protected areas, the viability of nature is little or hardly inversely proportional to its use by men. The most effective protected areas to preserve biodiversity are those in which men are excluded. Symmetrically, the viability of human groups in anthropized areas is partly based on the intensive exploitation of nature. Such a difference between the supposed viability of nature and the supposed viability of human societies may lead to an economic and ecological dead end. A solution to this problem could be found with the principle of coviability between men and nature. The promoters of the Millennium Ecosystem Assessment (MEA) consider ecosystem services a central element of this coviability since, as mentioned above, they bridge the ecosystems and the individuals and human groups who benefit from it (Fig. 12.1) and form a socio-system (Lapierre 1992). Defining reference values and valuing them in a monetary way seems a good approach for integrating the ecosystem services in the decision-making mechanisms of territorial planning and stewardship. It is currently the main objective of the TEEB initiative (The Economics of Ecosystems and Biodiversity), widely supported by the European Union[5] (TEEB 2010). Following this rationale, bringing to the

[5]The following text is particularly clear about the assumptions made by EU on ecosystem services and their monetary assessment: "*Human well-being is dependent upon 'ecosystem services' provided by nature for free. Such services include water supply, air purification, fisheries, timber production and nutrient cycling to name a few. These are predominantly public goods with no markets and no prices, so their loss often is not detected by our current economic incentive system and can thus continue unabated. A variety of pressures resulting from population growth, changing*

attention of decision-makers and the public the monetary value of services (Fisher et al. 2009) provided by ecosystems is a necessary and sufficient condition to generate virtuous practices among users of these ecosystems and thus limit their degradation. In a previous article (David et al. 2012), it has been shown that the application of this proposal to a coral ecosystem suffered from many conceptual and methodological inaccuracies that made it hardly operational. This may be extended to whole ecosystems if we refer to a recent study by economists and political scientists showing that among the 313 articles devoted to ecosystem services in the journal *Ecological Economics*, only eight of them mentioned the use of this concept in environmental management (Laurans et al. 2013). In this chapter, we assume that men/nature coviability is based on the existence of feedback loops between ecosystem services and sociosystems services. To test this hypothesis, we will take the ecosystem services provided by the coral reefs of the Reunion Island as an example.

Our chapter is structured in four parts. Firstly, we briefly present our methods. Secondly, we revisit the concept of ecosystem services and the rationale behind the classification of services operated by the MEA (2005) then we tackle the main criticisms made on this work (Fisher et al. 2009; Fisher and Turner 2008; Boyd and Banzhaf 2007; Wallace 2007). Thirdly, we will stress upon the identification and the classification of ecosystem services attached to the reef environment of Reunion Island. Previous studies about these reef uses, mainly (Mirault 2006), offer a quantitative and qualitative framework that allows to revisit ecosystem services in a spatial unit identified as the resource area. Fourthly, we will tackle the concept of sociosystem service, and conclude with the feedback loop between ecosystem services and sociosystems services as a main vector of coviability between nature and human societies.

12.2 Methodology

This work is part of the OREMSE project (Ontology, Coral Reefs, Mangroves, Environmental Services), included in GEOSUD, one of the projects selected as part of the call for proposals "Equipment of Excellence" in the Programme d'Investissements d'Avenir (large national bond issued in 2011). GEOSUD aims to develop a national satellite imagery infrastructure to serve research on the environment and territories and its applications in the management of public

---

diets, urbanization, climate change and many other factors is causing biodiversity to decline. As a result, ecosystems are continuously being degraded. The world's poor are most at risk from the continuing loss of biodiversity, as they are the ones that are most linking it the ecosystem services that are being degraded... The TEEB study evaluates the costs of the loss of biodiversity and the associated decline in ecosystem services worldwide, and compares them with the costs of effective conservation and sustainable use. "It intends to raise awareness of the value of biodiversity and ecosystem services and to facilitate the development of cost-effective policy responses and better informed decisions". http://ec.europa.eu/environment/nature/biodiversity/economics/index_en.htm

policies. In this framework, OREMSE focuses on coral reefs and mangroves, the two most emblematic coastal ecosystems of the intertropical zone. It aims to first analyze the services provided by coral reefs and mangroves to neighbouring populations and second the way to map them. The final goal is to provide knowledge to improve the use in satellite imagery for a better management and conservation of these two ecosystems. Two working papers were produced in this field (Cillaurren and David 2014a, b).

Our methodological approach is based on a simple observation. Ecosystem services provided by coral reefs have been little studied. As Hicks (2011) pointed out, most of studies are devoted to a reduced number of services. In fact, coral reefs provide a large number of services as many studies on the coral reef uses have shown. This discrepancy between the observed reality and the content of articles can be easily summed up: does the problem come from coral reefs or from the concept of ecosystem services? This question can be divided into two symmetric parts:

- Is the coral ecosystem so specific from a functional point of view that it makes it difficult to apply the concept of ecosystem services to it?
- Is the concept of ecosystem services sufficiently structured and rich to apply to coral reefs?

Responding to this last question led us to critically review the concept of ecosystem services, based on the scientific literature published in international journals. To answer the first, we have selected the coral reefs of Reunion Island as a pilot site. Our knowledge of this coast allowed us to follow a four-step method:

- Delimitation of the sociosystem, composed of beneficiaries of ecosystem services provided by coral reefs;
- Study of services linking the coral ecosystem to this sociosystem;
- Identification of sociosystem services provided by the coastal sociosystem to coral reefs;
- Analysis of the relationships between coral ecosystems, ecosystem services, coastal sociosystems and sociosystem services.

Located on a volcanic hotspot in the South Indian Ocean (55 ° 29 East and 21 ° 53 South), the Reunion Island is young. Coral reefs are poorly developed. They only cover 10% of the island's coastline (25 km) on its leeward side (Fig. 12.2). Despite its small size, the coral reef works as an "attractor" to human activities and urbanization. It can be seen as a social ecological system with many resource areas. Each resource area is a spatial entity that associates an area and the resources it houses as an object of use or representation. It can be distinguished from others by:

- two spatial components: geomorphology, bathymetry,
- two resource components: habitat, animal or algal populations,
- the ecosystem services it generates,
- the type and quantity of human uses.

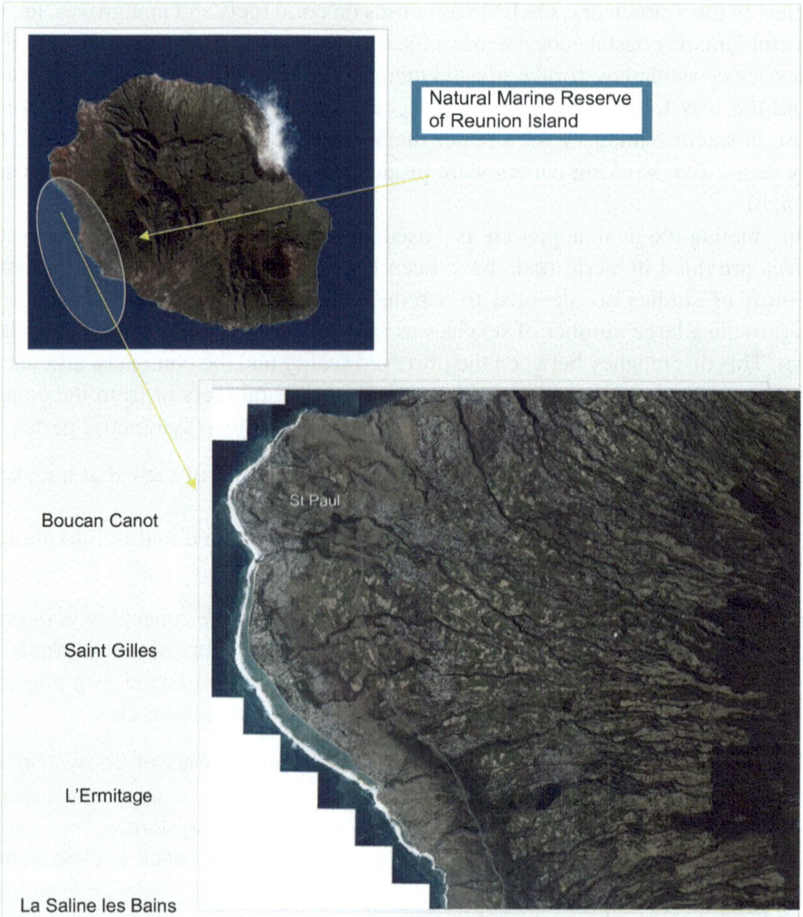

Fig. 12.2 The reef coast of Reunion Island from space (Pléiades 2013 in https://spatial.ign.fr); Licence Pléiades : Contains informations © CNES 2013, Distribution Airbus DS, all right reserved. Prohibited commercial use

12.3 From Viability to Ecosystem Services, When a Sociosystem Meets an Ecosystem

Viability is a prior concept to ecosystem services. In the 1980s, Aubin (1991) founded the basis of a mathematical theory of viability. At the same time, UNCTAD (the United Nations Trade and Development Conference) used the viability concept for planning the future of small island territories. The viability of a territory was seen as the combination of a static state, "*the meeting of necessary and sufficient conditions to exist and last*", and a dynamic state comparable to "*the conditions required to sustain island development, including the most comprehensive use of*

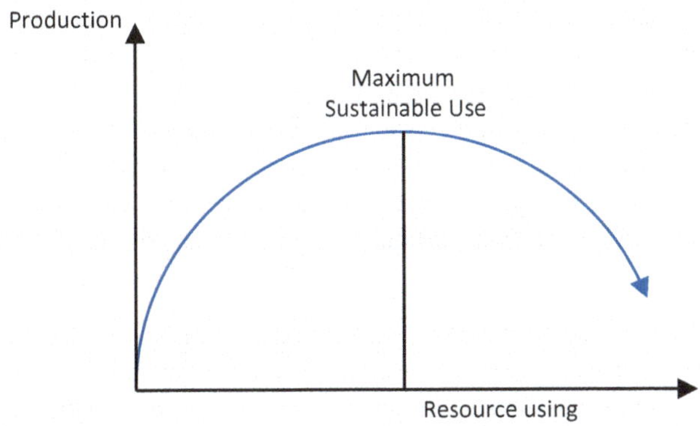

Fig. 12.3 The Graham-Shaefer model (Graham 1935) and (Schaefer 1954, 1957) applied to any resource

natural resources and the progress of social and economic living standards of populations" (Doumenge 1983, 1985).

Transcribed according to a simple mathematical formulation, this definition could be expressed as follow:

VT = f (MSUnr, GLS) where:
VT: viability of the territory,
MSUnr: maximum sustainable use of natural resources
GLS: growth in the social and economic living standards of populations

The concept of maximum sustainable use of natural resources is the maximum production extracted from a natural resource without altering its reproductive capabilities, in such a way that the resource quantity used remains sustainable over time and at its maximum level (Fig. 12.3). Coming from the Graham-Schaefer model for fisheries resources (Laurec and Leguen 1981), this concept leads to considering the used resources as a natural heritage, since their sustainable use allows their transmission to future generations.

In this sense, territorial viability works as true nature-society coviability. It requires both a viability of nature, i.e. an optimal functioning of the ecosystems making it up which results in maximum and sustainable productivity of these ecosystems, and a viability of the social or sociosystem, corresponding to optimal functioning of the society and the economy. Dependent on the ecosystems' production, this social system viability leads to optimum growth in the social and economic living standard of human populations. Beyond this optimum growth, the use of natural resources from is too great and breaks the balance between nature and society.

The concepts of maximum sustainable use of natural resources and viability as conceived by UNCTAD in the 1980s (Doumenge 1985) falls under the resources economy. The ecosystem services concept is part of the ecological economics, driven by the International Society for Ecological Economics created in 1989

(Gomez Baggethum et al. 2010). Ecosystem services structure an essential relationship of dependency between an ecosystem and a socio-system. This relationship is still considered to be unidirectional (Fig. 12.1). Daily (1997, p.3) has defined ecosystem services as "the conditions and processes through which natural ecosystems and the species that make them up, sustain and fulfill human life". For Costanza et al. (1997), ecosystem services are the benefits that human populations derive directly or indirectly from the functioning of ecosystems. They come in four types:

- supporting services as the cycles of nutritive elements within the biotopes which are the origin of the production of the other three services;
- provisioning services, such as fresh water, food, soil, breathable air, genetic resources, which allow life on Earth;
- regulating services, such as the reduction of wave energy and thus coastal erosion by the reef barriers;
- cultural services.

This typology was taken by the MEA (2005) which extended the concept of benefits to a set of material goods and services, intangible, present and future. Thus, the MEA defined ecosystem services in a very broad way, "the benefits people obtain from ecosystems" and shown as an essential component of the well-being of human societies (Fig. 12.4).

Two main criticisms were made of this definition by Boyd and Banzhaf (2007) and Wallace (2007). Firstly, it mixes ends and means. Thus, the two first authors have introduced the concept of final ecosystem services, which put in direct relationship the ecosystem which transmits services and the socio-system which receives them, in opposition to the concept of basic or intermediate ecosystem services. Supporting services are intermediate or services. Provisioning, regulating and cultural services are direct services. "*Final ecosystem services are components of nature, directly enjoyed, consumed or used to yield human well-being*". Being an integral part of nature means that "*ecosystem services should be isolated from non-ecological contributions to final goods and services. Once ecosystem services are combined with other inputs, such as labor and capital, they cease to be identifiably 'ecological*'" (Boyd and Banzhaf 2007). Secondly, the MEA puts services and benefits in a same set. For Boyd and Banzhaf and Wallace, it is clear that services are different from benefits (Table 12.1) Scuba diving is not a cultural service produced by the reef ecosystem but rather a benefit that divers obtain from the reef. The underwater landscape is the ecosystem service which provides this benefit.

Wallace's analysis is particularly interesting because it inserts the ecosystem services in an identifying framework named "human values" (Table 12.1), defined as "*... the preferred end-states of existence, including those required for human survival and reproductive success, which taken together circumscribe human well-being*" (Wallace 2007, p. 237). The sociocultural development is a key element of human well-being. This definition is close to the definition of viability given by Doumenge (1983), but applied at the individual scale. In this context, all ecosystem services are the material expression of ecosystem processes on a natural asset

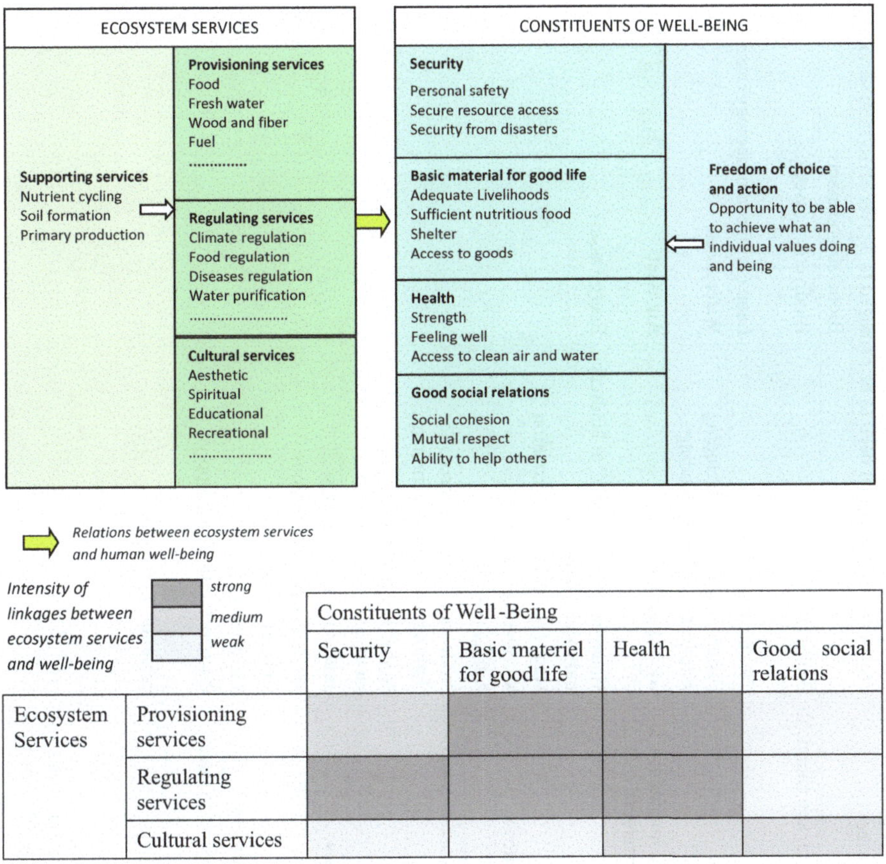

Fig. 12.4 Classification of ecosystem services according to the MEA (2005) and intensity of relationships between ecosystem services and human well-being

(Table 12.1). The author defines ecosystem processes as "*complex interactions (events, reactions or operations) among biotic and abiotic elements of ecosystems that lead to a definite result*". These processes work as transfers of energy and matter in natural assets, considered as resources and part of a natural capital.

Fisher and Turner (2008), Fisher et al. (2009) also brought constructive criticism to the concept of ecosystem services defined by the MEA. They define ecosystem services as "the aspects of an ecosystem utilized (actively or passively) to produce human well-being. Contrary to Boyd and Banzhaf (2007) who see services as only "*the directly consumable end points*"; they argue that "services must be ecological phenomena. They do not have to be directly used" ... "functions and processes become services if there are humans that benefit from them. Without beneficiaries they are not services" (Fisher et al. 2009, p.645) In this framework, these benefits are the outcome of services.

Table 12.1 Classification of the eco-system services suggested by Wallace (2007)

Category of human values	Ecosystem services experienced at the individual human level	Processes and assets that need to be managed to deliver ecosystem services	
		Ecosystem processes	Biotic/abiotic assets
Adequate resources	Food for organism energy, structure, key chemical reactions	Biological regulation	Biodiversity assets
	Oxygen	Climate regulation	Land (soil/geomorphology) assets
	Water (potable)	Disturbance regimes, including wildfires, cy-clones, flooding	Water assets
	Energy for cooking and warming component under physical and chemical environment	Gas regulation	Air assets
	Dispersal aid and transport	Management of "beauty" at landscape and local scales	Energy assets
Protection from predators/disease/parasites	Protection from predation	Management of land recreation	
	Protection from disease and parasites	Nutrient regulation	
Benign physical and chemical environment	Benign environmental regimes of:	Pollination	
	Temperature and energy (includes use of fire for warning)	Production of raw materials for clothing, food, construction, etc.	
	Moisture	Production of raw materials for energy, such as firewood	
	Light to establish circadian rhythms	Production of medicines	
	Chemical	Social-cultural interactions	
Social-cultural fulfilment	Access to resources for:	Soil formation	
	Spiritual and philosophical contentment	Soil retention	
	a benign social group, including access to mates and being loved	Waste regulation and supply	
	Recreational and leisure pursuits	Economic processes	
	Meaningful occupation		
	Aesthetical purposes		
	Opportunities values, capacity for cultural and biological evolution (knowledge and education resources, genetic resources.)		

An underlying link between ecosystem services and their use is implicit in the various given definitions (Boyd and Banzhaf 2007; Wallace 2007; Fisher et al. 2009). It seems important to detail this link and to put to the front the contribution of ecosystem services to the viability of societies. It is the topic of this third part, devoted to the reef coastline of Reunion Island.

12.4 From the Ecosystem to Its Uses and Services, the Case of Coral Reefs in Reunion Island

The two first typologies of ecosystem services provided by coral reefs were published by Moberg and Folk (1999) then Moberg and Rohnback (2003). They are more complete that the MEA typology (Table 12.2)

In the mid-2000s, some extremely detailed work was carried out on the uses of the Reunion island reef environment (Mirault 2006; Mirault and David 2006) in the VALSECOR project (Socio – economic values of the Reunion Coral reefs).[6] The first works on ecosystem services of the reef were published a few years later

Table 12.2 Comparison of three topologies of ecosystem services provided by coral reefs

Typology of ecosystem services according to		
Moberg et Folk (1999)	Moberg et Rönnbäck (2003)	Millenium Ecosystem Assessment (2005)
Physical structure services	Storm and flood protection	Supporting services
Biotic services within and between ecosystems	Nursery, feeding and breeding group	Provisioning services
	Maintenance of biodiversity and genetic resources	Regulating services
Biogeochemical services	Remineralisation of organic and inorganic matter	Supporting services
		Regulating services
Biotic services between ecosystems	Export of organic matter	Regulating services
Information services	Climate, pollution record	Cultural services
	Educational and scientific information	Cultural services
Social and cultural services	Support recreation	Cultural services
	Sustaining the livelihood of coastal communities	Provisioning services
	Cultural, spiritual and artistic values	Cultural services

[6]This research was carried out with a focus on the perimeter of the marine natural reserve (Lemahieu et al. 2013; Lemahieu 2015).

(Revillion 2009). This timeline logically led us to establish a strong link between uses and ecosystem services. We assume that ecosystem services are part of the ecosystem but only in a potential way. They become real ecosystem services when they are revealed when used by people. Users do not target a service but a resource in a resource area, which could be seen as the habitat housing this resource. In this framework, the ecosystem is not a whole but a set divided into many resource areas which have the potential to provide ecosystem services.

All uses can be identified according to four parameters: (a) the user who is the use producer, (b) the beneficiary of this use who is a use consumer, (c) the nature of the use and (d) the resource area which emits it, which can be seen as the morphological compartment housing the resource used. Transcribed according to a simple mathematical formulation, this definition could be expressed as follow:

$$U = f\ (Pu, Cu, Nu, Ra)$$

where

Nu: nature of use,
Pu: producer of use.
CU: consumer of use
Ra: resource area

All services can be identified according to four parameters: (a) the nature of the service, (b) the way by which the use reveals the ecosystem service, (c) the resource area which emits the service, (d) the beneficiaries of the service. Transcribed according to a simple mathematical formulation, this definition could be expressed as follow:

$$ES = f\ (N, R, Ra, B)$$

where

N: nature of service
R: way of revelation
Ra: resource area
B: beneficiaries

The two first parameters can be used to classify ecosystem services. Usually, the nature is the only way for classification. The way by which the service is revealed provides another valuable dimension of ecosystem services.

On Reunion Island, the link between the uses and the ecosystem services can divide the coral reef ecosystem into eight main resource areas from the beach to the open sea (Tables 12.3 and 12.4). Each of them can receive three types of use. Direct uses such as fishing, scuba diving, glass bottom boat cruises are directly driven by the coral ecosystem's functioning and processes. Semi direct uses such as swimming, beach practices, coastal housing are driven by the physical features

Table 12.3 Identification of main resource areas in Reunion Island's reef ecosystem according to their bathymetry, geomorphology and ecosystem

Bathymetry	Environments	Morphological units	Ecosystems	Resource areas
Subtidal area	Open sea		Pelagic ecosystem	Offshore area
	Outer reef	Deep fore reef slope	Fish and benthic ecosystems	Outer reef slope
		Fore reef slope	Fish and benthic ecosystems	Fore reef
Intertidal area	Reef flat limit	Reef crest	Algal ridge ecosystem	Reef flat
	Reef flat	Back reef	Coral ecosystem	Reef flat
		Reef pass channels	Coral ecosystem	Reef pass channels
		Back reef channel	Sand ecosystem	Back reef channel
			Coral ecosystem (Pinacles)	
	Shore	Beach	Sand and sedimentary ecosystems	Beach
Supratidal coastal area	Backshore	Beach	Sand and sedimentary ecosystems	Beach
	Terrestrial baseline	Upland	Terrestrial ecosystems with salt tolerant vegetation	Terrestrial coastline

Based on Mirault and David (2009)

generated by the presence of coral reefs (waves, coastal drift for example). Indirect uses are not related to the presence of the reef ecosystem or to its functioning, but to several direct uses, semi direct uses and indirect uses made of the ecosystem (Mirault and David 2009). Thus, scientific research, implementation of scuba diving clubs or kitesurf schools are indirect uses of a coral reef ecosystem.

These uses allow the related ecosystem services to be identified. Three major types of ecosystem services can be distinguished according to the way in which the services are revealed by uses.

(a) Matter or energy flows produced by a resource area become services only when they are put into use and/or revealed by the action of human beings. This can be labor which produces goods or a provisioning service. In the case of fishing, a resource, which is a stock of organic matter in the form of a fish population, is put into use when a fishing effort is carried out by fishermen (the resource users). As a result, this fish population becomes an exploited stock and then a fish production. The labor carried out on the ecosystem for providing ecosystem services returns to extract a resource from its resource area. So it can be described as an extractive use and the resulting service can be qualified as 'ecosystem service revealed by labor'. It provides identifiable and measurable well-being items (benefits according to the MEA) to the consumers of this resource, such as protein intake in the case of fishing. The resource area generating this can be called "a resource area supporting services revealed

Table 12.4 The reef ecosystem and resource uses in Reunion Island

Morpholo-gical units	Ecosystems	Resources	Uses	Use type
Deep reef slope	Fish and benthic ecosystem	Fish	Commercial fishing	Direct
			Recreational fishing	Direct
Fore reef slope	Fish ecosystem	Fish	Commercial fishing	Direct
			Recreational fishing	Direct
	Benthic ecosystem	Fish and benthic species	Recreational fishing	Direct
			Scuba diving (including illegal fishing)	Direct
	Reef ecosystem	Coral ecosystems	Spear-gun fishing	Direct
			Lobster fishing (including poaching)	Direct
			Filming and photo activities	Direct
			Glass bottom boat	Direct
Reef Crest and Back reef and Reef pass channels	Coral ecosystem	Coral environment	Foot fishing	Direct
			Spear-gun fishing	Direct
			Snorkeling	Direct
			Filming and photo activities	Direct
			Environment protection and nature conservation	Semi-direct

Backreef channel and Beach	Coral and sand ecosystem	Coral environment	Foot fishing	Direct
			Snorkeling	Direct
			Filming and photo activities	Direct
			Underwater trail	Direct
			Sliding sports	Semi-direct
			Canoeing, paddling	Semi-direct
			Swimming	Semi-direct
			Water sports school	Semi-direct
			Environment protection and nature conservation	Semi-direct
	Sand and sedimentary ecosystems	Beaches	Swimming	Semi-direct
			Sunbathing	Semi-direct
			Beach sports	Semi-direct
			Sport leagues or clubs	Semi-direct
			Nautical clubs	Semi-direct
			Walk	Semi-direct
Upland	Terrestrial ecosystems with salt tolerant plant species	Coast line	Walk	Semi-direct
			Picnic	Semi-direct
			Stores	Indirect
			Accommodation	Indirect
			Restaurant services	Indirect

Adapted from Mirault (2006)

by labor". The action of human beings can also be representational, which produces information. In this case, the service can be qualified as 'ecosystem service revealed by representation'.

(b) The services driven by the mere presence of a resource area require no human labor to be revealed and thus benefit human populations. The clearest example is the protection of the coast line against the energy of the waves breaking on the reef front. These services are referred to as 'non-revealed ecosystem services' and the resource area generating them are called "resource area emitting non-revealed services'. These services benefit the coastal population living near the emitting resource area (the sociosystem). They also play a role of induction towards economic stakeholders such as the CRPMEME (Regional Committee of Marine Fisheries and Aquaculture) and social stakeholders, such as traditional fishing associations, reef protection associations or the Group of Public Interest RNMR which manage the natural marine reserve in Reunion Island.

(c) In order to be revealed, some services require the presence of users on-site which means the use of the resource area by the simple attendance of people without any labor or representation to use the resource. They will be referred to as 'ecosystem services revealed by attendance'. The concerned resource areas will be called 'resource areas of attendance'. Non-extractive recreational activities, like swimming and surfing fall typically within such ecosystem services of attendance. It is also the case of the discovery of the environment, usually classified as cultural services (MEA 2005).

If uses reveal ecosystem services, there is no exclusive match between a use and a service. The same use can drive several services as shown in the case of fishing that can be classified as a supply service or as a recreational service. Table 12.5 connects resource areas, the ecosystem services they emit, the uses that reveal these services and the beneficiaries of these services, who are also the beneficiaries[7] or consumers of the uses that are made of the reef ecosystem resources areas. These beneficiaries are part of the reef socio-system. They are divided into four groups.

- public law stakeholders covering three groups of people: (a) local authorities, (b) State services and their civil servants (including researchers) and (c) the environmental managers studying, managing or promoting the reef ecosystem. All of them are involved in the management of the territory adjacent to the coral reef ecosystem and the maritime territory which hosts it;
- private law stakeholders who are users of the reef ecosystem and/or beneficiaries of its services. This group includes the islanders and the people not living in Reunion Island;

[7] All actors of the civil society can be seen as beneficiaries of ecosystem services whether they are (a) users of natural resources, (b) institutional or associative actors working on management, protection and information about the reef ecosystem, (c) resident population or tourist practicing recreation activities.

- private law stakeholders involved in a commercial activity based on direct, semi-direct and indirect use of the reef ecosystem resources areas;
- private law stakeholders involved in a public service mission (such as the CRPMEM) or in improving social well-being (such as environmental NGOs or NGOs of small traditional fisheries) concerned either directly or indirectly by the reef ecosystem.

Table 12.5 Matrix combining resource areas, ecosystem services, uses and service beneficiaries

a) Matrix of the services revealed by labor and non-revealed services

Ecosystem services	Resources areas	Uses	Services benificiaries
Provisioning	Back reef channel	Angle fishing	Recreational fishers
		Traditional net fishing	Commercial fishers
	Reef flat /	Net fishing	Self-subsistence fishers
	Fore reef / reef pass channels	Speargun fishing	Commercial, self-subsistence and recreational fishers
	Outer reef slope	Angle fishing	Self-subsistence and recreational fishers
		Drop line fishing	Commercial, Self-subsistence and recreational fishers
Regulation	Fore reef	Protection against erosion	Coastal population
Economic induction	All resource areas	Professional grouping (scuba diving, tourism, bottom glass boat)	Users of reef involved in commercial activities
Social induction	All resource areas	Customers associations	Association network and users
		Reef protection associations	Association network and users
		Traditional fishers associations	Association network and users
		Marine Natural Reserve	Reef users/Public institutions in charge of marine environment, coast inhabitants
Cultural services	Outer reef slope Fore reef Reef flat Reef Pass channels Back reef channel Beach Terrestrial coastline	Expertise and scientific research- Films and photo's production Education activities	Research agencies and consultants Professional and hobbyist photographers and video-makers, School population and public

NON-REVEALED SERVICES | SERVICES REVEALED BY LABOR

(continued)

Table 12.5 (continued)

b) Matrix of the services revealed by attendance

Ecosystem services	Resource areas	Uses	Services beneficiaries
Cultural	Outer reef slope	Expertise and scientific research	Researchers and consultants
	Reef Pass channels	Expertise and scientific research	Researchers and consultants
	Reef flat	Expertise and scientific research	Researchers and consultants
	Back reef depression	Environment discovery	School population and public
		Expertise and scientific research	Researchers and consultants
		Environment discovery	School population and public
	Beach	Expertise and scientific research	Researchers and consultants
		Environment discovery	School population and public
	Terrestrial coastline	Expertise and scientific research	Researchers and consultants
Recreational		Recreational boating and fishing	Recreational sailors
	Outer reef slope	Scuba diving	Local customers and tourists
		Scuba diving	Local customers and tourists
	Fore reef	surf	Local customers and tourists
		Snorkeling	Local population and tourists
	Reef flat	Recreational fishing	Local population and tourists
		Canoeing, paddle	Local customers and tourists
		Snorkeling	Local population and tourists
	Reef Pass channels	Canoeing, paddle	Local customers and tourists
		Snorkeling	Local customers and tourists
	Back reef depression	Sliding sports (WindSf/Kyte Sf)	Local customers and tourists
		Canoeing, paddle	Local customers and tourists
		Swimming	Local population and tourists
		Tanning	Local population and tourists
		Open air games	Local population and tourists
		Walking	Local population and tourists
	Beach	Picnic	Local population and tourists
		Walking	Local population and tourists
Economic Induction	Terrestrial coastline	Scuba diving clubs	Clubs employees and divers
	Outer reef slope	Surf clubs	Clubs employees and surfers
	Fore reef	Sliding sport clubs	Clubs employees and sliders
	Back reef depression	Canoeing and paddle clubs	Clubs employees and sport practicing people
		Itinerant sale	Shop keepers beach customers
		Equipment rental	Coastline population
		Residential housing	Hotel owners
	Beach	Commercial accommodation	Restaurant owners/tourists /local population
	Terrestrial coastline	Restaurant services	Local population and tourists
		Car parking	Local population and tourists

In Table 12.5, the services have been classified according to the two we have mentioned above (a) the processes characterizing their action, a distinction is thus made between services revealed by labor, non-revealed services (Table 12.5 a) and services revealed by attendance (Table 12.5 b); their nature, defined according to the typology of the MEA: provisioning services, regulation services and cultural services. This typology highlights the strong specialization of the beneficiaries of the supply services (all fishermen) and the importance of cultural services. Three types of fishermen have been distinguished. The less numerous are the food fishermen. They are all unemployed but they benefit from social welfare payments. Thus, they could live without fishing but this use provides them with a free source of marine protein and occupies their free time. Even if some are engaged in speargun fishing, most of these food fishermen are involved in on-foot fishing. They fish all coral reef resource areas in shallow waters. Fifteen years ago, seven types of fishing techniques were used (David and Mirault 2006).

Since then, the diversity of fishing techniques and boats has decreased. This food fishing is tending to disappear in favor of recreational fishing. Recreational fishermen show various profiles including jobless fishermen who angle in the reef passes and on the reef front at low tide and primary and secondary school teachers who practice boat fishing on the outer reef slope or speargun fishing. These fishermen are called recreational fishers because they are not involved in professional fishing. In fact, several of them fish on a regular basis, sometimes on a daily basis and because they sell their catches, they can be seen as informal and illegal professional fishermen. They practice speargun fishing and netting in particular.

The name "professional fishermen" reflects here two different realities. A first group of fishermen is composed of fishermen with a professional diploma. They are registered by the administration of Marine Affairs. A second group of 800 people receive each year a card of traditional fishermen awarded by the same administration. This card allows them to fish (*Mulloidichthys flavolineatus*) in shallow reef waters with a net of 10 m long and 3 m in height from February 1 to April 30, on Wednesdays, Thursdays, Fridays, Saturdays and Sundays, except holidays, from 5 to 9 h.

In the typology established by the Millennium Ecosystem Assessment (2005), the less well-treated aspect deals with cultural services. As stressed by Boyd and Banzhaf (2007) 'the MEA cultural services, including spiritual and religious values, aesthetic values, and recreation and ecotourism are particularly unsatisfying'. In Table 12.5 and 12.6, these services have been divided into four categories based on direct, semi-direct or indirect uses: cultural services, recreational services, economic induction services and social induction services.

- The cultural services are linked to four main types of uses:
  - The discovery of the reef ecosystem using either the underwater trail in the Hermitage or glass bottom boats (visiobulle) in St Gilles, or scuba diving, are direct uses of the reef ecosystem. They reveal a landscape service which belongs to the category of the attendance service and provides aesthetic wellbeing driven by the aesthetic value of the coral reef landscape.

- Scientific research is divided in three types: (a) visual observation during field trips, which means a direct use, the attendance of different resources areas in the reef ecosystem, (b) collecting samples in the field which means a direct extractive use of the reef, (c) data processing in the laboratory and then the writing of scientific articles about these results, which means an indirect use of the ecosystem.;
- The education of the public and schools via documentation that deals with the reef ecosystem is an indirect use, driven both by two previous uses: the discovery of the reef ecosystem and scientific research.
- The production of multiple photographs (including postcards) and videograms (Table 12.6a), which is a direct use of the reef revealing a cultural service.

- Recreational services are revealed by coral reef ecosystem attendance:(a) coastal and marine uses (sunbathing, swimming, snorkeling, picnicking), (b) board sports (surfing, parasailing, windsurfing) and learning how to do these through clubs and schools, (c) canoeing and 'paddle sports and learning how to these, (d) scuba diving and learning how to scuba dive[8];
- Economic induction services concern owners of main or secondary residences, restaurants, hotels and any other form of paid accommodation, stores specialized in beach activities, street peddlers selling cold drinks or swimsuits on beaches;
- Social induction services are so-called because the existence of the reef ecosystem and its uses, including conservation of habitats and biodiversity, generate the creation of association of users devoted to the promotion of small traditional fishermen or environmental protection. These services identified as social induction directly initiate socio-systems services (see below).

The economic and social induction services are non-worked services. They generate no extractive usage of the reef ecosystem (Tables 12.5 and 12.6).

Resource areas can emit several types of services and/or support several types of uses and categories of beneficiaries. The 7 resource areas listed in Table 12.6 have emitted a total of 116 services which means an average of 16 to 17 services per resource area. There is little disparity between the resource areas. The reef pass and channels are the least productive (14 services). The more productive (20 services) is the back reef channel. Its interface between, on the one hand, the beach (19 services) and the terrestrial coastline (15 services) and, on the other hand, the reef flat (15 services) and the deeper reef resource areas (total of 47 services) is probably favorable in terms of attendance. On the 116 services emitted by the reef ecosystem, 89% are cultural services, which is quite unusual in the literature (Hicks 2011; Failler et al. 2015; Schuhmann and Mahon 2015). A fifth of these cultural services are recreational services and economic induction services (Table 12.6a and 12.6b).

[8]Learning these sports are more socio-system services orientated than ecosystem services orientated, even if they are usually called cultural services.

Table 12.6 The ecosystem services according to their area-resource emission and their beneficiaries

a) Cultural and recreational services

Benificiaries	Resources Areas						
	Terrestrial Coastline	Beach	Back reef channel	Reef flat	Reef pass channel	Fore reef	Outer reef slope
Hobbyist and videomakers							
Professional photographers							
Researchers, consultants							
School population and public							
Recreational fishers							
Recreational sailors							
Surfers							
Snorkelers							
Board sport users							
Canoeing and paddle sports users							
Swimmers							
Beach goers							
Picnickers							
Walkers							

Cultural services (n =37)

Recreational services (n =21)

Revealed services (n=33)

Attendance services (n =25)

(continued)

Table 12.6 (continued)

b) Cultural services of economic and social induction

Benificiaries	**Resources Areas**						
	Terrestrial coastline	Beach	Back reef channel	Reef flat	Reef pass channels	Fore reef	Outer reef slope
Recreational fishers associations							
Scuba divers and spear guners associations							
Traditional fishers associa-ions							
La Réunion Natural Marine Reserve							
Reef protection associations							
Professional tourism groups							
Professional fishers groups							
Scuba diving clubs and divers							
Surf clubs and surfers							
Sliding sports clubs and sliders							
Canoeing and paddle sport clubs and users							
Beach clubs							
Shopkeepers							
Restaurant owners							
Hôtel and campings owners							
Coastline population							
Users associations							

Social induction services (n =24)

Economic induction services (n =21)

Nonrevealed services (n=32)

Attendance services(n =13)

(continued)

Table 12.6 (continued)

Benificiaries	Resources Areas						
	Terrestrial coastline	Beach	Back reef channel	Reef flat	Reef pass channels	Fore reef	Outer reef slope
Coastline							
population							
Drop line fishers							
Spearguners							
Anglers							
Net fishers							
Traditional net fishermen							

Regulation services (n =1)
Supply services (n =12)
Non-revealed services (n=1)
Revealed services (n =12)

Most of the recreational services (16 out of 21) are services of attendance. They are revealed by the presence of users in the practice area. The economic induction services are either non-revealed services (8 out of 21) or services of attendance (13 out of 21). The first group deals mainly with the implementation of professional organizations whose aim is partly to support the exploitation of the reef ecosystem through fishing or tourism. The second group deals with the creation of two kinds of businesses. Some aim to promote the practice of a sport (scuba diving for example) completely devolved to the reef ecosystem. The others are located on the terrestrial coastal strip or on the upper parts of beaches. All their customers are reef users.

Unlike the economic and recreational induction services, the social induction services and the cultural services are linked to a large number of resource areas. Cultural services are mainly information (photography, field data, writing, videos) brought to the attention of the researchers and the school population) or photographs and films shot by professionals and hobby photographers. As indicated in Table 12.6a, the same resource area can emit both non-revealed and services of attendance. In the latter case, services deal with information collected either in the field by researchers or consultants, either during field trips by school children aiming at discovering nature. The creations of the 'Réserve Naturelle Marine' of the Reunion Island or of environmental NGOs which aim to protect the reef ecosystem are the results of induction services which are both cultural services emitted by the reef ecosystem and socio-systems services aiming to protect this ecosystem. .

12.5 The Exploration of a New Concept: The Sociosystems Services

Sociosystem services are a set of services provided by human beings, as a social system, to ecosystems in order to secure and sustain ecosystem services. Under the frame of a social ecological system, the feedback loop between sociosystems services and ecosystem services is a necessary condition to nature/human society coviability (Fig. 12.5). For strengthening coviability between coral reef ecosystems and riparian human societies, four types of sociosystem services can be distinguished.

(a) The 'sanctuary of the ecological habitats' service' deals with the creation and implementation of Marine Protected Areas, including 'No take' zones, in order to restore habitats when they are degraded and to keep their biodiversity in good health status.
(b) The "ecological engineering" service has five distinct types:

   - The reintroduction of species whose natural numbers are too small to ensure their place in the ecosystem functioning, is the first one. Pacific Islands provide a good example with the reintroduction of both giant clams or mother of pearl shellfish such as 'trochus' (Teitelbaum and Freidman 2008). Planting mangroves for trapping sediments driven by the run off of eroded soils in watersheds is also an ecological engineering service. As it decreases the

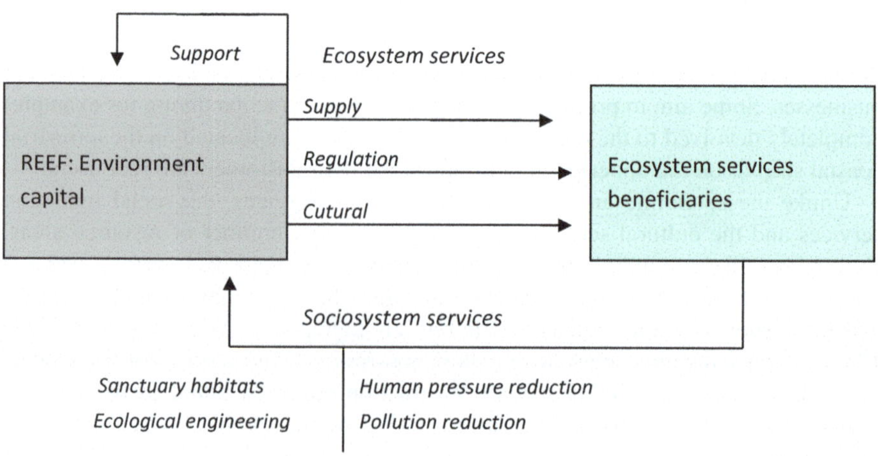

Fig. 12.5 The ecosystem and sociosystem services linked to the coral reefs

hyper sedimentation of coastal waters, it benefits indirectly to coral reefs. In any case, this engineering aims for the restoration of ecosystem services previously degraded by excessive anthropogenic pressure (Ronnback et al. 2007);

- The creation of artificial habitats is a second type of ecological engineering service. To put artificial reefs on shallow waters of reef lagoons is a good way to increase fish abundance on sand bottoms, where fish diversity is very low. Artificial reefs provide a new habitat to coral fish larvae where they can grow, sheltered from predators.
- The restoration of habitats so heavily degraded to be used as a sanctuary is a third type of service. The eroding dunes and the upper part of beaches offer a good example of this type of service. People plant herbaceous species such as the ipomoea family to reduce wind erosion and to slow down the wave's energy during high tides of major amplitude. Planting mangroves on seafronts where they naturally never or rarely grow is derived from such engineering. But here the result is often a failure.
- The fourth type of engineering service deals with the capture of larvae of reef species in the marine environment. Their breeding will produce juvenile and young adults for the reintroduction *in situ* or the aquarium fish sales. In nature, the life expectancy of these larvae is very low. Most of them will die, eaten by predators. So, their capture for breeding has negligible negative effects on the reef ecosystem. This service is upstream both to the 'reintroduction of species" service and to the "reduction of the anthropogenic pressure on ecosystems" service, to which it brings a quite valuable contribution, as it avoids to fish adults for sale to aquarium keepers.
- The fight against the proliferation of invasive species such as *Pterois volitans* in the Caribbean (Gonzales et al. 2009; Bouchon and Bouchon-Navarro 2010.) or the *Acanthaster planci*, a starfish that eats coral polyps[9] is the fifth type of ecological engineering service.

(c) The 'reduction of the anthropogenic pressure on ecosystems' service is one of the most common sociosystems services. It encompasses all of the regulations incorporated in the management of natural resources. This service takes three different forms:

- The regulation of the economic or social activities (including leisure) which have a direct impact on coastal or marine ecosystems.
- The regulation of economic activities that have an indirect impact on coastal or marine ecosystems.
- The integrated management of coastal areas, which requires regulation of all sectors of activities in this space.

[9]The coral reefs consist of animals, the polyps, which live in symbiosis with algae, the 'zooxanthellae', which provides much of the carbon they need via photosynthesis.

The regulation of economic or social activities can be carried out in two ways: first, the restrictions on the use of fishing boats and, second, the limitation of access to the resource areas. According to the geographical level of the decision-making process and the extent by which it is applied, two main types of limited access to areas can be distinguished: at both the local communities and the supra community scales, including national and the international levels. At the local scale, the limitation of access to the area is mainly carried out in the form of temporary closures of fishing and permanent closures on resource areas. At the supra community scale, an activity which is the subject to regulation is another parameter which can be used to classify socio-systems services.

For fishing, the three major limitations of access are respectively: (a) the permanent closure of fishing (fishing reserves), (b) the seasonal closure of fishing, for example in the spawning areas, in order to avoid the fishing of gravid females, (c) the fishing closures usually for a period of one to two or three years, under the context of rotating reserves.

For urbanization and tourism, regulation focuses on two points:

- The ban on construction in some areas. In the French Overseas Départements, the coastal Act (1986) governs urban planning and bans any new construction in a 80 m strip from the coast (except for exemptions) which corresponds to 50 geometric feet.
- The necessity for each coastal home or hotel to have an individual or collective sanitation system in order to reduce domestic effluents, sources of eutrophication of shallow coastal environments.

The regulation of economic activities that indirectly impact coastal or marine ecosystems deals mainly with activities which develop in watersheds, upstream the coral reefs, such as agriculture or urban development. This type of service is not yet common because few farmers, urban planners or public works contactors working on watersheds are aware of the potential impact of their activities on the coastal ecosystems. The example of chlordecone/kepone/in Martinique and Guadeloupe shows that chemicals used in watersheds for agricultural purposes can heavily impact the downstream reefs at range a of may kilometers away (Bertrand et al. 2010; Bodiguel et al. 2011). Coral is extremely vulnerable to pesticides and herbicides, hence the need to focus on sustainable agriculture that is more respectful of ecosystems in the islands, by minimizing the use of harmful inputs to the environment.

(d) The reduction of pollution in coastal ecosystems is a service provided mainly by water treatment plants. Their high cost is a huge constraint for the expansion of this service in poor islands. However, ecological engineering allows the implementation of biological plants using the remediation service provided by wetland vegetation, including mangroves.

12.6 Conclusion

Socio-system services aim mainly at sustaining the functioning of ecosystems when their health status is good or to restore them when they are degraded. Thus, they help sustain the quality of ecosystem services. Sociosystem services and ecosystem services are, therefore, linked by a feedback loop (Fig. 12.5). The preservation of such a loop to a high level of efficiency should be the top priority in the governance of coastal areas. It is the only way to ensure a sustainable and reasonable use of marine resources at both local and above local levels in order to promote nature/society coviability. In this perspective, the concept of a sociosystem service becomes central in the governance of coastal ecosystems. This relational interdependence between ecosystem and socio-system services leads to the combination of these two services under the name of environmental services. They are an essential key of nature/society coviability in coral reef areas as in any other coastal and marine environments.

References

Aubin JP (1991) Viability theory. Systems & controls. Foundation & application. Birkhauser (ed.), 543 pp

Bertrand G (1968) Paysage et géographie physique globale. Esquisse méthodologique. Revue géographique des Pyrénées et du Sud-Ouest 39(3):249–272

Bertrand C, Bertrand G (1992) Chap. 7: La géographie et les sciences de la nature. In: Bailly A, Ferras R et Pumain D (eds) Encyclopédie de la géographie. Economica, Paris, pp 110–127

Bertrand C, Bertrand G (2002) Une géographie traversière, l'environnement à travers territoires et temporalités. ARGUMENTS, Paris

Bertrand JA, Bodiguel X, Abarnou A, Reynal L, Bocquené G (2010) Chlordecone in the marine environment around the French West Indies: from measurement to pollution management decisions. In: Proceedings ICES science annual conference, Nantes 20–24 September 2010, Dvd ICES-CM 2010 /F07, 9 pp

Bodiguel X, Bertrand JA, Fremery J (2011) Devenir de la chlordécone dans les réseaux trophiques des espèces marines consommées aux Antilles (CHLORETRO). Rapport final de Convention Ifremer-ODE Martinique-DSV Martinique, 46 pp

Bouchon C, Bouchon-Navorro Y (2010) Invasion de la mer Caraïbe par *Pterois Volitans* et P. Miles. Rapport Scientifique. Université Antilles-Guyane/ DIREN Guadeloupe, 25 pp

Boyd J, Banzhaf S (2007) What are ecosystem services? The need for standardized environmental accounting units. Ecol Econ 63:616–626

Cillaurren E, David G (2014a) De l'écosystème vers le socio-système: une perception nouvelle de la notion de services éco-systémiques attachés au milieu récifal de La Réunion. Montpellier, IRD/UMR Espace Dev/GEOSUD, document de travail n°1 programme OREMSE, 17 p

Cillaurren E, David G (2014b) Formalisation ontologique de la structure et de la trame relationnelle de l'éco-socio-système attaché au littoral récifal de La Réunion. Montpellier, IRD/UMR Espace Dev/GEOSUD, document de travail n°1 programme OREMSE, 17 p

Corlay JP (1993) La pêche au Danemark, essai de géographie halieutique. Thèse d'Etat. Brest. 2 volumes. Université de Bretagne Occidentale, 1331 p

Corlay JP (1995) Géographie sociale, géographie du littoral. Norois, n°165, janv.-mars, pp 247–265
Corlay JP (1998) Chap. 2: Facteurs et cycles d'occupation des littoraux. In: Miossec A (dir.) Géographie humaine des littoraux maritimes. CNED-SEDES, Paris, pp 97–170
Costanza R et al (1997) The value of the world's ecosystem services and natural capita. Nature 387:253–260
Daily GC (1997) Introduction; what are ecosystem services. In: Daily GC (ed) Nature's services: societal dependance on natural ecosystems. Island Press, Washington, D.C.
David G, Mirault E (2006) L'épuisement de la ressource récifale de la Réunion: confrontation des perceptions et régulation de l'activité halieutique. In: Chaussade J et Guillaume J (eds) Pêche et aquaculture, pour une exploitation durable des ressources vivantes de la mer et du littoral: ouvrage en l'honneur de J.P. Corlay. Rennes, Presses Universitaires, Coll. Espace et Territoires, pp 127–140
David G, Mirault E, Pennober G, Révillion C (2012) "Unités Paysagères et services écosystémiques, l'exemple des récifs coralliens", VertigO – la revue électronique en sciences de l'environnement [En ligne], Hors-série 14 | septembre 2012, mis en ligne le 15 septembre, 2012, 15 p
Doumenge F (1983) Aspects de la viabilité des petits États insulaires, étude descriptive. CNUCED, Genève, p 39
Doumenge F (1985) The viability of small intertropical islands. In: Dommen E, Hein P (eds) States, microstates and islands. Croom Helm, London, pp 70–118
Ehrlich P, Mooney HA (1983) Extinction, substitution and ecosystem services. Bioscience 33: 248–254
Failler P, Pètre E, Binet T, Maréchal JP (2015) Valuation of marine and coastal ecosystem services as a tool for conservation: the case of Martinique in the Caribbean. Ecosyst Serv 11:67–75
Fisher B, Turner RK (2008) Ecosystem services: classification for valuation. Biol Conserv 141:1167–1169
Fisher B, Turner RK, Morling P (2009) Defining and classifying ecosystem services for decision making. Ecol Econ 68:643–653
Gomez-Baggethum E, de Groot R, Lomas PL, Montes C (2010) The history of ecosystem services in economics theory and practice: from early notion to markets and payment schemes. Ecol Econ 69:1209–1218
González J, Grijalba-Bendeck M, Acero PA, Betancur RR (2009) The invasive red lionfish, *Pterois volitans*, in the southwestern Caribbean Sea. Aquat Invasions 4:507–510
Graham M (1935) Modern theory of exploiting a fishery, and application to North Sea trawling. J Cons PermInt Explor Mer 10:264–274
Gunderson LH, Holling CS (eds) (2002) Panarchy: understanding transformations in systems of humans and nature. Island Press, Washington, D.C., p 554
Hénocque Y, Denis J, Gérard B, Grignon-Logerot C, Brigand L, Lointier M, Barusseau P (1997) Guide méthodologique d'aide à la gestion intégrée de la zone côtière. UNESCO, Paris. Coll manuels et guides de la COI, n°36, 47 p
Hicks C (2011) How do we value our reefs? risks and tradoffs across scales in "biomass-based" economies. Coast Manag 39:358–376
Lapierre JW (1992) L'analyse de systèmes, l'application aux sciences sociales. Syros, coll. Comprendre, Paris, 229 p
Laurans Y, Rankovic A, Billé R, Pirard R, Mermet L (2013) Use of ecosystem services economic valuation for decision making: questioning a literature blindspot. J Environ Manag 119: 208–219
Laurec A, Le Guen JC (1981) Dynamique des populations marines exploitées: concepts et modèles, n 45. CNEXO, Rapports scientifiques et techniques, Brest, 118 p

Lemahieu A(2015) Fréquentation et usages littoraux dans la Réserve Naturelle Marine de La Réunion, élaboration d'un suivi pour l'analyse des dynamiques spatio-temporelles et apports de l'outil à la gestion et la recherche interdisciplinaire. Thèse d'état de Géographie, Université Paris I, G., 333 p + annexes

Lemahieu A, Pennober G, David G, Lavigne F, Pothin K, Gérard L (2013) "Élaboration d'un protocole de suivi de la fréquentation au sein de la Réserve naturelle marine de la Réunion, France, Océan Indien", VertigO – la revue électronique en sciences de l'environnement [En ligne], Volume 13 Numéro 1 | avril 2013, mis en ligne le 16 avril 2013

Levêque C, Muxart T (2004) Anthroposystème, Hypergéo, 3 p

Lévêque C, Muxart T, Abbadie L, Weill A, Ven der Leeuw S (2003) L'anthroposystème: entité structurelle et fonctionnelle des interactions société-milieux. In: Lévêque C, Van der Leeuw S (eds sci.) Quelles natures voulons-nous? Pour une approche socio-écologique du champ de l'environnement. Elsevier, Paris, pp 110–129

Liu JC, Dietz T, Carpenter SR, Folke C, Alberti M, Redman CL, Schneider SH, Orstrom E, Pell AN, Lubchenco J, Taylor WW, Ouyang ZY, Deadman P, Kratz T, Provencher W (2007) Coupled human and natural systems. Ambio 36(8):639–649

Meral P (2012) Le concept de service écosystémique en économie: origine et tendances récentes. Nature Sciences Sociétés 20:3–15

Millenium Ecosystem Assessment (2005) Ecosystem and human well-being: synthesis. Island Press, Washington, DC

Mirault E (2006) Les fonctions et enjeux socio-économiques des écosystèmes récifaux: une approche géographique des valeurs de l'environnement appliquée à l'île de La Réunion. Thèse. Université de Paris X. IRD, 535 pp

Mirault E, David G (2006) Estimation des valeurs socio-économiques des récifs coralliens de l'île de la Réunion. Projet VALSECOR, DVD de restitution. St Denis de la Réunion, IRD/DIREN/IFRECOR/UE

Mirault E, David G (2009) Fonctions et logiques d'interface des récifs coralliens sur le littoral de La Réunion. Les Cahiers d'outre-mer 248:571–588

Moberg F, Folke C (1999) Ecological goods and services of coral reef ecosystems. Ecol Econ 29:215–233

Moberg F, Rönnbäck P (2003) Ecosystem services of the tropical seascape: interactions, substitutions and restoration. Ocean Coast Manag 46(2003):27–46

Revillion C (2009) Approche de la spatialisation des services éco-systémiques des récifs coralliens: le cas de la réserve naturelle marine de La Réunion. Université de La Réunion/IRD. Master II de Géographie, 59 p

Rönnbäck P, Crona B, Ingwall L (2007) The return of ecosystem goods and services in replanted mangrove forests – perspectives from local communities in Kenya. Environ Conserv 34: 313–324

Sauvé,L (1994) Pour une éducation relative à l'environnement. ESKA, Paris, p 361

Schaefer MB (1954) Some aspects of the dynamics of population important ot the management of the commercial marine fisheries. Bull Int AFK Trop Tuna Comm 1:26–56

Schaefer MB (1957) A study of the dynamics of the fishery for yellowfin tuna in the Eastern Tropical Pacific Ocean. Bull bit Am Trop Tuna Comm 2:247–285

Schuhmann PW, Mahon R (2015) The valuation of marine ecosystem goods and services in the Caribbean: a literature review and framework for future valuation efforts. Ecosyst Serv 11: 56–66

TEEB (2010) Mainstreaming the economics of nature: a synthesis of the approach, conclusions and recommendations of TEEB. UNEP/UE, Nairobi/Bruxelles, p 49

Teitelbaum A, Friedman K (2008) Successes and failures in reintroducing giant clams in the Indo-Pacific region. SPC Trochus Inf Bull 14:19–26

Wallace K (2007) Classification of ecosystem services: problems and solutions. Biol Conserv 139:235–246

Wilson CM, Matthews WH (eds) (1970) Man's impact on the global environment: assessments and recommendations for action. The MIT Press, Cambridge, ,MA

Texts

Déclaration finale de la Conférence des Nations Unies sur l'environnement. 5–16 juin 1972. Stockholm

Loi constitutionnelle n° 2005-205 du 1er mars 2005 relative à la Charte de l'environnement (JORF n°0051 du 2 mars 2005 page 3697)

Part II
Governance of the Coviability: Norms, Policy and Actors

Preview 2: The Governance and Environment and Humans

Governing is regulating. The first part of this volume positions regulation as a central element of coviability. The works undertaken by mathematicians (the space of constraints, the core of viability, etc.), jurists, biologists, geographers or philosophers, underline the place of the regulation in the viability of human and non-human systems. The contribution of this second part consists in formalizing the regulation through the governance of natural environment and humans.

Thinking about coviability at the time of decision-making, adoption of public policy guidelines and practices, and the implementation of standards is about designing and building social, economic, and ecological links. This notion of links brings us back to that of connecting local actors to their territory, to their living environment. The nature of these links determines the human/non-human nexus.

If the creation of protected areas aims to shape forms of reconnection to the biosphere, the economic issues rapidly compromise this ideal. The demonstration of this drift by commodification, referred to as the principle to pay a price to ensure this socio-ecological link, is already a practice (the example of pollination, yet so vital to human societies). Thus, the appropriation of the common reflects a form of regression. This is the case of pastoral practice, which is increasingly constrained by the pressure placed on the remaining common space. Whether it is food (herbaceous-wooded) or land, the common pastoral activity calls for the intervention of public policies to preserve its existence, "enhancing coviability" between the practice and the environment. The preservation and development of a practice that is both material and immaterial, yet locally rooted and declared patrimonial by the international sphere, will therefore depend on territorial regulation.

The issue of the coviability of social and ecological systems certainly requires man to reconnect to himself: how are we supposed to think 'biosphere' when the humans cannot think "humanity"? Does not assuming sociability start with a human diversity associated with biological diversity? Here again we are caught by the notion of the link associated to that of diversity … of the living.

Reconnecting humans to the biosphere is also reconnecting humans to humans. International environmental law is moving in this direction when it is in synergy with human rights. Maturation is underway, and while the concept of "sustainable" development is so well entrenched, its legal interpretation and implementation remain highly dependent on an exclusively Western economic model, with the system of values and representations that it entails.

Governance consists in regulating the relationships between humans and the relationships between humans and non-humans. At different scales and areas, the natural environment and humans interweave. Coexisting means living together. Governing includes integrating.

Four regulation rationales animate this part. Each includes two or three chapters dealing with the governance of protected areas, pastoral activity, human diversity and the environment:

1. **Environmental Regulation: governance of protected areas**

   Chapter 13 – Governance Of Protected Areas As A Tool For Coviability
   Chapter 14 – Social-ecological coviability of protected marine areas in Brazil
   Chapter 15 – Socio-ecological coviability confronted with the neoliberal system, the peace parks experience (Southern Africa)

2. **Territorial Governance: governance of pastoral activity**

   Chapter 16 – Coviability In The Governance Of Pastoral Systems, Permanence and Change
   Chapter 17 – Developing coviability through an eco-pastoral approach, the European project LIFE + MIL'OUV

3. **Regulation of human relationships: governance of human diversity**

   Chapter 18 – Reconnecting man to man: socio-cultural coviability ties and interculturality (Practical research in a sensitive neighborhood in Montpellier, France)
   Chapter 19 – Kinship as an Instrument for Coviability: Study Cases in Pará, Amazonia

4. **Regulation of human relationships to the biosphere: environmental governance**

   Chapter 20 – The price of coviability: pollination at all costs; legal approach of a new relationship between man and pollinators
   Chapter 21 – Can International and French environmental law accommodate coviability?

As in the previous introductory chapter, we propose the heuristic maps of each chapter that will support our discourse on the ontology of coviability.

Protected Area Governance

Chapter 13 presents the governance of protected areas as an example of coviability between anthropic and natural dynamics (Figs. Preview 2.1, 2.2, and 2.3).

Chapter 14 presents the case of marine protected areas in Brazil in which governance recognizes the role of traditional and ancestral knowledge in managing society and nature dynamics.

Chapter 15 is dedicated to the analysis of peace parks and the divergence of points of view within their governance.

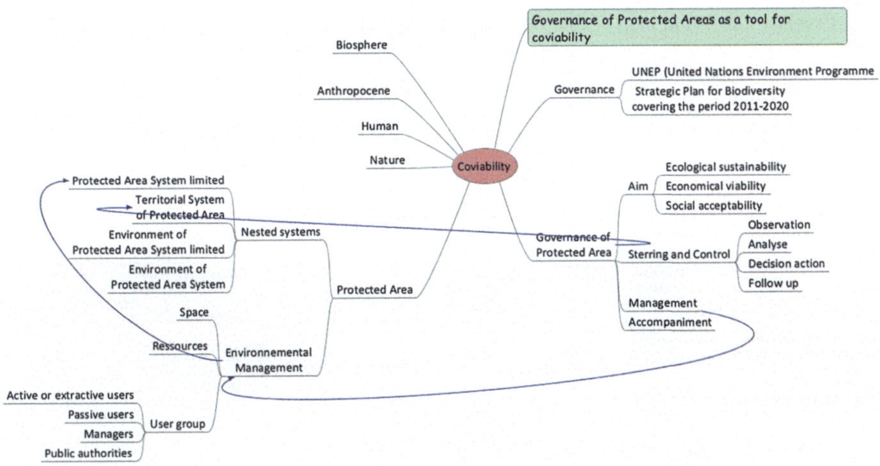

Fig. Preview 2.1 Mind map of Chap. 13 (®Thérèse Libourel)

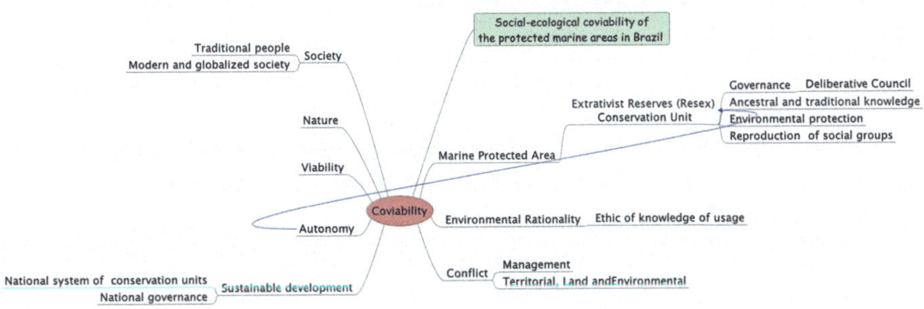

Fig. Preview 2.2 Mind map of Chap. 14 (®Thérèse Libourel)

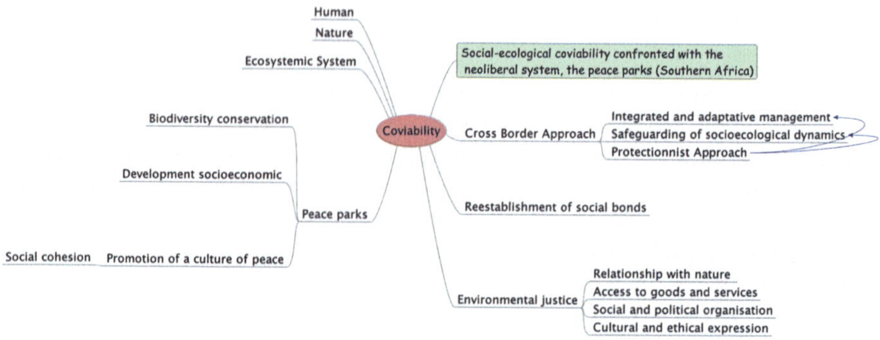

Fig. Preview 2.3 Mind map of Chap. 15 (®Thérèse Libourel)

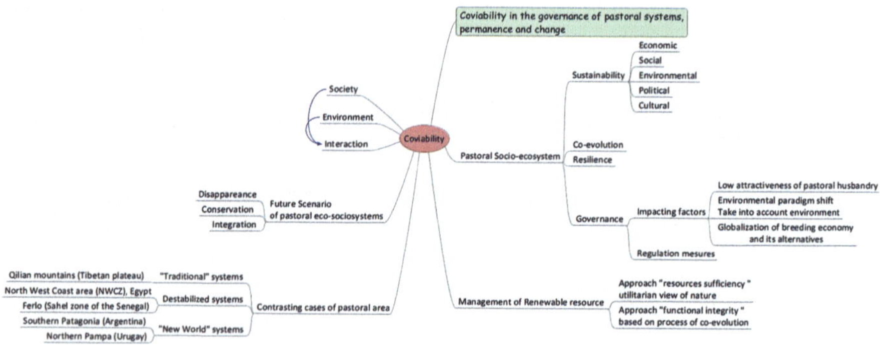

Fig. Preview 2.4 Mind map of Chap. 16 (®Thérèse Libourel)

The Governance of Pastoral Activity

Chapter 16 presents the governance of pastoral systems through the illustration of various situations (Fig. Preview 2.4).

Chapter 17 is dedicated to the approach implemented in the LIFE project which illustrates a cross-analysis of the viability of livestock farming and the natural environment (Fig. Preview 2.5).

The Governance of Human Diversity

Chapter –18 is dedicated to an experiment carried out within a multicultural society (Fig. Preview 2.6).

Chapter 19 analyzes the role of kinship in an Amazonian society in the State of Para, Brazil (Fig. Preview 2.7).

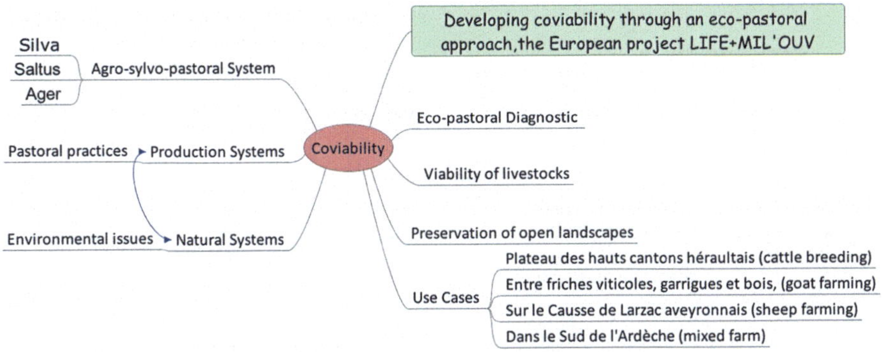

Fig. Preview 2.5 Mind map of Chap. 17 (®Thérèse Libourel)

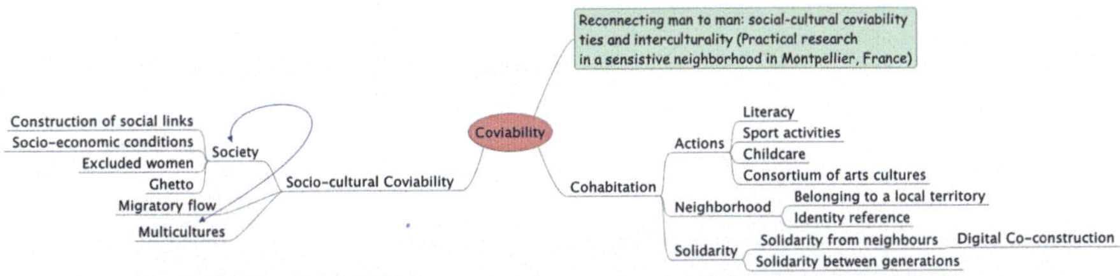

Fig. Preview 2.6 Mind map of Chap. 18 (®Thérèse Libourel)

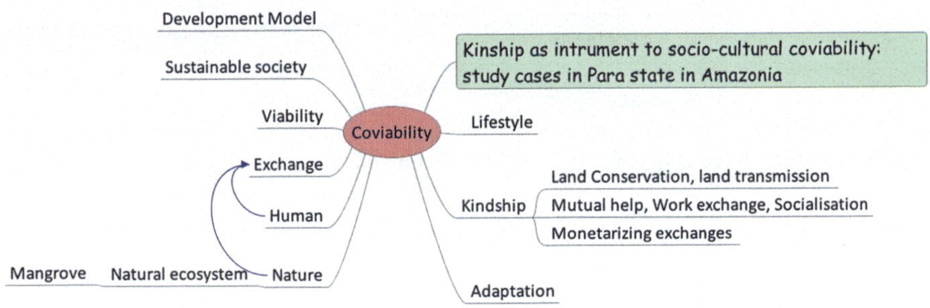

Fig. Preview 2.7 Mind map of Chap. 19 (®Thérèse Libourel)

Environmental Governance

Chapter 20 explains the approach in French and European laws of pollination by demonstrating the trend of coviability commoditization (Fig. Preview 2.8).

Chapter 21 questions the ability of international law and positive French law to mainstream the concept of coviability (Fig. Preview 2.9).

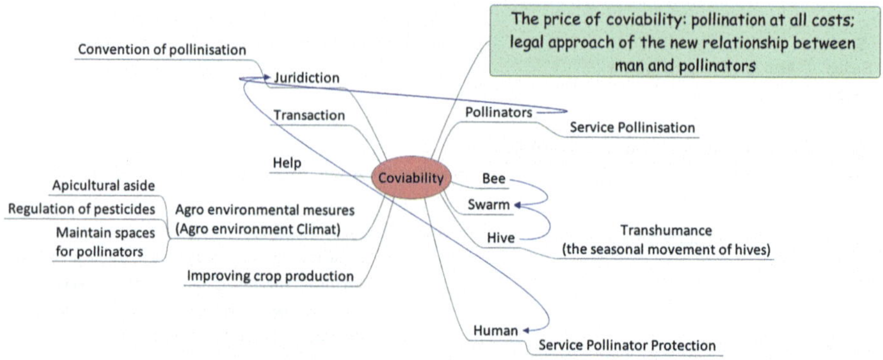

Fig. **Preview 2.8** Mind map of Chap. 20 (®Thérèse Libourel)

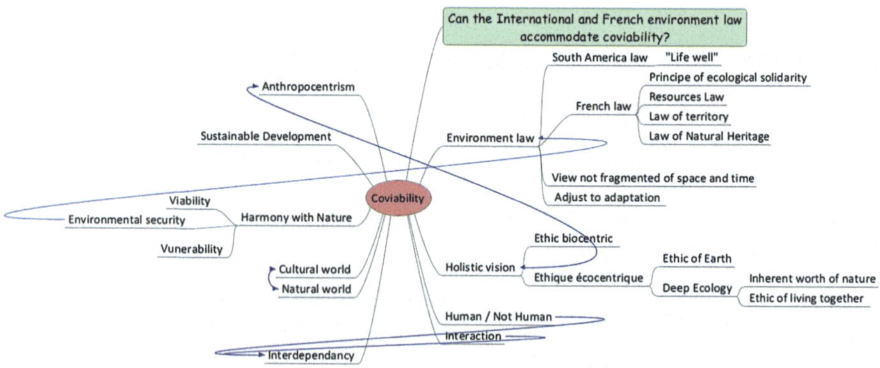

Fig. **Preview 2.9** Mind map of Chap. 21 (®Thérèse Libourel)

Chapter 13
The Governance of Protected Areas as a Coviability Tool

Gilbert David

Contents

13.1 Introduction

The intensification of ecosystems uses and the spatial expansion of human beings all over the planet are the two main characteristics of man/nature relationships through the time. This process has remarkably accelerated with the industrial revolution of the nineteenth century and then the demographic growth of the twentieth and twenty-first centuries. Faced with this evolution, a growing number of scientists questioned the future of our world during the 1960s and 1970s. "Nous allons à la Famine" (*we are heading for Famine*) wrote Dumont and Rosier (1966). Three years after, Commoner (1969) published "Science and Survival". In 1968, Erhlich and Erhrlich wrote "the Population Bomb". The year 1969 was a new stage in global awareness about our environment and the viability of humankind. The Apollo XI mission showed billions of people the geographical limits of our planet. In 1972, the

G. David (✉)
IRD - French National Research Institute for Sustainable Development, UMR ESPACE-DEV, Montpellier, France
e-mail: gilbert.david@ird.fr

O. Barrière et al. (eds.), *Coviability of Social and Ecological Systems: Reconnecting Mankind to the Biosphere in an Era of Global Change*,
https://doi.org/10.1007/978-3-319-78497-7_13

preparatory report for the first United Nations conference on the environment was entitled "We only have one Earth" (Dubos and Ward 1972). The final declaration stressed that: "A point has been reached in history when we must shape our actions throughout the world with more prudent care for their environmental consequences. Through ignorance or indifference we can do massive and irreversible harm to the earthly environment on which our life and wellbeing depend".[1] One of the major outputs of this conference is UNEP (the United Nations Environment Program). It aims at reversing this negative trend and sets out 25 principles to follow. The first two are the basis of what we call today coviability between man and nature. "Man has the fundamental right to freedom, equality and adequate conditions of life, in an environment of a quality that permits a life of dignity and well-being... The natural resources of the earth, including the air, water, land, flora and fauna and especially representative samples of natural ecosystems, must be safeguarded for the benefit of present and future generations through careful planning or management, as appropriate". Setting up natural parks and protected areas is an attempt to stop the man-induced damage on nature. UNESCO has played a leading role in their implementation. The MAB program (Man and Biosphere) organizes the creation of biosphere reserves. It has been in implementation since 1971. The World Heritage Convention was published in 1972. The conservation of the most remarkable natural sites on the planet is one of its two main objectives.

The 1975–1985 decade was marked by a dramatic increase in the number of protected areas and their area grew from 3 to 6 million sq km between 1970 and 1980. The growth was even more spectacular during the following decade. In 1990, the total area was close to 12 million sq km. In 2005, more than 15 million sq km of land was reserved for biodiversity conservation (Rodary and Milan 2008). After 1995, the average size of the new PA has tended to decrease. IUCN announced that 100,000 PA covered 12% of land area during the fifth World Congress of Parks held in Durban in 2003. Ten years later, by the sixth edition of the Congress (2014), the number of PA had more than doubled (209,000) with the protected areas covering 15.4 % of land.

Set in 2010 by the Convention on Biological Diversity, the objective of the "Strategic Plan for Biological Diversity 2011–2020" of having 17% of the planet's land and inland waters in protected areas, is getting closer. The ambition to protect 10% of marine and coastal areas (compared to 3% currently) will be more difficult to achieve. Setting up very large protected marine areas (PMA) seems to be a solution. Their number is growing fast. During 2014 and 2015 several MPAs of more than 1 million sq. km were established in the Pacific Ocean. This spectacular growth in the number of MPAs runs a high risk of reducing the overall effectiveness of protected areas for conserving biodiversity (Agardy et al. 2003), as a growing number of PAs becoming "paper parks". This risk was largely discussed at sixth IUCN Congress of

[1] http://www.unep.org/Documents.Multilingual/Default.asp?DocumentID=97&ArticleID=1503&l=en

Parks in Sydney in November 2014 (Di Minin and Toivonen 2015). Obviously, the question of the viability of a PA as a tool for the coviability of man/nature is raised.

13.2 Methodology

This chapter deals with protected areas as a central element of human/nature coviability. They count as one of the main services delivered by humankind to ecosystems in order to sustain the benefits people obtain from ecosystems.[2] Here, I assume that the viability of any protected area is primarily based on its governance that, as such, can be considered a coviability tool. Therefore, the reader is invited to explore the concept of governance of protected areas.

This work is based on a series of three pieces of research on PMAs. The first took place from 1999 to 2000 as part of the regional program for environment of the Indian Ocean Commission (David 1998).[3] The second was carried out from 2006 to 2008. It dealt with the socio-economic characterization of the initial state (before implementation) of the natural marine reserve of Reunion Island (Thomassin and David 2008). The third was devoted to indicators of governance in the Pampa project (MPA performance indicators of for the management of coastal ecosystems, resources and their uses) carried out from 2009 to 2011 (David 2011; David et al. 2010). This work is also based on ten years of participant observation by the scientific councils of the Reunion Island Natural Marine Reserve (2004–2014) and the Iroise Sea Marine Natural Park (2009–2011); by the steering committees of the protected marine areas network of the IOC (2006–2009) and the IFRECOR[4] programs devoted to protected marine areas.

The governance of a protected area is a new concept and appeared in 2003 during the fifth global Congress of Protected Areas (Cazalet 2004). This is not yet stabilized. First, it is highly polysemous by nature (Baron 2003). Second, the representations of PA are quite different when scientists are considered whose studies are upstream and support PA governance, and PA managers, in charge of developing and implementing such governance. Third, heterogeneity will also be found when the representations of scientific disciplines on what PAs are and how they work are considered (Chaboud et al. 2008). However, the majority of researchers studying PAs agree that the governance of Pas belongs to a broader category: environmental governance or resources governance. In their reference book on PMA, Pomeroy et al. (2004) define it as "*The way in which users and their*

[2]The notion of securing an ecosystem service deals both with the quality of the service provided and its sustainability in time (see chapter of Cillaurren and David on ecosystem services in this book.)

[3]The Indian Ocean Commission or the IOC is a regional organization based in Mauritius, which covers five territorial entities: Comoros, Reunion/France, Madagascar, Mauricius, the Seychelles.

[4]The IFRECOR or French Initiative for coral reefs is a program controlled by the French Ministries of Ecology and Overseas aiming at federating French research on coral reefs.

intentions are managed through a set of rights, rules and shared social norms and strategies. This includes enforcement mechanisms, such as policing measures and punishments, as well as incentives to direct human behaviour and use"... "*Resource governance can include: (a) formal and informal forms of resource ownership; b) use rights and the law that support these rights;" (c) the rules, rights and regulations that dictate how resources can or cannot be used*".

However interesting this definition may be, it does not remove two major ambiguities, due to the polysemous nature of governance:

- Is governance a purpose (as illustrated by the expression "good governance"), or a tool, or both at once?
- How can we distinguish the environmental governance from environmental management? Is the latter included in the first? In contrast, is governance a better way to manage the environment? Unless, of course, the two concepts merge? (Rey-Valette 2008).

The territorial nature of PA (David and Thomassin 2007) gives rise to a new level of complexity. PA governance is also part of territorial governance (Chia et al. 2010). Barrière (2005) gives a simple and operational definition of governance: "A process of decision-making and regulation of practices in terms of actions and interventions on a territory and of the implementation of public policies".

In this chapter, I aim to clear up the notion of PA governance. In the first part, the contours of this ill-defined object called "protected area" will be clarified. In the second part, the PA governance will be analyzed as a decision-making process dealing with information flow. The third part will focus on the steering and control module of this process and the social acceptance of PAs.

13.3 Governance and Protected Areas, The Need for Clarification

Any PA aims to preserve or restore habitats and their biodiversity (Dudley 2008; Dudley and Stolton 2008). This goal deals with environmental management, both defined as matter of relationships between human beings and nature, or as a matter of relationships between human beings with regard to nature (Weber, cited by Denis et al. 2001). It also deals with geography. Any PA is also a territorial system. Its implementation covers pre-existing territories (David and Thomassin 2007). We are here in territorialized environmental management, a specific form of environmental management which relies on the creation of territories to resolve conflicts of use and better manage natural resources. As such, a PA deals as much with the relationships between human beings and their territories than with relationships between human beings and nature.

These relationships between territories can also be expressed under the form of territoriality relations. They can be studied at the individual level according to a psychological perspective. Territoriality is defined as a "behavioral phenomenon associated with an organization of space in spheres of influence and distinct and defined territories, considered at least as exclusive by their occupants and designers" (Soja 1971). Territoriality can also be studied at a group level according to an ethnological perspective. It is defined as "the social and cultural relationship that a group maintains with the framework of routes and hierarchical and interrelated places, of which the figure on the ground is a territory" (Bonnemaison 1980, 1981, 1986, 1987). The use of these places and routes drive territoriality.

As a tool of both environmental and territorial governances, PAs are part of a complex dynamic between stakeholders[5] (human populations), their activities or uses, living resources and spaces (Fig. 13.1). Uses are represented as a relationship between the users and the resources they exploit or protect (Fig. 13.1a). Three categories of space can be distinguished: the support space, the resource space, which combines in a single entity the support space and the resources it houses as objects of use, and the produced space which is the social, cultural or economic production resulting from uses or representations on the support space. Territory is a produced space (Di Meo 1998a, b). Two groups of stakeholders can be distinguished: the users of the PA and its managers. Their objective is to regulate uses in areas under their control in order to improve the quality and quantity of resources. Finally, a PA is not only a set of resource space and stakeholders' space. It is also a regulated territory, in which the regulations of uses are implemented (Fig. 13.1b).

To move forward, four user groups can be distinguished (Fig. 13.1c):

(a) The resource users can be qualified as "active" or "extractivists" users because they carry out work (in a physical sense) which uses resources, as fishers allocate a fishing effort on fish stock to catch fish (Laurec and Le Guen 1981).
(b) The "passive users" benefit from a PA without any work to achieve this result. Thus, a hunting ban allows people to live be peacefully near a PA whereas people living near hunting sites suffer from noises and disturbance during the hunting season.
(c) PA Managers are a key group of stakeholders. They are at the heart of the PA governance.
(d) Public authorities play a major role as PA donors. We can distinguish national bodies, including ministries of environment and their different agencies, and international bodies such as UNEP, GEF (Global Environment Facility), FFEM (French Fund for the Global Environment) in France, which are major donors in the protection of biodiversity at a global scale.

Complexity of PA increases with the number of stakeholders but also with their heterogeneity in terms of territoriality, strategies and rationale behind action taken (within the PA as well as outside). Heterogeneity is also induced by the stakeholders'

[5]'Stakeholder' means any individual, group, institution, or eventually any social element endowed with capacity for action (reaction or initiative) and intervening, directly or not, in the process of managing" .the territory, here a protected area (Catanzano and Thébaud 1995, p. 38).

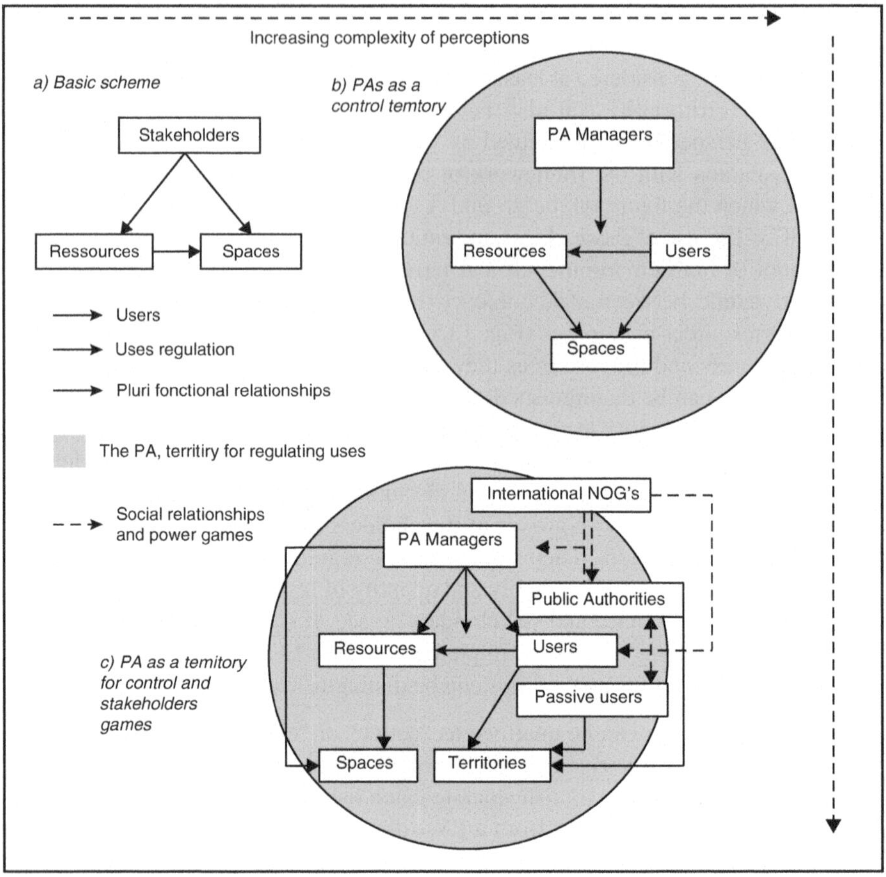

Fig. 13.1 The Protected Area, a territory for stakeholder games built around specific regulations

representations about PAs and their position of loser or winner in this new territory compared to their previous uses in terms of resources and space before the PA was set up. All these strategies and representations have a territorial anchor. PA is a territory of regulated uses. It can also be represented as a system of individual territories of representations, rationale and practices of users. This concept has the advantage of raising the question about the PAs' limits. As a legal and administrative object, PA is perfectly delimited. However, as a territorial system, this is not the case. The way in which natural bodies or institutions located outside the PA can influence the PA functioning is a crucial issue. We assume that any group which sends a flow of information towards the protected area but does not receive any flow of information from it does not belong to the PA system but to its environment. Thus any local political stakeholder who can sustainably influence the future of the protected area will only be considered part of the PA system on the express condition that his political career depends closely on the PA's future (which is rarely the case).

13.4 Governance As a Decision-Making Process

13.4.1 Deciding in Order To Act

Generally speaking, to manage, govern, or run a PA means above all taking action. Any action requires a decision which is a process based on the organization of information. As Le Gallou said (1992, p. 72): "*the decision, purpose and initialization of the action, is itself an action loop, but limited to the psycho-informative field. From the values and intentions, and with respect to stored information, it determines the objectives translated into concrete action*". This process is organized in three stages: to understand, to imagine, and to select (Fig. 13.2).

To understand means describe and analyze the problem requiring a decision. From a flow of information (which will be referred to as an objective reality) it is necessary to progress from the perception of this reality by the human brain (perceived as reality) to the projection of this real objective in the near future (desired reality). Designing consists in developing action plans or strategies leading to the desired reality from the perceived reality. Selecting means evaluating these action plans or strategies and then adopting one of the following three options: a) stop the decision to act; (b) choose to extend conceptual thinking by enriching it with new information, which requires returning to the two previous steps, to understand and to design; (c) change the first objectives of the decision-making process when no satisfactory decision for action can be planned. As quoted by Le Gallou (1992, p. 73), "*controlled and coherent action is the result of organization in*

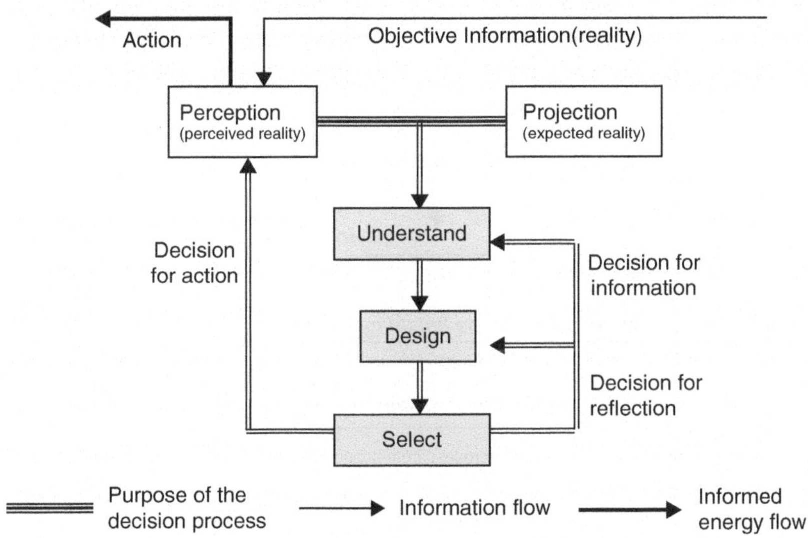

Fig. 13.2 The decision process, from decision to action (Le Moigne 1990)

space and coordination in time of four components: material, energy, information and willingness".[6]

The concepts of action and decision logically lead to see any AP firstly as a system intersecting nature and humankind and secondly the governance of any protected area as a decision-making steering/management module, where strategic or tactical decisions are made to enable it to function sustainably. Fig. 13.2 shows a decision-making process which is not specific to the management structure and management of PAs. It also applies to any stakeholder of the PA system or its environment for whom it filters the decisions and actions coming from the PA management structure, according to their mental patterns and past experience (what Bourdieu 1980 called "the habitus").

This discrepancy in the understanding of information flow between the PA manager who emits it in the hope that they will be properly understood by all and the PA stakeholders who receive them, is critical constraint in the governance of any PA. This leads to a distinction between the hoped management of PAs and their resulting management (Fig. 13.3). The first deals with the rationality of PA managers. It includes all the decisions for action needed to achieve their objectives. The second is the result of interactions between these decisions and hoped for actions for PA management and the representations of stakeholders on (a) the grounds targeted by the protection aims, (b) the way they used them before the PA's implementation, (c) the changes of uses driven by this implementation and the costs they will bear in terms of wellbeing and income. These representations are the basis for social acceptance.

One of the main problems in PA governance is the poor capacity of managers to properly design what PA acceptance is and to include it in their hoped management. Focusing on ecology and biology of populations by training, they mainly consider environmental management as a regulation of predator/prey relations. However, on a daily basis, this management deals more with speeches, representations and human

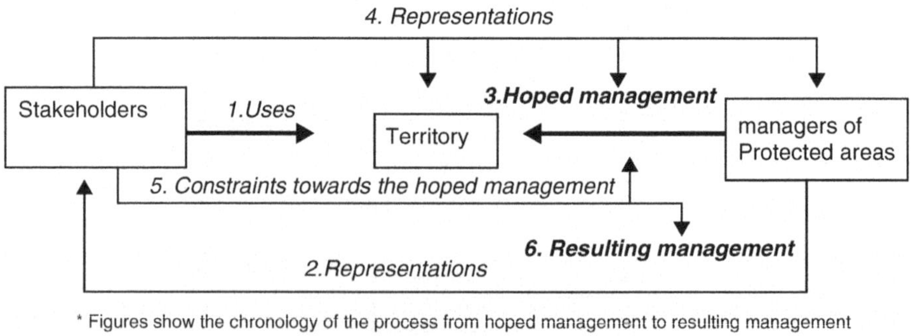

Fig. 13.3 From the hoped management to resulting governance of protected areas

[6]This last includes both energy and information and can be seen as informed energy.

relationships in regards to nature. The issue is less about understanding interactions between the prey and the predator (human) than internal dynamics of human societies which could drive the social acceptance of protected areas. Therefore, studying the functioning of PA ecosystems is not a priority in understanding the pressures of poachers on the preserved habitats, plant and animal populations. Detailed knowledge of these pressures is also secondary, except for purposes of a pitch at the decision-makers level in order to put new regulations in place or strengthen those that already exist, and for the environmental awareness of the public opinion. However, it is crucial to unsderstand (a) the reasons of users who are willing to defy the law to continue their use of protected resources; (b) The social, cultural, economic and political dynamics that drive these reasons.

Here, we are looking at dynamics of information with the notions of representation and values as central parameters. Any PA governance needs to properly manage information. But this management is becoming increasingly complex as the number of objectives assigned to PAs is increasing. This PA multi-functionality means a growth in the number and the diversity of stakeholders, as well as new issues in terms of PA governance.

13.4.2 Ecological Sustainability, First Purpose of Any PA Governance

Ecological sustainability is the first goal, or even the ultimate goal of any PA, regardless of the country concerned. It was designed to restore a degraded ecosystem or to preserve a rich biodiversity. Three types of resources are mobilized to achieve this goal:

- Rules for protecting biodiversity.
- A territory on which these rules applied and the zoning associated with it.
- A structure, known as 'management structure', responsible for administering, managing, and governing this territory as well as resources to be protected. This structure designs the internal rules of the PA, guarantees the respect of its zoning and implements the PA management plan. Therefore, it is located in the heart of PA governance.

The word 'rules' encompasses both the national regulations that apply to the ecosystems to be protected and the regulations locally implemented, including the PA management plan and the administrative measures that complement it. These local regulations concern the whole PA territory. They divide it into several zones according to the allowed uses of habitats and associated natural resources. Thus, even if the public imagination sees it as a natural area, any PA is primarily a territorial construction of legal nature and can be called 'territory of rules'.

This "territory of rules" applies to previous territorial constructions, mainly the territory of uses which are significantly restructured. This territory of uses embraces all areas used by stakeholders for improving their well being in terms of recreation, self-subsistence and income. The implementation of a PA results in the replacement of the former territory of uses by four new types of territories:

- "Territories with new uses", or uses whose frequency is significantly increased by implementing a PA. Marine Protected Areas (MPAs) provide good examples of this type of territory. Scuba diving benefits from the renewal of abundance of fish populations in areas with low economic interest before protection because their resources were severely overexploited by fishing;
- "Territories with recomposed uses". MPA still comes to mind in the case of a fishing area where this activity, henceforth prohibited, is replaced by scuba diving tourism.
- "Territories with lost uses. These uses are henceforth prohibited by the PA management plan.
- "Territories with unchanged uses". These territories are a residual part of the territory of uses previous to PA implementation. They can also be named as "unchanged territories of uses".

These territorial changes driven by the implementation of a PA have also generated changes among stakeholders. Three types of users can be distinguished:

- The actual winners, whose uses have increased;
- The potential losers or winners. These stakeholders are users of recomposed territories. Their win or lose character will be asserted over time according to the type of use practiced and the intensity of this use. Fishers are typical of this category. Any MPA implementation means they have lost in the exchange, unless important compensatory measures are put in place. However, it is hoped that after 10–20 years, they will become winners with (a) the replenishment of fish stocks in the MPA surroundings[7] where their will concentrate their fishing effort, and (b) the provision of a quality label associated with sustainable fishing, which should be inherent in all MAP and their surroundings.
- The actual losers whose uses have disappeared or have been severely regulated with the implementation of a PA.

These actual losers, and those among the potential winners or losers, who perceive PA implementation as negative can play a significant role in the first years of this implementation by opposing the project and shaping a part of public opinion as the latter considers this opposition as legitimate. This opposition challenges the pre-eminence of the territory of rules on the former territorial constructions. If zoning and their associated regulations on the uses of nature (which are the heart of any PA management plan) are no longer respected, the PA will no longer be operational. The habitats and their biodiversity will soon degrade again. The territory of rules is, thus, a territorial construction under two main constraints. First, the stakeholders' support

[7]This is the spillover effect (Buxton et al. 2014).

to the PA project and second, when this support is not effective, the repression of poaching and any criminal act against nature. This repression involves:

- Means of surveillance and control. In the case of MPAs, maritime surveillance needs a patrol boat, which is very costly in terms of equipment and operating. The implementation of these means follows two purposes: educational and repressive. Except in special situations when poachers are reported by observers or are the subject of denunciation, the controls are random. The probability of finding an offender is therefore low, unless poaching is generalized. But these outings at sea are essential to 'raise user awareness' about the risks of being fined if they do not respect the AMP rules. Often, PA ecoguards only have to show themselves to dissuade potential poachers.
- the right to fine offenders. Without this right, the PA ecoguards and managers have to call on security forces for any fining or organize joint patrols. But the low availability of security forces often reduces the frequency of these patrols and the effectiveness of surveillance at sea. The lack of sworn ecoguards also limits the educational controls if the potential offenders know that that the risk of sanctions is very low. To avoid this constraint, in many African countries, ecoguards have military status and enjoy the rights that police have.
- Legal assistance should be conducted in a way that offenders shall be sanctioned when their offense goes beyond a simple fine. This depends on two elements. Firstly, the personal convictions of the judge and the degree of information available to forge these. To this end, all PA managers should be in contact with all new judges in order to raise awareness of the roles of the PA and forge lasting relationships of trust with him/her. Secondly, the support of stakeholders to the PA project. The stronger this support, the more the judge will be inclined to be severer and the poaching will be less rampant. In contrast, if public opinion considers poaching as a minor offence, the probability of showing indulgence for the offender will be high, unless the judge's personal beliefs, partly nurtured and strengthened by information provided to him by the PA manager, urge him to override public opinion.[8]

Therefore, for any PA environmental sustainability cannot be achieved without minimum social acceptance. The same observation can be made with regard to economic viability. Without adequate financial resources, the management structure cannot be fully operational, particularly when concerning the surveillance, control and fining of offenders. PA governance aims to maximize or secure economic viability and the social acceptance of the protected area, so as to sustain its structure and functioning while monitoring both the ecosystem to be protected and the anthropogenic pressures on it. These are the tasks assigned to the steering and

[8]This view may seem a bit provocative because it puts forward the interpretation of the law more than its absolute normative character and integrates public opinion in the Court's decision. Therefore, rendering justice would mean to making a decision on a cost/benefit basis under a powerful constraint: respect for the law.

control module of any PA aiming at environmental sustainability. When the number of purposes increases, managing the PA becomes more complex.

13.4.3 The Growing Multi-functionality of PAs

In their guide for managers of protected marine areas, Pomeroy et al. (2004) identify 16 goals and 68 objectives to cover all of the purposes of any MPA. These 68 objectives fall within three areas: biophysics (5 purposes, 26 goals), socio-economic (6 goals, 21 objectives), governance (5 goals, 21 objectives). The conclusion is obvious: the ecological sustainability of the PA, to which are attached the 5 goals of biophysical nature of Pomeroy et al., is henceforth a minority goal compared to the governance and socio-economic goals. Together they cover two-thirds of the PA goals. Faced with the heterogeneity of these goals, which are sometimes very far removed from ecological sustainability, it is legitimate to ask if the PA is the most effective system to meet ecological sustainability.

The justification of Pomeroy et al., on the choice of these goals shows clearly their way in seeing protected areas: "*Experience shows that social, cultural, economic and political factors, more than biological or physical factors, shape the development, management and performance of MPAs (Fiske* 1992*; Kelleher and Rochia* 1998*; Roberts* 2000*). MPAs affect and are affected by people. For this reason, the goals and objectives of many MPAs include socio-economic considerations such as food security, livelihood opportunities, monetary and non-monetary benefits, equitable distribution of benefits, compatibility with local culture, and environmental awareness and knowledge. Understanding the socio-economic context of stakeholders involved with and/or influenced by the MPA (individuals, households, groups, communities, organizations) is essential for assessing, predicting and managing MPAs.*"

Two ways of interpreting this text are possible:

- The PA is a social ecological system which consists of three systems: the ecosystem, the social system (seen as a socio-economic system) and the governance system. These systems should be interconnected, but they are studied separately. There is no attempt to study how they can be combined into a single system and what the goal of this system could be (Fig. 13.4).
- The PA is an ecological system under human pressures driven by the populations of active or passive users of the PA. These users can be a threat to the sustainability of the PA if their acceptance of it is too low. It is important, therefore, to know the dynamics driving these human pressures in order to minimize a PA's vulnerability towards these human hazards and optimize a PA's ecological sustainability and viability. This social acceptance is at the center of the dynamics of a PA system. It is this social acceptance that motivated the type of six goals defined by Pomeroy et al. (2004). These choices reveal three keys elements which could structure the way of thinking of these authors:

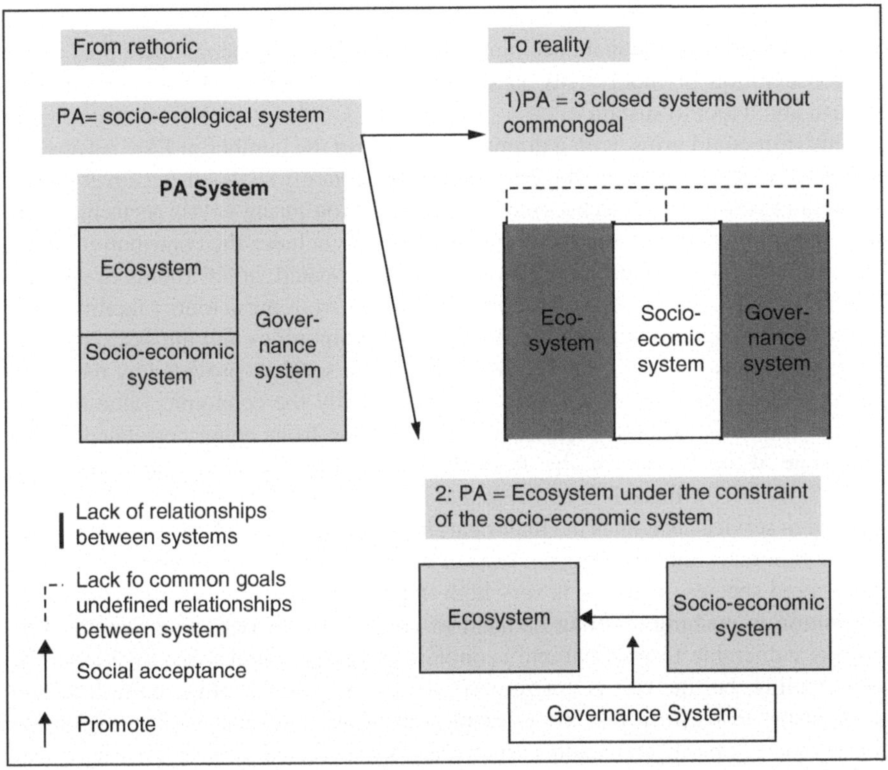

Fig. 13.4 The Protected Area as a socio-ecological system, from rhetoric to reality

- a significant portion of all PA users consider themselves as having lost out, the profits they were expecting from the PA is much lower than the costs they incur;
- this perception can be changed by acting both on the economy and society. Two economical changes are expected. First, maintaining livelihoods and food security. Second, guaranteeing an equal distribution of the potential benefits of the PA. The expectation in terms of society is to maintain the local culture and the non-monetary benefits for human communities living near or in Pas;
- environmental information can also contribute to reversing this negative perception of a PA which drives its low social acceptance.

Asking a PA manager to be a fully-fledged economic player is to consider that the original goal of all PAs – the conservation of biodiversity – is insufficient to justify the efforts made by the public authorities or users for the proper functioning of the PA. Adding an economic goal aims at increasing social acceptance of the PA. However, as soon as biodiversity conservation is no longer a goal, which is the subject of this social acceptance, but rather the combined goal of “biodiversity conservation – contribution to territorial development”, there is a clear risk of shift in the respective importance given to each component of this combination. A switch can be made from a dominant ecological goal to a dominant economical goal when the conservation of biodiversity is subordinate to an economic activity such as tourism and a tool to sustain it.

This shift could grow, with a dramatic increase in the number of PAs. As any PA is expensive, it is tempting to integrate it into the economy so that it can cover part of its operating costs, and even become "profitable". Monitoring a PA's performance is no longer limited to the impacts on biodiversity but includes the contribution of the PA to the economy, including the number of jobs created, hotel' rooms or shelters built on its edge, the number of people who benefit from these tourist facilities, the review of services provided by the PA to local communities and the assessment of the value of these services. Internalizing ecosystem services provided by the PA in the economy is an easy way to increase dramatically the economic value of PAs. Though the value of ecosystem services is still far from being correlated in the good state of the ecosystem, due to methodological problems of valuation (David et al. 2007, 2010), there is a real risk that the expected total economical value of ecosystem services becomes the major parameter in choosing whether or not to fund or to continue to fund PA projects instead of their capacity to preserve habitat and endangered species. This risk is very high in poor countries where the biodiversity convention is pushing a strong demand to create PAs. In rich countries, the PAs are less vulnerable to political and economic risks that could jeopardize a part of their funding, but the risk is not zero. To avoid any possible shift in the future, it is important to reaffirm that the primary goal of any PA must be the sustainable conservation of biodiversity and as such a PA is not created to be "profitable".

13.5 The Steering and Control Module and the Social Acceptance at the Heart of PA Governance

The steering and control module is the central element of the management structure of any PA. For the internal flow of information to the PA, it plays the same role as the heart plays in blood circulation in a human body. But its role goes beyond this, because it must also organize and manage the information flow emitted by the PA to its environment and those it receives from the latter in order to decide and act. However, notwithstanding this central role, governance cannot be reduced to a single, three-fold "management information/decision/action" approach. PA governance should also include laws, court judgments, police measures, legal and administrative procedures which drive PA functioning or which are likely to be applied to its users as well as to the management structure (Fig. 13.5).

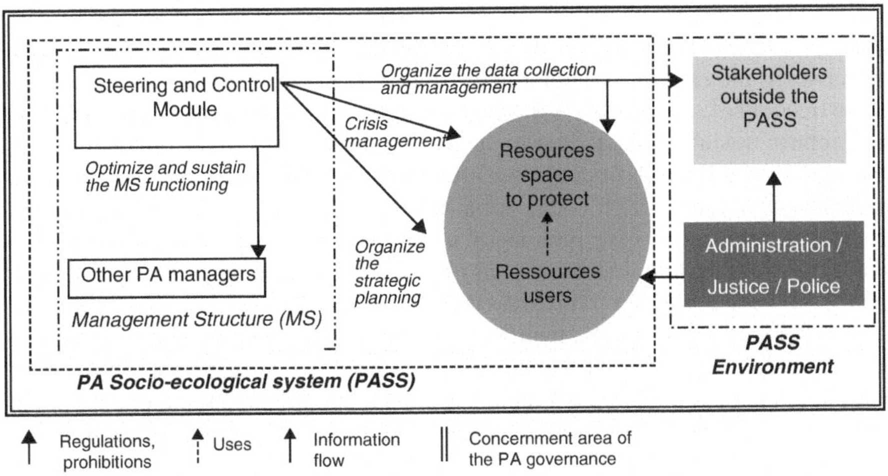

Fig. 13.5 The steering and control module in the governance of protected areas

For any PA, the steering and control module has four functions:

- To organize the collection and the management of information flow driving the decision for action;
- To develop the PA's strategic planning. This means the design and implementation of the management plan and its complementary actions.
- To optimize and sustain the operation of the management structure, particularly to organize every day actions of the latter;
- To manage the crises of the AP system, affecting the habitats and resources to be protected, or the stakeholders (Fig. 13.5).
- Structuring and maintaining information flow within the management structure constitutes the first function of the steering and control module. It is made up of two tasks:
- observing the status of different natural and anthropogenic elements of the PA system, objects and stakeholders of environmental management, as well as the dynamics that connect them and those that are driven by the environment of the system;
- following up the effect of decisions and actions taken by the management structure on the objects and stakeholders of the environmental management.

Here, we are at the heart of the PA steering and control module. Without this follow-up which is equivalent to a feedback loop in terms of information flow, it is impossible to efficiently redesign the decisions and actions for a better response to in-the-field constraints. Indeed, the daily operation of any PA management structure could be subject to many hazards coming from the PA system itself or from its external environment. Under penalty of losing its effectiveness and taking the risk of seeing the PA swept away by dysfunctions, the control module must be able to adapt and provide the appropriate answers, in accordance with Ashby's Law of Requisite Variety (1958). This law states that any regulation of an open system can only be effective on the express condition that the system control module has at least the same freedom of action as the disruptive environment. In other words, if the steering and control module wants to impose its willingness on the system, the number and the diversity of controls it could use should be at least equal to the number and diversity of hazards which could occur from the system's environment. It is not clear if all PA managers are sufficiently aware of the stakes of such follow-up. The description of the actions as listed in any activity report cannot be a substitute. This kind of document is designed to assess the conformity of actions undertaken with the strategic objectives of the PA, including its management plan. In a more secondary way, it also allows for inter-PA comparisons. However, even if the actions taken are consistent with the management plan, they may be ineffective or less effective than expected, if the context of their implementation has evolved from the prescription of such plan.

The social acceptance and acceptability are concepts which are not common in the world of PA managers. Yet in 1997, Gilmore (1997) put the social acceptance forward as one of the three pillars of durability of any protected area, but this proposal has not percolated in the scientific community. In France, it took until 2009 for first seminar to be held the on this subject (Laslaz et al. 2014). The topic was the social acceptance of protected areas and not their acceptability. Can these two terms be confused in the case of PAs? The answer is resolutely negative. Acceptability deals only with the PA steering and control module, while acceptance deals with the PA users and stakeholders. It corresponds to the effective support of PA users to information flow that the steering and control module emits in the form of decisions and actions in their direction. When it makes those decisions and reacts, any PA manager hopes that these measures will be socially accepted, meaning that they will generate adherence by the target public. Experience shows that this is rarely the case. In order to reduce the risks of failure, we should estimate the acceptability of these measures. That is to say, the potential adherence that they will generate among the target public, which is equivalent to estimating the probability of adherence to the decisions being made by the PA managers. By taking the terms in Fig. 13.3, it is, therefore, the estimation of the existing 'distance' between hoped management and resulting management, which is directly conditioned by its social acceptance. There are no accurate tools to estimate these distances or the probability of adherence. The estimation is generally made in an empirical way according to knowledge that the PA managers have about these target stakeholders or the representations they make. Table 13.1 formalizes this knowledge.

Taking the form of a matrix, it is based on a simple principle: the degree of potential public support to a decision/action from PA managers depends on the nature and the number of ties that the public has previously built with the space or resources targeted by the decision or action coming from the PA steering and control module. This type of matrix can drive a second type (Table 13.2) which allows the PA steering and control module to base its decisions according to an expected cost/benefits ratio and then to focus its efforts of communication and awareness on

Table 13.1 Matrix for assessing the social acceptability of the actions taken by protected areas managers (should be filled according to managers empirical knowledge)

Items involved in the decision	Nature of the relationship with the audience targeted by PA management			
	Economics[a]	Policy/law[a]	Identity/society[a]	Total
Resources			2	
Uses				
Population				
Material culture				
Territory/place				
Total				

[a]0: no link 1: normal link 2: very strong link

Table 13.2 The cost/benefit analysis and decision making in the control module of protected areas

Decision/action			Costs/benefits[a]	Response of the steering and control module justification
Urgency	Stakes	Acceptability		
Wlow	*Low*	*low*	*High*	*Abort action*
High	*Low*	*low*	*High*	*Abort action because the weakness of stakes is higher than the emergency action*
low	*High*	*low*	*High*	*Wait for action and put the emphasis on* increasing the social acceptability
High	*High*	*low*	Medium	*Immediate and risky action, measures for increasing social acceptability should be included*
low	*Lows*	*High*	*low*	*Decision aims to grow the acceptability of future decisions and actions*
High	*Lows*	*High*	*low*	
low	*High*	*High*	*low*	*Actions without risk but low priority*
High	*High*	*High*	*low*	*Immediate action without risk*

[a]Costs/benefits = (1/Acceptability)/(Urgency + Stakes)

the target PA stakeholders, whose social acceptance needs to be improved. Indeed, any decision or action performed by the steering and control module towards a target public is determined by two major criteria: the degree of emergency of the decision and the stakes associated with this decision. Two kinds of stakes can be distinguished: the benefits expected if the decision is made and the potential damage that could affect the PA system if the decision is not taken. It is in the light of these two criteria that we should consider the social acceptability and the costs/benefits ratio of the decision (Table 13.2)

When the stakes and the urgency are low, making or not making a decision will be driven by the social acceptability. When this acceptability is low, the steering and control module takes a significant risk in arousing negative feelings from users. The cost of a decision is then far higher than the expected benefits. Conversely, when acceptability is high, the decision aims more at satisfying the target public and thus strengthening the image of the PA in the hope to increase the acceptability of future decisions/actions than truly improving the functioning of the PA system (Table 13.2). When the decision is urgent but the stakes are low, the steering and control module is faced with a similar choice. In all cases, the low importance of stakes takes precedence over the urgency if the acceptability of a decision is low

However, when the stakes are high and social acceptability of the decision/action is low, the steering and control module faces a difficult choice. The most uncomfortable situation corresponds to a low urgency. It is then possible to defer the decision/action until the acceptability 'naturally' improves. When the decision needs to be taken urgently, it should not be postponed. The steering and control module must assess the threats that the low social acceptability of its decision causes on the PA system.

These threats are driven by three factors:

- The willingness of targeted stakeholders to be heard by the PA steering and control module and then to hinder the functioning of the PA by dramatic actions, or even violence. So, in high Islands, roads are usually structured according to the same model; a circular road that runs along the coast and a few crosscutting roads linking the sides East and West. Blocking this circular road is a classic way of putting pressure on public authorities;
- The vulnerability of the PA system to these actions. Let' us again use the example of a coastal road blocked by informal fishers who refuse a PA decision to ban net fishing. If the deadlock persists, transport and goods traffic could be seriously affected, causing damages to island's economy. The public authorities will, therefore, urge PA managers to negotiate with leaders of the demonstrations;
- The understanding of public opinion towards these actions and its empathy towards responsible stakeholders. If the island's population supports the claims of informal fishers, public authorities will be more inclined to invite the PA managers to discuss the issues and make compromises, than if these fishers are perceived by the public opinion as a corporatist group which blocks the island's economy. In this case, the police force will intervene to remove the roadblocks set up by activists and the PA managers will not change their position.

Ultimately, any PA has the same model. The degree of urgency in the decision, the stakes attached to it and the social acceptability of the decision made, act like feedback within the steering and control module which could modify the flow of information. This loop changes the control module's output flow. Tis flow of information can change in terms of quantity, which generally means delaying a decision or action. They also can change in terms of nature, which generally means adding to the decision a new flow of information whose purpose is to facilitate its social acceptance.

13.6 Conclusion

With the growing anthropization of our planet, the need to preserve the most remarkable habitats and the diversity of fauna and flora they support has been growing for several decades. The adoption of the Convention on Biological Diversity (CBD) at the Rio de Janeiro Earth Summit in 1992 was a clear acknowledgment of this need. It acknowledged a protected area as a tool for biodiversity conservation. Eighteen years later, the 10th conference of the CBD parties has set quantitative objectives. Five promising strategic goals were adopted. One of them covers the improvement of biodiversity by safeguarding ecosystems, species and genetic and fixed diversity with the ambition that. "*by 2020, at least 17 per cent of terrestrial and inland water, and 10 percent of coastal and marine areas, especially areas of particular importance for biodiversity and ecosystem services, are conserved through effectively and restructure managed, ecologically representative and well-connected systems of protected areas and other effective area-based conservation measures, and integrated into the wider landscapes and seascapes*". This 10th COP of the CBD raised, thus, the PA to the level of a major tool in humankind/nature coviability for land as well as the sea. But this tool is evolving. The spectacular increase in the number of PAs these past twenty years has been accompanied by a diversification of their objectives, transforming the PA into a more and more complex system but without actually matching its internal organization and operating rules with these new functionalities. A major and new risk is rising: the original purpose of any protected area (the preservation of biodiversity) is fading to other purposes, often of an economic nature and causing a loss in effectively conserving habitats and the species inhabiting them. This new context results in governance becoming a key factor in the future of PAs and, indirectly, that of humankind/nature coviability; effective governance constitutes a necessary condition so that a PA remains a powerful tool for this coviability (Fig. 13.6). PA "good governance" is all the more necessary because financial shortages and/or the low social acceptability can impact the functioning of many PAs across the world. However, the very polysemic character of the term governance leads to just as many interpretations of what "good governance" is.

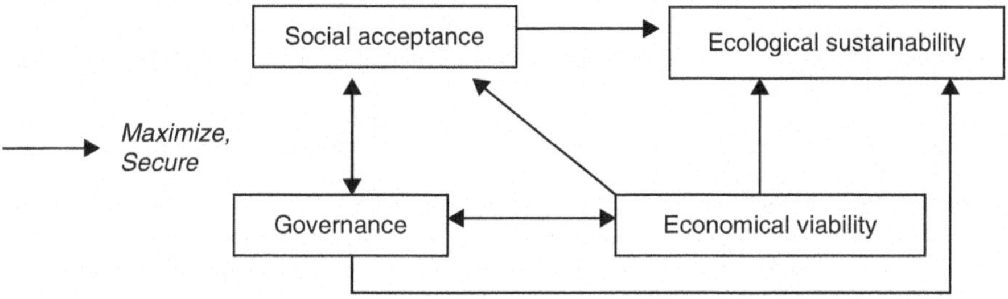

Fig. 13.6 Governance as a tool sustaining the structure and functioning of protected areas

In this chapter, it has been demonstrated that the governance is not only a decision-making process in terms of regulating the numerous uses of nature, it is also a system. A governance system is a subsystem of a PA system. In this context, four types of governance can be identified: a "purpose" governance related to the objectives of the PA; a 'PA control and monitoring tool' governance, which aims to provide managers with information on the performance of their actions; a "management tool" governance, which governs the relationships between users and resources to be protected, and an "accompanying governance of measures taken by PA managers" that focuses mainly on usage rights and all rules constituting the legal environment of the protected area system. On this basis, a distinction can be made between the management of the PA which is limited to the legal and administrative entity of the PA, and the governance of the PA that encompasses the whole territory of the PA's stakeholders and includes this management.

Considering the PA as a system interacting with its biological, socio-economic and institutional environment, results in new perspectives, particularly the central role played by the steering and control module within PA management (Fig. 13.5). Social acceptance is a three-dimensional concept that integrates at the same time: (a) the potential support of the target public to the decisions/actions issued by the PA steering and control module, (b) the anticipated sustainability of this support and (c) the degree of the public's empathy towards this targeted public. Increasing this acceptance/acceptability should be a key factor of PA governance because it largely determines the success of decisions and actions taken by PA managers for the PA system users and stakeholders concerned by its environment.

References

Agardy T, Bridgewater P, Crosby MP, Day J, Dayton PK, Kenchington R, Laffoley D, McConney P, Murray PA, Parks JE, Peau L (2003) Dangerous targets? Unresolved issues and ideological clashed around marine protected areas. Aquat Conserv Mar Freshwat Ecosyst 13(4):353–367

Ashby WR (1958) Introduction à la cybernétique. Dunod, Paris, 235 p

Baron C (2003) La gouvernance: débats autour d'un concept polysémique. Droit et Société 2(54):329–349

Barrière O (2005) Une gouvernance environnementale dans une perspective patrimoniale: approche d'une écologie foncière, in Christoph Eberhard (dir). Droit, gouvernance et développement durable, Cahiers d'Anthropologie du Droit, Paris, LAJP- Karthala, pp 73–98

Bonnemaison J (1980) Espace géographique et identité culturelle en Vanuatu. Journal de la Société des Océanistes 68:181–188

Bonnemaison J (1981) Voyage autour du territoire. L'Espace Géographique 4:249–262

Bonnemaison J (1986) "L'arbre et la pirogue", Les fondements d'une identité: territoire, histoire et société dans l'archipel de Vanuatu (Mélanésie). Coll. Travaux et Documents n° 201, vol 1. Orstom, Paris, 540 p

Bonnemaison J (1987) "Tanna, les hommes lieux", Les fondements d'une identité: Territoire, histoire et société dans l'archipel du Vanuatu (Mélanésie). Coll. "Travaux et Documents", n° 201, vol 2. Orstom, Paris, 680 p

Bourdieu P (1980) Questions de sociologie. Ed. Minuit, Paris, 268 p

Buxton CD, Hartmann K, Kearney R, Gardner C (2014) When is spillover from marine reserves likely to benefit fisheries? PLoS One 9(9):e107032. https://doi.org/10.1371/ journal.pone.0107032

Catanzano J, Thébaud O (1995) Le littoral, pour une approche de la régulation des conflits d'usage. Institut océanographique/Ifremer, Paris, 149 p

Cazalet B (2004) Les aires marines protégées à l'épreuve du sous-développement en Afrique de l'Ouest", *VertigO – la revue électronique en sciences de l'environnement* [En ligne], Volume 5 Numéro 3 | décembre 2004. [En ligne], mis en ligne le 01 décembre 2004. URL: http:// vertigo.revues.org/3274. Consulté le 14 février 2011

Chaboud C, Galletti F, David G, Ferraris J, Brenier A, Méral P, Andriamahefazafy F (2008) Aires marines protégées et gouvernance: contribution des disciplines et évolution disciplinaire. In: Aubertin C, Rodary E (eds) Aires protégées et développement durable. IRD, Marseille, chap 2, pp 55–81

Chia E, Mathé S, Rey Valett H, Michel L, Soulard L et al (2010) Comment étudier (analyser) la gouvernance territoriale? Mise à l'épreuve d'une grille de lecture. Colloque AISR -ASRDLF 2010. Identité, Qualité et Compétitivité Territoriale. Développement économique et cohésion dans les Territoires alpins, Sep 2010, Aoste, Italy. 23 p, <hal-00621169>

Commoner B (1969) Quelle terre laisserons nous à nos enfants? Paris, le Seuil, 206 p

David G (1998) Les aires protégées, laboratoires de la gestion intégrée des zones côtières: l'exemple des pays membres de la Commission de l'Océan Indien. In II rencontre Dynamiques sociales et environnement, Bordeaux 9–11 septembre 1998. Bordeaux, UMR-Regards CNRS-Orstom, t.2, pp 343–360

David G (2011). Petit voyage autour de la notion de gouvernance des AMP. Brest, PAMPA, WP4, rapport final n°1, 56 p

David G, Thomassin A (2007) *Les aires marines protégées, petit voyage autour d'un territoire en mal de reconnaissance.* 1er Colloque national Aires Marines protégées, Boulogne, 20–22 Novembre 2007, 14 p

David G, Herrenschmidt JB, Mirault E, Thomassin A (2007) Valeur sociale et économique des récifs coralliens du Pacifique Insulaire – Eléments méthodologiques. Nouméa, CRISP/IRD, 47 p

David G, Cazalet B, Barnay AS, Couture E, Ferraris J, Fleury PG, Malterre P, Pascal N, Pelletier D, Salaun P, Tessier E, Thomassin A (2010) Participatory approach to identify governance indicators for integrated coastal zone management, the case of marine protected areas. Nantes 20-24 september 2010, ICES Science Annual Conference. http://www.ices.dk/ sites/pub/CM%20Doccuments/CM-2010/B/B1510.pdf

Denis J, Hénocque Y, Antona M, Barbière J, Barusseau P, Brigand L, David G, Grignon-Logerot C, Kalaora B, Lointier M (2001) A methodological guide: steps and tools towards integrated coastal area management. UNESCO, Paris. IOC Manuals and Guides n°42, 42 p

Di Méo, G (1998a) Géographie sociale et territoire. Paris, Nathan, coll. Fac Géographie, 250 p

Di Méo G (1998b) De l'espace aux territoires: éléments pour une archéologie des concepts fondamentaux de la géographie. L'information géographique 62(3):99–110

Di Minin E, Toivonen T (2015) Global protected area expansion: creating more than paper parks. Bioscience:2 p. https://doi.org/10.1093/biosci/biv064

Dubos R, Ward B (1972) Nous n'avons qu'une terre. Paris, Ed Denoël, 360 p

Dudley N (ed) (2008) Guidelines for applying protected area management categories. Gland, IUCN, 106 p

Dudley N, Stolton S (eds) (2008) Defining proted area: an international conference in Almeria Spain. Gland, IUCN, 220 p

Dumont R, Rosier B (1966) Nous allons à la famine. Le Seuil, Paris, 279 p

Fiske SJ (1992) Sociocultural aspects of establishing marine protected areas. Ocean Coast Manag 18:25–46

Gilmore DW (1997) Ecosystem management – a needs driven, resource-use philosophy. For Chron 73:560–564

Kelleher G, Rochia C (1998) Lessons from marine protected areas around the world. Parks 8(2): 1–4

Laslaz L, Gauchon C, Duval M, Héritier S (2014) Les Espaces protégés, entre conflits et acceptation. Belin, Paris

Laurec A, Le Guen JC (1981) Dynamique des populations marines exploitées: concepts et modèles, vol 15. CNEXO, Rapports scientifiques et techniques, Brest, 118 p

Le Gallou F (1992) Activité des systèmes. In: le Gallou F, Bouchon-Meunier B (eds) Systémique Théorie et applications. TEC & DOC – Lavoisier, chap. 4, Paris, pp 70–90

Le Moigne JY (1990) La Modélisation des systèmes complexes. Dunod, Paris, 178 p

Pomeroy RS, Parks JE, Watson LM (2004) How is your MPA doing? A guidebook of natural and social indicators for evaluating marine protected areas management effectiveness. IUCN, Gland/Cambridge, 230 p

Rey-Valette H (2008) Gouvernance du littoral, revue des pratiques, axes et questions de recherche. MEDADDT, CS du programme LITEAU, Paris, 5 p

Roberts CM (2000) Selecting marine reserve locations: optimality versus opportunism. Bull Mar Sci 66(3):581–592

Rodary E, Milan J (2008) Extension et diversification des aires protégées: rupture ou continuité? In: Aubertin C, et Rodary E (eds), Aires protégées, espaces durables? chap 1, pp 32–53

Soja E (1971) The political organization of space. Ann Assoc Am Geogr LX:1–54

Thomassin A, David G (2008) Caractérisation socio-économique de l'état initial de la réserve naturelle marine de la Réunion. St Clotilde, la Réunion, IRD, 165 p

Chapter 14
Social-ecological Coviability of the Protected Marine Areas in Brazil: Contradictions in the Co-management of Protected Marine Areas of Brazil to Policies for the Coviability of Social and Ecological Systems

Catherine Prost

Contents

C. Prost (✉)
The Federal University of Bahia, Salvador, Bahia, Brazil
e-mail: cathprost@yahoo.com

O. Barrière et al. (eds.), *Coviability of Social and Ecological Systems: Reconnecting Mankind to the Biosphere in an Era of Global Change*,
https://doi.org/10.1007/978-3-319-78497-7_14

14.1 Introduction

From the early 1990s, in the midst of sustainable development, Brazil innovated with the creation of specific protected areas called extractive reserves in reference to the method of reasoned extraction of natural resources of either plant or animal predominance. This protected area status is distinguished by its sustainable use, maintaining local populations in their traditional cultures, as well as by the deliberative nature of the management council in which these populations are the majority. However, the growing institutionalization of environmental law is working to limit the power of local decision-making at the expense of the territorialities of local social groups and their mode of coviability with nature. As part of ongoing protected areas trials, the analysis of two marine extractive reserves of the Bahia State sets necessary factors to promote a real power of planning and management of their territories by traditional peoples and thus support the practices of sustainable use of natural resources.

With the growth of the environmental impacts of human activities, especially since the last quarter of the twentieth century, the ecological movement has launched alerts on the real and potential problems of certain activities and Earth Summits have multiplied. The creation of protected areas appears as one of the measures to ensure environmental protection. In Brazil, there are protected integral and permanent preservation areas and other areas of sustainable use, maintaining local populations in their midst in view of coviability.

Here, such a notion is\enlargethispage*{-14pt} understood through Bourgine (1996), who states that "the autonomy of a system is nothing other than its ability to maintain its viability in varied and changing environments through its self-organizing process. In other words, the test of autonomy is viability". But viability must also seek coviability since the system is connected to other systems.

Coviability between society and nature is thereby fundamental for the future of humanity because of the strict dependency of the first on the second to live and produce and thus ensure the basis of its viability. The Earth Summits show the concern of States and the pressure of societies as to the accelerated growth of the environmental problems in the context of globalization.

Nature, for its part, has been so deeply transformed by man over the centuries, that in the nineteenth century, Marx (1985) spoke of "second nature", a process significantly intensified since then. It becomes difficult to identify strictly natural processes at a time when human activities influence even the global climate. These two sets, society and nature, are intrinsically linked and it is not possible to think of the viability of one without the other.

The current environmental crisis, caused by the current process of growth, clearly shows its limits from the social point of view, excluding a growing number of individuals from the positive benefits of growth. Switzerland's Credit report reveals that in 2015, less than 1% of the richest people in the planet hold as much wealth as the rest of humanity. How did we get to here? As noted by Porto-Gonçalves (2006), the society – nature relationship has undergone great changes through the revolution of science and technology. Currently, these are presented by many as present or future solutions to environmental problems, whereas they are in fact part of them. But in addition, the author also emphasizes that the domination of nature through the technological revolution implies the domination (of some groups) of men over others. Also, the domination of nature is impregnated with power relations within a society and between different societies, a process aggravated in a global capitalist context.

The extractive reserves aim at ensuring environmental protection and reproduction of social groups. However, their mission takes place in a context of a capitalist market economy and, more importantly, of very recent and fragile democracy, so a context in which traditional peoples are historically marginalized. From the experiences of governance in progress, extractive reserves are providing elements for thinking about coviability, but they are not devoid of contradictions, as it will be analyzed.

14.2 The Extractive Reserves, a Brazilian Environmental Legislation Breakthrough

The extractive reserves belong to the category of conservation units (CU) for sustainable use, as opposed to those of permanent and integral conservation, as stated in the National System of Conservation Units (SNUC, law 9985/00), adopted in 2000.

14.2.1 A Protected Area Which Gives a Voice to Traditional Peoples

Like the other sustainable use CU, an extractive reserve is managed by a council, but it differentiates itself from the majority of the protected areas of sustainable use[1] by its deliberative nature and by the majority among which the targeted traditional populations[2] have. These two measures show acceptance in involving local people in such a way that the voice of "silent and invisible" individuals (finally) acquire resonance in decisions concerning them.

Functions of planning and management are assigned in the Deliberative Council, particularly: (i) the definition of the peoples carryout extractive activities, who will benefit from public policies, (ii) the development and implementation of a natural resources usage plan based on traditional environmental knowledge and (iii) the

[1]Excluding extractive reserves and sustainable development reserves, all protected areas of sustainable use are managed by management councils that are barely advisory.

[2]They have 50% of seats plus one in the composition of the Council.

development of a natural resources management plan, developed in the partnership of the scientific community.

This model of protected areas is the result of a social conquest. Indeed, in the 1970s, after a century of economic exploitation and political marginalization, the *seringueiros* – small rubber tree (*hevea*) tappers scattered throughout the forest – initiated a social movement with the help of the Catholic Church and basic ecclesiastical communities (CEBs[3]). They invested themselves in unions of rural workers, but their territorial identity substantially consolidated through the *embates*[4] strategy, facing the henchmen of large farmers and the aggressive intervention of the military police. In parallel, to better convince public opinion, the *seringueiros* declared themselves, along with the Amerindians, "people of the forest", the guardians of natural resources, the global discourse of which boasts how important this is.

The National Council of *Seringueiros,* created in line with the first national meeting in 1985, – the year where the military hand over the power to civilians, first demanded the guarantee of traditional usufruct of the land and environmental protection. In 1987, the Government accepted the request of extractive reserves of the social movement, supported by the international pressure against deforestation, intensified after the assassination of the historical leader Chico Mendes[5] and in 1990, President Sarney created four large extractive reserves.[6] With the institutionalization,[7] the reproducibility of the process is facilitated, what is very appreciated by the social movement.

Thus, the *seringueiros* received official recognition of their importance within national society for their role in nature conservation carried out through their traditional practices. Because of the great diversity of traditional peoples in the country, the decade of the 1990s, witnessed the creation of reserves in other environmental areas other than the Amazon forest, such as extractive reserves of babaçu coconut, in Maranhão State, and marine reserves.

14.2.2 Coviability Based on Traditional Environmental Knowledge

Here, the term tradition is understood as a process marked by a predominance of the preservation of values and know-how, with, however, the inclusion of technical

[3]Including religious and non-religious persons

[4]These are the mobilizations of peaceful resistance of a group - including women, children and elderly -in order to prevent, by the mere presence, the deforestation by wood operating companies.

[5]Chico Mendes (1944–1988) was the main leader of the seringueiro movement in the Acre.

[6]Cf. Decree 98.897/90.

[7]Institutionalization is defined as the set of rules and standards for the satisfaction of collective interests and, as a dialectical movement which, by guaranteeing the routinization of procedures allowing reproducibility and application, appears as the materialization of more general social forms and, of its contradictions, allowing movements in its interior (Cunha 2010, p. 73).

innovations and the evolution of values and social practices from dominant society, but at a more gradual pace, without imposing breaks.

In this perspective, the populations are considered traditional by their close relationship with nature, from which their environmental knowledge and all or most of their subsistence and/or their incomes have been derived.

These peoples have an in-depth knowledge of nature and the multiple relationships between its components thanks to the observation of natural processes and research relayed by intergenerational oral transmission. The richness and the complexity of social relations have also contributed to this knowledge (Brandão 1994; Castro 1997). Thus, the relationship with nature is not based on a utilitarian view of the exploitation of natural resources. It assumes above all an attitude of complicity and respect built from direct physical experiences, without a scientific approach. The representation of nature frequently joins the idea of unity with society, in contrast to the modern dichotomous view. Seen as a heritage or a gift from God (or gods), nature is full of symbolism, ranging from founding Cosmo-visions of a social or ethnical group to the toponymy of the people. In this sense, it deserves particular attention and respect (Boff 2000), built according to the techniques and values of each local traditional population.

In small-scale fishing, tradition and respect observed in the relationship with nature are subtly expressed by a variety of fishing systems, in which various equipment (and boats) are combined to capture multiple species, in different places and during specific seasons. We may note that the less capital a fisherman has, the more he tends to diversify fishing tools, more numerous but more artisanal. Therefore, diversified fishing practices and catches[8] avoid intensive extraction of a resource for a perfect strategy of sustainable management of natural resources. Thus, it is more likely to respect the rhythms of reproduction and/or migration of targeted species and better manage efforts on fish stocks, in a mechanism that does not maximize profits, but guarantees some resilience in the case of rarefaction or extinction of a specific resource. This mode of multi-diversified operation is accentuated when the fishing communities carry out several activities at the same time, such as family farming, plant collect, or even "small jobs".[9] In fact, many communities specialize in fishing due to land expropriation procedures, as shown herein (see Sect. 14.3.2).

The diversity of activities of artisanal fishermen contributes to their resilience. Multi-activity brings flexibility that helps overcome major changes, such as the implantation of a dam upstream to Iguape Bay, completed in 1985. With this infrastructure, the dynamics of water crossing the Bay was changed, with a higher

[8]Fishermen who have more capital can acquire a boat that allows them to browse further for longer, thus following schools of fish of high commercial value; and therefore, exercising more specialized fishing. Fishermen with less capital use other practices, such as gillnets – that can be fixed to the ground and trap fish-, the collection of shellfish, net fishing on the edge of rivers or mangrove areas, etc. The practices are chosen depending on the seasonality of some catches, climatic conditions, access or no access to a boat. The fisherman adapts to changing conditions throughout the year.

[9]They can include interim jobs in building, mechanics, informal trade, etc.

salinity and therefore important alterations in wildlife and fishing. Once a new ecological equilibrium was established, certain types of fishing were abandoned in favor of others. These difficulties have been circumvented through the level of environmental knowledge and fishing alternatives. This will not exempt fishing communities from claiming support from society to continue exercising, through their socio-cultural practices, environmental services.[10]

It is thus established that traditional practices of social use of natural resources, have generally low negative environmental impacts thanks to pretty artisanal means of production and sustainable management techniques of resources.

Even if it is accepted that some traditional populations tend to increase their pressure on the natural resources beyond their regeneration capacity, it is often because of the pressures of the market economy materialized by growing demands for products derived from fishing, the reduction of access to the commons areas, and the introduction of urban values. These give rise to new needs, which means major income and a greater exploitation of nature. However, public policies can avoid practices with high environmental impacts by basing themselves on existing environmental knowledge and maintaining sustainable exploitation of nature, following the rhythms and changes over time. It is the role of protected areas such as extractive reserves, whose usage plan must be submitted to a five-year review in order to reconsider the environmental changes as well as social and economic changes, expressed both by the adoption of techniques and accepted values.

In other words, regulation of the territories must be guaranteed in order to avoid ecological depredation. Two types of regulation are defended here: local communitarian and governmental. In the specific case of extractive reserves, the traditional regulation of resources justifies their creation. Nonetheless, historically, under an extremely confrontational context from a land point of view, the support of the state – or other local authority – is fundamental in enforcing rules of local use of natural resources, as well as in ensuring the territory of extractive populations, place of production, lifestyle, and culture.

14.2.3 Creation of Brazilian Marine Resex

In the 1990s, two marine extractive reserves were created in southern and southeastern regions of Brazil.[11] The process is thriving in the next decade in the North and Northeast regions, where artisanal fishermen predominate.[12]

---

[10]By temporarily interrupting fishing during the reproduction of species, fishermen contribute to the conservation of fish stocks through the preservation of female individuals.

[11]The Pirajubae extractive reserve was established in the State of Santa Catarina in 1992 and that of Arraial do Cabo in the State of Rio de Janeiro in 1997.

[12]Over 80% according to the Census of 2000 (IBGE 2000) or 77% according to the General Cadastre of fisheries of 2008 (MPA, 2003 apud Alencar and Maia 2011).

In the Bahia state, the first marine extractive reserve was founded in the Iguape Bay in 2000, that of Canavieiras and Corumbau in 2006 and that of Caravelas in 2009 which closed a phase of creation, with a total of 22 Brazilian marine extractive reserves in a decade.[13]

With the institutionalization of the protected areas, it would seem that Brazilian environmental policy is aware of the wealth and importance of its abundant human diversity, in social, ethnic and cultural terms, called socio-diversity by Neves (1995). The author claims that it reveals biodiversity and contributes, through the management of the commons, to the conservation of the latter.

Nevertheless, a more in-depth look shows that the challenge of making the extractive reserve an instrument for benefiting environmentalism for the poor (Alier 2007) remains to be launched. The author thus highlights a third ecological trend, as opposed to the two other better known ones. Indeed, some of the environmental slogans preach the absolute protection of a "wilderness", considering the human being as a predator of the nature. A second trend advocates sustainable development, what Alier (2007) called "Gospel of eco-efficiency", i.e. a belief in solutions made by science and technology. With environmentalism of the poor, the author denies the idea that ecology would be a luxury to be achieved only after reaching a certain degree of socio-economic development. The right for a healthy environment should be included in human rights in order to guarantee all the citizens of a country, favorable conditions of life.

14.3 An Analysis Resulting from The Social Movement

The effective application of laws (and the spirit of laws) depends on the context in which they have been designed and adopted. However, the National system of Conservation units was created under the influence of sustainable development, which significantly limits the possibility of real human and environmental coviability because of a vision of development too close to the economic growth. Here, it is understood that this problem stems from the same concept of development, proper to the West, where the industrial revolution was born. Although social sciences debated on development, defining it as a social welfare, it is understood as an increase in the level of comfort of a population expressed by a continuous and increasing consumption of goods. Economic growth is therefore introduced as a condition to achieve social well-being and presuppose a greater domination of nature.

Some examples of the limits of sustainable development explain the effects on extractive reserves.

[13]Three new marine extractive reserves were created in the State of Pará by President Roussef during the campaign of the presidential election in 2014.

14.3.1 Sustainable Development: Ecology and Participation of Civil Society in a Neoliberal Context

The Bruntland report, written in 1988 in preparation for the Rio de Janeiro Earth Summit in 1992, introduced the notion of sustainable development, commonly known by the general public as the fact of promoting today's development without prejudice to future generations.

However, despite the efforts of scientists to define a real concept of sustainable development, particularly by detailing the different dimensions to consider (cf. Cavalcanti 2012; Leff 2004, 2006), the Governments – and *a fortiori* the firms – take action in the name of sustainable development, quite insufficient as to the environmental and social dimensions.

A striking example is given by the suggested solution to reduce pollution caused by vehicles. Huge areas of land up until now intended for agricultural production to feed populations are henceforth exploited for the cultivation of bio-fuels to supply cars. In addition, the main species used are grown in monocultures cropping system at the expense of high biodiversity in tropical countries. Finally, in Brazil, with an average productivity of more than 10 tons/man/day, farm workers accumulate, after a few years of working, chronic pathologies in relation to intensive cultivation practices. Therefore, it has been established that bio-fuels demonstrate the perpetuation of the belief in infinite growth in a finite physical world thanks to science and technology.

In 1992, the second Earth Summit was organized in a neoliberalism context, introduced as the only possible solution by major international creditors, which imposed structural adjustment plans in all countries in need of financial liquidity. In this context, the successive Earth summits are struggling to renovate the discourse of capitalism since the choices of companies and states contribute to promote over and over again the market economy. The example of major agreements managed in international summits, such as carbon credits, or CFC gases,[14] also reflect economic interests always present and preferred.

However, in a world marked by information broadcasting, it becomes essential to legitimize the dictates of the market, whence the emphasis on the (so-called) ecological variable, often accompanied in local development projects by a so-called participatory methodology, as a form of adjustment of actions according to the needs of local people.

[14]In addition to the example cited in the introduction on the market of carbon credits (Porto-Gonçalves 2006), the work of Tomasoni (2011) provides information on concealed commercial interests regarding alternatives to the greenhouse gases (CFC – Chloro-Fluro-Carbone).

Certainly, the participation of the population to public action can potentially represent an opportunity for the democratization of decisions thanks to the existence of debate, greater transparency of projects, and the adherence to the rules generated by the population. Nevertheless, it is essential to analyze the nature of participation since the term has become polysemous (Pretty and Pimbert 2000).

Here, it is understood that the minimum level required for effective participation is the possibility to debate in a contradictory way on a proposed project and to make any changes in order to improve according to the common interests of the local population. Participation becomes stronger when the population is associated in the project management process – right from its design phase- and it reaches its maximum degree in self-management.

To ensure full participation, it must result in the individual and collective autonomy of interests, according to Castoriadis' precepts (1982), emphasizing the right to decide is achieved in full consciousness and freedom. Awareness is built through access to all necessary data to obtain full knowledge of the analyzed topics, for example, through consultants paid by the state and chosen by the people to provide assistance in the analysis of complex issues. In a context of historical marginalization of the disadvantaged segments of the Brazilian society, access to information is fundamental in order to advance citizenship by encouraging the various social groups to mobilize themselves for their rights. Without knowing their rights, the latter cannot be exercised, despite this does not mean however that they cease to exist. Consciousness is strengthened by the existence of critical and contradictory debate in which the local population should feel listened to and be reassured about its freedom of opinion, of expression and to taking action in response to other economic or political players, thus guaranteeing real democratic processes (Souza 2002).

Yet, in a context where the ideology of the minimum-State is advocated the participation of society can take a whole new meaning. In fact, the state disengages from one of its main functions: land use planning. So, the participation can be interpreted as the willingness of population to link themselves to local creditors – and to their requirements-, to decrease the action of the state in sectors that are not interesting financially, and encourage collaboration rather than conflict under the framework of "good governance". However, some divergence of interests is not soluble by consensual decisions and lead to relative social resistance that can turn into conflict.

Furthermore, the participatory nature is only applied to projects at essentially a local scale, thus to effects limited in space. To compensate for any public intervention, partnerships with organized civil society are encouraged. However, apart from religious organizations, the projects carried out with the third sector (non-governmental) are more fragmented because they are dependent on funds usually allocated for a few years. The example of the Brazilian extractive reserves is enlightening in this respect.

14.3.2 *The Institutionalization of The Social Conquest at The Expense of The Social Movement*

By incorporating an environmental variable, the movement of hevea rubber tappers (the *seringueiros*) developed links, bringing visibility and international support, made possible by the international character of the process for establishing a new model of ownership of natural resources in the Amazon.

By asserting their extractive reserve project, the *seringueiros* changed the geography of power and law (Porto-Gonçalves 2002), claiming their autonomy, without guardianship or coercion, and joining up with the State in the creation of a new land status "of communitarian usufruct". First of all, it was the *seringueiros* who were responsible for elaborating a usage plan of resources, with the official supervision of the environmental organization limited to referencing decisions and working for their application. But the objectivity of social agents was quickly altered.

The sectorization in the public management of the extractive reserves separates the two major platforms of the movement, namely, land reform and environmentalism. Yet, they are intrinsically linked since the guarantee of a differentiated method of production ensures the social group's own reproduction. The problem arose when the environmental variable took precedence over social issues, or were even split from them, moving away from Chico Mendes' recommendation as to the belief of the movement in "areas where the *seringueiros* more or less assume the reins" (apud Cunha 2010, p. 103).

First, the initial self-management of extractive reserves by the *seringueiros* was replaced by a system of co-management[15] (shared management), perpetuating the State territorial control of local land. Furthermore, the initial recognition of environmental knowledge has to be verified by technical and scientific studies, thus restoring the primacy of scientific-technical knowledge. However, traditional environmental knowledge is richer in terms of resources used and usages than those of modern society,[16] and more integrated into its analysis than scientific knowledge, albeit specialized but fragmented.[17] The importance of these knowledge for modern society is manifested in several forms: the payment of a basic income to fishermen during the months of reproduction of protected species (*salario-defeso*), to the cutting-edge research on natural resources in the domains of bio-technology, pharmaceutical industry, and cosmetics whose advances are accelerated thanks to information drawn from traditional knowledge. The exchanges with scientific and technical knowledge are desirable and desired by the traditional populations, but not in the form of a mandatory task to establish the legitimacy of ancestral knowledge.

[15]The Extractive Reserve Council is chaired by an employee of the federal environmental agency responsible for the management of protected areas, ICMBio.

[16]Porto-Gonçalves (2001) points out how traditional environmental knowledge are crucial for advanced science such as, inter alia, bio-technology, pharmaceutical and cosmetic industry, among others.

[17]Countless scientific references show this, of which barely a few examples will be quoted here: Adams 1994; Castro 1997; Diegues 2000; Souto and Martins 2009; Leff 2004, 2006.

Then, the association of the social movement with NGOs encouraged the transition from protest to a policy of results, with the incentive to set up projects to support local actions. With less militants and more volunteers and professionals, the explosion of NGOs – strongly linked to the propaganda for promoting government efficiency and de-bureaucratization – highlighted the passage of a process aiming at transforming the State into a transformation within the State (Cunha 2010). The original autonomy of the social movement then drifted towards an autonomy limited to specific requests and not to the construction of an alternative global vision.

Semantic changes also express the slide in the treatment of social groups. Social groups initially characterized by the Brazilian Institute of the Environment and of Renewable Natural Resources (IBAMA) by their type of extractive production are now defined by a cultural criterion (traditional population), reinforcing the idea of homogeneity within the extractivists, in contrast to reality[18] (Lobão 2006).

The National Centre of sustainable development and traditional population innovated, seeking to incorporate the vision of these populations in environmental management. Nevertheless, linked to IBAMA, it did not escape from the internal contradictions of the organisation mainly "preservationist" and subjected to various pressures for growth objectives. The Federal Government significantly reduced the action of the Centre by putting an end to the funding of the extractive reserves project[19] at the end of the 1990s.

So, the apparent progress of environmental legislation can also be read as a regression from the point of view of the social movement. The eroded autonomy, with the increasing control by the State and the reduction of the funds allocated, shows a search for restraining social struggles (Cunha 2010).

14.4 In the Field, Despite the Advances Made, Conflicts Are Multiplying

14.4.1 Potential and Actual Legal Regressions on Rights of the Rural Workers and the Environment

Successive governments promote large industrial projects or infrastructure, often at the expense of the local population, without sparing protected territories. The rights of traditional peoples thus face global constraints posed by environmental standards and fishing regulations.

[18]Brazil illustrates well that since the extractive reserves were also created among the babaçu coconut harvesters and fishermen, not to mention many other traditional peoples, as the descendants of maroons (called Bushiningues, people that escaped from slavery and settled in the forest) or Amerindians across the country, or even substantive pasture communities in the Northeast region, the geraizeiros of Minas Gerais States and Bahia, the pantaneiros of Mato Grosso and Mato Grosso do Sul, and so on.

[19]Largely funded through the PPG-7, commending program for Brazilian rainforest protection, whose major creditor was Germany.

Moreover, the evolution of political power relations in the country shows the permanent challenge faced by the populations concerned because of the repeated attempts to reduce the environmental achievements and labor rights.

In the field of the environment, the proposed reform of the Forest Code planned that mangroves should be removed from Areas of Permanent Preservation (APP), in order to facilitate the implementation of institutions of intensive shrimp breeding, having significant environmental impacts (MMA 2004; Machado 2007; Tahim 2009; Oliveira and Souza 2015). This particular measure was not adopted, but the threats of such decisions persist.

In terms of fishing legislation, one type of problem is related to the magnitude of some measures. During the period of reproduction of specific protected species, fishermen receive a type of "environmental wage" (*salario-defeso*) in exchange for the suspension of their activity. However, the fisherwomen often claim that the shellfishes they capture are not on the list of protected species.

Worse still, in 2015, in political turmoil and an economic situation of crisis, the government adopted a series of measures that reduced public support for small-scale fishing. In April 2015, the Government announced the suspension of this compensatory income for that year and Decree 8425 restricted access to this right by "artisanal fishermen" only. This definition was amended the same month by Decree 8424 of 01.04.05, in order to reduce the number of beneficiaries, by eliminating the category of professional fishermen who practice other activities in parallel (such as family farming or plant extractive operations) and the "support workers of small-scale fisheries".[20] Finally, the interim norms (MP 664 and 665 of April 2015) restricted also several rights of rural workers.[21] These measures are likely to cause greater strain on fish stocks in order to compensate the lack of support of sustainable small-scale fishing by public policies.

To these (draft) laws aiming to erode the environmental advances or small-scale fishing, are added examples of conflicts specifically related to the management of extractive reserves, as well as to numerous territorial conflicts, both possibly related.

14.4.2 Management Conflicts

Under the aegis of the National System of Conservation Units (SNUC), the Deliberative Council of the extractive reserve is chaired by The Chico Mendes Institute

[20]This category includes people who treat the catch (ex: cleaning and filleting fish, extraction of the flesh of crustaceans). The activity of these workers is closely associated with fishing and therefore to the seasonality of the latter. However, with Decree 8424, they will no longer receive the *salario-defeso*.

[21]Among the new standards, instead of being able to receive the *salario-defeso* after 6 months of reported activities, the fishermen will have to prove a year and a half in office before being covered.

for the Biodiversity (ICMBio).[22] However, training in participatory management is not a strong element of the continuous training of recruited employees, frequently affecting relationships with local people.

Thus, examples of authoritarian attitudes in the field from some leaders have been observed, with the omission of important information transmission to the Council (Oliveira 2012; Dumith 2012). Nevertheless, these officials are also pressured by their superiors in order to influence some decisions favoring the imperatives of economic growth rather than environmental protection.

Beyond these difficulties, the decision-making power of the extractive reserves is gradually being eroded, as revealed by a careful reading of the laws.

Its art.18, § 2°, the National System of Conservation Units (SNUC) states that "the extractive reserves will be governed by the Deliberative Council". However, Decree 4340/02, which regulates the SNUC, states in art. 20, § VIII that it is up to the Council to "manifest[23] itself on work or activity potentially causing impacts in protected areas, buffer areas, mosaics or ecological corridors". This semantic nuance translates an obvious disparity, with the reduction of the Council to an advisory function.

Not satisfied, the leadership of the Chico Mendes Institute for Biodiversity (ICMBio) enacts normative instructions (NI) that withdraw from Protected Areas Management Council the power to authorize (or not) projects with potential or significant environmental impacts. In 2007, Normative Instruction no 02/07[24] holds that the Council shall preliminarily (only) "consult" traditional peoples on the licensing of products or services potentially generating environmental impacts that cause a financial burden on these people or their organizing bodies. With Normative Instruction 05/09, preliminary approval is limited to the analysis of significant environmental impacts, determined by resolution 237/97 of the National Council of the Environment that defines when an impact becomes significant, a task which in practice is left to the Federal Environmental Body[25] or the federated bodies.[26]

Finally, generally reduced financial, material and human funds of The Chico Mendes Institute for Biodiversity, express lack of State priority in the management of protected areas, in favor of productive sectors, a process exacerbated by the current economic crisis.

[22]The ICMBio is a federal body for the environment protection, detached from the Brazilian Institute of the Environment and Renewable Resources (IBAMA) in 2007 and responsible for the management of protected areas of sustainable use. An employee of the ICMBio responsible for a extractive reserve is called, in a revealing fashion, "Chief".

[23]Underlined by me.

[24]Article 17 § 10.

[25]IBAMA – a Brazilian Institute of the Environment and Renewable Natural Resources.

[26]See: http://www.MMA.gov.br/port/CONAMA/RES/res97/res23797.html

14.4.3 Territorial, Land and Environmental Conflicts

In response to the increase in pressure on the land, the extractive reserves should act as a guarantee against the tragedy of the Commons described by Hardin.[27] This book is highly criticized because the author does not distinguish between lands of free access and lands for common usage. In a capitalist context, unregulated free access to areas can be the object of intense economic exploitation, substantially degrading local ecological conditions. The progress of pioneering fronts in the Amazon teems with many examples in this sense. In the case studied here, there is a specificity which is the aquatic environment, and in particular, the coastal environment. "The sea belongs to everyone" is a common sentence among Brazilian fishermen. Without effective regulations, the tragedy of the commons may be achieved, as shown by studies on the evolution of fishing in the North of the country for example.[28] This regulation can be exercised by customary law and at various levels of governments. In this perspective, the extractive reserves should theoretically be well-protected areas, thanks to the "common use of areas" rules based on traditional environmental knowledge (common usage lands), and to their legal status as Federal Conservation Units. However, many territorial disputes are taking place.

The extractivists of the four reserves in Bahia state denounce tensions or conflicts with large farmers practicing extensive farming around the Corumbau and the Bay of Iguape extractive reserve. In the latter, farmers have converted part of their land into fields of sugar cane, removing the historical activity of vegetal extrativism.[29] However, with the gradual reduction of access to commons, the variety of exploited resources is decreasing and the pressure on fish stocks is increasing, reducing the resilience of the communities. From an ecological point of view, the sugar cane monoculture causes a drastic reduction in biodiversity and in the conversion of forest to pasture areas contributes to erosion, as is observed in the Bay of Iguape.

Even though the mangrove is declared a permanent protected area (APP) in Marine Land, that is to say, under the jurisdiction of the Federal Union, farmers go beyond their rights by installing fences (some are electrified) in this ecosystem. Regrettably several deaths occurred during accidents that took place in the Bay of Iguape (Kuhn 2009). In the Corumbau extractive reserve, farmers installed in an area historically occupied by a fishing village have sent back their inhabitants to 2 km inland and make the access to the beach and the docking of fishing boats difficult.

---

[27]This author published in 1968 an article on the tragedy commons property. REfs = The tragedy of the Commons, Science, December 13, 1968, – or communal – thesis criticized for its lack of distinction between access lands and land of Common use, regulated by customary law, not written but based on traditional environmental knowledge.

[28]See Isaac, 2006; Santos et Isaac 2014.

[29]Mainly palm for the manufacture of oil and straw of the Palm *piaçava* for the manufacture of sweeping brushes.

To understand the proliferation of land issues, it is worth mentioning that Brazil distinguishes itself by the extent of the phenomenon of land usurpation. This reveals an extremely undemocratic national history. Since the beginning of Portuguese colonization, the huge administrative units distributed to the nobles to promote development were later listed as heritage areas. Under the Empire (1822–1889), the adoption of 'Land Status' in 1850 made the land a commodity, not accessible to the freed slaves. This movement of land concentration has been increasing up to today in a recent democratic country, perpetuating the process of primitive accumulation of capital by expelling small rural producers from their land.

Returning to Bahia, the two studied regions were affected by a regional economic downturn. Around the Bay of Iguape, plantations of tobacco and cigar industries employed thousands of people until the early 1990s. In the South of the State – where Canavieiras is situated-, thousands of rural workers contributed to the production of cocoa, whose production dates back to the XVIII century. Freitas (2008) highlights the effort of State intervention to increase the technical density of land for big landowners and capitalist companies, as well as to legitimatize the expropriation and exploitation of rural workers. The sector, which was a major export product,[30] underwent various crises since the XIXth century, particularly due to an illness affecting farmers. During the second half of the twentieth century, international competition strongly contributed to the decline of the activity, causing bankruptcy for many rural properties.

In both cases, rural workers had the choice of migrating to urban areas or of finding an area where they could meet their needs. They chose to settle on the coast, a communal area with free access. Canavieiras is an example, with the establishment of new neighborhoods converted to fishing activities. It is important to emphasize that this process of land concentration explained that some communities focus exclusively on fishing and do not practice multi-activities (as mentioned in the example of the Bay of Iguape).

Currently, traditional people of marine extractive reserves are faced with large-scale industrial or infrastructure projects, either public or private. New vertical rationale is needed to respond to exogenous targets.

Therefore, the extractive reserve of Iguape Bay is taken in middle between a hydro-electric power station upstream, in service since 2004, and a shipyard downstream, which started to establish itself during this decade. In the South of Bahia State (where three of the four marine extractive reserves are situated), companies are exploring for gas and oil in a gigantic underwater reservoir which have been discovered in recent years, situated off the coast between the State of Rio de Janeiro and the South of Bahia. However, mineral extraction has big potential environmental impacts, characterized by a critical contamination index in the mangrove areas due to the inability to extract pollutants covered of silt.

In the South of Bahia, the expansion since the beginning of the century, of industrial shrimp growing in the mangrove and apicum area – whose impacts are

[30]In the 1930s, cocoa was the first export product of Bahia and the third nationally.

in particular described in the report of MangueMar network (Red MangueMar Bahia, available online : http://goo.gl/KEsk4[31]) – was halted by the creation of Canavieiras and Caravelas extractive reserves. However, some conflict persists with the institutions located in the mangroves and communities vicinity.

Finally, thanks to the existence of long and beautiful beaches, and with relatively easy access by road or air, the Canavieiras and Corumbau extractive reserves attract tourists, whose financial contributions are mostly absorbed by outside investors, as occurs in many highly tourist places (see Virgens 2010; Coriolano and Lima 2003; Cruz 2003).

In response to all these challenges, the extractive reserves show management difficulties. However, a positive experience continues in the Canavieiras extractive reserve, in contrast to Iguape Bay despite being the oldest. It sheds light on certain conditions to be met to ensure coviability thanks to a rational and sustainable management of resources, discussed collectively and freely and revised when necessary. This form of governance aims also at consolidating the social group against pressures of economic players and who themselves are frequently supported by the State.

14.5 Lessons from the Bay of Lguape and Canavieiras Extractive Reserves

An in-depth analysis of management processes enables factors that limit or increase the autonomy of traditional peoples to be identified.

14.5.1 Canavieiras: When a Conflict Encourages the Autonomy

In Canavieiras, just like the marine extractive reserves of Pará, a parent association which brings together all basic associations, gathers systematically before each Council meeting in order to discuss the objectives. Thus, a real "locking" of the Council by fishermen is made possible by grouped votes of extractivists, who have the majority.[32]

This social cohesion owes its strength by remembering the conflict which existed when the reserve was created. The initial extractive reserve project, in a strictly marine territory, was expanded to include areas of land thanks to the mobilization of associations of fishermen and the support that was given to them by an environmentalist NGO, Pangea. In response to the project's evolution, various sectors of the local elite mobilized themselves and launched a campaign with ambiguous slogan,

[31]The report is undated, but the collected data are from September 2007.

[32]It is recalled that extractivists hold 50% of the seats, plus one.

"Extractive Reserve No, Nature Yes", defending the establishment of a protected area with a Management Council whose merely advisory role would reduce the limits on economic development projects.[33] During this period (2005–2006), the main social leaders were threatened with death. On the advice of the NGO Pangea, local associations coordinated themselves and successfully sought judiciary support at the regional scale. The parent association of the extractive reserve, the AMEX,[34] retains the lessons of the conflict that recalls how being unified is essential for the social movement.

14.5.2 The Bay of Iguape Fragmented and Trapped in the Middle of Major Industrial Projects

In the Bay of Iguape, the pattern is different. This extractive reserve did not benefit from encouragement by its successive heads, to socially organization of around tens of communities.[35] Certainly, the 'Pastoral Council of Fishermen (CPP), an NGO associated with the Catholic Church, brings technical and political assistance to the movement by fishermen and fisherwomen[36] (MPP). However neither it nor the MPP managed to establish themselves in the all of the communities. Finally, no parent association has been created to promote internal debate. In short, no body thus promotes collective discussion of all fishing communities, despite being a fundamental process in the debate of ideas for communitarian interests.

With this weak base link, investors establish themselves with the support of public authorities, from a municipal scale to a federal one. In order to win the appreciation of the local population, they carry out small social or environmental actions, or even co-opt too rebellious social leaders. The resistance to a hydroelectric power station upstream to the extractive reserve and a shipyard downstream is carried out with the participation of the CPP (Pastoral Council of Fishermen) and a few more engaged scientists, but in a disorganized way.

These major projects directly affect the local natural and social conditions.[37] The capacity of fishermen to be resilient has been undermined with, in 2005, the start-up of a hydro-electrical power station at the Pedra do Cavalo dam which had already caused some intense ecological alterations but on an ad hoc basis. With the production of electricity, variations in several water parameters (salinity,

[33]The proposed model was an Environmental Protection Area.

[34]Parent Association of the Extractive Reserve of Canavieiras (AMEX).

[35]The ICMBio had 21 communities forming the extractive reserve, but the study by Viviane Martins (BRASIL 2009) identifies communities that bring several communities together, with the total being more than 80.

[36]The MPP is a national organization, which brings together formed organizations across States.

[37]For more details on the effects of the hydro-electrical central, see Prost 2007, and as for the shipyard, see Prost 2010.

temperature, current...) are constant and irregular. They disrupt fishing activities when the valves are opened. Water flow does not match the natural river rhythm but the requirements of electricity demand instead, causing the death of mollusks and regular[38] escape of crustacea and fish. The hydroelectric plant management is achieved through a policy of opacity of information: without any warning or detailed report on the flow, nor compensation for loss in earnings. People do not receive the support of the State in this unfavorable balance of power. Worse, in the south of the extractive reserve, the ongoing expansion of a small shipyard, deactivated a long time ago in São Roque of the Paraguaçu, is funded by a pool of big companies and receives strong support from the Bahian government in the name of development.

Indeed, in the current international situation, there is a growing demand for the shipyards. Nationally, the discovery of huge reservoirs of oil *offshore* in the continental platform between Rio de Janeiro and Bahia also leads to prospects of growth in the oil sector, and thus, large equipment orders. With the encouragement of the Lula government, the Bahian government launched in 2010 the 'Speed up Bahia' program, including the Pronaval, an incentive program for the shipbuilding industry, relying on high economic results.[39] As for the town halls of the region, they welcome the project due to the jobs created and the expected royalties.

The intensity of the alliance between the public authorities and large companies' led to the adoption of Amendment No. 7 of the 462 Provisional Measure, confirmed by Act No. 12.058 of 2009. The Act changed the extractive reserve's[40] polygonal chain in order to exclude the area occupied by the shipyard, without discussing the location of the territorial compensation by the advisors of the resex (Prost 2010).

Therefore, the traditional extractivists saw their land, the basis of social reproduction and cultural reference, being taken away from them because of their immobility[41] and their economic and political weakness in the face of the companies' establishment, which express a national, and even a global rationale (Haesbaert 2004), which only targets intensive exploitation of natural resources as an exchange value. Certainly, fishermen also get exchange values through the sale of their catches, but in their eyes, nature holds an important part of values in usage, starting with food hygiene – rich in protein – and symbolic values that cannot be estimated in monetary terms. We have to remember that fishermen (like other traditional populations) build their own vision of development with is slower than the dominant mode. Their paradigm of development feeds, and is fed by, the values of democracy, social welfare, and community harmony, which are in contrast with the spirit of

[38]From several days to several weeks.

[39]The Government's interest is obvious when we know that the annual construction of two ships or platforms generate two billion Reals.

[40]The trace of both polygonal chains is accessible on the site of the pro-Iguape commission, in the following address: http://comissaoproiguape.WordPress.com/2010/05/19/34/

[41]In the two extractive reserves mentioned here, boats vary between motorized canoes (possibly to sailing boats if conditions are good) and small motorboats which make long expeditions on the high seas impossible. Fisheries occur mainly in estuaries and rivers, and to a lesser extent, in coastal zone.

competition that steers the current mode of production and consumption. They are in this sense "slow men", as qualified by Milton Santos (1996). This represents a new ontological category insofar as slowness qualifies mobility, as well as the intensity of production and its rhythm, close social relationships and local cultural values. This new category does not however exclude exchanges with other socio-economic groups. Public authorities possess the critical capability, through the enactment of standards and public programs, to make sustainable the maintenance and the slow development of harmonious relationships between traditional populations and nature.

14.5.3 Autonomy as a Way of Emancipation to Establish Coviability

The lack of cohesion in the Bay of Iguape not only play against the fishermen in terms of votes at Council meetings, but also against the necessary autonomy to withstand the pressures of the market economy. As conceived by Castoriadis (1982), without full autonomy, participation is limited to empty words.

Here, it is recalled that such autonomy is justified in many ways. Beyond the principles of citizenship and democracy, it is understood that local people possess knowledge about local problems and define their priorities based on objective conditions as well as locally constructed values. Furthermore, environmental knowledge constitutes a solid basis upon which the reasoning of traditional local populations is founded. Some question-raising examples about (draft) environmental standards by fishermen, armed with knowledge, are mentioned.

The Canavieiras crab fishermen disputed, for example in 2013, the fishing ban dates, during the period of reproduction that lasts for three months. In fact, in that year, the standard enacted two, one-month periods, which doubled the periods when it was banned to crab fish, but without financial compensation for environmental services rendered by crab fisherman. Furthermore, the standards by The Chico Mendes Institute for Biodiversity – ICMBio, which extend into a region that covers the Northeast region and the State of Pará, generalizes the reproduction phases, which does not always match local specificities. More recently, a list of endangered species led to banning catches.[42] Once again, the AMEX, without disputing the validity of the list, denounced the generality of measures. The Canavieiras extractive reserve is full of both species cited, namely the land crab (*Cardisoma guanhumi*) or the Blue Parrotfish- (*Scarus coeruleus*),[43] whose capture would be heavily sanctioned, without taking into account the traditional (and more sustainable) use

[42]Portaria 445 of the ICMBio.

[43]Portugal Em, popular names are respectively guaiamum and budiao azul.

of these catches, nor the protection of the species during reproduction.[44] During the month of August 2015, opposition pressure from Brazilian artisanal fishermen and Paraná ship-owners have forced the government to remove the list and admit that some species, caught during fishing, are not threatened throughout the national territory. These few examples illustrate the strength of traditional knowledge in strategies of social organization in Canavieiras.

For their part, fishermen of the Iguape Bay extractive reserve are suffering from the lack of important information from the ICMBio or companies located upstream and downstream to the Bay. This access to information is acquired to Canavieiras through the 'National Commission of Enhancement of Marine and Coastal Extractive Reserves (CONFREM[45]), which maintains regular relationships with the federal government. It is worth noting that important information is also systematically discussed within the parent-association in order to socialize and analyse data, and better define actions, at the local, regional or national scale. The existence of a free debate of ideas is asserted as a, certainly not sufficient, but essential condition so that the extractive reserves define their political action in accordance with their community's interests.

Enjoying strong social cohesion and environmental knowledge, the extractivists of Canavieiras do not hesitate to put pressure on the ICMBio so that it intervenes on the field. Therefore, regular monitoring enabled the deforestation of the mangrove tree to be halted, which led to an extension of the species. In addition, the shrimp farms stopped their activity due to environmental impacts and, from the 18 existing ones, less than five are currently precariously active, thanks to a temporary injunction of Justice. This partnership is not free of tensions and the extractive reserve has a history of several leaders whose replacement was demanded by the parent-association of the extractive reserve (AMEX) for reasons of authoritarian management. The AMEX repeatedly demonstrates its determination to keep functioning mechanisms democratic and exercise authority legally delegated by extractive reserve status.

Another factor in the experience of Canavieiras lies in the progressive renewal of social leaders at the head of local associations and advisorsat the conservation unit. Due to the multiplication of projects designed, developed and implemented in the extractive reserve through various partnerships with public funds or NGOs, some ex-presidents of association or ex-advisors are dedicated to new tasks.[46] The Costeiros[47] research team noted during field visits carried out, since 2011, the

[44]Crab collectors do not receive the *salario-defeso*, an income (corresponding to a minimum wage) distributed to fishermen in order not to fish, incorporating the rhythms of nature in public management of resources.

[45]Comissão Nacional de Fortalecimento das Reservas Extrativistas Marinhas e Costeiras.

[46]Ex: Coordinator construction project of working class houses, BAMEX Community Bank employees, head of the women network - project funded by the UN-women, Secretary of the CONFREM, etc.

[47]Environmental management group in coast region, registered at CNPq, coordinated by Catherine Prost.

political maturation of several social leaders insofar as they exercise new associative responsibilities. These examples illustrate the importance of the social and political praxis to promote the development of knowledge, skills, and abilities of reasoning and awareness of this process, i.e. an increased awareness through social movement.

Thanks to these factors, the fishing communities of the Canavieiras extractive reserve grouped together in the parent-association, AMEX, are building their history by their local political positioning, initiatives of local development projects, and their various social articulations, as much with the federal government as with NGOs, the scientific community, and gradually with other social movements.

A factor carrying weight in the socialization process concerns the carefully weighed choice in the battles fought by the AMEX before mobilizing the communities, a respected precaution since the conflict around the creation of the conservation unit, when the batles were selected because of the multiplicity of political opponents at that time. Thus, the polygonal chain of the extractive reserves travels along certain lines in order not to include shrimp farms already existing[48] or areas coveted by landowners who were neither occupied nor used by communities. More recently, in face of the liberation of the exploration of oil or gas reserves off the coast of the south of Bahia, and in particular a block located at almost 20 km of the extractive reserve, the AMEX chose not to engage mobilization against the activity of the oil company in question, but to carry out with it environmental compensation.[49] This decision was made because the AMEX heard that resisting to an economic sector of this magnitude would be futile due to the extremely unfavorable power balance against the traditional fishermen, and would cause popular demobilization. Without a horizon of victory, social leaders of the AMEX did not call upon the communities.

In parallel, to overcome local resistance – fruit of the intense anti-extraction reserve campaign cited previously, extractive reserve organizations opted for a work policy in favor of the protection of nature and traditional fish populations, hoping thus, to convince, in a biased and demagogic way, criticisms by the opposing camp.

14.6 Conclusion

The institutionalization review of the movement conquest of *seringueiros* expresses a step forward in Brazilian environmental legislation by acknowledging a relationship with nature differentiated from a dominant society, based on a non-dichotomous view by traditional populations with nature and on environmental knowledge built throughout generations by socio-cognitive mechanisms. Because of their lifestyle and by incorporating in a selective way modern elements, they

[48]However, by encircling the existing farms, the creation of the extractive reserve prevented thereby institutions from expanding their area.

[49]The company is responsible for funding public research on mangroves and rivers of the extractive reserve and financed the renovation or construction of seats of local fishermen associations.

contribute to nature conservation. The pressure of the market economy and the new values that it brings in its wake can be offset by rules which guarantee the land of traditional peoples, a place for living, working, and social and cultural reproduction. The presence of public authorities, to ensure that the rules of common use of relationships with nature are adhered to, remains fundamental. Therefore, it is up to modern society to redefine again the commons that natural resources are, which includes redefining users and usage rules using models other than the dominant ones.

However, in response to the productivity rationale of the Brazilian State, the race towards growth requires large-scale projects that end too systematically by negative social and environmental impacts for the local (economically) poor people. In this context, extractive populations have to pursue certain objectives in order to get a more favorable balance of power.

Firstly, following the advice of Chico Mendes, the extractive reserves should be located in areas crossed by conflicts – which means asking for public intervention to ensure the existence and the reproduction of social groups – and in which the population is preliminarily organized (Mendes in 1989 quoted by Cunha 2010, p. 103) so as to avoid the risk of increased State control. Once created, these Conservation Units must be managed for democracy and transparency purposes, ensuring freedom of expression and open debate.

The local governance must be worked in the perspective of popular education for which: "educational processes are unproductive for society if people who cross them have no access to collective bodies enabling them to participate in the public spheres to make proposals which can provide a solution to socio-environmental problems" (Peralta e Ruiz 2010).

The network of social articulations, from a local to the national scale (e.g.: MPP, CONFREM) or even international (e.g.: exchanges already carried out with the network MangueMar at the Latin American scale), social movements enriching the analysis at several scales of some problems, failures, successes and challenges, within a framework that promotes dialogue, learning, analysis and thus the (re) construction of the world in communion with the others (Freire 2005). Through these exchanges, social movements weave networks in a horizontally (Santos 1994, 1996) that consolidate the resistance to the vertical rational of States or big corporations.

It is interesting to add the AMEX recommendation on the choice of battles fought, seeking to reduce the risk of failure, which is a potential factor in popular demobilization. The renovation of social leaders and the sharing of community tasks also contribute to the development of skills and political awareness.

Other non-developed factors in this article may also be mentioned such as policy coherence of social leaders and the clear awareness of the value of their traditional knowledge to direct their actions. Finally, the fishermen are also protagonists of their fate by combining many local development projects and national campaigns to ensure or to advance their rights.

However, despite the apparently positive developments, such as the creation of extractive reserve status, the considered traditional populations, as are others who have legal lands like the Amerindians and descendants of slaves[50] – without mentioning those who don't enjoy the same rights-, are facing many obstacles. The (so-called) large development projects impose progressive limitations on the power of local management councils and cause environmental and territorial conflicts. They expose the common bias of the State, concerned with growth imperatives and thus benefiting the dominant sectors. Such contradictions reflect how the dominant ideology of economic growth is against the purpose of social justice and ecological conservation despite the discourse advocating sustainable development.

Society in its current capitalist stage is reaching its limits, caused by an unlimited ideology of growth in a finite world where nature is regarded through a very strict utilitarian viewpoint and private ownership. Environmental knowledge and the way of life of traditional communities can serve as valuable elements for the necessary revision of our modern civilization, as confirmed by scientists from various disciplines (Cavalcanti 2012; Leff 2006; Boff 2000).

For this to occur, the consideration of various matrices of rationality, external to science, acts as a condition for better rethinking our relationship with nature and within society since the "environmental issue" refers to both natural and social processes. In fact, it seems clear that the problem is not with a lack of knowledge but rather with the ethical sense to using it (Peralta e Ruiz 2010). Beyond interdisciplinarity, environmental knowledge must cover a complex range of knowledge, particularly the knowledge of the people, in order to define an "environmental rationality" supported by values that do not claim to be scientific (Leff 2004). Such a context allows a dialogue between science and knowledge, tradition and modernity. The opinion defended herein is that the stake lies in the recognition of endogenous forms of organization of land, relationships with nature and social relations, and in other words, diverse social and cultural practices, pluralistic and rooted locally. The challenge remains open because the adoption of new forms of coviability involves changes in power and the production of new meanings of civilization (Leff 2006).

References

Adams C (1994) As florestas virgens manejadas. Boletim do Museu Paraense Emílio Goeldi. Série Antr 10(1):1994

Alencar CAG, Maia LP (2011) Perfil socioeconômico dos pescadores brasileiros. Arquivos de Ciências do Mar 44(3): Labomar, Fortaleza, pp 12–19

Alier M J (2007) O ecologismo dos pobres. São Paulo: Contexto

Boff L (2000) Saber Cuidar: ética do humano. Compaixão pela Terra. Vozes, Petrópolis

Bourgine P (1996) Modèles d'agents autonomes et de leurs interactions coévolutives. In: Penser l'Esprit, V. Rialle et D. Fisette (eds), Presses universitaires de Grenoble, 1996. Vailable online: http://lefur.jean.free.fr/coviab/concepts/coviabil.htm. Accessed on 15 July 2016

Brandão CR (1994) Somos as águas puras. Papirus, Campinas

[50] *Quilombolas* em Portuguese.

Brasil – ICMBio (2009) Documento final de caracterização da unidade e temas complementares. Reserva extrativista marinha Baía do Iguape. Coord. de Vivian Martins. Maragogipe: ICMBio
Castoriadis C (1982) A instituição imaginária da sociedade. Paz e Terra, Rio de Janeiro
Castro E (1997) Território, biodiversidade e saberes de populações tradicionais. In: Castro E, Pinton F (orgs) Faces do trópico úmido: conceitos e novas questões sobre desenvolvimento e meio ambiente. Belém: Cejup-UFPA-NAEA
Cavalcanti C (2012) Sustentainability: a mantra or a moral choice? An ecological and economical approach. Estudos Avançados, São Paulo 26(74):35–50
Coriolano LNMT, Lima LC (org) (2003) Turismo comunitário e responsabilidade socioambiental. EDUECE, Fortaleza
Cruz RCA (2003) Introdução à geografia do turismo. 2° ed. Rocca, São Paulo
Cunha CC (2010), Reservas extrativistas: institucionalização e implementação no Estado brasileiro dos anos 1990. Thèse (Doctorat en Psycho-sociologie). UFRJ, Rio de Janeiro
Diegues AC (2000) Etnoconservação da natureza: enfoques alternativos. In: DIEGUES, A. C. Etnoconservação. Novos rumos para a proteção da natureza nos trópicos. Nupaub-USP, São Paulo
Dumith RC (2012) Dinâmicas do sistema de gestão na reserva extrativista de Canavieiras (BA): análise da robustez institucional e de possibilidades para o ecodesenvolvimento. Mémoire (Mastère en Géographie). UFRG, Rio Grande
Freire P (2005) A pedagogia do oprimido. 41ª ed. Paz e Terra, Rio de Janeiro
Freitas HI (2008) A questão (da reforma) agrária e a política de desenvolvimento territorial rural no Litoral Sul da Bahia. Mémoire (Mastère en Géographie). UFBA, Salvador
Haesbaert R (2004) O mito da desterritorialização: do "fim dos territórios" à multiterritorialidade. Bertrand Brasil, Rio de Janeiro
IBGE (2000) (Instituto Brasileiro de Geografia e Estatística) Censo de 2000. Online: http://www.ibge.gov.br
Isaac VJ (2006) Explotação e manejo dos recursos pesqueiros do litoral amazônico: um desafio para o futuro. Ciência e Cultura, São Paulo, v. 58, n. 3, set. 2006, pp 33–36. Available online: http://cienciaecultura.bvs.br/scielo.php?script=sci_arttext&pid=S0009-67252006000300015&lng=en&nrm=iso. Consulté le 26.05.2013
Kuhn ERA (2009) Terra e água: Territórios dos pescadores artesanais de São Francisco do Paraguaçu-Bahia. Mémoire (Mastère en Géographie), Salvador
Leff E (2004) Saber Ambiental: sustentabilidade, racionalidade, complexidade, poder. 3ª ed. Vozes, Petrópolis
Leff E (2006) Epistemologia ambiental. 5ª ed. Cortez, São Paulo
Lobão R (2006) Cosmologias políticas do neocolonialismo: como uma política pública pode se transformar em uma política do ressentimento. Thèse (Doctorat en anthropologie sociale) – Université de Brasília, Brasília
Machado RAS (2007) O meio natural na organização produtiva da população pesqueira tradicional do município de Canavieiras/BA. Mémoire (Mastère en Géographie). UFBA, Salvador
Marx K (1985) Le capital. Livre 1. Sections 1 à 4. Flammarion, Paris
MMA (Ministère de l'Environnement) (2004) Diagnóstico da Carcinicultura no Estado do Ceará. Mandado de Intimação n° 300/2004. Ação civil pública – Processo n° 2003.81.00.0024755-5, 240 p
Neves W (1995 Sociodiversity and biodiversity, 2 sides of the same equation. In: Clüsener-Godt, M., Sachs, I. Brazilian perspectives on sustainable development of the Amazon Region, MAB Series, v. 15, UNESCO, pp 91–124
Oliveira GC (2012) Identidade territorial das comunidades do entorno da reserva marinha Baía do Iguape. Mémoire (Mastère en Géographie). Salvador, UFBA
Oliveira DV, Souza SP (2015) Avaliação dos impactos gerados pela carcinicultura. Revista Ambiental. vol. 1(2), FPB, João Pessoa, pp 66–75
Peralta JE, Ruiz JR (2010) Educação popular ambiental. Para uma pedagogia da apropriação do ambiente. In; Leff, E. A complexidade ambiental, 3° ed. Cortez, São Paulo

Porto-Gonçalves CW (2001) Meio ambiente, ciência e poder: diálogo de diferentes matrizes de racionalidade. In: SORRENTINO, M. Ambientalismo e participação na contemporaneidade. EDUC/FAPESP, pp 133–161

Porto-Gonçalves CW (2002) Geografando nos varadouros do mundo: da territorialidade (o seringal) à territorialidade seringueira (a Reserva Extrativista). Ibama, Brasília

Porto-Gonçalves CW (2006) A globalização da natureza e a natureza da globalização. Civilização brasileira, Rio de Janeiro

Pretty JN Pimbert MP (2000) Parques, comunidades e profissionais: incluindo "participação" no manejo de áreas protegidas. In: DIEGUES, A. C. Etnoconservação. Novos rumos para a proteção da natureza nos trópicos. Nupaub-USP, São Paulo, pp 183–223

Prost C (2007) Efeitos da barragem Pedra do Cavalo sobre a pesca artesanal na baía do Iguape. II Encontro Nacional de Ciências Sociais sobre Barragens, Salvador, pp 1–17

Prost C (2010), Resex marinha versus polo na Baía do Iguape. Novos Cadernos do NAEA, v. 13, n. 1, Belém: NAEA-UFBA, pp 47–70

Santo RVE, Isaac VJ (2014) Desembarques da pesca de pequena escala no município de Bragança Pa, Brasil: I-esforço e produção. Boletim do Laboratório de Hidrobiologia (UFAMA. Impresso) 25: 31–48

Santos M (1994) Território, Globalização e Fragmentação. Hucitec, São Paulo

Santos M (1996) A natureza do espaço. técnica, tempo, razão e emoção. Hucitec, São Paulo

Souto FJB, Martins V (2009) Conhecimentos etnoecológicos na mariscagem de moluscos bivalves no manguezal do Distrito do Acupe, Santo Amaro-BA. Biomas 22(4): UFSC, São Carlos, pp 207–218

Souza ML (2002) Mudar a cidade. Uma introdução crítica ao planejamento. Bertrand Brasil, São Paulo

Tahim DF (2009) A carcinicultura e o meio ambiente: o desafio da sustentabilidade. Anais do III Congresso Brasileiro da SOBER. SOBER, Porto Alegre, pp 1–20

Tomasoni MA (2011) Mudanças globais: a problemática do ozônio e algumas de suas implicações. GeoTextos 7(2): POSGEO-UFBA, Salvador, pp 141–178

Virgens DA (2010) Turismo e transformações socioespaciais: o caso do município de Cairu – Bahia. Mémoire (Mastère en Géographie). UFBA, Salvador

Chapter 15
Socio-ecological Coviability Confronted with the Neoliberal System: The Peace Parks Experience (Southern Africa)

Nadia Belaidi

Contents

15.1 Introduction

Southern Africa is developing an innovative tool for transfrontier cooperation by creating conservation areas with the objective of biodiversity conservation, socio-economic development and the promotion of a culture of peace (Hanks 2003). Consequently, these peace parks appear as privileged sites of observation for the reconstructions underway in southern Africa, while giving consideration to the integration of environmental issues into the public policies of emerging states. From this viewpoint, they present a double interest.

On the one hand, peace parks constitute vectors for social bond (re)creation in all of its socio-cultural and socio-economic dimensions. They are consequently privileged sites permitting the analysis of the politico-legal mutations that are occurring in this geographical space, underlining the originality of such an experiment. On the other hand, these parks are also initiators of biodiversity conservation projects

N. Belaidi (✉)
Department of Human & Environment, UMR CNRS 7206 Eco-Anthropology and Ethnobiology, Paris, France
e-mail: nadia.belaidi@mnhn.fr

O. Barrière et al. (eds.), *Coviability of Social and Ecological Systems: Reconnecting Mankind to the Biosphere in an Era of Global Change*,
https://doi.org/10.1007/978-3-319-78497-7_15

along with a reorganization process for transfrontier social groups. This positions the sustainable development perspective in relation to its environmental, ecological, regional and social aspects, through the changing game of identities.

In this context, we are particularly interested in the Kavango-Zambezi Transfrontier Conservation Area (KaZa TFCA), located at the meeting point of the borders of Angola, Botswana, Namibia, Zambia and Zimbabwe (see map below).

Through these objectives, as emphasized by the Peace Parks Foundation – the initial promoter of these transfrontier areas in southern Africa – peace parks aim to achieve a co-existence between humans and nature.[1]

In this case, peace symbolizes the social cohesion and political co-operation dynamic sought in these formerly conflicted zones, during the period of apartheid. Social cohesion is determined by the social link. As a minimum, this link allows populations to live together. It targets all of the relationships that unite individuals who are part of a social group: collective solidarity mechanisms, norms, rules, the values that give this group a minimum sense of collective belonging (Cusset 2007). However, as a framework for the social link, ecosystems and their dynamics influence the type of social relationships that are established (Belaidi and Koubi 2015): for example, fishing societies significantly differ in their social relations from nomadic breeding societies or farming societies. In fact, in 2005, the Millennium Ecosystem Assessment (MEA)[2] confirmed this relationship even going so far as to consider that the damage done to the ecosystem violated the principle of equity and justice (See Fig. A, MEA 2005, pg VI).

Consequently, by recreating bonds through the transfrontier approach, peace parks seem to tend towards coviability by suggesting that human beings are an integral part of ecosystems and that a dynamic interaction exists between these human beings and other elements of these ecosystems. However, by subjecting Zambeze-

---

[1]"Peace parks are about co-existence between humans and nature, about promoting regional peace and stability, conserving biodiversity and stimulating job creation by developing nature conservation as a land-use option", Peace Parks – TFCAs, http://www.peaceparks.org

[2]The concept of the Millennium Ecosystem Assessment (MEA) was developed in 1998 and 1999 on the invitation of the World Resources Institute, the United Nations Environment Program, the United Nations Development Program, and the World Bank. The Millennium Ecosystem Assessment synthesizes information from the scientific literature and relevant datasets and models that have undergone the test of counter review. It incorporates knowledge held by the private sector, practitioners, local communities and indigenous peoples. More than 1360 authors from 95 countries participated in it. The evaluation focuses on the bonds between ecosystems and Man's well-being. It particularly targets "ecosystem services." The MEA deals with the full range of ecosystems (a dynamic complex composed of plants, animals, microorganisms, and the surrounding dead nature interacting as a functional unit). It extends from relatively intact ecosystems such as forests, landscapes controlled by human exploitation, to ecosystems under intensive human control and undergoing modifications due to man's actions affecting, such as agricultural lands and urban landscapes for example. The services that ecosystems provide consist in the benefits that humans derive from them. These include sampling services such as food, water, timber, and fiber. They also include regulatory services that affect climate, flooding, disease, waste, and water quality; cultural services that provide recreational, aesthetic, and spiritual benefits; self-maintenance services such as soil formation, photosynthesis, and the nutrient cycle.

Kavango's ecosystem, heritage and cultural resources management objective to the "best models of conservation and tourism", as planned by the Treaty establishing the KaZa TFCA (art. 6.1 a-j), we can wonder whether the very idea of coviability is not corrupted.

In effect, whilst the notion of coviability can designate the creation of operating system pathways that maintain both the viability of ecosystems in the long-term and the viability of the lifestyles they support (Weber 2013; see also Washington 2013), the submission of the latter to a model of production and consumption that largely depends on the dominant politico-economic apparatus, advocating a *micro* and individualistic vision of the social bond, makes them responsible for the viability crisis (Norgard 1994). In our example, this signifies that the coviability of lifestyles and socio-ecological systems, far from favoring *peace*, would ultimately contribute, through the destruction of the Man-Nature bond, to fueling conflict.

15.2 The Transfrontier Area, a Tool to Reconstruct Links

The transfrontier approach is introduced as an antidote to the inefficiency and/or absurdity of borders drawn by the colonizer. In many cases, these frontiers have in effect severed not only the unity and integrity of ecosystems but also the cultural and social relationships they underpin. Consequently, the new geographical framework introduced by transfrontier logic serves to rationalize the management of ecosystems by the public authorities while stimulating social relations (Sect. 15.2.1), transforming marginalized zones into new centers advocating the social link (Sect.15.2.2) and alternatives to the commodification of the *living environment* (Sect. 15.2.3).

15.2.1 A Renewed Geographical Framework

Transfrontier conservation has the advantage of preserving nature, of (re)creating a bond in the area concerned. It is also a driving force for development firstly thanks to the ecological and cultural integrity it helps to reconstruct and secondly, through the promotion of ecotourism.

Based on an inclusive method, the transfrontier approach advocates an integrated and adaptive management of land, water and natural resources on the scale of the ecosystem, consequently preserving the ecological and evolutionary processes that create and sustain biodiversity (Mace and Purvis 2007). Transfrontier areas as large-scale conservation initiatives therefore facilitate exchanges between populations and enable complete ecosystems to be studied. Similarly, the capacity of transfrontier areas to open borders, given that they often overlap, promotes exchanges between human communities separated by political frontiers. These frontiers have often been established under political and historical pressure regardless of population groups.

The transfrontier approach can therefore serve to rebuild cultural integrity. Moreover, the vast parks created offer tourists free and easy access to a region, which presents an investment opportunity. In fact, most transfrontier areas have been developed with well-defined ecotourism development objectives that represent strong economic justifications in favor of their establishment (Wolmer 2003b; Duffy 2007).

The transfrontier approach has the particular feature of creating a new space dedicated to safeguarding socio-ecological dynamics. This means that it takes into account interactions between social systems and ecological systems from the sustainability viewpoint, understood as a dynamic process and not a final objective. Within this context, the development sought corresponds to the needs of populations and local communities.

This vision is composed of socio-ecological logics which invite us to remember that the physical and chemical cycles of nature are important to sustain human and non-human life. Within this context, development and the ability to live take on a new dimension: it is (the maintaining of) all living systems which make them possible. The relationship to nature is defined not only by an environmental relationship but also by a relationship to a (complex) environment, summoning the notion of alterity. This is what synthesizes the concept of "social-ecological systems" (SES) by designating integrated systems linking societies and nature. This ultimately aims to redefine ecosystems by explicitly considering all of the actors, by integrating man as an active component of the system (Liu et al. 2007; Folke 2007; Lagadeuc and Chenorkian 2009).

This geographical approach also results in a specific management mode: bioregionalism (Wolmer 2003b). *A bioregion is a geographical space forming a homogeneous natural whole, whether for soil, hydrography, climate, fauna or flora. The population is also part of the bioregion insofar as it lives in harmony with these natural data and is sustained by it in the long term.*[3] It differs from an ecoregion which, whilst also corresponding to a geographical space that brings together different biological and geological criteria, does not refer to human societies (WWF 2009). The population is therefore an integral part of a bioregion, but on the condition that it protects and maintains its natural balances. This "*re-inhabitation*" consists in a relationship of interdependence and exchange with the ecosystem of the bioregion. It requires knowing the particular ecological relationships of and between each place in order to establish a socially and environmentally sustainable

[3]Berg P., 2001, "Aux sources du biorégionalisme" interview by A. de Benoist and M. Marmin, *Elément,* n°100; Berg P. and Dasmann R.,1977, "Reinhabiting California", *The Ecologist*, vol. 7, n°10, pp. 399–401; SALE K., 1985, *Dwellers in the Land: The Bioregional Vision,* Sierra Club, San Francisco; ALEXANDER D., 1990, "Bioregionalism: Science or Sensibility", *Environmental Ethics*, 12, 2, pp. 161–173; M. McGinnis (ed), 1999, *Bioregionalism,* Routledge, London and New York. For a review, S. Frenkel, 1994, "Old Theories in new Places? Environmental Determinism and Bioregionalism", *Association of American Geographers*, 46, 3, pp. 289–295; Meredith D., 2005, "The Bioregion as a Communitarian Micro-region (and its limitations)", *Ethics Place and Environment*, Vol. 8, n°1, pp. 83–94.

system (*Living-in-place*). In this sense, a bioregion has a unique cultural identity (a bioregional identity), whilst also constituting an area within which local populations can determine their own development.

Bioregionalism consequently pleads in favor of popular power and the decentralization of governance. Given that they are defined according to criteria of ecological, cultural and social coherence, bioregions form the basis of political organization.

15.2.2 A Space Based on a Dialectic of Law/Justice/Peace/Development

Considering transfrontier spaces as fully-fledged geographical entities superimposed on political frontiers opens up a new aspect of transfrontier conservation. In effect, these spaces could allow the gradual restoration of ecological, cultural and social bonds. However, protected areas are neither socially nor politically neutral (Hammill and Besançon 2007). This approach consequently constitutes a veritable challenge. Protected areas can take on different roles (for environmentalists, protected areas are an effective measure to protect biodiversity. For tour operators, they constitute a basis for the development of ecotourism. For neighboring local communities, protected areas can limit access to subsistence resources, or lead to relocations or generate income through tourism, etc.). All of these roles make protected areas political objects able to result in resistance and conflict.

This means that it is important to find the common denominator that would allow all groups to dialogue and reach an agreement in order to avoid or overcome potential conflicts. It is on the basis of the idea that nature conservation represents a common value to all countries, actors and populations involved that the concept of environmental peacemaking was born (Conca and Dabelko 2002; Ali 2007). The theory of environmental peacemaking is essentially based on the idea that cooperation with respect to a stake of common interest which nature represents, opens the door to a political dialogue around more contentious issues (political tensions, regional security, cultural diversity etc.) (Carius 2007) by fostering the trust of the parties involved.

Building on this trust, which would then help address more sensitive issues, requires the joint development and application of rules for environmental protection and natural resource preservation. Mutually negotiated environmental management policies and programs tend to be based on harmonized legislation. This legislation aims to maintain interactions between social systems and ecological systems in order to ensure ecological bonds and to maintain and/or re-establish the cultural and social link. This orientation of legislation underlines that inequalities are largely linked to the socio-economic structure and the socio-political organization of various societies (and more broadly to contemporary world order).

A legislation which focuses on ecological and socio-cultural bonds opens the way for social, environmental and ecological justice (Belaidi 2015a). This means that the benefits provided by the environment, as well as its constraints and nuisances, would be shared and distributed among populations (for an environmental social justice) irrespective of the racial, ethnic or social origin of groups and individuals. In addition, there is a rebalancing of the relationships between humans and the rest of the environment (defining ecological justice) with a view to ensuring the sustainability of life support systems and access to resources within the sustainable limits of the planet. (Low and Gleeson 1998; Kutting 2004). *Environmental justice* therefore encompasses *social environmental justice* (Sperber 2003).

Defined here as "*the ultimate objective that legal norms must reach*" (Shelton 2012, p. 20), *environmental justice* would articulate redistribution, social recognition, the sharing of environments and the maintaining of ecological processes (Low and Gleeson 1998; Shrader-Frechette 2002; Sperber 2003; Kutting 2004; Schlosberg 2007; Pezullo and Sandler 2007; Westra 2008). This presupposes two major changes: firstly, in the relationship to nature and secondly, in the relationship to others. On the one hand, the environment must be understood as the physical, chemical, biological set of bonds that exist in ecosystems between the various elements and processes that act within it. On the other hand, solidarity must extend to humanity as a whole, at the present time and in the future, so that everyone can benefit from fundamental human rights without suffering from exclusion or discrimination. On this basis, beyond simple procedural aspect (Jolivet 2012), it is a Law that could legitimize a worldview where the environment is understood in its complexity in order to give more thought to social relations (Belaidi 2014).[4]

The idea of "living together" consequently highlights the fact that the environment in which we live allows us to grasp and understand with whom – natural and human – we live. This conception is based on a holistic vision of man, social relations and the world (where man's survival depends on the survival of the social group and the environment in which he lives).

The transfrontier approach suggests a geographical framework that emphasizes the ability to cooperate not only across borders, but also between different stakeholders from governments, civil society and the private sector. This approach consequently favors other forms of normative, political and social production that would allow, as suggested by the IUCN (Sandwith et al. 2001), to aim for peace. It is a dialogue, or even better, a set of conditions for interactions that favor or even reinforce bonds between countries and/or groups of populations which are

[4]For a transfrontier application, Belaïdi N., 2009. "Le Great Limpopo Transfrontier Park: une gestion régionalisée de la biodiversité au service du développement?", Echogéo, n° 7/2008, décembre 2008-février 2009, "Les nouveaux enjeux régionaux dans l'océan Indien occidental", Thibaud B. (dir.), http://echogeo.revues.org/8523; BELAÏDI N., 2008. "Entre terre et eau, la gestion du delta de l'Okavango: un mécanisme d'ordre public écologique?", Cahiers d'anthropologie du Droit, Foncier et environnement en Afrique. Des acteurs au(x) droit(s), O. Barrière et A. Rochegude (dir.), pp. 189–214; Belaïdi N. 2008. "L'eau en milieu aride entre survie des populations et équilibre des écosystèmes" in Serfati C. (dir.), Une économie politique de la sécurité, Paris, Karthala, pp. 189–198.

sometimes opposed (Conca and Dabelko 2002), in the name of and on behalf of the living environment: coviability. Considered here as a *pre-requisite* of the normal performance of any social, economic and cultural activity, *this* peace would consequently constitute a true catalyst for development.

15.2.3 An Initiative Offering Alternatives

By proposing a social recognition of environmental dynamics, which tends to give priority to the use value, i.e., to the utility of a good or a service to satisfy the necessities of everyday life, as opposed to the exchange value (the transaction of goods or services), favored by market economies,[5] this "model" of conservation represents a change in the understanding of nature-society relationships.

In this respect, it echoes numerous social demands that highlight the need for a *justice* that seeks to preserve the rights of current and future generations by preserving the *environment* in which they live. This environment is not only the physical environment, in all of its complexity; it is also the social and cultural environment, which in reality, is intimately linked to the physical environment. Preserving the environment requires that economic and political "environments" do not jeopardize it, meaning that they should not be considered superior to these components (nature, society and culture) (Belaidi 2012).

These demands for justice are articulated around the four pillars of Man-Society-Nature co-existence: the "four vital requirements of every society" (Houtart 2001):

- the relationship with nature;
- the ability of all to access goods and services based on the maintaining of the material bases of physical, cultural and spiritual life – in sum, life and its reproductive capacity;
- the collective social and political organization, i.e., the participation of each individual or collective subject in the processes of socio-political organization;
- the possibility of individual cultural and ethical expressions.

These demands therefore tend to ensure that this management mode evolves. They also serve to combat the appropriation of "natural resources" by individuals or firms with the objective of maximizing their profits and the commodification of the elements necessary for the reproduction of life.

[5]For capitalism (the most developed form of commodity production), a good or a service that cannot be converted into a commodity has no value since it does not contribute towards the accumulation of capital, which is the objective and motor of the economy. Capitalism considers ecological destruction to be inevitable collateral damage (destined to be eventually reduced), or worse still, as "externalities" since they do not enter into market calculations and are therefore not taken into account in the capital accumulation process. Godelier M., "Transition" in Bensussan G. et LABICA G., Dictionnaire critique du Marxisme, PUF, Paris, 1982. See for an example applied to climate change, Stern N., The Economics of Climate Change. The Stern Review, Cambridge University Press, 2007.

In the case of southern Africa, the transfrontier initiative was immediately perceived as a way to heal the wounds of pre-and post-independence wars (Koch 1998) and render the reconciliation process effective. Reconciliation lies within certain African values such as the *Ubuntu*.[6] It aims for *peaceful coexistence* and seeks to produce *civil peace*.

However, even if in South Africa the process undertaken via the Truth and Reconciliation Commission (TRC) proved effective in rethinking a shared history, it has shown itself to be limited in terms of considering a more egalitarian future along with social justice in the long-term (Krog 2002). All the more the Reconstruction and Development Plan (RDP, the socio-economic platform to access democracy in 1994) was certainly presented in favor of the poorest populations but failed to question the structures of the South African economy. By privileging economic growth, the free market and rejecting nationalization, in South Africa redistribution constitutes the last link in a context marked by liberalism (Porteilla 2010). Emerging from South Africa, transfrontier initiatives are modeled on the same economic logic. This may explain why the promoters of peace parks have reversed the peace-development liaison in the dialectic supported by the transfrontier approach. It is the (sustainable) development that will seek to achieve peace, which is not without consequences for coviability.

15.3 The South African Approach, a Corrupted Coviability

In southern Africa, whilst conservation professionals have quickly adopted the idea that transfrontier dynamics may be a driver of peace by creating their own category of cross-border area (Sect. 15.3.1), peace is seen as a result. It is the creation and the maintaining of livelihoods through environmental conservation that leads to peace (Sect. 15.3.2). By subjecting the promotion of peace to the achievement of (sustainable) development, the transfrontier area is subjected to a purely managerial logic (Sect. 15.3.3).

15.3.1 An Appropriation of Transfrontier Dynamics

According to the Peace Parks Foundation, "by establishing conditions that are favorable to sustainable development," transfrontier parks allow a fair and harmonious society to be maintained – which is a category of the Soft peace (Ali 2007). The

[6] According to the Bantu philosophies of sub-Saharan Africa, the word UBUNTU refers to the term "humanity". It is better understood in the sentence: "umuntu ngumuntu ngabantu": man becomes man only by and with other men, Republic of South Africa, Truth and reconciliation commission, Report, vol. 1, chap. 5, p. 128.

Peace Parks Foundation bases itself on a broad understanding of the various forms of peace and stability which, according to the Foundation, can be achieved through a viable management of natural resources via the establishment of parks along national borders. This would alleviate international antagonisms and promote a spirit of international cooperation on the African continent, hence the "Peace Park" label[7]. On this basis, this organization justifies the use of the peace concept to describe all of the transfrontier areas in the zone.

The peace objective links these transfrontier areas to the specific category proposed by IUCN: Parks for Peace, defined as "transfrontier protected areas that are officially dedicated to the protection and conservation of biological diversity and the related natural and cultural resources, as well as to the promotion of peace and cooperation" (IUCN 1994).

The term "park for peace" refers to a particular subset of protected areas where there is a clear objective of conservation and protection of biodiversity, cooperation between at least two countries or sub-national authorities and a clear goal of peace.

Peace knows different gradients, from "soft peace" to "hard peace." These extremes range from habitual excellent relations to their incitation when tensions or hostilities are present. In the intermediate position, we find, on the one hand, the creation of interaction conditions where relations between countries are satisfactory but not excellent and, on the other hand, the reconstruction of a relationship of trust following a war. The Kavango-Zambezi transfrontier conservation area (KaZa TFCA) belongs to this last category.

During the apartheid regime, frontier parks and wildlife areas were integrated into the regime's defense system against opponents who might attempt to intrude from the outside. Relative isolation and strict access control sometimes turned these areas into paramilitary or infiltration training places in neighboring countries. In Namibia, for example, during the last years of their presence, the South African armed forces transformed the present Bwabwata National Park into a training ground. Large military bases were installed and used against the SWAPO[8] in Namibia; they were also used to support the Unita[9] anti-government movement in Angola. On this occasion, veritable poaching networks were created in some South

[7]It should be distinguished from the "Peace Park" of the United Nations University for Peace, which designates protected areas whose history has been marked by conflicts whether or not these areas are in a transfrontier location, such as the memorial monuments at the Nagasaki or Hiroshima Peace Parks.

[8]Historically, the South West African People's Organization is a Namibian union that has become an armed separatist movement. It has been Namibia's main political party since independence in 1990.

[9]Until 2002, the União Nacional para a Independência Total de Angola (National Union for the Total Independence of Angola) was essentially a military force that fought during the Angolan civil war against the regime of the People's Republic of Angola. It then renounced the armed struggle and showed itself in favor of the democratic process. It is now the 2nd biggest party in the country.

African force units, in connection with "friendly" movements, revolving around the traffic of ivory and rhinoceros horns.

Spurred on by these elements, since 1988 the KaZa TFCA has been identified by the IUCN as a *park for peace* on the basis of temporary classification criteria concerning these parks (Thorsdell 1990). This is not the case, for instance, of the Great Limpopo Transfrontier Park (South Africa, Mozambique, Zimbabwe), which was established by treaty in 2002. The treaty establishing the KaZa TFCA (signed on 18 August 2011 between the 5 States) is different from the other active transfrontier parks in the area (notably Great Limpopo Transfrontier Park) in that it explicitly aims to achieve peace – understood as a combination of solidarity and security via environmental issues[10].

However, although the conditions for peace, pluri-national cooperation and biodiversity protection are met, it is not a park for peace that the treaty has established, but a transfrontier conservation area (TFCA). The KaZa TFCA is indeed defined as a " component of a large ecological region that straddles the boundaries of two or more countries, encompassing one or more protected areas, as well as multiple resource use areas."[11] It constitutes a peace park which aims to establish cooperation for the economic development of the area.

15.3.2 *Cost-Effective Environmental Conservation*

By introducing the term *TransFrontier Conservation Area* (TFCA) as part of its initial project on transfrontier conservation in Mozambique in 1996, the World Bank supports the idea that, in terms of wildlife, transfrontier cooperation must zoom out of protected areas in order to emphasize the multiple use of resources, in particular those of local communities (Hanks 2000). This extension spreads beyond national parks and hunting reserves to include private lands, communal lands, forest reserves and wildlife management areas as well as biosphere reserves or conservancies (communal management of natural resources) (World Bank 1996).

This is the pattern adopted by the peace parks in southern Africa. Consequently the Kavango-Zambezi Transfrontier Conservation Area (KaZa TFCA) encompasses 36 different types of protected areas (national parks, nature reserves, forest reserves, conservancies, world heritage sites, Ramsar sites . . .) situated in the five jurisdictions that constitute it. (Fig. 15.1)

[10]"For the execution of the objectives expressed in this Treaty, the Partner States undertake to uphold the following principles: [. . .] b. advocacy for solidarity, peace and security within the KAZA TFCA;", Article 1 of the treaty on the establishment of the Kavango-Zambezi transfrontier conservation area, August 18th, 2011.»

[11]Article 1 of the treaty on the establishment of the Kavango-Zambezi transfrontier conservation area, August 18th, 2011.

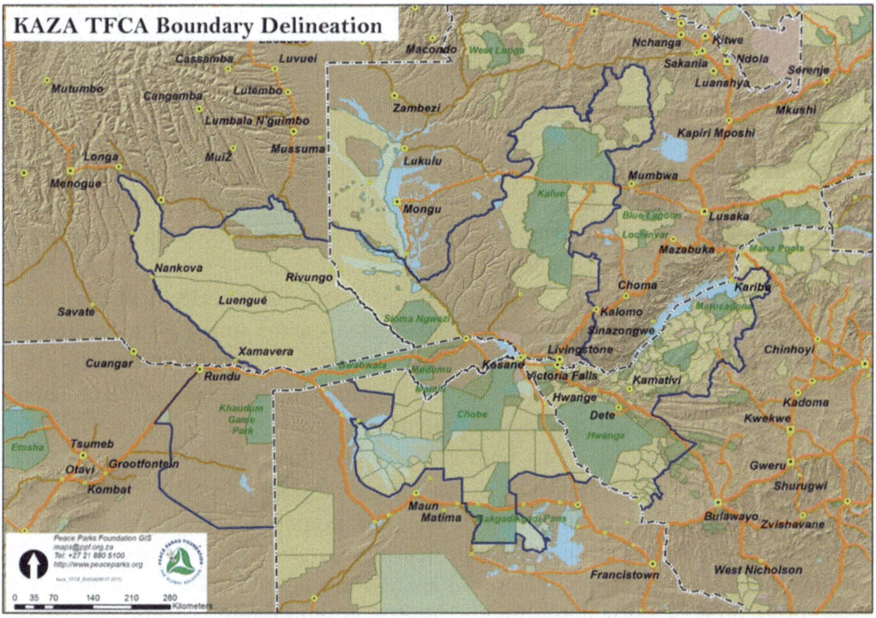

Fig. 15.1 Map: Kavango-Zambezi transfrontier conservation area (TFCA)

The transfrontier conservation area (TFCA) constitutes a category which is specific to southern Africa. The term originates from a biodiversity support program developed by the United States Agency for International Development (USAID) and is part of transfrontier cooperation processes that facilitate or improve the management of natural resources (Griffen 1999). These processes were adopted by the Southern African Development Community (SADC) in order to promote “regional co-operation in the development of common frameworks for conservation of natural resources and enforcement of laws governing sustainable use” (Protocol on wildlife 1999). These transfrontier areas strengthen the regional economic integration inscribed in the 1992 SADC Treaty. Furthermore, the SADC’s trade protocol, ratified in 1999 and which constituted the basis for the creation of a free trade area, ensures that the USAID provides *technical assistance* within the framework of the negotiation process (Wolmer 2003a).

These transfrontier conservation areas (TFCA) have rapidly become the focus of numerous new natural resource management initiatives funded by the region’s donors. Eighteen have either been established or are in the process of being developed in southern Africa. They fulfill the objectives of two important organizations in the region: the SADC, whose main goals consist in furthering cooperation, and the New Partnership for Africa’s Development (NEPAD), which advocates a developmental approach based on investment. As part of the peace objective, the elimination of non-tariff trade barriers (e.g. poor quality road infrastructure and long and complicated customs formalities) or the free movement of capital could be suggested, facilitated by the fact that they also favor the opening of borders and the promotion of transfrontier tourism.

In addition, since both the SADC and the NEPAD are dominated by South Africa, transfrontier areas reinforce South African political and economic domination throughout the region. International tourists arriving in Johannesburg will be able to visit the main natural parks of neighboring countries whilst remaining based in South Africa. The touristic hinterland has, to some extent, been expanded to encompass several neighboring countries.

With this shift from the conservation of specific sites to connectivity preservation, the Peace Parks Foundation (PPF) has capitalized on the image of a "boundless southern Africa" (http://www.peaceparks.org), where mega-fauna could freely traverse its natural territories in order to develop the lucrative potential of nature tourism and therefore peace and development.

The PPF can better define its own vision given that, at this moment in time, the procedure for classifying Parks for Peace proposed by the IUCN is not internationally recognized. In addition, after the preparatory work whose results were made public in 1988, in 2001 a project was launched on this issue with the involvement of several IUCN Commissions and the Peace Parks Foundation. The Peace Parks Foundation is therefore a co-author of the reference documents on the parks for peace meaning that peace can be defined according to its specific needs.

The links between peace and "sustainable development" have been largely recognized in international texts and instruments[12]. They include peace and security objectives within conservation initiatives now permitting the significant political status of these challenges to be taken advantage of, bestowing them with a greater visibility and an access to new funds such as humanitarian aid (and funds to alleviate conflictual situations). Moreover, the characteristics traditionally attributed to environmental issues are reversed when they are posed in a security context. This could help support governments and local communities for initiatives dedicated to environmental conservation (Matthew et al. 2002). However, with this top-down approach to conservation, bioregionalism is supplanted by a "technical eco-regionalism" where peace and the coviability logic that it carries, become tools of political and monetary domination.

[12]With all the ambiguities and contradictions that the term contains, see for instance, Hajer M. A., 1995, The Politics of Environmental Discourse: Ecological Modernization and the Policy process, Oxford University Press; Young S. C., 2000, The Emergence of Ecological Modernization: Integrating the Environment and the Economy? Routledge, London, 2000

15.3.3 Promoting Technical Eco-Regionalism

Transfrontier conservation suggests the promise of a transfer of power from central government to local communities through the reorganization of space into bioregions. In practice however, transfrontier areas represent an extension of the power of central government and international organizations, most often in distant and marginal regions which were previously more or less ignored. This distance leads to a "technical eco-regionalism" (Wolmer 2003a). It is a purely managerial approach where scientific constructions of space are used to justify imposing a top-down approach to conservation.

Moreover, the KaZa TFCA is part of an ecoregion (Article 1 of the Treaty of 2011 that established the KaZa) and not a *bioregion* as advocated by the transfrontier approach. Defined according to biological and geographical criteria (WWF 2009), this ecologically coherent region encompasses plots of protected areas and unprotected lands. Whilst it emphasizes the preservation of connectivity as opposed to the conservation of specific sites, it does not refer to human societies (WWF 2009).

Whilst the first phases of creating the conservation areas effectively focused on opening borders, they also lead to the launch of the "KaZa Uni-Visa" pilot project (a single visa to visit the KaZa transfrontier area). Launched between Zambia and Zimbabwe in November 2014 the objective of the project is to avoid obliging tourists to apply for multiple visas to cross the zone in order to make their journey through the partner countries of the KaZA TFCA zone more "comfortable", consequently fostering the growth of tourism for the region (http://www.kavangozambezi.org).

However, the policies and legal frameworks already in place in the party States enable the development of a common management plan to be envisaged (Jones 2008). This plan could help stakeholder countries to address, in a simple and rapid manner, the problem of elephant overpopulation and its status in the area[13]. More substantial common approaches could then be developed on the basis of this first cooperation. Questions such as the destruction of vegetation by elephants, disease control, illegal hunting, tourism control or the improvement of staff qualifications, equipment and infrastructure in parks would also be open for discussion. This dialogue would even possibly trigger exchanges on sensitive issues because they are subject to the specific historical and political contexts of each country, such as the recognition of forest authorities or the access of communities to resources and their participation in the process.

[13]In Zambia, the elephant is listed in Appendix I (Endangered Species) of the Convention on International Trade in Endangered Species of Wild Fauna and Flora (CITES). An easily achievable objective would be to move it to Annex II (species that are not necessarily threatened with extinction but whose trade in specimens should be regulated to avoid an exploitation which is incompatible with their survival) and thus to harmonize its classification over the entire area.

In the same way, the absence of any transfrontier area representatives in redefining (in progress) the management plan of one of these units (the Hwange National Park) raises questions. The KaZa TFCA "integrated development plan" was presented to all of the unit's "stakeholders", including village authorities, by the national coordinator (Zimbabwe) of the program in Hwange National Park (main camp) on May 9th, 2013. The discussions on the management plan are carried out in consultation with these same "stakeholders" but independently of the transfrontier area.

On the initiative of the Southern African Development Community and with the support of the Zambian and Zimbabwean governments, the World Bank and the KaZa TFCA Secretariat, regional cooperation has played a role in facilitating transfrontier movements of large mammals... and of tourists. It has therefore played a role in the commercialization of the largest contiguous natural area in southern Africa (approximately 287,132 km^2) as a tourist destination.

Within this context, large-scale conservation tends to shift away from community conservation even if it has been favored in the area in recent years (see, for example, Derman 1995; Child 1996). The tendency to marginalize the interests of local communities was denounced very early on. It was, for instance, demonstrated that the establishment of the Great Limpopo Transfrontier Park (35,000 km^2 between Mozambique, Zimbabwe and South Africa) on the Mozambican flank and the introduction of animals was carried out without the local community issue having been resolved. These populations were not fully informed or integrated during the park creation process and they have been the subject of a highly controversial "resettlement" process (Spenceley and Schoon 2007, Ramutsindela 2007, Chap. 6). In their haste to fulfill the project's objectives, namely to generate income thanks to the development of ecotourism, the project architects and sponsors (powerful donors, large NGOs such as the Peace Parks Foundation and government agencies) have neglected the needs and wishes of local communities. Yet, the Great Limpopo Transfrontier Park is the transfrontier area that served as a basis for reflection on the TFCA category (origins, http://www.peaceparks.org).

In conclusion, current interest in eco-regional planning and transfrontier areas seems to reflect a resurgence of the protectionist approach to conservation (Brosius and Russell 2003). We are witnessing a return to the top-down approach, where the involvement of local communities becomes minimal and where the particular interests of international NGOs and government agencies take priority.

Although claiming to promote peace by preserving an environment based on common value, the economic priorities of donors supplant the wishes of those living within conservation areas thereby creating fundamental conflicts of interest (Belaïdi 2008). Transfrontier conservation and the coviability objective it supports find themselves subordinated to development, owing to a prioritization which transformed the southern African experience into an initiative led by a neoliberal approach to conservation (Buscher 2013) as opposed to an original initiative supposed to present alternatives (Belaidi 2015b).

References

Alexander D (1990) Bioregionalism: science or sensibility. Environmental Ethics 12(2):161–173

Ali SH (2007) Introduction: a natural connection between ecology and peace? In: Ali SH (ed) Peace parks: conservation and conflict resolution. MIT Press, Cambridge, MA, pp 1–18

Belaïdi N (2008) Le Great Limpopo Transfrontier Park. Une gestion régionalisée de la biodiversité au service du développement? EchoGéo, n°7. http://echogeo.revues.org/8523

Belaidi N (2012) L'eau de la rareté économique aux considérations de "développement durable": une ressource pluridimensionnelle. In Belaidi N (dir) Eau et société: enjeu de valeurs. Les ambivalences du droit face à la complexité de l'environnement, Bruylant, coll. Droits, Territoires, Cultures, Bruxelles, pp 11–25

Belaidi N (2014) Identité et perspectives d'un ordre public écologique. Droit et Cultures, n° 68–2014/2, "L'environnement, une valeur sociale essentielle? L'Ordre public écologique, du concept à la juridicité. Exemple des polices de l'eau en droit français" coordonné par N. Belaidi, pp 15–49

Belaidi N (2015a) Théorie du droit et front écologique: apport à la (re)définition de la justice environnementale. Développement durable et territoires, vol.6, n°1, mis en ligne le 31 mars 2015

Belaidi N (2015b) Justice environnementale et dynamiques socio-écologiques: processus et perspectives à partir de l'exemple des Parcs pour la paix. In: Kolata A et Benest G (dir) Territoires écologiques: quelle écologie, quelle économie pour un territoire? L'Harmattan, Paris, 2015, pp 101–109

Belaidi N et Koubi G (dir) (2015) Droit(s) & Géographie(s). Développement durable et territoires, vol 6, n°1

Berg P (2001) "Aux sources du biorégionalisme" Propos recueillis par A. de Benoist et M. Marmin, Elément, n°100

Berg P, Dasmann R (1977) Reinhabiting California. The Ecologist 7(10):399–401

Brosius JP, Russell D (2003) Conservation from above: an anthropological perspective on transboundary protected areas and Ecoregional planning. J Sustain For 17(1):39–65

Buscher B (2013) Transforming the frontier: peace parks and the politics of neoliberal conservation in southern Africa. Duke University Press, Durham/London

Carius A (2007) Environmental cooperation as an instrument of crisis prevention and peace-building: conditions for success and constraints. Adelphi Research. http://www.adelphi.de/files/uploads/andere/pdf/application/pdf/us_503_carius_environmental_peacemaking_06-07-02.pdf

Child B (1996) The practice and principles of community-based wildlife management in Zimbabwe: the CAMPFIRE programme. Biodiversity & Conservation, March 5(3):369–398

Conca K, Dabelko GD (eds) (2002) Environmental peacemaking. Woodrow Wilson Center Press/John Hopkins University Press, Washington, DC/Baltimore, 264 p

Cusset PY (2007) Le lien social, Armand Colin, collection 128, Sociologie

Derman B (1995) Environmental NGOs, dispossession, and the state: the ideology and praxis of African nature and development. Hum Ecol 23(2):199–215

Duffy R (2007) Peace parks and global politics: the paradoxes and challenges of global governance. In: ALI SH (ed) Peace parks: conservation and conflict resolution. MIT Press, Cambridge, pp 55–68

Folke C (2007) Social-ecological systems and adaptive governance of the commons. Ecol Res 22:14–15

Frenkel S (1994) Old theories in new places? Environmental determinism and bioregionalism. Association of American Geographers 46(3):289–295

Griffen J (1999) Study on the development and Management of Transboundary Conservation Areas in southern Africa. USAID, Lilongwe

Hajer MA (1995) The politics of environmental discourse: ecological modernization and the policy process. Oxford University Press, Oxford

Hammill A, Besançon C (2007) Measuring Peace Park performance: definitions and experiences. In: Ali SH (ed) Peace parks: conservation and conflict resolution. MIT Press, Cambridge, pp 23–39
Hanks J (2000) The role of Transfrontier Conservation Areas in southern Africa in the conservation of mammalian biodiversity. In: Entwistle A, Dunstone N (eds) Priority for the Conservation of Mammalian Diversity. Has the Panda Had its Day? Cambridge University Press, Cambridge, pp 239–256
Hanks J (2003) Transfrontier Conservation Areas (TFCAs) in southern Africa: their role in conserving biodiversity, socioeconomic development and promoting a culture of peace. J Sustain For 17(1–2):127–148
Houtart F (2001) Des biens communs au Bien commun de l'Humanité. Fondation Rosa Luxemburg, Bruxelles
IUCN/WCMC (1994) Guidelines for protected area management categories. IUCN, Gland/Cambridge
Jolivet S (2012) L'équité et la coopération transfrontalière appliquées à la conservation de la nature. In: Michelot A (dir) Equité et environnement. Quel(s) modèle(s) de justice environnementale? Larcier, Bruxelles, pp 413–434
Jones BTB (2008) Legislation and policies relating to protected areas, wildlife conservation, and community rights to natural resources in countries being partner in the Kavango Zambezi Transfrontier conservation area. Windhoek, Conservation International
Koch E (1998) "Nature has the power to heal old wounds": war, peace and changing patterns of conservation in southern Africa. In: Simon D (ed) South Africa in southern Africa: reconfiguring the region. James Currey, Oxford
Krog A (2002) Country of my skull. Random House, 4th edn, 2006
Kutting G (2004) Review essay: environmental justice. Global Environmental Politics 4(1):115–121
Lagadeuc Y, Chenorkian R (2009) Les systèmes socio-écologiques: vers une approche spatiale et temporelle. Natures Sciences Sociétés, 2009/2 17:194–196
Liu J, Dietz T, Carpenter SR, Alberti M, Folke C, Moran E, Pell AN, Deadman P, Kratz T, Lubchenco J, Ostrom E, Ouyang Z, Provencher W, Redman CL, Schneider SH, Taylor WW (2007) Complexity of coupled human and natural systems. Science 317(5844):1513–1516
Low N, Gleeson B (1998) Justice. Society and Nature, An Exploration of Political Ecology, London, Routledge
Mace GM, PURVIS A (2007) Evolutionary biology and practical conservation: bridging a widening gap. Mol Ecol 17(1):9–19
Matthew R, Halle M et Switzer J (eds) (2002) Conserving the peace: resources, livelihoods and security. IISD/IUCN. http://www.iisd.org/pdf/2002/envsec_conserving_peace.pdf
McGinnis M (ed) (1999) Bioregionalism. Routledge, London/New York
Meredith D (2005) The bioregion as a communitarian micro-region (and its limitations). Ethics Place and Environment 8(1):83–94
Millennium Ecosystem Assessment (2005) Ecosystems and human well-being: synthesis. Island Press, Washington, DC
Norgard RB (1994) Development betrayed. The end of progress and co-Evolutionnary Revisionning of the future. Routledge, London
Pezullo PC, Sandler R (eds) (2007) Environmental justice and environmentalism, the social justice challenge to environmental movement. MIT Press Books, Cambridge
Porteilla R (2010) L'Afrique du Sud. Le long chemin vers la démocratie, Illico. Infolio, Genève
Ramutsindela M (2007) Transfrontier conservation in Africa: at the confluence of capital, politics and nature. CABI, Wallingford, 192 p
Republic of South Africa Truth an reconciation commission, Report, vol 1, chap 5, p 128
Sale K (1985) Dwellers in the land: the bioregional vision. Sierra Club, San Francisco
Sandwith T, SHINE C, HAMILTON L, SHEPPARD D (2001) Transboundary protected areas for peace and co-operation, gland. IUCN, Switzerland/Cambridge

Schlosberg D (2007) Defining environmental justice: theories. Oxford University Press, Movements
Shelton D (2012) "Préface" in Michelot A. (dir.), Equité et environnement. Quel(s) modèle(s) de justice environnementale? Larcier, Bruxelles, pp 17–24
Shrader-Frechette K (2002) Environmental justice, creating equality, reclaiming democracy. Oxford
Spenceley A, SCHOON M (2007) Peace parks as social ecological systems: testing environmental resilience in southern Africa. In: ALI SH (ed) Peace parks: conservation and conflict resolution. MIT Press, Cambridge, pp 83–104
Sperber I (2003) Alienation in the environmental movement. Regressive tendencies in the struggle for environmental justice. Capitalism, Nature, Socialism, volume 14(3)
Thorsdell J (1990) Parks on the borderline: experience in transfrontier conservation. IUCN, Gland/Cambridge
Washington H (2013) Human dependence on nature: how to help solve the environmental crisis. Routledge/Earthscan, London/New York
Weber J (2013) Gestion des ressources renouvelables: fondements théoriques d'un programme de recherche. In: Bouamrane M, Antona M, Barbault R, Cormier-Salem MC (eds) Rendre possible. Jacques Weber, itinéraire d'un économiste passe-frontières,Versailles Cedex, Editions Quæ, coll. Indisciplines, 2013, pp 35–52
Westra L (2008) Environmental justice and the rights of unborn and future generations. In: Law, environmental harms and the rights to health. Earthscan, London
Wolmer W (2003a) Transboundary conservation: the politics of ecological integrity in the great Limpopo Transfrontier Park. J South Afr Stud 29(1):261–278
Wolmer W (2003b) Transboundary protected area governance: tensions and paradoxes. Paper prepared for the workshop on transboundary protected areas in the governance stream of the 5th world parks congress, Durban, South Africa, 12–13 September 2003. http://www.tbpa.net/docs/WPCGovernance/WilliamWolmer.pdf
World Bank (1996) Mozambique Transfrontier Conservation Areas Pilot and Institutional Strengthenineg Project. Report No.15534-MOZ, Agriculture and Environment Division, Southern Africa Department, The World Bank
WWF (2009) Ecoregions: earth's most special places. http://www.panda.org/about_our_earth/ecoregions/
Young SC (2000) The emergence of ecological modernization: integrating the environment and the economy? Routledge, London, 2000

Chapter 16
Coviability in the Governance of Pastoral Systems, Permanence and Change. How Does the Governance of Pastoral Systems Appeal to the Coviability Concept?

Mohamed Taher Sraïri, Jean François Tourrand, Ruijun Long, Adama Faye, Hermes Morales Grosskopf, Fernando Raúl Coronato, Christian Corniaux, and Bernard Hubert

Contents

M. T. Sraïri (✉)
IAV Hassan II – Hassan II Institute of Agronomy and Veterinary Medicine, Rabat, Morocco
e-mail: mt.srairi@iav.ac.ma

J. F. Tourrand
DVM, Center for International Cooperation in Agricultural Research for Development (CIRAD), Montpellier, France
e-mail: tourrand@cirad.fr

R. Long
International Center for Tibetan Plateau Ecosystem Management (ICTPEM), Lanzhou University, Gansu Province, China
e-mail: Longrj@lzu.edu.cn

A. Faye
Animal Production Scientist, Université Cheikh Anta Diop, Dakar, Senegal
e-mail: adama.faye@hotmail.fr

H. M. Grosskopf
IPA – Institute for Livestock and Agriculture Planning, Montevideo, Uruguay
e-mail: paisanohermes@hotmail.com

O. Barrière et al. (eds.), *Coviability of Social and Ecological Systems: Reconnecting Mankind to the Biosphere in an Era of Global Change*,
https://doi.org/10.1007/978-3-319-78497-7_16

16.1 Introduction

16.1.1 The Coviability in the Governance of Pastoral Systems, Permanence and Changes

Pastoral systems involve both rangeland, as a natural open space enhanced and developed by a society and its herds, and to a long span of time, usually cyclical with seasons, seasonal migration and nomadic practices. Here, we will deal with a socio-ecosystem which can be regarded as exemplary stakes of coviability, which is developed in the introduction of this book. The pastoral farmer leads his herds. This is also and above all the generic term referring to the collective group made up of men, women and children, sedentary, migrating or nomadic, permanent and temporary, working for the sound functioning of their socio-ecosystem, that is to say for the control of its dynamics and its ability to adapt to uncertainties. This set made up of societies and pastoral resources and its environment has been studied by various authors all over the world (Costanza et al. 1997; Blench 2001; Wood and Porro 2002; Biber 2006; Gerber et al. 2010; Steinfeld et al. 2010; Dong et al. 2016). We know the stories, customs, rules and practices which differ mainly according to the conditions of the environment and culture, which is normal due to the diversity of these systems that extend over more than one third of the land surface of our planet. However, we are very little informed about the means of improvement since these socio-ecosystems are on the one hand varied, as previously mentioned, and on the other hand complex, especially the several and contrasting interactions between society and environment. But it is precisely these interactions, gradually built over time, that underpin the resilience of these socio-ecosystems (Moran 2006), in dynamics of co-evolution which it is necessary to anticipate.

In this particular situation of pastoral systems, the notion of coviability calls directly upon the process of resources co-production through the interactions themselves between a society and its environment occupied by domestic herbivores. Therefore, we should be equipped with the means to move on to *stewardship* of ecosystem functionalities through a frame in which an agro-ecosystem, pastoral in this case, is regarded as the sum of consumption/production relationships in

F. R. Coronato
CONICET/CENPAT – Patagonia National Center, Puerto Madryn, Chubut, Argentina
e-mail: coronato@cenpat.edu.ar

C. Corniaux
Center for International Cooperation in Agricultural Research for Development (CIRAD), Dakar, Senegal
e-mail: christian.corniaux@cirad.fr

B. Hubert
Agropolis International, Montpellier, France
e-mail: bernard.hubert@avignon.inra.fr

an environment considered stable (or predictable). The purpose is to facilitate "environmental services" and the acquisition of knowledge and skills such as the ability to adapt to changes (climate, policies, values, standards, etc.) based upon concepts such as dynamics, thresholds, resilience, viability core, learning processes and collective action, etc. Such an approach is based on co-evolution and system/environment interactions in a situation of uncertainty, in the spirit of the holistic approach developed by Bland and Bell (2007). To clarify what this changes in the approach, concepts, and methods, we could take Thompson's proposals (1997) or Hubert and Ison's (2011) who distinguishes two approaches about the management of renewable natural resources:

- A classical approach in terms of *resource sufficiency*, based on an utilitarian view of nature, on the notion of resources as a given capital (a stock), which can be considered abundant, renewable or critical and for which solutions to maintain their sustainability is to reduce their rate of consumption, or increase the effectiveness of their productivity, or eventually substitute other resources with the help of a technological change. This is what happens in a caricature way when a breeding system faced with the depletion of pastoral resources falls back on distributed foods, purchased from outside. According to this view, the assessment of the state of an agro-ecosystem is reduced to the sum of creating resources decreased by their consumption, in an environment considered as being stable enough to ignore its variations. This approach leads to public policies to improve the systems' efficiency, in a universal perspective, besides all things being considered equal;
- An innovative approach, which Thompson (1997) qualifies as functional integrity, based on a process of co-evolution, where resources emerge from interactions within a socio-ecosystem, whose critical points need to be identified – of a technical or social order – as they may endanger its sustainability, i.e. in this case its own transformation capacity. Thus, all activities and social organizational forms need to be examined in terms of their interaction and the way they allow – or block – the generation of what is considered as resources within the system. This approach results in policies adapted to local situations, which focus on securing these critical points and must be contextualized and adapted, and whose application must be monitored and regularly evaluated while avoiding prescriptive and centralized devices...

In the first session we will define what we mean by viability and coviability of pastoral systems. The five studied cases serving as references in this article will be introduced in the second session. The third session will analyze the four key factors of change affecting the governance of pastoral systems, namely: the low attractiveness of breeding, the new environmental paradigm, the globalization of the breeding economy and its alternatives, and finally the issue of functional integrity. Three scenarios will be described in the fourth session, first about the disappearance of the pastoral system, a second allowing its conservation and a third targeting its integration. In the conclusion, we will return to the necessary questioning of our own representation systems of pastoral activity and questions portending to the future, particularly the new Human-Animal relationship.

16.2 Viability and Coviability of The Pastoral System

The viability of a system refers to three components of sustainability: economic, social and environmental, and even two complementary ones, cultural and political, of eco-development (Sachs 1998). Coviability integrates more particularly the collective dimension, community of this viability in its relationship with its environment (biophysical, economic, social, political), and thus corresponds to the system's governance.

The multi-functionality of breeding is one of the viability pillars of herder families, especially in pastoral areas (Duteurtre and Faye 2009; Alary et al. 2011). In fact, as a tool of production, the herd covers a part of a family's needs, starting with the food (milk and meat), as well as wool, leather, hides and skins for clothing or equipment making, the use of droppings as fertilizer or fuel, etc. The herd also produces goods to be sold or exchanged (young animals, milk and dairy products, wool and hair, leathers and skins, etc.) providing income, monetary or barter, which allow the family to acquire other assets to cover their other needs. As capital, animals are regarded as available savings in case of accidents, and therefore can, in this case, finance or be used as collateral in any project. The originality of a herd is to integrate in the capital non-marketed young animals, the case of young females integrated into the reproductive herd or young males, possibly castrated for deferred sale. This is considered at the household or family level, which is the basic unit of management of a herd. Even when the herd is managed at the extended family level, there is almost always an allocation of animals to individuals and households by whom final decisions are made (Tourrand 2000).

Pastoral resources are managed either at the level of the household, when the latter is a beneficiary of the rangelands either at the extended family level, seen as a set of households connected to a common parent, or at the community level in the case of common ranges. Pastoral resource management is a major stake because it depends on the production of animals and thus the living conditions of the household, family, community... The resilience of pastoral ecosystems is in general high, in both dried and wet areas with a great adaptation of fauna and vegetation cover respectively to drought and flood. Archaeology shows changes at a geological scale, as for example the Saharan desert today but with much more green cover five or six thousand years ago (McGee et al. 2013). Changes happen also at the human scale, classic case of the tragedy of the commons (Hardin 1968) for many community rangelands. But this also applies to rangelands allocated to households, as in the case of the Patagonian steppe devastated in a few decades, between 1880 and 1930, by settlers' sheep flocks from the Falkland Islands and the Pampas (Coronato and Tourrand 2015), as will be shown later in one of the case studies. The impression is that facing the herds' pastoral pressure, the cover resists, adapts and changes its state dynamically, taking into account climate variations. However, it could collapse because reproduction would be impossible if the resource management remains that of mining, i.e. it does not take into account the necessary renewal of the resource.[1] In addition, unlike the herd, pastoral resources have no mobility as a factor of adjustment in time and space. Therefore, they remain at the mercy of management, the purpose of the major stake of such herd and resource management.

The viability of each system, meaning a tripod made of "pastoral farmers – herds – territories" (Lhoste 1984; Landais et al. 1987), also depends on external factors impacting different components and interactions of the system, within the socio-ecosystem as a set of united or grouped breeding systems. This is how drought, epidemic diseases, variations of the prices of products' sales, affect each breeding system at the same time, certainly in a different way in accordance with the proper characteristics of each of them, but affects the socio-ecosystem in its entirety as well. Coviability corresponds to this collective level, in the governance of the socio-ecosystem. We expect from this governance, in response to changes taking place or to come, to be able to maintain a favorable context for the emergence of sustainable forms of use of the resources by the herds. In the same way, the viability of a breeding system depends on its ability to adapt and manage risks at the family level, and therefore the governance must be able to anticipate and implement appropriate collective measures. In situations of shock or accident, such as an extended drought, each breeding system can cope using its own means, including networks. This will be even easier, quicker and successful if appropriate means to potentiate its own means are found at the collective level. Taking the extended drought as an example, the most frequent climate incidence,[2] resulting in a significant fall in rangelands' productivity, the breeder will mobilize its savings, that is to say selling some of the stock to finance the purchase of food in order to save the rest of his herd, most often playing on the variety of species he raises, large and small ruminants whose fragility to vagaries, patterns of reproduction and growth rates are not the same. Moreover, such food must be available, at an acceptable price, which depends either on market or political measures adopted to control prices, with many risks of perverse effects. In the same way, in response to a deterioration of the pastoral resource, any breeder would be at first glance ready to invest his resources, labor, capital/livestock etc., in view to recover this resource. However, he will only do that if he, on the one hand, estimates his means and those available for him are

[1]This is what happened in Southern Patagonia (Argentina) where the initial animal load at the end of the nineteenth century was around 1–2 sheep per hectare, or even 3 sheep per hectare, on rangelands that only supported 1 sheep per 2–3 ha, and once degraded they no longer supported more than one sheep for 8–10 ha, with no real possibility of recovery.

[2]For example, during the eruption of a volcano in Chilean Patagonia, 2011, Chilean herds were allowed to have shelter in Argentina; the border control was suspended though it strictly considered animal health issues.

sufficient, and on the other hand, if he will be a beneficiary of this recovery, two factors of socio-ecosystem governance. Finally, all breeders have medium-to-long term view, because of the management of his herds, his rangeland resources and his family. A medium-to-long term project will only make sense if there is a real chance of success. Otherwise, a breeder will consider exiting of the socio-ecosystem, a little like the exceptional seasonal migration and nomadic fleeing once resilience is no longer possible (Eldridge and Freudenberger 2004). He can transfer his lands to agribusiness or to a more favored neighbor or at least less sensitive to the risk. The construction of a medium-long term project, even stronger if straddling two generations or more, has meaning only if the context and prospects are favorable, and allow the establishment of some confidence in the long term, which is part of the modes of governance at different levels of interlocking organization whose interactions provide – or not – the coviability of the systems at these different scales.

The sustainability and coviability of pastoral systems will be analyzed based on the five following case studies; case studies which are simultaneously contrasting and in geographic areas far enough removed from a geographical, societal and cultural point of view as they have been subjected to significant interactions, strongly impacting their respective dynamics.

16.3 Five Contrasting Cases of Pastoral Areas

Five case studies were carried out in contrasting biomes, climate zones, geographical and cultural areas, but representing a gradient of types of pastoral situations and their coviability conditions:

- "traditional" systems changing with the property regime, in the mountains of Qilian in the center of China, whose coviability will depend on the consistency of decisions and multi-scale public actions;
- destabilized systems by the loss of part of their traditional territories due to their assignment to other activities and the reduction in their mobility on the Mediterranean coast of the Northwest in Egypt, and in the Sahel of Ferlo in northern Senegal, and whose sustainability was based on interdependencies rarely taken into account in decisions made based on other considerations and on other objectives;
- the 'New World' systems, for which speculations chase each other in the Northern Pampas of Uruguay and southern Patagonia in southern Argentina (Fig. 16.1) and whose viability at the local scale is completely determined by decisions taken at other levels without taking into account local dynamics and equilibrium.

The Mountains of the Qilian are located in the northeast of the Tibetan Plateau along the Silk Road. They are located above 2500–3000 m in altitude. This coupled with a nearby 36 °N latitude is the cause for the cold, the main factor limiting agricultural activities which is thus essentially fodder and animal rearing. According

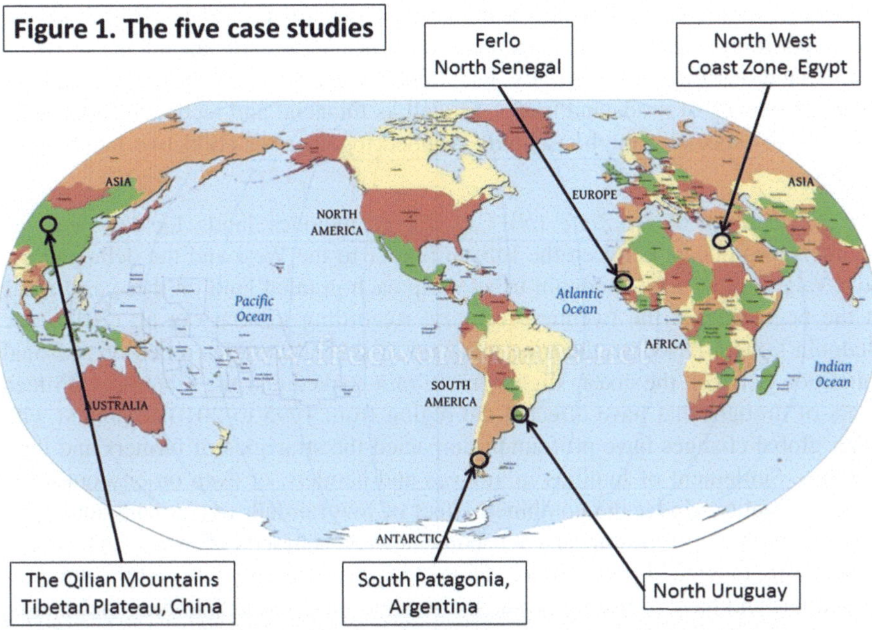

Fig. 16.1 The five case studies

to Long et al. (2014), livestock managed at the households' scale consist of yaks and Tibetan sheep, particularly suited to the cold, and conducted separately. Several sheep breeds including the Merino have been introduced, but they must be supplemented in winter and spring with forage produced in irrigated plots during the 3 months of summer, whereas supplementation of yaks and Tibetan sheep depends more on the strategy of the household. Resource management is based on seasonal migration. During the 3 months of summer, herds migrate to the pastures, above 3000–3500 m. Yaks stay there in the fall. Sheep flocks pasture over the course of fall, then move on to winter rangelands located near villages, whereas yaks remain on the autumn ranges for the best part of the winter period, then come back, in late winter and spring, close to the villages where lambing and calving occur. Sales of lambs are held in early fall whereas yak calf sales are more spread over the year. Resource management strategies depend on the villages. For some, all the rangelands are common. Only allocated land is devoted to the production of forage, given to the animals during scarce feed availability periods, mainly in winter and spring. Common forage fields also exist. In some villages, all lands are attributed. In others, only the winter pastures and forage fields are assigned. Governance is based on two entities dedicated to work together and discuss, namely: the leader of the village, selected for 1 or 2 years by households on the one hand, and the manager of the village, a territorial agent in charge of the application of public policies on the other hand. Households receive several subsidies for land, for the number of animals, equipment, repairs and the construction of housing facilities, etc. In addition to subsidies management, governance concerns the allocation of the use of land, equipment (electricity, telephone,...) and services (health, education), the maintenance of roads and tracks, as well as financial and technical support for innovation, especially the households grouped together for collective fattening of herds.

The North West Coast Zone (NWCZ), Egypt includes lands located near the Mediterranean coast between the Libyan border to the West and the delta and the Nile Valley to the East. Bedouin tribes were each granted land on these arid areas at the beginning of the twentieth century. According to Alary et al. (2014), the Bedouin system based on the tribe made up of several families relied on seasonal migration between the coast, slightly more rain-soaked and the hinterland. Fifteen years of drought that have affected the region from 1995 to 2010 combined with other global changes have profoundly impacted the strategies of farmers and their families. Settlement of families in villages and hamlets, or even on city outskirts, is now a reality. Under the combined effect of low rainfall, overgrazing and wind erosion, many steppe pastures have become nowadays deserts of stones. In response to declining rangeland productivity, the Bedouin have sharply reduced the numbers of animals and have to use by-products found on local markets to feed their herds, sometimes more than 8–9 months a year. Many have left animal production activities to migrate abroad, sometimes as shepherds, particularly in Libya, or any other job in the Gulf States or along the Nile Valley, and also along the Mediterranean coast where resort and seasonal tourism is growing strongly. Those who have had access to 'Oued' riverbeds, are engaged in the food production of figs, olives and vegetable crops thanks to the infrastructure built by the State and its landlords to use the little water available. Finally, others, with more capital, are engaged in intensive breeding such as poultry, for the national market and exportation, and in the production of local sheep for the Muslim religious festivities. Rangelands are the cornerstone of the development. In fact, in accordance with successive military government policies, national and international institutions in charge of the development of these areas have preferred habitat, education, health, irrigated agriculture and tourism, leaving the rangeland to the fate of low rainfall, overgrazing and wind erosion. The land issue is at the heart of this renunciation because the Bedouin tribes claim their land rights in many lands now in the hands of the army, dignitaries of the regime, tourism and urban development along the coast. Furthermore, tribes were little associated with public policies carried out by successive governments, with approval or not of their landlords. There is therefore a real problem in terms of governance that the Arab Spring exacerbated. However, tribal leaders are well aware of the urgency and interest to back the tribes at the center of public action, for the sake of their recognition as well as their ability to listen and operate.

The Ferlo is the Sahel zone of *Senegal*. Bound to the East, North and West by the Senegal River, populated mostly by the Fulani ethnic group, the Ferlo is a breeding zone for seasonal migrating ruminants. Once upon a time turning towards the Valley of the Senegal River, seasonal migration is today focused on the Center and the South of the country (Cesaro et al. 2010), due to the building of dams and wide irrigated plants dedicated to cropping and no more livestock free ranging. According to Touré (1986) breeding is based on the extended family with its household, herds of cattle, sheep and goats, with individual assignment of animals and transmission to descendants. Even if it relies on breeding, the Fulani cultivates millet in rainy seasons, especially for family consumption. In case of drought, the Fulani may migrate on very long distances, with their families. Herd management is a family affair while rangelands' management is a community one. Some family management initiatives of range resources have been implemented but with no convincing results. Building wells within the Ferlo in the middle of the twentieth century has profoundly changed the seasonal movements allowing pastoral farmers to stay on their land throughout the year, with some of their livestock, mainly cattle, even in times of drought as water supply is provided by the wells. From the 60s–70s, the agricultural development of the Valley of the Senegal River by irrigation and the advance of the groundnut to the South have also strongly impacted livestock systems by reducing drastically the pastoral space in some areas. As in the previous case, a part of Fulani pastoralists converted to irrigation agriculture, for those who had access to the water low lands in the Senegal River delta. Others have preferred to get jobs in the established agribusiness activities which emerged. Others finally take advantage of regional dynamics to sell their produce, particularly dairy products and meat. According to Homem et al. (2008) and Touré and Planchon (2008), some flagship measures in the 1980s gave the Fulani leaders real power at the local and regional scale. However, despite this the Fulani remain, with the exception of a few cases, overall excluded from the building of public policies, most often in the hands of persons from Dakar or agro-industrial operators seeking land and advantageous financial operations claiming a potential technological progress that the political discourse calls for, and has repeatedly done so for nearly half a century. Even if some of these operations are successful, the Fulani remain overall on the edge of the governance of the local land, a little like the hinterland which watches the coast becoming urbanized and developed.

The image of the hinterland matches also *The Northern Pampas of Uruguay*. A natural prairie known as *Campos*, dedicated to breeding since the establishment of European settlers in the sixteenth and seventeenth century, populated by gauchos, it is part of the Brazilian Pampas of the space initially colonized by Jesuit missions creating communities integrating migrants and native Americans, largely around the common management of the pastoral lands (Moraes 2008). Gradually the land has been attributed with areas of a few hundred hectares to a gaucho family (couple, parents and children) with its herds of cattle for meat and leather and sheep for meat, wool and leather. Very soon turned to the export of leather and skins, then meat with steam refrigerated transport, the country developed a good reputation, over time, of attractive meat to grass, natural, healthy and quality. As previously mentioned, breeding has often been associated with a bit of rain-fed agriculture, grain and oilseed, mainly for food to families and the farmyard. The necessary equipment and construction wood came from the small groves planted in all farms. The first big change comes with mechanization after the 2nd world war, with tractors used in the natural pasture to plant a more productive but less resilient forage grass, especially during summer and winter periods of drought. This was an attempt to copy and paste New Zealand's model, but as in the case of Argentine's Patagonia previously mentioned and detailed further, the context of the paste is far from that of the copy. In other words, the weather conditions of the North Island of New Zealand are much more favorable than that of the Pampas to single intensive forage practices because of the better distribution of rainfall and more uniform temperature throughout the year. But progressively, a few dozens and then hundreds of thousands of hectares of natural grassland of the 16 million hectares of the original area have been returned and used for growing crops.[3] The second big agricultural change comes from the succession of financial and economic crises at the end of the 1990s that ruined many small farmers forced to sell their land and move to Montevideo looking for jobs. Benefiting from a benevolent policy, a few international forestry groups have planted a large part of this land, then gradually developed so that today about 1.5 million hectares are handled (Arbeletche et al. 2012), planted with eucalyptus and pine trees, especially in North Uruguay. The result is interesting from an economic point of view by the added value and jobs created. However, there are no more wells less than 8–10 m in depth and the original natural grassland will not re-grow after this tree planting. The last big change is the wave of the soybean growing, also in the hands of a few groups, referred to as *Pools of Siembra* (Pool of seedlings) that swept from Argentina, in connection with the interest of the capital for this crop following the collapse of markets in 2007/2008. They cultivate today nearly 15% of Uruguay under a very special form because land is rented, using the services of local operators, and only managing the capital, the application of technology and the export of the production. The great social question is how many young gauchos will remain as breeders on their land, while the economic rationality is to rent to the *pools of siembra*, or even become operators of these same *pools of siembra*, and especially live in the city near services to which rural life only give partial access. The environmental challenges are the impacts of pine and eucalyptus plantations and the cultivation of GMO soybeans, pesticides and especially the erosion of soils.

Argentinian Southern Patagonia is one of the best examples yet unknown, on how breeding destroys, in less than half a century, a millennia-old natural steppe ecosystem. According to Coronato and Tourrand (2011) and Dong et al. (2016), not used to the herds of ruminants, in addition to the guanacos, wild camelidae of average size, the steppes of Patagonia have not supported the strong loads imposed by settlers at the end of the nineteenth century, in the order of 5–10 times what they could accept. The land system, based on private property at the fixed limits, has only increased the pressure on the resources. Therefore, in less than four decades, the less watered pastures and thus most fragile, i.e. the center and the east of Patagonia, have been destroyed and depopulated due to abandonment of properties. As in the Bedouin area of NWCZ, under the effect of low rainfall, overgrazing and wind

[3]Morales, H. 2007. Consequences evaluation of strategic decisions for plants based breeding in Uruguay. An approach by multi-agent systems (SMA). *Thèse de Doctorat, ABIES/AgroParisTech*, Paris, France, 159p.

erosion particularly strong in these latitudes, rangelands have turned into deserts made of stones, glacis and very sparse herbaceous vegetation and stunted shrubs. The western part, at the bottom of the Cordillera of the Andes, and the wetter southern area held up better and remained farming land. The family land is usually big, some thousands of hectares, but generally not enough for more than one to two thousand sheep, which give a much lower salary than a technician operating in oil or gas, less the urban living conditions. The characteristic remoteness of rural Patagonia settlements, in relation to the size of farms, combined with the abandonment of many properties, does not facilitate community life and various services, health and education in particular. As in the previous case of the Pampas, a property housed during the "Belle Epoque" a couple, one or both parents, children, as well as one or two employees. Today, very rare farms house the whole couple, as the wife prefers to live in the city, especially when she has children, for the reasons already mentioned. The center of southern Patagonia is mostly populated by guardians of farms rather than farmers, mostly single who dream to find an urban job in the near future, a situation farm owners have already been familiar with for the last decade or two.

16.4 Forces of Change Impacting Governance

In the previous five case studies, it appears that three sets of factors impact already the governance of pastoral systems and control measures that can act on their viability, and should continue. The first one is the low attractiveness of breeding, especially for young people. The second is the paradigm shift towards the environment. The third is the globalization of the economy and its alternatives.

The adoption of governance based on the principles of the functional integrity of pastoral systems seems relevant, because based on a systemic vision, means integrating the entities and their interactions to produce ecosystem services. Therefore, the functional integrity is based on the resolution of endangered specific critical interactions of each situation, but can also be found in the general considerations that are included in this part and which are at the heart of the interactions between social systems and ecological systems at different levels of organization, from the local up to normative considerations undertaken on a global scale.

16.4.1 Low Attractiveness of Pastoral Livestock Rearing

This is not recent. It dates back to the second half of the twentieth century when the search for modernity became the recurring theme of development and public policy. The Uruguayans locate it in the post-war years with the development of agriculture mechanization, when the Senegalese and the Argentines place it in 1960s–1970s. The Chinese experienced it a little later, at the very beginning of the 1980s, at the end of the Cultural Revolution. It would be most recent among the Egyptian Bedouins of the NWCZ and would coincide with the beginning of the drought of 1995–2010. In our case studies, pastoral breeding gradually became the effigy of tradition, the emblem of past values with its outdated side, somewhat unfashionable. It is worth noting that unlike Western Europe, the neo-rural life of 1960s–1970s has not affected the Southern hemisphere countries, or only very marginally (Tourrand et al. 2016).

The rural exodus has continued inexorably since the second half of the twentieth century, and has done so even if the conditions of living in the countryside have considerably improved and the rural population has increased slightly, as in the case of the Ferlo region (Faye 2016). Geographic isolation has long been advocated as one of the key factors for the low attractiveness of livestock rearing, the rest in the most remote corners of Patagonia or Ferlo, and to a lesser extent of the Pampas. This is not really the case elsewhere, especially with improved road infrastructure, mainly thanks to the new technologies of information and communication. In return, telecommunication antennas have become the standard-bearers of economic development in rural areas. Basic services such as access to drinking water and electrification in rural areas have significantly been improved, as well. However, health and education are still two areas requiring high public investment – even impossible to fulfill – given the low population density. So, they are always regarded as topical reasons of the rural exodus. It is clear that all these factors and indicators fall more within the community and thus the governance of the territory than the *sensu stricto* breeding activity.

The lower remuneration of livestock rearing appears as a second reason, especially compared to other agricultural activities or a remunerated job in the city. The figures show that the income of a couple of breeders in the Qilian Mountains or Uruguay is equivalent to that provided by a job in the city (Long et al. 2014; Saravia 2016). In addition, the breeder has a capital savings providing him with a guarantee and allowing him to invest. A simple Bedouin shepherd earns barely less than a university graduate. A Bedouin breeder meanwhile earns more money by keeping his own herd, in so far as it reaches a size estimated at 120–150 sheep and goats. The situation is different in Argentinean southern Patagonia because of the chronic shortage of labor in oil exploitation which already absorbed almost all young people of the rural world by offering them high wages. However, on the other hand, the cost of living in the city is also higher and compensates only partially for the income differential. On the other hand, access to health and education services is much easier. The Ferlo is a special case, especially the northwest. In fact, following the droughts of the early 1970s and 1980s, many Fulani are engaged in irrigated agriculture on the land they farmed in the dry season and took up paid jobs in agribusiness in the irrigated agricultural development. The pluri-activity at the family scale became the dominant model in some villages, less in others, but remained strong. Livestock rearing is also a way to enhance the savings of the families more quickly than in the banking system.

Unlike remuneration, the question of the workload within livestock rearing is of concern. Even if the saying "*breeding is all day, all week, all month and throughout the year*" should not be taken literally; it is clear that breeding is a demanding activity, with inevitable duties and productivity that much depend on the presence and attention of the breeder. In many situations pastoral breeding is based on the "custody" of the animals by shepherds (*sensu lato*) in order to often protect them from predators, particularly small species, but in particular to drive them to grazing areas to associate in a daily circuit a diversity of plants of different nutrients contents and manage pastures, the renewal of which is the most delicate. In other situations, even if the books state that a daily visit is enough for a herd at pasture, it is better to go there twice, once in the morning, once in the afternoon, even better three or four evenly spaced times over 24 h. This allows early intervention in case of accidents, illness or other untoward event, and thus limits losses. The story is similar in the calving period. The existing figures show that herd productivity suffers a lot. Thus, it is hard to believe that herds of yaks do not need a shepherd when they are in the mountains. Any breeder will say that it is better to go there once a day, and if not possible, once every 2 days or twice a week. This takes time because pastures are far away from homesteads, but the gain in productivity is significant. In traditional pastoral societies, herd management at the extended family scale, more intense community life as well as mutual assistance, were methods for addressing the high workload of breeding. Management on a household scale, or even on an individual scale, is inadequate against this constraint. So, it is actually expected that the level of mutual assistance, teamwork, rotating labor, quite similar to grouping together for the management of farm equipment. Incentives for new forms of work and work organization should also be carried out collectively and therefore in terms of governance.

16.4.2 New Environmental Paradigm

The need to take into account the environment to reduce the impacts of livestock breeding is fairly recent in the pastoral world. Certainly, farming has long been accused of being a key factor in the desertification of the Sahel, but "*have you ever seen a goat with an axe in the mouth or in paws?*" to use the words of a Senegalese researcher. However, the desertification as a result of an overgrazing is a frequent process as shown in our case studies, particularly in Southern Argentinean Patagonia (Coronato and Tourrand 2015) and the Bedouin area of NWCZ (Alary et al. 2012), following the introduction of domesticated herbivores in an environment not made for it or to the reduction of mobility as a result of the loss of a part of the pastoral territory and withdrawal into areas originally dedicated to domestic seasonal use. Environmental lobbies advance exorbitant figures about land degradation and desertification due to animal overload. What exactly is the situation? We have seen that in the Bedouin cases, as in Patagonia, and to a lesser extent in the Ferlo and on the mountains of the Qilian, the degradation of the rangelands, including the soil, is the result of a complex process in which various climatic, environmental, social, economic, political and cultural factors intervene. There are also greenhouse gases of which ruminants are large producers, nearly 18% of the global production

(Steinfeld et al. 2006; IPCC 2009), because of their digestive metabolism. This metabolism allows them in return, however, not to enter into competition with human beings for food, and to be complementary in the development of the plant resources, and to be one of few foods produced without consuming fossil fuels while promoting biodiversity. Moreover, pasture areas fix carbon into soils.

The environmental issue has become nowadays at the heart of our global society, especially for issues related to rural and agricultural topics. In fact, all public policies on agriculture in general and on livestock in particular now refer explicitly to environmental issues. This is true regardless of the country. This comes either from environmental lobbies, global players by self-imposition, or national and international institutions. It should also have a significant role in the funding of public policy. It is, therefore, clear that the environmental issue has gone beyond the breeding system experienced by the breeder and his family, as his almost only alternative is to play a little on the standard imposed, for example, such as Tibetan breeders who plan the maximum number of animals on their rangeland areas. Environmental issues fall within the domain of territory governance, the only level that can intervene on standards and their applications, and even only then within a scope granted by national and international bodies.

Furthermore, pastoralists have a landscape role through maintaining the areas open by reducing the overgrowth, which also prevents fires. In several European countries, the pastoral farmer receives bonuses for this function so as to control the impact of livestock on the ecology of pastoral resources, especially the change in biodiversity, the balance between grazed, non-grazed plants, or only grazed under certain circumstances, at certain vegetative stages or during periods of less resource availability or associated with others. The mobility of herds also plays a key role in biodiversity. Yet little is known about the processes and dynamics especially in the medium and long term. The Argentinean southern Patagonia is an example of the brutal and fast degradation of an ecosystem by breeding, but what can we learn beyond the fact that it is imperative to adapt animal loads to the pastoral resource or to radically change the management of the herd? We know that the *Campos* of the pampas in northern Uruguay evolved under the pressure of the herds, but in what proportion and how? This demonstrates the lack of information and data, and thus research on these ecosystems, a scientific challenge which also falls under governance. In the same domain, access to Internet has also become an asset because it allows information to be obtained remotely on foraging availability in remote areas (or the attendance by "competitors") as well as to access the weather forecast, which is crucial for resources' renewal, etc.

Another role of pastoralism consists in being a presence, a witness to human activity, and an icon of the land in some areas. Who would imagine an Alps landscape without a few cows during summer? The same concept applies to Mediterranean rangelands. Could someone think of them without a flock of sheep or goats? What about the Tibetan without his herd of yaks and sheep? What is a Fulani, a Bedouin or a Gaucho without his herd? In all these examples, there is an identity role of breeding combined with a landscape role. These roles and functions can be learned within the family, but are shared within the community during holidays, religious events, decisions related to the future of the land, etc.

16.4.3 *Globalization of the Breeding Economy and Its Alternatives*

The globalization of the economy has strongly impacted pastoral breeding in various ways. First of all, there is the largest movement of animal products over long and very long distances. However, these products, especially the perishable products such as milk and meat, were generally produced and consumed at a local or national scale, even if two of our case studies, Patagonia and the Pampas, exported meat and wool to Europe at the end of the nineteenth century. The recent scandals on the meat market (Tourrand et al. 2015) show circuits which were quite unimaginable barely two or three decades ago. Pastoral breeding is affected insofar as these products compete at very low prices in relation to their industrialization. In our case studies, these are the cases of milk powder and other dairy products that are sold in groceries in Ferlo, the Tibetan plateau and Bedouin zone of the NWCZ area. This is also the case for lyophilized preparation of meat-based products and other animal-based products, which are winning more and more shares in the market, especially in fast food, even in rural and pastoral area, mainly due to their low cost and ease of preparation.

The second factor, mentioned before, is the price of this industrial production. The cost of production has significantly declined through the provision of better technology and economy of scale. Transportation costs have increased relatively little and their environmental impacts are not yet taken into consideration. Moreover, while previously the price of industrial products was generally fixed based on the price of local production, the global economy tends to impose its price with little care of the impacts at the local scale. The result is greater vulnerability of local production, although alternatives to global systems are emerging everywhere.

Much like price, globalization has imposed quality criteria. In fact, industrial products are designed to be normalized and theoretically irreproachable from the point of view of their health status. Their standards have thus gradually been imposed on the market forcing local production to carry out controls and make investments, especially since public policies have taken over for the sake of public health. Even if contributions to public health are undeniable, it is interesting to note that quality concerns above all, health issues, in addition to the short and medium term. Thus, the organoleptic and symbolic quality of industrial products is only very rarely addressed, as are additives to products, mainly food preservatives and residues from production, which could be disastrous to public health over the long term. Clearly, standard measures are one-way playing in favor of industrialized products, and above all systematically at the expense of local production. In addition, it excludes from the market upstream and downstream, losses due to the imposed constraints.

In response to food uniformity or standardization, also called junk food, alternatives are emerging. These rely on the short, or even very short, circuits with direct sales from the producer to the consumer. They also have origin names, even if the latter are increasingly dominated by global players. They are called agro-ecology and organic farming strongly differentiating themselves from the concept of industrial production. These alternatives attract more and more consumers, and not only among the most favored social classes, who are apparently ready to pay a significant difference in price. With its natural side, pastoral production has a major role to play, as can be seen in our case studies. The Berki sheep of NWCZ is in demand in Cairo and Alexandria for the Muslim religious holidays (mainly Aïd El Adha) because of its quality, as in the case of the pastoral mutton in other North African countries during religious and festive ceremonies (Sraïri 2011). The same thing applies for the Patagon sheep for the *parrilladas*, as well as the butter and dried meat of the yak in China.

Among the positive points of the global economy should be mentioned its impact on labor and its management, as well as in many other activities of agriculture in general, and in breeding in particular, mainly by bringing an end to isolation and the lack of information. In fact, cell phone services provide stakeholders in pastoral areas with almost instant information about the employment market. For example, in a few phone calls, a prospective employer and a prospective employee know the pay grade practiced in the sector, the material and financial advantages associated with the position, the reasons for the departure of the previous employee, the actual qualification of the future employee, etc. A similar situation consists in the urgent need of labor. Mobile phones have become essential in collective labor, technical assistance, transfer of funds avoiding thus robberies, the marketing of products and the supply of inputs, etc. Having access to operational mobile telephone services in the area is much needed, and this is the responsibility of the community, or even the nation.

16.4.4 Governance and Functional Integrity

Clearly, we see many mechanisms and public actions acting on the collective scale that can reduce the vulnerability of breeders' families. Not all have the same function and operating mode, but all fall within what might be regulating measures for the increase of resilience of the pastoral socio-ecosystem, in the sense of maintaining the functional integrity of the pastoral system.

Some of these mechanisms do not focus directly on livestock rearing, but more on the territorial context in which it develops, as for example the development and maintenance of road networks and communication or those of social services in health and education. Although not being in direct connection with the pastoral production, they are, nevertheless, essential to the assessment of the conditions of life and work, to the prospects of local actors and thus the future of territories. Among our case studies, this concerns the Qilian Mountains of China, northern Uruguay and Argentinean southern Patagonia; three territories in which there is a significant effort of public investments at the request of the pastoral population supported by communities and systems of representation of stakeholders, professional and political systems. In contrast, other mechanisms directly target pastoral production at the family scale. They often operate through the provision of funding (for equipment, expansion, diversification, particularly in ecotourism, compensation in case of climate crisis, etc.), of subsidies (for inputs, animal, rangelands, etc.), to encourage, for example, the implementation of new practices so as to improve limiting points at the territorial scale. Among our case studies, the Qilian Mountains are more significantly engaged in these mechanisms averaging 20% of the budget of families (Long et al. 2014; Qi et al. 2014). They are wanted by the central ruling power and implemented by organized local society which previously took upon itself to raise local demands, according to a process proper to the Chinese model applied in pasture lands. These levers are less topical in the four other case studies due to the more liberal policy adopted towards breeding, in Uruguay and Argentina, and to the deficit of public funding specific for this sector in the Egyptian Bedouin area and the Senegalese Ferlo.

Another set of proactive-type mechanisms, apply directly to breeding activities by fostering the emergence of new forms of production and commercialization for economic and social added-value, in particular by keeping young people in the territory. This is the case of locally sourced produce systems which have been emerging in industrialized nations in recent years, but also in emerging and developing countries. It is also the case of intensive poultry breeding in the NWCZ in Egypt and sheep fattening workshops, also in the NWCZ, as well as in the Qilian Mountains in China. Among such mechanisms, some resolutely aim for no environmental impact and/or the greatest ecological efficiency. We think of organic farming that local governance measures could support by operations of general public promotion, in schools, public catering, etc., as well as by facilitating procedures, and organizing rangeland uses across the territory, etc.

Finally, there are mechanisms specific to the pastoral sector but interacting with other sectors with which they are involved in the territorial dynamics. Particularly concerned is technical assistance, information and training, forums of exchange among actors, best knowledge, evaluation and access to data especially economic, support to research on collective and territorial dynamics, etc. This type of mechanism is particularly visible in Uruguay where there is always a significant investment by local governance officials and national ones in general, in these areas, if only due to the livestock sector, which was, for a long time, the main provider of resources of the country thanks to its exports.

However, all these mechanisms are mainly human, animal and plantation-type agriculture, and very little natural resources, i.e. the rangelands, as if they were, if not unalterable, inexhaustible. This asymmetry is explained by both the little visible return on both the medium-long term of the actions on the resource while actions on the human and animal are more immediate, and by the lack of knowledge on the dynamics of the rangelands, particularly in the countries of the southern hemisphere. Therefore, it seems necessary to invest in research with a functional integrity approach in order to better understand animal-plant-human interactions, in the medium and long term, mainly on water, soil, biodiversity, and land. It is imperative to integrate local knowledge in this approach because numerous rangelands are not natural, and thus without sound human intervention, no sustainable management of resources would be possible, as some stakeholders have the impression of an infinite resource.

16.5 The Future Scenarios

The three scenarios (1/Disappearance, 2/Conservation, 3/Integration) introduced here come from the comparison, jointly conducted by the MOUVE[4] research project and the LIFLOD[5] network, of results obtained from various fields of research including the five case studies of this chapter (Fig. 16.1). Other scenarios can be developed according to the criteria adopted. These are the most relevant in terms of the five case studies.

The scenario 1 called ***disappearance*** is the gradual disappearance of the pastoral socio-ecosystem by the abandonment of one of the three entities, then successively of the other two, in the same way as to what has occurred over the past 30 years in the center of the southern Patagonia and that which has just occurred in the hinterland of the NWCZ. In both cases, the rangelands have gradually disappeared. Herds have gone the breeder families too, as well as other families who also lived through livestock rearing. In the case of Patagonia, we saw that a strategy of mining of a renewable resource without seeking to understand the determinism of its reproduction leads to desertification in the region. While in the case of the Bedouin area, it seems that the key factor was the drought of 15 years, from 1995 to 2010, with very limited rainfall (an average of 80 mm per year to more than 140 mm) and the loss of the coastal areas of seasonal use leading to a permanent focus on resources that did not support this type of use. The overgrazing and wind erosion, far from being only secondary factors, are worsening the effects of low rainfall.

The case of the Uruguay is very different. From the beginning of the 1990s, eucalyptus and pine were planted in the southern and eastern parts of northern Uruguay on permanent pasture lands. The dynamic strongly escalated in the early 2000s with the boom of agribusiness through the same forest group for timber and cellulose, to which *pools of siembra* with soybean as standard bearers were added. In less than a decade, nearly 1 million hectares of prairie (out of the 16 that the country

[4]MOUVE: Breeding and Territories interactions in the setting in motion of ecological intensification of ANR-SYSTERRA research.

[5]LIFLOD: Livestock Farming and Local Development Network (www.liflod.org).

counted) have been transformed, and 3 million in a little more than two decades (Arbeletche et al. 2012; Saravia 2016; Morales et al. 2015). In these cases the herds also have disappeared from the landscape. Not all humans have left because there are at least those in charge of the field work in agribusinesses, but the rural exodus is still significant with more and more single persons, while families leave to go and settle in the city.

These three cases of the disappearance of pastoral socio-ecosystems are entirely due to different factors, but lead to quite similar situations with regards to pastures, herds and families who lived from their interactions.

What can be said about the governance in these three cases? In the case of Patagonia, this concerned a colonization war led by the Argentinean army, stimulated and supported by the British agribusiness seeking to locate in the region, its herds cramped in the Falklands and, to a lesser extent, in the Pampas. In the Egyptian case, it is clear that from their own confessions, that even if the tribes participated in the planning and development of agriculture in the Oueds, they were not, or only very slightly, associated with the management of the pastoral zone for which the public authorities have never granted them a property right, in contrast to the army which was able to take over land. The case of Uruguay is more complex because the call of modernity affects the policy maker as city lights affect the young rural. The 2 million hectares of natural pastures transformed into plantations and soybean fields represent only one-eighth of the country's agricultural area. The economic benefits are significant, cellulose and soybeans have become two of the main export products. Environmental damage may occur, but it would be difficult to assess them precisely. Fortunately, in its political transition, Uruguay managed to implement environmental policies which will at least prevent generations of soy-soy *ad aeternam* until the complete degradation of soil fertility.

Scenario 2, called ***Conservation***, consists in setting aside land in order to maintain and protect the socio-ecosystem. Several options are possible, the prohibition of grazing of ecosystems with or without maintaining native populations, much like natural parks or reserves with populations largely under control. In both cases, governance is outsourced. In the case of natural parks, totally, with their management bodies, their Board of Directors, their own staff... and only partially in reserves where heads of clans and tribes are involved in the management of their territories. This right to participate is attributed to them by the operators of these projects – more or less closely with public authorities – who are also their major landlords. In return, those same authorities keep, as a last resort, their hands on the riches of the subsoil, which is a non-negligible compensation. Among the many natural parks existing in southern Patagonia, very few are in the degraded area, but they are already integrated in tourist circuits.

Such a scenario is ultimately possible for certain parts of the Qilian Mountains, especially if the rural exodus continues and if developing new forms of more community-based breeding and collective work is adopted, particularly in the bottom of the valleys. One can thus think of pastures' exploitation during seasonal migration where shepherds would alternately take care of herds, etc. This would look quite like what probably existed before the Chinese Revolution at a time herd-owning tribes and land and rangeland-owning monasteries!

A similar scenario can also be designed for the Ferlo, at least in the less populated areas and in which families are reluctant or will be reluctant to go to as it is far from everything, and especially as public authorities would not invest very quickly in the construction and maintenance of communications infrastructure.

Scenario 3, called ***Integration***, aims at maintaining the socio-ecosystem with its three components: Rangelands-Human-Animal, such as it is, such as it evolves, trying to better integrate these three components, reduce impacts, produce ecosystem services, potentiate complementarities in time and between areas, expectations, etc., far removed from a backward-looking vision. The public authorities desire this form for the Qilian Mountains in China, in which a lot of subsidies have been invested for nearly 20 years. We can find a comparable system in French regional parks where specific governance is implemented with, initially, a large place for local actors.

This is the scenario strongly desired by many elected officials and managers of territorial administration because it combines different actors with both consensual and cooperative governance. This scenario is scalable which allows proposals for improvements to be made, shared and discussed and then eventually to build and integrate new public measures. Two major disadvantages associated with this scenario are: firstly the impression of being unsuccessful, being in permanent change: "the constant search for improvement", which is normal because this is inherent in its definition; then the need for a strong dose of self-denial from leaders, a sinecure if they don't have any political ambitions.

This scenario is the one in which the governance of northern Uruguay is engaged for areas still little affected by cellulose and the soybean agribusinesses. Sharing arenas[6] of information and discussion between those in charge and local leaders can raise the requests and expectations of rural populations. The small number of intermediaries between local leaders and ministries, two or three at most, speeds up the process with quick feedback on the measures adopted.

A similar system is being investigated in a few specific areas of Argentinean southern Patagonia with promising results. As in Uruguay, the system is based on arenas of negotiation bringing together local actors with whom information is shared then discussed before formulating measures of public action to be implemented for local governance.

This is a scenario already underway for the central part of the Senegalese Ferlo, still highly populated and relatively less confronted to peanut growing in the south and to the irrigated agriculture in the north. In addition, these two recent agricultural activities, namely groundnut culture and irrigated farming, should not be regarded as competition. Indeed, they provide crop residues usable by cattle, especially in times of scarcity. Livestock fertilize the fields with their excrement even though this is often insufficient and poorly distributed.

[6]*Mesas de Desarrollo Rural* (MDR) Tables/Arenas of Rural Development.

16.6 As a Conclusion ...

Recently, a lot of pastoral systems (just like any other agricultural production systems, forestry, fisheries, etc.) are not sustainable. They even rely on counter-productive practices, because their logic is only based on how resources could be technically produced and operated, whereas it is necessary to see resources as dynamic and emerging from the society/environment interaction, transforming themselves due to their connections to complex ecosystems. It is time to produce the conceptual frameworks (theoretical, methodological, evaluative, etc.) allowing the development of other practices that respect both environment and societies (Hubert and Ison 2011). The sustainability and the viability of such socio-ecosystems should thus be seen as an emerging property of interactions between societies and their environment (biophysical, economic, and political) and not as an intrinsic and technical quality of an ecosystem, or even of society itself in a kind of stable equilibrium phase. The two are engaged in co-change dynamics which are very dependent on interactions with other systems having their own dynamics. Therefore, this is about the self-organizing capacities of these subsets that will allow them – or not – to regulate the fabric of interactions that ensure functionalities and interdependencies with the rest of the world at different scales and levels of organization, ensuring coviability. We have seen that pastoral systems are both very fragile – because it is easy to destabilize them – and very resistant – because they are deeply rooted in the culture of human-animal relationships, with their resilience factors still remaining poorly known to us. It is always easier to analyze a situation or define what its ideal functioning would be than to develop technologies and policies to be adopted, in order to ensure its coviability in a perspective of co-evolution that respects men, animals and their environment.

References

Alary V, Duteurtre G, Faye B (2011) Elevages et sociétés: les rôles multiples de l'élevage dans les pays tropicaux. INRA Productions Animales 24:145–156

Alary V, Daoud I Abdelzaher M, Salama O, Aboul-Naga A, Merveille N, Tourrand JF (2012) Adaptation des societies bédouines de la Côte Nord-Est de l'Egypte au changement global. Autrepart (62), France, pp 183–200

Alary V, Hassan F, Daoud I, Aboul-Naga A, Abd El Zaher M, Bastanelli D, Lescoat P, Moselhy N, Tourrand JF (2014) Bedouin adaptation to last drought in the North Coastal Zone of Egypt: continuity or rupture? World Dev 62:125–137. https://doi.org/10.1016/j.worlddev.2014.05.004

Arbeletche P, Perugorría A, Saravia A, Correa P, de Torres MF, Morales H, Franco R, Gedouin M, Poccard-Chapuis R, Bonaudo T, Capdevila L, Valarié P, Tourrand JF (2012) Norte del Uruguay: ¿cómo analizan los actores la dinámica de la ganadería? In: Estudios agrarios y agroindustriales. UBA, Buenos Aires

Biber JP (2006) Review of the literature on Pastoral Economics and Marketing: Europe. Report prepared for the World Initiative for Sustainable Pastoralism, IUCN EARO, 2006

Bland WL, Bell MM (2007) A holon approach to agroecology. Int J Agric Sustain 5:280–294

Blench R (2001) 'You can't go home again'-Pastoralism in the new millennium. FAO report

Cesaro JD, Magrin G, Ninot O (2010) Atlas de l'élevage au Sénégal. Cirad, Montpellier, 34 p

Coronato FR, Tourrand JF (2011) La Patagonie à la croisée des nations. In: Mobilités, Territoires et Identités en Amérique. De l'Indépendance aux années 1930. Presses Universitaires de Pau, Pau cedex

Coronato FR, Tourrand JF (2015) Mouton, dessine-moi un pays. Techniques & Cultures, n°63, Paris, pp 162–181

Costanza, R., d'Arge, R., de Groot, R., Farber, S., Grasso, M., Hannon, B., Naeem, S., Limburg, K., Paruelo, J., O'Neill, R.V., Raskin, R., Sutton, P., van den Belt, M., (1997). The value of the world's ecosystem services and natural capital. Nature 387, 253–260

Dong S, Kassam KAS, Tourrand JF, Bone RB (2016) Building resilience of coupled human-nature systems of pastoralism in seveloping world. An exploration of interdisciplinary strategies for sustainable pastoralism. Ed Springer (in press)

Duteurtre G, Faye B (2009) L'élevage, richesse des pauvres. Coll. Sciences & Technologies, Ed. QUAE, Versailles, 284p

Eldridge D, Freudenberger D (eds) (2004) People of the Rangelands. Building the future. In: Proceedings of the VI International Rangeland Congress. VI International Rangeland Congress Inc., Townsville, pp 819–824

Faye A (2016) Les politiques d'élevage du Sénégal de 1960 à 2012. Effets et impacts sur les systèmes pastoraux de la zone sahélienne. In: Waquil PD et al. Livestock Policies. (in press)

Gerber P, Mooney HA, Dijkman J, Tarawali S, Hann C (2010) Livestock in a changing landscape. Experiences and regional perspectives, vol 2. Ed. Island Press, 210p. http://www.fao.org/docrep/013/am075e/am075e00.pdf

Hardin G (1968) The tragedy of the commons. Science 162(3859):1243–1248

Homem V, Morales H, Champredonde M, Tourrand J (2008) The south-american livestock and the sustainability challenge. In: Hubert B, Kammili T (eds) A shift in natural resources management paradigm: from resources sufficiency to functional integrity? QUAE, Paris

Hubert B, Ison R (2011) Institutionalising understandings: from resource sufficiency to functional integrity. In: Kammili T, Hubert B, Tourrand J-F (eds). A paradigm shift in livestock management: form resource sufficiency to functional integrity. 28th & 29th June 2008, Hohhot, China. Cardère éditeur, Lirac (France), pp 9–22

IPCC (2009) International painel for climate change. 2009 Report

Landais E, Lhoste P, Milleville P (1987) Points de vue sur la zootechnie et les systèmes d'élevage tropicaux. Cirad-Inra-Orstom, Paris, 26 p

Lhoste PH (1984) Le diagnostic sur le système d'élevage. Cah Rech Dév 3–4:84–88

Long R, Qi XJ, Ding L, Yang TT, Bonaudo T, Hubert B, Tourrand JF (2014) Sur le toit du monde, les pasteurs tibétains face au changement. In: Bosc PM et al (eds) Diversité des Agricultures Familiales. Exister, se Transformer, devenir. Ed. QUAE, Versailles, pp 301–314

McGee D, de Menocal PB, Winckler G, Stuut JB, Bradtmiller LI (2013) The magnitude, timing and abruptness of changes in North African dust deposition over the last 20,000 years. Earth Planet Sci Lett 371–372:163–176

Moraes MI (2008) La pradera perdida. Historia y economía del agro uruguayo: una visión de largo plazo, 1760–1970. Ed. Linardi y Risso, Montevideo, 189 p

Morales H, Coronato FR, Carvalho SA, Saravia A, Schweitzer A, Bendahan AB, Tourrand JF (2015) Building new coupled human-natural systems for sustainable pasture management in South America. In: Dong S, Kassam KAS, Tourrand JF, Boone RB (eds) Building resilience of coupled human-natural systems of pastoralism in the developing world: an exploration of interdisciplinary strategies for sustainable pastoralism. Ed. Springer (in press)

Moran EF (2006) People and nature: an introduction to human ecological relations. Blackwell Publishing, Malden, p 52. ISBN: 978-1-4051-0572-9

Qi XJ, Ding L, Long R, Yang T Hubert B, Tourrand JF (2014) Coupling interdisciplinary researches and multi-scale approaches to understand the resilience of the socio-ecological system in the Qilian Mountains, Tibetan Plateau, China In: Global Fair and Workshop on Long-Term Observatories of Mountain Socio-Ecological Systems, Reno (NV), USA on 16–19 July 2014

Sachs I (1998) L'écodéveloppement. Editions Syros, Paris, 123 p

Saravia A (2016) Eleveurs et Société Gaucha face au changement global. Thèse doctorat, ED ABIES-AgroParisTech, Paris, France, 278p

Sraïri MT (2011) Le développement de l'élevage au Maroc: succès relatifs et dépendance alimentaire. Le Courrier de l'Environnement de l'INRA 60:91–101

Steinfeld H, Gerber P, Wassenaar T, Castel V, Rosales M, Hann C (2006) Livestock's long shadow. Environmental issues and options. FAO Report, 380 p. www.fao.org/docrep/010/a0701e/a0701e00.HTM

Steinfeld H, Mooney HA, Schneider F, Neville LE (eds) (2010) Livestock in a changing landscape. Volume 1: drivers, consequences, and responses. Island Press, Washington, DC. http://books.google.fr/books?id=lug9AF-wVR8C&redir_esc=y

Thompson PB (1997) Sustainabilty as a Norm. Phil Tech 2(2):75–93

Touré O (1986) Peul du Ferlo. Ed. L'Harmattan, Paris, 77p

Touré O, Planchon FKL (2008) La difficile intégration de l'élevage dans les politiques agricoles. In: Dahou T (ed) Libéralisation et politique agricole au Sénégal. CREPOS – KARTALA – ENDA GRAF DIAPOL

Tourrand JF (2000) L'élevage dans la révolution agricole au Waalo, delta du fleuve Sénégal. QUAE/Cirad, 165p. ISBN: 2-87614-404-2

Tourrand JF, Duarte LG, Valarié P, Capdevila L, Coronato FR, Faye A, Oman MA, Daoud I, Strankmann P, Long R, Ding L, Wedderburn L, Morales HG, Duteurtre G, Sraïri MT, Fleury P, Lasseur J, Waquil PD (2015) Livestock policy: progressive globalization led by market, focused on food quality and safety in the past, environmental issue currently and new human-animal relationship in the future. International Grassland Congress, Delhi, India, Nov. 2015

Tourrand JF, Neumeister D, Daoud I, Morales H, Bendahan AB, Long R, Faye A, Strankmann P, Sraïri MT, Basuno E, Hamadeh S, Wedderburn L (2016) Regards croisés sur l'attractivité du métier d'éleveur? Colloque Journées Rurales – La Renaissance rurale d'un siècle à l'autre? 23–27 mai 2016. Toulouse, France

Wood CH, Porro R (2002) Deforestation and land use in the Amazon. University Press of Florida, Gainesville, 386 p

Chapter 17
Enhancing Coviability Through an Eco-Pastoral Approach, the European Project LIFE + MIL'OUV

Jacques Lepart, Jessica Huron, and Sébastien Girardin

Contents

17.1 Introduction

In 1945, Georges Kuhnholtz-Lordat, Professor of Botany at the School of Agriculture in Montpellier,[1] published in the annals of this school, a long paper (82 pages) entitled "La Silva, le Saltus et l'Ager de Garrigues" (1945).[2] It was devoted to the extensive use of the garrigue (Mediterranean dry scrubland) through pastoralism, and to the characterization of a "good balance" between the silva ("anything that provides standing timber while maintaining itself by its own means"), the saltus ("all land that provides nutrition for livestock") and the ager ("all crops which ensure food for human and livestock"). As part of this agro-silvo-pastoral balance, the grazing of areas with serious ecological constraints was no longer considered as an abusive and harmful practice but was treated as one of the components of land

[1]In 1954, he becomes the first holder of the Chair of applied ecology and nature protection of the MNHN (Paris, Brunoy); see Heim (1965) for a biography

[2]This article is a very personal account of works of a garrigue committee (consisting of scientists and socio-professionals) which met in 1943 and 1944 under the framework of the 10th Economic Region

J. Lepart (✉) · J. Huron · S. Girardin
Conservatoire d'Espaces Naturels du Languedoc-Roussillon (CEN L-R), Immeuble le Thèbes, Montpellier, France
e-mail: jacques.lepart@cefe.cnrs.fr; life.milouv@cenlr.org

O. Barrière et al. (eds.), *Coviability of Social and Ecological Systems: Reconnecting Mankind to the Biosphere in an Era of Global Change*,
https://doi.org/10.1007/978-3-319-78497-7_17

development. This was a revolution compared to the positions of his predecessors: for instance, Charles Flahault (1891, 1926, 1927) stressed that grazing should be banned in mountains area and in environments with strong constraints like the garrigue. The majority of foresters stressed (with the exception of few "leplaysien" foresters[3]) the incompatibility between the pasture and the forest. The paper even bears witness of the change in Kuhnholtz-Lordat compared to the works that brought him, a few years earlier, to work with the "Nîmes school"[4] foresters such as Ducamp or Flaugère. It is most likely that the article went unnoticed, but (Kuhnholtz-Lordat 1938) had outlined the theme in a well-known book, *la Terre Incendiée* and again a few years later in another book, *l'Ecran Vert* (1958).

The notion of agro-sylvo-pastoral balance is taken up by other authors,[5] and the journal *Revue des Eaux et Forêts* dedicated in 1954 a special issue to the agro-sylvo-pastoral balance. Nevertheless, the interest did not last long and the notion, without being forgotten, is no longer a subject of research for agronomists and foresters. The notion goes too clearly against the intensification and the industrialization of agriculture which has been growing since the sixties. It continues to be mentioned by environmentalist geographers such as Bertrand (1978), and it is used today in promoting an agriculture of high natural value (Poux and Ramain 2009).

The word "balance" seems to be inappropriate. It refers to a point of balance, an ideal proportion between silva, saltus and ager which probably was not the intention of Kuhnholtz-Lordat. It would probably be better to use the notion of system which is, henceforth, used in ecology.[6]

What Kuhnholtz-Lordat (1945) highlights, is the possiblity of the sustainable development of the garrigue (taken as an example of a particularly fragile habitat) by ensuring the coviability of production systems and natural systems. It is possible to restore habitats or to keep them in good condition while having a pastoral use of the garrigue. However, this coviability should not to be taken for granted. As reflected in the history of the garrigue which, for Kuhnholtz-Lordat, is one of "all abuses leading from what was once forest to denudation [soil erosion and loss of vegetation]", it has to be enhanced.

More than half a century later, the situation has considerably changed. Almost everywhere in Mediterranean France, old pastures are slowly becoming covered by forest again, the landscape is changing by reducing itself to two components,

[3]The adjective *'leplaysien'* was forged by Kalaora and Savoye (1986); it refers to the sociologist 'Frédéric Le Play' who was devoted to forests and who replaced them in their economic and social context to derive the management rules from them. The 'Leplaysiens' foresters distinguish themselves from other foresters by being interested in rural mountain societies and taking into account their development issues in the management and reforestation of these areas.

[4]Foresters of the Nîmes School object to reforestation (artificial plantations of conifers) and urged their colleagues to be very careful about the natural dynamics (see Mure and Lepart 2006).

[5]Olivier Nougarède (1994) made a critical analysis of its use

[6]The notion of an ecosystem, which focuses on interactions without assuming equilibrium, was put forward as a theory by Tansley (1935) and the very first applications were carried out on natural systems during the Second World War

artificialized zones (ager and urban) on one hand, and the silva on the other hand. The huge decrease in the livestock is leading to the disappearance of the saltus, of the plant communities that compose it and of part of a natural heritage, namely the Mediterranean landscape, and the many species associated with open environments.[7]

For centuries, the Mediterranean landscape has maintained a close link with pastoral activities. Today, the sustainability of pastoral animal husbandry is becoming an essential part of its conservation. It is also the best way to avoid the damaging consequences of forest fires and to maintain water resources in the summer grazing period. However, breeding has become rare in Mediterranean areas (compared to the mid-nineteenth century, sheep stocks declined by a factor of at least ten[8]). The breeders which remained or settled, while adapting to local ecological conditions, had to adjust to land issues and to economic and legal opportunities/constraints. Rather than seeking to define in a normative way what coviability could be, we will take these experiences mainly by breeders and briefly analyze them from a technical, ecological and economic point of view.

We will use work carried out under the framework of a European program which brings together information on the strategies implemented by a large number of breeders in Mediterranean France. We will base our work on a few cases studies, focusing on the diversity of solutions adopted, and we will try to understand the rationale.

17.2 Method of Work

17.2.1 A Framework: The LIFE + MIL'OUV Project

The LIFE + MIL'OUV project aims at improving the use of natural resources through the breeding livestock and the conservation of agro-pastoral habitats in the Mediterranean and sub-Mediterranean regions, by implementing pastoral and ecological referentials. Promoting of the sharing of know-how, information, methods and skills to all agro-pastoralism actors (breeders, experts, students, teachers, decision-makers) is a key stake for this project.

In order to encourage breeders to optimize their use of their pastoral resource and and thus improve their self-sufficiency and thereby ensure the conservation of open landscapes and their biodiversity, forty voluntary farms have benefited, throughout the project from 2013 to 2017, from personalized support for pastoral and ecological management of their open environments.

---

[7]Lepart et al. 2011

[8]Lepart and Fonderflick (in preparation)

Fig. 17.1 Farm assessment with breeder and experts (Credit: Mario Klesczewski)

17.2.1.1 Three-Phase Approach to Support Breeders

(a) The Global Understanding of the Farm

The first phase of the fieldwork is to evaluate each farm. During this phase, the overall functioning of the farm is studied. It analyses how the herd feeds and what part of this food is provided by the natural resources available in open and semi-open environments (saltus). Therefore, different tools are used: Strat'Alim (OIER Suamme, unpublished) in order to describe the herd feeding strategy, a functional analysis (Moulin 2002) to describe the periods and types of grazed areas, as well as a general survey to assess possible developments on the farm (pastoral management, marketing strategies . . .).

This overall assessment takes place during a three-party dialogue involving the breeder, the pastoralist expert and the naturalist expert (Fig. 17.1).

(b) Eco-Pastoral Assessment of the Field

In a second phase, this breeder-pastoralist-naturalist trio attempts to establish links between pastoral practices and studied vegetation through field observations. This allows the ecological issues to be assessed on the present environment, the conservation of natural habitats based on a series of indicators (over/under-usage, control of woody encroachment, etc.) creating a permanent exchange between the experts and the breeder, a key factor for a successful assessment.

(c) Advice and the Follow-Up

Finally, the summarizing and the restitution of the assessment to the breeder allow adjustments to practices to be suggested and how to carry these out, and define related field indicators allowing the effects of these changes to be followed-up.

This method is detailed in chronological order (from top to bottom) in the following diagram (listing of steps /objectives/tools):

Therefore, the assessment phase allows the tandem of experts to familiarize themselves with the technical, ecological and socio-economic characteristics of the farm as well as the mobilization of the expert's skills by the breeder.

In another dimension, the exchanges between breeders, coordinated by the experts, enables a process of co-active search for solutions. This approach is even valid within for the naturalist-pastoralist tandem itself formed during the entire assessment phase (Fig. 17.2).

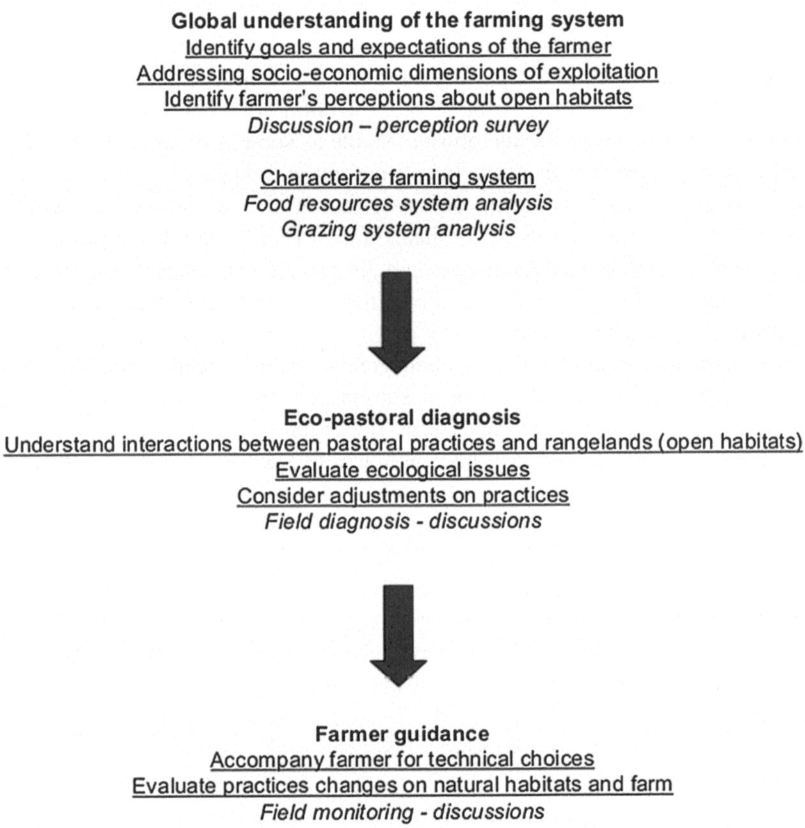

Fig. 17.2 A three-phase approach to support breeders

17.2.1.2 Pastoral Breeding Systems in Order to Combine Ecosystem Viability with Lifestyle Viability: A Few Examples

By looking for a balance between the use and renewal of resources, and that of the viability of an agricultural holding and the ecosystems on which it relies, many of the breeders met during the project illustrate the concept of co-viability through their practices. Below, we focus on a few pastoral systems linking conservation practices with food production for human beings.

Example 1
The high plateau districts in the Hérault area, France, cattle breeding Mr. W is transitioning towards more pastoralism to "produce with what the environment offers". After taking over a farm whose feeding system was essentially based on artificial grasslands and distribution of purchased feed, breeding is today moving towards food self -sufficiency of herds based on outdoor grazing. A decrease in the number of animals should also contribute to achieving this goal.

Today, the natural grasslands are grazed by the herd in spring and the enhanced rangelands, virtually year-round. In summer, the wooded rangelands offer an interesting, still green, resource (due to the protection of herbaceous plants by trees) and shade for herds. The animals return to stable as soon as the first snow falls, and are fed with hay harvested from the farm's natural grasslands.

Most of the farm's 250 ha are under agricultural lease, sometimes with constraints imposed by the owners for the installation of fences and water points, which are necessary for outdoor herd management. Today, a few hectares under precarious statutes (without a lease signed with the owners) are also cultivated to supplement the fodder ration and achieve autonomy.

To support this change of herd management, fences were installed and the overgrown environment, in addition to pastureland, reopened by mechanical means, in order to achieve a more pastoral system. Better hardiness of animals is sought by introducing the 'Aubrac' breed of cow, able to calve outdoors and more flexible in terms of feeding.

In order to ensure the economic viability of his breeding activity, Mr. W, is currently developing direct sales and marketing through a producers' store, allowing better promotion than through normal commercial channels. This essential added-value for the farm's turnover also depends on the proper functioning of local transformation units (slaughterhouses, meat cutting plants, etc.). Today, 40% of the farm turnover is obtained from aid from the Common Agricultural Policy (CAP). By clearing a salary, the activity is considered economically viable by the breeder and will be even more so in the future when everything is sold directly, decreasing the inputs further.

The CIVAM *Empreinte*,[9] a network of livestock breeders looking for more sustainable management practices of rangelands, is supporting this breeding in transition, allowing technical exchanges between peers. According to Mr. W, it will take another 3 years to reach the pastoral-type of functioning targeted with more diverse environments, offering, thus, a better feed potential and a reestablished quality landscape as well as savings on certain production costs: less purchase of equipment including a tractor, reduction of fodder purchase to zero or to the lowest level possible which is highly beneficial for the environmental quality of the rangelands.

Example 2

Between abandoned vineyards, garrigue and wood, goat rearing by Mr. X. is based on the diversity of Mediterranean habitats which are the farm's main resource.

The 80 goats which are of a hardy breed, are herded for 5–7 h per day all year round. The good management of the herd allows all kinds of environments to be exploited according to the seasons while meeting the needs of the herd.

During autumn, mattoral (scrubland) and oak forest (holm oak and kermes oak) are grazed, providing wood resources, acorns and young leaves, while aphyllante represents a very nutritious herbaceous resource in this season. The impact of goats on the kermes oak garrigue progressively contributed to the reopening of the environment and the reestablishment of an herbaceous layer. According to breeders, present for more than 20 years, bird of prey have returned, enjoying the reappearance of their prey in the mosaics of open and semi-open environments of high biological diversity. In the winter season, a period of high needs during gestation, the herd grazes the old vineyards and the garrigue near the building. When lactation begins in the Spring, food is supplied by heath: young shrubs, bushes, flowers of broom scorpions, etc. During summer, the more wooded rangelands offer shade, and also coronilla, brambles, the fruits of broom and wild rose, and only a complement of alfalfa is brought. Within the same day of herding, the mobilized resources vary. The herd is guided through different environments to meet the goats' nutritional requirements: rich fallow land at the beginning of grazing, and more wood resources in late afternoon.

Milking is carried out once per day, and cheese is produced in a small workshop on the farm, which requires a second full-time person. Therefore, the resources of the rangeland are economically valued by this transformation and their sale via an AMAP (*Association pour le Maintien d'une Agriculture Paysanne*[10]). With social aid and European CAP subsidies, two salaries can be paid enabling a family with children to be supported. Today, this frugal pastoral system (only three tonnes of hay are purchased) allows the cheese production as well as the fattening of some of the kid goats for direct sale in a context where the goat meat market is not profitable.

[9]CIVAM: Centres d'Initiatives pour Valoriser l'Agriculture et le Milieu rural, @: http://www.civam.org/index.php/agriculture-durable/agriculture-projets-locaux/471-le-civam-empreinte-un-groupe-d-echange-pour-valoriser-les-vegetations-spontanees

[10]The AMAP promotes a marketing system based on the principle of a direct contract between the consumers and the producers

Several networks support this livestock farming: the *CIVAM Empreinte*, and also the '*Terres Vivantes*'[11] association which prompted farmers to sell their produce through the AMAP, and stop going to the markets. This change, associated with a move to milking once daily, saves about 1.5 days per week which can be spent as a "time for children". Hence, the system is appropriate for breeders despite the daily task of herding.

Started in 1992, the pastoral activity continues and has been adapted resulting in this system that better values rangelands. Although agro-environmental aid is vital to the economic equilibrium of the farm and that the land control of the 200 ha of rangeland is not completely guaranteed (difficulty in obtaining written agreements from the owners), the breeding activity is integrated in this Mediterranean ecosystem which generates produce but also provides protection against fires and maintains the landscape.

Example 3

On the plateau of Larzac, Aveyron area, France, for more than 10 years, Mr. Y's farm has being undergoing adjustments in order to better use the surrounding environment for milk production, used in Roquefort cheese.

The farm, passed down from generation to generation in the family, consists of a flock of 570 sheep. In the center of the farm, the milking shed is today surrounded by 70 ha of rangeland, 80 ha of grasslands and 50 ha of cereal crops. A few kilometers away, 220 ha of rangelands are used.

Historically, the flock was shepherded and when the farm was taken over in 1997, a system of fenced areas around the building was implemented. The aid given for the installation of a third associate allowed the construction of a building close to the distant rangeland of 220 ha. The area was then grazed by a flock of sheep for meat production, in addition to the flock used for milk production that has remained nearer to the main farm. Six years later, the meat production workshop shut down and consequently the meat-producing flock no longer exists. The flock used for milk continued to grow and the rangelands were then used for the dry ewes (after the lactation period), causing extra work to feed this batch.

Subsequently, special attention is paid to these 220 ha of rangeland. They have been divided it into smaller areas of land in order to optimize the use of these additional resources while preserving remarkable dry grassland (maintenance of the open environment without causing degradation). The large investment for fencing is borne by a 'Mesure Agro-Environnementale' (MAE), measure established with the 'Grands Causses' Natural Regional Park (NRP) and the *Chambre d'Agriculture*. Thanks to the dividing up of the rangelands into small plots of land, instantaneous grazing pressure has become stronger as the animals select less the vegetation they eat and graze more optimally than when left to roam in larger areas. Overgrowth is better controlled, thereby ensuring the preservation of open grasslands. In addition

[11]*Terres Vivantes* supports the creation of agricultural activity and the animation on rural areas. It welcomes and offers support to people wanting to settle in agriculture, often in a progressive manner, sometimes with "atypical" projects. @: http://www.terresvivantes.org/

to these developments, the feeding of ewes on rangelands also allows the part of pasture in the diet to be increased and food requirements in the sheepfold to be decreased. Today, with this training of ewes to pasture, they prefer the rangeland at the expense of grasslands, a very interesting perspective!

Farming provides a livelihood for the whole family, and one third of the farm's turnover is made up of CAP aid despite the good earnings from milk for the Roquefort industry. In a dozen years, the system has decreased its purchase of forage by more than 80%, and increased the number of sheep and the volume of milk produced thanks to the use of semi-natural areas. The breeder promotes this evolution with his conception of rangelands "as a real resource". These innovations will be shared through the MIL'OUV project. By doing so, the breeder will host the project's next technical workshop in order to exchange with other breeders as well as pastoral and naturalist experts.

Example 4

In the South of the Ardeche, Mr. Z's mixed breeding was established 30 years ago on land colonized by brushwood and oaks. The reopening of the environment was made gradually by playing on the complementarity between goats, sheep and, at the beginning, a few horses. Henceforth, the grazing of goats and sheep is now possible under the oaks (consumption of acorns and leaves) with a herbaceous resource in clearings.

The effect of the animals on these environments has convinced reluctant owners to let the animals graze on their properties. In addition to 100 ha of green oaks, 400 ha of plateau located away from the main farm are, thus, mobilized by the herd and flock from September to January. The animals maintain the open dry grasslands of brachypode (Brachypodium retusum), a natural habitat of European interest with great biological diversity, where aphyllante grows "which helps to go through the winter".

The feeding of 90 milk producing goats and 70 sheep for meat is guaranteed by the resources offered by these different environments, through the harvest of 5 ha of natural pasture hay, 10 ha of alfalfa and a barley complement during lactation. During the summer, the sheep migrate to the Tanargue, a range well known for its ecological value in the Parc Naturel Régional d'Ardèche mountains, perpetuating this tradition of migration of herds and flocks, in search of grass.

The produce from this pastoral breeding provides three salaries with a farm turnover that is essentially made up of the direct sale of goat's cheese. The system benefits only slightly from CAP aid, partly due to the lack of long-term contracts with many owners (leases, etc.). The sale of sheep meat is economically in second place but represents little additional shepherding work for a high beneficial return linked to the management of grass growth and resources in general.

Despite implementing this in an environment abandoned and closed for several years and despite the skepticism of inhabitants, this varied pastoral breeding has been developing for more than 30 years. It is part of an area where pastoralism is no longer a priority, replaced by major touristic stakes, but which also requires services offered by pastoralism, such as the maintaining of the landscape's quality, biodiversity, and of course and providing high quality products (Fig. 17.3).

Fig. 17.3 Pastoral breeding in Ardeche (Credit: Diane Sorel)

17.3 Discussion

On the basis of these examples, a wide variety of strategies and organizations of private farm emerges. This is not only the result of a choice in presentation, favoring diversity. It is a reality. Farms whose functioning is based on an important use of the semi-natural environment are not organized on the basis of a standardized model. They must take into account the varied constraints and opportunities of the land in which they are part of. Firstly, the constraints and opportunities are ecological (climate variability, fragility of soils, nature of the pastoral resources, and forest dynamics), technical (land management, nature of the livestock, size of plots of land and proximity to the main farm, presence of drinking points, ability to store and manage fodder, etc.), economic (proximity to urban centers, installation of transformation plants cooperation, marketing, etc.), and often financial (start-up capital, etc.). In addition, there is also the choice of breeders, for example, regarding the family organization. These constraints and opportunities are related to each other, which leads to simultaneously dealing with the ecological viability and economic viability: an ecological constraint can be offset by an economic opportunity and vice versa.

Farm work is rarely routine. It changes in order to take into account the considerable variability over time of the ecological and/or economic context. The coviability of farm and its environment, is based on the exploration of resources and environmental constraints specific to each farm, and on the exploration of economic opportunities. There is a permanent need to innovate, usually in a minor way. Running the farm is not carried out to establish a balance but simply to correct the major imbalances in order to remain resilient in the system and take advantage of variability when it is favorable.

All these changes are not only the consequence of present ecological or economic determinism but also represent deliberate choices in management, the construction by the farmer of his work tool, often based on convictions reinforced by experience.

Nevertheless, within such diversity, there are factors that can bring farms to converge. On the one hand, in a region or an area, private farms are subject to similar constraints and opportunities. They often have a common history on which they rely on for their development, they share a similar heritage even if the owner is a new to the country. This is largely because a farm keeps renewing traditional practices that it helps in the conservation of the species present in the area. On the other hand, experiences are often shared, possibly by imitating, sometimes by creating common farming or marketing tools *(Coopérative d'Utilisation du Matériel Agricole* – CUMA, pastoral associations, economic interest groups), and of course by mutual assistance. Advisory structures which share models qualified as effective, contribute to this convergence.

The system's management is never total; it has its own dynamics which are only understood retrospectively. Kuhnholtz-Lordat (1938, 1945), who especially made reference to the "mattoral" and "garrigue" of the nineteenth century, stressed the risk of overexploitation of saltus and the importance of denudation and soil erosion. The herds and flocks which provided the transfer of fertility to the ager only had very insufficient fodder in their pastoral area. So, they used to encroach the Silva, contributing to reducing it to small coppices, and relentlessly moved around to graze, often too early, even the smallest source of fodder, contributing to erosion by trampling. It is by intensifying the agricultural production (for example, fodder plants in crop rotation) that it was possible to overcome this vicious circle.

Today, the situation is almost reversed, afforestation is progressing almost everywhere, and the focus is now on wood dynamics. Erosion has become rare and there is more concern about the risks of fire, the conservation of landscapes, types of vegetation, and natural heritage species. Coviability is based on forms of intensification of pastoral practices that have an impact on tree dynamics but also on the ability of breeding systems to adapt to such dynamics, taking thus into account long-term phenomena such as forest cycles (Vera 2000), and to find resources in woods; the development of sylvo-pastoralism meets this challenge. Therefore, these forms of pastoral intensification practices' must be thoughtfully developed, with regard to the environments' capacities to react, in some ways "to manage what is manageable", to focus efforts on zones of priority from a pastoral and environmental point of view in order to avoid carrying out practices disconnected from the systems on which they are based (do not open up or burn a piece of land only to 'clean it'). Coviability also depends on the ability of farmers in maintaining a certain amount of freedom, finding flexibility that allows them to cope with variations in climate. Farmers benefit from a better understanding of the functioning of the natural systems which they are part of and the stakes in biodiversity in order to use new management criteria adapted to the current situations and challenges. The link with advisory structures that have a good knowledge of these natural systems has become more necessary than ever.

References

Bertrand G (1978) Le paysage entre la Nature et la Société. Revue géographique des Pyrénées et du Sud-Ouest 48:239–258
Flahault C (1891) La question forestière. Bull de la Société de Botanique de France XXXVIII:39–45
Flahault C (1926) Trois conférences forestières. Société française des amis des arbres, pp 1–45
Flahault C (1927) Le boisement des terres incultes et des montagnes. Bull de la Société centrale d'agriculture de l'Aude, 1–33
Heim R (1965) G. Kuhnholtz-Lordat (1888–1965). Comptes rendus des séances de l'académie d'agriculture (28 Avril 1965), pp 505–517
Kalaora B, Savoye A (1986) La forêt pacifiée. Les forestiers de l'Ecole de Le Play, experts des sociétés pastorales. L'Harmattan, Paris. 134 p
Kuhnholtz-Lordat G (1938) La terre incendiée. Essai d'agronomie comparée. Editions de la Maison Carrée, Nîmes. 362p
Kuhnholtz-Lordat G (1945) La silva, le saltus et l'ager de garrigue. Annales de l'Ecole Nationale d'Agriculture de Montpellier 26(4):1–82
Kuhnholtz-Lordat G (1958) L'écran vert. Mémoires du Museum National d'Histoire Naturelle, IX, 276 p, XVI planches
Lepart J, Marty P, Fonderflick J (2011) Landscape dynamics and biodiversity of agro-pastoral landscapes in 'Les Causses'. Fourrages:343–352
Moulin CH (2002) Une méthode pour comprendre les pratiques d'alimentation des herbivores domestiques: analyse du fonctionnement des systèmes d'alimentation par enquêtes en élevages. Guide méthodologique. Institut de l'Elevage
Mure V, Lepart J (2006) L'Ecole de Nîmes. Les conceptions de la gestion forestière en région méditerranéenne de Roger Ducamp, Conservateur des Eaux et Forêts (1861–1938). Forêt Méditerranéenne XXVII:275–284
Nougarède O (1994) L'équilibre agrosylvopastoral, premier essai de réconciliation entre l'agriculture et la forêt. Revue forestière française, XLVI, n° spécial, pp 165–178
Poux X, Ramain B. 2009. L'agriculture à Haute Valeur Naturelle, mieux la (re)connaître pour mieux l'accompagner, EFNCP, 41p
Tansley AG (1935) The use and abuse of vegetational concepts and terms. Ecology 16:284–307
Vera FWM (2000) Grazing ecology and forest history. CABI Publishing. 509p

Chapter 18
Reconnecting Man to Man: Socio-cultural Coviability Ties and Interculturality (Practical Research in a Sensitive Neighborhood in Montpellier, France)

Catherine Barrière

Contents

C. Barrière (✉)
PACIM Association («Passeurs de cultures, Passeurs d'Images»/links between cultures), Montpellier, France
e-mail: passcultim@gmail.com

O. Barrière et al. (eds.), *Coviability of Social and Ecological Systems: Reconnecting Mankind to the Biosphere in an Era of Global Change*,
https://doi.org/10.1007/978-3-319-78497-7_18

18.1 Introduction

An anthropologist, working in the non-profit sector, working in a sensitive area of Montpellier, has been involved in "practical-research"[1] to spread knowledge in order to propose answers to emerging questions as part of fieldwork.

The neighborhood where the research took place is characterized by the coexistence of several communities, in this case a European community established in the 1970s with a low renewal rate and a North African community, immigrating massively in successive waves along with other more unnoticed Korean, Turkish and African migrants.

While this multiculturalism constitutes an undeniable cultural enrichment, the inhabitants are in fact confronted by a feeling of "social weight" tinged with relegation, due to urban density and the constant influx of migrant populations that erode the feeling of cultural diversity. Moreover, the social and economic difficulties encountered by many families, reinforced by the massive unemployment of young people, are generating increasing social precariousness[2] and reinforcing social inequalities.

The impression of living in a ghetto, often expressed by the residents, suggests a certain form of withdrawal of "common community cultural references". Indeed, «This discourse of withdrawal most often refers, in a pejorative way, to the noticeable grouping together of families of immigrant origin, who, by grouping together, are responsible for their lot» (Carlon and Dessis 2013, 23). It is therefore necessary to give voice to these social groups. They are often from the same geographical regions and confined to the same area. Whereas these families are considered as withdrawn within themselves, are they not rather in a search of openness and of a certain demand for social diversity that they have been denied?

The thread of our inquiry is to understand how social ties were built yesterday and today in this neighborhood that has now become sensitive. How have the changes in the structure, the immigration process and socio-economic conditions affected social relations? Are these ties growing at the level of individuals or socio-cultural

---

[1] The term "practical-research" is attributed to Lewin (1946), a German experimental psychologist, who argued that through practical research theoretical advances could be made simultaneously with social change. He described the phases of practical research as a spiral of research circles progressing from a description of the existing to a plan of action.

[2] "A lack of basic security is the absence of one or more factors that enable individuals and families to assume basic responsibilities and to enjoy fundamental rights. Such a situation may become more extended and lead to more serious and permanent consequences. Extreme poverty results when the lack of basic security simultaneously affects several aspects of people's lives, when it is prolonged, and when it severely compromises people's chances of regaining their rights and of reassuming their responsibilities in the foreseeable future" (OJ of 10 and 11 February 1987) definition by Wresinski Joseph, to the French Economic and Social Council and taken by Mr. Léandro Dsepouy in his report to UNO entitled;, Extreme poverty and Humans rights, Essays on Joseph Wresinski, available online: http://www.joseph-wresinski.org/wp-content/uploads/sites/3/2016/09/Attacking_Poverty_WBk_.pdf; and: atd-quartmonde.org

groups? What are the factors of assimilation and isolation? Can digital tools provide support and enliven social ties, including between individuals with distant socio-cultural backgrounds and in what ways?

Finally, what tools can the ethnologist invent to try to build intercultural bridges that would allow people to reconnect around common issues and interests in order to strengthen social cohesion and "living well together"? These questions will lead us to define the vectors of cultural and social coviability.

We will address these questions in three main areas: (1) the history of the construction of a district: from the wishful thinking of the "ZUP" (area to be urbanized as a priority) to the rise of a sensitive neighbourhood (2), the images of the residents, a desire to live together tarnished by a strong impression of social segregation, and the gamble taken on new sociabilities around solidarity thanks to an urban platform of mutual support (3).

18.2 History of the Mosson District: The Construction of a Ghetto

The history of Mosson is one of urbanization in response to migration. From vineyards and scrubland have emerged a priority urban area (1), which is densifying, autonomous and enclosed at the same time (2).

18.2.1 A Priority Urban Area (ZUP) Emerging from Rural Areas

In 1961, the mayor of Montpellier, the lawyer François Delmas, bought the estate Bonnier de la Mosson from the De Baroncelli family. This area represented a vast area of 225 hectares on which he wanted to establish a new neighbourhood to cope with the demographic surge, partly linked to returnees from Algeria and to the shortage of housing. This new district would have the advantage of "possessing both an internal balance and a certain coherence with the whole of the city".[3] The municipal council of October 2, 1961 validated the creation of a ZUP located at 6 km from the town center and gave it the name of "Paillade".[4]

The commendable municipal intentions were communicated to the residents of this new district via a booklet of 32 pages entitled the "White Paper of the Paillade". The first of these was "to create a new district according to a rational method (...)

[3]See: "Livre blanc de la Paillade", 1973, p.9.

[4]The word "Paillade" comes from Occitan "palhada" meaning straw or strewn with straw. This reference is explained by the fact that this area was since 1710 part of the dependencies of the castle of the "Bonnier de la Mosson" family.

a new city that would allow a certain return to nature with greenery, calm, clean air"; The second was to "create a harmonious living environment by managing the space and facilities, bringing together families of all origins and finally ensuring the coherence of the new district with the whole of the European agglomeration". Finally, it was a matter of "fostering a community lifestyle, just like villages (stimulating social relations, collective participation, integration of newcomers)".

The first bulldozers started up in January 1964 and 3 years later, on 15 April 1967, the first houses were delivered. The foreman for the first works at the Paillade, who still lives in the neighborhood, recounts: "At that time there was absolutely nothing at the Paillade, we had started to clear it.. In fact, it was not much, for it was not even like the "maquis" that one finds on the other side of the Mosson., It was more half cultivated and half fallow! My job was to first establish all the streets, all the platforms of the buildings" (P. Salamand, former foreman, survey 2011).

18.2.2 An Urbanization That Becoming Denser and Is Isolating the Residents of the District

The construction of this new district did not depart from the principles of construction, inspired by the movement of modern architecture in vogue in the 1960s: slab planning, standardization of buildings due to the system of standardization and prefabrication of building elements, orthogonality of the urban plan, verticality in order to minimize the footprint of buildings on the ground and to favor pedestrian traffic. This architectural scheme in which Nature is conceived as "the green zone", "the green protection" or "the antidote to the artificial environment" (Bechmann 1966) constitutes a trace of Le Corbusier's progressive utopia (inspired by the Charter of' Athens) by which open spaces were supposed to facilitate the encounter between all the residents and facilitate social relations. However, could the symbolic breaking down of physical barriers be enough to suppress the existence of social barriers and to lay the foundations for coviability between different socio-cultural groups?

As early as April 15, 1967, the first residents settled in the south of the district where some primary schools as well as the college of the Escholiers de la Mosson were ready to welcome the children. At the same time, towers continued to rise, notably the five towers of the Tritons (1968), the city Phobos (1969), the Belvedere des Garrigues, the tower Plein Ciel, as well as the Tower of Assas, the highest tower of the Languedoc-Roussillon, with 176 apartments.

The district grew rapidly since, at the beginning of September 1973, the southern part of the Paillade already had 79 primary classes, 28 kindergarten classes and 6 canteens, or about 2700 children. The teachers of the different schools did their utmost to integrate the many families coming in successive waves: "Sometimes we did not know in which classes to put the children who had never attended school. Some of them arrived in winter, frozen in their sandals and their light clothes ... The parent representatives managed to equip everyone" (Magali Bonnet, former director of Les Troubadours School 2011 survey). This period of the 1970s and 1980s is remembered by residents as being a pleasant social life, flowing, characterized by social exchanges and vibrant: "When we arrived in March 1976 I remember the spring, the height of the tower and the beautiful view ... And then very quickly we started a community-type life ... Between the floor above and the one below, we knew each other very well ... At the foot of tower 5, there was a plot that housed a Family of traders. The vegetable merchant was a Kabyle and he was a born animator, he organized couscous parties. There was a life ... From the fifteenth floor, I watched my son riding a bike. On the same level, there were 6 apartments from one bedroom to four-bedrooms, so there was diversity in terms of ages and families (...) if we opened our doors, children could ride tricycles on the landing and we helped each other out, we had an excellent relationship. When there was a concern, we talked ... "(M. and J. Valat 2011 survey).

Similarly, in Phobos, a newly built estate, life was full of life, but characterized by an atmosphere of solidarity: "At first it was very cosmopolitan, there were Spaniards, "Pieds noirs" (French national born in a North African colony), Moroccans who were arriving... But after all the Moroccans arriving were housed in Phobos, the gypsy families were housed there too ... It made us a ghetto! We were a little isolated, we had no grocery stores; street vendors who delivered to us! It was the school of life, this estate: solidarity, everyone helped each other in every sense of the word, we watched over the dwellings of others, we babysat each other's children, etc. No one was richer than the other. Some Saturdays we had a mobile disco in the courtyards of the estate ... We had a real quality of life, even if nobody came from outside the estate because such bad things were said about our estate!" (Dolores, former Phobos resident 2011 survey). This "positive" description of a successful mix is corroborated by other social actors in the neighborhood: "At that time, there was real dialogue, each one easily spoke what was on their mind, teenagers mixed and young North Africans came to the YCW[5] with the French youths. We also organized school fetes, lotteries, the popular world took power, and people mixed. This mix created good things; there is a style of meeting, of humanity, it's great, here you have human relationships that you cannot find anywhere else!" (Michel Peyre, Priest at the Paillade, survey 2012).

Finally, this distance from the city center provided the feeling of being cut off from the world and helped to build a "local amongst ourselves" (Billard et al. 2005) atmosphere, a strong local identity built on being distinct from the outside world. In the area thus delimited and defined, a form of familiarity emerged between the residents, a "local community"[6] where solidarity played a role of social cement.

[5]The YCW (Young Christian Workers) is the only national youth association of the working class. It is run and managed by the young people themselves.

[6]"The community is both a place, the people living in this place, the interaction between these people, the feelings arising from this interaction, the common life they share and the institutions that regulate this life" (Médard 1969).

Even if the heterogeneity faded, the coexistence between residents of different origins and ages worked all the better because they shared the feeling of isolation, the need for mutual help.

However, the neighborhood has continued to develop and the density has become worrying, so much so that in 1978, the residents demonstrated in order to stop the constructions being carried out: “We complained against the fact that they were still building the Hortus2, the Dourbi, the Aigoual, because we said: we cram and cram in the same place, we concrete everywhere . . . We were told that major investments had been made at the Paillade and that this had to be amortized” (Michel Peyre, Survey march 2012).

During a public inquiry carried out by the town in 1982, the residents were given the opportunity to express their concern about the risk of the disappearance of the socio-cultural mix and the harm that would be caused to neighborly relations and to the neighborhood’s auality of life in general. Despite the fears expressed, the policies did not stop the two processes of construction and segregation that continued simultaneously. In the early 1980s, Moroccan immigration experienced a strong surge, resulting in the construction of several additional residences and thus endorsed the feeling of crowding.[7] Little by little, the population of Moroccan origin became the majority. The tower of Assas called “the UN” and which housed 30 different nationalities was renamed “Ouarzazate” because its residents were all from the region near Ouarzazate. Despite the tram line service in 2000, the neighborhood continued to suffer segregation and self-defeat, not because of a desire to group together but because of a housing policy and strategies by other households, who had the option of choosing their neighborhood, to leave the area.

Today, the neighborhood (renamed Mosson since 2000) has 25,000 residents, of which 70% are French and 30% foreign. More precisely, «a tacit categorization distinguishes the “old”, “native French residents present since the construction of the district, the residents of North African origin, the “Catalan” or “Andalusian” gypsies,[8] and immigrants from other countries. In addition, “the” community from North Africa, brings together Moroccan, Algerian, Tunisian nationalities. Finally among the Moroccans, it is necessary to differentiate Berbers, Moroccans from the Rif or those from the South from the region of Ouarzazate» (Carlon and Dessis 2013).

The population of the Mosson neighborhood is precarious because half of it lives below the poverty line. The housing stock is composed of 68% of social housing (compared to 13% for the agglomeration, Carlon and Dessis 2013) and there are three times more inhabitants per dwelling than the rest of the municipality. The population is young and 45% of the population has no qualifications. Although the district has benefited from urban free zone status since 1997 to help create

[7]The neighborhood is part of a vast plan for urban renewal (ANRU): towers and blocks are gradually demolished for the construction of more modern estates.

[8]These denominations correspond to the identity claims of the gypsies themselves.

jobs, access to employment is particularly difficult and penalizes those under 26, of whom 46% are unemployed and have little access to employment are have hardly any qualifications. This status had not prevented the process of progressive segregation, accentuated by a settlement policy that gradually concentrated precarious populations in all aspects, financial, cultural and social.

18.3 Images and Expectations of Residents: A Desire to Live Together Tarnished by a Strong Current Impression of a Ghetto

Overall, a series of interviews conducted in 2014 show that the images are extremely contrasting, with residents expressing a more severe and worried view of the neighborhood's social climate, which varies according to the micro-neighborhoods where they reside.

18.3.1 A Double-Edged Cultural Diversity

One third of the residents consider the area to be very pleasant and "fairly quiet in some places". They highlight the following positive points: "diversity, multicultural richness, conviviality, solidarity". Many of them grew up in this neighborhood and have chosen to live there even now. They readily point out that "it is an enriching district where there are many cultures, several generations, several origins". However, if this cultural diversity is often perceived as an asset, it is also associated by some with a disturbing and troubled image: "When I arrived in the neighborhood, with the reputation that it has, I was not too reassured. And then by actually living there, you realize it's not Chicago. This image is also seen as disqualifying because it would have the contaminating effect of "rubbing off" on the residents: "When I have a professional interview, I do not say that I come from the Paillade".

Confirming their statements, stereotyped visions are projected from the outside and taint the image of the neighborhood. This bad "reputation" tarnishes and sets a pejorative image of the neighborhood, both in the eyes of the people living in Montpellier and the residents themselves, who suffer this poor appreciation: "People come here with fear in their stomachs because they do not know the Paillade, Nor the right people". Hence, the need to "work the image of the neighborhood to give it a more flattering identity, which would bring hope to young people" (O., man, 45 years).

Among the members of the voluntary community sector who have been involved in the neighborhood for many years, a rather positive vision of the neighborhood's resources emerges. They develop the overall idea that the neighborhood is "welcoming, alive" and "not such a dodgy area as the people of the town say it is". Their resolutely positive image of the neighborhood, anchored on their daily practices and exchanges, are nourished by a militant approach, aimed at implementing targeted actions to meet the particular needs identified. Nevertheless, local community associations are very aware of several problematic aspects linked to each other, in particular the massive unemployment of young people, leading them to display attitudes considered as "demanding, violent and problematic" and the social and economic difficulties faced by many residents and which generate a certain form of social misfortune characterized by great poverty, social isolation and the feeling of exclusion.

The local community associations indicate that the residents are experiencing significant economic difficulties linked to unemployment. Many families do not have the means to raise their children: "You wouldn't know it, but people do not have many resources and the lack these are not sufficiently compensated because of the precariousness of the community structures that assist them. It's impressive to see how the recreation center struggles to stay above water. Despite their local involvement, associations struggle to finance their projects" (woman, 36 years, 2015).

Moreover, young people lack training and 46% of them are unemployed, so that they feel both abandoned and stigmatized. This lack of employment deprives them of prospects and creates very problematic attitudes that hamper living together. Thus, young people are described as "noisy, disrespectful, hostile, engaging in trafficking (...) breaking the communal equipment and that of individuals": "it is a chic neighborhood here, but life is not good because of what happens with young people: with motorbike noise 24 hours a day, they do not let us sleep, they do unimaginable things, they race against the tram, burn trash cans" (man, 44 years old 2015).

18.3.2 Growing Concern

The anxiety of the adults is growing, between fatalism and anger. Parents wonder about the future of their children, their schooling, their circle of friends: So, should they register them in private schools to escape a doomed fate in the event they become close to some undesirable youths in the neighborhood? How can they be prevented from joining these groups that are being formed, that circulate in the vicinity of the St Paul shopping center, driving on the tram tracks? How, more generally, can they be prevented from becoming unemployed and the delinquency often associated with it?

Two-thirds of the respondents reported the disruptive behavior of young people who they consider as "disenchanted and discouraged". These residents point to the urgent need to remedy this by finding substantive solutions: to set up training courses, to find jobs, to create places where youths can meet and make areas their own and are thereby less likely to hang out on the streets: "Young people on scooters and who make noise in the street must be made to work because they are bored and bothersome" (man, 23 years old). The residents also blame the difficulties and attitudes of young people on "unemployment (. . .) and their lack of training": "It is unemployment and the absence of jobs for young people that are the cause of the neighborhood's problems and which makes them go adrift" (man, 38 years 2015).

18.3.3 A Feeling of Ghettoization

The residents evoke the coexistence of four facts that develop the feeling of ghettoization of the district:

- The first is a feeling of social weight linked to the density of the living space,[9] a living space designed in accordance with the of principles Le Corbusier and the supporters of the charter of Athens as having to respond to a city function. This living space developed so much that it no longer allows residents to imagine their living space; On the contrary, it generates a "feeling of crowding", of proximity, even of lack of privacy with too many neighbors described as "uneducated and uncontrollable". This image of social densification is associated with demographic expansion and has a negative impact on the quality of social relations. Moreover, the difficulty of immigrant families accessing housing outside the neighborhood causes overcrowded housing: It is not uncommon for three generations to share the same dwelling, which complicates family functioning. Many people feel that solidarity has declined over the last 20 years: "When I arrived, the neighborhood was less crowded. Families knew each other, and by knowing the problems faced by others we could help them, before we were more supportive! Today we have the impression that it is every man for himself"(A., 49). It also appears that intensification increases the feeling of insecurity. There is, of course, a potential for sociability[10] and solidarity between families, communities or neighborhoods, but they seem, according to the residents, to be more fragile than before.
- The second fact notes the permanent influx of populations from the North Africa, in this case from the south of Morocco. These numerous arrivals erase the mix and reinforce the feeling of confinement. They have a notable impact on the lives

[9]It is interesting to take the sociologist Henri Lefebvre's definition between "habitation" (the result of urban planning) and "inhabiting", which is no longer the result of a "good" housing policy, a "good" architecture, a "good" urbanism, but which must be "considered as a source, as a foundation", on which the quality of the private sphere depends: "Before habitation, inhabiting was a millenary practice, poorly expressed, ill-suited to language and concept, more or less alive or degraded, but which remained concrete, that is to say, functional, multifunctional, trans-functional" (Lefebvre 1968, p. 25 et s., our translation).

[10]Sociability is made up of "all relationships that we have with others" (Forsé 1991, 247).

of the residents and especially on the women who have chosen a western way of life and who aspire to a certain amount of autonomy. Linked to the ghetto aspect of the district, numerous testimonies highlight the town within a town phenomenon, reinforced according to them, by the presence of the State services and the necessary infrastructures, as well as by the profusion of shops. According to their interpretation, everything is done in such a way that the residents are encouraged to remain amongst themselves, which encourages the community withdraw into itsel: "What I feel is that they have put a town within a town, We were all housed in it, we were given the CAF (French family allowance) and national health service offices, lots of discount stores, lots of halal butchers, fruit and vegetable stores. So consequently, we do not need to go anywhere else and we remain amongst ourselves, we are crammed together. It looks like a ghetto» (Woman, 36 years old 2015).
- A third feeling is the impression of abandonment of the neighborhood, delivered to degradation and dirt: "In buildings, it is really degraded, we are invaded by cockroaches ... There are youngsters who break things, make the place dirty, who make noise with their music and motorcycles. It's a wild atmosphere!" (Woman, 60 2015).
- The fourth aspect that reinforces the feeling of living in a ghetto is the impression of the neighborhood's withdrawal into itself and its submission to the omnipotent gangs of youths who dictate their will and their law. In fact, day and night, some of the youngest, the most disadvantaged, wander in groups and occupy the area in a worrying way: they have left the abandoned school, are poorly trained, ill at ease in their families and within society, without rewarding jobs. Having experienced repeated failures, they arouse fear, rejection or distrust from other generations, totally baffled by a society whose only aspiration is consumerism. Their situation is precarious and their future totally uncertain, due to the lack of assistance.

This omnipotence of the youths, mainly male, is exercised in the form of verbal violence and intimidation. It is correlated with many acts of incivility and the exercise of a very strong influence on the neighborhood, reducing the freedom of the other residents. Moreover, it seems to influence younger children who wander at night. Several testimonies corroborate the idea that the gangs of youths take possession of the district, especially in the late afternoon, and that they impose themselves by through the sound of their motorized equipment (motorbikes, quad bikes) and by risky behavior: "our neighborhood is very inward-looking, distressing for the people who live here, scared of everyday life, in relation to these gangs that impose themselves and force noise, despair, and fear upon us, especially when one has children" (H, woman, 65 years).

If these juvenile attitudes generate the exasperation, fear, anxiety in other residents, who are silent and feel powerless, they nevertheless find in some eyes an explanation, sometimes even a legitimization. These attitudes are in fact analyzed as a need to exist in the eyes of others, in other words a need for social recognition in response to a feeling of abandonment in a relegated environment, to the impression of being repressed by this society. Their desire to mark their territory is a way of exercising a certain authority, of "taking control" in a context where their future eludes them: "Many young people are left to their own devices, lost and aggressive. There is nothing for them, we can understand this. What would be good is that there are a few more youth workers or educators. They have no work, they should be helped to regain self-confidence, they should be guided and their ambition restored to be able to move towards the better" (woman, 33 years old).

The spatial concentration of problems weighs on individual trajectories (Maurin 2004): "I, like my friends, are caught in a kind of curse. The impossibility to leave the neighborhood, to choose another, to find work. All this prevents us from reaching out towards the rest of the French population. It is like the tectonic plates: one is drifting, moving away from, and distancing oneself from the other French citizens. We needed to get out of this deadly trap, that of isolation, and join the crowd" (Nourdine Bara, journalist 2015).

A number of young women from Maghreb migration try to resist the pressure exerted by the youths and seek solutions to protect their children from the risk of delinquency: "As we cannot leave the neighborhood, we take our children out from neighborhood's, elementary schools and send them to the city center or into the private sector so that they can mix with other children" (woman 40 years, 2014).

Finally, the relational rules of the residents from the North African emigration reveal a rather divided and complex relationship between men and women, which are expressed in the form of a territorial appropriation and a differentiation of masculine and feminine areas: "Around 19 (Men) invade the neighborhood and the women go home, you don't see them anymore. Young women do not walk under the arcades, or in front of the Feuillade Hall in order to avoid the café. It's the looks from the men which disturb them. The elderly women who are divorced or widowed, do not apply this rule" (woman 74, 2014).

18.3.4 Recluse Women

The isolation and even seclusion of some emigrant women were mentioned several times: "Many women live under the control of men, some of them have been locked up for a long time, they are afraid to go out, to speak. They are locked up. We need to talk to the husbands, who do not want to let their wife go to the association, because they are afraid of having problems. Women need to learn to read, write, speak French, many don't know how to do that" (A, woman, 33 years old); "Those who could go out, do not dare to do so because of safety issues which prevents them from going out in the evening" (woman 43 years, 2014). Moreover, half of the women's testimonies converge on the idea that there are not enough structures open and sufficiently flexible so that women can freely find themselves among each other, at times that would suit them.

18.3.5 Living the Ghetto

As can be seen, the Mosson neighborhood's residents and associations describe a reality lived in 2015, that some of them assimilate to a situation of relegation (density of the ha dwelling space, impression of being a town within the town, lack of social mix, high unemployment, poverty, delinquency and juvenile violence, neighbors' incivility, etc.). However, all the residents do not share this image, to which Didier Lapeyronnie could retort that the ghetto "is not a homogeneous area". Many residents "can live in the neighborhood without living the ghetto (. . .). They are people who work, live their lives as a couple, raise their children, retirees, families" (Lapeyronnie 2008, 48).

Let us return to Loïc Wacquant's analysis of the definition of the ghetto: "For an urban district to become a ghetto, five mutually reinforcing characteristics would have to be found: increasing ethnic homogeneity, an increasingly complete enveloping of the target population, an increase in organizational density, the production and adoption of a collective identity, and finally, impermeable borders" (2012). The reports on the Mosson district are not sufficient to say that this district has become a ghetto, because, on the one hand, the borders of the district are truly open and Tram lines 1 and 3 make it possible to activate the going back and forth to the town's other districts. On the other hand, the phenomenon of the adoption of collective identity does not, except for the attitudes of certain gangs of youths, appear blatantly obvious. Yet the testimonies gathered show an often painful and anxious experience that must listened to, and even more so as the proposals of the residents to "reverse the process of ghettoization" are numerous.

18.3.6 Various Actions Advocated by the Residents

For families, it was first suggested to propose more outings outside the neighborhood and then to create squares and kindergartens so that the parents can meet there. In addition, the wish was expressed to draw the public from the other districts of the town to that of Mosson by organizing events such as the temporary artistic zone (ZAT) which was a great success in April 14, 2013.

As far as women are concerned, their emancipation could result from an increase in literacy[11] and French courses, but also from an increase in the sports activities offered at adapted times.

At the same time, mothers of below school age children should benefit from the setting up of solidarity childcare to welcome their toddlers and to involve them in

[11] Some residents explain how the lack of mastery of the language constitutes an embarrassment or even a "humiliation" that makes the elderly depend on their children or their daughter-in-laws. This situation deprives them of privacy insofar as any communication, however personal, must go through an interpreter.

a cyclical way in welcoming the children of others. Finally, create friendly places where exchanges can take place and would allow the women to find themselves amongst each other.

As for the young people, the residents suggest putting in place significant support and accompaniment services for students who are drop out of school as well as for young people in search of employment, by means of a structure that would be dedicated to them. They also advocate the hiring of young facilitators who would have a unifying role with young people.

Finally, the idea of creating a consortium of arts and culture, where all young people could come to meet artists, express themselves and debate freely, was raised.

The issue of the elderly has been little addressed by the residents as this category of resident is discreet about the neighborhood. However, these elderly people represent 17% of the local population. For their part, they are suffering more and more from geographical mobility linked to employment and the scattering of the family unit.[12] Children who are often geographically distant are less accessible and less available. They visit their parents at weekends, when shops and services are closed, so that they help them little in their daily lives. The social workers of the neighborhood mentioned certain situations of significant food insecurity and emotional and relational isolation, especially with regard to the elderly immigrants. A number of them are forced into force isolation due to the death of their spouse or their departure to the country of origin, which is also difficult to accept. This isolation is even more prevalent among elderly women who emigrated, many of whom do not have sufficient command of the French language (Fournier 2014).

In order to remedy the loss of the independence of the elderly and to support the other residents, an approach based on "humanitude" (Laroque and Pellissier 2007) needs to be developed in order to better understand old age and practice social supervision around the aging seniors. This psychosocial approach would enhance the quality of coexistence and provide answers to the loss of family and professional ties by building new links of solidarity.

However, more generally, seniors are not the only people who suffer from the individualistic and dividing change in society and who feel a sense of isolation. Our various surveys did indeed show that immigrant women found it difficult to integrate into the neighborhood and to find their independence. Finally, young people who have dropped out of school, are unemployed and stigmatized, are retreating into a "amongst one's own environment" and putting others at a distance through attitudes of defiance and provocation representative of their social suffering.

[12]According to a survey conducted by the Fondation de France in 2013, 27% of the elderly live alone (compared to 16% in 2010) and 41% have little or no contact with their own children (compared with 38% in 2010). 50% say they do not see their friends and 52% are not in contact with their neighbors. This trend is worsening among those over 75 years of age due to the death of the spouse, health problems and children living far away.

18.3.7 Towards What Type of Socio-cultural Coviability Are We Heading For?

Isolation, retreat, dependence, rejection, relegation are all significant features of the cultural, gender and generational divisions that weaken the social fabric and reinforce the harm done to the cohesion of the neighborhood, thereby constituting an attack on the socio-cultural coviability. This can be defined as the conscious and interconnected coexistence of several socio-cultural groups and their systems of meaning within the same territory, in a dynamic of reciprocal acceptance, guaranteeing a common future respectful of individual and collective identities. This socio-cultural coviability implies the maintenance of peaceful intergenerational and intercultural human relations, around a sharing of signifiers, values and stakes. By promoting semiotic, symbolic and dynamic interconnections, the comprehension of new experiences will generate new intercultural relations that will change the social fabric.

Socio-culturalcoviability is defined at the individual level in the concept of "active coexistence" defined by Grzybowski (2015) as "operational cooperation at the individual level between persons of different religions or convictions, freely and voluntarily engaged in a dynamic of mutual respect allowing the identification of a common project respecting and enriching individual identities" (Grzybowski 2015, 35).

Our practical research will therefore have to tackle sociocultural coviability as a dynamic-creating concept, from which social bonds can be observed and stimulated both as signifiers and as vectors of coviability. (Fig. 18.1)

Fig. 18.1 Fresco by the Nicaraguan painter Leonel Cerrato, which was offered to the Paillade neighborhood in 2011 as a testimony of friendship between peoples. The painting was performed here at the Mosson with the help of pupils of the Lycée Mas de Tesse. (Source: © Olivier Barrière/PACIM)

18.4 The Gamble Taken on New Sociability Around Solidarity Through an Urban Mutual Support Platform

In this worrying urban context, marked by divisions and individual or family survival strategies, it is indeed urgent to repair the social bond and to restore a certain level of conviviality and all means must be devised to achieve this. The joint activities and festivities organized by the local network of associations work on this in a "rationale of social transformation and emancipation" (Carlon and Dessis 2013) and help to regulate social relations. Similarly, the "Agoras" supported the journalist Nourdine Bara constitute "initiatives that go to the heart of the town to re-establish dialogue with other French people: in all these actions, there is the same urgency to share, exchange, to boost" (Bara Nourdine 2015). Finally, the residents put forward proposals and actions and they are ready to engage in active projects. They are all the more convinced when they are associated with the creation of these projects.

For its part, the PACIM ("Passeurs de Cultures, Passeurs d'Images") has been trying for several years to revive social ties by relying on a discovery of otherness and giving value to the life stories of residents. It is during ethnology initiation workshops (Ethnofil, Life Stories inheritance) that people can learn about methods of ethnological collection and become collectors of memory, engaged in a charter of confidentiality.

Following a collection of testimonies on the crosscutting theme of "Living together in the Mosson neighborhood", the PACIM association designed the project "Generations solidaires à la Mosson". This project is based on the republican value of fraternity and aims to stimulate community solidarity for the benefit of the elderly and to strengthen social cohesion. In order to establish the work on an anthropological basis and to ensure that it is adapted to the sensitive environment in which it is located, data have recently been collected[13] semi-directively from a panel of 40 residents; They found that in this urban setting, mutual support in terms of childcare, moral support, accompanying elders of the family or logistical assistance was most often exercised by the family unit or a close circle of friends. However, the geographical remoteness of the families makes this solidarity difficult. In order to compensate for this, while strengthening social cohesion, it would seem appropriate to stimulate neighborhood solidarity in this vast area of 5 km^2.

It is through an approach of intergenerational solidarity and under the auspices of the goodwill of the residents between them, of the indispensable support that must be provided to the oldest members in all the communities, that the PACIM

[13]The survey was conducted in 2014 on the theme of representations of the neighborhood and the role played by the Social Center CAF (Caisse d'Allocations Familiales / Family allowance).

association tries to re-establish links, by means of a digital intergenerational assistance platform. It was designed with groups of residents who identified potential needs and thought about what they wanted to share. In order for this solidarity to develop, the association involves young secondary school students in a process of collecting from the elders their experiences and stories in life in order to reconnect the generations to each other. By re-establishing links through ethnology workshops,[14] young people become aware of the concerns of the elders and return them to other generations in various artistic forms (travel diaries, audiovisual editing). Once the process of becoming aware of possible mutual encounters and support has begun, it is still necessary to invent acts of solidarity that are within the reach of all, which adhere to this social and cultural diversity. Concrete proposals can be formulated on the platform, then actions of dynamization relayed by the associative fabric and the citizen councils allow the different populations to be reached.

The question of solidarity and its repercussions on the development and consolidation of social ties in the Mosson district undeniably refers to an intercultural and intergenerational problem. It questions not only the conditions of aging, but also forms of today's sociability linked to the evolution of the neighborhood. It implies first taking stock of the cultural images of the neighborhood, the solidarity of the neighborhood, and the "possible relations" between the various residents of the neighborhood. Finally, it invites us to reflect on the triggers of neighborly relations, to what is likely to open up the prospect of a occasional and legitimate mutual support, i.e. what, in the eyes of the residents, makes their demand for service legitimate and what make their proposal for assistance acceptable?

18.4.1 The Notion of Neighborhood

"Being a neighbor" refers to a state of affairs which means "to be near something", "to be placed next to" or "to remain near another". If the notion of neighborhood indicates a more or less desired spatial proximity, the action of neighboring, that is to say, "the action of maintaining neighborly relations", refers to an intentionality and socio-cultural practices which are linked to the images of the space to be shared, of which, as Denis la Mache says, it would be naive to believe that it is "identically perceived by all and truly collective" (La Mache 1998). However, even if the shared neighborhood is a mental construction of images for oneself in a framework of life shared with the others, is it in fact "only a juxtaposition of distinct and sometimes antagonistic, parallel worlds"? Is there no convergent image?

[14]As part of these ethnology workshops, which take place once a week, adolescents are made aware of aging-related issues, learn to relate to elders and gather their memoirs, and then restore them in the form of a narrative.

Indeed, "neighboring" or "not neighboring" indicates a specific way of dwelling in space, which is based on a certain "model of dwelling", a way of conceiving relationships with the nearby residents are linked to "rationales of perception and appropriation of places that overlap, intermingle and sometimes oppose each other" (La Mache 1998). These rationales are related to the life histories and the expectations of residents, according to the way they imagine their life locally and whose neighborhood they invest emotionally Depending on whether the residents have settled for a long time, are waiting for to be relocated, or recently settled, their relationship with the neighborhood varies and does or does generate the "desire for solidarity" with the neighborhood. The action of "neighboring" is nourished by the desire to build relationships (between individual biography and collective history), in connection with the emergence of a sense of belonging to a local area or a group with shared identity references.

As a neighbor, the resident is thus in a form of "sociability", which allows him/her to maintain concrete ties with other occupants of the same space. Contrary to family or work relations, these neighborhood relations are essentially described by C. Bidart (1997) as "their non-structural, non-constraining dimension" in the sense that they require little coordination with others and thus little negotiation.

Confronted with an unconstrained sociability that allows invisibility in one's neighborhood, one can remain isolated or maintain a relational vacuum. Neighborhood relations are similar to the weak link, in that they are characterized by a form of lightness, respect for intimacy, and non-intrusion into private affairs. This action of neighboring can be rolled out in many ways, ranging from the good day-good evening, small verbal exchanges in the public area which means that the person is identified as a neighbor and thus distinguished from the mass of others, occasional mutual aid or friendly relations.

In the multicultural neighborhood of the Mosson, these neighboring cultures connected with different spatial and temporal imaginings need to be grasped in order to understand which relationships are the most elective. Are these relationships between people of the same age group, socio-cultural backgrounds, or similar life trajectories? What are the obstacles to the development of relationships? Are there rules for good neighborliness that are distinct according to the micro-spatial constituencies of the neighborhood, dependent on the local living conditions? Are certain neighborhood norms emerging, such as the standard of everyone mind's their own business?

18.4.2 The Notions of Neighborhood and Intergenerational Solidarity

For Edgar Morin (1989) our past was marked by the imperative duties of clan and family solidarity. Duties have more or less been abandoned because of "individual atomization and social protection (insurance, pensions, etc.)". The rapid evolution of society tends to widen the gaps between generations so that "we feel very far from our relatives and foreign to ours". However, blood brothers "have been substituted for brothers of choice or friends (. . .) with whom we maintain close relations, as much spatial as affective and moral" (Morin 1989).

If the notion of solidarity has fluctuated in history, whether facing the satisfaction of essential needs or defending the common interests, it always refers to the idea of community, territorial proximity and social affinities. Neighborly solidarity, on the other hand, implies the idea of a public space in which one recognizes others and one is recognized. This identification as a member of the same neighborhood is therefore based on a common geographical anchorage, referring to a sense of territorial belonging. Nevertheless, the feeling of belonging to the same group is not obvious, even if the individual identities of the residents of a same neighborhood overlap, each belonging exclusively to his town, neighborhood, family, but also to certain associations or bodies, sharing various strong interests etc. Within these groups, bonds arise from mutual acquaintance, based or not on a common praxis, involving or not mutual help or mutual support.

Mutual support only appears when an interest of care arises for others. It often results in informal gestures inscribed in a marked space, meaningful of a common belonging. Indeed, start-up of mutual help between neighbors is not automatic: it is often a particular event (a fall, a fracture, etc.) that will allow the neighbors to be solicited or which will legitimize mutual support and thus authorize others to cross the threshold of their own privacy. The concern for reciprocity,[15] donation and counter-donation, conveyed by the model of egalitarian exchange, may also act as an obstacle to these exchanges between neighbors of different ages.

Some acts of solidarity amongst neighbors rely on material help (occasional, regular or daily), others are translated by the setting up of moments of shared conviviality (a phone call, a snack, invitation to a family meal). A third trend concerns the reassuring presence through an informal watch (surveillance of the shutters, giving the alert, etc.). Finally, the existence of elements facilitating links (existence of shops or local public services, associative dynamics, digital network of mutual support) can help to stimulate links between residents giving them a reassuring and structured framework and offering pleasant key moments encouraging meeting people.

The specificity of the relational fabric of the neighborhood characterizes local solidarity. This will be based on ties of different intensities, i.e. on weak or strong ties (according to Granovetter 1973, 2000). Indeed, the strength of a tie is determined according to the amount of time shared, the emotional intensity, the

[15]A great part of human activities are subject to the principle of reciprocity, "which consists in obliging the one who acts on another to undergo the same act, and the one who is being subjected to the act to act. It reproduces in the opposite direction the situation of one in relation to that of the other and thus the perception of each one is redoubled from that of his opposite. For example, the donor (who loses what he gives) will feel that he is acquiring the value of humanity (prestige) while the recipient of the gift will feel the loss of face and hence for him the desire to regain prestige which is translated by the obligation of reciprocity, the obligation to give back" (Temple 1989).

mutual trust and the reciprocal services that characterize it. Strong bonds are thus frequent, long-lasting and emotionally intense bonds: they correspond to family or friendly ties. As for weak ties, they are associated with neighborly relations, distant friendly relations or professional relations. Insofar as weak ties are less restrictive, their solicitation brings flexibility and relational fluidity. As can be seen on social networks (such as Facebook), they enable information to be widely relayed (to friends of friends) and their scope increases the chance of obtaining answers. These ties are subject to fewer standards and emotional constraints than strong ties. They develop according to the curiosity and desire for openness of individuals who wish to develop networks for the exchange of information. We can therefore assume that these ties of weak intensity can contribute to activating neighborhood solidarity and to building bridges between people who are unlikely to get together, to dialogue together and who, at the same time, remain isolated.

Although social media has emphasized the process of ghettoization and acts of incivility that tarnish the reputation of the neighborhood, they do not provide a concrete answer to this phenomenon. This is why the PACIM association, which has been operating in the neighborhood for 9 years, wished to put in place this practical research which, on the one hand, offers a practical tool for reconnecting the residents with each other in order to encourage the development of weak ties and to strengthen e local solidarity and the others. In parallel, it is developing an anthropological study on the sociability of the Mosson residents[16] and on the cultures of the neighborhood in order to give tools to interpret the residents' behavior in the face of the implementation of the mutual support digital platform. This study will be a tool for evaluating the digital mutual support platform as a tool for supporting the socio-cultural coviability in the Mosson neighborhood.

18.4.3 A Digital Lever to Stimulate Neighborhood Solidarity in the Mosson Neighborhood

The "Générations solidaires à la Mosson" digital platform (generations-solidaires-mosson.fr), set up in the Mosson district, was conceived as part of an intergenerational project of participatory citizenship. It aims to jointly offer social mediation and digital support to trigger close solidarity and enrich social relations.

[16]This study is supported by the social and financial partners and by the network of local associations.

18.4.3.1 A Digital Co-construction in the Service of Solidarity Through Mutual Support

Its aim is to stimulate low-intensity social ties (we have just seen above that they can play a mobilizing role in community work) and, on the other hand, to build new, occasional or lasting relationships between residents who had not previously had the opportunity to meet. Given the innovative nature of this new tool, it was essential that its creative process proceeded through a bottom-up approach so that it really meets local needs. This is why it was progressively conceived and developed with a group of residents and associative partners involved in the life of the neighborhood Different one-year meetings were held to gather the views of the participants, make technical proposals and move forward until this supporting tool meets the expectations of all.

So that this digital space of mutual support is invested by the young people of the district, a partnership was put in place from the start with the "Escholiers de la Mosson" secondary school. A class of 15–16 year olds was asked to create the visuals and sound. The 14 students took advantage of the art lessons, in the presence of their main teacher and the PACIM team to discuss the graphic charter of the project. They created the pictograms identifying the welfare items and decorated the platform with local characters and symbols (trams, shops, etc.) representative of the district in their eyes. They also wrote texts and lent their young voices to the video presentation of the platform. In order to establish the project's indigenousness it was essential that these teenagers marked this platform of mutual support of a "local stamp" that symbolizes their adherence to this approach.

For their part, the various residents and associative partners who contributed to the development of this tool have "gamble" on an ethical and a social watch attitude shared by a community of neighbors that is supposed to be altruistic, caring and well-intentioned. This presupposes that these neighbors are ready to enter into a relationship and volunteer for one another as soon as their respective needs have been identified. The moral values underlying this approach thus seek to transcend pejorative images of neighborhood and proximity by means of "doing together" or "doing for others". The social assistance that is suggested is intended to contribute to the well-being of each individual by coming in different fields and in different forms.

Indeed, this platform invites us to open the field of possibilities in terms of the power of action, thanks to the diversity of the proposed mutual support items and the construction of new ties. Thus, the ties that were not built with potential neighbors for reasons of discretion, timidity or social convenience become spontaneously possible, since neighbors are identified as belonging to the same community and their request is made within a defined framework. By facilitating the establishment of new relationships between residents who had not yet integrated into the relational landscape, this digital medium gives material to read and therefore to hear new demands, but also offers of previously unheard of help.

18.4.3.2 The Modalities of Mutual Support

Mutual support registers were identified through a survey of a sample of 50 residents aged 14–90. In order to attract as many male and female users of different ages as possible, 15 types of mutual support have been proposed.[17] In order to meet the diverse needs of everyday life, these offers of mutual support can alleviate the lack of temporary autonomy of the elderly. They can also meet technical needs (administrative assistance, minor repairs, troubleshooting), ordinary needs (shopping, carpooling, assistance, etc.). In all cases, they are supposed to facilitate daily life and highlight the confidence in the competencies of neighbors. Volunteering is highly valued since it concerns the offering of a free and occasional helping hand, and in no way does this help replace the services provided to persons in need.

18.4.3.3 Exchanges Accompanied by a Socio-cultural Mediator

Use of the digital platform is accompanied, supervised and moderated by a socio-cultural mediator, who ensures respect for the rules of courtesy in exchanges at a distance and then in face-to-face meetings. Through regular contact with the members of the platform, she questions the exchanges that have taken place. She gives herself the means to temporarily or definitively "block" a resident whose attitudes could have been problematic, for example, by posting too often request, or by adopting disrespectful, conflicting behavior or financially self-interested. At any time, the users of the platform may report attitudes that they deem not to comply with the ethics mentioned on the home page.

If this new digital tool, designed with young people and adults, has an inter-generational and intercultural vocation, its access must made easy for all residents, including the elderly in the neighborhood and those for whom non-proficiency in the French language prevents them from integrating. However, the digital tool is generating great concern among the elderly, and sometimes even hostility. For example, elderly people consider that computers and tablets are not made for them, but for their children or grandchildren. They are also wary of the Internet, its supposed additive and intrusive effects in the lives of individuals. Of the 92 people over 65 with whom we had the opportunity to talk to, whether they are migrants or not, about 8% have Internet at home and a digital address. For those who are not Internet users the offer to have assistance in discovering the web or in digital workshops, is creating, for the moment, little enthusiasm.

Yet we think it is essential to break down these barriers and propose "Easy Digital" workshops that will enable seniors to become familiar with these new

[17] These categories of assistance are: Company for an outing or a visit, Shopping, Help with your trips on foot, Assistance with the preparation of meals, Lessons, Car-pooling and transport, Administrative procedures, Listening and conversation, DIY and troubleshooting, Babysitting, Advice, Sharing of activities, Loan of material, Gardening, Others.

technologies in order to receive the benefits in terms of maintaining cognitive functions and the enlivening of social interactions (which often diminish due to the gradual disappearance of strong ties). We are aware of the fact that electronic media are not enough to provide social relationships, but that maintaining them can largely be encouraged by a combination of interactions taking place in the virtual and real worlds.

18.4.3.4 "Easy Digital" Workshops for Seniors

Digital workshops adapted for older audiences begin to be conducted twice a week in different venues, sometimes at the association's premises and sometimes at the media library. These will allow seniors to choose their discovery media (fixed computer, laptop or tablet) and the diversity of related tools (software, applications, etc.). For isolated people who encounter difficulties in leaving their homes and going out, the project of a personalized digital accompaniment is set up at their home.

18.4.3.5 Community Access Points to Log In

In spite of the computerization of society resulting in digital tools imposing themselves in everyday life, the residents of the Mosson district remain disadvantaged concerning access to digital tools: the number of access points to computer stations remains restricted.: There is currently only one internet cafe and three multimedia access points for 25,000 residents. The challenge is to create new access points, offering adapted schedules and organizing a targeted reception for the most disadvantaged groups (due to language or age). In order to develop these access points, associations and partner structures that are already connected with vulnerable or specific persons, are being contacted.

We are working to implement actions to raise resident awareness of about the emergence of this dynamic of solidarity during events organized by the associations and during festivities, which provide appropriate opportunities to revitalize social connections and share and reconnect man to man. (Fig. 18.2)

18.5 Conclusion

It is premature to draw conclusions about this practical research on the social and digital ties, as it is in its intermediate phase. The tools were co-built and validated with the residents in order to create a locally recognized endogenous exchange tool. The official launch of the platform was carried out in June 2013 and registrations are progressing as communication actions are gradually being initiated. Digital interactions are booming. They are accompanied and encouraged by socio-cultural

Fig. 18.2 Human Constellation created by de Chen Zen in 2000; It consists of two giant plates attached back to back surrounded by 70 chairs of different styles. As a metaphor for both the terrestrial globe and the universal banquet, the Human Constellation symbolizes the modern fields of dialogue and fraternal exchange among peoples. (Source: © Olivier Barrière/PACIM)

mediation work that links up together an increasing number of associative and institutional partners in the neighborhood.

In this context of "a culture of depression, where everyone thinks of themselves and protects themselves" (Kader, a resident of the district), an anthropological survey related to the follow-up of this work proposes to explore how the notion of solidarity can be understood and integrated into everyday practices. This will require identifying the different meanings that are brought to it and the emerging common issues around which the residents can gather together and rally, beyond their usual circles of strong ties. We will seek to verify whether intergenerational solidarity is indeed a multicultural value, capable of deconstructing the ghetto from within. If so, what forms of traditional or reinvented solidarities may prove coviable at the crossroads of these different communities united in the same area? What will be their brakes and levers be?

In this period of technological change, where digital tools are becoming more and more important in everyday life, the "Générations solidaires à la Mosson" mutual support platform and the tools set up to accompany it (digital workshops, community access points, telephone hotlines, etc.) constitute a means of combining the reflexes of distance sociability (telephone, messaging, chat) in order to strengthen neighborhood ties and local solidarity.

It is a matter of reviving a local conviviality and mutual support, both crushed by the individualism, the anonymization or the stigmatization of the foreigner. If "identities are defined and renewed in relation to others" (Busino 2006, 185), this relational renewal will have an impact on the mental images of residents and enrich their social landscape. It will also allow the emergence of "new strategies for the management of otherness" (La Mâche 1998). It is a safe gamble that the activation of this neighborly solidarity, coupled with the establishment of a powerful digital lever, will enable us not only to live together and new expressions in the art neighboring, but they will also help in this process of "collective creativity" to ease tensions and gradually reduce the community's withdrawal insofar as they will encourage the progressive connection with neighbors (formerly invisible) belonging to the same neighborhood.

References

Bara Nourdine (2015) Nourdine Bara nous questionne, Midi Libre Newspaper of the 10/03/2015 https://www.midilibre.fr/2015/03/09/nourdine-bara-nous-questionne,1133046.php

Bechmann R (1966) Le Corbusier et la nature Textes recueillis et commentés in. http://documents.irevues.inist.fr/bitstream/handle/2042/50724/AetN_1966_2_9.pdf?sequence=1

Bidart C (1997) L'amitié, un lien social. La Découverte, p 403, halshs–00197849

Busino G (2006) Citoyenneté et démocratie providentielle: mélanges en l'honneur de Dominique Schnapper. In: Revue européenne des sciences sociales, tome XLIV, 2006, N°135. Librairie Droz, Genève/Paris

Carlon B, Dessis C (2013) L'évolution des territoires de la politique de la ville en Languedoc Roussillon, Cabinet Territori, Rapport 2, La Zone Urbaine Sensible de La Mosson à Montpellier (Hérault)

Casilli A (2010) les liaisons numériques. Vers une nouvelle sociabilité? Seuil, La couleur des idées

Forsé M (1991) Les réseaux de sociabilité, un état des lieux. l'année sociologique 41:247–264

Fournier L (2014) Les enjeux du vieillissement de la population immigrée à travers la situation de femmes immigrées du quartier de la Paillade à Montpellier, Mémoire master Développement Social, Université de Montpellier

Granovetter MS (1973) "The Strength of Weak Ties", American Journal of Sociology, 78(6): 1360–1380

Granovetter M (2000) Le marché autrement, Paris, Desclée de Brouwer, coll. "Sociologie économique", 239 pages, trad. de l'américain par Isabelle This-Saint Jean

Grzybowski S (2015) Manifeste pour une coexistence active, Les Editions de l'atelier, 109 p

La Mache D (1998) La conquête de l'espace, habitat et regards croisés dans un ilot sensible. Terrain 30:139–152

Lapeyronnie D (2008) Ghetto urbain, Ségrégation, violence, pauvreté en France aujourd'hui. Editions Robert Laffont, Paris

Laroque G, Pellissier J (2007) Humanitude: Comprendre la vieillesse, prendre soin des Hommes vieux. Armand Colin, Paris

Lefebvre H (1968) Le droit à la ville, Anthropos, Le Seuil (réédition 1974)

Lewin, K. (1946) Action Research and Minority Problems. Journal of Social Issues, 2, 34-46. http://dx.doi.org/10.1111/j.1540-4560.1946.tb02295.x

Libération (2015) Montpellier, la Paillade, sortir du piège mortel, 5 aout 2015, online: http://www.liberation.fr/societe/2015/08/05/montpellier-a-la-paillade-sortir-du-piege-mortel-de-l-isolement_1358885

Mathevet R (2009) L'invention du concept de solidarité écologique. Synthèse des occurrences du terme solidarité dans les rapports parlementaires, exposés des motifs, projets de loi, textes de loi, amendements et discours du ministre de l'écologie. Rapport CNRS CEFE, Montpellier, 10 p

Maurin E (2004) Le ghetto français. Enquête sur le séparatisme social. Seuil, Paris, 96 p

Médard JF (1969) Communauté et organisation communautaire aux Etats-Unis, In Cahier de la Fondation nationale des sciences politiques. Armand Colin, Paris

Morin E (1989) Pour sortir du XXe siècle, Nathan

Parks Malcolm R, Roberts LD (1998) Making Moosic: the development of personal relationships online and a comparison to their off-line counterparts. J Soc Pers Relat 15(4):517–537

Temple D (1989) Les origines anthropologiques de la réciprocité, online: http://dominique.temple.free.fr/reciprocite.php

Wacquant L (2012) "Repenser le ghetto. Du sens commun au concept sociologique" (Idees, n° 167)

Chapter 19
Kinship as an Instrument for Coviability: Study Cases in Pará, Amazonia

Voyner Ravena-Cañete

Contents

19.1 Introduction

The model of development adopted by the Western economy after the Second World War values the idea of economic growth. The latter has been transformed into a discourse of development and becomes a strategic principle for creating the basis to relationships of dependency between the old metropolises, now creditors of the Western world. However, this process of development, one which is limited to unmeasured economic growth, has promoted new economic actors, corporations and conglomerates. These have gained protagonist positions within the framework which forged them, that of economic, social and environmental exploitation. Global economic integration seems to dilute time and space (through the new communication technologies) and causes disastrous consequences for local populations, as they are less politically and economically involved in the global framework. This is true, even when such populations, within their own territories, own the natural resources essential to the machine of unbridled economic growth.

V. Ravena-Cañete (✉)
The Federal University of Pará, Belém, Brazil
e-mail: ravenacanete@gmail.com

O. Barrière et al. (eds.), *Coviability of Social and Ecological Systems: Reconnecting Mankind to the Biosphere in an Era of Global Change*,
https://doi.org/10.1007/978-3-319-78497-7_19

Technically transformed into an ecosystem,[1] nature loses a degree of its significance and its intrinsic value against the inexhaustible demands of established and emerging markets. Urban populations persist in pursuing an illusory possibility of unlimited consumption. This desire transforms the ecosystem: nature becomes a mere "natural resource" through "denaturation".[2] It is unviable, for both societies and ecosystems, to perceive one's access to nature in terms of resources only, without considering the cultural values that nature permeates and taking it as a mere strategy for unlimited economic growth. This is not about using the discourse of sustainable development encompassing economic, social and environmental viability, but rather about showing that the model of economic reproduction is in no shape or form, viable. A viable world implies societies exerting mutual sustainability with their natural environment. It refers therefore to societies that adopt a co-viable relation with the natural environment that hosts them. Coviability is thus a presupposition to thought about and to be incorporated as a response to the limits of sustainable development, in so much as this demonstrates such limitations given that it tends to wane when it breaks away from the idea of economic growth. Investigating a viable world then means finding coviability that engages various ecological and socio-cultural viabilities. However, it is important to emphasize that coviability hardly requires ecological and socio-cultural viabilities, but socio-cultural co-viability between them. Differently put, a social group may be ecologically and socio-culturally co-viable but not be co-viable in relation to other social groups, in that its lifestyle compromises that of other groups. It is, therefore, important to amplify the idea of coviability.

Understanding viability derives from the perspective that human beings are entities of nature (Adams 1994; de Castro 1996; Diegues 1993, 1997, 2008; Descola 2006), with reciprocal relationships *which* organize exchanges between humans and non-humans. However, these exchanges seem to have dissolved human relations through the artificiality promoted in the commodification of the mercantile exchange system (Simmel 1987; Polanyi 2000). It is nevertheless important to remember that all the processes of life are guided by exchanges. Indeed, everything is guided by exchange, from molecules, to cells, to ecosystems, to life in society (Guattari 2006). Life is exchange, but a sort of exchange that is not artificialized by money. Such artificialization has affected the Western civilization model which, while promoting the technological advances that the world has at its disposal, has emptied the

[1]The notion of nature could be understood from Latour (2004, 2013). (2004, 2013).

[2]The market transforms everything into goods. From the material to the non-physical, from concrete to imaginary, everything is converted into passive marketing. Nature does not escape this imperative movement. Disassociating the human being from its conditions as an entity of nature certainly consolidates nature as another element to be commercialized. Such a movement is born, organized and prevails in the system which today strengthens its dominion over the world. For this system, nature has ceased to be life, mother-earth, Pacha Mama, among so many definitions coming from ontologies that consider life as the imperative organizer of all relationships. As a commodity, nature ceases to be the organized and organizing expression of life, making a conceptual journey that technically *taxonomized* it to *subjugate* it as a resource.

planetary natural resources and compromised their quality, reinforcing thus an idea of the human being as an entity separated from nature and from other humans.

The global processes of trade commodification are precisely what is "enfeebling" the planet. And how non-*commodified* local forms of trade are instrumentalized to consolidate the new imposing global economic system draws attention. One may find plenty of examples of this in the Amazonian historical and social contexts. Kinship-oriented organizations are distinguished as native systems heavily used by the system that attracts them to the process of appropriation and involvement of local lifestyle. Simultaneously, kinship expresses itself as an organization capable of promoting dialogue in favor of future coviability between different lifestyles.

The Brazilian Amazon occupies more than 2/3 of the national territory. For a long time, this region has been at the edge of the traditional forms of occupation that marked the process shaping the Brazilian territory. Today, the Brazilian Amazon is under intense pressure because of its natural resources. This is manifested by a marked alteration of socio-diversity that characterizes the region to the point of transforming local lifestyles.[3] Brazilian public policies, notably developmentalist, exhaust natural resources of all types[4] by altering territories shaped by social and natural processes, over centuries of existence and that relied on a perspective of interaction between similarities and differences among the various groups found within them.

It is important to emphasize that although such policies focus on the idea of development as growth, towards the end of the twentieth century, this developer (developmentalist) discourse was already beginning to incorporate the idea of sustainable development, a concept that would dominate the last third of the twentieth century. Such a concept, combining ideas of technical efficiency and technological progress with environmental and social justice, seems to be struggling against the environmental crises that have ravaged and still ravage the planet. From classical economics, passing through environmental economics to ecological economics, discussions on sustainable development deserve special attention. Cavalcanti (2004) highlights the differences between classical, environmental and ecological economics in the discussion of sustainable development. For the author.

> Every human activity impacts the ecosystem, either as resource extraction or waste discharge in the form of degraded matter or energy. The economic process, which operates within an open subsystem surrounded by the global ecosystem, must respect limits, hence

[3]The discussion on lifestyle is broad and covers the different areas of knowledge. Given the Amazonian frameworks, this text still makes use of the classical definition of Wirth (1987). The author emphasizes that "... the transition from a rural society to a predominantly urban society taking place within the timespan of a single generation in industrialized areas such as the United States and Japan was accompanied by profound changes and in practically every stage of social life" (Wirth 1987: 90 free translation). The author considers lifestyle from a rural/urban perspective; that is to say, he takes as a reference, sociability and forms of interaction between individuals, individuals and the environment, and even between individuals and natural resources.

[4]Hydro-electric power stations, ports for distributing resources of all types, but especially minerals, transforming the country into a great exporter of commodities.

> the notion of sustainable development. From the perspective of sustainability, the type of process that matters is that which produces goods and services while at the same time taking into account all the costs (or evils) that are inevitably associated with it. This is the task of a new model of development and also of a science of economics with ecological foundations. Ecological economics come into play here. It brings with it a fundamental mutation in the perception of the problems of resources allocation and of how these must be treated. In the same way, a revision of the dynamics of economic growth is to be considered (p. 157 free translation).[5]

In addition to the environmental impact, acknowledged in every perspective, Brazilian public policies, based on "development," lead to erosion and consequently the destruction of traditional knowledge associated with a continually transformed or impacted environment.[6] In these processes of exploitation and incorporation of nature, local populations feel the consequences of their altered environment; they helplessly behold the scarcity or destruction of their natural resources. As the final blow, these peoples experience the transformation of their way of life and, consequently, the alteration of future expectations by the destruction of a logic of existence. Not only had these groups' specific ways of life guaranteed them socio-economic conditions of reproduction, it had especially guaranteed the construction of virtuous[7] processes of existence in the nature/culture relationship.

Pará is the second largest state in the Brazilian Amazon and covers a large share of the major natural resources exploitation projects in the region (de Castro and Hébette 1989). The traditional populations[8] living there suffer from the conse-

---

[5]For a deeper discussion on sustainable development and the approach to ecological economics, see Cavalcanti (2010, 2012), Daly (2002), Daly and Farley (2004).

[6]For more information on this theme, in Amazonian settings, consult Almeida and Ravena-Cañete (2015), Castro and Castro (2015), Castro et al. (2014), among others.

[7]We opted for the word "virtuous" because virtue is the quality of that which is considered desirable and commendable. However, we opted for this word specifically because of the relation of the term "virtuoso" with the skill of using a technique correctly.

[8]In Brazil, the definition of traditional population has penetrated the sphere of public management, summarized by the contributions of social movements and the academic context. The definition of the concept of traditional population occupies a significant position in the discussions of Social Sciences. The definition of traditional population that this text takes as a reference is taken from Ravena-Cañete and Ravena Cañete (2011). For these authors, traditional populations are those that "have a specific way of life marked by the strong symbiosis and relative harmony with the environment in which they live. They develop techniques of low environmental impact. They have feeble interactions with the market. They have a great knowledge of the biodiversity that surrounds them and a mode of production based on family labor. It should be noted that this term is under construction, as it has been created by national society to classify other societies. These populations do not always consider themselves traditional, but they are called so by social actors. Thus, "the so-called 'traditional' populations do not have to possess all these characteristics, nor do they have to always self-identify themselves as such. However, they see themselves as having a way of life different from the society which surrounds them in order to access rights inherentto this category" (Ravena-Cañete and Ravena Cañete 2011, online). The discussion on the concept, its trajectory and its uses within the Brazilian framework is broad; to narrow the debate, see Ravena-Cañete (2012).

quences of this perverse option of development. Within different environments,[9] but facing similar processes of knowledge erosion and destruction of local *ethos*, these populations still resist elimination by clinging to the social processes that shaped and guaranteed their ways of life for centuries. In this process, attention is paid to kinship relations as a social organization capable of maintaining the group, even in face of strong impacts, and especially on the necessity of adapting to the transformations experienced. Ultimately, it is the kinship relations that allow the reproduction of forms of incorporation imposed by the dominant society, which imposes foreign practices and rationales at the local level of ways of life by demanding the recreation of meanings, appropriations and adjustments for its functioning.

Kinship stands in fact as an exclusive study object of anthropology. It contributes to the constitution of this discipline, especially with regards to its original concern for the nature/culture relationship that has forged anthropology as a science. However, Schneider's publications (1972)[10] brought a new approach to this subject. As a consequence, kinship studies started to appear timidly in anthropological studies, especially when the subject studied finds itself inserted in a modern context as a result of the movements and dynamics of contemporary capitalist society.[11] Even if works on socially excluded populations still arouse the interest of anthropology and comparative sociology, problematizing and understanding the subjects are rarely approached by kinship as a theme; they are not even considered central to research interests. When one encounters kinship as a theme, it is challenging because of its classical approaches, such as the theories of lineage and alliances.

However, in Amazonian settings, theories of lineage and alliances appear to be important explanatory tools in the contexts of contact between traditional populations, present everywhere in the Amazon region and in the dominant society.

It is important to emphasize that the study of kinship is fundamental to understand how a society is structured in its different aspects (political, economic, among

[9]The State of Pará has a diversity of ecosystems. It would not be possible for this text to describe the environmental and therefore social diversity that characterizes its territory. For an introduction, see Morán (1990).

[10]Kinship studies explored new directions from the Schneider publications (1972). These inaugurated a dissolution of kinship ties from consanguinity, presenting subjectivity as the main component in the construction of these ties. In this way, from Schneider's viewpoint, kinship did not bind to biological questions, being limited exclusively by subjective choices. Such a resizing eventually transformed the classical approaches into an accessory often seen as a deformed mode, barely useful. However, such approaches, especially the theories of lineage presented by Radcliffe-Brown (1995) and those of alliances investigated by Levi-Strauss (1976), are still important instruments for analyzing and understanding traditional populations, especially in Amazonian cases. To broaden the discussion of kinship between the traditional Amazonian populations, see Ravena-Cañete (2005), Araújo and Schiavoni (2002).

[11]This perspective appears especially in Schneider's publications (1972) when kinship studies were gaining important. Marked by an interpretative approach, they began to see kinship relations as a social choice totally disconnected from biological constraints. Thus, the importance of these studies changed, as they became a new alternative to interpretive studies for modern society, where relations of kinship result exclusively from social choices.

others). In modern societies, kinship is generally less important than other aspects of social organization. However, this is not the case for traditional populations, to whom kinship represents an important aspect of social relations, and even the most important in some cases.

In Brazil, kinship studies emanated from research on indigenous populations undertaken by Nimuendaju and Levi-Strauss, in the first half of the twentieth century. However, such studies became popular only from the 1970s, particularly with the use of theories generalizing lineage and alliance, and their extensions (da Silva 2004: 650). This process has produced significant studies on indigenous Amazonian populations. On the other hand, in Brazilian urban settings, kinship approaches gained importance rather through individuation studies (McCallum and Bustamante 2012). Thus, kinship studies of traditional populations in an Amazonian framework *still await a future*.

This study presents three cases of traditional populations (riverine – watercourses -, farming, and fishing) that are subjected to the diktats of various public policies, provoking the same practices of erosion on local schemes. Our works utilize kinship as a strategy to preserve and adapt groups to the pressures of acculturation.[12] By choosing kinship studies to interpret the strategies used, the article presents three illustrative cases. The first describes a group of family farmers, typical of northeastern Pará. Access to the land in this region is subject to changes arising in response to pressures restricting the use of this natural resource and by redesigning relationships of work in agricultural activity. The second case presents a community of fishermen in the municipality of Curuçá, the salty basin (salgado) of Pará. The established public policy for fishing encourages the formation of cooperatives[13] regulated by the relevant legislation to this type of organization but which is not in harmony with local rationale. Lastly, the article investigates a community of rivers marked by the changes undergone by the artisanal fisheries with the arrival of industrial fishing in the locality. To conclude, the study presents kinship as an important input in the nature/culture relationship. Once understood, this theme can set a framework for investigating the coviability of various forms of mutual life between distinct societies within a broader global framework.

From a methodological viewpoint, the first case corresponds to the studies of an updated version of my doctoral thesis. This research required a 10-month stay in the field (from Feb to Oct. 2005). The stay was marked by an ethnographic practice oriented by processes of alterity and relativization of the perception on the subjects of the research. The other two cases emanate from the memoirs of Master theses overseen by myself, which show kinship as an adaptation strategy

[12]The first case corresponds to the studies of my updated doctoral dissertation (Ravena-Cañete 2005) while the other two cases emanate from Master theses (Diaz 2013 and Silva 2015) overseen by myself.They highlight kinship as an adaptation strategy against changes imposed by incorporation dynamics of areas and/or activities of the socio-cultural reproduction of the groups studied.

[13]In Brazil, cooperatives are organizations of small local producers, favored by national and regional public policies.

against changes, changes imposed by the dynamics of incorporating groups and/or socio-cultural reproduction activities of the studied groups. The case of the Lauro Sodré community in Curuçá is the result of 3 months of fieldwork (from Jul to Sep-2012), with questionnaires and community-based interviews,[14] as well as direct observation (Diaz 2013). In the case of Cajueiro, field research lasted 2 months, marked by the application of questionnaires and semi-structured interviews (Silva 2015).

19.2 Marriage and Succession: Social Organization and Kinship as a Strategy for Accessing the Land – The Case of Nova Redenção

Nova Redenção is a typical town of northeastern Pará, one marked by a rural lifestyle. It comprises 183 inhabitants, and it is located in the rural area of the municipality of Concórdia do Pará, about 200 km away from Belém, the state capital, in the northern region of Brazil. The economy of the municipality is mainly powered by activities linked to extensive livestock farming and agricultural production, dominated by subsistence crops. Pepper cultivation was successfully introduced by Japanese immigrants in the region,[15] which placed Concórdia do Pará in the 2000s among the largest producers of pepper in the state (Instituto Brasileiro de Geografia e Estatística[16] in 2011). This economy gradually yielded space to palm oil, nationally known as the *dendê* (Almeida 2012). Some of the small farmers in the region are still linked, however, to the production of cassava flour,[17] which is sold in the nearest towns or taken to Belém.

As a strategy to preserve Nova Redenção's agricultural activities and to guarantee access and use of the land, marriage and kinship, as well as the remembrance of these relationships, contribute to form the town's history. The land where Nova Redenção's inhabitants live and develop their planting activities corresponds to a division going back to a common ancestor, *Carlos Guimarães*, owner of these lands since 1903.[18] The heirs' rights follow an egalitarian division, as inheritance rights which extend to the men and women of next generations. Thus, kinship among the

[14]These are interviews without a preliminary question list, contrarily to semi-structured interviews that use a list of questions.

[15]In Pará, Japanese immigration is concentrated especially in Tomé-Açu town, 70 km away from Concórdia do Pará. This town has a large influence in that micro-region and energizes part of the local economy. Homma (2009) situates the arrival of the first Japanese in Amazonia.

[16]Brazilian Institute of Geography and Statistics.

[17]Flour is the main food for the traditional populations of the Amazon region. The process of its production marks the activities and sociability of these populations. Their primary resource is cassava which is an abundant tuber in the Amazon region.

[18]The land title was consulted during the fieldwork and, indeed, Carlos Guimarães held the title of the land he transmitted to his children as inheritance.

Guimarães obeys a cognate[19] principle which also governs land inheritance. In the first generation, the succession process divided the land equally. In the formation of the new domestic groups, obeying a principle of exogamy, marriages integrated three specific families that also live in the region: *Gomes da Conceição*, *Matos*· and *Batista*. The use of these lands was linked to various motives, such as *kinship* proximity to achieve the necessary exchanges for activities in agriculture or to have access to a forest covered land that guarantees an abundant harvest. In the second generation of heirs, marriages continued among the same families, maintaining the tendency to marry outside the main family.

The explicit right to use the land was directly linked to the constitution of a family by heir, accompanied by the obligation to be a resident in the locality. One-third of the second generation members (1950–1970) left the region. The alternative chosen by these individuals gave those who remained the possibility to arrange the land for the new domestic groups that *arrived*. Thus, in the second generation, the land made available was extended by *adding* domains of the *Gomes da Conceição*, *Batista* and *Matos* families. However, among the heirs of the third generation (1970/1990), the *Batistas and Matos* lands disappeared; only the *Gomes da Conceição* lands remained. Some of the heirs were able to rid themselves of the lands inherited from the *Guimarães* lineage because they possessed the inheritance of the *Gomes da Conceição* lineage, which they were able to use. The cognate principle, as we have seen, was also valid for land inheritance, but in this case the choices between the *Guimarães* and the *Gomes da Conceição* prioritized, through matrimonial unions, conserving the land among the *Guimarães*. The succession of the *Gomes da Conceição* remained an option for marriages outside the group.

Until the end of the 1960s, the majority of the families of familial farmers in the northern region of the country were able to possess public lands without having to legalize land ownership. However, since the 1970s, access to the land has begun to undergo greater restrictions due to the policies of the Federal Government for the occupation of the Amazon (Lima and Pozzobon 2005; Araújo and Schiavoni 2002; Ravena-Cañete 2005). In this way, the divisions of the inherited land on which the *Guimarães* settled began to intensify at a time when the conditions of land shortage, previously nonexistent, began to present its first signs in the region.

The third and fourth generations currently occupy the land of the *Guimarães,* and the land initially divided into plots of land still knows a similar division.

As already mentioned, to be *Guimarães* means to be the descendant of *Carlos Guimarães*. The established model may be defined as a group of cognate filiation (Radcliffe-Brown 1995) because all descend from a common ancestor. This lineage passes both through male and female sides. The right to inherit land also obeys the cognate principle. However, certain rules make it possible to maintain the land as an undividable property, kept in the hands of the group which, in obedience to the established cognate principle, divides the succession based on the right to it.

[19]Principle of a non-unilinear mode of filiation, which passes indifferently through men and women.

In order to gain property rights, without being a Guimarães, it is necessary to live in the locality because being an owner presupposes being able to play a social role until then played by the previous heir. Thus, a son takes over property rights after the death of a father and a mother. When there is no son, it is the woman who gains the rights and duties before the group and the family. This right may also be invoked when fathers can no longer carry out their economic activities satisfactorily. When it is not the case, children may use the land but not as owners. The inheritance can therefore be claimed by all children who have formed a family and who have remained in the locality.

As Fox points out (1986), inheritance expresses a principle of restricted cognate filiation,[20] which becomes effective with the death of parents. In this way, a Guimarães heir may leave the locality, but still be entitled to inherit land if he/she finds his/her parents still alive. However, this happens only when the social role which enables property rights, does not leave free space to be claimed by the heirs who already live there. Therefore, a son may leave and spend many years elsewhere, even start a family and, on his return, he would still be able to use the land if his parents are alive; and if he is able to guarantee his share of land usufruct when they die. The same rights are not guaranteed, however, if the parents are no longer alive when the group member returns. In this case, the death of the parents is necessary to activate the right to inheritance, as well as the right of land usufruct.

In view of the *Guimarães* and the *Gomes da Conceição* families, kinship seems to be the most categorical form of land access strategy within the local organization. If the cognate principle governs succession relations, the fact that men and women are also heirs makes marriages between members of these families possible, so that land can be maintained among relatives. This can be perceived in the third generation. Marrying another *Guimarães* makes available other inherited land for another brother. In other words, when a marriage between *Guimarães* takes place, the land used is the one added by the inheritance of this family. This leaves, for the other heirs, the inheritance land of the *Gomes da Conceição* family, for example. By dint of marriage, the *Guimarães* unite properties, leaving inheritance land for those who marry afterwards. The marriage between cousins occurs on a constant basis, and the partner's choice is directly related to the availability of land that the partner offers. Diagram 1 describes hist (Fig. 19.1).

The Fig. 19.1 shows three generations of heirs who characterize the manner by which the choices were made and how the arrangements were established. They reflect the typical arrangements between heirs.

In the first generation, members marry out of the group, but stay in order to use the *Guimarães* land. The *Gomes da Conceição* family appears as a choice for second-generation marriages, which also increased the availability of land for the

[20]It is worth recalling Fox's evocation of the concept of cognate filiation: "Those who opted for living in another place lose the right to belong to the lineage. If a classification of this type is enacted, the group would designate a group with 'restricted' cognate filiation. The cognate principle continues to be applied – all descendants of the common ancestor are entitled to the land of the group – but unless they exercise this right, they will lose the land" (Fox 1986: 183).

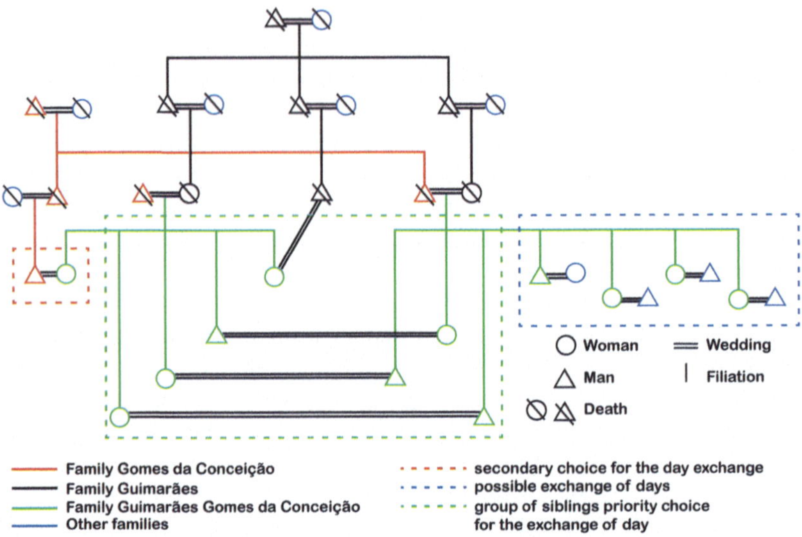

Fig. 19.1 Matrimonial arrangements and conservation of unity of land. (Source: Ravena-Cañete 2005)

heirs. In this way, the members are able to use and possess land belonging both to the *Guimarães* family and the *Gomes da Conceição* family, by extending thus available land. However, in the third generation, land availability returns to its original size. This happened because as the number of the group members grew, they married outsiders who did not possess land. Consequently, the division of the land returned to its original size, the one made available by the *Guimarães* family.

Of the ten families that currently constitute Nova Redenção, five settled in the locality by purchasing land. Marriages can be interpreted as a strategy to prevent that the division of land breaking the basic conditions for the reproduction of a group of family farmers (Moura 1978; Segalen 1986; Ravena-Cañete 2005). Among the *Guimarães,* it is clear that the land maintained its initial format because the rationale for the division gave priority to the integration of land in the format produced by the inheritance. In this way, when a *Guimarães Gomes da Conceição* woman marries an outsider, she uses the land inherited by the *Gomes da Conceição* descent. The *Guimarães* succession is then left to the sister who marries with a *Guimarães Gomes da Conceição*. Thus, the integrity of the inherited land is maintained as shown by the Fig. 19.2.

The land on the west side of Nova Redenção belonged to the *Gomes da Conceição*, *Batista* and *Matos* families. There, the only land that kept their original form are the additions of the *Gomes da Conceição* family, which makes it *Guimarães*'s preferred choice for exchanges, and therefore, for preserving the totally of the land of these two families. The land of the *Batista* and *Matos* families were sold, which added in the locality farmers coming from other areas of the State of Pará, especially the Bragantine area. This land was sold in time to a third party.

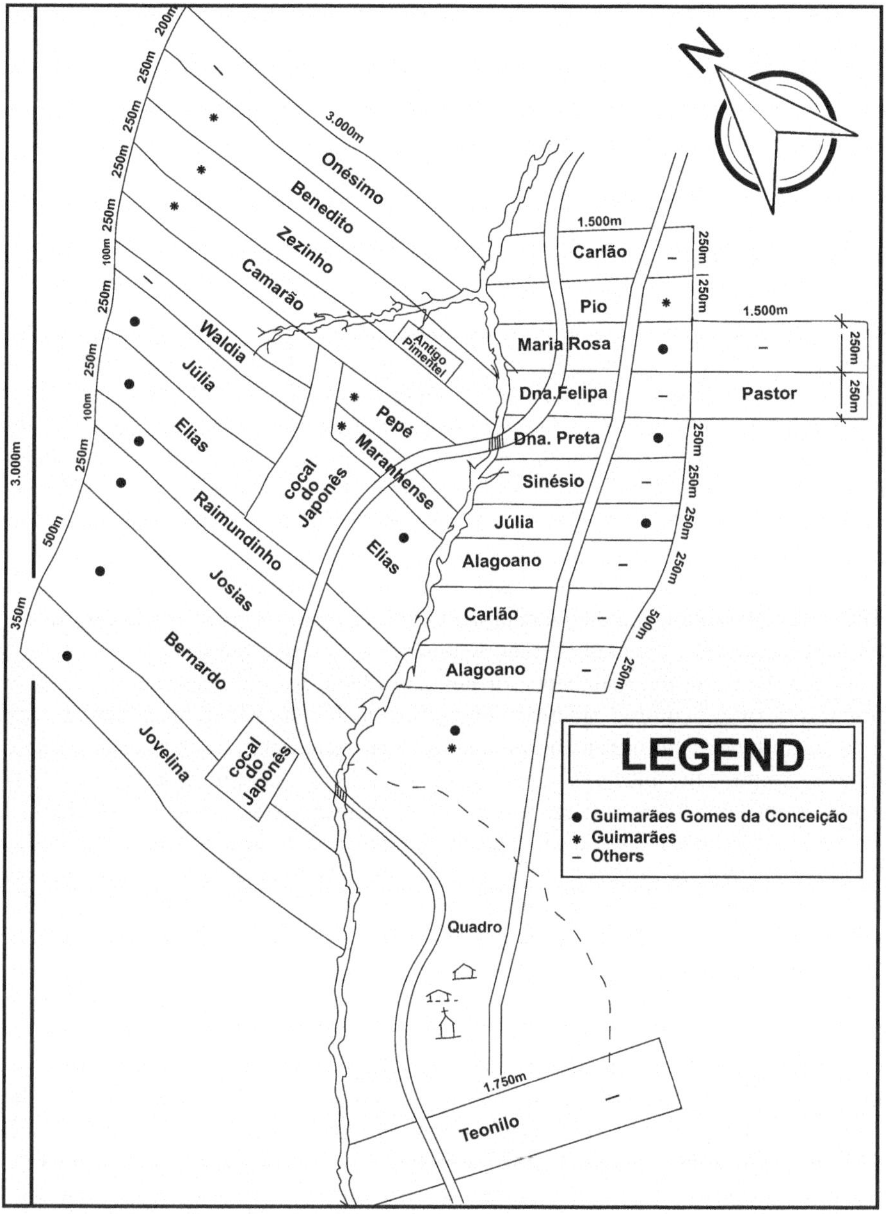

Fig. 19.2 Format and division of Nova Redenção's lands according to the successions. (Source Ravena-Cañete 2005)

More than 80 years have passed since the first inhabitants of Nova Redenção settled in the locality and their land remains similar to the initial form. Parenthood and marriage were the strategies that guaranteed such perseverance. They are also partly responsible for the labor relations that marked the history of the inhabitants and the agricultural practice of the locality.

19.2.1 Forms of Sociability in Agricultural Labor

Cutting and burning vegetation is a traditional technique that requires many participants when it is used for cultivation. A production and consumption unit of one family nucleus would not be sufficient to carry out all the tasks of the agricultural process based on this technique (cutting and burning). The solidarity of strong arms is necessary for this purpose and reciprocity is operated as a strategy that organizes work in the field.[21] Kinship-oriented relationships based on this type of exchange cover the history of Nova Redenção's agricultural daily life. However, just as the practices of this activity and the forms of land access are transformed into a frontier landscape,[22] the social relations that guarantee them also change. The commodification and the monetarization of the labor force, formerly exchanged through reciprocity, gradually extend within the agricultural process.

In the Amazon, the abundance of lands available for cultivation has created a framework of agricultural practices among its inhabitants which has given priority to the technique of cutting and burning in regional agriculture. Even if this technique removes some of the nutrients by fire, it still allows sufficient productivity if a fallow period is respected for soil recovery. The high availability of land that marked the regional agrarian context allowed this practice to continue. Thus, within 12–20 years, the vegetation *regenerated itself because the intense rainfall precipitation of the Amazonian equatorial climate contributed to* the recovery of soil and the re-composition of the forest. This was the general framework of the Amazon until 1966 when plans to integrate the region to the rest of the country were enacted. In the case of Nova Redenção, opening the PA-140 road during the 1970s helped in accelerating this scenario and stimulating a more vigorous contact between the region and the state capital. Despite the intensification of contact and the change in the availability of land, cutting and burning techniques have been maintained in the region to this day as an important agricultural practice.

In this Amazonian agrarian world, and specifically in Pará, ideas of abundance run through the common voice about the agricultural production of the past. The possibility of such production resulted in a high availability of forest-covered lands

[21]The field is the local definition for the space where agricultural work takes place; the term also designates the work occurring in that place.

[22]The term "frontier" refers to the classical definition in the Brazilian context, that is, a region with a weak presence of state action, where social agents impose themselves and exert dominance through economic power and violence (de Castro and Hébette 1989).

for the extension of *fields* in the region. With the burning ashes on the surface and using the nutrients left by the organic matter deposited on the soil by the vegetation covering it, the results of the plantations on these terrains were sufficient for the consumption of a domestic group. The remaining part of the production was to be sold in the markets of Belém. Transportation was carried out by river, starting the local creeks[23] up to the main rivers and from there to the capital. Families with a less developed transportation structure sold their production to a warehouse.

Until the 1960s, the fields' minimum extension amount varied between sixteen and twenty *tarefas,*[24] able to go up to forty. It equals a land area of approximately eight to ten hectares. Using the method of burning in an area of this dimension, an area covered with dense vegetation, required a large and intense work force. The *mutirão*, a local work organization based on reciprocity strategies, established itself as an important form for guaranteeing the quantity of laborers needed in an agricultural work. Present in the whole of the Amazon, the *mutirão* was characterized by a form of work regulated by mutual aid, organized by local rules and linked to the catholic liturgy of the locality. This was a strategy used to gather and organize the workforce. Without the facilities of modern agricultural equipment, many men were needed to work in the *field*. Similarly to other parts in the Amazon, *mutirão* was the common form of reciprocity in Nova Redenção.

Groups of 15–20 participants were structured, and by means of rotation the land of each and every one of them was clean within a few days. Domestic groups that worked together were related in some way by kinship as the initial occupation of the area was, as we have seen, made by the offspring of a common ancestor. Kinship regulated relations of reciprocity, although another form of organization was also present: a *society* or *brotherhood*. It presented itself as an organization based on Catholic traditions whose inhabitants were devotees at the time. Each locality had a patron to whom the religious vows were to be made by means of religious festivals. These were characterized as the culmination of a process initiated by agricultural work. Being part of a *society* meant counting on the help of others in the working the *fields*. The size of the land did not account as working hours, but as the certainty of help to be received.

However, after the arrival of the Protestant church in the area, around 1950, the religiously oriented festivals, as a form of devotion to saints, were slowly eliminated. The *society* or *brotherhood*, although organized on the basis of kinship relations, was founded on religious traditions. Once religious festivals disappeared, the *society* also disappeared leaving the *mutirão* as a form of agricultural work organization. The *mutirão*, as an organizing form of agricultural work, appears in the zones of cultivation which are regulated by criteria of kinship, neighborhood, and proximity. However, kinship relationships persevered as the most important regulatory principle, dominating agricultural labor until the 1990s.

[23]Local name given to small rivers, *igarapé* in Portuguese.

[24]Literally means task. By extension, it is the local name for a land with specific dimensions of approximately one-half hectare of land, corresponding to the area that a person can take care of individually during the day.

Currently, the *mutirão* as a form of reciprocity is no longer found in Nova Redenção's daily life. The *exchange of working days* has replaced it because men plan work in the *field* based on the possibilities of *exchanging days* between the nearest relatives. This exchange manifests itself in much more rigid and explicit accounting. A day's labor consisting of cleaning the land corresponds to the exchange of the same duration made available to clean the land of the one who worked. However, the practice that is spreading now corresponds to *empeleita.*[25] In this case, the purchase of labor force is explicit. A service is defined as well as the price for its realization. This form of work organization in the *field* enables different situations to occur, but it is important to note that such monetarization of work is only possible due to revenues circulation in Nova Redenção.

When accounting for these resources, originating from transfers realized through public policy, one can better understand the described framework.

Out of the twenty-four domestic groups in Nova Redenção's who reported having a fixed income, eight have incomes above two minimum wages. The same result is observed for those with incomes between one and two wages. Only eight remain with incomes below a salary. There is therefore a public transfer of income to Nova Redenção; it contributes in commercializing labor relations previously expressed only by reciprocity. In addition, there is the possibility of income, usually non-fixed, originating from members of the group who work as wage-earners in the capital or in other neighboring cities, and the old practice of alternating paid labor in the city with less intense work periods in the *fields*. Finding men who occupy variable jobs, 3–4 months, and who pay to have their land cleaned is a recurring situation.

Reciprocity is undoubtedly a well accounted form of exchange, as one does not work more than one has received from the work of the other, but the forms of the work in Nova Redenção are taking on a progressive commercial character.

The *mutirão* is replaced with practices such as the payment by the *day*, the *empeleita* and the *exchange of days*. The first two forms of an employment relationship are made effective on the basis of a cash payment. The *exchange of days* is a pale expression of the *mutirão*. This payment may be given for a specific activity or for taking care of the whole *field*. This situation is normally encountered when a domestic group is found without the main male Fig., such as the father. The latter may have obtained an income through occasional work and uses part of it for the payment of the *field* made by *contract*.[26]

The *empeleita* is considered a primary form of contractualizing the work. The payment of day labor, although it only appears in two domestic groups, may be in

[25] *Empeleita* comes from 'to undertake' (empreitar), i.e., to work according to a task. *Empreitada* or *empreita* consists of a contract in which somebody performs a certain task in exchange for a remuneration agreed on in advance.

[26] The manner by which the term "contract" is used between the inhabitants is specific; it should not be understood in the legal sense. It should rather be understood as an agreement between the one who pays for a service and the one who does it.

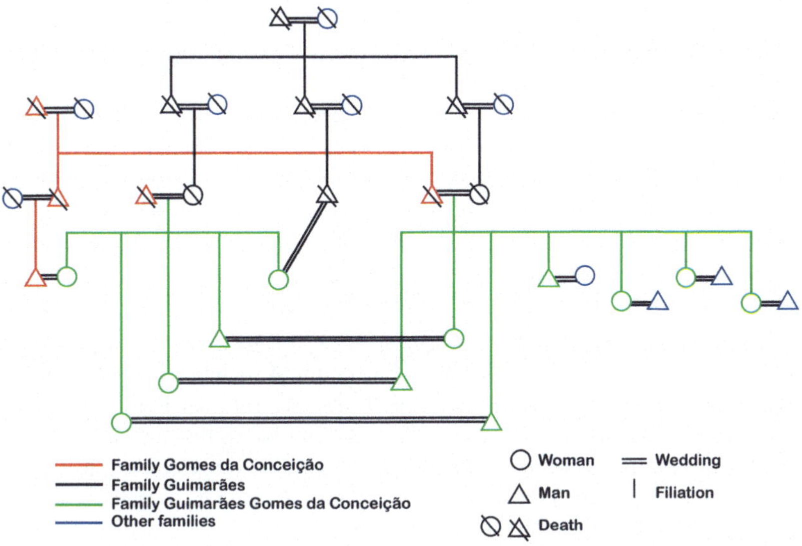

Fig. 19.3 Organization of the work in the fields according to the group of siblings. (Source: Ravena-Cañete 2005)

larger numbers when considering the activities of flour production.[27] In these, the presence of women, as day laborers, is common for grating the cassava. As for men, especially the younger ones, they may be day laborers to grill the flour.

Even if the *mutirão* has disappeared as a main form of work organization between domestic groups, *exchange of days* remains to this day the form of reciprocity that still orders these relations. The *exchange of days* is the most common form of reciprocity in Nova Redenção's agricultural sphere. There is a logic that organizes the choices of exchange partners. *A priori*, it is possible to establish an exchange with any domestic group, but the principle of kinship gives priority to certain groups. Thus, during the first preparations for planting, domestic groups favor exchanges with the nearest relatives. Considering that the inhabitants of Nova Redenção are mostly descendants of a common ancestor, the relationship between *siblings*[28] directs the choices for the exchange of days in the *field* labor. The exchange of work days is firstly chosen among brothers and sisters of the two parties. This group may be described as *siblings*, as shown in Fig. 19.3.

Radcliffe-Brown defines the group of *siblings* as "the body of brothers and sisters of common paternity and/or maternity, which has its own internal structure"

[27]Product derived from the cassava which makes up the population regime of the whole Brazilian Amazon.

[28]*Siblings* is the definition of the group of brothers based on a cognitive relationship given by Radcliffe-Brown (1995).

(Radcliffe-Brown 1995, p. 87).[29] Thus, agricultural work in Nova Redenção is based on the relationship between *siblings,*[30] since these are the groups that preferentially establish exchanges in agricultural activities.

Nova Redenção can be understood as an area of family farmers inserted in an environment which is subject to various changes. There are changes that result from seasonal economic activities referring to its components, and others experienced in the surrounding society and from the relations established with it. The group maintains its original family farmer characteristics as long as the changes that have occurred do not affect the structure of its production rationale. This rationale is directly related to the availability of natural resources for the group, especially the land. When the land is threatened, the group looks for mechanisms to maintain the rationale in which it is reproduced. The construction of such mechanisms is partially guaranteed by the possibility of contact with another form of logic present in the relationship with the whole of society. On the other hand, the surrounding society exerts pressure on the group by reducing the conditions for accessing the land, but at the same time it provides the group with cognitive elements to work with.

19.3 The Cases of Curuçá and Cajueiro

This section regards two cases of Amazonian fishing populations. The first is traditional, and the second is strongly inserted into the frameworks of the main society. Both use kinship as a strategy of permanence in the system, and especially for accessing and using natural resources.

Fishing in the Amazon has gradually undergone profound changes over the last two decades. As for coastal waters, the fishing potential at the coastline of Pará is scarcely known (Isaac-Nahum 2006). During the last decade, a complex picture of fishery-oriented social actors has been gradually emerging in the Amazonian region (Isaac-Nahum 2006). The coastline of the State of Pará is a typical example of aquatic of the overexploitation of resources, a process which affects the fishing sector on regional and national scales (Isaac-Nahum 2006). On Pará's coast, gillnet

[29]The definition of *sib* precedes the explanation of that which would become an organization of *siblings*. For Radcliffe-Brown, "the arrangement of parents by degrees of proximity or distance was based on the *sib* (in English *sib*, German *Sippe*). The sib of a man were all the individuals of the same origin to a certain degree" (Radcliffe-Brown 1995, p.76, free translation).

[30]Defining kinship from a Radcliffe-Brown perspective places this institution on a structural basis, where function and system represent basic concepts in the explanation of a given social fact. The theoretical and methodological orientation should be considered in a way to provide a better understanding of the approach to kinship: "A study of kinship systems, carried out worldwide with this method, reveals that while there is variation of its superficial characteristics, a few general structural principles may be applicable and may combine in various ways. Some of the first tasks of a theoretical study of kinship are to discover these principles through a process of summarized generalization based on analysis and comparison" (Radcliffe-Brown 1995: 60–61, free translation).

fishing predominates as an artisanal way to catch fish. The yellow *machoiran* (*Arius parkeri, Trail 1832*), the red acoupa (*Cynoscion acoupa, Lacepède 1801*), and the kingfish (*Scomberomorus brasiliensis, Russo e Zavala-Camin 1978*) are normally the species most captured in this way. Studies in biology and ecology, and even on the fishing of these species, have been gradually increasing in recent years (Barthem and Fabré 2004). However, much more is yet to be detailed about local actors' views on the depletion of fish in the region.

In representing the result of small-scale fishing, the dynamics of the actors that integrate this activity is not yet well understood and the seasonal variability, added to the spatial specificities, make this framework even less clear. The market and environmental peculiarities may contribute to the way in which this type of capture is described in various localities on the coast of Pará. The mangrove ecosystems, typical to the area, are characterized by the variety in fauna and flora, giving this framework an even larger socio-environmental complexity.[31] Diegues (2008) describes the importance of this ecosystem for maintaining species in the coastal zone, and its importance in the economy of the populations. In Pará, the mangroves represent approximately 2197 Km^2 of the totality (Filho and Martins 2005) as could be observed in Fig. 19.4.

19.3.1 The Community of Lauro Sodré à Curuçá

The mangrove located in the coastal zone of Pará is characterized by low plant diversity which contrasts with an intense variety of fauna (Mendes 2003; Isaac et al. 2003). Humans populations have been present in these areas for over 10,000 years (Figuti 1994) and the mangrove ecosystem is a fundamental environment for the survival of the populations living directly from plant collection and fishing (Maneschy 2003; Isaac et al. 2003; Almeida 2012). The knowledge of such an environment results from a long-established adaptation and from a social organization marked by its special relationship with natural resources. Among the eleven municipalities that form the Salgado area in northeastern Pará,[32] Curuçá stands out as one of the first to establish a Sustainable Marine Reserve (now called RESEX[33] marine), known as Resex Marine Mãe Grande de Curuçá. A RESEX is a type of environmental conservation unit (CU) created with the perspective of integrating human presence, their plans comprising the possibility of guaranteeing access and the use of natural resources to the traditional populations that already inhabit the area in a process

[31]The mangrove ecosystem zone on the Brazilian coastline is the second largest in the world. The country has approximately 13,400 km^2 of mangroves, second to Indonesia which has an area of 42,550 km^2 (Prost and Mendes 2013).

[32]Information available at: <http://www.cidade-brasil.com.br/microrregiao-de-salgado.html>.

[33]Resex refers to the Brazilian terminology of extractivist reserve. Extractivism is the extraction of natural resources. Although it can be applied to mining, this term refers rather to the plant collection activity or animal in the case of fishing.

Fig. 19.4 Mangrove ecosystem on the Brazilian coast. (Source: ICMBio 2013)

of generational inheritance. RESEX Marinha Mãe Grande de Curuçá computed 56 communities at the time of its creation in 2005, a number which increased to more than 90 communities today (ICMBio 2013). It results from the processes set out by national fishing policies and land regulation that integrate the processes following the creation of RESEX, initiating new territorial contexts often woven by conflict. It is important to emphasize that such conflicts result from the new model imposed by RESEX, which, in order to protect the area, redefines the forms of natural resources usage by imposing changes on the population who for centuries have occupied the region. Allocating a title to the land stands out as the main change introduced, as it conditions access to public policies, especially those relating to housing and financing for fishing and agriculture.

Although they value the protection of these populations, the RESEX impose a new way of life, one based on assumptions, rational and reasoning oriented towards the new values that are beginning to be introduced, while promoting the erosion of local forms of life.

Diaz (2013) demonstrates that in the RESEX Marine Mão Grande de Curuçá there is an activity brought about through the creation of this Conservation Unit: forming a cooperative for oyster production (*Crassostrea rhizophorae*) in the community of Lauro Sodré. It appears as a typical example in which forms of sociability and previous social organization are pressurised by new models which seek to impose themselves, even when they are not successful.

The RESEX region, particularly the community of Lauro Sodré, are known in the surrounding areas as suitable for o the production of seeds used in the creation of bivalve molluscs,[34] with the aim of creating and implementing a cooperative that produces oysters, and more particularly larvae. The cooperative, called Aquavilla, started its activity with 42 associates in 2006. The Fig. 19.5 shows the location of the community.

The agency that carried out the project, SEBRAE (Serviço Brasileiro de Apoio às pequenas e Micro-Empresas[35]), set up training courses for its cooperative workers, introducing seed production and cultivation techniques in addition to management training. However, 6 years after the agency was set up, barely 11 cooperator workers have continued the activity. The Fig. 19.6 shows the cultivation system used by the Aquavilla.

It is important to note that, from a legal point of view, cooperatives prohibit kinship between the associates of the same cooperative. If this legal measure were toin fact to be implemented, it would invalidate any cooperative in the Amazon, where traditional populations are the main objective of the intervention, because

[34]The region of Curuçá and particularly the community of Lauro Sodré present an ecosystem sufficiently favorable to the production of oyster seeds because they have an adequate salinity for this process. For more details on seed production in the community, cf. Diaz (2013).

[35]Brazilian Assistance Service for Small and Micro-Enterprises (free translation).

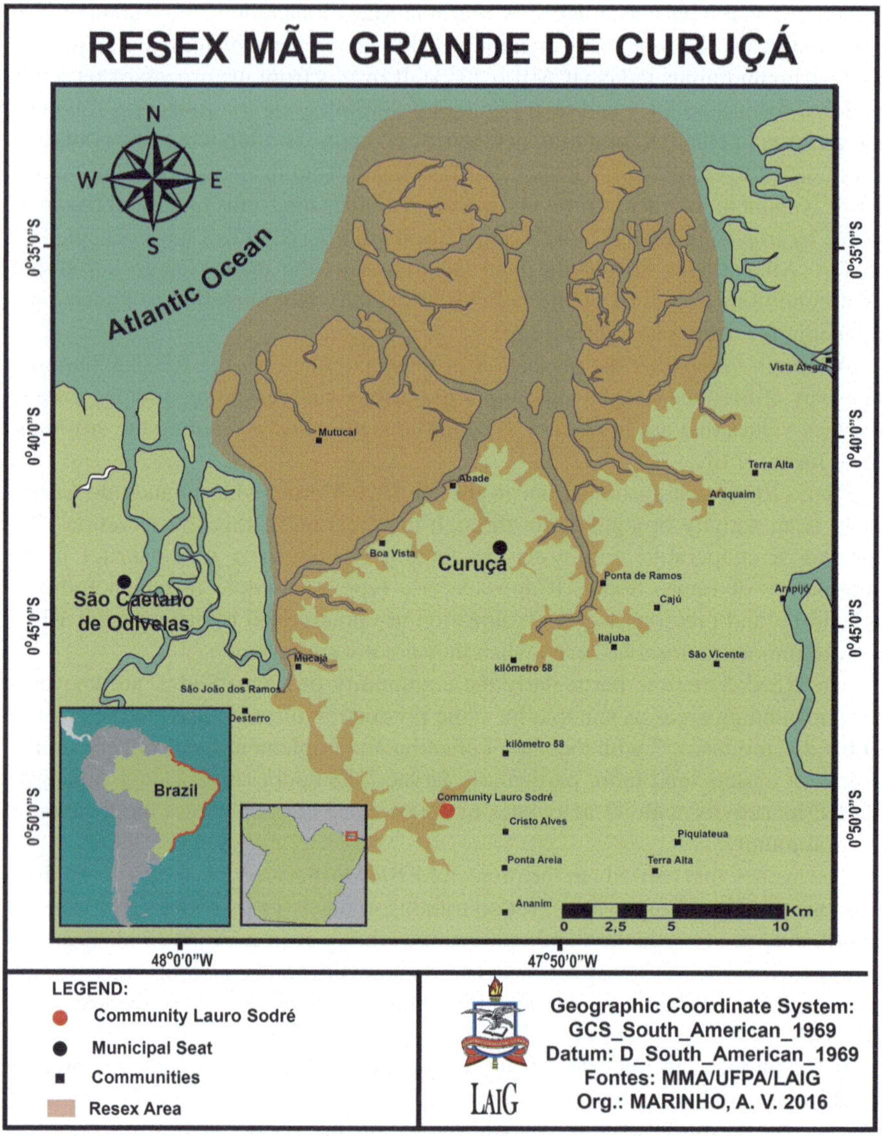

Fig. 19.5 The community of Lauro Sodré. (Source: ICMBio 2013)

Fig. 19.6 Oyster cultivation system. (Source: Diaz 2013)

they lead and guide their social organization mainly through kinship.[36] It should be noted, particularly in the case of Aquavilla, that after 6 years of the cooperative operation, it continued its activity through the social relations marked by the kinship that regulated the life of its members. Two families continued to develop the activity, Galvão and Pinheiro. This identifies the importance of kinship relationships in conducting the cooperative and the ways in which the activities had been developed (Diaz 2013).

The activities of the Galvão family were guided by a type of interaction marked by a hierarchical relationship and oriented by a system of descent that extends to power relations when interacting with social groups outside the community. As confirmed by Diaz (2013), the family's perception and action was oriented by an environmental rationale known as the "ecology of the poor" (Martínez-Alier 2014). Indeed, they possess and use their ancestral knowledge of the mangrove ecosystem by seeking positive results for the company proposed by SEBRAE, and by ignoring or modifying the procedures presented by the technicians. The Pinheiro family has a less rigid organization, with a less marked hierarchy. It executes and reproduces exactly the procedures indicated by the technicians of the SEBRAE, with little consideration to the local knowledge of cultivation. It turns thus towards a so-called eco-efficiency (Martínez-Alier 2014). It is remarkable that the Galvão family, precisely the one that uses local knowledge and is organized primarily by kinship

[36]Forms such as Brazilian law and legislation are found to be restrictive and, in some cases, inappropriate for frameworks marked by kinship relations and which deserve specific legislation. They can be found in Ravena-Cañete and Ravena Cañete (2011).

relations that are less altered by the relationship with the dominant society, ends up being the best model for the SEBRAE proposition, particularly in seed production. The Pinheiro family, in turn, definitely stands as a family that reproduces practices imposed by the technique, but practices which barely maintain the functioning of the cooperative or even the seed production activity. The relations of kinship in themselves, thus, allow the reproduction of local practices and knowledge that contribute decisively to the functioning of the cooperative.

19.3.2 The Community of Cajueiro

The case of the Cajueiro Community distinguishes itself by its artisanal fishing. Located in a continental area influenced by tides, this community was formed from its harbor, which is responsible for the second largest fish landing in Belém, the capital of the State of Pará. Although it is part of the territory of the State's capital, the community is located in a rural area. It offers its inhabitants thus both a lifestyle marked by rural life and a traditional feature expressed in fishing.

Silva (2015) shows in her study that, even though the inhabitants of Cajueiro narrate the difficulties of life in artisanal fishing activities, due to the desire of younger generations to abandon the activity, artisanal fishing is still maintained through kinship relations.

In Cajueiro, fishing structures the daily life of its inhabitants since it is the main economic activity, as everyone has either a direct or indirect relationship with it. The community adopts a division of labor based on gender. While men take the role of resource provider through fishing, women are mainly engaged in domestic activities.

In 1980s, industrial fishing was introduced to the community. This imposed a new order in the region's labor relations. Male work in particular underwent a tough move towards wage labor. However, after a period of destabilization, artisanal fishing consolidated itself as an activity that organized the life of Cajueiro. This framework was revealed by workers' constant return to this economic activity, after having experienced wage labor. In a context of confrontation between artisanal and industrial fishing, kinship becomes an important factor that guides choices. Silva (2015) shows that in Cajueiro, kinship orientates fishing according to three main factors: mutual assistance, similarities in occupational status permitted by this activity, and proximity of the fishermen's place of residence. The Fig. 19.7 draws the spatial configuration of the community, in which such proximity stands out.

According to Silva (2015), roundtrip in the world of fishing stem from the degree of the individual's connection with his family, the place where he lives and his occupation. These factors decisively contribute in preventing individuals from leaving their communities and artisanal fisheries. The geographical proximity of the relatives also gives encourages the exchange of favors.

Moreover, according to Silva (2015), the barely noticeable difference in the status of the profession, the similar financial earnings of fishermen, as well as housing and working conditions, function as an impetus for future generations to maintain the

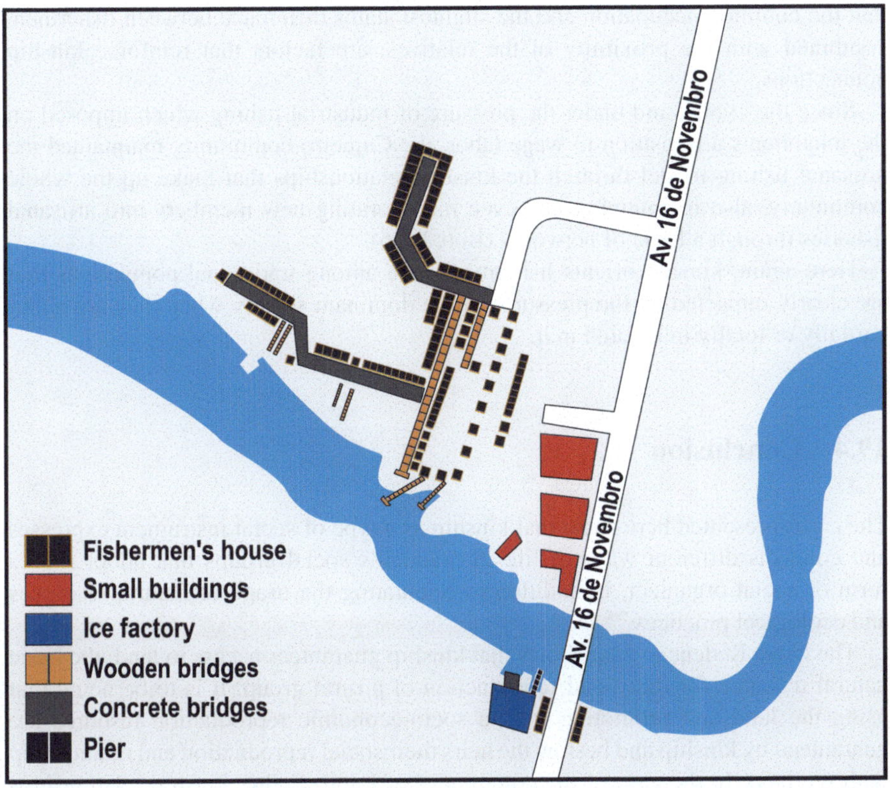

Fig. 19.7 The community of Cajueiro. (Source: Silva 2015)

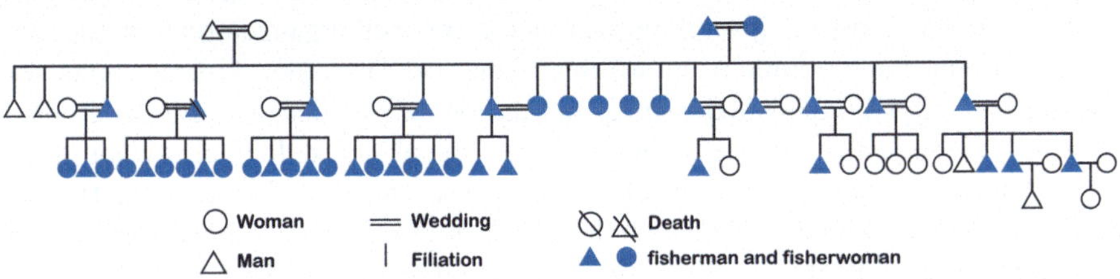

Fig. 19.8 Three generations of fishermen in Cajueiro. (Source: Silva 2015)

activity. Silva suggests (2015) that in Cajueiro individuals remain in the artisanal fishing domain less for personal reasons than for the effect of a tight knit community, woven by the kinship existing in the activity. This is a priority among younger generations despite a common discourse on an activity characterized by exhausting work. The Fig. 19.8 shows the presence of artisanal fishing over three generations of one of the Cajueiro families. This scenario is repeated in the other families that populate the place.

Through the figures and by Bott's reference (1976), Silva (2015) stresses that every society is formed by networks. In Cajueiro, this scenario is repeated. It is clear that the common occupation and the slightest status difference between fishermen, combined with the proximity of the relatives, are factors that reinforce kinship connections.

Since the 1980s, and under the pressure of industrial fishing which imposed on the inhabitants a transition to wage labor, the Cajueiro community maintained the artisanal fishing model through the kinship relationships that make up the whole community, also maintaining and even incorporating new members into artisanal fisheries through a logic of networks (Bott 1976).

Here again, kinship orients fishermen's life among traditional populations that are clearly impacted by the pressures of the dominant society when they are either partially or totally integrated in it.

19.4 Conclusion

The cases presented here show that kinship as a type of social instrument expresses and connects different ways of life. It maintains social groups that adopt it as a form of social organizer, in addition to orientating the usage of natural resources and ecological practices.

The Nova Redenção case shows that kinship guarantees access to land, the basic natural resource for the social reproduction of a rural group. It is to be noted that using the land and promoting it as a socio-economic reproduction resource are guaranteed by kinship and bestow the heirs their social reproduction and relationship with the land. In the case of the community of Lauro Sodré, kinship relationships are responsible for a certain type of agriculture and a practice in relation to fishing resources marked by a particular local knowledge that feeds on practices. Finally, in the Cajueiro case, kinship is an important factor of social organization with the role of maintaining the community in artisanal fisheries. Therefore, kinship organizes and guides both daily life and the social and ecological practices of different groups. Kinship could be understood as an instrument of social and ecological coviability, as defined at the beginning of this article. That is to say, kinship should be interpreted as an instrument maintaining the viability of social groups against the pressures of a dominant society, one that depletes common natural resources. As indicated, it is not possible to be coviable only from the ecological and social perspectives. Coviability should be created between societies and their ecological practices. Kinship appears as an adequate instrument for this purpose.

Coviability denotes life processes that are suitable for different societies. A society may be considered viable only when its socio-environmental reproduction processes do not excessively compromise other groups. In social life and in the forms of life and insertion in nature, kinship stands out as a human form marked by infinite possibilities.

After five centuries occupying the Amazon, questions of development are still the same: the exploitation of natural resources in frontier regions, exploitation relations, and the poverty of the populations. In the cases discussed here, the transformations in the Amazon over the last 30 years have imposed changes in farming practices in Nova Redenção. They have also altered the access and use of the mangrove ecosystem in Curuçá; and the transformation restructured artisanal fishing in the community of Cajueiro. In all three cases, kinship is an important strategy in the relationships of these groups with the dominant society; while it curiously appears as an important agent of change contained in the new economic practices imposed by the economy of the surrounding society. As a social organizer, kinship proved itself capable of defining different alternatives of socio-environmental interaction. It would be beneficial to consider it a type of representative or mediator in processes of social, economic and environmental transition, or even an organizing instrument for coviability.

References

Adams C (1994) As florestas virgens manejadas. Boletim do Museu Paraense Emílio Goeldi. Série Antropologia. 10(1)

Almeida NJR (2012) Saberes e práticas tradicionais: população pesqueira extrativista da Vila Sorriso – São Caetano de Odivelas/PA. Belém: Núcleo de Meio Ambiente. Programa de Pós-Graduação em Gestão dos Recursos Naturais e Desenvolvimento Local na Amazônia

Almeida NJR, Ravena Cañete V (2015) Recursos naturais e cotidiano: população ribeirinha e a gestão da RDS Alcobaça – UHE Tucuruí/PA. Tessituras: Revista de Antropologia e Arqueologia, Pelotas 3(1):191–207

Araújo R, Schiavoni G (2002) A ilusão genealógica: exercício do parentesco e vínculos locais na fronteira agrária da Amazônia. In: Albadejo C, Veiga I (ed) Agricultura Familiar. Pesquisa, Formação e Desenvolvimento / Universidade Federal do Pará. Centro Agroecuário. Núcleo de Estudos Integrados sobre Agricultura Familiar. UFPA/CA/CNRS, Belém, pp 15–39

Barthem RB, Fabré NN (2004) Biologia e diversidade dos recursos pesqueiros da Amazônia. In: Ruffino ML (Org.). A pesca e os recursos pesqueiros na Amazônia Brasileira. Provárzea, Manaus, pp 11–55

Bott E (1976) [1957]. Família e rede social. Francisco Alves, Rio de Janeiro

Castro RA, de Castro ER (2015) As monoculturas e a sustentabilidade: análises de Três Regiões do Brasil. Sustentabilidade em Debate, Brasília 6(2):228–248

Cavalcanti C (2004) Uma tentativa de caracterização da economia ecológica. Ambiente & Sociedade VII(1):149–158

Cavalcanti C (2010) Concepções da economia ecológica: suas relações com a economia dominante e a economia ambiental. Estudos Avançados 24(68):53–67

Cavalcanti C (2012) Sustentabilidade: mantra ou escolha moral? Uma abordagem ecológico-econômica. Estudos Avançados, São Paulo 26(74):35–50

da Silva MF (2004) Parentesco e Organização Social na Amazônia: um rápido esboço. Anuario de Estudios Americanos 61 (2), 649–679. Recuperado em 25 de novembro de 2015, de http://estudiosamericanos.revistas.csic.es/index.php/estudiosamericanos/article/viewArticle/136

Daly HE (2002) Desenvolvimento sustentável: definições, princípios, políticas. Cadernos de estudos sociais, Recife 18(2):171–184

Daly H, Farley J (2004) Ecological economics: principles and applications. Island Press, Washington, DC

de Castro EV (1996) Os pronomes cosmológicos e o perspectivismo ameríndio, Mana, Rio de Janeiro 2(2):115–144

de Castro ER, Hébette J, (Orgs.) (1989). Na Trilha dos Grandes Projetos. Modernização e Confronto na Amazônia. NAEA/UFPA, Belém

de Castro ER, Alonso S, Nascimento S, Carreira L, Correa S d A (2014) Hidrelétricas na Amazônia e dilemas políticos postos à sociedade no século XXI. Papers do NAEA (UFPA) 1(343):1–26
Descola P (2006) As lanças do crepúsculo: relações jivaro na Alta Amazônia. Trad. Cosac Naify, Dorothée Munch, São Paulo
Diaz RPO (2013) As populações pesqueiras e a maricultura: um olhar sobre os processos de diminuição dos recursos pesqueiros no litoral paraense – Resex Mãe Grande de Curuçá. Belém: Programa de PósGraduação em Ecologia Aquática e Pesca. Universidade Federal do Pará
Diegues ACS (1993) Populações Tradicionais em Unidades de Conservação. In: Vieira, Paulo Freire; Maimon, Dália (Org.). As Ciências Sociais e a Questão Ambiental: rumo à Interdisciplinaridade. NAEA/UFPA, Belém
Diegues ACS (1997) O mito do paraíso desabitado nas florestas tropicais brasileiras. In: Castro, Edna; Pinton, Florence (orgs). Faces do trópico úmido: conceitos e questões sobre desenvolvimento e meio ambiente. CEJUP, UFPA-NAEA, Belém
Diegues, ACS (2008). O mito moderno da natureza intocada. 6 ed. HUCITEC e NUPAUB, São Paulo
Figuti L (1994) Ecossistemas costeiros e homens pré- históricos. In: III Simpósio de Ecossistemas da Costa Brasileira – Subsídios a um Gerenciamento Ambiental, vol. I, 1993, Serra Negra. Manguezais e marismas. Aciesp, São Paulo, no 87, pp 212–217
Filho S, Martins PW (2005) Costa de manguezais de macromaré da Amazônia: cenários morfológicos, mapeamento e quantificação de áreas usando dados de sensores remotos. Revista Brasileira de Geofísica *23*(4): 427–435. Recuperado em February 18, 2015, from http://www.scielo.br/scielo.php?script=sci_arttext&pid=S0102-261X2005000400006&lng=en&tlng=pt
Fox R (1986) Parentesco e Casamento. Editora Veja, Lisboa
Guattari F (2006) As três ecologias, 16ª edn. Papirus, Campinas-SP
Homma AKO (2009) Os japoneses na Amazônia e sua contribuição ao desenvolvimento agrícola. Somanlu (UFAM) 9(1):113–133
Icmbio. Instituto Chico Mendes de Conservação da Biodiversidade (2013) Estudo Socioambiental Referente à Proposta de Criação de Reserva Extrativista Marinha no Município de São Caetano de Odivelas, Estado do Pará. ICBMBio, São Caetano de Odivelas
Isaac VJ, Paul US, Schneider H (2003) Programa MADAM: Cooperação internacional para a compreensão dos ecossistemas de manguezal do litoral paraense. In: Aragón, Luis. E. (Org.); Conservação e desenvolvimento no estuário e litoral amazônicos. UFPA/NAEA, Belém/PA
Isaac-Nahum VJ (2006) Exploração e manejo dos recursos pesqueiros do litoral amazônico: um desafio para o futuro. Cienc Cult [online] 58(3):33–36. ISSN 0009-6725
Latour B (2004) Políticas da natureza: como fazer ciência na democracia. Trad. Carlos Aurélio Mota de Souza. EDUSC, Bauru
Latour B (2013) Jamais fomos modernos. Trad. Carlos Irineu Costa. 3ª ed. Editora 34, São Paulo
Levi-Strauss C (1976) As Estruturas Elementares do Parentesco. Vozes, Petrópolis
Lima D, Pozzobon J (2005) Amazônia socioambiental. Sustentabilidade ecológica e diversidade social. Estudos Avançados 19(54):45
Maneschy MCA (2003) Sócio-Economia: Trabalhadores e Trabalhadoras nos Manguezais. In: Fernandes MEB (Org.). Os Manguezais da Costa Norte Brasileira. Fundação Rio Bacanga, Maranhão, pp 135–165
Martínez-Alier J (2014) O Ecologismo dos pobres: conflitos ambientais e linguagens de valoração. Tradução: Mauricio Waldman. 2 ed. 1. reimp. Contexto, São Paulo
McCallum C, Bustamante V (2012) Parentesco, gênero e individuação no cotidiano da casa em um bairro popular de Salvador da Bahia. Etnográfica 16(2):221–246. Recuperado em 15 de dezembro de 2015, de http://www.scielo.mec.pt/scielo.php?script=sci_arttext&pid=S0873-65612012000200001&lng=pt&tlng=pt
Mendes AC (2003) Geomorfologia e Sedimentologia. In: Fernandes, Marcus EB (Org.) Os Manguezais da Costa Norte Brasileira. Fundação Rio Bacanga, Maranhão, pp 13–31
Morán E (1990) A ecologia humana das populações da Amazónia. Vozes, Petrópolis
Moura MM (1978) Os herdeiros da terra. Hucitec, São Paulo

Polanyi K (2000) A grande transformação: as origens da nossa época, 2ª edn. Elsevier, Rio de Janeiro

Prost MTRC, Mendes AC (2013) Ecossistemas Costeiros: Impactos e Gestão Ambiental. 1 ed. Museu Paraense Emílio Goeldi, Belém

Radcliffe-Brown (1995) Sistemas africanos de parentesco e casamento. In: Julio Cezar Melatti (org), Radcliffe-Brown. 2ª ed. Grandes Cientistas Sociais. Ática, São Paulo

Ravena-Cañete V (2005) Os colonos de Nova Redenção: estratégias de permanência e dinâmica do campo rural da zona Guajarina do Estado do Pará. Núcleo de Altos Estudos Amazônicos. Universidade Federal do Pará, Belém

Ravena-Cañete TM (2012) Direito e populações tradicionais no Brasil: da revisão à crítica de aplicabilidades e definições acadêmicas/jurídicas/legais. Dissertação (Mestrado) – Universidade Federal do Pará, Instituto de Ciências Jurídicas, Programa d Pós-Graduação em Direito. Belém, p 125

Ravena-Cañete TM, Ravena-Cañete V (2011) Por uma sociologia do campo jurídico na/da Amazônia: as populações tradicionais amazônicas em foco. Revista Sociologia Jurídica 13(1):26–54

Schneider D (1972) What is Kinship all about? In: P Reining (ed) Kinship studies in the Morgan Centennial Year. Anthropological Society of Washington, Washington, DC, pp 32–63

Segalen M (1986) Historical anthropology of the family. Cambridge University Press, Cambridge

Silva IMC (2015) A rede que pesca gente:o caso da comunidade cajueiro na ilha de mosqueiro/Pa. Programa de PósGraduação em Ecologia Aquática e Pesca. Universidade Federal do Pará, Belém

Simmel G (1987) A metrópole e a vida mental. In: Velho OG (ed) O fenômeno urbano. Guanabara, Rio de Janeiro

Wirth L (1987) O urbanismo como modo de vida. In: Velho OGO fenômeno urbano. Guanabara, Rio de Janeiro, (X)

Chapter 20
The Price of Coviability: Pollination at All Costs. Legal Approach to the New Relationship Between Man and Pollinators

Philippe Billet

In memory of Jacques Weber, A tireless communicator.
To our fruitful exchange of views
Interrupted all too soon

Contents

P. Billet (✉)
University Jean Moulin – Lyon 3, Lyon, France

Institute of Environmental Law (CNRS, UMR 5600, EVS-IDE), University of Lyon, Labex IMU, Lyon, France
e-mail: philippe.billet@univ-lyon3.fr

O. Barrière et al. (eds.), *Coviability of Social and Ecological Systems: Reconnecting Mankind to the Biosphere in an Era of Global Change*,
https://doi.org/10.1007/978-3-319-78497-7_20

20.1 Introduction

Pollinators are part of the history of humanity: bees and honey are mentioned in the oldest writings ever discovered, from Sumer to Babylon, passing by the Hittite laws which penalized the theft of beehives. We also find their traces in the hieroglyphs of Ancient Egypt (Harissis and Harissis 2009; Pundyk 2010). There are numerous myths and legends that, on all continents and in all eras, have been dedicated to these relations (Ransome 1937). However, for a long time, honey and beeswax remained the only noticeable and visible benefits of beekeeping or of the search for so-called "wild" honey. The role of insects in pollination was not perceived at all, what was invisible was left to chance, dependent on the wind. As Pliny the Elder stated, summarizing the majority of ancient treaties on agriculture and breeding: "The following is the order which Nature observes throughout the year. First comes fertilization, taking place when the west wind begins to blow, which is generally from February the 8th. This wind impregnates the creatures that derive life from the earth-indeed in Spain even the mares, as we have stated: this is the generating breath of the universe, its name Favonius being derived, as some have supposed, from fovere, 'to foster'. It blows from due west and marks the beginning of spring. Country people call it the cubbing season, as Nature is longing to receive the seeds; and when she brings life to all the seeds sown, they conceive in a varying number of days and each according to its nature, some immediately, as is the case with animals, while some do so more slowly and carry their progeny for a longer period of gestation, and the process is consequently called 'germination' (Pliny).

The auxiliary role of insects was only understood much later: in fact, it wasn't until the Age of Enlightenment in Europe that the principle of sexual reproduction for the majority of plants was confirmed, bringing with it the possibility of the fundamental "insect-plant" relationship and laying the foundations for the entomophilous reproduction of angiosperm plants (Abrol 2011). Darwin then marked a decisive stage with his observations on the reproduction of orchids, speculating on the existence of an interdependence related to successive "plant-insect" adaptations (or "coevolution"): "(...) I have further shown that even Lepidoptera are able to penetrate other and tougher tissues. It is an interesting case of co-adaptation that in all the British species, in which the nectary does not contain free nectar, the viscid matter of the disc of the pollinium requires a minute or two in order to set hard; and it would be an advantage to the plant if insects were delayed thus long in obtaining the nectar by having to puncture the nectary at several points." (Darwin 1862).

These observations will fundamentally change the view of pollinators and modify the empirical relationships traditionally established until then. The transhumance of beehives intended to improve honey production (Lemeunier 2006; Fanica 2006), the only objective sought at the time, was gradually replaced by a more functional approach: to improve plant production (fruits, vegetables, plants). The domestic bee (Apis mellifera) became the precious auxiliary of the plants' fruit production whilst the displacement of beehives became a financial investment. This new idea reflected the economic appropriation of a natural function: domestic pollinators had already been the focus of economic transactions, but on a more modest family or local scale. Such transactions had now taken on an industrial scale. This also permitted a better perception of the "pollination crisis" characterized by the collapse of pollinator colonies (Colony Collapse Disorder) under the influence of various factors: depending on the regions of the world, beekeepers record losses of between 30 and 70% of their bee populations, without counting attacks on non-domestic pollinators, on which we have little data but for which the harmful effects are real[1]: substitutions of species, regression of the associated flora ... etc (Chagnon 2008; European Food Safety Authority 2013; United States Department of Agriculture 2013; Deguines et al. 2014). This awareness of Colony Collapse Disorder and the risks for biodiversity served to justify an institutional intervention in order to preserve the eco-systemic service provided by pollinators, and in particular by domestic bees.

The massive decline in pollinators was taken into account within the framework of the United Nations Convention on Biological Diversity (1992), during the third Conference of the Parties in 1996: The Conservation and sustainable use of agricultural biological diversity (COP 3 – decision III/11) consequently noted, concerning the "Impact of Biological Diversity on Agriculture", that: "a large proportion of crops depend on insect pollinators for good yields". It prioritizes them within the context of "Initial issues for conducting case studies": "1. Pollinators, including consideration of the monitoring of the loss of pollinators worldwide; the identification of the specific causes of pollinator decline; the estimation of the economic cost associated with reduced pollination of crops; the identification and promotion of best practices and technologies for more sustainable agriculture; and the identification and encouragement of the adoption of conservation practices to maintain pollinators or to promote their re-establishment". Following The Sao Paulo Declaration on Pollinators (International Pollinators Initiative 1999), the Fifth

[1]"Wild pollinators have declined in occurrence and diversity (and abundance for certain species) at local and regional scales in North West Europe and North America. Although a lack of wild pollinator data (species identity, distribution and abundance) for Latin America, Africa, Asia and Oceania preclude any general statement on their regional status, local declines have been recorded. Long-term international or national monitoring of both pollinators and pollination is urgently required to provide information on status and trends for most species and most parts of the world" (IPBES 2016).

Conference of the Parties of the Convention on Biological Diversity promoted an "International Initiative for the Conservation and Sustainable Use of Pollinators"[2] (International Pollinators Initiative 1999), based on the fact that "native pollinators need to be protected and sustainably managed for the pollination service they can provide and that agricultural practices be designed to incorporate the protection and sustainable management of bee populations. "The pollinator crisis exemplifies the intimate relationship existing between the welfare of natural environments and their biodiversity and the needs of sustainable agriculture". The aim of this is to promote coordinated action worldwide to: "(a) Monitor pollinator decline, its causes and its impact on pollination services; (b) Address the lack of taxonomic information on pollinators; (c) Assess the economic value of pollination and the economic impact of the decline of pollination services; (d) Promote the conservation and the restoration and sustainable use of pollinator diversity in agriculture and related ecosystems". It asks the FAO "to facilitate and coordinate the Initiative in close cooperation with other relevant organizations and to consider establishing a coordination mechanism, with geographical balance and with leading relevant organizations, to prepare a proposal for a plan of action taking into account the recommendations in the Sao Paulo Declaration on Pollinators, as well as on contributions submitted by countries and relevant organizations, for submission to and review by the Subsidiary Body on Scientific, Technical and Technological Advice and consideration by the Conference of the Parties at its sixth meeting" (COP5, decision V/5). This plan was adopted by the Sixth Conference of the Parties in April 2002 (decision VI/5) and is based on the networking of five regional initiatives, coordinated by the FAO: in addition to the Brazilian initiative for South America,[3] the African Pollinator Initiative (FAO 1999), established to promote pollination as a vital service for the survival of human populations and the conservation of biodiversity in Africa (FAO 2007); the European Pollinator Initiative (European Commission 2004), which aims "to integrate and co-ordinate local, national and international activities relating to pollination into a cohesive network in order to safeguard the services provided by pollinators across the continent"[4]; the North American Pollinator Initiative (Committee on the Status of Pollinators in North America 2007),[5]which aims to "raise public awareness and education and promote constructive dialog about pollinators' importance to agriculture, ecosystem health, and food supplies", to "promote conservation, protection, and restoration of pollinator habitat" and to "document and support scientific, economic, and policy research – creating the first international data bank (library) of pollinator information"; the Oceanic Pollinator

[2]Known under the name of IPI (International Pollinator Initiative) (decision V/5, section II). http://www.internationalpollinatorsinitiative.org

[3]http://www.webbee.org.br/bpi/english/linha_tempo.htm

[4]www.europeanpollinatorinitiative.org

[5]http://pollinator.org/nappc/index.html (NAPPC: North American Pollinator Protection Campaign).

Initiative (OPI 2008),[6] for which "Pollination is an essential ecosystem service, and a prerequisite to all the other essential services provided by plants, yet in Australia, New Guinea and on the Oceanic Islands". The aim is to develop information "about the distribution and ecology of pollinators, taxonomy of insect pollinators, whether or not they are in decline, the ecosystem service role of native and introduced pollinators, the economic value of pollination services by unmanaged pollinators. These regional actions are reinforced by the FAO's Global Action on Pollination Services for Sustainable Agriculture (2010): it's a full-sized project, working together with partners from seven countries: Brazil, Ghana, India, Kenya, Pakistan, Nepal and South Africa (2009–2014). The development objective of the project is "improved food security, nutrition and livelihoods based on better conservation and sustainable use of pollinators" (UNEP 2010a, b).

These services have been the focus of monetary appraisals. According to a study conducted on one hundred or so plants used for human consumption, they would reach an approximate value of $153 billion euros, or 9.5% of the agricultural production value used for global human food in 2005 (Gallai and Co. 2009). This estimation is based on a negative value, the "shortfall" due to the absence of pollination and the costs to replace the failed service. We are now better aware of the importance of pollinators and of what their disappearance could cost: a multitude of reports and books assess the services they provide (Ruhl and co. 2007; Government of Ireland 2008) or more generally, propose an analysis of these services' (Braat and Brink 2008). As noted by the TEEB, "The reason for this characterization is that the value of these services often remains invisible until it is no longer provided by 'ecosystems and biodiversity'. (The Economics of Ecosystems and Biodiversity 2014). However, is the service really free, "offered by nature", as we have considered it to be until now? Its scarcity reveals the hidden costs, and consequently, its economic value in the absence of having appreciated it until now for its environmental contributions. The "environmental good" it has become has the value of its contribution to procurement and support services: it can consequently be measured economically (Chevassus-au-Louis 2009; Deffairi 2015) from multiple points of view, whether in terms of promoting the labor of bees, as understood in French law (1), or remunerating the eco-systemic service provided through pollination agreements (2), or funding the eco-systemic pollination service (3) or the environmental pollinator protection service (4).

20.2 The Valuation of the Work of Bees in French Law

Bees visiting flowers has traditionally been valued in terms of honey and beeswax, sometimes resulting in agreements between hive owners and a third party, most often paid in kind. Such agreements characterize an ownership relationship based on the

[6]http://www.oceanicpollinators.org/

immobilization of the bee colony within a hive: the "hive-bees" consubstantiality consequently creates a close dependence on the beekeeper. This "servitude" on the part of bees transforms them into production animals, within a renewed perspective, from the time that the pollination service takes precedence over honey production. It raises the question of the legitimacy of beekeepers' rights with respect to the work of bees and leads us to ask whether it would not be possible to identify a common and collective right regarding this particular function of bees, in the form of a 'transpropriation'.

20.2.1 The Contractual Valuation of the Work of Bees

The contractual valuation of the work of bees is long-standing. From the fourteenth century, historians noted the existence of livestock leases concerning swarms of bees (Fournial 1976), ancestors of the current livestock lease agreement in the Civil Code. A simple livestock lease "is a contract by which one person gives to another animals to be kept, fed, and cared for, on condition that the leasee will profit from one half of the increase in stock and will also bear one-half of the loss" (art. 1804). If the Civil Code mainly concerns traditional "farm" animals (sheep, cattle, goats ...), jurisprudence admits its application to domestic bees, since this lease "may be made for all kinds of animals which can increase or be of profit for agriculture or trade." (art. 1802).

A contract model by private deed of 1806 consequently illustrates the profit-sharing regime of the hive in terms of beeswax and honey with half going to the leaser and half to the leasee: when the hives are withdrawn after beeswax is harvested if they weigh more than a certain weight in relation to the hives initially made available, the leaser must pay the excess weight just as he must pay half of the value of any additional hives resulting from his industry in comparison to the initial hives to the leasee. For his part, at the end of the lease, the leasee must pay the leaser half of the missing weight in relation to the initial weight of the hives; he must also abandon as many hives as necessary in order to meet the weight agreed upon in the terms of the contract, keeping any excess hives for himself (Beaunier 1806).[7] The weight therefore enables an assessment of the growth of the bee population in relation to the population initially made available, whilst the sharing of beeswax and honey during the life time of the contract as well as the sharing of additional hives at the end of the contract constitute the payment "in kind" of the industry and the care of the leasee. The gains, like the losses, are therefore easy to assess.

[7]The hypothesis described here is that of the simple livestock lease agreement of articles 1804 and following of the Civil Code, "contract by which we give to another cattle to keep, feed and take care of, provided that the leasee will benefit from half of any increases, and that he will also bear half of any losses".

20.2.2 "The Immobilization" of the Colony, the Basis of "Hive-Bees" Consubstantiality

Within the context of the livestock lease agreement, remuneration is justified for one party by the ownership of the hives and the colonies they are home to; for the other party, remuneration reflects the work to maintain and develop the bee population at his disposal and to encourage production. By nature, bees are movable, since they are able to move by themselves, are difficult to catch and consequently difficult to own. Their ownership is based on a legal fiction formalized by the Civil Code: immobilization by destination. As stated in article 524 of the Civil Code, "Animals and things that the owner of premises places thereon for the use and working, are immovable by destination. Thus, the following are immovable by destination when they have been placed by the owner for the use and working of the premises: (...) Beehives ..." (Michallet 2013). Civil law refers to the "hives" and not the "bees": at no time is the ownership of bees expressly asserted by regulations, which only refer to the hives that contain them. On this point the beehive is "unwavering" as regards ownership rules. It is considered to be an accessory to the "stock" to which it belongs as article L211-8 of the Rural and Maritime Fishing Code illustrates, according to which "in cases where the beehive could be separated from the stock to which it is attached, it can only be displaced during the months of December, January and February". The attachment is such that the Civil Code must consider a dissociation of stock and the beehive, during the time when the bees "can be displaced conveniently, bees being locked in their cell and numb with cold "(Gadar, Year VIII).

The "hive-bees" consubstantiality means that the bee colony cannot be conceived without the hive which immobilizes it and permits its appropriation. This property of the immobilized colony is corroborated by the right to pursue that which swarms and escapes this legal fiction owing to its natural movement: as stated in article L. 211-9 of the Rural and Maritime Fishing Code, "The owner of a swarm has the right to claim and recover it, as long as he has not stopped following it; otherwise, the swarm belongs to the owner of the land on which it has settled" (Prétot 2011). The attachment to the stock, the immobilization, therefore equates to ownership, but this ownership concerns the swarm as opposed to the bee. The latter is not legally (and biologically) individualized and is appropriated and exploited collectively, due to the impossibility of identifying it as an individual and of its attachment to a particular hive, and as such, to a particular owner. Domestic bee colonies, bred for the production of beeswax, honey and other products, are consequently immovable properties appropriated by the owner of the hives housing them. The latter consequently only has a right of claim to them if the swarm is immobilized in one of his hives.

20.2.3 *Bees as Production Animals*

The Rural and Maritime Fishing Code classifies bees as "production animals", echoing the provisions of articles 547 et seq. of the Civil Code relating to "The right of accession to what is produced". The traditional production of honey and beeswax is consequently appropriated by the owner or his co-contractor in exchange for work to maintain the hives, placing them near to favorable sites for feeding and conserving them during the winter period ... etc. A more realistic approach may consider that these procedures are simply the counterparty of what is extracted to the detriment of the colony, which must be compensated in order to ensure its survival until the next floral season: in its natural state, a colony of bees does not need man but this changes when the colony is immobilized in the hive and the fruit of its work intended for its single life-cycle is removed. Maintaining the bees' work force and guaranteeing the tranquility of their winter rest are consequently offset by the gains from beeswax and honey to the benefit of the beekeeper. The bee consequently upsets the traditional regime of animal ownership, since it is not recognized as an individual (swarm) and it constitutes an object of atypical ownership (as a hive content and not as an individual or a colony). Swarm ownership is only evoked when the latter leaves the hive, whereas the hive by itself benefits from the consideration of civil law when bees are found in it. These particularities benefit from an unusual prolongation with the valuation of a natural function of bees, pollination.

20.2.4 *Remunerating the Pollination Service*

The new issue of the remuneration of the pollination service can be questioned from a legal viewpoint. For a long time, the benefit of the bee's production was limited to its material components: beeswax, honey and other substances. For the past 50 years, it has exceeded this traditional approach to integrate the fruit production of certain plants, and therefore, food security and the resulting preservation of biodiversity, whilst ensuring the genetic diversity of plants. There are consequently a multitude of eco-systemic services provided by the "visit" made by bees to flowers. The pollination service provided by these particular worker bees shares this characteristic which is common to all pollinations in that it is not sought by the animals that perform it: it is seen as "accidental", even if in reality, it is the result of a complex biological process. However, this process may be sought by the owner of the colonies who can artificially guide the bees: this leads to distinguishing between "a global and passive service" carried out with no particular human intervention and "a targeted and active service" requiring human intervention to the benefit of a specific territory (Gerster 2012a, b). Such service has the cost of the intervention of the beekeeper and the equipment implemented, but nothing guarantees the pollination process in the desired conditions, nor that it will be carried out in its entirety by the bees in question. Therefore, the benefit is uncertain with regard to both the implementation process and the bees performing this pollination, even if there is though a strong presumption that the bees present in the hives effectively pollinate nearby plants. The payment made within the context of the contract between the beekeeper and the producer consequently serves to remunerate a random service., This can justify the fact that it is not conditioned by the result of making the bees available, or even that this service is totally absent from the stipulations of so-called "pollination" agreements.[8]

If a service remuneration can exist for providing hives during a certain time, given the fees and expenses incurred (handling, transport, immobilization, loss of livestock ...), the remuneration of pollination and associated services, considered separately, may surprise and raise questions as to their "appropriation" by the owner of the bee colony responsible for pollination. First of all, there is no service equivalence: the pollen and nectar which feed the colony[9] are harvested on flowers that do not belong to the hive owner, and the hive owner does not remunerate the plant owner for this "feeding" service, these plants being visited "naturally" by his bees and consequently "naturally" made available to the benefit of the beekeeper. Legally speaking, something is taken from something belonging to someone else, certainly with his implicit agreement, but it is still taken. In this perspective, given the reciprocal gains related to "plant/bee" mutualism, pollination and the resulting fruit production should appear as compensation for feeding, without any other compensation being called for. As a consequence, nothing justifies remuneration while the feeding of bees is not due to the plant owner any more than the resulting honey and beeswax produced. If feeding and the production of honey are certain, pollination is much less certain and plant fertilization even less so, consequently calling on intricate service assessments at every level. There is, to a certain extent, an exchange of services: the provision of plants with a view to feeding the colony and apicultural production is "remunerated/compensated" through pollination, as the "making available" of bees is "remunerated/compensated" though feeding the colony and the profits related to honey and other products produced. In contrast, within the context of "guided" pollination with the express placing of hives, two services are likely to be involved and are even likely to justify financial intermediation: on the one hand, if transportation is involved, there are the costs incurred for making hives and colonies available and, on the other hand, there are the potential costs of the lack of or reduced pesticide treatment of plants during the plant visiting period, in order to avoid killing the bees, costs which are linked to the risk of a reduced or lost crop production. In this last case, however, there is a single risk and no expenses actually incurred, contrary to those related to the provision of hives. A reasoned crop management during the flowering period necessarily involves limiting pest control treatments on plants in order to prevent the mortality of pollinators and allow them to work for the benefit of the vegetable or fruit producer. Ultimately, the profit for the beekeeper is linked to the payment of the service of making the

[8]V. infra.

[9]Carbohydrate intake by nectar and protein, lipid and vitamin intake, mainly by pollen.

hives available and, secondly, the profit related to pollination and subsequent fruit production. Whatever the case, remuneration depends on the agreement between the two protagonists, the value of the service that can be disrupted by market forces: the scarcity of colonies on the one hand and the increase in pollination needs on the other, almost naturally leading beekeepers to search for profitability and to provide the service to the highest bidder. Scarcity and need determine the price, in keeping with market forces, whilst honey production becomes a somewhat secondary concern.

20.2.5 The Legitimacy of Rights on Bees' Labor

Aside from the provision of hives and the associated work, what legitimacy can such remuneration have? In other words, does the hive owner have rights to the eco-systemic pollination service and the associated services carried out by "his" bees? Or, on the contrary, should he be excluded from any profits from this point of view, with this eco-systemic service being considered as a common good over which he has no rights, even if "his" bees participate in it? This contractual management of bees in fact isolates one of their particular qualities, that of providing eco-systemic services. This function queries the legal regime of the services provided and their management in terms of property law and questions the relationship between bee colony ownership and these natural functions. However, there should be no confusion: it is necessary to distinguish between the "pollination service", ensured by the provision of hives to a contractual partner in order to assist the fruit production by his plants, which effectively characterizes a service provision to the benefit of the identified beneficiary, and the "eco-systemic pollination service" provided by the bees for a common and collective profit with no pre-constituted legal relationship or the idea of carrying out a "service" on the part of the beekeeper. The beekeeper's rights are undoubtedly better guaranteed in the first case, since they are linked to the contractual provision of hives, but the question arises from the point of view of the bees and the rights that the beekeeper has over them. Their appropriated "work force" outweighs the growth that ordinarily characterizes accession, given that fruit production benefits the plants' owner. The "eco-systemic pollination service" is the result of a mutual benefit, the "visiting" of flowers by bees allowing them to be fed and the production of honey for "private" purposes, while ensuring the fruit production and reproduction of plants which are as much of general interest as of personal interest. The inability to determine to which hive a bee belongs does not however allow the beekeeper to claim any rights; "his" bees participating as much as any other bee to this service.

20.2.6 The Transpropriation of the Eco-Systemic Pollination Service

Beyond the question of the financial evaluation of the eco-systemic services consequently provided (Angel 1998; MEDDTL 2009; Maris 2014), several clues lead us to believe that ownership of this capacity to produce services is not as sovereign or more particularly, as personal as a first analysis could imply. In effect, we can wonder about the legal consubstantiality effective which would exist between the ownership of pollinator colonies and the services they can provide. This oscillates between the very strong link which would make the owner the sole beneficiary of their work, calling for a payment for the services provided, to a more stretched distant link allowing us to consider that the owner has no rights and cannot claim no any payment, aside from for his work as regards making hives available (displacement of hives ... etc.). Like François Ost, we can wonder whether there would not be a possible transpropriation, namely a legal detachment that could operate between the ownership of the colony and this ability to provide services, due to a "a multiple use concession [on a same good] to a plurality of concession holders" (Ost 2003). Undoubtedly, the colonies belong to the hive owner, but their capacity to provide eco-systemic services would, in this perspective, belong to the collectivity. Just like a historical monument whose building belongs to its owner but whose historicity belongs to the community. At the very least, the collective and undifferentiated service of pollination, given the production of honey and other substances, is individualized and appropriated. Consequently, the owner "does not necessarily have the same level of control intensity over each aspect of his property" (Ost 2003). If the beekeeper is "satisfied", as he always has been, with the traditional products of beekeeping (honey ...), the sole goods he considers as being goods and which he effectively sought, he is, a priori, indifferent to other services, unless they are related to feeding and maintaining his colonies from which he makes a profit. There would therefore be a 'patrimonialization' of this pollination capacity in a more collective sense, that cannot be appropriated, justifying the intervention of the collectivity to preserve this service, support its management or regulate it to take account of general interest. The only certainty is that "ownership of a good, either movable or immovable, gives a right to everything it produces and to what is accessorily united to it, either naturally or artificially. That right is called a right of accession".[10] For its part, pollination is more difficult to grasp: the property of what grows or fruit which is produced as a result of pollination is excluded, since it concerns the owner of the plant and not the beekeeper, who is simply the owner of the hive that houses the colony.

Pollination only serves to evaluate the exchange: the arborist or the horticulturist requesting the service is motivated by the profit derived from the plant, by the price that he's willing to pay for the service, based on the expected gain or the loss

[10] C. civ., art. 546.

avoided. When the bee pollinates it provides a service as we have seen in article 524 of the Civil Code: the honey hive (and consequently the colony of bees that it houses) has been moved by the stock owner "for the use … of the premises". The beekeeper doesn't have particular right over the pollination, but rather over his bees which pollinate, which justifies the remuneration of his service consisting in the provision of bees to a third party (plant or fruit producer) within a voluntary and sought framework. His rights over the bees' service is based on his ownership of the colony. However, it is not the pollination which constitutes the market service, but rather the provision of hives with a view to pollination. Remuneration is therefore founded when the service is individualized, provided to an identified third party which has expressly requested the service and the fruit of particular work on the part of the beekeeper (displacement of hives … etc.) according to a contract, whatever its form. For his part, the beekeeper is motivated however by the survival of his colonies: should he refuse the pollination service offer it, his good may decline and eventually disappear. He is therefore dependent on this service request, unless he considers that his bees are simply a "pollination machine" which he can replace once they are destroyed, without worrying about their sustainability.

On the other hand, what is the situation when the pollination service is provided to an anonymous collectivity, a non-individualized service beneficiary, with no advanced solicitation, no contract or particular constraints for the beekeeper (particularly when his hives are fixed) or with the simple purpose of honey production (transhumance of hives according to the flowering calendar)? Can the beekeeper claim a right over the service provided by the bees for the benefit of the collectivity and ask to be paid in this capacity? Furthermore, if pollination is clearly in the general interest, can the collectivity intervene to facilitate or perpetuate the achievement of this service by conducting actions to the benefit of pollinators (financing of floral fallow …) or of the beekeeper (financing the placement of the hives …) or by imposing ownership constraints? The development of the financial valuation of this eco-systemic service in the form of economic incentives – more than direct payment – shows that we have surpassed the question of legitimacy to wonder about the modalities of this valuation.

20.3 Pollination Agreements, Financial Valuation of the Eco-systemic Service of Bees

Because of the scope of action of these insects (3 km), the bees' domestication due to their placement in hives does not permit the pollination of certain sectors to be ensured, much to the discontent of producers (vegetable farmers, horticulturists, arborists …), since the success of crops depends on this service. In the face of the regression in numbers of pollinators severely threatening the durability of these activities, an "industrial" transhumance has been organized in certain countries (United States, Canada): tens of thousands of hives have been moved over distances of up to 20,000 km per year (Bruneau and Burget 2008),[11] (as opposed to an average of 100 km per year in Europe). "The commercial bee" (Downing 2007) has been born, whose service is subject to negotiations and contracts, in the form of Pollination Agreements (or Pollination Contracts) . Though non-written, the pollination agreement still constitutes a contract between a beekeeper and a farmer, though which the first undertakes to put at the disposal of the second a number of hives during the flowering season of a particular crop and to install the hives in the area specified by the farmer or, "if no area is specified, to install them in a place the beekeeper considers the most appropriate to ensure maximum pollination coverage".[12] The agreements are generally concluded before the beginning of winter, in such a way as to enable the beekeeper to prepare his season and ensure the availability of pollinators. An "official" definition of this activity has been issued by the French Ministry of Finances in order to submit it to an agricultural tax regime and avoid the service provision tax regime which is less favorable: This activity consists "in the provision of hives for a fixed-term to farmers in order to increase the production value of their plantations through pollination work carried out by bees".[13]

20.3.1 A Tried and Tested Agreement Mechanism

The existence of these agreements is quite ancient: the Insect Pollination of Cultivated Crop Plants manual, which already forty years ago, presented pollination and service contracts (MacGregor 1976), referred to the prices of making hives available and the correlation of the strength of colonies with the potential rental rates as they had been studied during the twenties (Farrar 1929). It is the pollination service itself which is concerned: "Beekeepers have an opportunity and a challenge to contribute more fully to the development of our agricultural resources by supplying the bees needed for pollination and making honey available to more people. The pollination of fruit and seed crops will be proportional to the increase in honey production,

[11]"Wintering in Florida to move on to California for almond trees and then to Oregon for fruit trees (apple, pear, cherry,...); then transhumance for small fruit such as blackberries, blueberries, strawberries, and then the production of vegetable seeds (carrots, onions, cabbage), leguminous plants (clovers...); finally, they go down south for wintering".

[12]According to the pollination contract model made available to beekeepers by the Regional Directorate of Quebec Capital of the Ministry of Agriculture, Fisheries and Food of Quebec (Revised Version of April 2016).

[13]"For tax purposes, this pollination activity can be regarded as exercised within the framework of a beekeeping operation. It is therefore accepted that the revenues derived from this activity should be taxed in the category of agricultural profits [CGI, art. 63, al. 3], the same way as other products from beekeeping" (Bull. officiel des impôts – BA-Champ 10-10-10-20140306 du 3 mars 2014. The site "Bulletin Officiel des Finances Publiques-Impôts" (BOFiP-taxes) brings together, in a single and consolidated database, the set of comments concerning tax legislation published by the Office of the General Directorate of Public Finance).

because both depend upon the number of blossoms visited by the bees" (Farrar 1947). Through experience, since the eighties, several tens of contract models have been proposed in the United States and Canada, by official agricultural organizations as well as by beekeeping or entomology journals, and have served as sources of inspiration for the models available in France.

20.3.2 A Simple Obligation of Means

This agreement imposes a reciprocal obligation of means but not a performance obligation. The contract model proposed in France by Apiservices[14] imposes for example on the beekeeper "to place his bees on the crops for the time required for effective pollination, estimated (. . .) at approximately x days, with a maximum duration of 15 days"; "to agree to show the strength of his colonies following random sampling conducted by the arborist"; "to ensure that the colonies remain in good pollination conditions for the duration of the contract"; "in the event of persistent bad weather, to feed his colonies to maintain them at good population levels" In return, the arborist is committed "during the rental period, to refraining from using any phytosanitary treatment involving products which are toxic to bees or whose use is prohibited during the flowering period (including on weeds"; "in the event of mandatory phytosanitary treatment posing a danger to to bees (eg: end of flowering period), the beekeeper should be informed three days in advance to enable him to remove or enclose his colonies".

20.3.3 Agreement Concerning the Availability of Hives

If pollination constitutes the purpose of the contract, it is lacks, however absent from the service remuneration conditions: the beekeeper must provide an agreed quantity of hives, "in exchange of which, the arborist will pay him x Euros per hive, for the displacement and maintenance fees of these hives, for a duration of X days". In reality, the contract is much more focused on the provision of hives and the associated bee colonies than on pollination, which can be explained by the fact that "successful pollination does not imply successful fertilization (. . .). The price reflects the distances required to move the colonies, the type of crop, the amount of manipulations and any additional costs" (Skinner n.d.). Pollination itself, whilst it appears in the title of the contract, is in contrast ignored in the stipulations of American and Canadian contracts on which the contracts proposed in France are based. Whilst the latter mention pollination, they do not make it a performance obligation: in this way, concerning remuneration, "the producer undertakes to pay

[14]Society that promotes the Bee "as a pollination agent".

the following fees for the pollination services: x colonies at a cost of X per colony for a total of X". However, provided they meet these obligations, "the beekeeper cannot be held responsible for any loss or decrease in production as a result of the pollination carried out within the context of the present contact". Pollination therefore constitutes the objective of the provision of hives, but it is not formally the objective. The beekeeper is not bound by any productivity guarantee since he has no control over the behavior of bees, the climate or any other events likely to affect flowering or honey production. Agricorp, a body attached to the Government of Ontario (Can.), has in fact compiled an insurance contract to the benefit of arborists for whom certain fruit crops would be affected due to "unavoidable pollination failure due to adverse weather conditions" (Agricorp 2016), but without referring to "guided" pollination as a means of preventing crop decreases or losses.

20.3.4 The Success of Pollination Agreements

This particular exploitation of bees' labor[15] is in the process of replacing remuneration related to the production and sale of honey. It sometimes reaches nearly half of beekeeper income, depending on the periods of the year and the plant species to be pollinated. Consequently, in the United States, "one out of every two beekeepers does not live from the honey trade but rather from the transhumance of his hives. Unlike what is happening in Europe, it is a veritable industry where beekeepers load several hundred colonies into trucks and travel the length of the country to sell their pollination service to large fruit and vegetable farms. The reduction in bee populations can already be felt: in the past, beekeepers use to rent bees colonies for between 45 and 65 dollars (32–46 Euros). This year, the price paid by almonds producers is in the region of 170 dollars (120 Euros) per colony. Overall, the cost of pollination has increased for all types of producers" (Downing 2007). The pollination industry, which only uses bees, is estimated at $16 billion for the United States alone (Gallai and Vaissiere 2009). To a certain extent, this increase is related to the structuring of the offer in relation to demand, with the consolidation of the itinerant beekeeping activity. It is also explained by a decrease in pollination service offers due to Colony Collapse Disorder, which needs to be taken into account in relation to the increase of spaces to be pollinated (Morse and K. Flottum 1998). The contract consequently corresponds to market forces: pollination has a market value or, more precisely, a value has been conferred on pollination through pollination contracts (Sagili and Burgett 2011). Needs, like pollination practices, differ according to states and continents: in France, therefore, pollination

[15]We will use the term "labor" concerning the bees' production activity in reference and equivalence to article L.211-10 of the Rural and Maritime Fishing Code that uses this term to refer to silkworms' activity: "production animals" like bees, they "cannot be seized during their work".

income only represents 2.3% of total global turnover as opposed to 90.7% for hive products (honey, pollen, propolis, royal jelly, beeswax) (Gerster 2012a). However, hive provision price is difficult to determine because it cannot be set according to a reference of what manual pollination would cost, at the risk of making the natural service prohibitive; therefore, the sole reference remains the cost of making bees available (Bastide 1993).

20.3.5 The Price of the Industrialization of Pollination

A price is omitted in these contracts: the price which is "borne" by the pollinators consequently made available and the potential risks to biodiversity. As noted by Professor Dave Shutler within the context of his interview with the Standing Senate Committee on Agriculture and Forestry: "Honeybee losses in recent years can be largely ascribed to increased industrialization of the bee industry. To make large-scale apiculture profitable, honeybees are sometimes transported long distances one stress of transportation might be being locked inside a hive for days while in transit, unable to get rid of waste, dead bees, et cetera". "In sum, industrialization of apiculture stresses honeybees by limiting their access to sufficient quantities and qualities of nutrition by exposing them to multiple pesticide residues, probably other contaminants (. . .), and to a suite of parasites with which they have not co-evolved. As is the case for pesticides, these stressors have multiple potentials for their own set of synergistic interactions" (Parliament of Canada 2014).

We also have to take account of the risks to biodiversity: the assignment of these pollinators to well-defined sectors leads to the other sectors they would have been able to visit being neglected. However, the neglected sectors are not necessarily visited by other species due to their specialization, which can lead to a depletion of plant diversity and associated species. In order to limit this decline and encourage the protection of bees, various measures have been adopted to ensure the financing of the eco-systemic pollination service, irrespective of the pollination service contractually provided to the benefit of a producer, which the latter pays to the beekeeper.

20.4 Financing of the Eco-Systemic Pollination Service

Can the hive owner whose bees pollinate be considered as "a provider of the eco-systemic pollination service" able to benefit from "payments for eco-systemic services"? A priori no, given that it is the bees who are responsible for pollination. It is therefore not the beekeeper who "provides" the service in the true sense of the term. On the other hand, the maintenance of these hives and the bees that they shelter characterizes a service on his part, contributing towards the good realization of the eco-systemic pollination service. In this regard, he accomplishes an environmental service that can constitute the basis for payment to encourage him to realize the service or pay him for his action. As Karel Mayrand and Marc Paquin stress, the payment for environmental services is a mechanism "which aims to promote positive environmental externalities through the transfer of financial resources between the beneficiaries of certain ecological services and the providers of the services or the managers of environmental resources" (Mayrand and Paquin 2014). Nevertheless, an uncertainty persists: do these aids characterize a remuneration or, on the contrary, compensation for the person who provides this service? The context in which the service occurs permits an answer to be given, as in the case of the transhumance of hives or that of the agri-environmental subsidies dedicated to bees. It is therefore by intermediation that the eco-systemic pollination service is the focus of a payment, by financing the set of actions aimed to preserve this service.

20.4.1 The Principle of Payment for the Preservation of Eco-Systemic Services

Some economists believe that "contrary to what the "payment for eco-systemic services" label can lead us to imagine, the purpose of the transaction within the context of this mechanism is not the eco-systemic service itself, derived from one or several ecological functionalities, which by nature, cannot be appropriated, but the adoption of particular uses of resources (mainly land), or specific practices likely to maintain or restore one or several eco-systemic services". In this hypothesis, what we are remunerating is "the human interaction permitting the preservation of eco-systemic services to be facilitated". Here, "the transaction namely focuses on an environmental service, in the sense of the services that people can render for one another by the intermediary of eco-systems", which justifies the use of the notion of "payment for environmental services". As noted by these economists, "certain authors tend to define environmental services solely from the perspective of the services provided by man to eco-systems, consequently reversing the logic and concealing the second part of the equation supposed to be taken into account by the mechanism, namely the dependence of human activities on the services derived from nature. Therefore, in order to remove such uncertainties that hinder the debate and the clarity of the mechanism, we suggest using the terminology "payment for the preservation of eco-systemic services (PPES)" (...)" (Guingand A. and co 2014).

We will retain this last approach that places the emphasis on the objective of the maintenance or the restoration of eco-systemic services. Unlike other identified services which are rendered by appropriate elements of the environment (soil, plants ...), the pollination service cannot be reduced to the mere intervention of the domestic bee (Apis mellifera), an insect appropriated in its collective dimension (swarm) or the bumblebee (Bombus terrestris), which is used for the pollination of certain plants in greenhouses. They must necessarily be associated with the cohort of non-domestic pollinator insects, otherwise the approach of payment for eco-systemic services related to pollination would be partial: it would in effect neglect all of the actions undertaken to maintain populations of non-domestic pollinator insects. Two distinct services must therefore be envisaged: the eco-systemic service, provided by pollinators for men and the environmental service that man provides for pollinators. However, the distinction is not really straightforward, because the complementarity of the two actions scrambles the reference points: hence, the environmental service which consists in maintaining flowering plants during the blooming period has the objective of maintaining an eco-systemic service while guaranteeing the durability of populations of certain pollinators, whereas the provision of the eco-systemic pollination service closely depends on the environmental pollinator protection service.

20.4.2 The General Interest of Pollination as a Basis for Subsidies

The European Parliament has suggested an allocation key for beekeeper/bee services. After having noted that "beekeeping and biodiversity are mutually dependent", it stresses that "via pollination, bee colonies provide important environmental, economic and social public goods, thus ensuring food security and maintaining biodiversity, and whereas, by managing their bee colonies, beekeepers perform an environmental service of paramount importance, as well as safeguarding has sustainable production model in rural areas".[16] Within this perspective, beekeepers are more than intermediaries, more than "simple" owners of pollinator bees: the good management of their colonies determines the good service provision. This management is in itself a service that justifies a financial reward in order to encourage them or help them to provide this service. However, do they have any choice but to ensure good management, as anything but would run the risk of losing their property or seeing a decrease in their honey harvest? They therefore strongly depend on a reasoned management of their bee populations whose productivity potential is conditioned by their behavior towards them and are therefore "captives" of the pollination to which they are finally dedicated. This dependence limits their means of action to claim interventions by the public authorities, because it is difficult to imagine a "pollination strike", which would result in the permanent loss of their "work tool".

Nonetheless, such a claim is not necessary: pollination being immediately recognized as being of general interest, their good offices with their hives making them service providers to the benefit of the community, ready to intervene as part of a collective interest in order to sustain the provision of eco-systemic services. Such a situation permits the material aid which beekeepers can request for the maintenance

[16] European Parliament resolution of November 15, 2011 on honeybee health and the challenges of the beekeeping sector (2011/2108(INI)).

of their colonies to be legitimized, and the public community participates in this management because of the benefits it gains from it, collectively. This compensation and the resulting reciprocal interests bring us beyond the logic of aid which prevailed until now, although the terminology has barely evolved. There is an environmental service on the part of the beekeeper to allow the implementation of an eco-systemic service, the realization of one favoring the realization of the other without determining its existence: the bee does not need man to pollinate. However, human intervention contributes towards improving the service by strengthening the number of pollinators and/or by placing them in the vicinity of the plants requiring pollination.

20.4.3 Uncertainty on the Classification of Paymentss: Remuneration or Compensation?

European Union law is relatively silent in terms of qualifying the payments made in favor of beekeepers: remuneration or compensation of a service, compensation of a constraint, financial incentive, subsidy, sui generis revenue ... it is difficult to rely on the fiscal qualification retained by the States to attempt to define its nature, because of the pragmatism they have shown in this regard, or the policy that they want to adopt. If the revenues derived from pollination activity are, in France, taxed according to the agricultural profits category,[17] the state of Washington subjects the income derived by the beekeepers to the "Business and Occupation Tax", under the heading 'Service and Other Activities', as a "farmer": "Farmer means any person engaged in the business of ... providing bee pollination services". Nevertheless, in order to take into account the economic impact of Colony Collapse Disorder on the agricultural sector, the activity had been exempted from this tax until the 1st of July, 2017, insofar as the beekeeper is eligible and registered (Washington State Department of Revenue 2013). In these two cases, the state aid classification is called for.

The direct support to apiculture, accepted by the European Union, is not in the sense of the recognition of a particular service: in fact, it is about intervening "in order to improve the production and marketing of apiculture products in the Community". "Member States may draw up a national program for a period of three years (hereinafter referred to as the 'apiculture program')": "in order to improve the production and marketing of apiculture products in the Community, national programs should be drawn up every three years, comprising technical assistance, control of varroasis, rationalization of transhumance, management of the restocking of hives in the Community, and cooperation on research programs on beekeeping and apiculture products with a view to improving the general conditions for the production and marketing of apiculture products. These national programs should

[17] V. supra.

be partly financed by the Community". In addition, the member states may grant "specific national aids for the protection of apiaries disadvantaged by structural or natural conditions or under economic development programs, except for those allocated for production or trade".[18] Community co-financing, provided by the European Agricultural Guarantee Fund (EAGF) stands at 50% of the expenditure incurred by the Member State under the National Program.[19] The production of honey and other hive products are the focus of these subsidies, or at least they were initially, and it is only recently that pollination has found its place in the mechanism, first of all allusively, then more explicitly. The subsidies developed in favor of pastoral apiculture are highly symptomatic of the extension of the approach.

20.4.4 The Particular Case of Hive Transhumance

Transhumance is a particular hive exploitation mode, which is different from sedentary apiculture: hives are moved in order to follow honeydew, increase production and diversify honey varieties. This is not a new practice (Lemeunier 2006). It should not however be confused with the controlled pollination linked to pollination contracts, since the first objective of transhumance is honey production.[20] In its

[18]Council Regulation (EC) n° 1234/2007 of October 22, 2007 establishing a common organization of agricultural markets and specific provisions for certain agricultural products (Single CMO Regulation): OJEU n° L 299, 16 nov. 2007, p. 1. The latter rationalizes the common organizations of the previous market each covering different products or groups of products on the basis of a basic regulation proper thereto, and in particular the Council Regulation (EC) No 797/2004 of April 26, 2004 on measures improving general conditions for the production and marketing of apiculture products (EUYO No. L. 125, 28 April 2004, p.1).

[19]The measures financed by the European Agricultural Fund for Rural Development (EAFRD) are excluded from the apiculture program: In conformity with the Council Regulation (EC) No 1698/2005 of September 20, 2005 on support for rural development by the European Agricultural Fund for Rural Development (EAFRD) (OJEU No. L. 277, 21 oct. 2005, p. 1), a same action may not be the subject of a payment under the framework of the apiculture program, and under the framework of another Community aid regime at the same time. This solution has been confirmed by Regulation (EU) n° 1305/2013 of December 17, 2013 on support for rural development by the European Agricultural Fund for Rural Development (EAFRD) and repealing Council Regulation (EC) n° 1698/2005 (OJEU n° L. 347, 20 déc 2013, p. 487). As indicated in the decision of the Director General of France Agri-Mer of September 13, 2013, with regard to the implementation of the French three-year bee program 2014/2016 "The Community funding allocated to each Member State is determined on the basis of its relative share in the community bee census. In France, this proportion is fixed at 10.42% which generates 3.52 million euros per year the Community co-financing for 7.05 million euros of planned spending in the French program" (v. also D. No. 2013-820 of 12 seven. 2013 related to the national aid program of the beekeeping sector for the 2014 to 2016 receipts: JO 14 sept. 2013, p. 15485).

[20]The Council Directive of July 22, 1974 on the harmonization of the laws of the Member States relating to honey (OJEC No. L. 221, 12 August 1974, p. 10) is only interested in honey, without worrying about the conditions of its production. The following guidelines are heading the same direction (Council Directive N° 2001/110/EC of 20 December 2001 relating to honey: OJEU n°

Resolution of October 9, 2003 on the difficulties faced by the European beekeeping sector, the European Parliament firstly focuses on the fate of insects before worrying about the effects on pollination and biodiversity: "the health of domestic bees is not just a matter of concern to beekeepers; it is also indicative of the state of the environment in general and of pollinating fauna in particular". "This bee mortality is indicative of problems in current beekeeping which are multi-factorial in origin and is concerned at the loss of biodiversity caused by the destruction of insects, in particular pollinating insects". In line with this resolution, the Council Regulation (EC) No 797/2004 of April 26, 2004 on measures improving general conditions for the production and marketing of apiculture products considers that one of the main functions of beekeeping is "the maintenance of ecological balance" and the promotion of the "rationalization of transhumance" within the context of subsidized apiculture programs. The French circular of November 8, 2004, which sets out the conditions for granting Community subsidies to beekeeping, never evokes the interest of pollination within the context of transhumance. Nor is it mentioned within the more general framework of the beekeeping activity. Taking stock of the implementation of measures relating to the beekeeping sector from Council Regulation No 1234/2007 (Single CMO Regulation) which codifies and replaces Council Regulation No 797/2004, the European Commission reminds that "The overall goal of the program is to improve the general conditions for the production and marketing of apiculture products in the Union". However, it insists for the first time on the fact that "in several Member States, transhumance is crucial to meeting the nutritional needs of bees and ensuring the pollination of plants" and that some Member States "support transhumance because of its importance for pollination".[21] In the meantime, in its Resolution of November 15, 2011 on honeybee health and the challenges of the beekeeping sector, the European Parliament went even further in its beekeeping interests analysis: beekeeping "provides an important ecosystem service via pollination, which contributes to the improvement of biodiversity by maintaining the genetic diversity of plants". The Parliament calls on the Commission to consider the possibility of creating a special scheme for assistance to beekeepers within the framework of the direct aid scheme, for example through bee colony payments, which will (...) ensure bees continue to act as pollinators". It particularly notes "a decrease in the number of bee colonies has been reported in both the EU and other parts of the world" and which "in the event of a marked intensification of this trend, farmers in the EU, as well as those in other parts of the world, may have to resort to human-assisted pollination, which would entail a twofold increase in expenditure on pollination". It is then concerned about the fact "that increased mortality among honeybees and wild pollinators in Europe would, if left unchecked, have a profound negative impact on agriculture, food

L. 10, 12 January 2002, p. 47 and Council Directive n° 2014/63/EU of 15 May 2014: OJEU n° L. 164, 3 June 2014, p. 1).

[21]Report from the Commission on the implementation of the measures concerning the apiculture sector of Council Regulation (EC) n° 1234/2007 (COM/2013/0593 final).

production and security, biodiversity, environmental sustainability and ecosystems". He calls on the Commission "to clarify, in the forthcoming reform of the CAP, the support measures and aid to be assigned to the European beekeeping sector, taking account of the environmental and social public goods that honeybee colonies provide via pollination and of the environmental service performed by beekeepers in managing their bee colony".

20.4.5 Agri-Environmental-Climate Measures for Honeybees

The protection of bees and their pollination function is the subject of particular financial subsidies within the framework of the new Common Agricultural Policy (CAP) developed over the period of 2014–2020. However, this "greening" of agricultural payments is nothing new: the first "local agri-environmental operations" defined within the framework of the CAP, though not specifically targetting pollinators, permitted this objective to be met by encouraging the development of plant diversity. It consisted in promoting the "ways of using agricultural land which are compatible with the protection and improvement of the environment, the countryside, the landscape, natural resources, soil and genetic diversity".[22] Nevertheless, the measure focused more on an environmental service than an intervention in favor of a particular eco-systemic service, since it consisted in ensuring that pollinators had access to a healthy and varied "food source".

It has howeverpre-figured in the agricultural policy developed since 2007 on the basis of which the Member States were able to introduce "Agri-environmental measures" (AEM), then "Agri-environmental-climate measures" (AECM) under the aegis of the new CAP, with specific components. France has consequently established a mechanism specifically dedicated to bees, the AEC "EHBPP (API in french)" (Enhanced Honey Bees Pollination Potential). This measure seeks to ensure that beekeepers modify their practices, namely by extending pollination zones in order to "improve the pollination potential of honeybees by modifying beekeeping practices in order to better implement this activity in such a way as to contribute towards biodiversity". "A part of the hives must be placed in areas which are labeled "interesting for biodiversity". As it is reminded when each campaign is launched "The EHBPP measure is a devolved measure, based on national specifications, with regionalized options".[23] For the beekeepers concerned,

[22]Council Regulation (EEC) n° 2078/92 of June 30, 1992 on agricultural production methods compatible with the requirements of the protection of the environment and the maintenance of the countryside: OJEC n° L 215, July 30, 1992, p. 85. This regulation was repealed by Council Regulation (EEC) n° 1257/1999 of May 17, 1999 on support for rural development from the European Agricultural Guidance and Guarantee Fund (EAGGF) and amending and repealing certain Regulations: OJEC n° L.160, 26 June 1999, p. 80.

[23]For each DDT or DDTM, v. "Notice of information Amelioration of the pollinator potential of honeybees for the preservation of the biodiversity (API). Campaign 2015".

the mechanism imposes "an increase in the number of locations being used, with the localization of a minimum proportion" in these areas, the effectiveness of the measure is assured "by minimum distance obligations between two locations, a minimum number of colonies per location and a minimum occupation duration". The specifications, which are the basis of the commitment, impose a doubling of the number of locations per batch of 100 colonies (4 as opposed to 2) "in order to ensure a better territorial coverage by pollinator insects". It also imposes a minimum number of locations in areas of interest to biodiversity (Natura 2000 areas ...): These areas are defined on a regional scale "and represent a minimum of 25% and a maximum of 50% of the regional territory". The mechanism only concerns professional beekeepers as, unless a specific regional measure of derogation is in place, they must engage at least 72 hives (colonies). The minimum distance between two locations must be 1000 meters, except in cases where natural obstacles are present (ridgelines and collars in a mountain area, groves), in which case this distance is reduced to 500 m. There must be at least one location per 24 colonies over 1 year; these locations that can be fixed or transhumant hive locations. Colonies must remain in place for at least 3 weeks on each location. As a reward for compliance with specifications, beekeepers receive a 21 € subsidy per engaged hive which is paid on an annual basis in exchange for a 5 year commitment.[24]

Like directed pollination, this mechanism has to deal with the health rules which determine hive displacement: the beekeeper must namely report "every transportation of bees outside of the department of origin (...) in the days which precede or follow the displacement"[25] to the veterinary services. Other States have adopted even stricter measures: some Swiss cantons oblige hive owners wishing to practice pastoral or floral beekeeping to apply for an authorization until May 1st of each year. This authorization is only granted after checks which namely certify the good health of colonies, and that the area is not sequestered for health reasons[26]. Others, more generally, impose that all displacements of bees from other provinces or regions are subject to prior authorization, "setting out the conditions permitting this displacement."[27]

However, these measures do not characterize a veritable payment for the eco-systemic service provided by domestic pollinators. The beekeeper is supported in his work to follow honeydew or put in place and maintain his hives, whereas the collectivity seeks to benefit from an increase in pollinator potential: It is, to a certain extent, paid "in kind" for the environmental service provided to pollinators, and more generally, to biodiversity.

[24]V. Instruction DGPE/SDPAC/2015-1070 of December 10, 2015 concerning the Technical Statement 2015 Agri-environmental and Climate measures (AECM) and subsidies for biological agriculture, p. 52 s.

[25]Amended order of August 11, 1980 concerning the health mechanisms to combat disease among bees: JONC 1st oct. 1980, p. 8684.

[26]Ruling December 9, 1997 concerning the execution of federal legislation epizootic diseases and the elimination of animal by-products, art. 27 (Canton of Jura – Switzerland).

[27]Ontario (Canada), Bees Act, RSO 1990, c. B.6, art. 12.

20.5 The Environmental Service to Protect Pollinators

The protection of pollinators with the objective of benefitting from their services is the result of an environmental service whose implementation depends on numerous and various tools, but whose main characteristic is to maintain a "healthy" environment. This is particularly true of agri-environmental and climate measures, certain aspects of which help to combat bee mortality. It is a question of maintaining pollinator-friendly spaces as well as developing various forms of fallow land or, more generally, good agricultural and environmental conditions, which may be accompanied by pesticide regulation.

20.5.1 Agri-Environmental-Climate Measures and Beekeeping

In its Resolution of November 15, 2011 on honeybee health and the challenges of the beekeeping sector, the European Parliament (mentioned above) (calls on the Member States "to lay down agri-environmental measures geared towards apiculture in their rural development programs and to encourage farmers to engage in agri-environmental measures supporting 'bee-friendly' grasslands on field margins and to employ an advanced level of integrated production, taking a holistic approach to farming and using biological control where possible". It thus complements a previous recommendation: the Parliament calls on the Commission "to improve coordination of the various research programs carried out in Member States with a view to establishing an action plan for tackling bee mortality; it points out that this should include mainstreaming sustainable, pollinator-friendly farming practices by avoiding monocultures without rotation".[28] According to the recommendation of 2011, these pro-biodiversity actions "are also vital in the non-farm sector: that green spaces along roads, verges of railway lines, forest cuttings for energy transmission networks and public and private gardens cover substantial areas where rational management methods can considerably increase pollen and nectar resources for bees and pollinating insects". The maintaining or recovery of plant diversity is therefore crucial to improve nectar and pollen potential in order to maintain and/or to reconstitute pollinator communities. Various measures have been developed for this purpose, such as rotational agri-environmental measures: the practice of growing different crops in succession on the same land chiefly to preserve the productive capacity of the soil is allowed to alternate the food for the bees (protein crops). The AEC "agri-environmental pasture premium", however, tends "to preserve prairies and encourage an extensive management of these surfaces through respectful practices toward the environment". Its specifications prohibit plowing permanent grassland and impose that the fixed components of biodiversity on the farm are

[28]European Parliament resolution of November 25, 2010 on the situation in the beekeeping sector (P7_TA(2010)0440).

present on at least 20% of the surface (permanent wet grasslands, protection of grazing areas and their withdrawal from production ...) (Saddier 2008; Heidsieck and Allier 2013). Other measures concern lowering the use of pesticides which affect the biotope of pollinators and the pollinators themselves, in order to maintain the areas which are beneficial to them.

20.5.2 Maintaining Favorable Spaces for Pollinators

The survival of wild and domestic pollinators is conditioned by the quality of their environment as well as by the availability of food. Anthropogenic changes in land use lead to the gradual disappearance of bee-forage plants, due to their artificialization, the replacement of plants by crops or the use of herbicides for the upkeep of certain surfaces. Various proposals were developed to ensure the coexistence between the human occupation of soil and pollinators and, more specifically, between agriculture and pollinators. Advocating for "Recovering the necessary biodiversity for beekeeping and beyond that, for all agriculture", the French plan for the sustainable development of beekeeping (Prev.) speculates that "the surfaces said to be of 'ecological interest' (...) would indirectly once again become of 'agricultural interest' thanks to the strengthening of the pollination service". This paves the way for farmers and beekeepers to "find a common interest for an intelligent application of the environmental objectives of the European Union for the CAP reform": "fallow land set aside for apiculture, strips of grass and flowers, intermediate crops with bee-forage plants, floral INTC (intermediate nitrate-trap crops), flowered hedges and slopes and acacia groves, cultivation of fodder legumes, growing methods favoring apiculture (alfalfa flowering), cultivated varieties of interest to apiculture (for example, in the case of sunflowers, consideration of attractiveness criteria for nectar and the quality of pollen in the variety selected)" (Gerster 2012b). The idea is not new, but it presents the particularity of promoting its integration within the Common Agricultural Policy (CAP). This policy will be developed with the new CAP: the eco-conditionality established in 2013[29] maintains the per-hectare subsidy along with a "green" payment (30% of the sum of direct payments) subject to certain agricultural practices being respected. These conditions may include maintaining an "ecologically valuable area" (EVA) "on the agricultural surface" of at least 5% of arable land (on increased to 7% from 2017 for farms that exceed 15 hectares), to which the surface of EVA excluding arable land is added. These surfaces namely concern land lying fallow, terraces, buffer strips, ("including buffer strips covered by permanent grassland, provided that these are distinct from adjacent eligible agricultural area"), areas with catch crops, or green cover

[29]Regulation (EU) n° 1307/2013 of December 17, 2013 establishing rules for direct payments to farmers under support schemes within the framework of the common agricultural policy: OJEU n° L 347, 20 Dec. 2013, p. 60.

established by the planting and germination of seeds and the areas with nitrogen-fixing crops (peas and field beans, lupins, lucerne, soya, clover ...), which offer a multitude of opportunities for pollinators. The new CAP also establishes "greening" equivalences which integrate existing practices with equivalent or higher benefits. We can namely mention the ecological set-aside, "the management of buffer strips and non-cropped field margins (cut regime, varieties of local or specified grasses and/or seeding regime, reseeding with regional varieties, no use of pesticides, no spreading of livestock effluents and/or mineral fertilizers, no irrigation and no soil sealing)" and the "borders, bands and parcels in the managed field for certain types of wild flora or fauna (herbaceous borders, protection of nests, bands of wild flowers, mixture of local seeds and non-harvested crops).[30] Some of these techniques have been developed within a non-agricultural framework (roadside maintenance), while others have been the subject of agreements in parallel to agricultural subsidies, mainly for the farms unable to benefit from these subsidies (lower surface area than the required surface area, area concerned outside of the agricultural surface area ...).[31]

20.5.3 Contrasting Effects of Beekeeping Fallow

Maintaining vegetation during a period which is favorable for flowering can even be the objective of the fallow land developed for apicultural purposes (wild-flower set-aside, bee-flora set-aside). The aim is to create plots composed of different floral mixtures chosen for their apicultural interest, with the spreading of the blossoming period to compensate the impoverishment and banalization of flora and provide food resources to pollinating insects (Decourtye A. and co 2007; Sebillotte 1993 M. and co). Maintaining this co-viability is controversial in the scientific community because "what benefits domestic bees may not necessarily benefit wild species". A high density of honeybees may "discourage other pollinators in the immediate vicinity" and lead to their regression. "However, most wild plants need wild pollinators and whilst the domestic bee Honey bee is a generalist, it does not feed on all plant species. Therefore, if if the domestic bee is overly encouraged, it, increases the risk of reducing wild, cohabitating species, and consequently, wild plants". In addition, using horticultural varieties can initiate competition among plant species

[30]Regulation (EU) n° 1307/2013 of Dec. 17, 2013, appendix IX.

[31]"Parallel", and not necessarily "complementary" as soon as farmers are not eligible to the actions listed in the framework of non-Agricultural – non-forestry. Natura 2000 contracts, and related to the maintenance of open environments by mowing: A mowed plot must be notified within the framework of the common agricultural policy and cannot therefore be the subject of a non-agricultural – non-forestry Natura 2000 contract. In addition, these actions are financed within the framework of the AECM (Instr. DGPAAT/SDDRC/C2012-3047 of April 27, 2012 relative to the contractual management of Natura 2000 sites largely terrestrials in application of articles R. 414-8 to 18 of the Environmental Code, § 3.1.2.2.2).

and lead, in the long term, to a depletion of wild varieties (Gadoum 2007). In other contexts (pollination of a same floral variety), other observers, highlight a functional synergy between pollinator communities once inter-specific interactions can modify the behavior and increase the efficiency of pollination: "In orchards with non-Apis bees, the foraging behavior of honey bees changed and the pollination effectiveness of a single honey bee visit was greater than in orchards where non-Apis bees were absent" (Cl. Brittain and co 2013).

20.5.4 Good Agricultural and Environmental Conditions

The wild-flower set-aside and bee-flora set-aside participate in the "good agricultural and environmental conditions" (GAEC) regime set up pursuant to the Council Regulation n° 1782/2003 of September 29, 2003 establishing common rules for direct support schemes under the common agricultural policy and establishing certain support schemes for farmers.[32] From the time that "the land set aside shall be maintained in good agricultural and environmental condition" and "it shall not be used for agricultural purposes and shall not produce any crop for commercial purposes", farmers may be required to comply with these good conditions set out in Appendix IV of the Regulation. This includes "Protecting soil through appropriate measures", by a "Minimum soil cover" and a "Minimum land management reflecting site-specific conditions" as well as "Ensuring a minimum level of maintenance and avoiding the deterioration of habitats", which involves "Protecting permanent pasture" and the "Retention of landscape features". It is a far cry from the protection of bees, but in its special report 8/2008 "Is cross compliance an effective policy?" the European Court of Auditors notes, "With respect to cross compliance, Member States have considerable room to define the obligations imposed on farmers, especially for GAEC standards. Where cross compliance and agri-environment apply to the same objects (landscape elements, biodiversity, etc.) this means that the level of agri-environmental obligations, and consequently its effects, are determined by the level of cross compliance obligations". (The European Court of Auditors 2008). This latitude is specifically illustrated in the "specific fallows" (wildlife set-aside land, wild-flower set-aside or apicultural set-aside) which, under the rules of good agricultural and environmental conditions, are included in "landscape features".[33] On the other hand, non-cultivated lands (i.e.

[32]OJEU n° L. 270 of oct. 21, 2003, p. 1.

[33]Rural and maritime fishing Code, art. D. 615-50-1; Ministerial order of July 13, 2010 related to rules of good agricultural and environmental conditions (GAEC): JO 17 July 2010, p. 13257, at last amended by Min. order of April 15, 2014 related to the eligibility of some surfaces: JO 20 of April 2014, p. 6974. See also technical instruction of June 3, 2014 with regard to surface payment for 1st pillar of CAP campaign 2014: for the floral fallow, 1 ha is declared equivalent to 1 ha of topographical elements while for the bee fallow, 1 ha is declared equivalent to 2 ha of topographical elements.

the non-producing areas that are beyond the areas benefiting from subsidies for fallow or arable crop area payments for land set aside) may not be the basis of an agricultural activity, which includes the establishment of hives, which is prohibited in this context.[34]

20.5.5 The Pesticides Regulation

However, this environmental service to pollinators can only be effective if it is accompanied by measures to make their environment healthier, through the regulation of the use of phyto-pharmaceuticals (Safety Authority 2013). If the placing on the market of these products is the focus of prior tests on their lethal effects for pollinators,[35] various provisions have also been adopted to promote a reasoned use for the benefit of these pollinators. This is the case for numerous environmental services "paid for" in this capacity and which are, in any case, compensated. Various AECM therefore include provisions on the reduction or the prohibition of the use of chemicals and their financing implicitly integrates their consequences for farmers. Therefore, specifications of the "Agri-environmental grass premium" AECM ban chemical weeding on agricultural plots under contract and impose a non-chemical control of brushwood. However, there is no correlation between a subsidy granted in this regard and the extra cost for the farmer related to the use of other techniques. Territorial farming contracts better corresponded to this approach: according to Regulation No 1257/1999 of May 17, 1999, they provided that the level of subsidies "is adopted for each action or agri-environmental measure, set out in a standard measure, based on the loss of income incurred and additional costs resulting from the agri-environmental commitments and the need to provide a financial incentive beyond the duly justified losses and over-costs. This additional financial incentives may not exceed 20% of the loss of income and over-costs, with the exception of that related to specific commitments duly justified and specified in the description of the measure, for which a higher rate can be set when this is essential for effective implementation of the measure. The calculation of income loss and additional costs resulting from the commitments is based on the level of reference for usual good farming practices in the area where measures or a combination of measures apply".[36] The new sustainable agriculture contracts are inspired from similar data and characterize a contractual regulation of the use of

---

[34] For example, Order n° 164 DDAF of 25 April 2006, setting the rules related to the good functioning of agricultural and environmental lands, and the maintenance of fallows of Côte d'Or departement.

[35] V. Regulation (EU) n° 1107/2009 of 21 October 2009 concerning the placing of plant protection products on the market: OJEU n° L 309, 24 Nov. 2009, p. 1.

[36] Order of 8 November 1999 on aid granted to holders of territorial farming contracts by the financing fund of territorial operating contracts, art. 3: JO 9 Nov. 1999, p. 16684.

pesticides[37]. The incentive, which implies a spontaneous approach, has therefore been preferred to a more binding, non-compensated regulation.

20.6 Conclusion

The anthropogenic approach to pollination clearly illustrates the interdependence of social and ecological systems. If pollinators can do without man, the contrary is far from certain because nature is difficult to imitate. This situation is far from simple: the natural relationship between insects and plants is thwarted by the influence of human activities on ecosystems and imposes the regulation of these activities to enable it to continue. This regulation is not easy because of the complexity of the relationships that the law maintains with biodiversity in general, and with bees in particular.

20.6.1 Thwarted Legal Relationships

These relationships are indeed particularly disturbed by a series of natural factors that require legislation to adapt since it struggles to leave behind its traditional patterns. The property law, which governs relationships between humans and animals, is consequently subject to the peculiarities of "domestic" bees. The beekeeper does not formally own bees, but rather the hive: the insect cannot be domesticated in the classical sense of the term and remains 'wild', as proved when the swarm escapes. The link to 'home' that characterizes domestication is therefore a legal fiction, just like immobilization – that of the hive – which fictitiously absorbs its content – the bees – remains movable mobile. Without the hive, there would be no appropriation; without the swarm, there would be no appropriated bees: given that, as a social insect, the domestic bee has no individual existence. Bees cannot be individualized, because they cannot be naturally individualized – they have no distinctive signs and no possibility of surviving without the colony. The law must therefore be adapted to these natural contingencies and invent a specific legal regime. The same difficulties are faced when tackling pollination, which largely escapes the rules of accession: pollination is not a product of the hive in the traditional sense of the word, but it is produced by the occupants of the hive, insofar as it is possible to be certain of the effective action of the bees in a given in hive. The invention, of the pollination "service" by economists, in order to give a value and, as a result, a price to a natural function, questions the adaptability of the law. There is a great temptation to classify this service within a known category in order to benefit from a tried and tested

[37]Order of October 30, 2003 related to the aid granted to holders of contracts for sustainable agriculture, art. 2: JO 7 Nov. 2003, p. 19043.

regime plan, such as accession, the fruit of the labors of the appropriated bees. However, if the work of the possessor is the basis for their ownership of fruits, the beekeeper cannot claim rights on what results from the labor of bees. He can only claim rights on the labor itself, that of the bees whose labor he exploits, and his own labor, linked to maintaining his colonies and the handling of his hives. This industry certainly has a heritage value, but it is not a good that can be owned.

20.6.2 A Necessary Coviability

The initial autonomy of pollination is followed by a necessary mutualism, supported by a strong anthropization: pollination now has a price and several costs. It has become a 'service', which gives rise to commercial exchanges: the price of making available insects due to the costs incurred for this purpose; the price of the provision of food resources for pollinators by the allocation of agricultural surface areas for purposes other than agricultural production, and the implementation of alternative techniques, with uncertain productivity levels. These costs must be seen in relative terms and must be related to the costs of the disappearance of these insects. These costs are individual for farmers whose crops depend on pollinators, when they need to offset the service deficit by a massive importation of pollinators; they these costs become collective when of the consequences of the reduction or even disappearance of the service could affect an entire entire human community. Paradoxically, the collectivity becomes dependent on farmers whose activities and growing methods account for a significant proportion of Colony Collapse Disorder. Communities have to help farmers to maintain bee colonies and their services, while the damages caused by their production activities to a collective service could have justified obliging them to meet the costs of their protection instead of turning it into gain (Weber 2012).

The pollination service illustrates how the economy has appropriated nature. Henceforth, this appropriation imposes a value on to a biological function, which until now has always been free of charge. We have since become aware of the cost of its possible disappearance. The premise on which payments for eco-systemic services are based is rather sobering and serves to reinforce the reasons to protect suppliers. At any cost.

La Rindyà, July 2016 – July 2017.

References

Abrol D-P (2011) Pollination biology: biodiversity conservation and agricultural production. Springer, New York, pp 25–26

Agricorp (2016) Fruit crops insuring agreement – part. III, (ed: Agricrop). Ontario

Angel M (1998) La nature a-t-elle un prix? Critique de l'évaluation monétaire des biens environnementaux. Les Presses de l'Ecole des mines de, Paris

Bastide P (1993) Production of cocoa hybrid seeds in the Cote-d'Ivoire. Various hand pollination trials. Cafe Cacao The. 37(4):295–302, annexe

Beaunier S (1806) Traité-pratique sur l'éducation des abeilles. Vendôme, p 18

Braat L, Ten Brink P (eds) (2008) The cost of policy inaction, the case of not meeting the 2010 biodiversity target. Alterra, Bruxelles/Wageningen

Brittain Cl and co (2013) Synergistic effects of non-Apis bees and honey bees for pollination services. Proc R Soc B 280:20122767

Bruneau E, Burget M (2008) Le dépérissement en Europe et aux Etats-Unis, des approches différentes. Abeilles et Cie 127:25

Chagnon M (2008) Causes et effets du déclin mondial des pollinisateurs et les moyens d'y remédier, Fédération canadienne de la faune (publisher). Bureau régional du Québec, Québec, p 5 s

Chevassus-au-Louis B (dir.) (2009) Approche économique de la biodiversité et des services liés aux écosystèmes. La documentation française, Paris p 237 s

Committee on the Status of Pollinators in North America (2007) Status of pollinators in North America, Board on Life Sciences, Board on Agriculture and Natural Resources, Division on Earth and Life Studies, National Research Council. National Academies Press, Washington, DC

Darwin C (1862) On the various contrivances by which British and foreign orchids are fertilised by insects, and on the good effects of intercrossing, 2nd edn. William Clowes and sons Limited, Londres. p 267

Decourtye A. and co (2007) Introduction de jachères florales en zones de grandes cultures: comment mieux concilier agriculture, biodiversité et apiculture? Courrier de l'environnement de l'INRA 54:33

Deffairi M (2015) La patrimonialisation en droit de l'environnement. IRJS Editions, Paris, T 61

Deguines N. and co (2014) Large-scale trade-off between agricultural intensification and crop pollination services. Front Ecol Environ. https://doi.org/10.1890/130054

Downing J (2007) On achève bien les abeilles. Le Courrier International n° 851, Paris

European Court of Auditors (2008) "Is cross compliance an effective policy?", report 8/2008, p 25

European Food Safety Authority (2013) EFSA guidance document on the risk assessment of plant protection products on bees (Apis mellifera, Bombus spp. and solitary bees). FSA J 11(7):3295, 268 p

Fanica PO (2006) Un exemple d'apiculture transhumante: le gâtinais, in P.-Y. Laffont (dir.), Transhumance et estivage en Occident. Des origines aux enjeux actuels, PU du Mirail, coll. Flaran, p 327 s

Farrar CL (1929) Bees & apple pollination. Massachusetts Cooperative Extension Service, Amherst

Farrar CL (1947) More honey from bees, Yearbook of agriculture, United States Department of Agriculture, Washington, p 680

Fournial E (1976) Les investissements urbains dans l'élevage: les baux à cheptel dans le Forez médiéval. Etudes foréziennes VIII:39

Gadar C-H (an VIII), Le nouveau Ferrière, ou Dictionnaire de droit et de pratique, Paris, T. 1, rubr. "Abeilles", p 6

Gadoum S. and co (2007) Jachères apicoles et jachères fleuries: la biodiversité au menu de quelles abeilles? Courrier de l'environnement de l'INRA 54:57

Gallai N. and co. (2009) Economic valuation of the vulnerability of world agriculture confronted with pollinator decline. Ecol Econ 68(3):810 s

Gallai N, Vaissière BE (2009) Guidelines for the economic valuation of pollination services at a national scale. Pollination Services for Sustainable Agriculture, FAO, Rome

Gerster Fr (2012a) Plan de développement durable de l'apiculture, rapport CGAAER n° 11 174-01, p 22

Gerster Fr (2012b) Plan de développement durable de l'apiculture, rapport CGAAER n° 11 174-01, p 6

Government of Ireland (2008) The economic and social aspects of biodiversity. Benefits and costs of biodiversity in Ireland. Government of Ireland, Dublin, p 39 s

Guingand A. and co (2014) Les paiements pour préservation des services écosystémiques comme outil de conservation de la biodiversité. Cadres conceptuels et défis opérationnels pour l'action, Les cahiers de Biodiv' 2050, févr 2014, p 2

Harissis V, Harissis AV (2009) Apiculture in the prehistoric Aegean. Minoan and Mycenaean symbols revisited. British Archaeological Reports, Oxford

Heidsieck H, Allier F (2013) Propositions pour une prise en compte des pollinisateurs dans les politiques agricoles nationales et européennes. ITSAP - Institut de l'Abeille, Cahier technique, Paris de

International Pollinators Initiative (1999) The Sao Paulo declaration on pollinators. Report on the recommendations of the workshop on the conservation and sustainable use of pollinators in agriculture with emphasis on bee

IPBES (2016) Summary for policymakers of the assessment report of the Intergovernmental Science-Policy Platform on Biodiversity and Ecosystem Services on pollinators, pollination and food production, p 3

Lemeunier G (2006) Aux origines de l'apiculture rationnelle: la transhumance des ruches (France et Espagne, V. 1750-V. 1850) in P.-Y. Laffont (dir.), Transhumance et estivage en Occident. Des origines aux enjeux actuels, PU du Mirail, coll. Flaran, p 263 s

MacGregor S-E (1976) Insect pollination of cultivated crop plants, agriculture handbook 496. U.S. Department of Agriculture, Agricultural Research Service, Washington

Maris V (2014) Nature à vendre. Les limites des services écosystémiques (ed: Versailles Q)

Mayrand K, Paquin M (2014) Le paiement pour les services environnementaux: Etude et évaluation des systèmes actuels, Unisféra International Centre, Montréal sept., p ii

MEDDTL (Ministère de l'écologie, du développement durable, des transports et du logement) (2009) Evaluation des services rendus par les écosystèmes en France: Etude exploratoire. Synthése, éd. MEDDTL, Paris

Michallet I (2013) L'existence des insectes pollinisateurs dans le droit, in V. Gratpain et I. Michallet, Protection des organismes pollinisateurs et des populations face aux incidences sanitaires et écologiques, Rapport EnvitéRA, p 84 s

Morse RA, Flottum K (eds) (1998) Honey bee pests, predators, and diseases, 3rd edn. Wicwas Press, Cheshire

Ost F (2003) La nature hors la loi. L'écologie à l'épreuve du droit, éd. La découverte, p 325

Parliament of Canada (2014) Proceedings of the standing senate committee on agriculture and forestry, Issue 8 – Evidence – Meeting of April 3, 2014

Pliny, Natural History, with an english translation in ten volumes, by H. Rackham, London, William Heineman Ltd, 1960, L. XVI, Chp. XXXIX-XL p 449

Prétot X (2011) L'essaim infidèle, le maître des ruches et le seigneur justicier. La propriété des abeilles dans l'ancien droit. Revue semestrielle de droit animalier 2:247 s

Pundyk G (2010) The honey trail. In pursuit of liquid gold and vanishing bees. St. Martin's Press – Macmillan Publishers, New-York

Ransome HM (1937) The sacred bee in ancient times and folklore. George Allen & Unwin, London

Ruhl J-B et al (2007) The law and policy of ecosystem services. Island, Washington DC

Saddier M (2008) Pour une filière apicole durable. Les abeilles et les pollinisateurs sauvages, Rapport au Premier ministre, 64 p

Sagili RR, Burgett DM (2011) Evaluating honey bee colonies for pollination: a guide for commercial growers and beekeepers. Pacific Northwest Extension Publication, Oregon State University, Corvallis

Sébillotte M. and co. (1993) La jachère et ses fonctions agronomiques, économiques et environnementales. Diagnostic actuel. Le Courrier de l'Environnement de l'INRA 20:11–22

Skinner JA (n.d.) Making a pollination contract, The University of Tennessee, Agricultural Extension Service, BP1516

The Economics of Ecosystems and Biodiversity (TEEB) for Agriculture and Food (2014). Concept note, 27 febr 2014, p 10

UNEP (2010a) Emerging issues: global honey bee colony disorder and other threats to insect pollinators. UNEP, Nairobi

UNEP (2010b) Conservation and management of pollinators for sustainable agriculture, through an ecosystem approach. UNEP, Nairobi

United States Department of Agriculture (2013) Report on the National Stakeholders Conference on honey bee health

Washington State Department of Revenue (2013) Tax exemptions for honey beekeepers: special notice, 18 nov

Weber J (2012) Biodiversité et production de richesse, une érosion inéluctable, in J.-Fr. Toussaint et al. (dir.) L'homme peut-il s'adapter à lui-même?: Marges d'adaptation de l'espèce humaine face aux changements environnementaux, ed. Quae, p 150 s

Chapter 21
Can International and French Environmental Law Accommodate Coviability?

Aline Treillard and Jessica Makowiak

Contents

Since the 1992 Rio conference, the protection of the environment has been crystallized around the concept of sustainable development which aims to ensure a balanced development by taking into account social, environmental and economic stakes. Initially, the issue covered by this concept is to give future generations the similar legacy to the one we have.[1] From the outset, the environmentalist doctrine has criticized the content of such a concept, highlighting the absence of a cultural pillar. However, sustainable development has become the major reference for many regulations in international environmental law (Convention on Biological Diversity, Convention of the United Nations on Climate Change, Kyoto Protocol). This wide consecration hides another reality. The inherent philosophy of this principle is far from being shared by the entire international community. The concept of

[1]The concept of sustainable development was used for the first time in the Meadows report of 1972. It was spread on an international scale by the Brundtland Commission's Report "Our Common Future" in 1986.

A. Treillard (✉) · J. Makowiak
CRIDEAU-OMIJ (EA 3177), Université de Limoges, Limoges, France
e-mail: aline-treillard@hotmail.fr; jessica.makowiak@unilim.fr

O. Barrière et al. (eds.), *Coviability of Social and Ecological Systems: Reconnecting Mankind to the Biosphere in an Era of Global Change*,
https://doi.org/10.1007/978-3-319-78497-7_21

sustainable development broadcasts an approach of a purely Western nature, in other words a very anthropocentric approach. Anthropocentrism is defined as "the idea according to which the interests, goods and human values, are at the heart of all moral evaluations of environmental policies" (Callicot 1984, 299) and that "these interests, goods and values are the basis of any justification of environmental ethics" (Callicot 1984). He sets the purpose as being the satisfaction of human needs. Its main features are the superiority of the human being over nature, the immediate usefulness, the market economy, and self-regulation (Apostolidis 1997). Sustainable development fully meets these characteristics insofar as it advocates development before anything else. Besides, for Serge Latouche, sustainable development is above all an example of verbal diplomacy that consists in changing the words in the face of our incapacity to change things (Latouche 1995). This philosophy of Northern Hemisphere countries is in total contradiction with the ethical conceptions of non-Westernized countries. Indeed, the countries of the Southern Hemisphere, Latin America for example, support a much more holistic view of the world. In this respect, we must differentiate between two major perceptions. The most radical is the biocentric perception. This is characterized by the attribution of a moral value to the individuals of the living world (Schweitzer and Trueblood 1970). Legally, biocentric ethics leads to the attribution of legal traits to the elements of nature. Nonetheless, the biocentric current suffers from the same objections as the anthropocentric current insofar as it is also based on an individualistic approach of living creatures, humans and non-humans (Frias 2006, 61). The second ethical aspect is the ecocentric one. It offers a median vision according to which any form of life deserves to be respected. This perception implies that nature conservation must be a moral imperative and not a vital necessity. The World Charter for Nature, 1982, perfectly illustrates the content of this approach.[2] Therefore, the formulated criticism against the concept of sustainable development, which determines the vision of environmental law, exceeds its only content. This is an ethical conflict. Some authors do not even hesitate to consider sustainable development a slogan that "allows, in the imagination, two opposing aspirations to be met: the indefinite continuation of a form of economy based on domination and the destruction of nature as well as the preservation of a healthy environment" (Lavieille 2010, 46).

Regulating activities harmful to the environment is no longer enough to protect it, the threats weighing on mankind are multiplying. Accordingly, the right to the environment is currently used by a part of the doctrine to drive the idea that building a new relationship between human and nature is needed. This movement is part of an approach similar to that of anthropological research that promotes the paradigm of coviability. Coviability is defined as "the balance between the needs of societies for the reproduction and the maintenance of ecological, biotic and abiotic processes, and their components", in other words "the union of cultures (social systems) and the natural world (ecosystems)" (Barrière 2008, 220). Therefore, can environmental

[2]World Charter for Nature adopted by the General Assembly of the United Nations, Resolution 37/7, 48th plenary meeting, 28 October 1982. In particular see the preamble.

law integrate coviability? The theoretical preconditions for receiving coviability seem to be gradually fulfilled (1), even if embracing it by French environmental law (2) is still unquestioning but limited.

21.1 The Theoretical Conditions for Receiving Coviability into Environmental Law

Receiving coviability into environmental law requires the recognition of the interdependency of humans and nature. We propose to introduce the ethical prerequisites of interdependence (Sect. 21.1.1) as well as its manifestations in international law, comparative law and the internal law of the environment (Sect. 21.1.2). Lastly, we wish to highlight the analogy between many concepts of environmental law and coviability (Sect. 21.1.3).

21.1.1 A Prior Ethic: The Recognition of the Moral Unity of the Living World

As it seeks to restructure the legal relationship of humans and non-humans, coviability requires a special relationship with ethical law (Sect. 21.1.1.1), an interpenetration which will allow the law to seize the central notion of intrinsic value (Sect. 21.1.1.2).

21.1.1.1 The Ecocentric Approach of Law

Each society determines its fundamental rules of ethics by culture, religion (etc.) and formalizes them into its legislation. As a branch of philosophy which scientifically studies the foundations of good and evil, ethics and law cross paths despite having totally different features, the first being a flexible regulatory instrument whereas the latter is a rigid one. The global environmental crisis that exploded from the 1970s onwards and which finds particular resonance during global threats, is the source of many philosophical questions which have "led to the search for new ethics, a global redefinition of our relationships with nature" (Larrere 1997, 14). The ecocentric approach briefly described in the introduction contains this idea of balance assuming that human beings are members of nature and not at the top of it. It contains in essence the idea of harmony and excludes the utilitarian criterion (Berros 2013). Two currents are founded around this approach. The first, *the ethics of Earth* is supported by Aldo Leopold (Leopold 2013). According to him, the Earth is formed by a biotic community characterized by the interdependence of its individuals. Applied to our societies, this ethic assumes that human beings take their responsibilities in regards to other members of the community.[3] This current is distinguished from the second one which radicalizes this egalitarian theory. "*Deep ecology*" has the particularity of including non-living natural elements in the moral field. (Naess 2005). Therefore, basing the theory on the biological equality criterion (i.e. the idea according to which any form of life has an intrinsic value) and the self-realization (which is, for its part, the guarantee by man that every member of an ecosystem will have the ability to develop according to its own needs), this current advocates attributing a moral consideration to natural objects such as forest, rivers and mountains.

By attributing a value to any form of life, ecocentric ethics recognize by essence the intrinsic value of nature, the necessary consecutive step for environmental law to receive coviability.

21.1.1.2 The Recognition of Intrinsic Value

The recognition by environmental law of the intrinsic value of nature, means admitting that it has an objective quality, something it would possess on its own (Larrere 1997, 23). This concerns the recognition that "any form of life is unique and deserves to be respected, regardless of its usefulness to humans".[4] However, "all modern thinking considers that moral qualities are subjective qualities, given by a subject, not properties of the object" (Larrere 1997, 23). This premise is in conflict with another concept of nature, according to which it would only have one instrumental value. According to Kant, the foundations of this dominant value are to be found in the distinction between the purposes and the means, between people and things (Kant 1985). According to his theory, natural objects are things because they "are devoid of reasons" (Kant 1985, 302). Only people who "should be considered as means in themselves" (Kant 1985, 302) can have intrinsic value, the rest of living beings are excluded from the circuit of moral consideration. Intrinsic value is by principle absent from dominant Western ethics that distinguish the "person-subject"-and the "nature-object"; nonetheless, this is strongly revealed in the ecocentric ethics (Frias 2006). In this regard, it is conceived as the "best way to contradict the utilitarian point of view in terms of ecology" (Benoist (de) A, 1993, 128).

Finally, the ecocentric approach of the law suggests "an ethic of living together" (Mathevet et al. 2011, 426). This concerns requiring that the responsibility of human beings is reinforced, when they intervene in biotic communities.

[3]This current is closer to the theory supported by Hans Jonas and Courtine-Denamy (2000) and Hans Jonas (2013).

[4]Preamble of the World Charter for Nature, 1982.

21.1.2 The Legal Demonstrations of the Ecocentric Perception of Law

The main demonstration of the ecocentric perception of law is analyzed in the evolution of the legal status of nature (Sect. 21.1.2.1). This movement, which disrupts Western concepts, is especially supported by the countries of South America, and non-exhaustively by New Zealand, India and Philippines (Boyd 2012). In French law, the main change is characterized by the emergence of the ecological solidarity principle (Sect. 21.1.2.2).

21.1.2.1 The Change in the Legal Status of Nature

Ecocentric ethics have the effect of repositioning living beings in the center of concerns. Legally, the normative center of gravity moves from Man to all living beings. This legal shift is particularly manifested in the status of nature which, at the beginning of twenty-first century, underwent profound changes by becoming subject to law Latin America States are forerunners in this area. Without using the precise term of coviability, the law applicable to nature and the duties imposed on natural and legal persons coincides with the essence of this paradigm. To illustrate our point, we will take the change in the legal status of nature in Ecuadorian and Bolivian law, as an example. Ecuadorian law undoubtedly reflects the most significant development. The Ecuadorian constitution, approved by referendum on September 29, 2008, contains a chapter, numbered 7, dedicated to the rights of nature. This chapter contains four articles of innovative content.[5] Article 71 gives rights to nature, rights which are rolled out in the following articles. This particularly concerns the right to fully respect its existence, the right to the maintenance and the regeneration of its vital cycles and the right to restoration. In order to ensure the respect of these rights, the Ecuadorian constituent has implemented legal obligations towards the State, individuals, peoples and communities. These include the duty to defend the "Pacha mama" (nature) by exercising litigation appeal on behalf of nature, in their capacity as human representatives. As to the Bolivian law, it was strongly influenced by article 7 of the Cochabomba Declaration which stipulates that "Mother Earth has the right to live and to exist, the right to regenerate her biocapacity and to continue her vital cycles and processes, free of human alteration".[6] Subsequently, the legislator intervened by adopting two laws that guarantee the right to life, the right to diversity, the right to water, the right to clean air, the right to equilibrium, the right to restoration and the right to live

[5]Articles 71 to 74 of the Ecuadorian Constitution, approved by referendum on September 29, 2008.

[6]Universal Declaration on the Rights of Mother Earth, World Conference of Nations against Climate Change, Cochabamba, 19-22 April 2010.

without pollution.[7] The Ecuadorian constitutional law and the Bolivian law are both based on the concept of good living (*buen vivir*) and the concept of *sumak kawsay* which makes a point of honor to equally mix human development and natural development (Haidar and Berros 2015). Without distorting the content of these founding concepts, this construction of law, as introduced, can be considered a construction of a legal system of coviability.

However, before concluding on this point, it is should be noted that these legal developments are at the border of ecocentric and biocentric concepts of nature, in which the limit is inevitably porous. In the West, and particularly in French law, this change is manifested by the recent construction of the ecological solidarity principle.

21.1.2.2 The Principle of Ecological Solidarity

For the first time, the principle of ecological solidarity was introduced into French law during the law reform on the National Parks Act in 2006.[8] This marks a qualitative step in the recognition of the interdependence of living beings connected to a same community with the same fate. As a result, the literature and environmentalist doctrine rushed to propose a definition for the principle. The most detailed one considers that ecological solidarity is "the close interdependence of living beings, among themselves and with natural areas or landscapes of geographical space" (Mathevet et al. 2011, 421). This definition presupposes that man is the main topic of the definition at the same level as all other living creatures. Two types of solidarity are distinguished by these authors. On the one hand, an ecological solidarity which in fact "emphasizes the community of fate between man, society and the environment" (Mathevet et al. 2011, 424) and on the other hand, an ecological solidarity of action reflecting a willingness to live together based on the recognition by human beings that they are included in this community. This type of solidarity requires a positive action from human beings, a helping hand to non-human creatures. In such a way "it allows the establishment of a pragmatic compromise between ecocentrism and anthropocentrism" (Mathevet 2012, 76).[9] To date, this definition is supplanted by the one adopted in the law for winning back biodiversity, nature and landscapes.[10] The legislator would like to include in article L110-1 of the French Environmental Code (the list of founding principles of environmental law), a 6° stating thus: "the principle of ecological solidarity, which

[7]Ley Derechos of Madre Tierra No. 071 of 21 December 2010 and Ley Marco of the Madre Tierra y Desarollo Integral para el Vivir Bien no. 300 of October 15, 2012.

[8]Law No. 2006-436 of 14 April 2006 related to national parks, marine parks and regional parks.

[9]For some authors, this provides "ecocentric ethics of leopoldien filiation" (Mathevet et al. 2011, 424).

[10]Law No. 2016-1087 of 08 august 2016 for the winning back of biodiversity, nature and landscapes.

requires taking into account the interactions of ecosystems, of living creatures and natural or developed environments, in all public decision-making with significant impacts on the environment of the territories directly or indirectly concerned". The principle of ecological solidarity as defined by the legislator proposes a new interpretation in time and space of environmental law. As a result, this consists in basing environmental law and more particularly the right to the protection of nature on spatial criteria able to go beyond the idea of a border. In this sense, it underlies a non-fragmented vision of time and space. Furthermore, by imposing the consideration of ecological solidarity in public decisions affecting the environment, the legislator connects this principle to those of information and the participation of the public and thus assures social solidarity in an indirect way. This connects individuals, the public, groups and public institutions. This assumes that local society feels "concerned by the fate of non-humans and/or aware of the interdependencies and complementarities" (Mathevet 2012, 89). Therefore, this reflects an ecocentric perception of law and contributes to the paradigm of coviability.

The entanglement between coviability and the ecological solidarity principle defended either by the environmentalist doctrine or by the legislature seems to have been proved. These analogies between the content of environmental law and coviability are multiplying.

21.1.3 The Existence of Similar Legal Concepts

The concepts of harmony (Sect. 21.1.3.1), of environmental and human security (Sect. 21.1.3.2), and the new legal category of non-humans supported by the associations for the protection of animals all around the world (Sect. 21.1.3.3) also show that environmental law may initially receive the principle of coviability.

21.1.3.1 The Concept of Harmony with Nature

The principle of coviability induced by essence the notion of sustainability. Currently, the ecological crisis can only be resolved through the questioning of individualism and irrational productivism and through the redefinition of our relationship with nature and other human beings. The technical and the economic measures cannot provide a viable answer on their own. Viability requires peaceful, sustainable and territorialized lifestyles. Combined with an ecocentric approach, this position recalls the recent interventions of the UN General Assembly on the concept of harmony. The Rio Declaration initiated the process by adopting its first principle: "human beings are at the center of concerns for sustainable development. They are entitled to a healthy and productive life *in harmony with nature*". Criticized quite rightly for its ambiguity, this principle, however, has the merit of having initiated discussions in international forums. However, it was necessary to wait until 2009 for the General Assembly of the United Nations addressed the issue. Three resolutions have been taken at this level.[11] The main report highlights the strategies and sustainable development initiatives promoting life in harmony with nature, while providing recommendations. For this purpose, a historical approach is proposed, highlighting the fact that "the most important lesson we can draw from the wisdom of sacred traditions is perhaps the most straightforward: honoring creation while respecting the relationship that unites us with nature." This statement perfectly explains how ancient civilizations perceived the relationship between the human body and Mother Earth and how, for them, being in harmony with nature guaranteed good health".[12] Since then the international community has shown interest in indigenous peoples who have a special relationship with the land and nature. In terms of legal demonstrations, this concept of harmony clearly highlights the interdependence between human rights, biodiversity rights and cultural rights. This also reminds us that the Western model of ownership is not universal. Such awareness of the existence of viable alternative models at the environmental, social and economic scale, creates new life for environmental law, especially in its ecocentric filiation. Under the influence of this new thinking, the international community meeting in Rio in June 2012 declared that it was convinced that "to achieve a fair balance between economic, social and environmental needs of present and future generations, it is necessary to promote harmony with nature".[13] This promotion is guaranteed by environment law through education on the environment, the principle of information and participation.

21.1.3.2 Human and Environmental Safety

The viability of the relationship between men on the one hand, and the nature on the other, inevitably requires a spirit of mutual trust, responsibility towards each other, and justice in order to put an end to the vulnerabilities that weaken the bonds of solidarity. The construction of the paradigm of coviability is born from observing the existence of a threat to humanity and the biosphere. This is the same for environmental law which, now subjected to the speed of global changes, is trying to identify alternatives by the mobilization of the security concept. Firstly,

[11] Resolution 64/196 Harmony with Nature, adopted by the UN General Assembly on December 21, 2009, 54th session; Resolution 65/164 Harmony with Nature, adopted by the UN General Assembly on December 20, 2010, 65th session. Resolution 66/204 Harmony with Nature, adopted by the UN General Assembly on December 22, 2011, 66th session.

[12] Report of the UN Secretary General A/65/314, 65th session, general distribution, on August 19, 2010, paragraph 23.

[13] "The future that we want", paragraph 39 of the "statement" Rio + 20. For indigenous peoples, see mainly article 8j of the Convention on Biological Diversity, which is extended by the Nagoya Protocol.

in respect of environmental security (Barnett 1997),[14] the right to the environment intervenes in the aim of ensuring sustainable use of natural resources, protection of natural elements and the maximum reduction of damage caused to the environment at the same time.[15] Secondly, as part of human safety, environmental law strives to promote a trans-temporal and trans-generational dimension. This dimension is prevalent thanks to the reference in multiple international conventions, in the expression "future generations".[16] Coviability can be analyzed through its ability to stand against vulnerabilities. The institutionalization of coviability can find resonance in the struggle led by the environmentalist doctrine that proposes to renew the distribution of power in international forums, particularly in a future World Organization of the Environment in order to guarantee a fair position for most threatened States. Legal visibility is a constant struggle for the most vulnerable, human and non-human. Environment law actively pursues such a path.

21.1.3.3 The Emergence of a Non-human Category

The terminology of "non-human" appeared in legal vocabulary as a result of the actions of environmental protection associations that use this name before courts to justify an extended legal protection of certain animals. For the most radical doctrine, the aim is to obtain from judges the recognition of fundamental rights of non-humans.[17] This movement is particularly important since 2010 and is spreading on an international scale. The idea discreetly gained consistency in India in 2013 when the Environment Ministry proposed to recognize in dolphins the quality of non-humans having, from that moment on, specific rights. Then, it was the turn of the South American continent which had been the subject of attention. After the first failure in the Supreme Court of New York State,[18] the Non-Human Rights Project Association managed to obtain from the Federal Chamber of Appeal of

[14]The Environmental safety is defined as "the pro-active minimization of anthropogenic threats on the functional integrity of the biosphere and its symbiotic human component" (Barnnet Barnett 1997).

[15]This objective is found particularly in recent international environmental law agreements: the Convention on Biological Diversity entered into act on 29 December 1993, the Protocol of Nagoya on access to genetic resources and the fair and equitable sharing of benefits derived from their use related to the convention on biological diversity, entered into act on 12/10/2014, the UN Convention on climate change entered into act on 21 March 1994.

[16]Cf., for example, the preamble to the Convention on Biological Diversity and the UN Convention on climate change.

[17]The starting point of this movement is attributed to Christopher D. STONE (1972) Should Trees have standing? Toward legal rights for natural object, Southern California Law Review, pp. 148-157. Peter SINGER has continued in this sense, P. SINGER (1993) the animal liberation, B. GRASSET, 382 p.

[18]State of New York Supreme Court, December 4, 2014, Tommy c / Patrick C. Lavery, n ° 518,336, Olivier LE BOT comment, "Pas d'habeas corpus pour un chimpanzé"," bi-annual animal law review, 2014, no.2, p.131.

Argentina, recognition of a non-human person for a female orangutan as well as her right to freedom.[19] For the authors of the appeal, this was not about extending the full range of human rights to animals, but only a limited number such as the right to life, the right to physical integrity, the right to security or freedom (the Bot 2010). Several statements follow these lines, including the Universal Declaration of Human Rights filed by the Lepage Commission to the President of the Republic, Francois Hollande, within the framework of the COP21, which states in article 5 that "humanity, like all living species, has the right to live in a healthy and ecologically sustainable environment".[20] Nevertheless, the difficulties are crystallized on the legal capacity of this emerging category. The dissenting doctrine seems to disregard the legal developments in this respect, and many lawyers recall with determination that the institutional capacity, on the model of the legal capacity of legal entities, is an interesting and applicable solution. Once again, the example of the Ecuadorian law is relevant to the subject. An appeal has been filed by a private owner on behalf of and exclusively forthe River Vilcabomba-Quinara in respect to article 71 of the Constitution[21] in order to defend the river which flows through his plot of land =.[22]

It is yet to be seen whether or not the (positive) law of the environment promotes the integration of such a notion, beyond the preconditions theoretically allowing coviability to be mainstreamed.

21.2 The Obstacles to Coviability Reception in French Environmental Law

If we can detect, particularly in international and comparative law, a change or – at least – a variety of legal concerns for the environment (ecocentric approach, emergence of concepts such as ecological solidarity), it is no doubt too early to embrace a new principle of coviability. This is what a review of French environmental law revealed. The law is based on foundations rejecting, in theory, social-ecological interdependence, reflecting instead a relative disconnection from man and his socio-ecosystem (Sect. 21.2.1). However, there are legal instruments to understand interdependence (Sect. 21.2.2), that some emerging principles could support (Sect. 21.2.3).

[19]Federal Chamber of Appeal of Argentina "Orangutana Sandra / recurso casacion S/HABEAS CORPUS" of December 18, 2014, no. CCC 68831/2014 / CFC1.

[20]See, Ms. Corinne LEPAGE and editorial team, "Universal Declaration of Human Rights", final report delivered on Friday, September 25, 2015, 133 pages.

[21]Article 71 line 2 of the Ecuadorian Constitution states that (our translation): "any person, community, people or nationality can require from public authority the fulfillment of the rights of nature". Paragraph 3 states: "The State will stimulate people and communities so that they protect nature".

[22]Court of Justice of Loja Province (Ecuador), March 30, 2011 Rio Vilcabomba against Governor of the Province.

21.2.1 *Foundations Apparently Less Favorable to Embracing Coviability*

Coviability calls for "the union of cultural (social systems) and natural worlds (ecosystems), in other words a reconciliation between man and the biosphere" (Barrière 2008, 584). Since the environment means in essence "the surrounding", its right should foster the understanding of the interdependence of its constitutive components (understood as including humans). Yet, we can consider that environmental law is first built materially as a law of nature, the latter being conceived in an exteriorized and objectified way. This distance from nature (whose design is necessarily anthropocentric) can be seen from article L. 110 – 1 of the French Environmental Code (originating from the law on the protection of nature of 1976) which defines as part of the common heritage of the nation "the open areas, resources and natural areas, sites, diurnal and nocturnal landscapes, air quality, living beings and biodiversity". Immediately we perceive the difficulty of defining the environment (necessarily changing, dynamic and dependent on cultural determinations) and, thereby, the scope of its law. Such a definition is not the reflection of the interdependencies, the legislature adopting a "restrictive" and "historic" environmental concept (Naim-Gesbert 2014, 6). It leads, as expressed so well by Jean-François Malherbe, in man being considered as both "inherent and dehiscent with regard to the biosphere. He detaches himself as a chestnut does from its bark and falls to the ground (...)" (Malherbe 2004, 590).

Furthermore, it appears at first glance that French environmental law has adopted a relatively compartmentalized approach of the objects which it encompasses. While coviability expresses a relationship or even a regulation (Barrière 2013, 220),[23] French environmental law makes a distinction between the right to resources (water, air), open areas (essentially protected and little developed), and "natural heritage" (originally understood as including species, and then gradually extended to habitat). This partitioning is obviously accentuated by the codification which took place in 2000[24] and which makes it immediately visible. However, it is conceivable that this corresponds to the historical rationale of the construction of French environmental law, and not only to a formal presentation of its various components. Indeed, the first time in the history of nature protection, that we can also see at an international level, is marked by the protection of species (wild fauna and flora), then by the environment and habitats, before the recent emergence of the willingness to understand ecological continuity (Bonnin 2008a, b). A sectorial approach in the prevention of pollution, risks and nuisances, i.e. perception of human activities separately from the protection of nature, accentuating the dichotomy of human-nature, is added to this relative compartmentalization. Once again, it is the historical

[23]The sustainability of resources depending on the viability of the community and the environment.

[24]Order No. 2000-914 of 18 September 2000 related to the legislative part of the environment code, JO Sept. 21.

factors (related to concerns in hygiene at the beginning of the nineteenth century) which encouraged such compartmentalization of approaches, that the book 5 of the French environment Code just sort to reproduce.

Does this mean that environment law is a "distorted fiction" (Naim-Gesbert 2014, 5), a fiction devoid of true adherence? We do not believe so. As stressed by E. Naim-Gesbert, the links between the two original cornerstones (nature and pollution) are becoming tighter so as to include a human dimension (through human health and development), and are witnessing a "progressive stretching of environment law" (Naim-Gesbert 2014, 6). This human dimension is found especially in the formulation of the human right to the environment, defined as the right to live in an environment that is "balanced and healthy" (article 1 of the French Charter for the Environment which is a part of French constitutional law). Above all, the gradual recognition of a social-ecological interdependence offsets the first impression of partitioning. In this respect, we affirm that "the resources and natural balances have conditioned the emergence of humanity" and that "the future and the very existence of humanity are inseparable from its natural environment", the 2005 preamble of the environment Charter is particularly revealing. From the point of view of its foundations, French environmental law is also based on a principle of prevention, whose legal instruments of implementation today allow the interdependencies to be understood. Therefore, by reforming environmental impact assessment (the main instrument of prevention), the Grenelle II Act of July 12, 2010 is no longer content in just addressing the initial state of areas and environments likely to be affected by a project as a "collection" of environmental elements (fauna, flora, natural habitats, cultural heritage . . .), but requires that "the interrelations between these elements"[25] s is analyzed. Analysis of project effects on the environment obeys the same rationale; it has to include not only the "natural" components of the environment (water, ecological continuities, species, etc.), (cultural, archaeological) heritage but also, and above all, "the addition and the interaction" of items between each other.[26] If coviability is supposed to consider the overlapping between the level of resources and the quality of the environment then an environmental impact assessment allows – at least in theory – to legally translate such a requirement.

Furthermore, the areas covered by environmental law must include human environments. On this point, we will identify, here or there, what falls under an attempt of reconciliation of interests rather than a principle of coviability. Therefore, in order, for example, to stop the loss of biodiversity, the "green and blue belts" established by the French Grenelle law, participate in the conservation, management and rehabilitation of areas necessary for ecological continuity, "while taking into account human, and particularly agricultural activities in rural areas".[27] Always in the field of biodiversity, the implementation of the European ecological network Natura 2000 proceeds along a similar rationale. If the designation of sites by

[25] Article R. 122 – 5 – II – 2 ° of the Environmental Code.

[26] Addition of the Grenelle Act pre-mentioned, v. art. A. 122-5-II-(3) of the Environmental Code.

[27] Article L. 371 – 1 of the Environmental Code.

the Member States of the European Union is based solely on scientific criteria (ecological),[28] the conservation measures of sites and the appropriate preventive actions "take into account the economic, social, cultural and defense requirements" as well as "regional and local specificities".[29] Moreover, these measures are defined "in consultation, particularly (. . .) with representatives of property owners, operators and users of lands and open spaces" included in the Natura 2000 site. The National Park Charter includes, in addition to a general section recalling the fundamental principles, a section specific to each Park, with measures "established from its territorial, ecological, economic, social or cultural specificities".[30] As to the national strategy for the sea and coastline, it still relies on the delimitation of maritime fronts, "defined by the hydrological, oceanographic, bio-geographical, socio-economic and cultural characteristics of the areas concerned".[31]

In the fight against pollution, the standard of "specific local circumstances" also tempers the abstraction of the legal rule, which stems directly from the scientific dimension of environmental law. The adaptation of the general rule to local circumstances can be seen as a factor reintroducing anthropological considerations, an essential element to coviability. This applies to the enactment of the general rules and technical requirements applicable to facilities classified under the "Classified Facilities for Environment Protectional"[32] regulations, which can be adapted to local circumstances by a prefectural authorization Order,[33] while assuring the protection of extremely various environmental interests: dangers for neighborhoods, health, safety, public health, agriculture, nature, environment and landscape protection, conservation of sites and monuments or even elements of the archaeological heritage . . .[34] Substantially rewritten by the law on the energy transition, the national waste policy includes a proximity principle that consists of "assuring the prevention and the management of waste in a area as close as possible to its production" and especially – going further than a simple adaptation to local circumstances – which allows to " environmental stakes to be met while contributing to the development of professional businesses, local and sustainable".[35] Social and ecological systems seem, thus, more nested.

---

[28]See in this regard the constant jurisprudence of the European Union Court of Justice, including CJCE, July 11, 1996, Regina c / Secretary of State for the Environment "Lappel Bank", aff. C 44/95, Rec. CJCE I-3805.

[29]Article L. 414 – 1 – V ° of the Environmental Code.

[30]Article L. 331 – 3 of the Environmental Code.

[31]Article L. 219 – 1 of the Environmental Code.

[32]The Classified Facilities for Environment Protectional is an administrative regime. It refers to any depot, yard, workshop and facility in general operated or owned by a public or private individual or corporate entity, involving hazards or inconvenience for the neighbourhood, health, safety, agriculture, or the protection of nature and the environment.

[33]Article L. 512 – 5 of the Environmental Code.

[34]Article L. 511 – 1 of the Environmental Code, which lists the various interests protected by the law of July 19, 1976 on the installations classified for the protection of the environment.

[35]Article L. 541 – 1 of the Environmental Code.

The rationale of environmental law seems to be evolving towards greater consideration of interactions between human activities, resources and the environment. But these instruments must be able to translate (and set) the interdependency. There is a qualitative leap that lies well beyond the reconciliation of environmental, social and economic interests.

21.2.2 Legal Instruments to Understand and Grasp Interdependency

As part of this contribution, it is not possible to exhaustively consider all the instruments which, in environmental law, allow more or less effectively understand the interdependency of socio-ecological stakes. However, we can focus thinking on the first notion of coviability, and on the ability of our normative system to "maintain, by means of its processes (...), its viability in varied and changing environments" (Bourgine 1996). Therefore, the analysis will be focused on the adaptability of legal rules, and on new interdependencies.

First, it is important to clarify that the order of validity of law is based on the existence of both spatial and temporal limits. Being in need of stability, law is not generally adapted to what is happening. Also the requirement of viability, which goes beyond the notion of durability, necessarily questions the traditional instruments of environmental law, certainly acclimatized in the long term, but not necessarily to "shifting reality" (Naim-Gesbert 2012, 300).

The thought of the legal rule adaptability (that applies social-ecological interdependence) will be considered from a double point of view, time and space. In the first case, this is about rethinking the content of the legal standard, which must understand and grasp the movement (Sect. 21.2.2.1). In the second, this is more about the form of the legal standard, which must understand and grasp the change (Sect. 21.2.2.2). In one case as in the other, it is the viability and the interdependency of human activities, species and environments that are at stake.

21.2.2.1 Understanding and Grasping the Movement by a New Material Approach of the Legal Norm

It is probably the domain of biodiversity protection, in a new context of climate change, which best illustrates the necessary adaptation of existing legal instruments. Today, we know that species conservation policies have been conducted according to a fixed model, based essentially on the conservation of a state of a species compared to a previous state (central criterion of the "Habitats" Directive of May 21, 1992, for example) or on the nativeness criterion of a species (as if the distribution ranges are frozen). The restoration of a species was, then, to find a past situation (by correcting the damage inflicted onto the species' habitats or by limiting the pressure on them).

Legal instruments are part of this real myth of a stable environment (Rochard and Lassale 2010). However in the context of climate change, areas such as parks and reserves which are today interesting for the conservation of biodiversity, will not necessarily be so tomorrow. Therefore, there is a real paradigm change, the adaptation of some species to climate change particularly dependents on their ability to disperse. Henceforth, this "third phase "in nature conservation is well-known (Bonnin 2008a, b), and is based on the protection of ecological continuity, translated into French law by the institution of the "green and blue belts".

Moreover, already in 2008, some French authors suggested the legislative recognition of a connectivity protection "principle", aiming at ensuring the movement of species or an even larger functionality principle for the protection of ecosystems functionalities (Bonnin 2008a, b, 167). In this respect, the law for winning back biodiversity in France[36] amends article L. 110 – 1 of the Environmental Code by adding, for open spaces, resources (...) and species, not only the need for management, but also the need to preserve their "ability to develop" and "safeguard the services they provide".

Nevertheless, we must add that this conservation strategy by the creation of corridors (belts) is sometimes scientifically questioned, given the weak capacity of dispersion of some species (Conseil de l'Europe 2007). As opposed to landscape policies, it would not allow all the interactions between the networks and the scopes to be conceived either (Luginbühl 2009). Therefore, in addition to the protection of ecological continuity would it be suitable, for example, to create new protected areas where species are likely to move i.e. at higher latitudes and altitudes (Halpin 1997; Scott and Suffling 2000)?. For environmental law, this means "adapting to the adaptation" (Naim-Gesbert 2012, 301), also assuming procedures which contribute to the changing the current limits of protected areas. Indeed, such limits are usually mapped and endorsed by a legislative or regulatory act and thus frozen in time and space. Some authors suggest that these limits, like those allowed for hunting or fishing, can fluctuate through appropriate legal means in order to adapt them to the change in the distribution of fauna and Flora (Godet 2015). This idea of fluctuating limits allowing the withdrawal or rather the addition of zones, however seductive it is, should, nevertheless, be in agreement with other imperatives, such as the requirements of legal safety.

21.2.2.2 Understanding and Grasping Change Through a New Formal Approach of the Legal Standard

From a temporal point of view and in a context of interdependence, legal instruments must also adapt to the urgency and to change. Currently marked by a certain normative rigidity (cumbersomeness of procedures for creating protected areas in

[36]Law No. 2016-1087 of 08 august 2016 for the winning back of biodiversity, nature and landscapes., article 2.

particular), nature conservation law should provide for regular adaptation of the instruments created (zoning, easements), requiring a follow-up of their effects and possible opportunities of adaptation. For example, we could imagine a hierarchy of procedures based on the stakes, following the example of what is already practiced in planning law (simplified change, common law, or review procedure). Here again, such a development must be reconciled mainly with the requirements of effective participation of the public.

It is precisely in the principles of the environmental law that coviability, quite implicitly, seemed to find resonance, during the reformulation of these principles by the French law on winning back biodiversity, nature and landscapes.

21.2.3 Emerging Principles in Support of Coviability?

Insofar as French environmental law prefers the first notion of nature that of, henceforth, biodiversity, it is not surprising that the legislature wishes to give a definition for this in the law which would be dedicated to this. In the recent law on wining back biodiversity, nature and landscapes, this definition dedicates the social-ecological interdependence, while understanding biodiversity or biological diversity as "the set of living organisms as well as the *interactions* which exist, on the one hand, between the organisms themselves, and on the other hand, between these organisms, their natural habitats and their living environments". The text on this point has clearly evolved, since its initial drafting, modelled on the definition of the 1992 Rio Convention, ignored the term interaction.[37]

Still regarding principles, and in a similar rationale of interdependency, the French law on winning back biodiversity has added a new "principle of complementarity" between the environment, agriculture and forestry, according to which "agricultural and forestry activities can be vectors of ecosystem *interactions* ensuring, on the one hand, the conservation of ecological continuity, and on the other hand, environmental services that use ecological functions of an ecosystem in order to restore, maintain or create biodiversity".[38] Fortunately, this new principle was inserted in the Rural and Sea Fishing Code, and not within the General principles of the Environmental Code, as initially contemplated. In fact, if the role of some human activities in the management of biodiversity is undeniable, such recognition would not meet the definition of "general principle" (in environmental law), because it is not likely to lead to particular legal effects.

---

[37]Original text (translated): "We mean by biodiversity, or biological diversity, the variability of living organisms from all sources including terrestrial, marine and other aquatic ecosystems, and the ecological complexes of which they are part. It includes the diversity within species and between species, as well as those of ecosystems". The first reading of the Bill adopted by the National Assembly and transmitted to the Senate on March 25, 2015, article 1er.

[38]Article 2 -I-8°.

Finally, supporting the notion of coviability, ecological solidarity would make its entrance into the General principles of environmental law. We have already shown that such recognition was part of law's ecocentric perception.[39] However, it is still necessary to make it clear that as a "connection that requires this from us" (Mathevet 2012), the ecological solidarity principle puts into perspective the issue of environmental liability. In fact, it involves an extended perception of it, understood in a more preventative sense than a curative one. This is about taking into account, before any public decision, "the interactions of ecosystems, living creatures and natural or developed environments". This concept of responsibility coincides, by the way, with that of the European Union, embodied in a 2004 directive, and which concerns both prevention of and compensation for, environmental damage.[40] Likely to guide a judge in assessing damages, the notion of ecological solidarity has the advantage of understanding all the stakes related to the use of an ecosystem: i.e., the possible damage suffered by the (natural or developed) areas, to be distinguished from damage suffered by those protecting or using the ecosystem (which may be material or moral). Besides, it is now in the same law on winning back biodiversity, that a long-awaited change in the French Civil Code,[41] is envisaged, aiming at introducing accountability for environmental damage, stating that "anyone responsible for abnormal damage to the environment must compensate it" (art. 1386-19). The recognition of such responsibility is in continuity, as we know, with the case law setting a legal precedent about the sinking of the Erika, sanctioning the notion of ecological damage.[42]

21.3 Conclusion

Is it not precisely on the side of the judges that the paradigm of coviability can be sought? From 2012, the French Supreme Court spoke in these terms: "the habit in simplifying the premises of an argument to make it easier, led to consider man, disconnected from his natural environment, disregarding the constant interaction of man with nature and forgetting that nature is part of man, just as he is part of it; (it) stems from this interdependence that any significant breach of the natural environment is an attack on the community of men who live in interaction with it and that this attack must be compensated (...)".[43] From such interdependence

[39]V. supra I, B) 2°.

[40]Directive 2004/35/EC of 21 April 2004, JO L 143 of April, 30th 2004.

[41]V. Draft of reform of the law of obligations and the prescription, P. Catala (dir): doc. fr. 2006; Retailleau Bill proposition may 2012; L. Neyret and G.J. Martin (dir.), Nomenclature of environmental damage, LGDJ, 2012; Report Jegouzo "for the repair of environmental damage", September 2013.

[42]V. Court of Appeal, crim. Sept. 25th, 2012, no. 10 – 82.938, published in the bulletin.

[43]Abovementioned decision.

necessarily stems the existence of prejudice "objective, autonomous", which refers to "any significant damage to the natural environment namely: air, atmosphere, water, soil, land, landscape, natural sites, biodiversity and the interaction between these elements in particular, that is, without repercussion on a particular human interest but affecting a legitimate community interest".

At an international scale, it is under the particular prism of human rights that another coviability figure emerges. A petition is currently pending before the Inter-American Human Rights Commission on behalf of the Athabaskan people, affected by the "black carbon" emissions from Canada (Canal-Forgues and Perruso 2015). For applicants, the failure to act by Canada violates their rights to culture, property, livelihood and health, rights in the American Declaration of Human Rights and Duties of 1948. In a very enlightening way for our purpose here, the request is based on the close link between native peoples and their environment, on international law as well as on case-law developed by the Inter-American Court of Human Rights.

The latter has indeed considered that it was imperative to "prevent environmental damage in the traditional collective indigenous territories, given that they depend on their physical survival, their social organization and their cultural integrity" (Canal-Forgues and Perruso 2015).[44] Required along with such case law developments is a new form of ecological solidarity, this link which further "requires this from us". These developments also contribute in specifying the constituting elements, or at least the outlines of what could be, in law, coviability: the recognition of the intrinsic value of nature, the sustainability and the interdependence of socio-systems and ecosystems. In other words, a notion placing the living beings in the center of concerns.

References

Apostolidis C (1997) La protection juridique de l'humanité, l'homme face à la mondialisation, droit des peuples et environnement, L'Harmattan, Paris, p 229

Barnett J (1997) Environmental security: now what?, Séminaire au département de Relations Internationales de la Keele University

Barrière O (2008) In: Lee C, Schaaf T (eds) Legal aspects of the coviability of social and ecological systems in African arid zones: an anthropological approach to environmental law, The future of drylands. Springer-Verlag & Man and the biosphere series, Unesco Publishing, Dordrecht, pp 583–597

Barrière O (2013) Repenser le droit de l'environnement dans une conception renouvelée du développement durable. Prospective d'un "droit de la coviabilité" des systèmes sociaux et écologiques, in S. Blaise, C. David et V. David, (ed.) Le développement durable en Océanie, vers une nouvelle éthique? Presses Universitaires de Provence

Benoist (de) A (1993) La nature et sa valeur intrinsèque, Revue Krisis, pp 113 à 126

[44]IDH Court, 31 August 2001, series C, no. 79, Mayagna community (Sumo) Awas Tingni c/Nicaragua, merits, reparations and costs. – Court IDH, June 27, 2012, series C, no. 245, Sarayaku Kichwa c / Ecuador, merits, reparations and costs).

Berros V (2013) El estatuto juridico de la naturaleza en debate (meulen en el mundo del derecho), Revista de derecho ambiental, n°36

Bonnin M (2008a) Les corridors écologiques. Vers un troisième temps du droit de la conservation de la nature?, L'Harmattan, Paris, p 586

Bonnin M (2008b) Prospective juridique sur la connectivité écologique, Revue Juridique de l'Environnement, n° spécial, pp 167–178

Bourgine P (1996) Modèles d'agents autonomes et leurs interactions co-évolutionnistes. In: Rialle V, Fisette D (eds) Penser l'esprit: des Sciences de la Cognition à une Philosophie Cognitive. Presses universitaires de Grenoble, Grenoble

Boyd DR (2012) The environmental rights revolution: a global study of constitutions, human rights, and the environment. UBC Press, Vancouver, p 443

Callicot B (1984) Non anthropocentric value theory and environmental ethics, Am Philos Q, n°21, pp 299–309

Canal-Forgues VE, Perruso C (2015) La lutte contre le changement climatique en tant qu'objet juridique identifié?, Énergie – Environnement – Infrastructures n° 8–9

Conseil de l'Europe (2007) Conserver la diversité biologique européenne dans le contexte du changement climatique, p 60

Frias LE (2006) Responsabilidad y sostenabilidad ecologica, una etica para la vida. Universidad Autonoma de Barcelona Facultad de Filosofia y Letras, Barcelone, p 321

Godet L (2015) La conservation de la biodiversité dans un contexte de changements globaux, in R. Mathevet et L. Godet (dir.), Pour une géographie de la conservation. Natures, biodiversités et sociétés. L'Harmattan, Paris, pp 113–128

Haidar V, Berros V (2015) Entre el sumak kawsay y "la vida en con la naturaleza": disputas en al circulation y traduccion de perspectivas respecto de la regulacion de la cuestion ecologica en el espacio global, Revista Theomai Estudios Criticos sobre sociedad y desarollo, Buenos Aires, n°32, pp 128–150

Halpin PN (1997) Global climate change and natural-area protection: management responses and research directions, Ecol Appl, vol. 7, n° 3 828 843

Jonas H (2013) Le principe responsabilité: une éthique pour la civilisation technologique, traduit par Jean Greisch. Flammarion, Paris, p 468

Jonas H, Courtine-Denamy S (2000) Une éthique pour la nature. Desclée de Brouwer, Paris, p 159

Kant E (1985) Fondements de la métaphysique des mœurs, trad. Franç., Œuveres Philosophiques, t. II, Gallimard, "Bibliothèque de la Pléiade", p 210

Larrere C (1997) Les philosophies de l'environnement. PUF, Paris, p 124

Latouche S (1995) La mégamachine, Editions la Découverte, Série Bibliothèque du mouvement antiutilitariste dans les sciences sociales, p 243

Lavieille JM (2010) Droit International de l'Environnement, 3ème édition. Editions Ellipses, Paris, p 368

Le Bot O (2010) Des droits fondamentaux pour les animaux: une idée saugrenue? RSDA 1/2010, et suivantes p 131

Leopold A (2013) A sand County almanac: & other writings on ecology and conservation. Etats-Unis, New York, p 931

Luginbühl Y (2009) Biodiversité, changement climatique et paysage, in Humanité et biodiversité, Manifeste pour une nouvelle alliance. Descartes & Cie, Paris, pp 212–239

Malherbe JF (2004) L'humain, l'animal et l'environnement: prospectives et défis futurs, éthiques et juridiques, in L'être humain, l'animal et l'environnement: dimensions éthiques et juridiques (dir. Leroux et Létourneau), Thémis, p 595

mathevet R (2012) La solidarité écologique, ce lien qui nous oblige. Acte Sud, Arles, p 205

Mathevet R, Thompson J, Delanoe O, Cheylan M, Gil-Fourrier C Bonnin M (2011) La solidarité écologique: un nouveau concept pour une gestion intégrée des parcs nationaux et des territoires. Nature Sciences Sociétés, volume 18

Naess A (2005) The selected work of Arne Naess. Springer, Oxford, p 688

Naim-Gesbert E (2012) Biodiversité et changements climatiques: la méthode et le discours. Revue Juridique de l'Environnement, 2-2012, pp 295–303

Naim-Gesbert E (2014) Droit général de l'environnement, 2ème éd. LexisNexis, New York, p 269
Neyret L, Martin G (2012) Nomenclature des préjudices environnementaux. LGDJ, Paris, p 456
Rochard VE, Lassalle G (2010) Conservation de la biodiversité et changement climatique: un nécessaire changement de paradigme. Le cas des poissons migrateurs amphibiens, Sciences Eaux et Territoires, n° 3
Schweitzer A, Trueblood E (1970) Reverence for life, London, Royaume-Uni, S.P.C.K, p 154
Scott D, Suffling R (2000) Climate change and Canada's National Park System: a screening level Assesment, Toronto, Adaptation and Impacts Research Group, Environment Canada
Singer P (1993) La libération animale, B. Grasset, p 382
Stone CD (1972) Should trees have standing ? Toward legal rights for natural object. Southern Calif Law Rev 45(2):148–157

Part III
The Paradigm of Coviability, a Future Challenge

Preview 3: The Urgency in Reconnecting

The first three parts of the volume lay down the foundations of a thinking that contributes to the definition of coviability through disciplinary or interdisciplinary approaches and multiple and varied experiences. The important number of researchers contributing to this undertaking opens the perspectives of a new paradigm for a humanity in regeneration.

The following four chapters outline the challenge to be dealt with by societies to reconnect with the biosphere. This link of humans to non-humans, and consequently of humans to each other, is situated between fiction and reality and at different scales, both global and local. It depends largely on the ethics of an 'obligation of solidarity' between systems for a joint viability, supported by a moral and a universal consciousness of global socio-ecological coviability.

The relationship to the other, the latter being non-human, is expressed by the 'co' of coviability. We thus enter into a relational complexity that is perceived differently according to socio-cultural patterns. The association generated by the 'co' reconstructs a link defined by a solidarity for an 'environmental peace', which not only transcends the consumerist aspect of resource use, but also the contemporary violent aspect of access to resources.

Chapter 22: Climate Change, a Catalyst for Coviability and for a New Utopia
Chapter 23: Approaching the Human-Environment Nexus Beyond Conflict: A Peace and Coviability Perspective
Chapter 24: The link to the biosphere: humanity condemned to otherness and coviability for its existence
Chapter 25: Tracking the origin of Western's man-biosphere disconnection, opening a view to a change

The last two chapters of this part present a first summary of the research works presented in the four parts of the book (two volumes). It highlights both the meaning and determinants of coviability and attempts to forge a definition of this emerging paradigm.

Chapter 26**:** Transversal Ontology Analysis: What Coviability Means
Chapter 27**:** Coviability as a Scientific Paradigm for the Ecological Transition, from an Overview to a Definition

Chapter 22
Climate Change, a Catalyst for a New Utopia Towards Coviability

Anne Coudrain

Contents

22.1 2015, a Tipping Point Year

After 1950, major Earth System changes became largely related to a global economic system. Human population has tripled, but the global economy and material consumption have grown many times faster (Steffen et al. 2011). The speeding up of numerous indicators such as fertilizer consumption, telephones and motor vehicles, among others, is known as the Great Acceleration.

In the twenty-first century, we face scarcity in critical resources, the degradation of ecosystem services, and the erosion of the planet's capability to absorb our

A. Coudrain (✉)
IRD - French National Research Institute for Sustainable Development, UMR ESPACE-DEV, Montpellier, France
e-mail: anne.coudrain@ird.fr

O. Barrière et al. (eds.), *Coviability of Social and Ecological Systems: Reconnecting Mankind to the Biosphere in an Era of Global Change*,
https://doi.org/10.1007/978-3-319-78497-7_22

waste (Steffen et al. 2011). According to calculations in the Living Planet Report (WWF 2014), it would take almost four planets to provide all humans with a North American lifestyle.

Climate change (CC), in this context, has a peculiar place. CC played a role of awareness-raising and warning since the 1980s thanks to the work of the Intergovernmental Panel on Climate Change (IPCC) and to the mobilization of scientific communities. CC reveals and catalyzes the link between the human and the nonhuman, whatever the scale from small territory to the planet one. "Most greenhouse gases accumulate over time and mix globally, and emissions by any agent (e.g., individual, community, company, country) affect other agents. Effective mitigation will not be achieved if individual agents advance their own interests independently." (IPCC 2014, p 17).

The present section tries to explore how 2015 is a tipping point and what the cultural and social consequences of a new paradigm such as coviability might be.

Coviability is a concept of sustained life of human societies with and in ecological systems, integrating respect for others, whether human or non-human. It is coherent with the terms of the report by Lepage et al. (2015) entitled "Universal Declaration of Rights of Humankind" where science is grasped by the unprecedented imperative to preserve the future of humanity and nature in an interdependent relationship, report that will be submitted in 2016 to the United Nations General Assembly.

2015 was a watershed year for the climate movement. Moreover, it concentrated a large number of world events related to the true nature of the behaviour and activities of human beings with regard to development and to the environment. State parties, Non-State parties, scientific, business and faith communities, and the United Nations proclaimed, on a world scale, their support in profoundly changing our references.

The State parties met in November–December for the 21st Session of the Conference of the Parties to the United Nations Framework Convention on Climate Change (UNFCCC) for a universal climate agreement (COP21 2015).

The Non-State Actor Zone for Climate Action (NAZCA 2014) launched at the COP20 in December 2014 recorded in October 2015 more than 4400 commitments to address CC by companies, cities, subnational regions, and investors.

The Business & Climate Summit (2015), lasting 2 days in May 2015, brought together 2000 international business leaders, policymakers and investors in Paris. In the Summit's press release, business calls upon policymakers, among others, to introduce carefully designed, robust and predictable carbon pricing and to eliminate fossil fuel subsidies.

The scientific community mobilized in July 2015 in the 4-day scientific conference "Our Common Future Under Climate Change" that gathered 2200 persons in Paris (CFCC15 2015). The outcome statement is clear "CC is a defining challenge of the 21st century. Its causes are deeply embedded in the ways we produce and use energy, grow food, manage landscapes and consume more than we need. Its effects have the potential to impact every region on Earth, every ecosystem, and many aspects of the human endeavour. Its solutions require a bold commitment to our common future."

The Roman Catholic Church issued in June 2015 an encyclical letter "Laudato Si" on care for our common home (François 2015) that places most of the blame on fossil fuels and human activity, while warning of an "unprecedented destruction of ecosystems, with serious consequence for all of us" if corrective action is not taken swiftly.

The Conference of Cult leaders in France brings together the bodies responsible for Buddhism, Islam and Judaism and Catholic churches, Orthodox and Protestant. Their common call (CRCF 2015) in favour of the adoption of a binding agreement at the COP21 is also a common voice on the climate crisis, described as a spiritual and moral challenge. For them, "solidarity and the common good must be our compass".

Islamic leaders have called on the world's 1.6 billion Muslims to play an active role in combating CC and have urged governments to conclude an effective universal CC agreement in Paris at the end of the year at an International Islamic CC Symposium in Istanbul (Islamic Declaration on Global Climate Change 2015).

A last but not least fact illustrating the deep evolution is the 2030 Agenda for Sustainable Development adopted by the United Nations General Assembly (2015). Instead of addressing the dimensions of sustainable development separately (as previously addressed) the Sustainable Development Goals (SDGs) are expected to adopt an approach that fully integrates the social, economic and environmental dimensions of sustainable development.

Conjunction of such events force the question: are we facing a tipping point from "business as usual" to "coviability – human and non-human"? This paper attempts to review some studies and reflections on the root causes and possible consequences of this eventual forthcoming coviability upheaval.

22.2 Climate Change, Initial Global Awareness and a Catalyst for Coviability

The obvious key root of the undergoing upheaval is CC whose term first appeared in the media and in conversations in the 1980s. In the IPCC's last report (2014), the first highlight is that the "Human influence on the climate system is clear, and recent anthropogenic emissions of green-house gases are the highest in history. Recent climate changes have had widespread impacts on human and natural systems."

Everyone is invited to consider his part of responsibility on CC. CC is a trigger for raising awareness of a common future to be considered together.

Scientific communities stress the urgency in limiting greenhouse gas emissions and changing our consumption behaviour and way of living. To maintain greenhouse gas (GHG) concentrations in the atmosphere at about 450 ppm CO_2-eq or lower (IPCC 2014) implies that the total cumulative emissions of CO_2 since 1870 is kept below 3000 Giga tons (Gt). With 2000 Gt already emitted from 1870 to 2010, the window remaining is 1000 Gt. Taking into account the annual global emissions from fossil fuel and industry, which amounted to 36 Gt in 2014 (Global carbon project 2015; Jouzel and Debroise 2014, p 184), the window will end before 2050.

The scale of the challenge lies here: achieving a balance of emissions reaching zero by around 2050.

The Great Acceleration is an expression that summarizes the increasing rates of world changes in human activity since around 1950 (Steffen et al. 2011). It allows sketching how climate change is intrinsically connected to biodiversity loss, and, among others, to ocean acidification that affects living beings.

Humans are presently a global geophysical force of the Earth System (ES) and the term of Anthropocene has emerged in scientific papers to define a new geological era during which the human impacts on the ES become important (Steffen et al. 2011).

Meadows et al. (1972) already clearly documented the limits of growth. In continuity, the Planetary Boundary (PB) framework, introduced in 2009, aims to define the environmental limits, including CC, within which humanity could safely operate. It is based on the intrinsic biophysical processes that regulate the stability of the Earth system and may be substantially altered with respect to the Holocene-like state by human activities (Steffen et al. 2015).

Of the original nine proposed boundaries, four are already transgressed and thus create a substantial risk of destabilizing the Holocene state of the ES: climate change, biosphere integrity, biogeochemical flows and land system change. Three should be considered at risk: freshwater use, ocean acidification and stratospheric ozone depletion. Two boundaries cannot yet be quantified: aerosol loading and novel entities (chemicals and other new types of engineered materials or organisms).

This extension to several PBs strengthens the alert and the urgent need for a new paradigm that integrates the continued development of human societies and the maintenance of the ES in a resilient and accommodating state.

Dirzo et al. (2014) present how Anthropocene defaunation is both a pervasive component of the planet's sixth extinction and also a major driver of global ecological change. For McCallum (2015) the magnitude of vertebrate extinction has exploded since 1980 and it is several thousand times faster than during the Cretaceous mass extinction: it is blatantly obvious that biodiversity losses must be reined in.

Facing these challenges is still possible, through reducing consumption and limiting waste. All of this will need to involve aggressive moves to decarbonise economies worldwide using new technologies, appropriate financial instruments, and changes to patterns of consumption and energy use (Pidgeon 2012).

Humanity has to engage in a new way of life immediately, taking into account consequences of its own activities on the environment and the rest of the biosphere. A subsequent question is: how are such a commitment and upheaval possible? What makes people want to act in such a way?

The most difficult step is the first one: sharing coping or in other words changing references from "business as usual and addiction to material consumption" to "coviability", human beings taking into account the other – human and non-human inhabitants of Earth.

22.3 Obstacles to Coviability

Despite the general consensus on CC, despite the serious threat of acidification of ocean, the loss of biodiversity and land grabbing and degradation, despite the claims from world actors such as the IPCC, United Nations, World bank and others, despite more than 25 years since the first powerful alerts, despite the narrowing of the window for taking action, despite the symbolic threshold of 400 ppm of carbon dioxide in the atmosphere reached globally by 2015, despite the occurrence of impacts on all continents and the oceans, greenhouse gas emissions still continue to rise. According to the Global carbon project (2015), these amounted to 22 $GtCO_2$ in 1990 and 35.9 $GtCO_2$/year in 2014. Despite the IPCC (2014) report and the scientific community's (CFCC15 2015) demand to urgently lower greenhouse gas emissions, these continue to increase according to the IPCC scenario in which atmospheric CO_2 concentration could reach more than 1000 ppm in 2100.

Facing this acknowledgement of failure, Oreskes and Conway (2013), in their fiction book "The Collapse of Western Civilization" imagine an external observer from the future saying: "the most startling aspect of this story is just how much these people knew, yet how little they acted upon what they knew." They broaden the subject in the register of emotion.

Why has so little action has been engaged thereby still allowing emissions to increase; what are the actual obstacles?

WBGU (2014) in their contribution to the SDG Debate identify three obstacles on the road to an international agreement on ambitious and universal SDGs: Resistance to restrictions, 'Unholy alliance' against goals that go beyond direct eradication, and Political barriers to global cooperation. WBGU argues in detail that development and environmental protection must be considered together and do not contradict each other.

Aykut and Dahan (2015), in their reference book on 20 years of international negotiations on climate, analyse what lay at the root cause of the major failure in Copenhagen (2009) and the durable regression of the environmental issues on the international political agenda. For these authors there was a growing gap between the reality of the world and the sphere of negotiations – they call it "schism shift of reality". The clash of events such as the US wars in Iraq and the financial crisis of 2008, in a context of market globalization and economic competition of states, led to the growth of unbridled exploitation of fossil energy resources. The sphere of negotiations and governance which pictures a large central regulator capable of defining and distributing emission allowances is less engaged with this reality.

Here, we extend this exploration of the obstacles to the roots of "willing/not willing to act" behaviour or "desire/no desire to commit" behaviour. The register is more social or psychological and we consider: (Sect. 22.3.1) present references of priority and (Sect. 22.3.2) sceptics and "merchants of doubt".

22.3.1 Present References of Priority Impede Moving Forward

Billaudot (2006) explores the question of justification that directs the fulfilment of human activities and concludes on the necessity of coherence between these justifications at individual and social levels. Later, Billaudot (2011) defends the theory that to achieve sustainable development, we must change the mode of justification of human practices in reference to the philosophy of Enlightenment as prevailed for two centuries in the "modern societies". These societies emerged in Europe and proposed their model to countries throughout the world. They only take into account humans and consider that nature is only a social backdrop. For this author, during the first modern period, their exclusive justification is based on the "priority of fairness" (fairness between humans). This explains why the modern society over the last two centuries has not been worried by the fact that the actual development was not sustainable.

The naive concept of progress as a triumph of enlightenment on a mythical past was criticized by Adorno and Horkheimer, who wrote their book during Second World War. A sentence summarizes the heart of the criticism "animism had endowed things with souls, industrialism makes souls into things" (Belaval 1990; Schmid Noerr 2002). Undertaking the conquest of instrumental reason produces its opposite; humans became alienated to what they produce.

According to Latour (2015c, p 248), when facing ecological change, most of us remain cold, indifferent, and cynical, as if, deep down, nothing could happen to us. For this author, one can of course appeal to the inertia of habits, fear of, the heady profits of consumption in the iron cage of capitalism; one can point the finger at the influence of lobbies actively engaged in misinformation; or take into account the work of psychologists on fear that paralyzes instead of provoking reactions. All this is probably true but, for this author, these reasons cannot explain why we quibble when the threats are overwhelming.

Exploring the emotional components of the lack of commitments, as illustrated in the pedagogical cartoon by Squarzoni (2012), Latour (2015a, b) describes Western civilizations fleeing archaism for centuries in a movement of emancipation. During this escape, science and myth progressed along different paths – or so it seemed. Admiring the magnificent remnants of numerous civilisations that are now lost forever, we may now look at them in an entirely new way as our own civilisation is in danger of being lost and because, like them, we are facing the unknown. Hence, the root of tergiversation preventing action in the face of climate change is not the lack of certainty (never has the degree of certainty has been so strong). Instead, humans are ill-equipped emotionally, intellectually, morally, politically and culturally. The root is in the stupor and in the inability to act in the face of what Western civilizations have fled for centuries.

Another image for our collective refusal to act would be that we are anti-revolutionaries (Latour 2015c, p 55), trying to minimize the consequences of a revolution that was made without us, against us and, at the same time, by us. In other words, we have been the blind instigators of this revolution and stay blind in face of its consequences.

Going further, Latour explores the historicity of these behaviours in the book entitled “Face à Gaïa” (2015c). Our summary of his conclusions is as follows.

Originally, there would be, from the twelveth century, a progressive misunderstanding of the apocalyptic message of the Christian religion. With the addition of a new era, that of the kingdom of the Spirit, monks have dreamed of achieving it on Earth. The certainty of the realization of this kingdom of the Spirit was strengthened, somewhere between the thirteenth and eighteenth century by the new format offered by the incontestability of science. This author does not call into question religion itself. For him, “being religious” is to listen to what others believe in, i.e., partly to behave as a diplomat (Latour 2015c, p 201). What is called into question is what some humans do in the name of religion.

Telling Westerners, or those who have just recently been Westernized more or less violently, that times are over, that their world is over, that they must change their way of life, can cause a feeling of complete misunderstanding because, for them, the Apocalypse has already taken place. They have already passed onto the other side. The world of the Beyond has already become a reality.

Hence, it would be necessary to take action to unravel the “certainty” that the Apocalypse has already taken place and instead develop what is “questionable”. It would be necessary to restore their place for politics, religion, art and science, they have distinct virtues and none has the monopoly of providing here bellow the “truth”.

If the key obstacle to engaging actions in favour of ecological questions is tied to Western civilizations, it is interesting to see what happen so in developing countries.

A cross-national study about public opinion on climate change (Kim and Wolinsk-Nahmias 2014), concerning 82 countries, was based on a comprehensive dataset that integrates 12 major international surveys over 5 years (2005–2009). It shows that citizens in developed countries tend to be less concerned about climate change and less supportive of certain climate policies than those in developing countries. The belief that global warming is a “very serious issue” shows a significant negative correlation with Gross Domestic Product (GDP). People in developing countries were significantly more likely to say that climate change was causing damage currently or would, within the next 10 years. Levels of “strong” public concern about climate change are generally higher in developing countries.

Concerning international negotiations (Aykut and Dahan 2015), at the time of the discussion on the Kyoto Protocol (1994–1997) the main political issue concerned the allocation of emission reductions. Developing countries were not affected by these objectives. They refused to consider CC as a global problem and saw it only as a Northern-hemisphere consumption problem. Even, the climate alert was considered as having been created by scientists and Northern countries to prevent the development of Southern hemisphere countries.

At that time the debate focused on the future scenario from a 1990 baseline. This “initial state” concept was particularly unacceptable for the South: for two centuries the North had grown, become industrialized and equipped and polluted. When it comes to defining a socio-economic future for the planet, developing countries play an important role in the methodology that must necessarily encompass the various political, economic, social and historical legacies. According to Aykut and Dahan (2015), these points of view were gradually included in the IPCC reports and made a very large place to these questions of vulnerabilities facing CC, land use, agriculture or forestry issues.

In conclusion, the Western references, linked to immeasurable supreme authorities, Spirit, Truth or Science, appear as obstacles on environmental issues. Understanding their historicity is an advantage for overcoming these. Developing countries, which have integrated to a lesser extent these references, played an important role in the last decade in promoting green issues.

22.3.2 Sceptics and "Merchants of Doubt" Hinder Changes

Sceptics are opponents of the scientific consensus on CC defined as the agreement that (a) the Earth is warming up and (b) that warming is due to human greenhouse gas emissions over the last few decades. In a population, there is always a certain rate of disbelief on any apparently reasonable topic. We try here to follow what the processes could be in making people doubt CC and the Great Acceleration impacts.

A cross-national study was performed for 14 advanced industrialised nations with 19,991 samples from the International Social Survey Programme based on surveys administrated in 2010 and 2011 (Tranter and Booth 2015). After this work, the proportion of sceptics ranges between 17% and 12% in Australia, Norway, New Zealand and USA (decreasing order), and between 4% and 2% for Germany, Switzerland and Spain. Analysing the data from existing research on climate scepticism, this study tries to find the possible predictors of climate scepticism that could apply across nations. The conclusion is instructive: "Climate scepticism persists despite overwhelming scientific evidence that anthropogenic climate change is occurring. The reasons for this are varied and complex."

Many studies and polls are documented on who the sceptics are and what makes them sceptical. However, like the study by Tranter & Booth, they conclude on the apparent complexity of causes.

Public opinion surveys suggest that doubt is increasing (Matthews 2015). According to Oreskes and Conway (2010a) one reason that the public is confused is that people have been trying to confuse them.

A book, "Merchants of doubt" by Oreskes and Conway (2010b), recounts the strategy of intentionally waging campaigns of doubt against climate science. This strategy of creating doubt was applied to combat the ideas that cigarette smoking causes cancer, that acid rain or the ozone hole is caused by man-made pollution, that the pesticide DDT should be banned, that the world is warming or, if it really is warming, that we ought to be worried. Doubt on the reality of CC and the role of humans in this change was organized during the 1980s onwards. The approach of these "merchants of doubt" is related to their social organization convictions against communism and in favour of free enterprise.

The "climategate" story illustrates the efficiency of waging campaigns of doubt. Shortly before the Copenhagen Summit (COP15 of UNFCC in December 2009), approximately 1000 e-mails from the Climatic Research Unit (CRU) were made public without authorisation. This small research unit plays an important role in climate science, in particular with developing global temperature trends. Important allegations were made following the e-mail release and the work of CRU and to the conclusions of IPCC were challenged, fuelled by these e-mails. Eventually, a number of separate independent reviews have supported the honesty and integrity of scientists in the CRU in 2010, just after the Summit (Russel et al. 2010; Deutsche Bank climate change advisors 2010). But the damage had been done; this story was extensively referred to during the Copenhagen Summit which was considered a fiasco.

The reasons lying upstream to the merchants of doubt again seem coherent with the references: nostalgia of leaving a world they knew or that made them powerful and recognized in the records of wealth, power or fame and, more significantly, to the references presented in Sect. 22.3.1: the deep forces that drive denying the existence or danger of CC, and focusing all the spotlights on economic life and consumption, refusing to take into account any other aspect of existence, running them down as mere "poetry", or any other term that has become pejorative in modern times.

22.4 Human Essence Is Social

This section, entitled "Human essence is social" after Freud notes (1921) explores the question of coviability at the scale of each human being and his interdependence with the others.

22.4.1 Collective Initiatives Preserving the Identity of Each One

To commit together towards a common objective, it is necessary to be self-confident and to preserve the identity of each other. In this respect, Kahan (2015) provides interesting elements. He presents the example of the Southeast Florida Regional Climate Change Compact where four counties have formally adopted climate action plans, an exceptional endeavour in the USA. This means that the collective initiatives were therefore identified as being worthwhile to pursue by city planners, business groups, resident associations, conservation organizations, civic associations and religious groups – who all took part in the public and highly participatory process that generated the Plan.

People's answers to whether or not one "believes in" human-caused global warming does not measure what they know; it expresses who they are. It is important to take into account that the answer to the question "whose side are you on?" has a much bigger impact on the life of a person than his answer to the question "what do you know?". As ordinary individuals have a bigger stake in maintaining their status within their defining groups than in expressing correct understandings of science on societal risks, they will, in such circumstances, use their powers of reasoning to give information that protects their identities as members of these groups.

Bombarding citizens with information such as "97% of scientists agree" doesn't diminish polarization but instead aggravates it by amplifying the association between competing identities and competing positions on CC. Effective communication of climate science has persistently been defeated. Effective communication on CC should have the aim to identify "Who knows what about what?" Empirical study shows that good instructors can impart genuine comprehension of modern synthesis in a high school classroom filled with culturally diverse students. If the question is addressed in a manner that disentangles identity and knowledge, there is no "consensus" gap between the scientists and the public.

Each person may face during his life numerous changes (CC – nanotechnology, etc) and the role of training and information of all, at school, and throughout one's life is of course particularly important nowadays.

22.4.2 *Subject and Social Link*

In the face of CC and the aim of lowering consumption, one may ask what makes individuals consume more and more. After Sauret (2009), a line of force emerges in the analysis of contemporary society by psychoanalysts. The capitalist functioning impacts on collective life: proliferation of consumption "pathologies" (anorexia, bulimia, addiction, nausea, disgust), of enjoyment (risk behaviours, extreme sports, arson for fun and other forms of violence); of desire (melancholy, sadness, loneliness, boredom); of ego (fatigue of being oneself, egoism).

According to Sauret (2009) in all media, scientific literature, blogs or other, these behavioural changes are discussed, explored and analysed. A relatively large number of classic depressive illnesses appeared in the seventeenth century and link depression to the contemporary social formation of symptoms. For this author, depression is a major psychopathological epidemic of our time.

The pioneering work of Durkheim (1930) in modern sociology establishes a causal relationship between unbalanced forms of social ties and the suicide rate. Chu et al. (2010) present a comprehensive analysis of literature describing the relationship between cultural factors and suicide. Included among the factors they identify are alienation and acculturation.

The World Health Organization (2014) claims that suicide is a global phenomenon. All countries are affected. Over 800,000 people die by committing suicide every year and it is the second leading cause of death in 15–29-year-olds. There are indications that for each adult who has died by committing suicide there may have been more than 20 others attempting suicide. In France, suicide is the first cause of death among 25–34 year olds with 20% of all deaths in this age group (Ministère des affaires sociales, de la santé et des droits des femmes 2015).

When the scope is extended to developing world, this shows that depression is not universal. Findings presented in the book edited by Carr and Schumaker (1996) imply that people living in the "developed" world, in their quest for modern living, may have been to some extent neglecting their own psychological development.

We found it important to explore in more depth, the link between markets and capitalism with the disintegration of social life and the proliferation of psychological pathologies. After Sauret and Askofare (2011): the subject is stubborn. He is made in a way that makes him constantly ask the Other for this "real" being, for which the Other substitutes a being of signifiers (similar to his own). No subject exists without other speaking beings that receive him, without the logical preexistence of language (figured by the Other): as Freud notes (1921), human essence is social.

Psychoanalysis, according to Lacan (quoted by Sauret 2009), appeared around 1940 as a practice precisely because some part of the subject (the individual) was threatened by capitalism.

Language is the habitat of the speaker (Sauret 2009). It refers to the power of representation, not only to represent "something" in its absence, but also to represent "something" non-conceived and hitherto unimaginable: language brings this "something" to life and thereby as "actually said". Language confronts the subject with his own representation: it asks him the question that science poses about its objects "What am I?" In doing so, language requires the subject (the one who says "I") to confront the fact that he is also represented in language, he answers the question of what he is by speaking i.e., by using words that represent him, where he is and where he is not, in which the reality of his being is not a language designation, except in the form of a statement. Language confronts the subject into agreeing his lack in being.

According to Sauret (2009), the coupling of capitalist discourse with science, exploiting the structure of the subject, makes him believe that science produces the object he is missing, that he may help himself on the market and that he has no need to establish social ties. The subject abandons himself to the suggestion that speech has no reason to appeal to being cu-off from the market. It follows symptomatology familiar to us: "rejection of the things of love", "claim for immediate gratification", "allergy to frustration", "violence", "sexual facelift on consumer goods", "psychosomatic and hypochondriac phenomena", "appearance of phobic and perverse behaviours". It is as if such a subject is deprived of the support of his own structure and cannot develop the type of solution that allows him to accommodate his uniqueness.

Dubois and Ceron (2015) highlight the role of citizens in climate policy through their consumer choices. Along with the present need of leaving the supremacy of market/consumption, is the challenge in the way in which everyone can be singular and may search for meaning in his life, a justification of his "inability to know everything" that would enable him to accommodate his uniqueness whilst "living together".

Sahlins (1976) shows that abundance is unknown to contemporary society as it corresponds to a situation in which one consumes only what he needs – what he wants. If primitive man does not make his activity a paying one, it is not because he does not know how to, but because he does not want to. The challenge would be to find abundance again.

Witzel (2012) in a book on the origins of the world's mythologies expresses in the epilogue that "Pan-Gaean humanity, with our most ancient tales and their meaning, is very close to us, and the Stone Age way of thinking is still akin to ours"; and that "Whatever belief system or religion people now belong to, they try to find meaning in their lives so as to provide some assurance that their stay here is not altogether accidental, brief, and futile, that they have a prospect to look for".

22.5 Precursors to the Development of Coviability

In this section, we try to investigate some social mechanisms that could sustain a possible future constructed along the paradigm of coviability.

A reason for hope in our future is that the world's population growth rate began to decrease in 1967 and the projections by demographers announce that the population could quasi-stabilize at around 10 billion by the end of the century. The number of children per woman has been globally falling since 1964 (Pison 2011). It was 5 in 1964; 2.46 in 2016; it should decrease to 2.1, the replacement level fertility, by 2073 and should continue to drop reaching1 by 2100. This fall is coherent with the widespread demographic transition that corresponds to the declining death rate followed, with some delay, by the subsequent declining birth rate as humans chose to have fewer children. Thus, for the demographer Pison (2011) the main current concerns focus on the capacity and the choice of humans to react in sufficient time to stop or roll back their influence on the planetary boundaries.

At the climate conference (COP20) in Lima in late 2014, it was decided that each of the 195 States would have to set out their roadmap to limit the effects of global warming to less that 2 °C by 2100. The UN Climate Change Secretariat has published an update to its summary report on the collective impact of national climate action plans (Intended Nationally Determined Contributions, or INDCs), submitted by governments as contributions to global climate action under the Paris agreement. The updated report (in May 2016) captures the national climate plans covering 189 countries which covers 95.7% of total global emissions.

Hence, at the level of international negotiations, the reduction of carbon emissions should be in progress. However, these are still not enough to keep the global temperature rise since pre-industrial times to below 2, or preferably 1.5 °C. The Paris Agreement is only a bridge between today's policies and climate-neutrality before the end of the century.

22.5.1 New Climatic Justice

On 24th June 2015, for the first time ever, a court legally required a state to take precautions against CC. Based on current government policy, the Netherlands will achieve a reduction of emissions by 17% at most by 2020, which is below the 25–40% reduction below 1990 levels by 2020 according to the target for developed countries, necessary to create a 50% chance of avoiding a dangerous 2 °C rise in global temperatures. The Urgenda Foundation and 886 citizens involved in the class action against the Dutch government aim to force it to take more robust action to reduce emissions. They also hope to offer a legal solution to the political impasse on international CC efforts. This verdict might provide support to all the other climate cases around the world.

In the absence of explicit treaties up to now, states have no legal obligations to curb their greenhouse gas emissions. However, if emissions continue on their present trajectory, the harm they cause will put the human rights of billions of people in jeopardy. As international human rights law is legally binding on states they are, therefore, not free to continue business as usual. But how much do human rights and other sources of law require each state to reduce emissions, even in the absence of a specific treaty?

The paper "Oslo principles on climate change obligations" lists a series of legal arguments to facilitate redress against states and companies that do not engage enough. It sets out existing obligations regarding the climate, along with a detailed legal Commentary that draws on the best joint interpretation of international law, human rights law, national environmental law and tort law.

22.5.2 New Foundations of Organizations, Social Responsibility and Business Law

The international standard of social responsibility, ISO 26000, launched in 2010, provides guidance on how businesses and organizations can operate in a socially responsible way. This means acting in an ethical and transparent way that contributes to the health and welfare of society.

Billaudot (2011) comments about this standard, saying that being responsible is no longer about not deviating from the legislation in force but rather to cause no harm to oneself by not respecting the rules imposed by one's personal ethics. Thus, the advent of ISO 26000 represents an important step in shared values compared to the previous ISO 9000 standard (launched in 1987) that answered the underlying issue "What is a good quality product?" without reference to any social or environmental responsibility.

The relationship between businesses and organizations' and the society and the environment in which they operate is a critical factor in their ability to continue to operate effectively. It is also increasingly being used as a measure of their overall performance.

However, if progress is undeniable, Segrestin and Hatchuel (2012) draw attention to the limits of its development. The first limit is a legal one. As an example, in France, the article 1833 of the Civil Code is: "Any company must have a lawful purpose and be incorporated in the common interest of the shareholders." In other words, social and environmental objectives cannot go beyond the interest of the shareholders. As shareholders have the possibility of revoking the officers or allowing the company to be bought back, concerns on corporate social responsibility (CSR) commitments have little weight.

However, Segrestin and Hatchuel (2012) mention an example of evolution from the state of California, in USA, who ratified in 2011 the creation of a new type of company. It encourages and allows companies to pursue one or more objectives in addition to the creation of economic value for shareholders. Furthermore, Segrestin (2015) suggests new contractual formats such as a "commitment contract" instead of an "employment contract" in order for workers to be engaged in the building of desirable futures instead of being subordinates.

If the principle of shareholder financial interest remains unchallenged, there is a risk that many social, environmental and innovative goals will not be pursued. On the contrary, a new corporate paradigm could result from a shift towards the integration of the concept of impact. According to the Impact Investing Task Force, this would have the potential to transform our ability to build a better society for all (Social Impact Investment Task Force 2014).

22.5.3 New (Eventual) Era for Geopolitical Negotiations

For 20 years, international climate negotiations have stalled on action to address the urgency of climatic degradation. This powerlessness is a result of the huge complexity of the issues in play and the format of the negotiations. The original framing from scientific expertise, through concepts such as the global thresholds or carbon budgets, has conveyed the illusion of the possibility of centralized control of the problem; but, organizing negotiations between states stumble upon issues of national sovereignty (Aykut 2015). Analysis by Aykut and Dahan (2015) of the 20 years of negotiations under the auspices of the United Nations Framework Convention on Climate Change shows that the solution to a low-carbon mode cannot be developed under the single pulse of such negotiations. While the agreement of the COP21 is a gateway to various favourable country commitments under no circumstances do the procedures of the COP prefigure a world government of world solutions. For Aykut and Dahan (2015), the routes to low carbon societies will necessarily be ones of multiple and mutual learning: India can be a model because of its very low consumption of meat, Europe because of the density of its cities, America for its capacity for innovation, China and Germany for supporting renewables.

Currently, efforts are driven bottom-up. States, corporations, and individuals are gradually engaging on the slow road to a paradigm shift. Among others, one can cite the understanding of the development of capitalism, the Sustainable Development Goals, the Human Development Index, or the concept of Gaïa.

The documentary series Ziv (2014) restores the historical debates and provides keys to understanding capitalism. It shows that trade can be driven and justified by something other than seeking short-term profit and that alternatives exist in which solidarity between individuals and groups and shared common goals can dominate trade.

Sustainable Development Goals (SDGs) launched by United Nations in 2015 replace the Millennium Development Goals (MDGs) and offer a new orientation towards the key message of the 1992 Earth Summit: that development and environmental protection must be considered together. The SDGs should not be reduced to poverty eradication, but must address all dimensions of sustainable development.

Environmental and social considerations must be incorporated: poverty eradication will become impossible with environmental degradation; people lose the feeling their lives are meaningful while receiving consumer goods in overwhelming abundance.

However, although many reports mention the concept of planetary guard rails or planetary boundaries, they do not back this up with specific targets. The WBGU (2014) presents recommendations on how guard rails for global environmental problems should be incorporated in the SDGs catalogue and operationalized by means of corresponding targets.

The Human Development Index (HDI) is also a world initiative that may help. This index is a measure of average achievement in key dimensions of human development: a long and healthy life, being knowledgeable and having a decent standard of living. As introduced in Duflo's book (2010), it was created in 1990's by United Nations Development Programme to replace the GDP as a measure of the development of a nation as the notions of freedom and well-being have no sense without the "capabilities" essential to human development. However, we must recognize that the HDI is not yet the driving concept of states.

In the book "Face to Gaïa", Latour (2015c) considers that the concept of Gaïa allows the limits of the Earth to be accepted or at least to be explored. Gaïa is a concept based on works that show how organisms not only adapt to their environment but also change their material environment as well, so that the whole evolves as a single, self-regulating system up to the scale of Gaïa. After Lovelock (2003), this metaphor of a living Earth reminds us that we are part of it and that human rights are constrained by the needs of our planetary partners.

Could a world state be a solution?

Latour (2015c), analysing the question in depth, concluded that we should not count on the mirage of having a world government that could, by a miracle of coordination and good governance, assign each (person or organisation) its share of CO_2 or financial compensation, or face sanctions. Moreover, a world government would be dangerous, giving way to autocracy under the law of "Global Nature" or "Natural Science". Additionally, the dream of a world government is just another extension of the vision that the world must be administered by man, as a large and single machine, to be able to work at all.

For Latour (2015c, p 349), to find the common world – and perhaps also the meaning (of) common – the solution is not to appeal to a God of Totality. The solution is to learn to represent differently the territory in which one belongs. This is basically the internalization of the innumerable encroachments from other territories and which we discover, little by little, how much we depend upon to survive. In geopolitical terms, the question is therefore to have several superimposed authorities on the same land. He proposes that the negotiations be not only between nation States but also include other entities such as "Oceans", "Atmosphere" and "Endangered Species". To remind humans about the necessities of these entities, it is necessary that they are represented by spokespeople. As the risks of acidification of oceans, of the composition of the atmosphere, of biodiversity loss are existential, humans (who are earthlings) should establish their representation in negotiations – and give them their full role.

When comparing seizure of brain epilepsy and the present behaviour of societies, Naccache (2015) detects a parallel. Seizure of brain epilepsy is a phenomenon in which several regions of the brain begin to communicate with each other too much ending up sharing poor and stereotypical information. The functioning of these brain regions loses in complexity and the characteristics that enabled them to be distinguished, dwindle.

The three key fields for the success of consciousness in the brain (and by analogy, the three recommendations at the macroscopic level of society) are: communicate a lot, maintain the complexity and, preserve the differentiation between the various system components.

This approach leads the neurologist Naccache to strongly recommend keeping in our societies, simultaneously, communication, differentiation and complexity in order to save or restore a conscious society.

Keeping separate virtues of science, politics and religion is essential; they become poison when we confuse them (Latour 2015c). Inviting the incontestability of religion or of science into politics is extremely dangerous. On the contrary, interactions between politics and science, with each one being kept separate, is fruitful, as for example (Latour 2010), in Antiquity, with Archimede's relationship (Greek scientist third century BC) with King Hiero. present researchers to be more assertive in their analyses (CFCC15 2015; Klein 2008; Oreskes and Conway 2010a).

Bringing together the above cited insights, leads to a multifaceted method and route. The nation States should keep their sovereignty but must transform their practices, interacting with other non-state delegations that represent existential entities. To this list of entities attempted announced by Latour (2015c), including atmosphere, ocean, biodiversity, indigenous people, we propose adding a coviability delegation.

22.6 Conclusion – Key Messages

In the second half of the twentieth century, climate change was the trigger for a new awareness in the link between humans and between them and their earthly environment. Gas emissions generated by human activities affect the composition of the atmosphere of the whole Earth and affect all beings, human and non-human.

The acceleration of human activities and their consequences since 1950 threatens not only the climate but several planetary boundaries, defined as the resilience threshold of the Earth's envelopes: loss of biodiversity and extinction of species, soil degradation, increases of nitrogen fluxes, and ocean acidification (Steffen et al. 2015).

Another serious threat is the development of individuals and societies functioning with a proliferation of pathologies, from anorexia-bulimia, to risk behaviours, fatigue of being who one is, egoism (Castoriadis 2005; Sauret 2009; Naccache 2015). The supremacy of market/consumption does not offer a satisfactory answer to the universal question "from where, how, why?"

The new paradigm to be constructed is, for the first time, bound to human-driven changes on the environment.

With the decrease in the world's population growth rate since 1967 and the perspective of quasi-stabilisation of the population at around 10 billion by the end of the century (Pison 2011), the main current concerns focus on the capacity and the choice of humans to react in sufficient time to stop or roll back their influence on the planetary boundaries. The challenge, already announced in 1972 by Meadows et al., is the combination of different possible transformations of aim and way of life in order to reach, before the end of the century, a new equilibrium at a mastered, limited and acceptable impact on humans and other species, and on climate, oceans and soils.

Coviability or the respect for others, human or non-human, in a relationship of interdependence would be the foundation for a consciousness of mankind, entangled with his earthly home. It benefits each individual, humankind and the other species.

After the present review, the obstacle or brake of the advent of coviability appear rooted in societal references prevalent in the world today. Modern society over the last few centuries was not concerned by the fact that actual development was not sustainable, having fled for centuries the archaism in a movement of emancipation.

Step by step, it is possible to understand the historicity of the Western-type references through work by Oreskes and Conway (2010b) on the question of "merchants of doubt", by Ziv (2014) on the development of capitalism, by Adorno & Horkheimer (cited in Belaval 1990; Schmid Noerr 2002) or Latour (2015c) on the links to the history of politics, science and religion, or finally on the ill-being of individuals and societies (Castoriadis 2005; Sauret 2009; Naccache 2015).

The ways and means of developing a coviability society are to be invented and are certainly complex and multiple. Humans, facing with stupor what Western-type civilization has fled for centuries, are presently ill-equipped emotionally, intellectually, morally, politically and culturally (as illustrated by Squarzoni 2012).

This review leads to the conclusion that the main issue is to develop an intelligent and conscientious society able to react and adapt and have a projective capacity in its behaviour, decisions and impacts. The keys for this are to communicate a lot, to maintain the complexity and, to preserve the differentiation between the various components of humanity and territories.

For various reasons, establishing a world government is not the solution. The experience gained over the last 20 years of international climate negotiations shows that this route is, at least, largely insufficient (Aykut and Dahan 2015). The risk of totalitarianism by a centralized system is contrary to an intelligent society capable of reacting. Moreover, there are no universal laws that prescribe the route. "Nature" itself is a concept built as being complementary to the culture of the seventeenth century (Latour 2015c), it cannot be the reference for a higher authority.

We retain as an important working hypothesis the innovation proposed by Latour (2015c) to add non-state delegations (e.g. climate or ocean) to those of the nation states. Their role is to have the floor on the political scene and to intervene on what is acceptable or not. These delegations should include several members among which scientists. And finally, we propose to add beside the State delegations a "coviability" delegation – as "coviability" has become essential.

References

Aykut SC (2015) Comment se construit la confiance dans les sciences et les politiques du climat? retour sur un colloque international. Nat Sci Soci Supp 3:102–110

Aykut SC, Dahan A (2015) Gouverner le climat? 20 ans de négociations internationales. Presses de Sciences Po, Références-Développement durable, Paris

Belaval P (1990) Une présentation: La dialectique de la raison de Max Horkheimer et Theododor Adorno, Germanica Accessed on 18 July 2016

Billaudot B (2006) Economique, justification marchande et démocratie. Paper presented at the 4th colloque international PEKEA "Démocratie et économie", Rennes, 4–6 nov. 2005, pp. 32 <halshs-00099473>

Billaudot B (2011) La norme ISO 26 000: une norme-définition qui a le statut d'un compromis. Chapitre 11 in ISO 26 000: une Norme "hors norme"? sous la direction de Capron M, Quairel-Lanoizelée F & Turcotte MF: 195–214

Business & Climate Summit (2015) Working together to build a better economy, Paris May 20–21, 2015 UNESCO Headquarters, http://www.businessclimatesummit.com/. Accessed 15 Sep 2015

Carr SC, Schumaker JF (1996) Psychology and the developing world. Edited book. Paeger Publisher, Westport

Castoriadis C (2005) Une société à la dérive: entretiens et débats, 1974–1997. Seuil, Paris

CFCC15 (2015) Conference "Our common future under climate change" at UNESCO Headquarters and Université Pierre et Marie Curie, Paris, July 7–10, http://www.commonfuture-paris2015.org/. Accessed 15 Sep 2015

Chu JP, Goldblum P, Floyd R, Bongar B (2010) The cultural theory and model of suicide. Appl Prev Psychol 14:25–40

COP21 (2015) 21st session of the conference of the parties to the United Nations framework convention on climate change (UNFCCC) for a universal climate agreement. Paris, Nov 30th to December 11th, http://www.cop21.gouv.fr/en. Accessed 15 Jan 2016

CRCF-Conférence des Responsables de Culte en France (2015) Statement on the climate crisis. http://lacrcf.fr/upload/20150701-CRCF-statement%20on%20the%20climate%20crisis.pdf. Accessed 16 Jan 2017
Deutsche Bank Climate Change Advisors (2010) Climate change: adressing the major skeptic arguments. https://www.uea.ac.uk/documents/3154295/7847337/Deutsche-Bank-CRU-report.pdf. Accessed 13 Oct 2015
Dirzo R, Young HS, Galetti M et al (2014) Defaunation in the anthropocene. Science 345(6195):401–406
Dubois G, Ceron J-P (2015) Consommation et modes de vie: une autre perspective sur les politiques d'atténuation du changement climatique. Natures Sciences Sociétés, Supp. 3:76–90
Duflo E (2010) Le développement humain. Lutter contre la pauvreté (I). Ed. Seuil, La république des idées, 106 p
Durkheim E (1930) Le suicide: Étude de sociologie. Ed. Presses universitaires de France et Félix Alcan, 462 p
François P (2015) Laudato Si!. Saint Siège, Vatican, Lettre encyclique, http://w2.vatican.va/content/vatican/it.html. Accessed 30 Nov 2015
Global carbon project (2015) http://www.globalcarbonproject.org/carbonbudget/index.htm. Accessed 7 Oct 2015
IPCC (2014) Climate change 2014, Synthesis report, Summary for policymakers. Fifth Assessment Report (AR5), 31 p., https://www.ipcc.ch/pdf/assessment-report/ar5/syr/AR5_SYR_FINAL_SPM.pdf. Accessed 27 Oct 2015
Islamic Declaration on Global Climate Change (2015) Declaration adopted at international Islamic climate change symposium, August 2015, Istanbul, http://islamicclimatedeclaration.org/islamic-declaration-on-global-climate-change/. Accessed 30 Nov 2015
Jouzel J, Debroise A (2014) Le défi climatique – objectif: 2°C ! Dunod. Quai des sciences, Paris
Kahan DM (2015) Climate-science communication and the measurement problem. Adv Polit Psychol 36(suppl. 1):1–43. https://doi.org/10.1111/pops.12244
Kim SY, Wolinski-Nahmias Y (2014) Cross-national public opinion on climate change: the effects of affluence and vulnerability. Global Environ Polit 14(1):79–106
Klein E (2008) Galilée et les Indiens. Flamarion, Café Voltaire
Latour B (2010) Cogitamus, six lettres sur les humanités scientifiques. La Découverte, Paris
Latour B (2015a) La civilisation mise au défi par l'anthropocène. Le Monde, 11 Mar 2015. http://www.bruno-latour.fr/sites/default/files/downloads/2015-3-LEMONDE-ANTHROPO-GB.pdf. Accessed 30 Sep 2015
Latour B (2015b) Anthropocène, anthropocène, est-ce que j'ai une gueule d'anthropocène! Show conference at Branly Museum in Paris (France) with Mathieu Protin and Frédérique Ait-Touatti as part of a series "L'ethnologie va vous surprendre", 14 mars 2015. http://www.bruno-latour.fr/fr/node/629. Accessed 30 Sep 2015
Latour B (2015c) Face à Gaiä, huit conférences sur le nouveau régime climatique. La Découverte – les empêcheurs de penser en rond, Paris
Lepage C et al (2015) Déclaration universelle des droits de l'humanité. Rapport à l'attention de monsieur le président de la République. Rapport, 133 p, http://www.sentinelle-droit-international.fr/?q=content/le-rapport-lepage-sur-les-droits-de-lhumanit%C3%A9-et-le-concept-de-lhumanit%C3%A9-indivisible. Accessed 3 Nov 2015
Lovelock J (2003) Gaia, the living earth. Nature 426:769–770
Matthews P (2015) Why are people skeptical about climate change? Some insights from blog comments. Environ Commun 9(2):153–168. https://doi.org/10.1080/17524032.2014.999694
McCallum ML (2015) Vertebrate biodiversity losses point to a sixth mass extinction. Biodivers Conserv 24(10):2497–2519
Meadows DH, Meadows DL, Randers J, Behrens WW (1972) The limits to growth. A report for the club of Rome's project on the predicament of mankind. http://www.donellameadows.org/wp-content/userfiles/Limits-to-Growth-digital-scan-version.pdf. Accessed 29 June 2016
Ministère des affaires sociales, de la santé et des droits des femmes (2015) http://www.sante.gouv.fr/etat-des-lieux-du-suicide-en-france.html. Accessed 26 Nov 2015

Naccache L (2015) L'homme réseau-nable, du microcosme cérébral au macrocosme social. Odile Jacob, Paris
NAZCA launched in (2014) Non-state actor zone for climate action showcases commitments to action by companies, cities, subnational regions, and investors to address climate change. United Nations framework convention on climate change, http://climateaction.unfccc.int/. Accessed 27 Oct 2015
Oreskes N, Conway EM (2010a) Defeating the merchants of doubt. Nature, Opinion 465(10): 686–687
Oreskes N, Conway EM (2010b) Merchants of doubt: how a handful of scientists obscured the truth on issues from tobacco smoke to global warming. Bloomsbury, New York
Oreskes N, Conway EM (2013) The collapse of western civilization: a view from the future. Daedalus 142:40–58
Pidgeon N (2012) Climate change risk perception and communication: addressing a critical moment? Risk Anal 32(6):951–956
Pison G (2011) Sept milliards d'êtres humains aujourd'hui, combine demain? Population & Sociétés, https://www.ined.fr/fr/publications/population-et-societes/sept-milliards-etres-humains-aujourdhui-combien-demain/. Accessed 1 Fev 2016
Russel M, Boulton G, Clarke P et al (2010) The independant climate change e-mails review. Report. http://www.cce-review.org/. Accessed 13 Oct 2015
Sahlins M (1976) Âge de pierre, âge d'abondance: L'économie des sociétés primitives. Bibliothèque des Sciences humaines/Gallimard
Sauret M-J (2009) Malaise dans le capitalisme. Presses univ Mirail, coll. Psychanalyse &, Toulouse
Sauret M-J, Askofare S (2011) The effect of liberalism on the development of clinical professions. Recherches en psychanalyse Journal of Psychoanalitic studies. Recherches en psychanalyse 2011/2 (n° 12), 115–123. https://doi.org/10.3917/rep.012.0115
Schmid Noerr G (eds) (2002) Max Horkheimer & Theodor W. Adorno. Dialectic of enlightment. Philosophical fragments. (trans: Jephcott E). Stanford University Press, Stanford
Segrestin B (2015) Subordonner le travail ou l'ordonner à un futur commun? Pour de nouveaux contrats d'engagement dans l'entreprise. Rev Française de Socio-économie 2(hors-série):73–88. https://doi.org/10.3917/rfse.hs1.0073
Segrestin B, Hatchuel A (2012) Refonder l'entreprise. La République des Idées/Seuil, Paris
Social Impact Investment Task Force (2014) Impact investment: the invisible heart of markets, Harnessing the power of entrepreneurship, innovation and capital for public good. Report, http://www.socialimpactinvestment.org/reports/Impact%20Investment%20Report%20FINAL%5B3%5D.pdf. Accessed 6 July 2016
Squarzoni P (2012) Saison brune. Delcourt, coll encrages, Paris
Steffen W, Persson A, Deutsch L et al (2011) The anthropocen: from global change to planetary stewardship. Ambio 40:739–761
Steffen W, Richardson K, Rockström J et al (2015) Planetary boundaries: guiding human development on a changing planet. Science 347:1259855. https://doi.org/10.1126/science.1259855
Tranter B, Booth K (2015) Scepticism in a changing climate: a cross-national study. Glob Environ Chang 33:154–164
United Nations General Assembly (2015) Transforming our world: the 2030 agenda for sustainable development. Resolution adopted on 25 Sep 2015. http://www.un.org/ga/search/view_doc.asp?symbol=A/RES/70/1&Lang=E. Accessed 3 Feb 2016
WBGU (2014) Human progress within planetary guard rails. A contribution to the SDG debate. German Advisory Council on Global Change, Policy paper 8. Berlin. http://www.wbgu.de/en/policypaper/policy-paper-8-2014-sdg/. Accessed 1 Feb 2016
Witzel EJM (2012) The origin of the world's mythologies. University Press, Oxford
World Health Organization (2014) Preventing suicide: a global imperative. http://apps.who.int/iris/bitstream/10665/131056/1/9789241564779_eng.pdf?ua=1&ua=1. Accessed 26 Nov 2015
WWF (2014) Living planet report. http://wwf.panda.org/about_our_earth/all_publications/living_planet_report/. Accessed 25 Jan 2016
Ziv I (2014) Capitalisme. a six episode film (about 53 mn each), directed by Ilan Ziv; Coproduction: Arte France, Zadig Productions, Filmoption International, T.A.M.I. Média.

Chapter 23
Approaching the Human-Environment Nexus Beyond Conflict: A Peace and Coviability Perspective

Mohamed Behnassi

Contents

23.1 Introduction

According to the Western paradigm, the natural environment is mostly seen as a separate entity from human systems, and often perceived in terms of a mere provision of resources to meet anthropogenic 'needs'. In this respect, natural resources often represent different interests and values related to their use, availability, and market value. When politicized, especially in times of scarcity, access and control of these resources and related benefits may stimulate competition among actors, which may eventually lead to conflicts. These are the key reasons why an increasing number of studies have investigated the environment-conflict nexus in the past. With the advent of securitization theories in the mid-1990s, environmental issues have increasingly been formulated in terms of security, thus giving birth to the concept of 'environmental security' (Wæver 1995; Westing 1986).

Many researchers have questioned the relevance of the environment-conflict thesis and related assumptions, whether there is a direct link between the scarcity or abundance of natural resources and violence, and whether environmental factors

---

M. Behnassi (✉)
Center for Research on Environment, Human Security and Governance (CERES), Ibn Zohr University of Agadir, Agadir, Morocco

O. Barrière et al. (eds.), *Coviability of Social and Ecological Systems: Reconnecting Mankind to the Biosphere in an Era of Global Change*,
https://doi.org/10.1007/978-3-319-78497-7_23

lead inevitably to violent conflicts.[1] While disputes involving natural resources have a higher chance of transforming into more sustained forms of conflict, it was argued that natural resources alone are rarely sufficient to initiate an outbreak (Homer-Dixon 1999; Kramer et al. 2013; Maas et al. 2013). In other terms, environmental factors may, under specific circumstances, increase the risk of violent conflict, but not necessarily in a systematic way and unconditionally. Hence there is, to date, no scientific consensus on the impact of environmental factors on violent conflicts, and further research and empirical investigation on the subject are vitally needed. For instance, one of the key issues according to Bernauer et al. (2012) is that the effects of environmental factors on violent conflicts are likely to be contingent on a set of societal conditions that determine adaptation capacity.

Inspired by these theoretical and methodological criticism, some scholars set out to test, validate or disprove assumptions made in the environmental conflict literature. The use of quantitative models in large cross-national tests allowed them to ponder the relative weight of various variables, and thereby to refine existing environmental conflict models. New ecologic and socio-political variables, identified as 'missing links' between the environment and violent conflict, have been considered. While these contributions innovated the empirical analysis of environmental conflicts, they failed to generate new theoretical insights or innovative concepts.

Currently, the dichotomy between humans and the biophysical environment as a passive, external entity is increasingly challenged by many researchers (like Descola and Pálsson 1996; Moran 2007; Robbins et al. 2013) especially in the Anthropocene era. Moving away from the conceptualizations presenting natural resources as a driver of conflict, this new wave of researchers have begun envisioning environmental change and natural resources as an incentive for cooperation and rather than for violence (Dresse et al. 2016). In this perspective, the concept of 'environmental peacebuilding' has evolved into a broader framework connecting mainly environment and peace studies as well as investigating the evolution of environmental cooperation into a conflict transformation instrument and how environmental cooperation can effectively contribute to peace, and if so, under what circumstances and conditions. The concept emphasizes the human-environment nexus and the cooperation that can emanate from this nexus. In this connection, nature is often perceived as constitutive of cultural identities, since it is deeply intertwined with anthropogenic activities, making it a possibly highly politicized issue which may open potential for peacebuilding when the key stakeholders are involved in a participatory and inclusive way (Dresse et al. 2016).

To prevent the conflict scenario, it has been recently suggested to switch to socio-ecological systems as the referent object of securitization (Zikos et al. 2015), in

[1] *Violent conflict* is defined as deliberate violent acts perpetrated by an organized or semi-organized group against state forces, other organized or semi-organized groups or civilians within the territorial borders of a recognized state (Van Baalen and Mobjörk 2016).

order to envision environmental issues (including resource scarcity) as a common concern, therefore inspiring cooperation rather than conflict between stakeholders. In addition, it was argued that environmental cooperation represents an opportunity to stimulate dialogue and confidence-building on environmental issues and to create strategic social spaces for regular interaction and systematic negotiations between actors on different scales (Maas et al. 2013).

Despite the attractiveness of environmental peacebuilding, more systematic research is still needed to make it a robust framework. In this perspective, this analysis suggests the coviability of social-ecological systems as an alternative to properly perceive the human-environment nexus. Given the fact that the viability of human societies depends intimately on the living components of natural and managed systems, the coviability approach has the potential to adjust our perception with regard to the position of humans in the biosphere. A position which should be mainly oriented towards ensuring solidarity between humans in order to maintain viable ecosystems instead of conflict or pragmatic cooperation driven schemes. This will raise hope that future targets – such as Sustainable Development Goals (SDGs), mitigation and adaptation strategies, biodiversity conservation schemes, environmental justice, etc. – can be achievable and that human societies are resilient and better prepared for a world of universal ecological change.

Against this background, the overall perspective provided by this chapter is built around four points. *Firstly*, the analysis begins by reviewing and discussing the environmental conflict literature.[2] *Secondly*, the relevance of the environment-conflict thesis is questioned with the aim to identify its main shortcomings in approaching the human-environment nexus as a theoretical and policy-making referential. *Thirdly*, the analysis highlights the evolution of the natural environment from a vector of conflict to one of peace; a discussion on the limits and opportunities of 'environmental peacebuilding' as an emerging analytical framework, with the identification of related contentious issues and research challenges, is provided. *Fourthly*, the analysis concludes by exploring the relevance of coviability as an alternative to properly perceive the human-environment nexus.

23.2 Environmental Conflict Literature: A Review

The divergent conceptual approaches, methodologies, and levels of analysis make a coherent presentation of the environmental conflict literature often difficult. Adding to this difficulty is the literature's division into specific sub-themes (such as energy, water, land or mineral resource conflicts) (Hagmann 2005). Previously, the state

[2]According to Hagmann (2005), 'environmental conflict', 'environmental security', or 'eco-violence' are often used interchangeably in the literature. Environmental conflict literature designates research contributions that portray or discuss the natural environment as a cause of violent conflict. There is no widely accepted definition of what constitutes an environmental conflict or environmental security, nor is there agreement on whether environmental conflict exists as a distinct type of violence.

of the art had been based on consecutive 'generations' of environmental conflict research (Rønnfeldt 1997), noted differences and shared aspects in methodology and research design (Gleditsch 2001a, b), or stressed underlying normative underpinnings and epistemology (Barnett 2000).

From a historical perspective, the evolution of the environmental conflict school has gone through various stages since the mid-1980s, and specifically from the time when the 1987 Brundtland Report identified environmental stress as a potential cause of conflict. The development of the 'environmental security' concept, at the juncture of global environmental politics and security and conflict studies, marked the beginning of this school. This concept, as a subset of broader concerns over human security, views ecological processes and natural resources as drivers or catalysts of conflicts, barriers or limits to human well-being, or conversely, as the means to mitigate or resolve insecurity (Scott and Thapa 2015).

This interdisciplinary and largely conceptual debate mobilized academic and political spheres alike, thus dominating and influencing the security discourse in the post-Cold War era, particularly in the United States (Dabelko and Simmons 1997). Subsequently, it was expanded and exemplified the search for alternative paradigms in international affairs and security studies through the focus on whether and under what circumstances environmental concerns represent a threat to national and collective security. The emergence of the 'securitization' paradigm in the mid-1990s (Wæver 1995) and the discourse on environmental security – as a potential threat to stability or a policy goal that needs to be achieved – had been part of an epistemic community that critically advocates the broadening of conventional security thinking to include other non-military issues (Westing 1986).

According to this trend, which is amply documented, it was widely assumed that environmental factors can play a key role in triggering violent inter-community and inter-state conflicts. Depending on the respective theoretical premises, some scholars have even argued that *scarcity* of renewable natural resources inevitably leads to violence (Matthew et al. 2009), especially in countries of the global South. Others, like Gleditsch (2004), have tried to show that it is not scarcity, but *abundance* of natural resources which has the potential to generate conflicts. In all cases, major contributions on empirical tracing of the environment-conflict nexus were characterized by a strong emphasis on experimental evidence and a 'process-tracing' methodology applied to numerous case studies.

Two research groups were at the forefront of the endeavor to demonstrate and typify the relevant causal mechanisms: conflict researchers at the University of Toronto directed by Thomas Homer-Dixon, usually referred to as the *Toronto Group* (Homer-Dixon 1991, 1994, 1995, 1999; Homer-Dixon and Levy 1995; Percival and Homer-Dixon 1998; Schwartz et al. 2001); and scholars associated with the *Environment and Conflict Project* (ENCOP) of the Swiss Federal Institute of Technology in Zurich and the Swiss Peace Foundation in Bern (Baechler 1998, 1999; Baechler and Spillmann 1996). Both groups used different terminology and concepts but aimed to reveal empirically how and under which circumstances, environmental factors and resources (especially renewable ones that are key for food production such as cropland, freshwater, and forests) cause violent conflict.

Both projects operated exclusively on the basis of *ex-post* analysis of cases where scarcity had actually led to conflict. Consequently, both attempted to define conflict typologies and theorize on the socio-political processes that led to violent conflicts (Hagmann 2005).

The subsequent research trend was inspired by theoretical and methodological criticism of the Toronto Group and to a lesser degree of ENCOP. Many researchers associated with the International Peace Research Institute (PRIO), Oslo, Fig. prominently among this strand of environmental conflict research (Gleditsch 1997, 2001a, Hauge and Ellingsen 2001; de Soysa 2000, 2002b). This heterogeneous group of scholars initially set out to test, validate or disprove assumptions of previous research (namely the alleged determinism between resource scarcity and violent conflict). Consequently, their contributions provided a clearer picture of geographic and diachronic frequency distributions of environmental conflict cases. The use of quantitative models in large cross-national tests allowed them to ponder the relative weight of various variables, and thereby to refine existing environmental conflict models. New ecologic and socio-political variables – such as poverty (Smith and Østreng 1997), political regime type (Gleditsch and Sverdrup 2002) or cultural variables (De Soysa 2002a) identified as 'missing links' between environment and violent conflict, were considered. Nonetheless, scholars in this phase remained attached to the idea of investigating causalities and correlations between environmental variables and domestic violent conflicts. While these contributions of PRIO-associated researchers innovated the empirical analysis of environmental conflicts, they failed to generate new theoretical insights or innovative concepts.

In this vein, Dresse et al. (2016) claim that the environment, natural resources, and human actions interact in different ways. During a violent conflict, environment/natural resources can impact on (when they cause or contribute to a dispute) or be impacted by human violence (when they suffer from the violence as a direct or indirect target). Therefore, environment/natural resources may act either as an irritant or as a unifier (when they help mitigate tensions between conflict parties). These interactions remain generally determined by the inherent characteristics of a conflict and the local context.

Taking into account the fact that our understanding of conflict and peace is challenged in a changing, 'globalized' world, where the types and shapes of conflicts are mutating (Wolf 2007), Dresse et al. (2016) mention the existence of several *degrees and types of conflicts*, whether violent or non-violent, internal or transboundary, in which a multiplicity of actors can be involved, each with their own interests. Accordingly, several arguments, concerning the possible interactions between environment/natural resources and violence, have been distinguished by the authors:

- *First argument*: Even if the biophysical environment is rarely the sole cause of a conflict, environment/natural resources may indirectly contribute to the escalation of events into a violent conflict or be part of a wider political strategy. In this respect, it was shown that natural resources play a key role in determining and shaping conflict and its development. When coupled for example with political instability, scarce natural resources can cause conflicts over access to, and ownership of, shared transboundary resources between competing states (Giordano et al. 2005). In some cases, it is the political use of natural resources by state or non-state actors which leads to an escalation of the violence.

- *Second argument*: Environment/natural resources may become a direct or indirect target during violent conflicts, either as a weapon[3] (when environment is used by one of the conflict parties as a direct means to wield violence against the opposing party) or victim (the case of negative environmental externalities generated by a violent conflict).
- *Third argument*: Violent conflicts may entail beneficial effects on environment/natural resources such as the creation of 'de facto ecological havens in demilitarized zones', which contribute to biodiversity conservation[4] (Jarraud and Lordos 2012). These zones, referred to as peace parks, can in turn be used as conflict resolution tools (Ali 2007). Research by Lelieveld et al. (2015) also establishes a correlation between armed conflicts and regional air quality improvement in the Middle East (i.e. a decrease in nitrogen dioxide emissions since 2010 as a result of the Iraqi and Syrian crises has been noticed).

The interactions between environment/natural resources and conflicts are also determined by the *types of natural resources* involved. Dresse et al. (2016) believe that different resources have different effects on the development of a conflict and/or cooperation; however, despite some attempts no clear typology exists so far to conceptualize these interactions.[5] In an effort to categorize natural resource-related civil wars since the end of the Second World War, Ross (2003) and Matthew et al. (2009) focus, in their studies, on the *high financial value* of natural resources that may be put forward as a cause of many conflicts – such as metals (gold, copper, tin, cobalt or iron), minerals, timber, gemstones (i.e. diamonds), and consumption goods (i.e. cocoa and coffee), oil and other fuels. Therefore, depending on the kind of natural resources involved, the types and levels of violence observed in conflicts are variable. In addition to this, Le Billon (2001) differentiates between

[3]The Israeli-Palestinian conflict is a well-known example where water scarcity plays a role in a conflict and has been used by different parties as a means of applying pressure. Gleick and Heberger (2014), for instance, mention the 2011 destruction of water tanks, pumps and wells by the Israeli army in several Palestinian villages. The Vietnam War is also a good example: the US Army used the thesis of P. Gouro, who had drawn up a map of the dikes in northern Vietnam, to bombard these dikes and cause massive flooding. Yves Lacoste then wrote a famous book on this subject stating that: *Geography was primarily used to make war*! showing that the fine knowledge that geographers can provide about the relations between environment and society, can be used for the purpose of damaging this binomial. Here, we are in a process of the deliberate destruction of coviability.

[4]The geographer of the ORSTOM M. Benois has shown that among the first Protected Areas in Africa, several were carried out in areas that had been depopulated by slavery and inter-ethnic conflicts.

[5]For instance conceived a typology to classify water-related conflicts, but these categories are generally ever-evolving, adding conceptual confusion to the existing body of environmental peacebuilding framework literature.

natural resources involving *extraction* and *production*, and argues that extracted resources (e.g. minerals) tend to lead to *physical* violence over territorial control, while produced resources are more likely to cause *structural* forms of violence.

A recent addition to these diverse types of relations between environment/natural resources and conflict is *climate change*. Given its severe implications for the availability of natural resources and ecosystems' balance, many scholars (such as Matthew et al. 2009; Gleditsch 2012; Behnassi 2016, 2017a, b; Behnassi and McGlade (2017) and policy-makers have expressed concern over the climate-conflict nexus; therefore presenting climate change as a potential security issue that is capable of escalating latent conflicts into violent outbreaks.[6] In its latest assessment report, the Intergovernmental Panel on Climate Change (IPCC), as a scientific authority, states that some of the factors that increase the risk of violent intra-state conflict are also sensitive to climate change and variability (IPCC 2014). For Das (2015), this is a potentially vicious cycle as violent conflict could, in turn, negatively affect environmental and human systems, thereby leading to a new cycle of destruction.

From a regional perspective, Blondel (2012) considers climate change, when combined with poor governance and development and security challenges, as the largest and most global environmental variable to peace and security and its impacts are set to fall disproportionately on the world's most vulnerable populations,[7] especially in the Asia Pacific Countries. The climate effects have the potential to compound existing low-intensity conflicts,[8] spill insecurities into neighboring countries, and create new bases for insecurity. Due to inequalities and inadequate social protection programs, the most vulnerable sectors of society are likely to be the most exposed to the adverse effects of climate change. These are primarily the poor, indigenous peoples, the displaced, women, children, and the elderly. If improperly addressed, climate effects threaten to further impact on these vulnerable social groups; a dynamic that is likely to have humanitarian and security ramifications which may undermine progress toward the national and international development goals.

[6]Kelley et al., by reference to the hypothesis of climate change-induced conflicts, argued that the Syrian conflict was partially caused by a preceding drought.

[7]Focusing on Asia Pacific Countries, Blondel (2012) claims that low-and high-intensity resource-based conflicts in the region have thus far been primarily intra-state and this trend is likely to continue with climate change. Resource-based conflicts are likely to be both within and between sectors and user-groups with causes and effects also crossing state boundaries. Resource-based conflicts in the region already also stretch across borders and this has affected sub-regional stability.

[8]According to Blondel (2012), several overarching conflict risk factors are present in the Asia Pacific Countries and relate primarily to supply and demand dynamics. This is true particularly in regards to water. For example, seven of the world's greatest rivers, fed by glaciers in the Himalayas and the Tibetan plateau, supply water to roughly 40% of the world's populations. Yet as these glaciers decline several countries, primarily China where many of Asia's key rivers originate, may divert waters as their own needs increase. This could adversely impact on other countries, however, and low-intensity conflicts already exist over several proposed dam projects.

Similarly, Van Baalen and Mobjörk (2016) report that a large body of scholarly literature on climate-related environmental change and violent conflict in East Africa shows that environmental changes – such as changing rainfall patterns, droughts, changes in vegetation cover and increasing resource scarcity – have contributed to various types of violent conflict. The link is particularly evident for conflicts involving livestock herders. Case study research also shows that local resource conflicts are sometimes drawn into more intense power struggles related to civil war, for example those in the Sudan[9] and Somalia. This does not mean that climate-related environmental change automatically causes violent conflict since the political, economic and cultural context is often key.

For Blondel (2012), studies that chart the inter-linkages between environmental change and security are based primarily on the resilience, or lack thereof, of social systems to change and not on a direct 'cause and effect' dynamic between, for example, climate change and conflict. In this respect, Lee (2009: 24 cited in Blondel 2012) compares the effects of climate change on security to the dynamics present in 'slow wars', highlighting that "the chances for conflict will gradually accumulate and appear as a long-term process...[where the] prolonged effect may do more to destroy the social fabric for many years to come...[and] be more dangerous and more volatile".

Yet, despite this potential, the climate concern remains critically peripheral to national, regional and international peace and security strategies (Blondel 2012). The links between climate change, security, and conflict are not always clearly established because the climate aspect is rarely the only contributing factor in many conflict situations even if it is increasingly perceived as a 'threat multiplier'. The arguments supporting the securitization of climate change or the hypothesis of climate-induced conflicts are still weak and need robust empirical evidence. For this reason, it is still challenging for decision-making processes to perceive climate change judiciously as a potential security threat and to deliberate accordingly. This seems to indicate that the gap between what is necessary from the scientific community's point of view and what is possible from a political point of view is still wide (Behnassi 2016, 2017a, b).

23.3 Relevance of the Concept of Environmental Conflict from the Perspective of Human-Environment Nexus

As shown above, the argument that environmental factors may lead to conflict has dominated environmental security studies. However, numerous criticisms have been addressed to the environmental conflict school by scholars (like Barnett

[9]A causal link between climate change and armed conflict was clearly explained by the United Nations Environment Programme (UNEP) in its 2007 report entitled Sudan: Post-Conflict Environmental Assessment, http://postconflict.unep.ch/publications/UNEP_Sudan.pdf

2000; De Soysa 2002a; Dalby 2003; Matthew 2002) who attempt to explore other alternatives to overcome its different shortcomings and contradictions. It was argued that the field's value has been depressed by its simplified renderings, the shakiness of its core concepts and assumptions (Diehl and Gleditsch 2001) and that the school has reached a theoretical impasse unhelpful for decision-making processes. Therefore, most of these scholars subscribe to a fairly pessimistic assessment of the state of the field.

In this respect, Barnett (2000) claims that the causes and consequences of resource-driven conflicts are traditional concerns of International Relations, and these powerfully inform the environment-conflict thesis. The supply of resources problem is fundamental to neo-Malthusian theories and is commonly (but arguably mistakenly) thought to have been the central environmental problem advanced in *The Limits to Growth* (Meadows et al. 1972). Regardless of this aspect, what is of concern here is the way in which these longstanding resource issues are reinterpreted under the label 'environment'. A pervasive difficulty with this literature is the conflation of *resources* with *environment*. With respect to the question of environment and violent conflict, the environmental conflict school is by and large concerned with *resources of economic value*, rather than environmental issues *per se*. With respect to the question of scarcity for instance, which is central to this school, it should be noted that it is a relative phenomena since the problem of scarcity comes in most cases from the expectation of abundance which is denied for structural economic reasons rather than natural ones (Bookchin 1982: 71).

For many scholars, however, 'resource' and 'environmental issues' are one and the same, they are of interest only in as much as they relate to security, and the key to understanding them lies in the study of Realism's traditional geopolitical texts which have the tendency to resurface under the rubric of 'environment' (Barnett 2000). A notable function of this conflation of resource scarcities with environmental issues is that it offers strategic rationality a beachhead on the environmental agenda, because resources and conflict are part of strategists' stock-and-trade. It is important, then, to make the distinction between resource scarcity and environmental concerns clearer to provide a membrane against the inappropriate colonization of environmental issues by the resource/strategy agenda. The environmental conflict school is mainly dominated by *resource-conflict research* and does not seriously consider the substantive difference between resource and environmental problems, therefore allowing their misleading conflation (Barnett 2000).

Barnett (2000) also claims that the most complex, uncertain, and potentially disruptive problems lie not in the realm of environmental concerns but in silent, apolitical and pervasive processes which are overloading the planetary 'sinks'[10] (oceans, atmosphere, marine sediment, etc.). Accordingly, a rule of thumb is that in most of the cited instances of 'environmental' resource scarcities, where the scarce resource can be costed, its price altered according to the balance of supply and

[10]The capacity of the planet to absorb pollutants, including carbon dioxide.

demand, and if necessary substituted, then the problem is more *economic* than it is *environmental*. Environmental problems are those effects or externalities that cannot be costed or reasonably substituted – such as increasing rates of pollutant-induced cancer, biodiversity losses, and climate change effects. These issues, perceived as the essence of environmental insecurity, are already discernible in declining human security, felt mostly by the already insecure. Water and land seem two basic resources that defy this classification, having both economic and ecological functions; however, the issue of land degradation has yet to be seriously considered as a cause of conflict, and the arguments that there will be 'water wars' are also unconvincing.

In this respect, Barnett (2000) believes that the most striking difficulty of the 'water wars' thesis is the impossibility of clearly distinguishing among the many factors which contribute to warfare. When one examines this thesis, it seems that few violent conflicts have been induced solely by water scarcity, and that the broader political context is often more relevant. Nevertheless, there appears to be sufficient evidence that water is an important variable in violent conflict within, if not always between, states (Homer-Dixon and Percival 1996; Oktav 2017[11]). Therefore, the argument about water wars, which is overstated, reflects a particular product of strategic rationality and undervalues the historical and contemporary evidence that water is as likely to 'cement peace' as it is to induce violence (Cooley 1984).

In the same vein, many scholars have raised a number of theoretical and heuristic questions in order to challenge the core assumptions of the environmental conflict literature. It was argued that the inconsistency of this research is not limited to methodological weaknesses and theoretical shortcomings, but concerns also the concept of 'environmental conflict' which is: fundamentally flawed, as it neither allows for convincing empirical substantiation nor for sound theory-building; relies on preconceived causalities; intermingles eco-centric with anthropocentric philosophies; and neglects the motivations and subjective perceptions of local actors. Therefore, the concept of environmental conflict represents an inappropriate research strategy in our quest to understand human-nature interactions:

- ***The inconsistence of the one-sided fixation on causality***

Environmental conflict literature is mainly characterized by a one-sided fixation on causality based on which environment/natural resources may trigger violent conflicts. The basic dilemma here is that, in most cases, research first labels conflict as a single-issue (environment/natural resources play a decisive role in the emergence of conflicts), and subsequently adds other intervening non-environmental variables. This causal paradigm is not useful in explaining the environment-conflict nexus and often prevents a more holistic view on the complex and non-linear interactions between natural and human systems (Hagmann 2005).

[11]Focusing on Turkey's water policy and its impact on bilateral relations with Syria and Iraq in Euphrates Tigris basin, Oktav (2017) claims that water can be used both as a weapon of war during hostilities and as a source of cooperation.

Accordingly, van Baalen and Mobjörk (2016) claim that the interactions between climate-related environmental change[12] and violent conflict does not exist in a political and social vacuum. Political processes, for instance, permeate every link in the causal chain from environmental change to an increased risk of violent conflict. A group's access to natural resources or vulnerability to climate change is determined by both biophysical and political processes. Political institutions are critical for understanding why some local resource conflicts turn violent, while most do not. Analysis of the case study literature provides many examples of this; one example concerns East African pastoralists, who face increasing resource scarcity as a result of longstanding political, social and economic marginalization, in combination with more frequent and longer droughts. Acknowledging the political and social nature of the climate-conflict link is important, since it highlights the political manoeuvrability that exists for easing vulnerability and preventing violent conflict.

- ***The intermingling of eco-centric and anthropocentric philosophies***

Environmental conflict literature often assumes that environment/natural resources have the capacity to modify human behavior by 'causing' conflicts. For Hagmann (2005), the questions as to where this capacity stems from and why and how environment possesses the capacity to stimulate and transform human behavior remain unanswered. This challenge cannot be resolved through definitional exercises, methodological innovations, or large empirical samples; it depends fundamentally on whether one adopts a more eco-centric or more anthropocentric philosophy. However, the environmental conflict literature often amalgamates eco-centric and anthropocentric conceptions of agency that are incompatible.

At first sight, the idea of environmentally-induced conflict reflects an eco-centric assumption, as the environment is portrayed as being capable of modifying human behavior. However, a closer examination of the literature reveals that collective action, mostly inter-group conflict, is represented as the outcome of the interaction of environmental stresses with non-environmental factors. Thus, scholars ultimately fall back on an anthropocentric argument to explain human behavior. As long as research intermingles eco-centric and anthropocentric philosophies that are not made explicit, coherent knowledge on the subject of environmental conflict will be difficult to achieve.

- ***The non-inclusion of actors' motivations and perspectives***

Lederach (1995) claimed that conflicts often emerge due to humans' perceptions of related actions and events.[13] Yet, the environmental conflict literature has failed

[12]*Climate-related environmental change* is defined as a change in biophysical conditions that are or will be affected by a change in the state of the climate or by variations in the mean state of the climate (van Baalen and Mobjörk 2016).

[13]Insular Oceania shows that, in small islands, wars have far more serious consequences than those occurring in continents because defeated parties can only escape by sea and have little chance to survive. Also, the war is either ritualized or avoided by measures of space management that prevent

to consider how social actors contribute to, perceive, and cope with environmental stresses. Often concerned stakeholders are reduced to functional categories (such as ethnic groups, marginalized groups, pastoralists, peasants, women, etc.) or casualty numbers. This shortfall is partly due to the somewhat biased conception of conflict that guides numerous studies on environmentally-induced conflicts. Conflicts are identified exclusively when manifest in violent inter-group relations that result in a significant number of casualties.

For Hagmann (2005), focusing research agendas on the explanation of violent conflict is unquestionably a legitimate strategy. However, it bears the constant risk of disregarding the social dynamics that produce and shape collective mobilization and action before the outbreak of violence. Although the natural environment represents the physical matrix in which human interactions are inscribed, the simple reason why the environment matters to people remains unspecified. The literature typically regards ecological concerns as independent variables in modeling the ‘causal pathways’ to environmental conflict. Paradoxically, the selection of explanatory variables – such as desertification, freshwater availability, or arable land – is often accompanied with vague explanation as to *why* they are relevant in the causation of conflict or instability.

Similarly, minimal indications on the actors’ theory underlying environmental conflict analysis are occasionally given. Nonetheless, most of the literature fails to come up with an unambiguous statement on the sociological rationality of actors and a theoretical account of why humans behave the way they do (Bourdieu 1998). Many scholars, like Homer-Dixon or Gleditsch, have circumvented this admittedly challenging question by adopting methodologies that include a large number of variables supposed to *explain* violent human behavior. Again, these variables tell little about the ontology of these conflicts in the local context of many Southern countries, home to the vast majority of today’s violent conflicts (Hagmann 2005).

- ***The importance of temporal and spatial dimensions***

Van Baalen and Mobjörk (2016) have recently highlighted the importance to take the temporal and spatial dimensions into account, especially when analyzing the linkages between climate-related environmental change and violent conflict. The reason is that climate change involves boundary crossing and delayed impacts (even if many of them are already palpable), as do the dynamics of violent conflict. Mere correlation-based analyses limited to short periods or limited spatial units risk overlooking the complex and non-linear relations that shape the causal pathways between environmental change and violent conflict. It is thus essential to incorporate these dimensions into the relevant research.

---

potential belligerents from interaction, and thus develop animosity towards each other. Rather than having lines of contact between populations, as are the borders in Europe, non-contact zones are created with crossing points (an Oceanic version of the Berlin Wall with checkpoints, except that instead of the wall there are spaces inhabited by evil spirits). For centuries, people have lived with little contact with their neighbors, which has led to a linguistic speciation with no earthly equivalent (nearly 1000 languages in Melanesia which are real languages, not dialects). Therefore, from the most severe natural constraints for the survival of humans, adaptive measures may be created within viable human societies.

In addition to the above, many scholars criticized the environmental conflict literature in terms of robustness of research designs, the conceptual value of core variables, its neo-Malthusian assumptions, and the epistemology of Northern-driven discourse on environment and conflicts:

- ***Robustness of research designs***

The research designs proposed by environmental conflict scholars have been repeatedly challenged. The Toronto Group, for instance, has been accused of violating important principles of research design, such as lacking control groups, offering imprecise variables, and neglecting variation on the dependent variable, all of which make the falsification of the hypotheses impossible. Likewise, ENCOP has been accused of lacking *an ex ante* formulation of research hypotheses (Hagmann 2005). The field's tendency to propose "overly complex models which offer only very general conclusions" (Rønnfeldt 1997) or "to refer to future crisis as empirical evidence" (Toset et al. 2000) has been considered as a key limitation.

- ***Conceptual value of core variables***

The core concepts and variables of environmental conflict research reflect misleading assumptions and definitions. This holds true for seemingly technical vocabulary relating to the status of non-human entities such as 'resource scarcity' or 'environmental degradation'. It also applies to terms embracing social phenomena such as 'environmental discrimination' (Hagmann 2005). The concept of resource scarcity – a core independent variable in many studies – raises serious criticism, as virtually all natural resources are or can become scarce, and as scarcity by definition leads to conflicts of interest (Gleditsch 2001a). Dalby (2003) rightly points out that certain non-renewable resources – such as diamonds or oil – are by definition a resource, precisely because they are not ubiquitous. Furthermore, the complexity and multitude of intervening variables weaken the explanatory power of theoretical models proposed (Levy 1995). Thus, the operational measurement of many concepts of environmental conflict research leaves considerable room for interpretation.

- ***Neo-Malthusian assumptions***

Considerable attention has been paid to the links between demography, the environment, and conflict. The standard argument is that demographic growth will overextend the natural resources of the immediate environments, leading to deprivation which, it is assumed, will lead to conflict and instability either directly through competition for scarce resources, or indirectly through the generation of environmental or climate refugees (Barnett 2000). For example, according to Myers (16) "so great are the stresses generated by too many people making too many demands on their natural-resource stocks and their institutional support systems, that the pressures often create first-rate breeding grounds for conflict".

One of the main theoretical criticisms addressed to the environmental conflict school concerns this conception which assumes simplistic theoretical relationships between resource availability, demographic growth and conflict (Barnett 2000; Dalby 2000; Haas 2002; Gleditsch and Urdal 2002; Peluso and Watts 2001). For Barnett (2000), the ways in which population growth leads to environmental stress are reasonably well known; however, the particular ways in which this leads to conflict are difficult to prove. In the absence of proof, there is a negative style of argumentation, and there are blanket assertions and abrogations. Empirically speaking, it is possible though, that rather than inducing warfare, overpopulation and famine reduce the capacity of a people to wage war. In addition, from this outdated neo-Malthusian viewpoint, resources become scarce, thus exacerbating conflict once they have been overused, depleted, or degraded to a certain threshold. However, these thresholds can only be determined through inductive reasoning based on *ex-post* analysis and a selection of cases on the dependent variable (Hagmann 2005). In many cases, the literature perceives resources either in terms of scarcity or abundance, and thereby neglects that these two situations are themselves the outcome of societal processes (such as cultural patterns and market dynamics) that vary across time and space (Lipschutz 1997; Le Billon 2001).

For Barnett (2000), there are three principal features of the population-environment-conflict literature. *First*, by scripting demographic growth in Southern countries as a threat to the interests of the Northern countries, it presents the phenomenon as a problem which requires management by the North. However, this is rarely seen to involve the relinquishment or adjustment of economic power. *Second*, it assumes that the number of people is absolutely indicative of environmental impact. This totally ignores the question of what kinds of lifestyle these people lead. Overall, the environmental footprint is not merely a function of numbers, but also a function of the resources people use and the types of wastes they generate. So *lifestyle* is as important as the *number* of lives. In this respect, the most overpopulated country in the world should be the United States rather than India or China whose populations consume less natural resources and energy and emit less carbon per capita. Hence, overemphasizing the demographic aspect turns a blind eye to the complicity of Northern nations.[14] Similarly, the same inequalities may be found within countries composed of classes with differentiated consumption capacity and ecological footprint. *Finally*, by viewing demographic growth as a threat, by indicating this threat through impersonal statistics, and by seeing this from a 'global' perspective and in Malthusian terms, this literature ignores the positive social and biological aspects of population growth.

[14]This population-environment-conflict rationale is captured in an early pronouncement by Robert MacNamara (former US Secretary of Defense and former President of the World Bank) who said in 1984 that: "short of thermo-nuclear war itself, population growth is the gravest issue the world faces over the decades immediately ahead". Barnett (2000) warns against this suspicious reasoning likening population growth to nuclear war since it comes from key figures in the Northern world order such as MacNamara; whose 'world' is MacNamara referring to? If MacNamara the philanthropist is talking here about the plight of those who are adversely affected by rapid population growth and famine, then the 'world' in question may be that of the Southern people at the receiving end of the exploitative, poverty-making global economy. This 'world' is at risk from those very institutions with which MacNamara is so familiar – the World Bank, the Pentagon, and Global corporations. More probably, MacNamara is referring to the world of the wealthy and powerful and the possibility that the growth in the number of *Others* might undermine the stability of (Northern) world order. In environmental security discourse, claims to the 'global' often mask the pursuit of the Northern powers' interests (Dalby 1999).

- ***Epistemology of the Northern-driven discourse on the environment-conflict nexus***

Environmental conflict scholars have been accused of 'securitizing' environment and natural resources while disregarding other dimensions of the human-environment nexus. Dalby (2000: 173), for instance, rejects the idea of "militarizing the relationships between the poor and the rich in the face of rapidly growing disparities, and turning the poor into a military threat to the affluent, looms over this whole literature (...)". In the same vein, Buckland (2007), who is skeptic about the securitization discourse, argues that climate-related security implications may lead, among other things, to the military responding to issues for which a military response is unnecessary and potentially even detrimental.[15] For Barnett (2000), the environment-conflict literature is theoretically rather than empirically driven, and is both a product and legitimation of the North's security agenda. This literature is almost entirely premised on the ethno-centric assumption that people in the South will resort to violence in times of resource scarcity – this assumption acts as a smokescreen that diverts attention from the fact that Northern countries consume and extract most natural resources worldwide (Hagmann 2005). Rarely, if ever, is the same argument applied to people in the North. There is continued scripting of people from the South as barbaric, strongly implying that those in the North are more 'civilized'.

It is perceivable indeed that the Northern societies are characterized by a certain level of institutional/social resilience that often prevents large-scale violence. This fact offers hope as a meaningful research agenda for environmental security. Nevertheless, this resilience is the outcome of at least three possible reasons: *First*, as the Northern economies partake in the global division of labor, they bring about a global division of environmental degradation as well, thereby transferring negative environmental externalities abroad. Given this, practicing environmental security seems to be the practice of securing the ecological health of the North by transferring undesirable externalities. *Second*, the levels of wealth in the North – which is mainly gained through the exploitation of cheap materials and labor abroad – allow for institutions that provide stability and resilience to environmental change. The market, well-financed governments, the insurance industry, transport and communications infrastructure, a degree of democratic participation, and a base level of personal affluence all seem to help hedge against turmoil in the face of environmental stress. *Third*, trade between similarly affluent liberal democracies assists in the transfer of the necessary food and technologies that help enhance resilience and decrease the likelihood of conflict. Underwriting all this, however, is the ability to pay and to participate in the domestic and global economy without great disadvantage. This ability, of course, is limited to the few and powerful by the exploitation of the many (Barnett 2000).

[15]It is due to this suspicious scenario that the Non-aligned Movement and the Group of 77 objected to the securitization of climate change during the two Security Council debates on climate change (2007, 2011); in part because of the perceived potential, and even risk, that Council members may abuse this approach by relying on it as an excuse for using military intervention to enforce legal obligations in respect of climate change (Scott 2012).

Based on this, the real irony of the environmental conflict literature is that it is the North which assumes that the South will threaten; the North creates its own fiction, based on little or no evidence[16] (Barnett 2000). In this literature, the Northern strategic vision projects onto the South its own violent rationality. It assumes that the 'South' will behave as the North would, that is with aggression and force; yet this is merely *an assumption*. There may be indeed 'rogue' states, but these few are exceptions and do not represent the vast majority of Southern states. Hence, the threat to peace and security, which is fully apparent to some scholars (like Gleick 1990), is by no means apparent. The peace and security being referred to is the peace and security of the North, not the positive peace and security to which the majority of the world's people are entitled. This Northern 'peace' is a negative peace, and its 'security' is a resistance to change.

23.4 Environmental Conflict Versus Environmental Peace-Building

According to Barnett (2000), the more telling question about the linkages between environment and conflict is not '*is environmental degradation likely to lead to violence?*' nor even '*how might environmental degradation lead to violence?*', but rather '*why are we interested in the environment-violence nexus?*' In short, *why this literature*?

In the previous section, it was argued that the environment-conflict thesis is generally unconvincing and is mainly a reflection of Northern theoretical and strategic interests. Therefore, the first two questions are by and large irrelevant. The answer given to the latter question is that the environment-conflict literature is the discursive primer to legitimate the permanence of *status quo*. Thus the obsession with only one of the possible effects of environmental dynamics (*conflict*) at the expense of other effects and at the expense of taking seriously the root causes of these dynamics (such as environmental degradation and stress). The net effect of the environment-conflict thesis is, then, the justification of a state response that maintains the legitimacy of the security and military elite, and the justification for impending military and economic defense of Northern lifestyles.

In this vein, many scholars (such as Smith and Østreng 1997) have repeatedly called for consideration of null cases in which environmental dynamics (such as scarcity) do not lead to conflict. According to Hagmann (2005), little is known about the causes and processes that foster, for instance, cooperation rather than conflict over resources. The analytical shift towards an appreciation of natural resources as a source of cooperation – or rather as a source of conflict *and* cooperation – has only occurred sporadically in the recent past. Limited empirical evidence has been presented with regard to the environment-cooperation nexus. This happened

[16]The same logic applies to the current international counter-terrorism agenda.

mainly to refute overly deterministic 'eco-violence' assertions or arguing for the 'peacebuilding' potential of environmental policies such as conservation (Canter and Ndegwa 2002; Conca and Dabelko 2002; Matthew et al. 2002).

Nevertheless, the environmental cooperation hypothesis has recently and increasingly gained momentum as a complementary approach to explain human-environment interactions (Dresse et al. 2016). A growing trend, furthering the fact that conflict and cooperation can coexist and focusing on the transformative potential of environmental cooperation, has been recently observed (Pachauri et al. 2009; Giordano et al. 2005). Conca and Dabelko (2002) believe in this respect that the complex, uncertain, and long-term nature of environmental dynamics seems to create 'functional interdependencies'. These interdependencies, when coupled with the cost-effectiveness of cooperation over conflict, often create an incentive that may bring former, current, or possible future conflict parties to dialogue and eventually cooperation. For Matthew et al. (2009), the ability to find arrangements to manage shared natural resources can even provide new income sources, thus supporting post-conflict economic recovery and the peacebuilding process. Beyond stabilizing interstate relations, environmental cooperation can contribute to fostering trans-societal relations and lay the ground for a shared collective or social identity (Zikos et al. 2015). The critical nature of environmental problems for human survival is key to this potential and renders environmental cooperation an important potential component of peacebuilding. Natural resources that are shared by conflicting parties are, thus, a good entry point for dialogue and negotiation, which can later extend beyond environmental issues, laying the roots for peace and reconciliation (Fig. 23.1).

Conca and Dabelko (2002) were behind the initial development of the 'environmental peacemaking' framework. It has somewhat mutated since then and is commonly referred to as environmental peacebuilding, peace ecology, or environmental peace. From a conflict resolution tool, this framework has gradually evolved towards a more comprehensive, transformative peacebuilding approach. Researchers supporting this paradigm (such as Maas et al. 2013; Conca et al. 2005; Carius 2007) commonly classify related activities into three types:

- *Firstly*, activities which aim at preventing environmental conflicts. As pointed out by the environmental conflict or scarcity school, when natural resources are not sufficient for all groups exploiting them, tensions and violent conflicts might emerge. Therefore, limiting human pressure on these resources, coupled with their appropriate institutional management, is one way to alleviate these pressures and the risk of associated conflicts. This is especially true in situations of power asymmetry between groups, where access to natural resources and their economic benefits is determined by ethnic, economic or other socio-cultural differentiations (Carius 2007; Conca et al. 2005).
- *Secondly,* activities which aim at building peace through cooperative responses to shared environmental concerns (Conca et al. 2005). This approach tries to bring conflicting parties together to stimulate dialogue through environmental cooperation, in order to foster trust between former conflicting parties, paving

Fig. 23.1 Environmental peacebuilding and the conflict cycle. (Source: Dresse et al. 2016)

the way for conflict de-escalation, political cooperation, social transformation, and eventually reconciliation (Carius 2007; Wolf 2007; Matthew et al. 2009; Jensen and Lonergan 2013). Through regular interaction and dialogue, competing parties gradually evolve from a narrative of resource scarcity, often characterized by uncertainty and security concerns, in order to identify sustainable, win-win solutions to shared environmental concerns. The benefit of environmental cooperation is special because it has the potential to bring together conflicting parties even while violence is ongoing.[17]

- *Finally*, environmental cooperation can lay the foundations for future cooperation in other domains through the creation of opportunities for interstate bargaining. It can lead the way to political and institutional forms of cooperation between conflicting parties and, in turn, lasting peace by promoting conditions for sustainability (Conca et al. 2005; Carius 2007). Ultimately, by treating the root-causes of conflicts, environmental peacebuilding has the potential to prevent future conflicts between competing parties, rendering the use of violence as unthinkable (Conca and Dabelko 2002).

[17] Water negotiations between Israel and Jordan illustrate this (Wolf et al. 2005; Jägerskog 2013).

Despite the attractiveness of environmental peacebuilding, there is still no unified model or definition of this concept, and the available empirical evidence to corroborate the existence of a direct relationship between environment/natural resources and either conflict or peace is either insufficient or contradictory (Ide and Scheffran 2008). Following this line of thinking, Maas et al. (2013) claim that environmental peacebuilding is not yet a "coherent theoretical school", nor a "distinct set of practical activities", but is instead "an umbrella term that covers a wide range of aspects on the relationships between environment, conflict, and peace". For Dresse et al. (2016), the difficulty in testing this link between environmental dynamics and peacebuilding is partly due to the absence of adequate indicators to measure environmental cooperation, in addition to the fact that several elements of this emerging framework are still unclear. In order to move forward and address these shortcomings, Dresse et al. (2016) focus in their research on the following particular aspects of the environmental peacebuilding framework:

- ***The need to clarify the terminology of the environmental peacebuilding concept***

It was argued that the interchangeable use of environmental 'peacemaking' and environmental 'peacebuilding' by many researchers is problematic because the *timeframe* to which each of these terms refers is unclear. While environmental cooperation is predominantly implemented during periods of low violence intensity according to some researchers, there is no clear-cut separation between conflict and peace, especially in the case of prolonged conflicts where periods of acute violence alternate with latent phases. This might explain the confusion between peacemaking, which is traditionally seen as activities that are implemented to end a conflict, and post-conflict peacebuilding.

Meanwhile, these two terms refer to two distinct sets of *objectives* pursued by peacemaking and peacebuilding and the types of *activities* they imply. While peacemaking aims at deescalating the violence level (achieving the negative peace or the absence of violent conflict), peacebuilding aims to secure lasting peace (positive or sustainable peace and reconciliation between stakeholders) by solving the root causes of the violence in a peaceful and cooperative way.

Based on this, and since conflicts are complex, multifaceted processes, there is a need for a more comprehensive approach to conflict transformation than that which is envisioned by the environmental peacemaking framework. This does not mean that environmental peacemaking does not exist, but simply that it refers to a more limited framework than that of peacebuilding. When environmental cooperation is used as a means to foster trust and dialogue between communities, thereby preventing future conflicts and impacting sustainability, this corresponds rather to environmental peacebuilding. Accordingly, environmental peacebuilding measures should be implemented when relevant: in the pre-conflict phase to prevent an escalation of latent violence; during a conflict to support a smooth transition to peace; and in the post-conflict phase to ensure sustainable peace (Conca et al. 2005).

- ***The need to consider the specific qualities of the environment in contrast to other areas***

Natural environment is just one of many other issues around which peacebuilding can be articulated. However, the environment has distinctive qualities which potentially strengthen peacebuilding efforts and offers a broad range of types of actions – such as transboundary water agreements, joint research projects, education, or peace parks which promote biodiversity conservation and eco-tourism. It appears from the literature, though, that some types of environmental issues are more likely to result in cooperation than others, and that some types of cooperation are easier to implement than others. Depending on the local needs and non-environmental factors, the environment and natural resources can thus be a more or less suitable peacebuilding tool. However, more detailed quantitative and qualitative data are needed to further investigate which type of environmental cooperation is best suited and in which contexts.

In addition, it has also been noted that environmental cooperation can contribute to durable peace, regardless of whether the environment caused the conflict in the first place (Ali 2007). The environmental peacebuilding framework can also be seen as an entry point for broader cooperation in other areas, ultimately restoring peaceful relationships (Matthew et al. 2009; Amster 2015). Environmental issues are often lower on the political agenda, and may as such provide a good entry point for dialogue and cooperation (Maas et al. 2013).

23.5 The Human-Environment Nexus from Peace-Building to Coviability Perspective

Taking into account the outcome of this assessment, peace and conflict researchers should endeavor to develop alternative approaches. In order to circumvent the weaknesses of existing literature, Hagmann (2005) proposed, for instance, a shift from environmentally-induced conflict to natural resource-use conflicts through at least three major analytical changes. *First*, one must not assume that resource scarcity or environmental degradation predispose violent conflict. Rather resource use should be viewed as a contested process that inscribes itself in cooperative and conflictive relations between different resource user groups. Hence, natural resource management strategies and conflict management practices should gain importance and become new research themes. *Second*, the analysis of resource use patterns and conflicts requires a thorough understanding of institutions that shape the rules and rights of resource use. Different layers of environmental governance at local, national, and international levels need to be incorporated into the analysis of resource-use conflicts. For instance, the overlap of customary and modern state rules for resource and conflict management in Southern countries deserves more attention. *Third*, a shift from a purely objectivist analysis to one taking into consideration the intentions, meanings, and logic for action by local communities is imperative. Dedicating more interest to the rationale of actors in resource-use conflicts is also a precondition for formulating conflict transformation strategies.

In the same spirit, the dichotomy between humans and the biophysical environment as a passive, external entity is being challenged by many researchers (like Descola and Pálsson 1996; Moran 2007; Robbins et al. 2013; Barrière 2017). These researchers increasingly believe in the inseparability of humans and their natural surroundings, especially in the Anthropocene era, which has been triggered by anthropogenic actions, resulting in numerous global risks to both humanity and ecosystems. Moving away from the conceptualizations presenting environmental dynamics as a potential driver of conflict, many researchers have begun envisioning environment as an incentive for cooperation rather than for violence. This was relevant especially as conflict and cooperation often coexist and are not mutually exclusive (Dresse et al. 2016).

While cooperation often presents the best chances to implement equitable solutions to apolitical environmental issues, many critical natural resources – such as water, land, energy, and climate – are still securitized in many scientific and decision making spheres. This often fosters distrust and division between competing groups, leading to unilateral, detrimental decisions to perceived scarcity threats and ultimately increasing tensions and conflict escalation. To reverse this trend, some researchers – like Zikos et al. (2015) – suggest switching from the biophysical environment, as defined by political borders, to socio-ecological systems as the referent object of securitization. This paradigmatic shift provides the opportunity to envision environmental dynamics as common concerns, therefore inspiring cooperation and solidarity rather than conflict between stakeholders. Thus, the focus will not be so much on the environment and natural resources as such (from a purely consumerist and utilitarian point of view), but on the linkages between social and ecological sub-systems with the prospect of broader peacebuilding, as well as the subsequent institutions which regulate the interactions between these two sub-systems (Zikos et al. 2015).

As discussed above, the environmental peacebuilding paradigm does not focus on the environment and natural resources as such, but rather emphasizes the human-environment nexus, as well as the cooperation that can emanate from it. However, to ensure an inclusive, equitable and just process, all stakeholders involved in this nexus should be considered when exploring, analyzing, and/or applying the environmental peace-building framework. The mainstreaming of gender equity and rights-based approaches, for instance, seems relevant in this perspective. Failing to do so could result in the exclusion of some groups along with their interests and expectations, thus creating additional conflicts or tensions (Dresse et al. 2016).

Environmental peacebuilding has also the potential to stimulate dialogue and confidence-building between grassroots actors (academia, civil society actors, local authorities, etc.) on environmental issues by creating strategic social spaces for regular interaction and systematic negotiations on different scales. This may provide the opportunity to exchange insights, knowledge, practices and more on shared concerns while deconstructing mutual stereotypes (Maas et al. 2013), thereby preventing future conflicts. As such, these grassroots actors can be seen as 'de-securitization' actors that are capable of restoring peaceful relations between stakeholders (Coskun 2009). Nevertheless, cooperation, when limited to grassroots initiatives and scientific or technical collaboration, is insufficient because these initiatives need to develop into institutionalized forms of cooperation in order to have a sustainable impact on peacebuilding. Otherwise, conflicts are more likely to arise. Yet, most environmental cooperation initiatives are not formally institutionalized, and researchers often fail to identify which institutional forms are best suited to ensure peace and in which context.[18] It is thus important, when considering, describing and implementing environmental peacebuilding to not only understand what this concept implies but also to ensure an institutional process, including a transition from mere technical to political cooperation (Dresse et al. 2016).

In the same vein, according to Blondel (2012) the decisions that states and the international community take today directly impact on the vulnerability of populations, sectors, peace and security. However, when it comes to some kind of environmental dynamics, they are still being primarily framed within scientific and environmental discourse and less so within contexts of development and peace. The case of climate change is specific in this respect since it represents the most global environmental variable in international peace and security and today's climate governance choices will be central in determining future prospects. Yet, not only are the human implications of climate change serious, but also the global climate regime is still not sufficiently shaped to reduce them and a large part of response mechanisms may even exacerbate environmental damage and human rights violations. Also, regional and domestic climate governance frameworks do not systematically refer to justice, equity, and human rights approaches. Therefore, the significant challenge currently being faced is how to ensure that human rights are widely recognized and genuinely mainstreamed in the global climate regime and response mechanisms at all scales (Behnassi forthcoming).

The mainstreaming of human rights may lead to a focus on the right to life, to diversity, to a health environment, etc. and this in turn may result in a focus on the imperative need to ensure the viability of humans on planet Earth. In other terms, the need to maintain solidarity between humans and non-humans on this planet according to a certain balance which may condition the mutual existence. In this perspective, environmental peace-building will be guided by this ultimate objective (i.e. with regard to resource use, everything will be managed to ensure the viable permanence of both the biophysical environment and human systems). All existing regulatory frameworks, for instance, should be adjusted within this direction, therefore preventing conflicts and creating favorable conditions for inter- and intra-states peace and solidarity.

[18]Transboundary agreements on shared natural resources are an often-used example of institutionalized environmental cooperation, which is understood in the literature and among many stakeholders as a positive indicator of political will to cooperate on environmental issues. Political cooperation, based on pre-existing environmental cooperation initiatives, has similarly been shown in a number of cases to be a successful way to restore dialogue between states (Dresse et al. 2016).

In this sense, the coviability of ecological-social systems – based on the premise that the viability of human systems is conditioned by the viability of ecological systems – can be perceived as a paradigmatic framework within which new conceptual alternatives with regard to environment-conflict-peace nexus can be developed. The global human society's viability is currently at risk due, among others, to environmental/climate risks. Therefore, the ecological imperative is increasingly embodied in the links that connect man to the biosphere by simultaneously questioning human and environmental security. Within this context, coviability, as an emerging paradigm, can be perceived as another neutral and objective approach to perceive the human-environment nexus, not under the ideology of sustainable development or anthropocentrism, but according to a viewpoint which repositions humans as part of the biosphere. Such an approach, which is still in its elementary or embryonic stage, has the potential to foster the notions of resilience and adaptive capacity, therefore informing future works on human-environment nexus.

Achieving co-viability depends primarily on a theoretical, cultural, and political shift. From a pragmatic point of view, it depends on the adequacy of institutions and ecosystems (Ostrom 2009) which requires an adapted governance system with different scales of decision-making and supported by a socially-accepted regulation for an effective implementation (Barrière 2017). Blondel (2012) believes that addressing these issues requires a marriage of strong international and regional coordination and state-based actions with peace, governance and coviability at its core.

At national levels, several measures exist to address key issues – such as policies related to climate mitigation and adaptation, disaster risk reduction and SDGs – yet, while there is no one-size-fits-all solution, these are mostly ad hoc and unlikely to prevent conflict. In many cases, information with which to form comprehensive and targeted peace and security-based approaches on existing and possible resource conflict fault-lines is insufficient and at times contradictory. This is particularly true of smaller-scale tensions and conflicts that easily go unnoticed but could increase with environmental stress. Increasing information on resource-based conflicts is, therefore, central to developing sound mitigation and adaptation policies (Blondel 2012). Curbing the effects of climate change that could undermine security at the national level means, therefore, increasing information of conflict and natural resource fault-lines, prioritizing good governance, and advancing the involvement of local communities and vulnerable groups in the identification of both problems and solutions. This could also mean: prioritizing natural resource management; and adjusting domestic demand and trade to alleviate vulnerability in regards to international market fluctuations and local availability. Policies related to climate mitigation and adaptation and disaster risk reduction should also incorporate conflict prevention and resolution strategies and ensure comprehensive participatory processes.

At the regional level, countries must advance coordination and cooperation based on information sharing and through the development of a regional perspective to countering the effects of environmental/climate change from a coviability perspective. Regional agreements and mechanisms should also been seen as a coordination of common regional concerns and as an opportunity to increase the voice of regional bodies within the international fora. The growing political and economic strength of some regional powers should be mirrored in their influence in other areas that promote active peace. Regional resource management should be considered as a basis for peace and stability within and between states, and this has been true particularly in regards to shared water management. Climate-friendly policies, for instance, not only represent a mechanism to ensure long-term peace and sustainability, they also present opportunities for building bridges across communities and countries through comprehensive representation and cooperation.

At the international level, countries must advance the issue of environmental/climate change and peace. Address should span across the relevant multilateral bodies. Support should be provided to vulnerable countries to incorporate resource-conflict mitigation and resolution strategies into national and regional policy and legislation.

23.6 Concluding Remarks

Over the past decades, the conceptualization of the human-environment nexus has evolved in the scientific literature from environmental conflict to environmental peacebuilding. In this work, key findings and remaining gaps were identified for each research trend.

Regarding the literature on environmental conflict, the analysis showed that its core assumptions remain questionable and its empirical and theoretical conclusions are contested. More precisely, the concept of 'environmental conflict' has failed, due to many shortcomings and contradictions, to provide an appropriate framework to perceive and manage the human-environment nexus. It is unable as to analyze well the heterogeneous trends within ecosystems and the multitude of existing natural resource management practices. Its one-sided fixation on causality and attempt to produce causal chains between a specific state of the environment (preferably a degraded, depleted, and overpopulated one) with a specific type of inter-group relationship (violence, warfare) have proved empirically controversial and theoretically unsound. To this day, the concept of environmental conflict represents a global paradigm in search of a local reality. It hinders rather than improves our understanding of the interrelations between ecology, politics, and violence. In sum, it represents an inappropriate research strategy in our quest to understand the human-environment nexus (Hagmann 2005).

The focus on environmental conflict literature has been made in order to show the full potential of the concept of environmental peacebuilding. However, despite the attractiveness of this emerging concept, and by reference to the literature review upon which this work is based, there is still a disparity and lack of consensus regarding this concept and how it is to be applied. A main challenge is to demonstrate the effectivity of environmental peacebuilding, and to identify the circumstances under which it can be a successful conflict resolution tool. While several case studies explore the linkages between the environment, conflicts, and peace, more systematic research is needed to understand if and how environmental cooperation can contribute to peace. Indeed, little empirical evidence substantiates the causal relationship between environmental interdependency and either violent conflicts on the one hand and cooperation and peace on the other hand (Dresse et al. 2016). Although recent research strongly suggests that shared environmental concerns can effectively contribute to sustainable peace-building, the question of whether and how this can be achieved remains open. Therefore, the conditions of its success in contexts that are often very different demand further analysis and systematization.

The analysis concludes by suggesting the coviability as an alternative to appropriately perceive the human-environment nexus. This approach has the potential to adjust our perception with regard to the position of humans in the planet Earth, which should be mainly oriented towards ensuring solidarity between humans to maintain a viable biosphere. This perspective may substantially increase the sense of belonging to the same planet, thus reducing eco-centric perceptions, which often lead to tensions and conflicts, and promoting cooperation and solidarity between different stakeholders to serve common objectives.

Human society, in order to survive, will have to organize itself in a different way. It has yet to appreciate the implications of unprecedented environmental changes for life on Earth, including for human lives. Meeting these challenges requires a paradigmatic and societal shift, enhanced awareness, supported by appropriate governance frameworks that can anticipate and adapt to changing conditions, as well as minimize negative consequences. Our knowledge, lifestyles and management schemes need to be recalibrated to reflect the realities of environmental/climate change impacts on our natural systems. At the local level, a range of responses may be needed to enable affected places and communities to survive or thrive under new conditions (Pecl et al. 2017).

References

Ali SH (2007) Peace parks: conservation and conflict resolution. MIT Press, Cambridge/London

Amster R (2015) Peace ecology. Routledge, Oxon/New York

Baechler G (1998) Why environmental transformation causes violence: a synthesis. Environ Change Secur Rep 4:24–44

Baechler G (1999) Violence through environmental discrimination: causes, Rwanda arena, and conflict model. Kluwer, Dordrecht

Baechler G, Spillmann KR (eds) (1996) Environmental degradation as a cause of war. Vol. 2: regional and country studies of research fellows and environmental degradation as a cause of war. Vol. 3: country studies of external experts. Rüegger, Chur/Zurich

Barnett J (2000) Destabilizing the environment-conflict thesis. Rev Int Stud 26:271–288

Barrière O (2017) Human relationship to the land from a legal perspective as a human and environmental security challenge. In: Behnassi M, McGlade K (eds) Environmental change and human security in Africa and the Middle East. Springer, Berlin/Heidelberg

Behnassi M (2016) The 'securitization' of climate change: relevance and implications for the global climate regime. In: Lewis M, Couzens E, Honkonen T (eds) International environmental law-making and diplomacy review. University of Eastern Finland – UNEP Course Series 16. University of Joensuu, Joensuu, pp 55–87
Behnassi M (2017a) Climate security as a framework for climate policy and governance. In: Behnassi M, McGlade K (eds) Environmental change and human security in Africa and the Middle East. Springer, Berlin/Heidelberg, pp 3–24
Behnassi M (2017b) Mainstreaming a rights-based approach in the global climate regime. In: Behnassi M, Himangana G, Pollmann O (eds) Human and environmental security in the era of global risks. Springer, Berlin/Heidelberg (Forthcoming)
Behnassi M, McGlade K (eds) (2017) Environmental change and human security in Africa and the Middle East. Springer, Berlin/Heidelberg
Bernauer T, Tobias B, Vally K (2012) Environmental changes and violent conflict. Environ Res Lett 7:1–8
Blondel A (2012) Climate change fuelling resource-based conflicts in the Asia-Pacific. Asia-Pacific Human Development Report Background Papers Series 2012/12
Bookchin M (1982) The ecology of freedom. Cheshire Books, Palo Alto
Bourdieu P (1998) Practical reason: on the theory of action. Polity Press, Cambridge
Buckland B (2007) A climate of war? Stopping the securitization of global climate change. International Peace Bureau. Available at: http://www.ipb.org/uploads/tbl_contingut_web/176/documents/paper.pdf. Visited 24 June 2015
Canter MJ, Ndegwa SN (2002) Environmental scarcity and conflict: a contrary case from Lake Victoria. Glob Environ Polit 2:40–62
Carius A (2007) Environmental peacebuilding: conditions for success. In: Scholars WWICF (ed) Environmental change and security report
Conca K, Dabelko GD (2002) Environmental peacemaking. Wodrow Wilson Center Press, Washington, DC
Conca K, Carius A, Dabelko GD (2005) Building peace through environmental cooperation. In: Institute TW (ed) State of the world 2005: redefining global security. Island Press, New York/London
Cooley J (1984) The war over water. Foreign Policy 54:3–26
Coskun BB (2009) Cooperation over water resources as a tool for desecuritisation: the Israeli-Palestinian environmental NGOs as desecuritising actor. Eur J Econ Polit Stud 2:97–115
Dabelko G, Simmons PJ (1997) Environment and security: core ideas and US government initiatives. SAIS Rev 17:127–146
Dalby S (1999) The threat from the south: geopolitics, equity and environmental security. In: Deudney D, Matthew R (eds) Contested grounds: security and conflict in the new environmental politics. State University of New York Press, Albany
Dalby S (2000) Jousting with Malthus' ghost: environment and conflict after the cold war. Geopolitics 5:165–175
Dalby S (2003) Resources and conflict: contesting constructions of environmental security. Paper presented at a conference in Kathmandu, Nepal
Das O (2015) Climate change, the environment and armed conflict. Working Paper No. 6, Centre for Legal Research
de Soysa I (2000) The resource curse: are civil wars driven by rapacity or paucity? In: Berdal M, Malone DM (eds) Greed and grievance: economic agendas in civil wars. Lynne Rienner, Boulder/London, pp 113–135
de Soysa I (2002a) Ecoviolence: shrinking pie, or honey pot? Glob Environ Polit 2:27
de Soysa I (2002b) Paradise is a bazaar? Greed, creed, and governance in civil war, 1989–99. J Peace Res 39:395–416
Descola P, Pálsson G (1996) Nature and society: anthropological perspectives. Routledge, London/New York
Diehl PF, Gleditsch NP (2001) Controversies and questions. In: Diehl PF, Gleditsch NP (eds) Environmental conflict. Westview Press, Boulder/Oxford, p 6

Dresse A, Nielsen JØ, Zikos D (2016) Moving beyond natural resources as a source of conflict: exploring the human-environment nexus of environmental peacebuilding. THESys Discussion Paper No. 2016-2. Humboldt-Universität zu Berlin, Berlin, Germany, pp 1–22. edoc.hu-berlin.de/series/thesysdiscpapers

Giordano MF, Giordano MA, Wolf AT (2005) International resource conflict and mitigation. J Peace Res 42:47–65

Gleditsch NP (ed) (1997) Conflict and the environment. Kluwer, Dordrecht

Gleditsch NP (ed) (2001a) Armed conflict and the environment. In: Diehl PF, Gleditsch NP (eds) Environmental conflict. Westview Press, Boulder/Oxford, pp 251–272

Gleditsch NP (ed) (2001b) Environmental change, security, and conflict. In: Crocker C, Hampson FO, Aall P (eds) Turbulent peace: the challenges of managing international conflict. US Institute of Peace, Washington, DC, pp 53–68

Gleditsch NP (2004) Beyond scarcity vs. abundance: a policy research agenda for natural resources and conflict. In: United Nations Environment Programme (ed) Understanding environment, conflict, and cooperation. UNEP, Nairobi, pp 16–18

Gleditsch NP (2012) Whither the weather? Climate change and conflict. J Peace Res 49:3–9

Gleditsch NP, Sverdrup BO (2002) Democracy and the environment. In: Page EA, Redclift M (eds) Human security and the environment. Edward Elgar, Cheltenham, pp 45–70

Gleditsch NP, Urdal H (2002) Ecoviolence? Links between population growth, environmental scarcity and violent conflict in Thomas Homer-Dixon's work. J Int Aff 56:283–302

Gleick P (1990) Environment, resources, and international security and politics. In: Arnett E (ed) Science and international security. American Association for the Advancement of Science, Washington, DC, pp 501–523

Gleick NP, Heberger M (2014) Water and conflict: events, trends, and analysis (2011–2012). In: Gleick PH, Institute P, Ajami N, Christian-Smith J, Cooley H, Donnelly K, Fulton J, HA ML, Heberger M, Moore E (eds) The world's water volume 8: the biennial report on freshwater resources. Island Press, Washington, DC

Haas PM (2002) Constructing environmental conflicts from resource scarcity. Glob Environ Polit 2:1–11

Hagmann T (2005) Confronting the concept of environmentally induced conflict. Peace, Conflict and Development: Issue Six

Hauge W, Ellingsen T (2001) Causal pathways to conflict. In: Diehl PF, Gleditsch NP (eds) Environmental conflict. Westview Press, Boulder/Oxford, pp 36–57

Homer-Dixon T (1991) On the threshold: environmental changes as causes of acute conflict. Int Secur 16:76–116

Homer-Dixon T (1994) Environmental scarcities and violent conflict: evidence from cases. Int Secur 19:5–40

Homer-Dixon T (1995) The ingenuity gap: can poor countries adapt to resource scarcity? Popul Dev Rev 21:587–612

Homer-Dixon T (1999) Environment, scarcity, and violence. Princeton University Press, Chichester

Homer-Dixon T, Levy MA (1995) Correspondence. Environment and security. Int Secur 20: 189–198

Homer-Dixon T, Percival V (1996) Environmental scarcity and violent conflict: briefing book. American Association for the Advancement of Science, Toronto

Ide T, Scheffran J (2008) Climate change: source of conflict or promoter of cooperation? Working paper. CLISEC, Hamburg

Intergovernmental Panel on Climate Change (IPCC) (2014) Climate change 2014: synthesis report. Contribution of Working Groups I, II and III to the fifth assessment report of the Intergovernmental Panel on Climate Change [Core Writing Team, Pachauri RK, Meyer LA (eds)]. IPCC, Geneva, Switzerland, 151 pp

Jägerskog A (2013) Water-sharing among Israelis, Jordanians and Palestinians. In: A world of science. UNESCO, Paris

Jarraud NS, Lordos A (2012) Participatory approaches to environmental conflict resolution in Cyprus. Confl Resolut Q 29:261–281
Jensen D, Lonergan S (2013) Assessing and restoring natural resources in post-conflict peacebuilding. Taylor & Francis, Florence
Kramer A, Wolf AT, Carius A, Dabelko GD (2013) The key to managing conflict and cooperation over water. In: A world of science. UNESCO, Paris
Le Billon P (2001) The political ecology of war: natural resources and armed conflicts. Polit Geogr 20:561–584
Lederach JP (1995) Preparing for peace: conflict transformation across cultures. Syracuse University Press, Syracuse
Lee JR (2009) Climate change and armed conflict: hot and cold wars. Routledge, New York
Lelieveld J, Beirle S, Hörmann C, Stenchikov G, Wagner T (2015) Abrupt recent trend changes in atmospheric nitrogen dioxide over the Middle East. Am Assoc Adv Sci 1
Levy MA (1995) Is the environment a national security issue? Int Secur 20:35–62
Lipschutz RD (1997) Environmental conflict and environmental determinism: the relative importance of social and natural factors. In: Gleditsch NP (ed) Conflict and the environment. Kluwer, Dordrecht, pp 35–50
Maas A, Carius A, Wittich A (2013) From conflict to cooperation: environmental cooperation as a tool for peace-building. In: Floyd R, Matthew RA (eds) Environmental security: approaches and issues, 1st edn. Routledge, Oxon/New York
Matthew RA (2002) In defense of environment and security research. Environ Change Secur Rep 8:116
Matthew RA, Mark H, Jason S (eds) (2002) Conserving the peace: resources, livelihoods and security. International Institute for Sustainable Development, Winnipeg
Matthew RA, Brown O, Jensen D (2009) From conflict to peacebuilding: the role of natural resources and the environment. United Nations Environment Programme, UNEP/Earthprint
Meadows D, Dennis M, Jorgen R, William B (1972) The limits to growth. Universe Books, New York
Moran EF (2007) The human-environment nexus: progress in the past decade in the integrated analysis of human and biophysical factors. In: Hornborg A, Crumley CL (eds) The world system and the earth system. Left Coast Press, Walnut Creek
Oktav ÖZ (2017) Turkey's water policy in the Euphrates- Tigris Basin. In: Behnassi M, McGlade K (eds) Environmental change and human security in Africa and the Middle East. Springer, Berlin/Heidelberg, pp 239–258
Ostrom E (2009) A general framework for analyzing sustainability of social-ecological systems. Science 325:419–422
Pachauri RK, Gnacadja L, Afes-Press HG, Cutajar MZ, Steiner A, Behera NC, Briceno S, Kameri-Mbote P, Ogwu J, Dimas S (2009) Facing global environmental change: environmental, human, energy, food, health and water security concepts. Springer, New York
Pecl GT et al (2017) Biodiversity redistribution under climate change: impacts on ecosystems and human well-being. Science 355(6332)
Peluso NL, Watts M (eds) (2001) Violent environments. Cornell University Press, Ithaca
Percival V, Homer-Dixon T (1998) Environmental scarcity and violent conflict: the case of South Africa. J Peace Res 35:279–298
Robbins P, Hintz J, Moore SA (2013) Environment and society: a critical introduction. Wiley, Chichester
Rønnfeldt CF (1997) Three generations of environment and security research. J Peace Res 34: 473–482
Ross M (2003) The natural resource curse: how wealth can make you poor. In: Bannon I, Collier P (eds) Natural resources and violent conflict: options and actions. The World Bank, Washington, DC
Schwartz DM, Deligiannis T, Homer-Dixon T (2001) The environment and violent conflict. In: Diehl PF, Gleditsch NP (eds) Environmental conflict. Westview Press, Boulder/Oxford, pp 273–294

Scott SV (2012) The securitization of climate change in world politics: how close have we come and would full securitization enhance the efficacy of global climate change policy? RECIEL 21:220–230

Scott CA, Thapa B (2015) Environmental security. environmental science, ISBN: 9780199363445, January

Smith D, Østreng W (eds) (1997) Research on environment, poverty and conflict. PRIO, Oslo

Toset HPW, Gleditsch NP, Hegre H (2000) Shared rivers and interstate conflict. Polit Geogr 19:978

United Nations Environment Programme (UNEP) (2007) Sudan: post-conflict environmental assessment. http://postconflict.unep.ch/publications/UNEP_Sudan.pdf

Van Baalen S, Mobjörk M (2016) A coming anarchy? Pathways from climate change to violent conflict in East Africa, Stockholm University, Stockholm International Peace Research Institute & The Swedish Institute of International Affairs. Available at: http://www.diva-portal.org/smash/get/diva2:928237/FULLTEXT01.pdf

Wæver O (1995) Securitization and desecuritization. In: Lipschutz RD (ed) On security. Columbia University Press, New York

Westing A (1986) Global resources and international conflict: environmental factors in strategic policy and action. Oxford University Press, Oxford

Wolf AT (2007) Shared waters: conflict and cooperation. Annu Rev Environ Resour 32:241–269

Wolf AT, Kramer A, Carius A, Dabelko GD (2005) Managing water conflict and cooperation. In: Institute TW (ed) State of the world 2005: redefining global security. Island Press, New York/London

Zikos D, Sorman AH, Lau M (2015) Beyond water security: asecuritisation and identity in Cyprus. Int Environ Agreem Polit Law Econ 15:309–326

Chapter 24
Link to the Biosphere: Man, Condemned to Alterity and Coviability

Auguste Eyene Essono

Contents

A. Eyene Essono (✉)
Ecole Normale Supérieure de Libreville, Libreville, Gabon
e-mail: eyenessono@yahoo.fr

O. Barrière et al. (eds.), *Coviability of Social and Ecological Systems: Reconnecting Mankind to the Biosphere in an Era of Global Change*,
https://doi.org/10.1007/978-3-319-78497-7_24

24.1 Introduction

24.1.1 Man and the Anthropocene Epoch: From Excess to Recklessness, a Veritable Rupture

Humanity has definitively entered a new era known as the "Anthropocene Epoch" (Bonneuil and Fressoz 2013). This is the direct consequence of a "calculated" and programmed grip man has gained on nature, as desired by certain philosophers from past centuries: "it is indeed this mission to conquer nature and to dominate the world that Bacon, Descartes, Buffon and Marx have, in the name of science, entrusted humanity with" (Gould 2000, 60). We have in fact achieved this since the first industrial revolution, given that "man, becoming the sovereign of the cosmos," (ibid.) has colonized "everything". The domination he has today over the world now goes beyond all conceivable limits.

Humanity's grip on the planet thus marks a rupture (a) characterized by the disappearance of links between the biosphere and the technosphere (b). Reconnecting man to the biosphere calls for both an ethical awareness of the ecological danger (c) and a scientific basis to put in place a joint obligation ethic (d).

It is this race with no landmarks, this "madness men were exposed to in the industrial era" that characterizes the Anthropocene Epoch (Guery 1989, 106). In this way, humanity (western and westernized) is still running, constantly carried along by the exploitation of materials, without any concern for the future. This is clearly a consumerist teleology which rejects humanity: a culture that obligates us to accept progress as being synonymous with perpetual overshooting. The purpose of surpassing each level of performance achieved in the geostrategic race is to control the decision-making levers of the planet's resources. Thanks to progress in science and technological innovation, the most advanced societies are mastering the laws of nature upstream. They also control the possibilities for the use of natural or genetic resources, etc. Consequently, all, or almost all of the World's nations are constantly looking to establish their hegemony as regards production flows and exchanges of goods. To that s end, each nation only knows how to grow by increasing its level of control over its energy consumption and resource levers.

However, nations do not always take enough time to distance themselves and ask themselves about the meaning and the ultimate consequences of their actions. Moreover, "the will to tackle major global environmental problems, those that will affect the habitability of the planet for years to come, is failing everywhere" (Bourg 2014). We excessively consume the resources and benefits bestowed by nature but we remain insufficiently alert to the major risks and threats posed. In other words, "we exploit goods, which have been forever pulled from the Earth" (Lombroso 1931, 225), but we remain insensitive to the impact of our actions on nature. Ultra-consumerism, a secular dogma of a culture which harms and of this reckless Western civilization,[1] of the excesses that carry it along,[2] is at the same time a mode of thought and a mode of production and consumption. In short, it is a way of being, a dominant cultural paradigm. The model of thought,[3] which, at the basis, consolidates the paradigm in force is being rooted in the Cartesian dogma that elevates man to the status of "master and possessor of nature". Goethe understands and clearly presents – through his fascinating discussion with Johann Peter – the characteristics of this model: "henceforth in France and in the study of nature, the mind will dominate and will govern nature. We will take a look at the great laws of Creation and peek into the Lord's secret workplace" (Goethe, quoted by Saint-Sernin 2012).

In brief, thought, science and the prowess of leading-edge technology are taming the physical world, elaborating its laws and deriving the greatest industrial profit from it; by excessively[4] exploiting natural resources[5]:

> Man, in this technological age, is provoked into a showdown, firstly at the level of nature which is the chief storehouse of our energy reserves. The "offensive" behavior of man, in a corresponding manner, is firstly revealed in the appearance of the modern, exact science of nature. The representational manner characteristic of this science pursues nature, which is considered as a calculable set of forces. Modern physics is not experimental because it uses apparatus to study nature. Rather, the reverse is true. Because physics – already as pure theory – calls on nature to exhibit itself as a set of forces which can be calculated and anticipated, which experiments are called on to question, in order to discover whether and how nature responds to this request. (Heidegger 1958, 28–29)

In this context, nature becomes a "*Stock*: a standing-reserve or supply store that leads up to a declassified waste cycle that can only be reused in a degraded form to give rise to a new entropic product whose quality will be diluted into infinitesimal proportions" (Guery 1989, 107–108). The ability to totally tame matter, a resource and a source of energy to be consumed at will, immediately endangers the very foundation of the human being in a constantly-evolving world.[6] As a result, life in its diversity and ecosystems are seriously affected. This crisis, is deep and multifaceted: "the environmental crisis affects all levels of biodiversity and the

[1]Civilization characterized by a sort of "industrialist orgy" and the obsession concerning systematic and teleological progress.

[2]Gina Lombroso describes this well (1931). She affirms that "the goods of which we are proud are not eternal," "by exploiting them as we do, we provoke problems for the future whose seriousness we are only beginning to measure" (p 232).

[3]That of the ultra-liberal economy that structures the dominant technological civilization of the moment.

[4]The extreme exploitation of nature by "mechanical civilization" is limitless. To this end, François Guery asserts that "in the ultimate form of an affront to nature, industry appears capable and therefore responsible: both stupid (it wastes) and wicked (it is violent)" (Guery 1989, 109).

[5]According to Olivier Godard's definition, "'Natural resources' or 'natural assets' refer to non-produced and non-producible goods by humans but which still respond to a demand from human agents." (Godard 2015, 114).

[6]These are co-evolutionary relationships between species and their common biotic link with the biosphere.

interrelationships between them: ecosystems and natural habitats are destroyed, fragmented or polluted by human activity, and genetic and intraspecific diversity suffer the most devastating destruction" (Delord 2005).

24.1.2 *The Rupture Between the Biosphere and the Technosphere*

The same crisis leads to a rupture of the original link between the biosphere[7] and the technosphere.[8] Since the advent of the first industrial revolution, technology is no longer thought to be in harmony with nature. At best, technology is tending to move away from nature, whilst, in the worst cases, it is becoming its antinomy. Yet, the unity of benchmark values and the multimodal and internal interdependence bond that shapes each sphere is no longer a matter of controversy among researchers. If, since Charles Darwin, evolutionary theory "provides a means of thinking out the unity of the constitution and the history of "living organisms" (Saint-Sernin 2012), it is necessary to recognize that "since the middle of the 20th century, the globalization of the economy renders more evident the "cross-linked" character of the technological world (Saint-Sernin 2012). On the one hand, according to Bertrand Saint-Sernin, the biosphere is "the entire living world, including the human race, as well as the inorganic and organic supports that condition its existence" (Saint-Sernin 2012). On the other hand, the technosphere is "all of the technical actions of humanity on nature" (Saint-Sernin 2012). Apart from the internal coherence of each universe named in this context and the interdependence between the artifacts that compose them and the networks of interrelations that constitute the framework of each of these systems, technology and nature were not always so discordant. On the contrary, the two phenomena were in fact originally linked. They have been interwoven through a unity of principle and a co-evolving history since the invention of manufacture by *Homo habilis*. Indeed, the first hominids invented technology not to separate themselves from the biosphere, but to better adapt to nature and build an ontological bond with it.

Unfortunately, since the Industrial Revolution of the eighteenth century, the advent of the Anthropocene Epoch has misled man by setting him against nature, and in doing so, opposing the biosphere and the technosphere. We are referring to the avatars of an era, our era, which is chained to the frenzy of unnecessary endless growth and caught up in the vertigo of a blind exploitation process[9]: the downward spiral of an ontology of excess, with the planetary anxiety and real anguish that it generates within modern societies.

---

[7]This term was defined in 1926 by a Russian scientist Wladimir Vernadsky 1863–1945 (Vernadsky 1926). One of his major works, translated into French in 1928, is entitled *La biosphère*.

[8]Two systems between which hiatus could cause significant damage.

[9]It is this phenomenon that Gina Lombroso analyzes in her book *La Rançon du machinisme*.

In view of all of the above, we believe that the unease is profound, that the crisis is permanent, and that the future is marked by stops and starts. Consequently, the hominization process may undergo serious substantive mutations. In any case, man has, in this way, dismantled the relationship of necessity with the biosphere (Raphael Mathevet (2012) calls this "Déliance" [10]) and has instead established a mere contingency relationship. The environment consequently appears to be no more than an accidental, or even random, artifact.

This technological civilization carries within it an unwavering sense of unease. An unease inherent in the modern world, which is constantly deepening our "shortfalls" and paradoxically scuppering our aspiration to happiness. This happiness which also corresponds to a "desire for eternity" among a humanity addicted to an innovation, which lacks reference points. A humanity, which is in the grip of the linear asymmetry of an infinite progress curve. A humanity, which pushes man's imagination and creativity to the extent that he/she no longer has the means to contain the excesses (in terms of negative impacts on the biosphere); and is even less capable of dispelling the preoccupations generated by the catastrophes of our time. To that end, our inability to safeguard the viability of all biotopes to last raises the red flag, which shows that the human species is perhaps more threatened than ever before. As evidence, if it was still needed, we are less and less certain, given the feverishness of governments and the growing complexity of the challenges ahead, that we can ensure the quality of life required for the well-being of our species. Worse still, we are no longer able to guarantee equity in the distribution of available resources, not only between societies and nations, but also and especially between generations.

In the end, it seems urgent to establish a joint obligation ethic for future generations. With that in mind, Edgar Morin speaks of "prolonging, in the future, the ethics of responsibility and solidarity with our descendants" (Morin 2004, 185; Hans 1995).

24.1.3 The Obligation to Put in Place an Ethical Awareness of the Ecological Danger

The evidence of the "ecological emergency" imposes on modern society the imperious necessity to build the basis for a new culture. More precisely, it is necessary to lay the foundations for a new mode of operation, appropriation and consumption of energy. Developing a joint obligation ethic consequently seems essential for the future of humanity. Hence, defining the major principle of another vision of "alterity," one which is both endogenous and exogenous, becomes, for all human beings, the primal requirement of what we refer to as the awareness of

[10]The term "Déliance"(a French neologism for disconnection) is used here to describe the rupture or distance between humanity and the parent strain, its "living matrix."

risks and dangers. Within this context, alterity can be endogenous when it describes an internal relationship, that is, within the limits and the economy of a particular system. Examples include the intramural relationship between the elements of the same system, either organic or non-organic. The relationships between individuals within a population or a species are referred to as "endogenous Alterity." Conversely, the interaction or interrelation between two systems, two different species or two entities from two different orders is referred to as "exogenous Alterity."

Hence, a new type of bond between individuals, nations, systems and world civilizations would allow us to transvalue the Anthropocene Epoch. This could be achieved through a real awareness of risks and dangers whose basis would be the vision of an "alterity which is both endogenous and exogenous." The latter advocates the joint obligation between societies, species, systems etc. within a context of coviability (term defined below).

Such an ethical pre-requisite has become an absolute necessity in order to gradually reduce the permanent risks of a definitive break with equilibrium within the biosphere. The joint obligation in the name of "ecological emergency" requires that we are already constantly held responsible via a relationship of interdependence between societies, species, systems and so forth. The aim here is to create a link, not only "the principle of non-nuisance against others" (Bourg 2014), but a fresh ethic based on the new vision of alterity. This joint liability is individual, collective and universal. Such an "ethics of planetary understanding and solidarity" constitutes a vital emergency. Edgar Morin outlines its content, clearly highlighting our relationship as humans with the Earth, the Cosmos and the Biosphere, through a simple notion, which is perfectly adapted to our time, the notion of "realization": "[. . .] man comes to the realization that humanity is finite in the cosmos. This leads us to understand that, for the first time in its history, humanity must define the limits of its material expansion and correlatively undertake its psychic, moral and spiritual development." The "(. . .) ecological realization of our earthly condition, including our vital relationship with the biosphere". Earth is not the mere sum of a physical planet, a biosphere and humanity. Earth is a physical-biological-anthropological complex whole, where Life is the emergence of the history of Earth and man is the emergence of the history of life. The relationship between man and nature cannot be conceived in a reductive or disjointed manner. Humanity is a planetary and biospheric entity. The human being, both natural and supernatural, must revitalize himself in the living and physical nature from which he emerges and in which he distinguishes himself through culture, thought and consciousness. Our consubstantial link with the biosphere leads us to abandon the Promethean dream of the mastery of nature to aspire to conviviality on Earth" (Morin 2004, 185), especially since "the future of any society is intimately linked to the natural and anthropized environments within which it evolves" (Roué 2006).

We are consequently bound by a duty of ethical awareness not only *vis-à-vis* ourselves but, and above all, *vis-à-vis* "alterity." Contemporary Alterity designates the neighbour, the neighboring nation; a system other than one's own; the ecological order and not only the order of the Living; other species and not only human species. Alterity also refers to future generations and not merely to those that are contemporary to us.

Our actions as individuals, communities, governments or States, on the heritage of humanity and of the entire planet, put lives of Others at risk as well as the lives of other communities or nations, those of other species and the ecological reign. The planet is a common Heritage. Therefore, in the same way and in the same capacity, we are all bound to a requirement of ethical awareness in order to rationalize resources, safeguard the "vital Link" and preserve equilibriums. Within this context, our personal interests always depend on general interest. From this moment, our responsibility (individuals, communities and nations) is to see to the preservation of all terrestrial and marine biotopes that form the ecosystems throughout the biosphere, for "species are not the only ones that deserve protection. Rather, all biodiversity and the entire biosphere deserve protection, considering the current phase of extinction" (Norton, quoted by Delord 2003).

It is only on this ultimate condition that a "rapprochement at all levels between man and nature" remains definitely conceivable, with a view to "reconciling societies and people with their environments" (Delord 2003).

Ultimately, the question to be dealt with as top of priority is to understand that humanity has been rejected by a civilization of mistrust and an ontology of excess. It is imperative to properly assess the scope of the danger and above all, to consolidate the horizon of the movement to "re-anthropize the planet, to move towards "a paradigm of "reconciliation" between man and his environment" (Rosenzweig 2003; Teyssedre quoted by Delord 2003). The future, the serenity and the viability of our societies depend on it.

To the question posed by the meaning and significance of the issues raised here, we answer that man has been "ordered" to develop a culture of emergency in his relationship to all ecosystems; for mankind urgently needs a more ambitious environmental ethic.

24.1.4 Laying a Scientific Basis to Establish the Project Concerning the Joint Obligation Ethic

This project is possible in the light of knowledge that jointly derived from the sciences of heredity and evolution. Firstly, molecular biology in particular is helping us to discover the full genetic potential of the individual. Secondly, the theory of evolution and natural selection shows that living beings in general (and the human species in particular) is an offshoot of evolution, a product of the natural environment and of the environment in which they evolve. Therefore, our study will mainly focus on the forthcoming project of scientifically positioning the principle of a joint obligation as regards an ethical awareness of risks and dangers. This constitutes the foundation for a new environmental ethic, based on the concept of "coviability" in terms of hereditary sciences (Mendel's theoretical heritage) and the favourite theories of natural selection and evolution of Darwin (Darwin 1992).

The twofold endogenous and exogenous alterity of a system or a species may then be explained through heredity, adaptation/natural selection and evolution. For Leigh Van Valen, "the driving force of the evolution of every living species is represented by the other species with which it shares resources. Any progress in the adaptive value of any species modifies the environment of the surrounding species and forces them to adapt. In turn, this adaptation causes a change in the environment of the first species, forcing it into a new episode of selection and so forth" (Combes 2000). In this case, each species is always dependent on both the environment and on other species.

On another level, the genetic inheritance of an individual, or the karyotype of a population, of a species, is provided through a nucleus surrounded by a membrane. All of the elements within a cell, namely acids, molecules, chromosomes, sequences, genes or alleles, etc., function as a system. In this way, every element is systematically and *de facto* related to the other elements. In other words, all the elements of the mechanism interact, or relate to each other. The environment described herein belongs to the endogenous medium which is internal to the cell.

However, a systemic description of the chemical and genetic potential of a cell cannot suffice to explain the development of an individual's DNA or genotype, or the karyotype of a group or a population. We must also take into account the external factors that influence the development of heredity on two levels. This namely consists in explaining, firstly, how the different cell types in the individual's organism interact through ontogenesis, and secondly, how the external environment of the organism participates in the biological development of the being in the making.

It is within this context that the concept of "coviability" becomes relevant to explain the matrix-interaction between the endogenous environment of the cell and the external environment in which the individual or the population evolves. In this case, "viability" refers to the potential for the vitality and preservation of a system, an organization or a structure. "Coviability" consequently means joint and interdependent "viabilities" between two phenomena, two systems, two organizations, two structures, etc., or even between two natural environments. In other words, guaranteeing the "viability" of an entity or a system always depends on the principle of "exogenous alterity." Consequently, asserting that the external surrounding environment of an organism influences its development immediately raises the question of whether the quality of the relationships between the individual and his external environment[11] partly depends on the viability of the individual's genetic inheritance.

More clearly, does the viability of the genetic structure and inheritance (of an individual or a population) not originally depend on the quality of the environment? On the other hand, are the relationships between the individual and his environment not dependent on the genetic inheritance and resources of the individual? Answering these questions may enable us to lay the foundations, on a scientific basis, for a joint obligation ethic in order to respond to the "ecological emergency."

[11] In terms of evolution.

The argument presented here consists in two parts: The contribution of a new vision of the environmental ethics project for a humanism which is in the process of being re-founded (2); and the need to scientifically establish the joint obligation to put in place ethical awareness (3).

24.2 A New Vision of the Environmental Ethics Project for a Humanism, Which Is in the Process of Being Re-founded

As a product of evolution (1) and in its intraspecific relationship, humanity depends on its relationship with the biosphere (2). A co-evolution link consequently appears between man and his environment (3).

24.2.1 Man as an Offshoot of Evolution

Living together, through a social bond, is an absolute evidence, given that the reality, which is characteristic of the human condition is based on *alterity*. But before being a social reality, the human being within a community depends on the way in which each society imparts a certain vision of existence within the world, a way of inhabiting the Earth. However, far from the avatars of modernity, the living world in general and the Earth as a habitable place constitute a space shared between all living beings and the ecological artifact that extends into the biosphere. This applies to the entire physical world in the form of ecosystems (biotopes and biocenosis). It remains valid in all cases, at every level of cultural relativism, beyond social reality and the human race as a "product" of evolution, especially given that we are all of the same, indivisible origin. Consequently, since the appearance of life on Earth, "it has seemed clear that microbes, fungi, plants, animals, humans, in short, every living entity, descends from an initial proteo-bacterium" (Jacob 2000, 34).

Basically, we all have the same genetic inheritance, "the genes that form the structure of a human being are the same ones that can be found in a fly or a worm ... It must be assumed that all of the animals, which currently exist on Earth, descend from the same organism" (Jacob 2000, 33). The human being is therefore not a unique being, separate from other beings and ecosystems, far from it. His life wholly depends on other living beings and on nature in general. To that end, "the fundamental properties of living organisms" may be summarized as follows: "species evolve, influence each other within ecosystems, within a complex dynamic of emergence and feedback" (Thomas 2011).

Consequently, is the condition of man firstly a combined biological, ecological and geophysical fact?: "the history of man is part of the history of life which is part of the history of the Earth that is part of the history of the universe. We are here, by the same token, projected into a history of between twelve and thirteen billion years … Less than five billion years ago, the solar system appeared and, of course, among our Sun's planets, the Earth was born. At least four billion years ago, the so-called inert matter of the Universe, found in Earth's water, in turn became more complex and organized, transforming a part of itself into living matter (…) Several hundred meters of layers of calcium carbonate concretion, known as stromatolite, built by photosynthetic cyano-bacteria in many parts of the world, testify to this" (Coppens 2006).

We can therefore clearly understand that, as such, the human being is the product of phylogenesis, the evolutionary history of living organisms within ecosystems, a process which has spawned all of the living species that have populated the hydrosphere and the lithosphere for billions of years. For Yves Coppens, "over its four billion year journey, a part of living matter has never ceased to complicate and organize itself…" (Coppens 2006, 18). The French paleoanthropologist portrays a richer and more varied picture of living matter through relevant and abundant scientific information, and more precisely, from a category and gradual scale on the unwavering hominid's allegiance to the only living matrix that gave life to all living beings. As Coppens elaborates "man is evidently a living being and he appears as such at the end of one of the multiple branches of the immense phyletic tree of the living world. When we organize beings, which all naturalists have done to try and find their place in the amazing creativity of nature, it is within the Eukaryotes, the Metazoa, the Chordates, the Vertebrates, the Gnathostomata, the Sarcopterygians, the Tetrapods, the Amniotes, the Synapsids, Mammals and Primates that man is placed" (Coppens 2006, 20).

It therefore appears that mankind stems from and depends on both biological fact and the animal world. Whilst "our neurons have a short memory (…), reminds Michel Serres, DNA has a much better one". It remembers the common inheritance shared by all living beings, for "living species are places of memory". This memory unites us; it educates us (**we** humans) on our common origins with the other living beings. Unfortunately, "men are leaving these places behind" (Serres 2003). Humanity is consequently tending to **become** emancipated **from** its only place of origin: the inseparable Life and Earth, even though the essence of its condition and of its integral being are still linked, and even more so today, to this biological and ecological determinism, for being one species among many others, the human being is part of biodiversity, the "living fabric of the planet." Through this fabric, genes, individuals, species and ecosystems interact and co-evolve. And this in the true sense of the term *biogenesis,* which in this instance means that man, is naturally **born of the same live strain (issu de la meme souche vivante)** as all other species that exist within the biosphere. We consequently refer to the "living world" because "given that we ourselves are living beings, we are, as humans, along with our activities, part of biodiversity, of which we are both the product and a determining player. The consequences are not only scientific,-; they have repercussions of a philosophical nature and even of a 'political' nature in a broader sense" (Thomas 2011, 129–132).

24.2.2 The Relationship Between Humans Depends on Their Relationship with Ecology

What is at stake here, from the viewpoint of the human condition, which emanates from our primary inheritance, is the assurance that we do not need to worry for future generations. Above all, the challenge concerns being able to safeguard the balance of community life. Certainly, a priori, the human being can only survive the misery of his own finitude through alterity, the relationship to others, the link between communities, between nations and so forth. "Evolution has made us the determined physical beings that we are; as it has also made us the determined social beings that we are" (Ruse 1993). Edgar Morin speaks of a possible "improvement in human relationships to which every individual, deep down, aspires" (Morin 2000). However, "humanity needs nature so that it can survive as a species. This is why the destruction of the natural environment is detrimental to our interests. Consequently, in order to contribute to man's evolution (to preserve his survival), Edward O. Wilson believes that it is necessary to favor the good state of the environment and ensure that moral actions are directed towards this end" (quoted by Ruse 1993, 38–39). In other words, the ultimate condition for humans to live together depends first and foremost on the ecological phenomenon (biotope and biocenosis) and not on the social phenomenon.

Moreover we have long been accustomed to a certain model of ethical awareness and of representation of our relationship to the world (both socio-centrism and anthropocentrism). This relationship considers that the bond of humanity (in society, between communities or nations) does not depend on any external referent but on a simple endogenous intra-human alterity. As it happens, we are accustomed (even unconsciously) to thinking that the inter-human link only depends on the "intra-mural" relationship between individuals of our own species, i.e., the relationship that humans maintain among themselves within a society or which nations develop among themselves. Yet, in reality, the foundation of human alterity (or the human bond) depends on the viability of ecosystems. Why? "We are humans together," undoubtedly, and this is irrevocable. However, the bond between humans, on all possible scales, owes its viability to the co-evolution and coexistence relationship that they maintain with individuals of other species and with habitats. In fact, beyond human beings, interdependence and interrelationships constitute an unsurpassable ontological norm, which for billions of years has structured the "biotic community." Humanity has been part of this community since the appearance of the hominid, hence the "ecological urgency" to return to our roots, to the state of nature: the need to reconnect with primary origins, to re-tie the original knot, to reconnect with the elementary or original matrix. In other words, the need is formalized in the "reconciliation of man with the biosphere" (Delord 2005): it is a matter of "renewing our links with living beings, becoming aware of our interdependence" (Mathevet 2012).

However, to bring about the advent of this "return to nature," it is necessary to break with current models of thought, production and consumption. It is a major pre-requisite in order to found a community life ethic, widely shared among all communities and societies at a global level. The establishment of the "new Alterities" developed here would firstly require dismantling the foundations of the culture characterized by the uncontrolled consumption of fossil energies. Such a project requires establishing a new resource exploitation grid. It consequently forces us to modify the map of "useful" economic stakes. This consists in re-founding a type of humanism for the human being of the twenty-first century for a better understanding of the requirements of coviability. The Leitmotiv of re-founding humanity, which is essential for our viability and therefore for our survival, is urging us to reconnect with nature.

Re-founding humanism has structured (in part and implicitly) the history and evolution of environmental ethics for decades. According to a particular environmental philosophy our societies were disconnected from nature. This philosophy therefore attempted to rethink our relationship with ecology judging consumer society and advocating the transcendence of the simple utilitarian relationship of modern man with the environment. In other words, "environmental philosophers based themselves on the unanimously shared observation that evil fed on the ignorance of modern societies, which only attributed instrumental values to nature. (. . .) Environmental ethics were consequently split between three canonical forms (. . .): for anthropocentrism, it is all about man; for bio-centrism, it is all about every individual organism; finally, for eco-centrism, it is all about the "biotic community" (Delord 2005).

Moreover, eco-centrism perfectly fits in with what we refer to here as "a joint obligation for ethical awareness".[12] It has been able to lay the foundations for a universal ethic, which considers that the environment is not merely a tool to be used, a pure artifact or a simple "piece of equipment,"[13] but a dimension of the being which is indispensable to life. The environment is a dimension which is just as essential as man, if not more so, as it integrates the whole of the biosphere. To that end, eco-centrism ties in with the urgent need for alternative thought. This alternative considers that the relationship to Others within a human community, for example, far from being guaranteed by social fabric, bases its viability on an authentic, more global link to the biosphere. It is a matter of writing the odyssey of an eco-phylogenesis[14] of which man would be a mere witness to the original imprint, which is the bearer of a memory, and that of the place and the consciousness of time. This memory also bears the imprint of manufacture such as the emergence and the authentic development of technology through the blossoming of the being constantly evolving. Conversely, anthropocentrism places man at the "center of gravity," making him the *alpha* and *omega*, just as it turns nature into a mere domain, which provides raw materials and energy. Its ethics consequently become obsolete

[12]Instance of legitimate anxiety and incessant questioning.

[13]In Martin Heidegger's words (Heidegger 1958).

[14]A development of the human species always in connection with nature.

and completely inappropriate. Bio-centrism also proves itself to be an insufficient response, for it does not take account of the integral and systemic dimension of the biosphere as a whole.

24.2.3 The Co-evolution Link Between Man and the Environment

It would therefore be a matter of redefining a solid foundation for a revolutionary vision[15] of the future of humanity with respect to the danger and ordeal of the Anthropocene Epoch, by taking reference from the original narrative of the hominid. The new paradigm, "the joint obligation for an ethical awareness with respect to risks and dangers," would be set up by a sort of universal memory. The memory of places of origin, that which reminds all humans of the invention of technology, speech, narrative and art. But which would also recall the development of a sedentary culture, and the establishment of the first forms of breeding and agriculture and so forth. All of these inventions bear the trace of phylogenesis. They show future generations the co-evolutionary bond between human beings in the making and the environment. They also reveal the memory of the hominid in his relationship with nature, and delineate the sediment or the framework of the memory of the human being's evolution and that of his bio-cultural environment. In addition, the development and knowledge of the first cultural rudiments[16] transmit the memory of co-belonging between nature and culture through phylogenesis. They relay the indelible trace of the unity of the world, between the biosphere and the technosphere, but also that of the unity of place through myth as an awareness of time and consequently, the legacy of a shared bond with the biosphere. Such a vision is given through an ethical awareness whose filiation clearly considers itself eco-centric. Through it, we are certainly aware of the fact that man experiences a visceral need for intra-human alterity, and he consequently develops a secular fiber that naturally orients him towards society. However, "the joint obligation for an ethical awareness of risks and dangers" reminds the hominid that he is a product

[15]In the sense in which Thomas S. Khun (1962/1983, 30) understands it, a shift in paradigm: "paradigms, a term that is related to that of normal science. In choosing it, I want to suggest that certain recognized examples of real scientific work - examples that include laws, theories, applications and experimental devices - provide models that give rise to particular and coherent traditions of scientific research, for example that which historians describe under the headings of "Astronomy of Ptolemy" (or of Copernicus), "Aristotelian Dynamics" (or Newtonian), "Corpuscular Optics" or "Wave Optics," and so on. It is the study of paradigms, many of which are much more specialized than the ones I have just enumerated, which mainly prepares the student to become a member of a particular scientific community with which he will work later (. . .). Men whose research is based on the same paradigm adhere to the same rules and standards in scientific practice. This commitment and the apparent agreement it produces are necessary prerequisites of normal science, that is to say, of genesis and the continuation of a particular tradition of research."

[16]Since it is about this when we talk about invention.

of phylogenesis. As such, he owes his survival to the ingenuity of nature. The homonid's existence in the world is renewed in the confines of places of memory are referred to as (sont definis par) the Cosmos, Earth, Life (RNA[17] and DNA[18]), Ecosystem, Biotope, cosmological Time, geological Time, biological Time and so forth. Memory is consequently seen as a phyletic memory of the human being in the making. A human being which has been rethought through co-evolution owing to which no species, no biotope, no ecosystem owes its viability solely to its endogenous reality. Within this context, the hominid is a sort of "chronic little runt (avorton chronique)"[19] with a relative guarantee of viability, one he owes not to a social reality but to an ecological reality (as we have already pointed out). Ultimately, alterity must extend itself and take on a more global form, and this, through a dynamic of co-evolution and interrelations between all beings and all ecological realities.

Aldo Leopold, an American environmentalist, affirmed as early as 1948 that "ecologically speaking, ethics is a limit imposed on the freedom to act in the struggle for existence (and) finds its origins in the tendency of individuals or interdependent groups to develop modes of cooperation" (Léopold 2000 quoted by Delord 2005). The future of humanity can no longer be written according to an anthropocentric thought, "which only recognizes man as a moral subject and which, according to a utilitarian philosophy,, places no value on biodiversity depending on the economic, scientific, medical, aesthetic, religious interests that humanity takes in it and wishes to protect" (Delord 2005). Within the Anthropocene Era, both the survival and the future of species depend on ecology, whose phylogenesis has determined conditions for a long time. Bio-centrism also seems to be questionable, essentially because it "bases its ethics on respecting the intrinsic value that every living being possesses insofar as it demonstrates vital fundamental goals: to survive and to reproduce are goals which possess a value in themselves (intrinsic or inherent), irrespective of human judgments" (Delord 2005). Such a doctrine does not sufficiently take account of the idea of a "biotic community", which includes all individuals, species and non-organic realities, not because of what they are individually, but because they cannot survive when taken individually. In effect, both visions are interdependent, through a common ecological destiny, which consolidates the viability of an individual or a system. "According to the eco-centric position of American philosopher Baird Callicott (1989), and mirroring the thoughts of Aldo Leopold, all of supra-individual entities must be considered from a holistic viewpoint as exhibiting irreducible emergent properties. Consequently, for an eco-centrist, it is the 'biotic community'

[17]RNA or ribonucleic acid "used primarily for information transfer" (Jacob 2000, 25).

[18]DNA or deoxyribonucleic acid "which ensures the preservation and reproduction of cellular information" (Jacob 2000, 25).

[19]It is a being which has been born too early but survives in the long term. A term of Arnold Gehlen used by Jacques Poulain in *De l'homme, éléments d'anthropobiologie philosophique du langage* (2001), edition of Paris, P. 12, quoting Arnold Gehlen, in *Der Mensch, Francfort, Athenäum Verlag*, (1939); *Urmensch und Spätkultur, Francfort, Athenäum Verlag*, 1956 and *Zeitbilder, Francfort, Athenäum Verlag*, (1965).

(…) that must be considered as the intrinsically valued object. This ethic, inspired by Darwinism, attempts to extend the idea of cooperation to the entire biotic community" (Delord 2005).

A question then arises: based on this ethical foundation, how can we re-invent a secular anteriority of the human species (which seems beyond our reach today), with a view to renewing the link to the biosphere? Is this even possible? "What world could we form together, humans, non-humans, species and populations?" (Latour 2001, quoted by Delord 2005). The re-founding and the lasting security of this link, our original link, requires the establishment of a true coviability[20] aesthetic in order to reinforce the new vision of the re- anthropization of the human. In this case, a coviability aesthetic means another way of existing in the world, another way of feeling among other living beings, another way of living on Earth. Such a state is only possible on the basis of an ethical foundation that takes account of the human being, of course, but above all, the phyletic anteriority of the relationship between life and the biosphere. Consequently, these are certain pre-requisites, which need to be defined in order to establish the conditions for a viable solution:

- Getting rid of the dominant consumerist ideologies that stem the world's shared culture of unbridled wear and tear that is currently threatening our life balance. In this context, "the aim is to preserve the Earth's ability to promote life in all of its diversity, to respect the limits of natural resources and to guarantee a high level of protection and the improvement of environmental quality. It is also about preventing and reducing pollution and promoting "sustainable" consumption and production modes in order to break the link between economic growth and environmental degradation" (Augier 2012, 29).
- **Transvaluing (transvaluer en profondeur)** (basically) the Anthropocene Epoch and literally rethinking our relationship to the Earth, to the biosphere not only as a protected place, the ultimate dwelling, but as the "ultimate Alterity", which is the supreme possibility for the survival of the hominid.
- Establishing (consequently) new forms of Alterity and building more solidarity-based approaches to secure the original link, by increasing the awareness of major risks and future dangers.
- Informing, communicating, training and educating.
- Promoting and better adjusting three major elements: the vision of securing vital balances, the widely shared culture of sustainable production and consumption, finally, a sound and equitable management of the world's resources.

Taking into account these pre-requisites would imply a global vision of secure impregnation. This vision simultaneously includes the social being, biocenosis and all biotopes. The said vision is looked on as a division of a more objective basis

[20]Coviability implies a relationship of joint viability between two systems or joint models via interdependence and interrelationships. This would include interdependence between the human species (in its bio-cultural diversity and geo-diversity) and the set of biotopes that constitute the architecture of ecosystems.

for the promotion of a new link for humanity. In this case, this link of humanity is defined in the emergence of a universal moral value of planetary socio-ecological coviability.

24.3 Scientifically Basing the Joint Obligation for Ethical Awareness

Among man, solidarity results from a process that is biological (Sect. 24.3.1), genetic and adaptive to his living environment (Sect. 24.3.2). As a result, heredity associated with evolution is at the origin of a joint obligation, specific to the human species (Sect. 24.3.3).

To make the concept of coviability clearer within the context of this study, we suggest a work based on the theoretical legacy of Charles Darwin[21] and Gregor Mendel.[22] This enables us to lay solid foundations for an introduction to a future theory on the phyletic memory of the hominid, as it has the potential to give us an objective basis for founding "the joint obligation for ethical awareness." We will consequently begin with a simple principle, according to which the sciences of heredity and evolution offer a veritable scientific basis to the concept of coviability. Let us be guided by the healthy debate between the heirs of Lamarck and the custodians of Darwinism: "For more than one century, controversies concerning evolutionary processes have been based on caricaturized conceptions of Lamarck's and Darwin's works. The first focused on the importance of internal factors in evolution, and the second considered external factors. Current research attempts to take both into account. Internal factors concern genetics, developmental genetics, embryology, processes of ontogeny and traits inherited from ancestors. Their combination associates ontogeny and phylogenesis in that they originate from our evolutionary history (phylogenesis) and are inscribed, for some, in our genetic inheritance (genome). External factors are related to the environmental conditions surrounding the organism: other individuals within social systems (competition for food and sexual partners), other species (competition for the same resources, predators, parasites), regular variations related to seasons and climatic cycles, disasters (volcanism, plate tectonics, meteorites). These constitute the processes of natural and sexual selection" (Picq 2003, 209).

The viability of the being in the making depends on two major bases, both of which are considered to be "internal processes" and "external modalities."

[21]Charles Darwin explains his theory of evolution and natural selection through a book entitled *On the Origin of Species by Means of Natural Selection or the Preservation of Favoured Races in the Struggle for Life.*In addition, Michel Morange recalls the relevance of Mendel's works in a book entitled *La part des gènes* (Jacob 2000, 21).

[22]For his part, Gregor Mendel has lifted the veil on the mechanisms of heredity.

24.3.1 A Brief History of the Birth and Development of Biology

In the nineteenth century, biology was still in its infancy. Certain researchers were only beginning to take account of the "properties of living beings unlike inanimate objects and use the word biology" (Jacob 2000, 23). However, development was rapid, especially towards the end of the century. During that period, great theories emerged and went on to play a key role in the history of biology in general and in the sciences of evolution and heredity in particular. This prosperous period, that was the end of the nineteenth century was an exceptionally fertile period for biology. It was the epoch of great theories" (ibid., 24). A period that encouraged the emergence of great theorists and established the foundations of biology. In this line of scholars, Pasteur was one of the major pioneers in his discipline. His "theory of germs" profoundly marked this century of "explosion of knowledge" (Foucault 1969). According to François Jacob "micro-organisms were discovered at the end of the 17th century, thanks to the invention of the microscope (. . .). Pasteur brought to light the role of these small living creatures in the diseases of man and animals, as well as in certain industries such as wine and beer. Moreover, Pasteur demonstrated that microbes originate from other microbes and that spontaneous generation does not exist" (Jacob 2000, 24).

Asserting that "microbes are born from other microbes" implicitly suggests that there are specific properties characteristic of each type of individual. Each individual belongs to a well-defined type, to a species, presenting with the same general characteristics and carrying the same karyotype as his congeners. One can then already attempt to understand the scientifically underlying elements of Pasteur's theory: "microbes are born from other microbes." In other words, how can we explain that a microbe has the ability to engender another microbe? The idea that an organic material exists, and that the latter would be transmitted from generation to generation through a lineage, which thrives within a species, was becoming an increasingly plausible thought. It is here that the discovery of an organic matter, essential for the future of biology, takes on its full meaning: the living cell. It is in fact a membrane within which life develops, namely, genetic material. The cell theory developed by Schleiden for plants and by Schwann for animals consequently constituted a major step towards the definitive birth of the sciences of heredity. Schleiden and Schwann remarkably discovered that "all organisms are composed of cells (. . .). The cell is the smallest element which possesses all of the properties of living beings" (Jacob 2000, 24). They go further still, elaborating pertinent hypotheses and developing new knowledge in order to better understand and explain the mechanisms of reproduction, cell division and the formation of the embryo: "reproduction occurs through fertilization, i.e., the fusion of two sexual cells: the spermatozoid and the ovule. The development of the embryo occurs from the egg which is subsequently formed, by the multiplication of cells and their differentiation into specialized cells" (Ibid.). Following these discoveries, two fundamental and connected disciplines emerged at the dawn of the twentieth century: biochemistry and genetics (Jacob 2000, 25).

Biochemistry has revealed entire sections of the living universe, promoting the development of research on hereditary. It "seeks to analyze the constituents and reactions of the cell. It is thanks to biochemistry that experimentation gains access to life chemistry. It analyzes a considerable number of relatively simple reactions. It follows transformations permitting energy reserves to be constituted and construction materials to be elaborated" (Ibid.). Consequently, thanks to biochemistry, we discovered a new dimension of the organism, which was not visible to the naked eye: a domain which is still inaccessible. A new scientific discipline was then born, whose furthering(approfondissement des connaissances) of knowledge contributed to the advent of another discipline that was equally decisive in terms of knowledge of the living world, genetics. Genetics emerged and experienced unprecedented development thanks to the discovery of two nucleic acids.[23]

Indeed, genetics was properly born during the twentieth century. However, it was the discovery of the three properties of the living world (conservation, reproduction and information transfer) that enabled Gregor Mendel[24] to inaugurate a new field of knowledge for scientific research in the nineteenth century: heredity transmission mechanisms. He was far from imagining the scientific revolution that was taking place before his eyes at the time: "the official story attributes the discovery of the laws of heredity in 1865 to the Moravian monk, Gregor Mendel, with evidence showing that the characteristics of living organisms (. . .) are due to the action of two determinants – now known as genes – one given by the father and the other by the mother. It was shown that these determinants do not mix but are transmitted unaltered and randomly to offspring. Finally, it was proven that the determinants, which correspond to different characteristics, are transmitted independently" (Morange 1998, 21).

Learned society hesitated a long time before realizing the value of Gregor Mendel's works and understanding that he had just succeeded in opening up an immeasurable horizon of possibilities for science and humanity. At the very beginning of the twentieth century, the role of four researchers was to become decisive in this respect. They re-exhumed the fruit of research, which had been ignored for decades by the scientific community: "Ignored, Mendel's results were simultaneously rediscovered by Carl Correns, Erich Von Tschermak and Hugo de Vries in 1900. Nonetheless, genetics only experienced a real development at the end of the first decade of the 20th century with Thomas H. Morgan who chose to study the organism that was to be identified with this discipline, the small drosophila fly. In just a few months, Morgan showed that genes were carried by chromosomes, consequently establishing the very first genetic maps" (Ibid.). Subsequently, varied

[23]"Nucleic acids are polymers of the elements chemists call purine and pyrimidine bases; they are four in total. There are two types: deoxyribonucleic acid (DNA), which preserves and reproduces cellular information and Ribonucleic acid (RNA), which is mainly for information transference" (2000, 25).

[24]Michel Morange reminds us of the relevance of Mendel's works (the illustrious predecessor of all the great geneticists of the twentieth century) in a book entitled *La part des gènes* (Jacob 2000, 21).

and complementary scientific domains bloomed. Among them, we can mention molecular biology, population genetics, molecular anthropology, embryology and so forth. François Jacob specifies that Mendel's works, when conducted and published in the 1860s, had not attracted much attention." (...) They lead to the idea that "character," what we see, is underpinned by a "particle" that we do not see and which is hidden at the heart of the cell. This particle is called a "gene." Since then, tireless research was pursued in genetics with the purpose of understanding what a gene was, how it functioned and its properties, and the more scientists learned, the clearer it became that genes are situated in the heart of every cell, of every organism, that genetics underlies all biology" (Jacob 2000, 29). Later on, knowledge on genes grew and, research rapidly developed around important issues.[25] From this point onward, genes were no longer a mere idea, they were increasingly becoming a physical, chemical and biological reality: "with works showing that it is deoxyribonucleic acid, DNA, which carries hereditary traits in bacteria and viruses,, the gene, which until this time, was a pure mental construction, began to gain in consistency" (Ibid.).

Gregor Mendel had made his discovery too early. The individual is the repository of a genetic inheritance that is transmitted from generation to generation. Therefore, the phenomenon of heredity is no longer a mere conjecture, unrecognized by the scientific community. Mendel's discovery marked a real turning point in the history of science, especially sciences that deals with living organisms. However, it was necessary to await the discovery of the structure of DNA, which controls the hereditary transmission of characteristics from one generation to the next, in order to definitively lay the foundation of genetics: "the structure proposed by Watson and Crick served to resolve, in the properties of a molecule, one of the greatest questions posed to mankind, heredity" (Ibid., 27). Science giving "chromosomes the role of supporting" heredity,[26] served to give real content to the concept of the gene and shed light on the importance of genetic information by lifting the curtain, with Watson and Crick, on the molecular structure of DNA.[27]

[25]For example, "the demonstration that a given gene occupies a precise position, which could be assigned a place on a particular chromosome, dates from 1910. The linear arrangement of genes on a chromosome and the first genetic map with several markers were published in 1913" (Jacob 2000, 26).

[26]The chromosomal support with Thomas H. Morgan (in 1910).

[27]"With the deciphering of the genetic code" we understood that genes have as a major role to provide information that "allows the synthesis of a protein. The latter fulfills, in all or a part of the cells of the organism, one of the many structural or enzymatic functions that this class of macromolecules is capable of accomplishing" (Morange 1998). DNA, a genetic material, plays an essential role here: its function is to "code information" for both the genetic identity of the species and the hereditary characteristics of the individual. Hence the name of the genetic code, which is the "relationship between the succession, alongside the DNA molecule, of its basic constituents, nucleotides, and the chain of amino acids that form proteins" (Ibid., 28).

24.3.2 The Individual Is Under the Influence of Both Heredity and the Surrounding Environment

What interests us more than the progress of biology[28] is the idea and the inexorable proof that biological filiation and heredity are scientifically confirmed,[29] with all that this implies in terms of consequences (the models whose vocation is "to account for heredity phenomena") (Morange 1998, 22), especially on an epistemological level. Whether speaking of nuclear DNA or mitochondrial DNA, science has just revealed stakes of a markedly superior dimension and opened up new fields of research and formulated new problems, from which interesting hypotheses are resulted. The latter are related to innovative themes such as genetic lineage, lineage convergence (coalescence) and so forth.[30]

However, knowing that hereditary signs come from our most distant ancestors arouses old concerns in the Homo Sapien Sapien species: "many believed that flaws rather than gifts, handicaps rather than benefits, were what passed through generations, gradually leading to a degeneration of the human race" (Morange 1998, 23). Such a feeling is quite legitimate; even more so when science asserts that gene mutation, variation, diversity and distance can, at first glance, conceal disturbing phenomena. Through an interval of one hundred thousand years, for example, between two generations, the mutation of DNA or genotypes may have tolerated and recorded events that are detrimental to the human species. However, what if the genetic odyssey of the hominid was in fact the true vector of progress for individuals, groups and species?

It is evident that science is progressing, but at the same time it is troubling, worrying and upsetting the benevolent clichés, preconceived ideas and certainties of society; for man instantly worries when novelty threatens to oust a long-standing traditions or prejudices. Heredity can undoubtedly transmit "genetic anomalies," such as germs causing serious pathologies. However, it is both remarkable and more than obvious that the individual, within a population or a species, is not the perfect copy of his congeners even if they have the same karyotype; just as he is different from individuals of other species. The mutations of "Life" [31] on the scale of individuals, groups or populations through the phyletic history of beings and species, generate mutants. The intergenerational transmission of traits within a

[28]"Namely, molecular biology put into perspective and the epistemological turn of the "rediscovery of the laws of heredity".

[29]Notably the fact that the genetic basis of individuals, groups, sub-species and species, in the form of what is now called genotype (in relation to the individual), the genetic program of the person, and the phenotype, the physical manifestations of this program, are transmitted from generation to generation, through reproduction within the limits of a species).

[30]The ancestral DNA, the disclosure of the genetic origins of the Hominid, the genetic distance and kinship between populations, DNA mutations, genetic variability and difference, genetic diversity of a population, and so forth.

[31]Through the great phylogenetic odyssey of transformations.

population is therefore subject to mutations or re-combinations. In a word, it is a question of all phenomena which are likely to provoke genetic variations.[32]

Fortunately, because of these phenomena, all genetic inheritance is not reproduced in exactly the same way. As a result of reproduction, individuals, taken on a global scale within a species, are therefore highly likely to escape "the genetic anomalies" transmitted by heredity.[33] This is thanks to natural selection. Evidently, there is still a risk of *genetic strangeness* in genetic inheritance. However, without being a panacea, natural selection always rewards the individuals who adapt better than others to the conditions of existence and consequently ensures humanity evolves. According to Charles Darwin, "no organ is formed with the intention of causing pain or prejudice to its possessor. If we equitably establish the balance of good and evil caused by each party, we shall perceive that, in short, each of them is advantageous. If, in the course of time, under new conditions of existence, any part becomes harmful, it changes. If not, the being dies out, as so many millions of other beings have died out before it" (Darwin 1992, 254). According to the formula of Marc Kirsch (1993, 23), "nature fosters the survival of the fittest." Minor changes, sometimes important and decisive for the future of species, occur through the prism of evolution. However, when an individual is unable to adapt because a distant, harmful mutation causes him a disability, he is purely and simply eliminated by natural selection. For this purpose, individuals or species, depending on adversity or hostility, become more or less happy with their surrounding environment. They consequently transcend their conditions of existence by improving their resilience, their defense and survival mechanisms. Otherwise they disappear purely and simply because of natural selection (Cavalli-Sforza 1997). "Mutation happens by chance, fostering innovations that can be either useful or detrimental. Natural selection chooses automatically, promoting advantageous mutations, and eliminating those that are unfavourable in particular living conditions of a population. The selection consequently helps the population to adapt to environmental conditions, depending on the environment. "(Luca and Francesco Cavalli-Sforza 1997, 123). So, a species improves both its constitution and its destiny through the abundance of those mutations that happened in the history of its evolution, and that are under the influence of the environment. It has been" formed for a long time. And it is generally spit into several other subspecies. The latter correspond to the environment adaptations through mutations and are under the action of natural selection" (Jean Chaline 2000, 19–20).

[32]"The genetic diversity of a species is due to mutations, re-combination, deletions, and so on, in genetic transmission from one generation to another" (Eccles 1994, 18).

[33]If the gene carries any anomalies.

24.3.3 Heredity and Evolution as Scientific Bases for the Joint Obligation for Ethical Awareness

Abundant information shows that the viability of hereditary mechanisms[34] does not just depend on the "support of heredity." Consequently, genetic mutation is of paramount importance. Depending on nature and the state of the surrounding environment, dislocations and re-combinations occur. In turn, these lead to divergences and expand the horizon of genetic variability and diversity,[35] – variability between individuals, groups or species, in terms of natural selection. In other words, genetic processes[36] generate different individuals within a single species, as well as new species, thanks to three main factors: mutation, natural selection and chance" (Cavalli-Sforza 1997, 130).

The interactions that develop between heredity and an individual's (or a population's) living environment involve, in terms of coviability, a new type of alterity, which is both endogenous and exogenous. The aim is to bring together two systems: one consisting in preserving and transmitting genetic traits, the other concerning the social or bio-cultural environment of the individual. Scientifically understanding the phenomenon of a joint viability between genetic inheritance and the living environment, could make the formulation of a new ethic possible. This would be reflected by a joint obligation between the individual and his environment, between a group and its bio-cultural environment, between nations, and also between generations.

Every individual would have a joint obligation vis-à-vis himself and vis-à-vis others to protect the environment and to guarantee its safety and viability. The optimal development of our genetic resources depends on the state of the environment, whose relationship with the living being is partly determined by genetic inheritance. It is in the same way that the interactions that the individual develops between his genetic resources and his environment either jeopardize the future of next generations, or guarantee their viability.

Depending on the context or situation, alterity refers as much to a third party, a nation, a society, a generation, an ecosystem, a biotope or even the biosphere in general. Consequently, when we are dealing with intra-organic or intra-human relationships, we speak of "endogenous alterity;" and when we are dealing with the relationships between an organic or a human system and a non-human, animal or ecological system, we refer to "exogenous alterity." The joint responsibility of the individual, within the framework of an ethical awareness obligation, therefore requires safeguarding the living environment. Not only because the blossoming of

[34]Hereditary mechanisms such as endogenous processes of the development and evolution of the individual, a group or a species.

[35]This happens over the course of births, minor or major events that occur within a population or species and through the transmission of roots between generations.

[36]This is based on the "molecular support of heredity" and on mutations and re-combinations.

our genetic resources depends on the quality of our environment, but also because this is the price to pay for the survival of the social link or the link of humanity.

It is only via this principle of solidarity between systems, from the acceptance of the principle of joint viability, that we will be able to better manage major risks and preserve the life balance. Systems have been interacting and cooperating with one another since the beginning of life on Earth, owing to the fact that living beings have always evolved at the same time as matter. Co-evolution is therefore a principle that helps us to understand the coviability phenomenon.

We believe that no reliable environmental ethic[37] is possible without the paradigm of coviability as its ultimate foundation. It is on this basis that we can reconstruct links with the biosphere. On the other hand, the anthropocentric monist vision of the viability of our social and bio-cultural systems is today completely inadequate. It is equally difficult[38] to think up the viability of a system from the sole internal parameter of its operation; for alterity is always considered from a double endogenous and exogenous viewpoint. The future may only be immediately envisaged through a common vision of the viability of systems, in this case, either between the biosphere and the technosphere, between humans and ecosystems, or between heredity and the environment (whether organic, natural, social, cultural or bio-cultural).

This phenomenon of interdependence has structured the relationships between man's heredity and the cultural environment for millions of years. It has also been able to modify the content of the human genome. More clearly, it is now attested that "cultural practices modify the genetic characteristics of human populations" (Cavalli-Sforza 1997, 130). We are consequently speaking of the culture-genome coevolution because "by their cultural traditions, transmitted from one generation to the next, men have increasingly influenced the natural selection of genetic information" (Combes 2000, 97).

However the culture of man has influenced the transmission of the genome as much as heredity has impacted our relationship with the environment. The dynamic and dialectic of interdependence between living beings[39] and their environment has structured the being in the making since the dawn of life. More precisely, the reciprocal influence between hereditary inheritance and cultural inheritance through phylogenesis has been central to coevolution between the genetic material of humanity and cultural artifact. According to Claude Combes "if the evolution of genomes in hominids' lineage has led to the emergence of the brain of modern man and consequently to that of culture, it has, in turn, profoundly modified the selective pressures exerted on genomes" (Combes 2000, 98).

The paradigm of coviability highlighted here defines and describes how an individual within a given population or the group to which he belongs adapts

[37]For a humanism which is in the process of being re-founded according to the theory of the hominid's phyletic memory.

[38]The complexity of the problems faced by modern societies in the Anthropocene Epoch.

[39]Taken at the level of the individual within a group, population or species.

to his living conditions. This occurs under the double influence of his genetic inheritance and of his environment, but also, in view of the interaction between his heredity and his biotope. In short, the individual or the group owes its social and bio-cultural viability to the dynamics of coviability between their genetic inheritance and their environment. Consequently, the relation to life is neither linear nor merely synchronous; it is diverse, entangled, multiple and involves relations of differentiation. It is made up of ruptures (i.e. endogenous ruptures) inside and outside the system.

Our paradigm of coviability, as defined above, allows a better study of the potential for the safe impregnation of life balances for a population or a species which, by definition, is changing. This insofar as it prospers or it doesn't, depending on the general conditions of existence. Other elements should also be taken into consideration, namely the direct or distant influence of the modifications undergone by the biocenosis, or the effects of the formation and the ruin of biotopes on population movements. The inter-systemic approach that founds the joint obligation for ethical awareness to create a viable environment[40] takes on its full meaning here Namely via the idea that a being in the making could only flourish under the double sign of genetic processes and environmental conditions and modalities, owing to hereditary variance and the random nature of what the individual can later become. The interaction between the two internal and external factors can cause the death, extinction or survival of an individual, a group or a species.

Consequently, the environment (including in its random components) plays a leading role. We observe, for example, that the environment in which the embryo, and later the fetus, develops has a non-negligible influence on the behavior of embryonic cells and on cell differentiation. A cluster of interactions and interrelationships is developed between heredity and all that report to external factors: "changes in the environment exert a dramatic influence on the construction modalities of bodies (. . .). The environment does not change the nature of genetic information. On the other hand, it modifies the nature of the information that the cells of the embryo, in the process of being constructed, will be able to consult and therefore the proteins that they will be able to produce. In turn, these proteins influence the behavior of other cells, gradually modifying the configuration of the entire cellular society (. . .). These are cell differentiation phenomena. Nevertheless, they are not only born from the dialogue between the cells that compose the body in the making. The external environment in which the body of the embryo is immersed also places its mark on cell differentiation phenomena" (Ameisen 1999, 296).

Fukuyama agrees with this in his comments on the works of geneticist Richard Lewontin: "the individual's interaction with his environment begins long before birth. The characteristics that we tend to attribute to nature are, in this argument, the product of a complex nature/environment interaction" (Fukuyama 2002, 204). To escape genetic determinism, the" salvation" of heredity would come from factors (natural selection and evolution) external to molecular biology. This is all about

[40]In the context of eco-centrism.

coviability, for emancipating individuals and species, between the endogenous parameter of DNA and modalities related to the environment. Ultimately, mutation would not be the only cause of genetic variability. Indeed, "it is the Danish geneticist Wilhelm Johanssen who succeeded in showing that this variability has a dual origin. There exists a variability related to the environment, to everything that surrounds character formation. There is also a variability of hereditary origin due to the transmission of factors of slightly different natures" (Morange 1998, 24–25). In this context, the struggle for survival selects the "best" and ensures not only the viability of individuals and entire generations, but also that of species.[41] We can even say that the genetic identity of a species paradoxically owes its viability and longevity to the significance of its individual genetic variations, according to their development through interactions with the outside world.

As regards the viability of the being constantly evolving and the viability of its species, we can therefore retain a dual influence: internal biological mechanisms and environmental factors. In short, individuals and species are the custodians of the genetic program, which they develop/fine tune via a contact with the environment. This increases adaptation to conditions of existence while permitting increased resistance to the hostility of the surrounding environment and avoiding extinction.[42]

24.4 Conclusion

After Mendel and Darwin, most geneticists, anthropologists and evolutionists agree that the genotype alone cannot determine the future of an individual, a group, a population, or the future of a species.

"DNA"[43] supports heredity and represents, in its viability, the endogenous factor of natural selection and evolution. Its particular trajectory and viability nucleus remain strongly determined by physically external, geographical, socio-cultural and bio-cultural factors. Consequently, every living organism, every human system, and every species can only be viable from a double internal and external phenomenon. The evolution of the human being, as much in terms of phylogenesis as in terms of ontogenesis, is consequently driven by an ambivalent development.

Human beings carry a genetic inheritance which they share with the rest of the living world. The rate of genetic variability, from one individual to another within a same group, from one community to another, or from one species to another,

[41]"Some individuals better adapt to changes to the environment or when they occupy a new environment. These individuals will produce more children over several generations, who will be able to use the resources available to them in order to survive and reproduce" (Edelman and Tononi 2000, 102).

[42]"Natural selection only tends to make every organized being perfect or somewhat more perfect than the other inhabitants of the same country with which the being is in competition." (Darwin 1992, 254).

[43]A macromolecule.

may only be beneficial and promising for the future when it respects its eco-environmental, bio-cultural or biogenic status. In a word, our genetic inheritance, as an internal factor of evolution, may only help us to better adapt to life if we respect our relationship with the environment, as an external factor of evolution. Our species and the groups and societies which compose it, and individuals within communities may only evolve and make the most of their genetic potential if they have a "healthy relationship" with the environment.

Furthermore, on a purely ethical level, the possibility of belonging to a community and a generation, whilst being able to lay the foundations for the future and the survival of the community, does not necessarily depend on the social link. It remains profoundly dependent on the viability of a more or less stable living environment, a bio-cultural reality of a "healthy and reliable" environment. In this sense, we must rethink our relationship with the world, lay the foundations for the future. Science cannot succeed alone, just as ethics cannot sustain any viable project without basing itself on knowledge, and more specifically, on the link between different types of knowledge. On that subject, Rabalais said "Science without conscience is but the ruin of the Soul".

This early alliance between science and ethics would allow man to re-hominize by returning to the biosphere and developing a greater complicity in his relationship with nature. Man would, *de facto*, undo this rupture between himself and his environment. The concept of exogenous Alterity may offer us a reliable comprehension tool in order to establish, from the intramural domain of a community, models of coviability and coevolution. Developing the concept of coviability would enable us to undermine the hiatus between the different spheres of the living world as a whole in the most appropriate manner, and consequently fully measure the original link between man and the biosphere.

References

Ameisen JC (1999) La sculpture du vivant, le suicide cellulaire ou la mort créatrice. Seuil, Paris

Augier H (2012) Le développement peut-il être durable? Pour le meilleur et pour le pire. Sans de la Terre, Paris

Barde, JP (2003) (n.d.) Développement durable et devenir de l'Homme, un enjeu pour la paix; Economie et politique de l'Environnement

Berg P Singer M (1993) Comprendre et maîtriser les gènes, Le langage de l'hérédité, Trad. Nicolas Glansdorff, Paris, Vigot

Bonneuil C, Fressoz J-B (2013) L'Evénement Anthropocène. Éditions du Seuil, Paris

Bourg D (2014) Droit et défis écologiques du long terme, Revue Juridique de l'environnement. Lavoisier 3(39):413–416

Cavalli-Sforza, Luca et Francesco (1997) Qui sommes-nous, Trad. Françoise Brun. Flammarion, Paris

Chaline J (2000) Un million de génération, aux sources de l'humanité. Seuil, Paris

Combes C (2000) *La coévolution, in qu'est-ce que la vie*, volume 1, sous la direction d'Yves Michaud, Paris, Odile Jacob, p 97

Coppens Y (2006) Histoire de l'homme et changements climatiques. Librairie Arthème Fayard, Paris

Darwin C (1992) L'origine des espèces, Trad. Edmond Barbier. Flammarion, Paris
Norton (2003) "La sauvageté": un principe de réconciliation entre l'homme et la biosphère. Natures Sciences Sociétés 3(13):316
Delord J (2005) "La sauvageté": un principe de réconciliation entre l'homme et la biosphère. Natures Sciences Sociétés 3(13):316–320
Eccles JC (1994) Evolution du cerveau et création de la conscience, Trad. Jean-Mathieu Luccioni. Flammarion, Paris
Edelman GM, Tononi G (2000) Comment la matière devient conscience, Trad. Jean-Luc Fidel. Odile Jacob, Paris
Foucault M (1969) Archéologie du savoir. Gallimard, Paris
Fukuyama F (2002) La fin de l'homme, Les conséquences de la révolution biotechnique, Trad. Denis-Armand CANAL. La Table Ronde, Paris
Godard O (2015) Environnement et développement durable, Une approche méta-économique. De Boeck, Bruxelles
Goethe, cité par Bertrand Saint-Sernin (2012) Biosphère et technosphère, Canal Académie–Les Académies et l'Institut de France
Gould SJ (2000) Quel avenir pour l'espèce humaine? In: Les clés du XXIè siècle. UNESCO. Seuil, Paris
Guery, François, (1989) La Société industrielle et ses ennemis, ed. Olivier Orban
Hans J (1995) Le principe responsabilité. Une éthique pour la civilisation technicienne. Cerf, Paris
Heidegger M (1958) Essais et conférences, Trad. André Préau, Paris, tel Gallimard, n 52 pp 28–29
Jacob F (2000) Qu'est-ce que la vie? in Université de tous les savoirs, Volume 1, sous la direction d'Yves Michaud. Odile Jacob, Paris
Khun TS La structure des révolutions scientifiques (1962, 1983, Flammarion pour l'édition actuelle)
Kirsch M (1993) *Introduction, Fondements naturels de l'éthique*, d irected by Jean-Pierre Changeux, p 23
Latour B (2001) Quoted by Delord J "La sauvageté": un principe de réconciliation entre l'homme et la biosphère. Natures Sciences Sociétés 3(13):316–320
Léopold A (2000) Quoted by Delord J "La sauvageté": un principe de réconciliation entre l'homme et la biosphère. Natures Sciences Sociétés 3(13):316–320
Lombroso G (1931) La rançon du machiavélisme. Payot, Paris
Mathevet R (2012) La solidarité écologique. Ce lien qui nous oblige. Acte Sud, Paris
Morange M (1998) La part des gènes. Odile Jacob, Paris
Morin E (2000) *Quel avenir pour l'espèce humaine, in Les Clés du 21è siècle*, Seuil, Editions Unesco 64
Morin E (2004) La méthode 6 – Ethique. Seuil, Paris
Picq P (2003) Au commencement était l'homme, De Toumaï à Cro-Magnon. Odile Jacob, Paris
Rosenzweig (2003) Teyssèdre, (2004), quoted by Delord J "La sauvageté": un principe de réconciliation entre l'homme et la biosphère. Natures Sciences Sociétés 3(13):316–320
Roué, Marie (2006/1) Introduction: entre cultures et natures, Revue internationale des sciences sociales, 187, 11–18.
Ruse M (1993) Une défense de l'éthique évolutionniste, in Fondements naturels de l'éthique à la base, Jean-Pierre Changeux (dir), Odile Jacob
Saint–Sernin Bertrand (2012) biosphère et technosphère, Philosophie des sciences, Canal Acadèmie – Les Acadèmies et l'Institut de France, Online: http://www.canalacademie.com/ida9087-Point-de-vue-Philosophie-des-sciences-Biosphere-et-technosphere-29-05-2012.html
Serres M (2003) L'Incandescent. Le pommier, Paris
Thomas F (2011) Cosmologies, diversité bioculturelle et préservation de l'environnement. Natures Sciences Sociétés 19:129–132
Verlag A (1956) Quoted by Poulain J (2001) in De l'homme, éléments d'anthropobiologie philosophique du langage, eds du Cerf, Paris p 12
Vernadsky W (1926) La biosphère, Nauchno-techn. Izd. Leningrad. Translated in French in 1928

Chapter 25
Tracing the Origins of Western Disconnection from Nature, to Envision a Change

Vincent Douzal

Contents

V. Douzal (✉)
IRSTEA, National Research Institute of Science and Technology for Environment and Agriculture, Montpellier, France
e-mail: vincent.douzal@teledetection.fr

O. Barrière et al. (eds.), *Coviability of Social and Ecological Systems: Reconnecting Mankind to the Biosphere in an Era of Global Change*,
https://doi.org/10.1007/978-3-319-78497-7_25

25.1 Introduction: A Feeling of Disconnection, Dissociation and Divergence

25.1.1 A Cultural Gap

In a 1959 conference that was to become famous, Snow (1961) stated that the intellectual life of Western society was split into two cultures, mutually unintelligible and deaf: the sciences and the humanities, unable to communicate through ignorance of the most basic tenets of the other's thinking. A few decades later, Rupert Riedl goes one step further about the academic milieu: "The division of these philosophies is firmly established. There is no negotiation across its boundaries, and whoever tries it must be prepared for the wrath of both sides." (Watzlawick 1980).

But it is not enough to remark that the two cultures are incommensurable, or excommunicate one another as soon as they feel in a position of strength: notice above all that sciences are the 1st culture, overwhelmingly dominating the 2nd, and determining the course of society. It claims to subsume all of reality, and disqualifies any of the 2nd's statement, as if nonexistent. As the 2nd spends a fair amount of its energy proclaiming that there are notions that escape the formal hold of the 1st, not only does it admit the full power of physical sciences in their native realm, but it feels like living on borrowed time in its own terrain, that it strives to fight against cut down, yet in terms that everyone feel inaudible to the hegemonic dogma. Worse still, it applies to adorns itself with the regalia of sciences.

Hence everyone acknowledges at least implicitly the pre-eminence of the sciences, even though everyone groans or rebels here and there against it. Thus at a closer look, the demarcation line reproduces its irreconcilable split in every discipline of any side. Economy is the best example, with microeconomic theories (physics in disguise) asserting the dogma of a perfectly rational agent, whereas in the next room, marketing boasts of being able to sell anything, on purely imaginary grounds. Though contradicting one another, they live happily together; it seems that every generation of professors smoothly teaches their students how to part their mind to harbour the incompatibility. Hard core sciences are not devoid of their inconsistencies, as we shall see, suddenly pulling out of a hat notions that they exclude as a postulate, and they stubbornly keep denying as they progress. Eventually the split is internalised, tying every mind in labyrinthine twists and turns.

The typical taboo mentioned by Riedl must be broken, if we are to have any hope for our Titanic civilisation veering off course from its promised iceberg. We know the inevitable disaster—a film whose scenario everyone knows could not be an international hit without secretly saying something very important, that we do not conceive of properly. Such being the case, I invite the reader to this sacrilegious transgression.

25.1.2 The Picture of an Integral, Endogenous Crisis

The current state of integral crisis—not only ecological, psychosocial, economic, moral... it grows its effects in every functional compartment of our world—is not the first example of a major crisis on earth, a crisis endogenous to the living—as opposed to exogenous cataclysms such as a collision with a large meteorite, or extensive volcanic eruptions such as those of the Permo-Triassic Siberian traps: one can remind the great oxygenation event, which 2.4 billion years ago radically transformed the physico-chemical conditions of the atmosphere, the oceans, mineral genesis and all earthly living conditions, when cyanobacteria, having invented photosynthesis, grew huge stromatolites colonies that saturated water and atmosphere with their metabolic waste: oxygen, a powerful cytotoxic agent, forming free radicals that bomb biochemical pathways. Its high reactivity allows combustion, from which the highly energetic metabolic pathways of cellular respiration benefit, putting to use this overflowing explosive waste that interweaves power and destruction.

But the current crisis has something unprecedented. No-one would hold stromatolites *responsible* for the great oxygenation event, whereas man would feel guilt for causing such tremendous effects. Hence the current crisis is made unique through what is specific to man in the living world; it stems, *radically*, that is, at its root, from man's consciousness of his responsibility for his acts, it stems from the fact that man is conscious. It is therefore a crisis in consciousness, a crisis of conscience. *We* are in a state of crisis, and at its base is the range of ways and the magnitude in which we interact with our surroundings. The subsequent total transformation of our world makes it less and less comfortable for us, unravels legions of species and dislocates their living systems; mirroring our internal state of crisis. Our consciousness remains partial. There seems to be no sign of awareness that this pressure increases despite the full range of actions we undertake to curb it. Even more so: it increases through these actions. In terms of symmetry, the increase is left unchanged by all the actions of change to counter it: it is invariant under all these actions (Rosen 1995). We do all what is conceivable, and the process accelerates. This means that we should be doing something *else than* "everything conceivable." Something which implies questioning our behaviours from their most basic postulates.

On initial examination, this unprecedented crisis seems intimately connected to the appearance of human consciousness, which we are to investigate. It is also the largest crisis witnessed by man, infinitely more extensive than the last deglaciation, which however, provoked sudden rises in sea level of tens of meters, still recounted in Aboriginal legends, and glacier debacles that swept along entire European and Near Eastern regions, echoed in many deluge myths (Capart and Capart 1986). Then, just as now, most population concentrated in low-altitude lands, and once flooded, they surged back onto other's territories. But however cataclysmic these events were, they do not compare with present time's systemic effects.

25.1.3 In Search of a Connection to Restore

The prefix *co-* for coviability which is the subject matter of this book suggests there is a link to be restored, but which link, how should we represent it? The demarcation line we should bridge the gap over lies in the level of self-awareness that differentiates man from other animals, which causes him to experience himself as distinct. Man experiences a feeling of responsibility when he discovers an extended range of choice, a new-found freedom. We shall track this novelty.

This remark takes us back to the origins of humanity. This specific origin which is a threshold of consciousness: the birth of man, a birth of the world in the consciousness of man, and of that man to himself. Consciousness must not be seen within an isolated individual only. It is settled and maintained in a human group and as wild children cases have shown, it must be socially inducted in a subject; without proper activation within the developmental slot of competence for the acquisition of language, they were never capable of elaborate language if any, or capable of having a life rich with relationships. Hence consciousness is a social enactment and construct. Now, before we plunge into origins, let us make for a panorama of the present.

25.2 Tracking Down the Divide, in the External World Described by the 1st Culture

We now hunt down this disconnection within reality as established by our science.

25.2.1 The Crisis: Dislocation of Space, of Established Relationships

The present crisis that we chose to consider from a crisis in consciousness is above all a generalised feeling of discomfort, loss of meaning, of references. It is a resounding theme throughout all the written works from the end of the 19th century onwards: press, essays, novels. A definite sign is that of complaints about words losing sense, syntax loosening, and even more, reported facts in political speeches being overtly disconnected from actuality. As language reflects a culture's view of reality, a dislocating language has always been a sign of deep turmoil.

Space is being dislocated in a very concrete sense: suddenly all the stable elements that furnish our space, be they values, practices, *things*, seem to give way with the distraught terror of a trap opening under one's feet. Close witnesses of natural catastrophic events, say, a volcanic eruption, frequently report gazing in awe and wonder —quite in contrast to animals: they flee. However, among all these events major earthquakes are special, causing a 'metaphysical fear', as

Haroun Tazieff calls it (Prévost 1984), because it is not something *happening*, but the blurring of all the elements that furnish our space into a muddle, an amorphous, primordial chaos. Hence by contrast, with this disappearance of space the possibility of any *thing* happening vanishes; there is nothing to be the witness of, except the intimate experience of the implosion of the self losing any reference it can lean on. Such collapse into a closed monad is the most extreme psychological regression that can be. This occurrence of an archaic terror of the psyche returning to the limbs, into near non-existence, teaches us that our individuality is in fact leaning on the perceptual surface from which we build our world —those stable things each of us manages to elaborate from his world—, and that any state of crisis is experienced as a space dissolving. Again this echoes matters of origin, as expressed in both creation myths and scientific cosmogonies.

But before we break apart space, let us build it.

25.2.2 Building Up a Space, Perceptively: Meeting a World That Resists

Contrary to the mould imposed onto western minds by a millennial tradition, strongly maintained by school and academic teaching, space is not primarily an empty abstraction, that would be secondarily peopled by objects. Space can only be apprehended by what fills and *furnishes* it, obstructs it, *hinders*. One cannot perceive a vast void; this is the shock when, in a summer sky, you see a lit up storm cloud, and suddenly notice before one of its swelling cauliflowers the tiny arrow of a jetliner: at once the whole spectacle of the sky expands to a gigantic stage, the cumulonimbus 'just there' within reach, is pushed miles away and expands to prodigious proportions; now the world is a dazzling, towering vastness. Not only is space perceptible only through what furnishes it, but it only *exists* in any definable sense through the obstacles we can interact with. Only when an exploratory movement encounters something which opposes movement —an invariant in the variant— can we account for something existing. Something we can *engage a relationship* with. All our universe of phenomena, any object tangible at any degree is aroused by interaction. Thus space, in the intuitive sense of openness, gains its extension and structure by a resistance that contradicts its exploration. You only see what blocks your sight. You sense *something*, not the nothing. Perception has no grasp on uniformity.

It supposes meeting an otherness, a contrast; it is intrinsically a 'self' meeting a 'non-self' —perception is an encounter, at any scale: a subject and an object, a sensory cell and a stimulus, a sensory receptor and an activation signal. Remove one of the sides, perception disappears: it is an intrinsic relation. The uniformly blue sky, a few minutes before the storm was absolutely flat and depthless: void. Just outlined by a roof, a tree, the hill.

Hence we confirm that a disintegration of space —the ground giving way under our feet— takes the generic form of the non-self failing to offer resistance, leaving the self without a relation: all the objects that build up our world, that we lean and rely on, and that actually delimit our bodily envelope, blurring into uniformity, plunging into chaos, leaving the self siderated, collapsed into a closed monad. A seclusion in a self left without a space. People who work in anechoic rooms, hearing no response from the surroundings whatever noise they emit, filled with the dreadful noise of their breath, their blood flowing at each heartbeat, are much watched over, they easily become unnerved and disoriented.

Perception is an encounter. The self, in all radial directions, meets a non-self as a surface of contrast between him and the world, the locus where the dive along the perspective line is blocked. This *surface of horizon* is itself perceptible because under a circular scan it displays transversal contrasts. In the visual field, this surface may be seen as multiple successive tiered planes; a few centimetres away is the side of my hand, because my fingertip touches the corner of my mouth, figuring out the next sentence, the fountain pen sticking out from my fingers, the page a cubit away masks part of the desk, itself covering most of the floor, then in the wall a window opens a few metres away, looking onto a house, behind a line of trees in the distance emerges a hill kilometres away and in the darkening blue sky of indeterminate distance, near the huge evening storm cloud and its jetliner lies Venus that I locate at hundreds of thousands of kilometres. These planes are where my sight comes into contact with the world, my visual bodily envelope, where up my body extends through vision. Objects and theirs depths are models that I speculate: a space built by experience from flat colour patterns.

Hermann von Helmholtz (1985, p. 220–221) gives a striking account of how visual space must be instructed, and objects formed by intersecting various sensory afferences, eventually taking them for granted in one's guts —their *incorporation, embodiment:*

> I am inserting here two reports [...] on two persons born blind, whose vision was only restored later in life by an operation. Cheselden operated on a boy, 13 years old who had a very pronounced congenital opacity of the crystalline lens (grey cataract, as it is called). Concerning his ability to distinguish forms, Cheselden's report is as follows:
>
> "When he first saw, he was so far from making any judgement about distances that he thought all objects whatever touched his eyes (as he expressed it) as what he felt did his skin; [...] We thought he soon knew what pictures represented, which were shewed to him, but we found afterwards we were mistaken: for about two months after he was couched he discovered at once, they represented solid bodies; when to that time he considered them only as party-coloured planes, or surfaces diversified with variety of paint; but even then he was no less surprized, expecting the pictures would feel like the things they represented, and was amazed when he found those parts, which by their light and shadow appeared now round and uneven, felt only flat like the rest: and asked which was the lying sense, feeling or seeing?
>
> Being shewn his father's picture in a locket at his mother's watch, and told what it was, he acknowledged a likeness, but was vastly surprized; asking how it could be, that a large face could be expressed in so little room; saying, it should have seemed as impossible to him, as to put a bushel of any thing into a pint."

This boy initially has no perspective, no depth of field. This sense is learnt, as is every other. This rare case teaches us that *the essence of perception* is not going over hackneyed things in everyday life, but *the building of a space*, the sudden appearance of something new (a new object) in the perceptual field. Which seldom shows up daily.

Nevertheless, by (mainly) visual afferences, the limits of our body extend quite far away. But neither by touching does our bodily envelope confine to our skin. The stick that sounds the river extends our perceptive surface to its end, so do the whiskers (vibrissae) of many animals, or the spider's web, or the bow and violin in virtuoso hands, or any mastered hand tool. And of course the white stick of the blind.

Note also that it is as wrong to separate each sensory modality as it is to count only the traditional five of them —just consider the sense of verticality for which the inner ear plays a major, but again not exclusive rôle. The virtuoso's playing involves sensitivity to touch and muscular sensations, but of course through the ear, both by aerial and inner transmission through the bones, the whole body is fully committed in the performance. A peal of thunder has just crossed through all these layers of perception to impress a new shape upon me and make me look up. Perception is profoundly multimodal, it is *only* relating things.

To sum up, it is sound and consistent to define the limits of our self as our projected sensory surface, a complex construct whose edification is what perception is really about. The self knows itself by this panoramic horizon of non-self. It spreads tentacles in varied reaches. It loses its integrity the moment this surface of contrast with the non-self blurs into uniformity, because its envelope disappears when it expands by radiating in all directions without meeting limits. Losing its usual buttresses, it scatters, dissociates. Having nothing left to relate to —the world is taken apart—, this explosion is equivalent to a solipsistic collapse.

> The earth teaches us more about ourselves than all the books in the world. Because it resists us. Man discovers himself when he confronts the obstacle.
>
> —Antoine de Saint-Exupéry *Terre des hommes (Wind, sand and stars)*, 1939.

25.2.3 Constructing the Structure of a Space, Physically: Relativity

The tautology that we sense *something*, not nothing, is reflected up to the formalisation of elementary measurement operations at the basis of *geo-metry*, the measurement of the world, the building up of a physical space.

One cannot survey a completely uniform space, that is, invariant under any motion —that displays no difference, whatever your motion about it, or at least you *think* you move: for to begin with, you cannot even account for motion, you "believe you move," but nothing is changed that could report it. *There is no contrast between before and after*. To account for a motion, one must stick ranging poles that serve as reference points. Motion is then expressed relative to these landmarks. (Note again the apparent paradox that their breaking the uniformity of space is necessary to conceive of a uniform space; a grid and an origin pole are necessary to claim that "all places are identical.") One's *relation* to reference points changes during motion. It is improperly said that the reference points are not invariant under the action of motion. The sentence is faulty, because motion is not a thing *per se*, but a *relative* notion, it requires both sides of the relation: myself, on the one hand, and the reference points, on the other hand; motion is about neither myself or reference points, it is only about *the relation* between the two of them, and this relation is symmetrical—neither of them is fixed, while the other would be moving. Movement is a characteristic of a *link* between two objects, neither side of the relation undergoes motion, only the *relation* is a displacement. Technically it is wrong to talk about *one* object moving. Of course everyday language abounds in such statements. We did, at the beginning of this section, talk about moving oneself.

Notice the logical circularity: in order to plant landmarks, you need an already well-established stable ground; there would be no need, no *possibility* to stick references in a bubbling magma. You need to have beforehand at your disposal what you pretend to establish by measurement: to "move about" and plant references, means you *already* have the means to establish relative motion, and to ensure that you place each reference at a distinct, stable position. Building-up a physical, *geo-metrical* space is not made *ex nihilo*, it is not invented, it is made by accentuating the type of an already available perceptive space. This migration is made by clearly setting measurements and keeping the record of every manipulation. Then one may discover that these operations can be idealised into a geometry, e.g., Euclid's, for what goes on a flat table, or a flat field, when one considers ideal points and straight lines and the figures made out of them. Given these conditions (from a few centimetres to a few tens of metres), a key property of Euclid's geometry, to wit, that the sum of the angles in a triangle is a flat angle (180°), is *experimentally* realised with a good approximation. The physical space made out of this process is not established in a strong sense, it crystallises as an idealisation, from an already available material. In the idealised geometry, the angles in a triangle add up *exactly* to a flat angle.

25.2.4 Euclidean Geometry as an Experimental Science, a Physics

Euclidean geometry must therefore be regarded as what it is natively: an experimental science, a physics, made by an observer interacting with his world. But what about measurements, this fuel for physics? They are commonly considered as 'objective', as though they provided us with an absolute bedrock, a delivered truth about the world: the foundation of ontological materialism which is the unchallenged ideology of the 1st culture.

Measurement operations do not lead to absolute truths any more than perceptual processes they are a specialisation of. Ancient Greeks had already clearly stated that senses deceive us —at times—, and every wisdom has recognised it. Senses do not deliver anything ultimate, one can never attain certainty. Simply put: *there is no first order truth about the world.* Direct statements about reality are not liable to truth. Everything is approximation-stricken, irreducibly imponderable.

There is nothing absolute in measurement processes. The fundamental type is a measurement of length (and similarly of time, mass, electric intensity . . . all amount to the same principle of extensive measurement). It consists in putting an arbitrary standard stick end to end, counting how many times it fits in the length to be measured. More precisely, and though usually omitted, the result is an interval $[n, n + 1]$, $n + 1$ standards going beyond the measured length. The practical key operations are putting end to end and comparing two lengths; both imply already having the concept of length, which does not exist in every culture (the fact that *we* may *translate* many things that occur in such cultures in *our* notion of length, does not make it a concept of *these* people). Besides, it is acknowledged from the onset that a length is to be measured, hence it preexists. The idealised putting *end to end,* exactly and without a gap, allows to establish a correspondence between the length category of the object to be measured and the *scale of measurement,* which is a numerical set, *end to end* matching addition on numbers. The result of measurement, i.e., a category —length— combined to a numerical value —actually an interval— is a length approximated in the arbitrary standard unit, it is *relative* to that standard. The standard is the bootstrap of all measurements, it cannot measure itself (this is true for the initial Revolutionary standard metre kept in the *Pavillon des poids et mesures* as well as for all its ulterior replacements), it is the 'primary cause,' all the measurements made around the world descend from it. It closes the otherwise infinite regress of relativities. It is an arbitrary, conventional choice, decided by men.

No component of physical measurements is absolute. Any measurement is relative, thus anything ensuing from measurement is relative. The building up of a physical space, anchored in the world, is relative.

25.2.5 Constructing a Space, Mathematically: The Elements of Euclid, and the Euclid Myth

25.2.5.1 The Euclidean Synthesis

In the third century B.C., Euclid rearranges all known geometry, starting from elementary terms given as self-evident, and proceeding by definite logical and mathematical manipulation, he establishes all geometrical properties as a deduction resulting from mechanical transformations of the basic elements. He 'encyclopædiases,' 'refounds,' synthesises a new geometry. It is the first axiomatisation in mathematics, and will remain the sole one until the nineteenth century. The concept of demonstration that stems from this system is still valid today.

25.2.5.2 Euclid's Feat: A True Machine, Conservative, Deterministic, Consistent and Disconnected

Euclid, from evident notions that in his day *no-one educated in his culture* would challenge —straight line, points, and operations applied to them—, rigorously built geometry as an abstract space, a pure system of relations. These notions are evident —more precisely, *ideal*—, only to those who have grown in a social-historical context that have brought them into existence, that inducts them culturally (you can try the experiment out in Papua). They were cultivated long before the ancient Greeks, whose heirs —Europeans, notably— also take for granted. But they are all Greek to, say, a Pirahã, and totally devoid of interest for them (Everett 2008).

The property of being ideal is that *they remain perfectly unaffected, unchanged* through the manipulations applied to them. The notions are the exact invariants of these operations, and both jointly crystallise and mutually reinforce, they *contrast* one another. The whole forms a perfectly laid-out, well-oiled machine. Even more: it is perfectly exact and wear-free, thus eternal. Definite and deterministic, without escapement or room for error, it is conservative: "nothing is lost, nothing is created, things are transformed" during its operation, as stated early on by Anaxagoras (c. –510 to –428) (Voilquin, 1941, p. 136). One can replay at will its correct operations, always confirming that it is eternal in our mind.

It is a *true machine*. We have already mentioned of machines, without special care, but what is a machine, and why add the extra term of *true?* The machine par excellence of modern man may seem to be the car and its heat engine or the computer. But these are material artifacts, and as such cannot be true machines. There is always a day when pressing the button, they do not run: at which point, they are no longer the machine they are supposed to be. A true machine does not wear out, there is no limit to iterating its steps. A true machine is a mental object. Its two major properties are *determinism and conservativity*. It is completely definite (completeness), it is the realm of exactitude, and it provides with the *rules of production* or generation of a space.

But let us return to Euclid: with this machine, he created an autonomous domain whose main characteristic is its consistency, closed onto itself. Its internal consistency derives from it being constructed by its own rules of organisation: *auto-nomos*, it defines its own rules. *Disconnection from reality is perfected.* Euclid has lifted geometry of the ground, brought it into the realm of Ideas (Platonic), of cognitive machines. He shifts from physical to mathematical space.

Of course Euclid's geometry matches —approximately— the experimental geometry as known before. It is much better than measured geometry, since by contrast, it is exact. Hence the conviction of having captured the essence of reality, truth about reality. Between the near-abouts of physical space and the fascinating exactitude of mathematical space, the choice is promptly made: the second is the right one, the *true* one. But here is the additional step: this system, disconnected from the world, is proclaimed to be truth *about the world*. Henceforth, if reality deviates from exactitude, reality is wrong!

To gauge the extent of this reversal, note that until the end of the twentieth century, the *International vocabulary of metrology* (ISO Guide 99:1993) would define error of measurement as the difference between a 'true value' and the measured value: there *is* a true value, but the world refuses to conform to it; each time a measurement was made, one would cold-bloodedly state that the world is mistaken. A bizarre inversion of what an experimental science is supposed to be. This reputed exact value is a pure creation of the mind. It exists nowhere, and no one has any means to reach it. It is a property of an idealised model, (only in part) necessary, because calculation requires definite values. As a mathematical object, it complies with constraints intrinsic to our descriptions. The postulate of a true value indicates a confusion between what goes on in our minds with its own constraints and what is a matter of reality, that is, what is seen as being outside our mind.

25.2.5.3 The Emergence of the Euclid Myth

Thus the Euclid myth was born. As usual, it does not bear the name of its instigator who happened to potentialise a collective fabrication nurtured over centuries, and that blossomed in his posterity. Its name was given only recently by Davis and Hersh:

What is the Euclid myth? It is the belief that the books of Euclid contain truths about the universe which are clear and indubitable. Starting from self-evident truths, and proceeding by rigorous proof, Euclid arrives at knowledge which is certain, objective, and eternal. Even now, it seems that most educated people believe in the Euclid myth. Up to the middle or late nineteenth century, the myth was unchallenged. Everyone believed it. It has been the major support for metaphysical philosophy, that is, for philosophy which sought to establish some a priori certainty about the nature of the universe (Davis et al. 2012 p. 363).

To be more precise, until the middle of the nineteenth century, believing in the Euclid myth could not be an issue, no one was conscious of it being a belief, it was absolute fact.

Euclid invented axiomatised geometry, what we now call a *formal system*. Certainly this term does not convey much to most readers, yet it is essential to have a manageable image of it for the sequel.

25.2.5.4 The Concept of a Formal System: A Board Game

A Formal System Equals a Board Game

For our concern, we can identify a formal system with a game, thinking of a board game such as chess, go, awalé, or the very ancient royal game of Ur, found in Mesopotamian excavations, dating from 2600 years before our era. That the equivalence is well-founded is apparent in a standard formalist position about mathematics, as exemplified by Davis and Hersh: "The whole thing doesn't really mean anything, it's just a game, like chess, that we play with axioms and rules of inference" (Davis et al. 2012 p. 72).

The rules of chess are not found anywhere in nature, it is a domain entirely fabricated by man, their sole instigator, defined 'freely', arbitrarily—not without meeting internal constraints, though: a demiurge is not all-powerful. In this axiomatic system, the basic elements are the pieces and chessboard, the logico-mathematical operations are the valid moves—the rules, in the narrow sense, including the initial configuration. One usually learns with wooden board and pieces, but this materialisation is not of necessity, pieces and board are respectively abstract categories and structure (and that rules are immaterial is probably obvious for the reader) —in Greek *u-topos*, without a place, nowhere just as Davis and Hersh point to the question "Where is the place of mathematics?" (Davis et al. 2012, p. 8). This place cannot be shown, it is a new space, created *ex nihilo*. This domain is born in our cognition.

The Rules of Chess: The Only Place of Exactitude and Truth

Whoever can learn the rules of chess —which, we must emphasise, is *not absolutely everybody*—, knows how to operate the machine that plays legal moves. And those who can, believe with absolute certainty that they play *by exactly the same rules* (playing *well*, however, does not belong to the formal system; it is a totally different matter). This *complete and exact communication* is a characteristic property of formal systems, they are the only objects liable to exact and complete communication of information; even more precisely, they are the only systems where completeness and exactitude of a given state is meaningful. Equally, truth is exactly defined only in a formal system: a move at chess is legal or not, is a truth. Exactitude, completeness, and truth, out of these three pillar notions of our culture, none is applicable as first-order statements about present reality.

Formal Language: Algebraic Notation for Chess Games

These properties are made explicit in the algebraic notation of chess games. A game is entirely expressed in the succession of moves encoded in the form departure square-ending square, denoted by the 8 × 8 Cartesian coordinates, a1 to h8. The pieces involved need not be specified, only one can be on each square. The resulting text is very concise (actual notations are varied and differ, the involved piece is named and its origin square omitted if redundant, etc., but this is secondary); good players read games fluently, just as they can play without a board. The game is abstract, it needs no materialisation, a game is formally expressible in a sequence of symbols taken from an alphabet; and of course the rules are abstract and primarily given in a language. The game is a formal system.

The autonomy of the system is obvious on the case of a problem, e.g., checkmate in two moves: absolutely nothing can happen in the world that could alter the solution in any way. No event, however cataclysmic, has any effect on the answer, nothing has any ability to interact with it, it is unaffected, even were we in the middle of a volcanic eruption. It does not depend on it. It does not trade any sort of relationship with the world —except that man has made it materialise in it. That is the reason why we talk about a new, autonomous domain, that most do not hesitate to term absolute, eternal, relieved from any of the world's vagaries: Platonists.

Formal systems are *insensitive*, indifferent to any thing. To put it plainly, they are dead. Which was already clear from their being eternal. A living being cannot fit in with these characteristics.

Overall, formal system equals board game, true machine, and alphabetic text.

A Formal System Excludes Semantics: The Myth of Artificial Intelligence —Of Exoteric Knowledge

No event can *bear any meaning* to a formal system, since it cannot *interact* with it; without a possibility of relationship —it is closed in isolated, pure relations—, a formal system is deprived of any ability for semantics. A statement which has resounding consequences, because it denies the possibility of artificial intelligence (a computer in its *normal* function does formal operations, and is a good approximation of a true machine): the helplessness of a machine in establishing a relationship to reality, to other subjects, to understand them, to reply, to live with them. Formalism excludes semantics. It yields the *disconnection* we pursue. This comes as a shock in today's context. But read back to "the whole thing does not mean anything..." There, formalist mathematicians rashly told us that formulæ *per se* have no meaning; that meaning resides in the interpretation a person endows the text with, by connecting it to experienced reality. Mathematics singly tells nothing about the world. *We* do put it to use by connecting it to reality.

Hubert Dreyfus (1992) made strong points about unspoken postulates of artificial intelligence, virtually putting the field to an end by mid 1990s. It is reviving, stronger than ever, in this early twenty-first century. Which will force us to return to it further on, to wreck it not on abstract, but on practical, solid grounds. And again to track it down to its deepest roots, and tear it up.

But while we are at it, let us recall that one of the postulates of artificial intelligence is that all knowledge can be expressed formally, as a piece of text —that all knowledge is exoteric. This belief is so ingrained and pervasive that I challenge the reader to raise a counterexample, till I disclose it further on.

A Formal System Is What Has an Alphabetic Floor

The algebraic notation leads us to another factual definition: is formal what can be turned into an alphabetic text. With rules of manipulation.

More precisely, is formal what has an alphabetic floor. In an alphabetic writing, there is nothing deeper once you have read characters, no mystery behind the *a* you read. Nor on the chessboard is there anything to be found beyond the Knight, once identified: it is known *in being*, totally. Just as the word is naked to its bones, in its sequence of letters, a chess composition is known all out. Formal means total transparency, full knowledge, every item in bright light.

Contrast this with an ideogrammatic writing: a sign stands as a block for a thing or concept in its whole depth, woolly, unfathomable. The sign shows but still veils. Reading never experiences reaching a firm, definitive floor to rely on. Mystery and life remain palpable under the thickness of every sign.

With the skeleton of words exposed in alphabetic writing, the mystery behind each word surfaces only when one buries oneself in a dictionary, seeking to strengthen a meaning, unfailingly emerging clouded in ever so many sliding usages: the word cannot be focused on. The repeated disappointment of these journeys is evidence that we cannot dispel the conviction that words do carry a sense as clear as suggested by the sequence of their letters. We tend not to believe it, although every interaction with reality shows the same effect: perfectly outlined categories seen from a distance are blurred at a closer look; the initial pure uniformity reveals many inside structures, and neat edges vanish: look at a sharp four-colour magazine photograph through a magnifying glass, and you shall see.

By Contrast, Our Relationship With the World Creates a Horizon of Reality Along a Perspective Line

Now we are lead to properly defining the very important notion of a horizon: look again at the cumulus swelling in the sky, it has a perfectly neat boundary. Its edge on the sky background is its horizon, it generalises the edge where the earth meets the sky. Now climb up the hillside, the cloud swallows you, and you are unable to say precisely where it began. A horizon is a border of an object, or the object itself, as experienced in any sort of approach. Each perspective line of approach will make for a distinct, proper horizon; objects come and go according to the context of interaction —the horizon in the sky is not found from the slope of the hill. Generalising a step further, a horizon is not only a line, but the whole surface of what obstructs your view: we again run into perception as the self meeting a surface of resistance, inducing a horizon by interacting with a non-self. We live by constantly shifting points of view and assembling the resulting horizon facets into 'objects.' We never access an ontological core, only horizons of reality. There is no 'alphabet of senses.' The generic experience is an indefinite perspective dive: a point at the horizon unfolds unexpected structures as we near it, the sharp edge of the cloud extends in many shades, a flat plate of colour reveals a world of contrasts —as with magnified magazine photographs. Perception never delivers ultimate categories, there is no alphabet for the world. We are tricked into believing we have found the alphabetic floor, only when we stick to prompt things in the same stubborn way, thereby always obtaining the same answers. Compulsion brings the delusion of absolute categories.

The Horizon of Reality Against the Myth of First Order Reality

Ferdinand Gonseth made these points clear in 1949, in a contribution held in high esteem by prominent physicists (Pauli 1948): "The previous results extend beyond the framework of geometry; They span over any knowledge, we mean how any knowledge presents itself to us, at a given moment: nothing allows us to think that our knowledge, even at its ultimate frontiers, is any more than a *horizon of knowledge;* that the last 'realities' we conceive are more than a *horizon of reality."* (Gonseth 1949, p. IV-46 —310) In modern physics, Gilles Cohen-Tannoudji has developed it in a lucid, but singular view:

> Essentially, the answer I propose [. . .] is the thesis that universal constants would express *limitations in human knowledge*, limits as unavoidable, inalienable, but also moveable as *horizons* are. [. . .] They have even been termed 'constants of nature.' Today, we are forced to leave that behind, [. . .] constants are not physical constants of the universe, but universal constants of physics. [. . .] Universal constants express in principle limitations imposed upon man in its cognitive relationship to nature. [. . .] Physical concepts, our concepts, do not claim any more to describe the world directly, but only *horizon lines* that practical or experimental sensory knowledge draws upon the world. A horizon is virtual and inaccessible, but is easily moved. Its movement is easily expressed in mathematics. Whatever its movement, it occurs *on* the world. Thus the world can be thought of as *the geometrical locus of all possible horizons* (Cohen-Tannoudji 1992).

Try once to meet the rendez-vous where the rainbow ends, and you shall feel what a horizon of reality is. The myth of reality described at first order, which dominates current physics, swears to meeting the alphabet of the world —where the rainbow ends. In the same vein, we routinely hear of *laws of nature* —whereby a speculation made by humans is ontologically attributed to nature. A typical case of mistaking the map for the territory (Bell 1933, p. 135). Mature physical theories are formal, axiomatic at the core —just as chess is. Who would claim that chess rules reveal the deep nature of reality? A theory informs us about the world not by its being formal, but by its semantic bond to it, its approximate predictive match with observed categories.

Out of this muddle rises a large figure. The Euclid myth is just one of many faces of a universal trait of Western 1st culture: the claim for absolute knowledge.

Assigning Categories: Measurement, Convergence and Exclusion

In a normal reading, all the variations of letter shapes instantly *converge* to a symbol, say 'a.' Over-distorted letters will block reading: *measuring* varied shapes to categories crucially involves *excluding* unacceptable forms. The dual action of convergence and exclusion permits building a local alphabet in our cognition, e.g., see a board and pieces of wood as a chess situation. The assignment or *measurement* of categories is the elementary apparent step of the building of a formal domain.

This process has a notable side-effect: once trained to form specific objects, cognition, at all stages, tends to reject every sign that does not fit the expectations.

We stick to a theory, well-after it is refuted by events. If we cannot let go of the handrail of deep-rooted conducts, we are captive to the orbit of compulsion, reinforced in the belief of ultimate building blocks.

Seduction of the Alphabet and the Unique Enjoyment of Play

The need for a stable ground cannot be avoided. Biological life calls for stable terms; the same extends to psychology. However, there is something more in the unique pleasure experienced in playing a board game, mixing a relief with freedom, glee and empowerment. They surely stem from the uniqueness of a situation known exactly and entirely, whose outcome depends only on the operations of our mind. We feel invested with the autonomy of a formal system, protected from any outside event —in an invented world, a speculative bubble. Such instant rising into a new space is found also in listening to a storyteller by a bonfire, readily entering a new world, truer than any other, as soon as he starts "once upon a time", or, in a more recent history, burying into a book. This particular flavour of minting worlds has been discussed under the label of 'suspension of disbelief' in literary theory.

But in a board game, the mind is in full power. It is longing for itself, it revels in its own pure operation as a true machine. The alphabetic transparency. Playing is thus a reflexive experience. The mind knows itself as independent. Men —from Africa to the Eurasia at least— have felt it invaluable, cultivating it over centuries. It would be a full-blown experience, if it were not only to marvel at playing, but to see it consciously as an occasion to know rational thinking in intrinsic operation. To sight in isolation the map that serves to frame our representations of reality —precisely because it *contrasts* with it— instead of the usual superimposition —and confusion— of map and territory, whence the Euclid myth springs. As we shall see it comes in many flavours —artificial intelligence gave us a foretaste.

25.2.5.5 Debunking the Euclid Myth

At this point it becomes urgent to explode the Euclid myth.

Of Large and Tiny Triangles

Euclidean geometry is a broad match in flat earthly conditions at common scales. Now take a bold step and join Hawaii, Vanuatu and the Galapagos: the sum of angles exceeds a flat angle. Go broad scale and join one pole to two points on the equator: in both of them is a right angle, and the one at the pole comes in surplus. Move one of the points along the equator, until it comes opposite to the first, your triangle has become a *biangle* —the quarter of an orange—, with still two right angles plus a flat one at the pole. Euclidean properties no longer apply.

Draw the smallest triangle you can, under the microscope you experience once again the magazine picture effect: no more lines —not to speak of straight ones— all notches and nicks and pieces. The uniform sheet of paper has become tangled spaghetti. The basic Euclid elements have vanished. Their horizon has dissolved.

Practical Euclid's geometry, *just as any category we raise*, is grounded only in a limited range of interaction. Its basic elements vanish in a blink of an eye out of this context.

Still Objecting

This should be enough to demolish the Euclid myth in any literate reader. But what about the general one? Every Westerner keeps hearing of '3D' space, and his gut feeling about it is further fuelled by every film and video game: each is built-in evidence, that cannot be escaped, since it is unsaid. As Marshall McLuhan put it, *the medium is the message:* the strongest message distilled by any medium, is in the carrier itself, *esoteric* (McLuhan et al. 2008). Thus our myriad screens glimmer from the shooting painful Euclid myth. The Western world is under this permanent narcotic suggestion.

In the City

How do you give directions to a place in town, how do you find your way? You proceed by lines of buses and underground, stops, stations and connections; by streets. In Euclidean space, you should go straight. You follow streets because there is no way out of these corridors. The concrete geometry of your city is that of a graph: look up a map of transportation, a map of the city —that is: *of streets*. Not *one* graph, many: try a wheelchair once, then a mere step turns into a wall. Or, if you happen to be bound to a slow pace, a busy street can no longer be crossed. Look up Euclid's *Elements* (1966): no trace of walls, doors, trees, no sort of obstacle: the space is completely empty, transparent to the infinite —*it cannot be sensed*. Look up: if you see something, it means you are not in Euclidean space. You may persist and dream up your city as Euclidean —but you may dream *anything you like*, dreams are not bound by facts. You still shall not be able to move by Euclidean operations.

By the same token, there can be no single geometry, total and exclusive, that would be true to our world. "A geometry cannot be truer than another, it can only be more convenient," says Henri Poincaré (1968, p. 76), and we agree, but he goes on: "Now Euclidean geometry is and shall remain the most convenient," which makes us wonder if he ever had to give directions to his laboratory. Or, to be more faithful, *he meant another interpretative path*, he had in mind another connection to reality, another semantics. He was on other horizons.

How we can mentally exclude any feature that does not fit with Euclid's frame, is just a case of *exclusion* noted above. This is to *abstract*, to rise over the earth, in a detached space.

25.2.5.6 Debunking Generalisations of the Euclid Myth

Debunking the Myth of Artificial Intelligence

Since I promised, there it is. Just as a reminder, the project of artificial intelligence, worthy the golem or Frankenstein, is nothing less than the synthesis of a fully-fledged subject from within a machine, a formal system. It would amount to having by some magic burst, the Euclidean void teeming with life. The crux is with semantics, of course. Relating to *some thing*, to the world. As the root *sem-* expresses, a sign, pointing to something *outside*, which is the *meaning* of the sign.

To this day, machines are (of course) unable of understanding a sentence, in the very concrete following sense: Vinton Cerf, Vice-president and Chief internet evangelist at Google, confesses that their indexing engine for corporate servers does not work well: it returns loads of unranked documents. Why? because these documents refer to one another in terms clear enough for human usage, such as 'last annual report' or 'the letter from ... ' Of course machines do not understand sentences: they cannot even extract those names and resolve them to actual documents. To the indexer, these documents all have similar content with no relationship between them: a big amorphous set. Which is what it returns. How then is the same indexing engine able to provide us —often— with satisfying answers, *over the whole web?* Because there, millions of human subjects have typed in character by character, explicit machine-readable links to other pages together with their text. By scanning the whole web, the indexer gathers statistics of which link is found close to which words in pages, and should therefore be returned when a user queries these keywords. Hence a machine can rank the pages, *without any understanding*. In this case, humans are slaves to the service of machines. Until all employees are under duress to make explicit links to other files each time they allude to something, the corporate indexer shan't work.

Please note that in all of these situations, machines have full access to the exoteric text. It does not help, because the meaning *is not in the text*. It is in connecting to the outside. No formal manipulation can do this. *Formal manipulations are by nature internal laws of composition, they strictly stay in the confines of the formal domain, thus without any means to connect to anything else; this is in fundamental contrast to living systems*. If you have to *make* sense of a text, to *give* it meaning, it is perfectly clear that the meaning *is not in the text*. If you want a hint about how a machine sees text, take a look at the Rosetta stone, or a Maya codex, or a page of runes —whatever script you have no clue about. Or, for computer-goers, read your text in binary code. And try to picture what it means: nonsense, isn't it?

Now, could you translate what you do not understand? Google does this, apparently. Of course with a mere dictionary, a machine can send a flow of the words that appear on the Rosetta stone, which does give you an idea of what it is about. But this is not translating, or *I* would translate Mandarin to Quechua. Mind the illusion of *sometimes* getting decent translations: it is not for having bright machines, but simply large stores of multilingual corpuses, translated the hard way by plain humans, where picking a corresponding phrase is easily automated.

Where is the artificial intelligence if it is so easy to block any 'robot' on the web with a captcha? Last, do take a look at Donald Knuth's *Art of computer programming* (1997, p. v), whose preface requires on the part of the reader "an ability to put the solutions to problems into such explicit terms that a computer can "understand" them. (These machines have no common sense; they do exactly as they are told, no more and no less. This fact is the hardest concept to grasp when one first tries to use a computer.)" Today, he keeps maintaining that he sees no sign of any artificial intelligence, and for instance David Lorge Parnas (2017) is even more categorical.

Gödel's Debunking of Hilbert's Deadly-Laden Program

A look into the origins of artificial intelligence leads us right to the *crisis in foundations* in mathematics, from the late nineteenth century to the first quarter of the 20th. It is best expressed in Hilbert's words:

> The goal of my theory is to establish once and for all the certitude of mathematical methods. [...] The present state of affairs where we run up against the paradoxes is intolerable. Just think, the definitions and deductive methods which everyone learns, teaches and uses in mathematics, the paragon of truth and certitude, lead to absurdities! If mathematical thinking is defective, where are we to find truth and certitude? (Davis et al. 2012 p. 374–375)

Hilbert's *programme* —don't laugh, it is no joke— was to turn every mathematical statement into a formal expression —an alphabetic sequence of symbols—, then the *ideal mathematician* was turned into a machine processing symbols.

But sacrificing the living subject on the altar of certitude proved useless. Gödel's 1930 theorems broke the machine from inside. No formal system strong enough to include integers could prove its own consistency. Autonomy, self-sufficiency of a formal system is illusory and hopeless.

In spite of the wreck of Hilbert's arch on Gödel's shelf —laugh, it is intended— the Western twentieth century has rushed in the wake of Hilbert as if his quest had been crowned. A generalised chase for the same rainbow, from the sciences to philosophy and humanities. The least not being artificial intelligence. Why this torrential, turbulent flow was so irresistible, where it originates from, is going to be our quest from now on.

Debunking the Myths of Modern Physics: Non-existence of Time, Ultimate Theory

Physics is distinctively *defined* by the successful completion of a fully formal, detached theory, secondarily connected to observations and proving useful (connection to observations *alone* is just plain life, usual sensorimotor loops). They form respectively its formal, and its meaningful, semantic side. From what precedes, it is no wonder that no actual notion of time appears in physics: in classical mechanics the machine-theory connects in one coherent block all the states along a trajectory; they are all given at the onset; nothing new can occur. Time is at best an index, trajectories can be travelled in any direction; in fact all 'times' are simultaneous: the world is completely still, dead, just as a formal system is. It is dead *because* it is regarded as a formal system. Relativistic or quantum mechanics make no difference in this regard, though a more complex mathematical paraphernalia keeps the bystanders out of the temple. As we can see, the neutralisation of time is a property of our map —a formal system—, certainly not of the territory —where all things pass. However, let us listen to Ilya Prigogine:

> I like to quote the correspondence between Einstein and his old friend Besso. In the latter years Besso comes back again and again to the question of time. What is time, what is irreversibility? Patiently Einstein answers again and again, irreversibility is an illusion, a subjective impression, coming from exceptional initial conditions.
>
> Besso's death only a few months before Einstein's own was to interrupt this correspondence. At Besso's death, in a moving letter to Besso's sister and son, Einstein wrote: "Michele has preceded me a little in leaving this strange world. This is not important. For us who are convinced physicists, the distinction between past, present, and future is only an illusion, however persistent." (Speziali —Einstein, Besso 1972, after Prigogine 1982)

First, the reader can see that I drew physics to a true likeness. Second, as Einstein quietly denies the existence of time, he denies everything he has lived by. All his life is illusory, including his publishing his theories. He denies that his life was quite different after 1905, that it makes sense to write a letter; he denies that to live is to sum and carry experience over a lifetime: an integration in memory. How on earth can a living being insist, in the name of a theory his healthy body gave him leisure to elaborate, that his theory proves his biological existence is null and void? How can you *live by* such a belief? Thus Einstein is an illusion to himself *—however persistent—*even though, *he* claims to know something absolute about the universe that proves his non-existence. Paradoxes pile-up.

Einstein was not born with the view that he strives to convince Besso of, he *acquired it* over his lifetime. The self-contradiction is obvious: he gained the conviction that gaining experience is illusory. Since time is only measurable by gaining information, we see the intricacy of denying time and denying the living, experiential subject.

Einstein's decree is issued from within the bubble of a speculative theory, inside a mental bubble, itself in a biological bubble. The decree requires these three nested contexts to exist —it is relative to them. Now it pretends to look at the world sideways, from outside any context, from nowhere. Although Einstein has extended physics more than anyone else by bringing in relativities, he is caught in the sin of absolutism —anti-relativity. Einstein makes his theory a first-order, absolute truth about the universe. Such an ontological statement, as the Euclid myth, is unacceptable because non-relative. And to dispel possible misunderstanding, stating apparently 'in the absolute' that absolute statements about reality are unacceptable is not self-contradicting: it is about *representations* —second-order—, not first-order, about reality.

Without a "distinction between past, present and future," *there is no life*. This case of cognitive self-denial, a cognitive suicide that we shall have to return to later, however remarkable, is a mere replica of Hilbert's doing away with the living mathematician for the sake of certitude. Time and the living subject are expelled together, replaced by a true machine. Both cases purport to a denial of reality, as if it were possible to pump oneself into the formal bubble and withdraw in that invented domain, the nowhere land of mathematics, making it our home and universe, clean and exact, remote from the dirty reality. Both are just variants of the Euclid myth, since there is no thing, no living being in Euclidean space. We cannot be there. It is empty.

The dream underlying Einstein's conviction is that a theoretical bubble can not only be detached, but completely liberated from its originating substrate, and grow until it encompasses the whole universe. As we know quite well by now, it results from the hypnotic effect of seeing a formal system detached from reality.

So great is the fame of Einstein —a legend much alive—, that I cannot hope to be heard unless I balance him with another icon of physics. Werner Heisenberg is one of the few lucid ones, who does not forget the native substrate of a theory:

- P B How does quantum mechanics deal with time flow or does it in fact say anything at all about it?
- WH I would have to repeat what C. von Weizsäcker said in his papers: that time is the precondition of quantum mechanics, because we want to go from one experiment to another, that is from one time to another. But this is too complicated to go into in detail. I would simply say that the concept of time is really a precondition of quantum theory. (Buckley, Peat 1996).

But these words cannot go against the overwhelming 1st culture tide. Its inexorable machine reigns supreme.

Scientists, physicists never unveil their beliefs in scientific communications, formal, impersonal and devoid of meaning, as we should expect. Where can these be found? Apart from direct personal exchanges, or from peering into their correspondence —but that comes many years later—, the best place to look for, is in their popularisation books. Let us pick the number one bestseller of the genre, Stephen Hawking's *Brief history of time* (1996). Notice: vulgarisation is no excuse for talking insanely. On the contrary, responsibility is heavier when you address a helpless, trustful audience who believes and cannot belie you.

Beginning Chap. 1, Hawking seems to agree with our relative view of a theory —*pardon him the nature of the universe:*

> In order to talk about the nature of the universe [. . .] you have to be clear about what a scientific theory is. I shall take the simpleminded view that a theory is just a model of the universe, or a restricted part of it, and a set of rules that relate quantities in the model to observations that we make. It exists only in our minds and does not have any other reality (whatever that might mean). [. . .] Any physical theory is always provisional, in the sense that it is only a hypothesis: you can never prove it. No matter how many times the results of experiments agree with some theory, you can never be sure that the next time the result will not contradict the theory.

But soon further down in the chapter, no more going through the rituals, etiquette is jettisoned, Hawking claims an absolute, ultimate theory, and he goes on throughout the entire book:

> The eventual goal of science is to provide a single theory that describes the whole universe. [...] And our goal is nothing less than a complete description of the universe we live in. —Chap. 1.
>
> A complete, consistent, unified theory is only the first step: our goal is a complete understanding of the events around us, and of our own existence. —Chap. 11.

Since a theory is a formal system, he unavoidably believes that the universe has an alphabetic floor:

> So the question is: what are the truly elementary particles, the basic building blocks from which everything is made? —Chap. 4.
>
> We do have some theoretical reasons for believing that we have, or are very near to, a knowledge of the ultimate building blocks of nature —Chap. 5.

And for the 10th anniversary edition, the devil take good manners: the foreword already conflates formal systems and reality; laws, speculated by reason become laws of nature:

> ... but it would not change the most important point: that the universe is governed by a set of rational laws that we can discover and understand.

With the rational map confused with the territory, one wonders why there is any difficulty in understanding that territory.

The ultimate theory is by definition completed, and single —there can be only one. Nothing can appear that is not already contained in it, everything is spread out in text, completely still and fully transparent. We are in still land: a dead place. There is no room for life there. We are lost in the absurd. No wonder Albert Camus made it a major feature of the twentieth century.

Hawking gently mocks Einstein's naïveté

> Einstein spent most of his later years unsuccessfully searching for a unified theory, but the time was not ripe. [...] As I shall describe, the prospects for finding such a theory seem to be much better now because we know so much more about the universe. [...] I still believe there are grounds for cautious optimism that we may now be near the end of the search for the ultimate laws of nature—Chap. 11.

Hawking follows in the footsteps of Einstein; he knows but he is not aware, he learns nothing from history. A deluge of absolute. A permanent rush to the rendezvous where the rainbow ends. The promise that we are closing in. This portrays our 1st culture.

Physics Number Zero Hypothesis: Repudiation of the Observer

We have seen artificial intelligence pretends to make a subject out of a formal system. Now for physics to repudiate the subject in flesh and blood, for a bit of symmetry. Not only physics: it is the core of the 1st culture, it spreads throughout our societies.

All was fine with Isaac Newton's contribution. *Hypotheses non fingo*, he did not fashion hypotheses about the essence of things, he resolved to describe simultaneous apparent differences occurring during motion, by equating them. Explanations tend to speak in terms of forces and causes, but the equation merely says that synchronous differences occur 'magically' on each side of the equal sign, none of them acts as a cause, nor of course is there any intent anywhere.

Let us turn again to Hermann von Helmholtz, renowned physicist, but also a huge contributor on the physiology of perception and philosophy of science; he has exposed the underlying attitude in the clearest possible terms (1985):

> In my opinion, therefore, there can be no possible sense in speaking of any other truth of our ideas except of a *practical* truth. Our ideas of things *cannot* be anything but symbols, natural signs for things which we learn how to use in order to regulate our movements and actions. [...] Not only is there *in reality* no other comparison at all between ideas and things —all the schools are agreed about this— [even Hawking is, at the beginning of his first chapter] but any other mode of comparison is entirely *unthinkable* and has no sense whatever. This latter consideration is the conclusive thing, and must be grasped in order to escape from the labyrinth of conflicting opinions. To ask whether the idea I have of a table, its form, strength, colour, weight, etc., is true *per se*, apart from any practical use I can make of this idea, and whether it corresponds with the real thing, or is false and due to an illusion, has just as much sense as to ask whether a certain musical note is red, yellow, or blue. Idea and the thing conceived evidently belong to two entirely different worlds, which no more admit of being compared with each other than colours and musical tones or than the letters of a book and the sound of the word they denote.

A scientific description is made explicitly *relative* to the observer exactly as we have regarded the perceptive build-up of a space —a subjective relativity, in a technically exact sense (Douzal 1998, 2015).

Of course, Newton connects what appears *to a subject*, but he never mentions this observer, and all is framed in a Euclidean space. In fact nearly all classical mechanics —kinematics, Galilean relativity— is entirely decided by choosing a Euclidean space. Thus mechanics inherits the global, absolute view implied by Euclidean space. On top of that, observations —measurements— are supposed to be made without any mutual influence between object and subject, as if separated: no interaction, no *relation*, no perception. The perfect observer was insensitive —a 'formal observer'. Both effects result in a dematerialisation of the observer in a sort of immanent witness, at once nowhere and everywhere, that was felt as the key to the success of physics, because Newton's attitude also removed all sorts of intents and misconceptions that hindered previous attempts at a theory. His pragmatic, tight connection to actual appearances was thus understood as disconnection from reality. And as the way to follow.

This became physics' number zero hypothesis: there is no observer. And the observer must be hunted down, expelled, since his *subjectivity* sullies the purity of science. The fear of pollution extended to all earthly, dirty matters, changing and unreliable, while the formal theory was eternal and unmoving, the realm of certainty. By a mean pun, the outcast subject insured the end of subjectivity, and by contrast guaranteed the 'objectivity' of science, without the need to define what that meant. Heinz von Foerster lifts the ambiguity: "objectivity is the delusion that observations could be made without an observer" (von Glasersfeld 2001 p. 37).

By the twentieth century, the observer has become a taboo in all academic fields, and avoiding to mention him, a sort of magic spell to good science. Not only the ideal mathematician, who tries to conform to Hilbert's machine: "His writing follows an unbreakable convention: to conceal any sign that the author or the intended reader is a human being" (Davis et al. 2012 p. 67). Most academic editors in all fields fanatically required —and many still do— writing impersonally third-person, and passive (Williams 1990 p. 40): you should conform to this magic spell so as to show you grovel before the idol of the zeroth hypothesis. You should annihilate yourself. The fashion did not stop in academics; William Zinsser (2009) denounces it nearly everywhere.

Interviews in streets or at work of social workers, police officers, city clerks... show this awkward style, as if going straight to the facts meant dirtying one's hands, and sounded unsuitable. If you believe that languages, or modes of expression, registers of language, do express deep cultural feelings, you are bound to conclude that impersonalisation has become an internalised command in society at wide, as soon as an utterance is felt of a certain public character.

We already know this scene of the cognitive subject trying to heave himself up into the formal bubble, hoping to be liberated, away from earthly vicissitudes, servitudes, dirt and filth.

But before we proceed to the extent of these ideas in society at large, let us finish with physics and see where the obstinate hunting down of the observer leads to.

With Arthur Eddington (1920) talking about the knowledge provided by physics, we feel at home again, as with Poincaré, Heisenberg or Helmholtz, and back to a reasonable man-made science, an observer confronted to a horizon of reality:

> And yet, in regard to the nature of things, this knowledge is only an empty shell —a form of symbols. It is knowledge of structural form, and not knowledge of content. [...] And, moreover, we have found that where science has progressed the farthest, the mind has but regained from nature that which the mind has put into nature. We have found a strange foot-print on the shores of the unknown. We have devised profound theories, one after another, to account for its origin. At last, we have succeeded in reconstructing the creature that made the foot-print. And Lo! it is our own.

One physics theory in particular (physics are many, essentially non-unified —disconnected— theories), has assumed the zeroth hypothesis —still without ever making it explicit—, until it washes up on the shore of the observer: quantum mechanics. This horizon is inevitable, since a perfectly rigorous theory is conservative; no subject at the onset, no subject till the end.

By strictly transporting the hypothesis in an axiomatics (formulated by John von Neumann), and by reducing measurements to their tiniest minimum, quantum mechanics blocks against the quantum of interaction implied by any measurement, and is forced to account for the act of observation, a perception, the meeting of two entities. Any observation must pay at least a quantum, and it becomes unavoidable to integrate in the theory that the observed phenomenon is inescapably modified. The banished subject is not reintegrated though. He is nowhere in the axioms. The formalism describes how an elementary light wave propagates, for instance when emitted from the centre of a room. However, it cannot say where, when the wave reaches the walls, it is going to materialise as an observable photon. The theory cannot produce a definite result, something that *happens*. This is called reduction, or collapse of the wave function. Since the measurement is made with a macroscopic device, completely unwieldy for quantum description, it is described in classical terms and one often reads that the transition to classical provokes the irreversible event happening. We cannot accept this idea: quantum mechanics does say that there is no way to know all the infinitesimal states involved in the device —an attempt to measure them would change them all—, but still it is claimed to work, entirely and in totality from the bottom up —as a good theory en route for the ultimate one. To accept the transition from quantum to classical, one would need accept an inaccessible horizon, which contradicts the dogma of full reconstruction from elementary bricks. This ontological materialism excludes the notion of a horizon. But if all is quantum mechanical *as in the theory*, nothing can happen in the world, wave functions keep on composing more possible issues. Now however, things do happen, this is what our lives are made of. The scientistic ideology has but two options once in this dead end.

The first maintains that nothing happens. Since this is a bit too rich, here is the trick: for each interaction, wave functions compose and the universe splits into as many offspring-universes as there are possible events; we just happen to be in one of them. This is plain dupery because this branching-out *does not belong in the theory*. What *happens* is entirely accounted for by the bifurcating multiverses. The problem has been pushed onto the neighbour.

The second is more instructive. Since the theory cannot have a thing happening, von Neumann and Eugene Wigner have proposed an *interpretation* of quantum mechanics in which the consciousness of the observer reduces the wave function. Calling it an interpretation is apt, because the observer is outside the formalism —he appears nowhere in the axioms— and, as we have seen, an interpretation *makes sense, gives meaning* to a formalism by connecting it to reality. We can even say that a formalism is *precisely built* to be devoid of meaning, to serve as a substrate structure, perfectly manageable. With this sudden convocation, the observer is airdropped through the horizon to give a helping hand to the lost physicist. Although this interpretation is non-scientific, it is treated as first-order, as if resulting from the theory, and given in addition an ontological status. The overall result is a sort of sophisticated solipsism, with the claim that in the absence of a 'consciousness', nothing happens in the universe. Since the physics-based scientific cosmology tells us *a history* of various structures building up (do not miss that history contradicts the non-existence of time), and geology, palaeontology and evolutionary biology take over from there leading to the late birth of man, one wonders how all these things occurred, before a conscious man arrived, allowing anything to happen.

Once again, there would be a way out, only if we knew how to build a consciousness from scratch, out of a formal system. Physicists are no more able of doing it than computer scientists. From the point of view of mathematical physicists, this interpretation is a rout. Not only have they lost sight of reality, they are lost in the maze of their own formalism.

Since a theory is man-made, if it outlaws any account of an observer, there is nothing else it can block on as an ultimate horizon; at the end of the journey, *there can be only one*, none but the observer himself, obstinately refusing to reach a higher level of consciousness.

What Eddington said, von Neumann-Wigner did: staring wide-eyed at the night-horizon, and failing to reflect on their own mind, they do not realise that the features they see in the deep are that of their own face.

> A man sets out to draw the world. As the years go by, he peoples a space with images of provinces, kingdoms, mountains, bays, ships, islands, fishes, rooms, instruments, stars, horses, and individuals. A short time before he dies, he discovers that the patient labyrinth of lines traces the lineaments of his own face.
>
> –Jorge Luis Borges, *Aleph and other stories* (2004).

The Whole Society Resonating in Unison with Variations on the Euclid Myth

To picture how pervasive and extensive this 1st culture faith is, a few more mentions are apt. In 1972, the 70-year old much renowned epistemologist Karl Popper publishes the quintessential enlightenment of his philosophy of science, *Objective knowledge* (1979). With evident pride he proclaims "an epistemology without a knowing subject", he marvels about *knowledge without a knowing subject*. I do not want to spoil more paper on commenting such nonsense. Just note that this is precisely the promise of artificial intelligence. The reader should be shocked that Popper brandishes his find as if it were new in any way. In 1986, when Thomas Nagel publishes *The View from nowhere*, the fate is sealed: since he tries to reconcile this 'objective view' with a subjective one, he wholly recognises it. First, we must greet the feat: not anyone can fit an absurdity in a title; he is in good company with Popper. A *view from nowhere* is literally a view from Utopia. The oxymoron is apparent if you express a view as a relativity, that is, how to transform a view from a frame of reference into another. Since from nowhere means no frame, you cannot transfer any information from or to it. It has no existence. But . . . does not that tune sound familiar? "He's a real nowhere man/Sitting in his nowhere land/Making all his nowhere plans for nobody/Doesn't have a point of view/Knows not where he's going to/Isn't he a bit like you and me?/Nowhere Man, please listen/You don't know what you're missing/Nowhere Man, the world is at your command/He's as blind as he can be/Just sees what he wants to see/Nowhere Man can you see me at all?/Nowhere Man, don't worry/Take your time, don't hurry/Leave it all till somebody else lends you a hand." John Lennon (1965) expressed in plain words the distress of the disconnected believers in the 1st culture dogma. It is not new that artists' antennæ capture early signals and warn us much in advance of what is coming. And what is more, they express it in a form that gains acceptance, whereas many a prophet or free-thinker has been scapegoated for having revealed the same sort of things, but all too clearly, instead of telling a fable. In 1516, Thomas More depicted the peninsula of Utopia, where all things were arranged mathematically, with perfect equality and symmetry. Utopia had managed to cut the isthmus that used to connect it to the continent, and thereafter live in full autonomy and perfection. Hence, the ideas that blossomed in the twentieth century were floating around much earlier.

The claim for universal knowledge in the manner of Popper has wormed its way into the French Constitution, since it has rashly included the *Charte de l'environnement (Environmental charter)*. The latter indirectly refers to scientific knowledge as the ultimate criterion for truth that should motivate any decision. Decisions should be machines. Only when truth is still to come should we resort to a precautionary principle. Without notice, this postulates that any decision to be made is totally informed: exactly the phantasm of microeconomic theories mentioned in the introduction —real-world situations are like a chess problem. Such a head-on denial of the irreducibly imponderable character of life amounts to a state of madness; in a Constitution it reveals a mass psychosis.

Hence science should rule. It is a complete reversal from the Athenian democracy. Science —*technè*— was denied any intrusion in the public debate, where values —*doxa*— confronted strongly, but on a common ground. For this very reason suggests Paulin Ismard (2015), all the specialised tasks of supply to the public life —to operate the machine of the city—, implied exclusion from the democratic debate. They were allotted to slaves, be they roadmen, accountant, master of the standards of weights and measures, or controller of coins. All denied a democratic voice, not for lack of knowledge, but for excess in the matter: in the realm of his *technè*, you cannot debate with a specialist. Hence the Greeks had clearly split apart the exoteric operation of the city from the sphere of decision and values, the connection to reality. Note that likewise, until the nineteenth century, the Ottoman Empire had their slaves, *Mamelukes*, bought from the Mediterranean area, some of whom highly educated, carrying out duties up to ministerial level.

The total overturn completed in the twentieth century, excludes any idea of *commons;* the dogma is that decision is founded on technè, making for *a society of distributed specialities*. A vast machine where each specialist holds sway over a micro-domain whose local programme he runs impersonally and exclusively. There is absolutely no place for debate. Uttering any specific viewpoint is violently repressed, disqualified in advance: truth belongs to nowhere, not here, and only truth is allowed existence. No one can claim the view from nowhere, or, he who can must always be elsewhere. Values, which used to be the foundation stone of political decisions, are obsolete. The ethics committee is but remains, for want of actual scientific responses.

The consequences are huge and appalling. The ideology of a subsuming knowledge 'from nowhere' denies the existence of individual persons, and as a by-product, of any cultural specificity —everything melts into one single overall pot—, while dressing it in universalist praise. Beneath the dress we unmask the spectre of automatic translation again, that connects all humans together in a continuous medium, in total denial of Babel's reality. The Western culture sees only itself, since it holds the whole; it sits on the universal vault. Cultural idiosyncrasies exacerbate, as hypersensitivity reactions to this planing down into uniformity. Thereby a specific viewpoint is admitted, if and only if mediated by some sort of lobby. We now live in a society of lobbies, says Cornelius Castoriadis (2005). Between lobbies there are no debates, no thinking, only struggles for influence. The place of the agonal debate has deserted the society of men.

25.2.6 To the Sources of the Euclid Myth: The Sequence of the Greek Revolution

We have done our best to debunk the Euclid myth and its myriad generalisations. But the mystery where it originated from remains.

25.2.6.1 The Stage of Philosophy: Parmenides, Plato

Before Euclid, Plato

When Euclid writes his *Elements*, he realises in rigour *what everyone knows, what everyone feels*. (We shall find where the sentence comes from further on.) It is said that one century before, Plato's Academy gate addressed: "Let no one who is not a geometer enter." His *Republic* explains:

> [...] the real object of the entire study [of geometry] is pure knowledge. [...] That it is the knowledge of that which always is, and not of a something which at some time comes into being and passes away. That is readily admitted, he said, for geometry is the knowledge of the eternally existent. Then, my good friend, it would tend to draw the soul to truth, and would be productive of a philosophic attitude of mind, directing upward the faculties that now wrongly are turned earthward.

The picture needs no further comments, it is exactly the frame of the 1st culture. No-one should enter who is not unswerving in his belief in the eternal truths of geometry. Plato always turns to geometry to prove the existence of certain, timeless knowledge. In the *Phaedo*, the soul inherits this eternity, because to know geometry, it must be of the same nature, and for the same reason it must have always known it. In the *Meno* however, Socrates leads a young slave to *discover* that a square inscribed with a vertex in the middle of each side of a larger one (a "diamond in a square"), has half its area. The logical steps seem so inevitable that Meno seems to have always known it. There is no *learning* in the process, nothing new. And though Meno could not figure it out at the onset. Plato justifies the contradiction of having to rediscover what you have always known by saying that the soul always knew but forgot, and now has recollection.

This comes as no surprise to us: a conservative machine cannot account for any novelty —its raison d'être is to ban it. Hence learning is impossible. But, by the same token, forgetting is impossible: the theory of recollection is a trick. If you stick to conservative reasoning, there is no way to explain why anything happens. Sticking to the 1st culture hard core, you are condemned to never apprehend novelty, creation, becoming, nor learning, nor perception, proper. They are all beyond the horizon of a machine. You live on the fringes of the happening world. Hence, by Plato's time (c. -428 to c. -348), the fate was sealed.

Before Plato, the Parmenides Symmetry Breaking, and Pythagoras

Plato evidently builds on established foundations. A few passages from Parmenides (c. -515 to c. -440) will reveal them (Burnet 1920): "Being is, and Not being is not". This being is "uncreated and indestructible, alone, complete, unmovable and without end". There is no room apart from it, no otherness —'not being.' Nothing can "come into being or pass away," it is eternal, and change is impossible.

At this point in our journey, this mysterious *Being* which to naïve ears sounds as a senseless tautology, is obviously *the formal domain*. It has all its characteristics. Besides, Parmenides states it: "The same is thinking and being". This being is a pure figment of the mind, a proper figure of rational thinking, it is rational thought itself. Since it works only by conservative operations, there can be no question of creation, destruction; all its elements must be eternal and unchangeable. Or they cannot be. Time vanishes —were you looking for the source of Einstein's denial of time? This interpretation makes perfectly sense, since the only things that can be known in their being —exactly, totally, to the (alphabetic) floor, ontologically—, are formal. Another cross-check: we have seen that nothing can happen in quantum mechanics (nor classical, nor relativistic), for mere formal reasons. Parmenides already made the point quite strongly: change is impossible.

By Parmenides' time, thinking rationally was old practice throughout the Eurasian continent. But to practice and to become conscious are quite different things. Parmenides was first in our history to see consciously and reveal to his contemporaries the cogs of the rational machine in their minds, and all these pure beings it operates on. The vast stage of a sort of chess game appears in the mind. It sparked the Greek revolution, the birth of rational thought. Rational thought emerges by watching and analysing itself through its own structures; delighting in the exact figures and the machines it uncovers in itself; this closure in an autonomy bubble concurs with an explosive inflation of the space generated by playing its production rules in all their combinatorial possibilities. A cosmic inflation. It is a chain reaction, a crystallisation, a shockwave of transparency whose light dazzles all minds, enlightens everything —but what is not rational, which becomes nil, untrue, sent into non-being. Rational thought tolerates no alterity, there is only its own being. It makes senses why the ancient Greeks were interested, and even eager to know about science, customs, culture of all barbarians, *provided they came to be expressed in the logos*, the language of the universal, of rational thinking. Anything resisting translation was non-being. We see the root of universalism, and automatic translation, and denial of a real *other* culture.

Parmenides makes a radical partition between the way of truth, where the being is —the only deserving attention— and that of opinion, of appearances, of perception, which is repudiated —all that comes through the senses is imperfect, unreliable. The way he rejects is that of reality, of what is felt, that of life! Of the opinion on the *agora*, where the life of the city takes place. Parmenides rejects any dealings with reality. The mind is in love with a desert world.

Parmenides is credited for introducing logic in philosophy. For the first time, the truth of a statement is checked, not by confronting it to reality, but by disconnecting from it, based on its internal consistency. The picture is clear, since we have seen that truth cannot apply first-order to reality, and has substantial meaning only in a formal context. Many find his writing obscure; for Aristotle, it is close to madness. And so should it be, as it strives to come as close as possible to pure logic, that it disconnects from any tangible thing: the poem loses all *meaning*, it revolves on itself, solid-consistent but inevitably looping in tautologies to express what is a connected formal block. Once you grasp this point, Parmenides' discourse is the exact contrary: crystal clear. And we see why the Greek revolution is best seen as having made space transparent. The earlier mythical world is opaque: the souls that animate each thing —grow the wheat, flow the water, move the leaves in the wind . . . are hidden beneath the surface, closing a horizon that veils the forces at work. Each one knows he must pace his actions —where his bodily envelope meets the world— at the risk of triggering unforeseeable events, arousing the wrath of some soul. A world swarming with innumerable agents. The contrast could not be more striking with the emptied rational space.

While Parmenides dismisses perception, he describes it as "the same perceiving the same". Which comes as no surprise: otherness does not exist. We have found the origin of contemporary science being unable to deal with perception, proper: the meeting of subject and object cannot take place by staying in the confines of formal operations —without admitting otherness.

In Parmenides we find the most basic form of the Euclid myth, and the origin of the disconnection we have been tracking so long. He makes a split for the way of being, immutability, stillness, eternity, against that of life the Occident would never recover from. Alfred North Whitehead (1979) summed up European philosophy as "a series of footnotes to Plato" and in turn, Plato's writings seem like footnotes to Parmenides.

Hence Parmenides sealed the schism between the two cultures. He committed a major brain damage to the Western world: a commissurotomy —a bisection of the *corpus callosum* that connects the two brain hemispheres. This operation was practised as a last resort —successful— treatment to cases of epilepsy, refractory to any medication. Subjects with a split brain live nearly normally, but presented an opportunity to observe that several functions are mostly localised in one of the two hemispheres. For the typical right-handed subject, when an object is located in the left visual field, thus seen only by the right side of the brain, the patient proves to identify it by handling it correctly but cannot name it aloud: the left brain is usually that of language, of rational thought. The Western world lives with a split brain, and has invested all its energy on the left side —that of the machine, of Parmenides' being, of the 1st culture. A mirror paraphrase of Betty Edwards' *Drawing on the right side of the brain* (2012) —in which she releases spontaneous drawing abilities in students by training them to enter the receptive state expressing the right hemisphere activity—, would be the West's *Being on the left side of the brain*. The cultural gap actually splits each Westerner. Each one must carry *The Unbearable heaviness of being* (Kundera 1984) left to us in Parmenides' will.

Parmenides of course did not go all the way by himself. He inherited from the school of Pythagoras (c. -580 to c. -495), which originated the first recorded fully-fledged physical law in Western history: that of musical intervals. This was a leap from just surveying surfaces and lining up square angles in flat fields: it relates length, weights and intrinsic vibrations of strings of different diameters. And thus the sounds of speech in language and the melodies of music —the vehicle of all the ancient myths sung by the Greek bards—, the world of art and that of technique, were all intimately connected together by number ratios. This earthshaking accomplishment blew up Pythagoras' mind, who altogether declared that numbers are the first and last things in all of nature, the eternal essence. In the words of his successor Philolaus (c. -470 to c. -385): "All things which can be known have number; for it is not possible that without number anything can be either conceived or known." (Bell 1933 p. 84).

It is now clear that Parmenides generalised this conception by discerning the formal objects of rational thought. While Pythagoras conflated numbers and reality, at a deeper level of abstraction formal thought begins with Parmenides.

25.2.6.2 The Stage of Politics: Athenian Democracy

At a certain point, rational thought invading all the compartments of human life, crosses a threshold of collective awareness. In its inflationary expansion, rational thought reverberates on all the components of life; it suddenly strikes that it can apply to the circle of men in the city. In this unique event in history, men explicitly express that their laws do not come from any sort of transcendence —a god, possibly through the medium of a king—, they are entirely decided by men, here and now (Castoriadis, 1986, p. 383). At this stage of social consciousness, the society breaks off from heteronomous —laws coming from elsewhere— to an explicit autonomous self-institution, as Cornelius Castoriadis has named it. The rules can no longer be absolute, they are *relative* to actual choices in a social-historical process, depending on men only, and nothing else. They realise they are alone and responsible on earth.

The Greeks see fit to project their mental structures on their city, and build it anew. They feel the command to do so; since they envision perfection, they want it in actuality (Watzlawick et al. 1974 p. 48). The Greek phenomenon singles out in the entire Eurasian continent, buzzing in intellectual activity from end to end, with evident cross-fertilisation along ancestral migration, then trade roads. China lived the period of the hundredth schools of thought, India knew similar effervescence. But in these civilisations the society stood unflinching. The art of the Indus valley shows Hellenic influences following Alexander of Macedonia's conquest, but the structure of the society remained unconceived, not consciously addressable in India whose caste system, abolished in the twentieth century, is still enduring in the 21st. The quake of the Greek revolution did not shake China either.

25.2.6.3 The Stage of Mathematics: Euclid

Ending a process which ran over three centuries, rational thought born as an awareness in philosophy, put into action to rule human affairs in the *polis* —the 'easy' part, which only needed to be decided—, finally yields a perfect mathematical form with Euclid's tour de force. This sequence is interesting because it has another occurrence, and because it may inspire a way out of the current crisis.

25.2.7 The Sequence of Enlightenment

25.2.7.1 Descartes Ignites Enlightenment

Parmenides had shown to all the spectacle of rational objects; Descartes reveals to all how to use rational thinking to experimentally question reality: a representation must be a true machine, then it can be communicated, and put to the test. This may seem trivial in today's culture, but is a huge feat. In retrospect, the Greek thought appears to have mainly exerted its action by itself, in a closed circuit, and this is true through early in principle reasoning in philosophy, in politics, and in mathematics. Of course there were also experimental theories (Pythagoras' musical intervals, Archimedes works; and as we have seen, geometry started anchored as a physics) but how they connected to reality was not fully articulated. Or, more precisely, it went one way from reality to formal, and never were formal systems used to investigate reality.

Descartes (1983) makes it very clear that one never obtains certainty about reality, only a temporary adequation, always hanging on a possible refutation. Refutability as key to science, and the pragmatism readily apparent in Descartes, would only acquire formal status in the twentieth century (Popper 2005). The key concept that permeates all the *Discourse*, though not explicitly stated, is relativity: how the *cogito* elaborates theories relative to its experience. By neatly telling apart reality and the true machines we use to question it, Descartes is able to establish a relation with each other. Now back up to Knuth's quotation about machines in the artificial intelligence section: His words could be Descartes': machines are helpful not because they are clever, show common sense or have intent, but because they are totally deterministic, hence you know when they *do not work as you want*. Hence you select those machines which 'work, up to now'.

Descartes' mechanistic device as a way to experiment with nature is a rule of production that immediately radiates in all directions and inflates a new space of science. It sparks the shockwave of Enlightenment. A larger universe is made transparent at once. This time, not only are gods expelled from the earth —Greek revolution's deed, with the cities mathematically arranged as in Utopia— "Gods are expelled from the firmament," sings Georges Brassens. Once again the process started in philosophy.

However, Descartes' relativity is quickly buried under layers of absolute convictions and forgotten: Western thought was never to recover from millennia of gathered momentum since Parmenides' time.

25.2.7.2 The French Revolution

Friedrich Nietzsche (1973 p. 114) aptly makes Descartes "the father of rationalism, and hence of the Revolution." The Revolution implements fanatically and from top to bottom the reorganisation of society according to rational plans, and another extensive real-size experiment in autonomous self-institution. Alexis de Tocqueville's (1856) words are unsurpassable to express the process, and there we find every point that we have made up to now —including going astray in writing style issues:

> Now when it is borne in mind that this French nation, [...] was, at the same time, the most literary of all the nations of the world, it may be easily understood how writers became a power in the state, and ended by ruling it. [...] in France, the political world was divided into two separate provinces without intercourse with each other. One administered the government, the other enunciated the principles on which government ought to rest. The former adopted measures according to precedent and routine, the latter evolved general laws, without ever thinking how they could be applied. The one conducted business, the other directed minds.
>
> There were thus two social bodies: society proper, resting on a framework of tradition, confused and irregular in its organization, with a host of contradictory laws, well-defined distinctions of rank and station, and unequal rights ; and above this, an imaginary society, in which every thing was simple, harmonious, equitable, uniform, and reasonable.
>
> The minds of the people gradually withdrew from the former to take refuge in the latter. Men became indifferent to the real by dint of dwelling on the ideal, and established a mental domicile in the imaginary city which the authors had built.
>
> Our revolution has often been traced to American example. [...] In other European countries the American Revolution was nothing more than a strange and new fact ; in France it seemed a striking confirmation of principles known before. It surprised them, it convinced us. The Americans seemed merely to have carried out what our writers had conceived ; they had realized what we were musing. [...]
>
> The student of our revolution soon discovers that it was led and managed by the same spirit which gave birth to so many abstract treatises on government. In both he finds the same love for general theories, sweeping legislative systems, and symmetrical laws ; the same confidence in theory ; the same desire for new and original institutions ; the same wish to reconstruct the whole Constitution according to the rules of logic, and in conformity with a set plan, instead of attempting partial amendments. [...]
>
> Still, setting details aside, and looking only to main principles, it is readily discerned that all these authors concur at least in one central point that each of them seems equally to have invented by himself, that seemed to exist in his mind before all particular ideas, and to be their common source. [...] They all started with the principle that it was necessary to substitute simple and elementary rules, based on reason and natural law, for the complicated and traditional customs which regulated society in their time.

To finish convincing us that the Revolution pursued nothing less than axiomatising the society, here is how the Marquis de Lafayette addressed the assembly on the 11th of July, 1789:

> You have already been presented a work plan about the constitution. This plan, so justly applauded, presents the necessity of a declaration of rights as the first object of your attention. Indeed, [. . .] your ideas must first focus on the declaration that contains the first principles of any constitution, the first elements of any legislation. [. . .] The merit of a declaration of rights consists in truth and preciseness; it must say what everyone knows, what everyone feels. [. . .] Nature has made men free and equal [. . .] (Galois and Ray 1840, p. 148b–149a)

25.2.7.3 The General Axiomatisation of Mathematics

The three-century sequence ends again by the end of the nineteenth, with the next blossoming of mathematical axiomatisation to occur since Euclid; an enduring pursuit of certainty and of an oxymoron-like ‘absolute frame of reference,’ or *view from nowhere*.

25.2.8 Upstream from the Two Sequences: Alphabetic Induction

We have seen Parmenides can serve as the explanatory source of the great Western disconnection, and how the Enlightenment was a striking replica and merely an amplified version of the sequence of the Greek revolution. Without diving into infinite regress, can we hint at how it clicked? Usage has it to call what we have named revolution the Greek *miracle*. When a 1st culture-dominated society speaks of miracle, we face a von Neumann-Wigner effect: it gives up all its own rules, and abandons any possibility of explanation. It wants to be fooled. As described above, rigor *mortis* rationalism is bound to wash up on the shore of its intrinsic inability, to block on its own horizon. It necessarily ends up in bursts of total irrationality. This *coincidence of opposites* that we have recognised on building a space from perception under the notion of contrast was at the centre of Heraclitus’s thought. It becomes inconceivable following Parmenides, although Plato mentions it, but as a concept *per se*, disconnected from his otherwise articulated philosophy, and without mentioning Heraclitus.

The gist is in the previous sections; the central concept of a true machine coincides with alphabetic writing. Indeed Harold Innis (1950) points out that the ancient Greeks had their world mutating from mythical to theoretical, without their understanding of how it took place. How, basically, from an opaque world, teeming with entities, full of will and creative potency, where nothing lasts and everything flows in becoming (*panta rhei* says Heraclitus), did they suddenly conceive of a transparent world, with a limited number of immutable, eternal beings, all the *appearances* being only combinatorial arrangements of these ultimate building blocks (as Hawking will say some time later)? Where did any of them ever see anything like that? Marshall McLuhan expands on Innis's view that the suggestion came from the adoption of alphabetic writing (McLuhan et al. 2008; McLuhan 2011). They most probably borrowed from the Phoenicians an alphabet, whose purely consonantal structure matched well Semitic languages, but could not accommodate Greek dialects as is. Thus they completed it with vowels, yielding a full, unparalleled transparency of the *logos* on the written line. This contribution went back to Semites, who added diacritics to express vowels in their alphabet. The original Semitic writing continued to require from the reader —and still does in today's most printed forms of languages such as Arabic or Hebrew— to mentally complement the characters to obtain actual words, and sentences, hence an enduring opacity. The Greek alphabet saw the birth of a complete, absolutely transparent, writing, saying the *logos* without any gap.

The idea of the letters generating the whole world was present before in the Semitic legacy, and would be repeated later in various forms in its descent. But it would take another dimension while, as seen above, the medium would distillate the hypnotic message that everything that can be thought neatly, thus said, and written to enter the realm of eternity, must reduce to the combinatorial arrangement of the limited set of characters of the alphabet. Now, bear in mind that any representation first and foremost presents itself as reality, and only secondarily do we have a chance of recognising it as relative —thus *learning consists only in learning relativities:* the collapse of the world and of writing is complete. The beings of Parmenides reflect exactly alphabetic letters, and the mystery, the opacity is gone. Alphabetic writing displayed openly and hypnotically the operation of a true machine that would soon be revealed in Greek's, minds by Parmenides. The reader should backtrack to Hilbert's programme, to find a clear embodiment of this machine. And see why so many physicist talking today about the universe being nothing but a computer —ontologically, as ever— is but the disappointing naïve and blind, mechanistic repetition of the same scheme, *that each of them seems equally to have invented by himself*, for lack of a conscious view of what is at stake. Conservativity, the key property of this machine —inseparably conjoint to immutable beings—, was consciously laid down by Anaxagoras, a contemporary of Parmenides: "nothing is born or dies; on the contrary, everything is assembled out of existing things and then dissolved." He was only to cursorily watch the written line for the idea to occur in his mind. In passing, science would have to wait until Antoine Laurent Lavoisier (1743–1794), following Descartes' path, would expose how to use this sentence as an experimental principle, in the study of chemistry.

The Greek revolution was contemporary to a disenchantment of the world, it lost its mystery and its magic. Gods were expelled from the surface of the earth, from within all the things surrounding men. As said above, the alphabetic induction also met something special in the Greek mould that remains behind the horizon of what can be grasped: the alphabet spread to India, but this society stood stiffly. Even more so in China, standing firm behind the Himalayas, which stuck to its ideogrammatic writing. Notably however, China developed an independent, combinatorial model of the world, directly based on a binary code: the Yi king (Knuth 2011). But without ever taking over all minds. And Parmenides has no equivalent in the Orient.

25.2.9 *Ninth Circle*

25.2.9.1 The End of the Search

Our Ideal 1st culture subject does not clearly tell apart what originates from himself proper, and what results from meeting the outer world. He is not fully conscious. He does not know properly his bodily envelope. Thus he literally pours and spreads into space, becomes an all-immanent presence —such is his phantasm. At the end of the Odyssey, just as Ulysses, he is bound to end up in himself —also Eddington's realisation.

In particular, the standard Westerner, scientists included, is fascinated, bewitched by mathematical structures —all sorts of formal systems. They are conceived as an external realm of Platonic Ideas. Here is a much more fruitful definition of my own: *mathematics is the constructive structures of cognition*. They are made by men, and only in *some* cultures. The keystone of Western 1st culture, with its many facets, is now clear: formalism-worship.

The world is proclaimed ontologically a machine, it is without a mind, without a subject. It has no means to form a frontier —such as an individual's inside and outer world—, no relation —and perception cannot be properly accounted for—, no others —to *lend you a hand*—, no alterity, all as a result of the shockwave of transparency.

The 1st culture dogma's key property, conservativity, means nothing new can happen. If something new happens, it means there has been a trick. Our wizards von Neumann-Wigner pulled consciousness out of the hat. The general audience does not know it is being tricked, and does not want to grow disillusioned. It is so fond of absolute promises. Quite in contrast to *real* magicians' audience: it knows there is a trick but demands the illusion to be perfect (Mannoni 1969).

There is no magic in still-land says quite profundly Edward Tufte (Tufte 1997, p. 63). Still-land is the written line. He means that a trick, explained in a book, conveys no magic. We know why: alphabetic writing makes everything transparent. Nothing new or unpredictable can occur in a formal system.

25.2.9.2 The Self-Sacrificial Origin of Our 1st Culture

Socrates threatened democracy at its core, more than *flesh and blood* could stand, enough to sentence him to death, although the reason is nowhere apparent in the traces of his trial, nor as far as I know in any of its comments to date. The threat was felt radical, but could not be thought over consciously, even less voiced.

It is easy to see why today with open eyes —I admit to be making now and then outrageous hypotheses of this sort—: in every of Plato's dialogues, Socrates claims that every rôle in society must be accomplished by its best specialist, politics included. It is easy to see because Socrates' dream was therefore *the society of distributed specialities* whose advent we in Westernland —the left side of the brain— both celebrate and strive after even harder, with all our faith and will and strengths. In Socrates' dream, there was only one true man left, sat enthroned at the top of society, the thinking philosopher—himself apparently. All the others were but exoteric operators of the machine's supplies—life is wiped out from the city's *map*.

By doing so, Socrates dissolved the place of the agonal in the Athenian democracy: the agora where *doxa* confronted, and the commons took place, shaping the decisions of the city. Debate was sent nowhere. The very existence of the society was impossible, save under the iron rule of a tyrant. By the same token, Socrates' dream annihilates human consciousness: no one thinks any more in his dream. Again the spectacle is before our eyes: France, where intellectuals held the traditional lead expounded by Tocqueville up to the 1980s, is now devoid of any prominent figure.

Socrates managed a much greater exploit through his trial. He consecrated his victory over the democrats by forcing them to sentence him to death. He deliberately chose the option of self-sacrifice on the altar of the absolute to sanctify the eternal glory of his Ideas. He caught the democrats in their own trap, turned them into mere instruments to commit suicide. In their wake the whole Western world still finds no way out of the trap, not even conscious of swimming in downright paradox by revering at once Plato and democracy.

To us this plan of 'dying for ideas' as sung by Georges Brassens is déjà vu, well exemplified in Einstein's denying his own existence on intellectual grounds—ceasing to be grounded, in fact.

Plato's *Phaedo* (1966, 65a—67e) sums up many facets of this whole section on our 1st culture, and should now take on its full meaning:

> Now, how about the acquirement of pure knowledge? Is the body a hindrance or not, if it is made to share in the search for wisdom? What I mean is this: Have the sight and hearing of men any truth in them, or is it true, as the poets are always telling us, that we neither hear nor see any thing accurately? [...] Then, said he, when does the soul attain to truth? For when it tries to consider anything in company with the body, it is evidently deceived by it. [...]—65a—65b

At first sight these remarks seem congruent with the impossibility of first-order truths about reality. But then Socrates takes up the torch lit by Parmenides, that truth lies in a formal system—but remember, only as far as mathematicians find a practical agreement, pragmatically, as emerged from the crisis in foundations—and he strides right into the mad idea that truth about the world is to be found by disconnecting from the world:

> Do we think there is such a thing as absolute justice, or not?—We certainly think there is. —And absolute beauty and goodness. —Of course. —Well, did you ever see anything of that kind with your eyes?—Certainly not, said he. —Or did you ever reach them with any of the bodily senses? I am speaking of all such things, as size, health, strength, and in short the essence or underlying quality of everything. Is their true nature contemplated by means of the body? [...] Would not that man do this most perfectly who approaches each thing, so far as possible, with the reason alone, not introducing sight into his reasoning nor dragging in any of the other senses along with his thinking, but who employs pure, absolute reason in his attempt to search out the pure, absolute essence of things, and who removes himself, so far as possible, from eyes and ears, and, in a word, from his whole body, because he feels that its companionship disturbs the soul and hinders it from attaining truth and wisdom? Is not this the man, Simmias, if anyone, to attain to the knowledge of reality?—65d—66a

Socrates completely repudiates the body:

> For the body keeps us constantly busy by reason of its need of sustenance; and moreover, if diseases come upon it they hinder our pursuit of the truth. [...] We are slaves to its service. —66b—66d

Do keep in mind this unfriendly relationship, of repudiation of one's body —schizophrenic self-repudiation— which makes clear Socrates' feeling regarding the analogue supply of the city —equally assigned to slaves.

And finally, Socrates concludes, quite in line with what stands out of the hard core of the 1st culture, that the ideal man is a dead man:

> Then, as I said in the beginning, it would be absurd if a man who had been all his life fitting himself to live as nearly in a state of death as he could, should then be disturbed when death came to him. Would it not be absurd? —67d

Do not misunderstand: this is not acceptance of being mortal. It is promising oneself immortality. Socrates is in complete denial of his earthly body. Thereby he seals the fate of the Western world, in a totally absurd cosmogony.

To the Greek people who had fearlessly stepped into relativity by admitting none of their norms or laws came from any transcendence, the ultimate ill omen lurking in Socrates' dream was his implying to impose another overhanging 'absolute reference frame' —which the readers knows to be an oxymoron.

25.2.9.3 The Undivided Domination of the 1st Culture

The Athenian society was entirely organised to stage the place of the debate, to master and maintain it. Western 'democracies' have put through Socrates' grand plan. The debate is *nowhere* —where it is dreamt to be! Their individual members have surrendered any autonomy of judgement, thus any responsibility, which endlessly percolates through the network of the global machine. Nowhere in that network shall you meet a subject. We should be appalled to have built the frame of Nuremberg's defence, to the scale of our whole society. Moreover, what better official repudiation of democracy could we imagine than the European Union and international bodies putting an elected leader under the guardianship of a troika of technocrats, *in Athens*, the historical and symbolic birth place of democracy? There have been neither popular nor intellectual protests, which means unanimous agreement. Consciousnesses are ready. Or, more accurately, *unconsciousnesses* are ready. Even more chilling is the result of a 2015 poll by Ifop, revealing that a majority of the French —67%!— favourably regarded the advent of a technocratic, non-elected government (Fourquet and Dubrulle 2015). It would of course be selected for its highly specialised qualification, by a competitive market process —the optimal solution to *any* problem, in the received dogma; refer to previous mentions of microeconomic theories.

The Western world trampling underfoot the cradle and principles of its cherished 'universal' democratic regime is no accident. Nor is the systematic bombing of the Fertile Crescent over nearly three decades now. All express an unconscious but imperious desire to disown the origins of our culture, to erase any of its traces. A civilisational refusal to know one's own history, to know oneself: a resignation from self-awareness, consciousness. Hence arduous is the quest I am taking the reader along.

What it amounts to has been spotted many a time; by George Santayana "those who cannot remember the past are condemned to repeat it" (Santayana 1906 p. 284)— what about those who dedicate themselves to erasing it?, by Edmund Husserl "tradition is the forgetting of the origins" (Davis, et al. 2012 p. 36). By the way, anyone who has told fairy tales to children knows it: they keep asking for the same story while they still wonder about it. Once understood, they move on to something *new*. Compulsive repetition implies locking unconscious the mythical figure that fuels it. It means enduring lack of understanding. Just like an unsaid postulate shall be the unavoidable eternal horizon of a theory, whatever is ignored or rejected from the self, repressed in the unconscious, is bound to appear repeatedly as an inescapable fate. "We cannot change anything unless we accept it," says Carl Jung accordingly: to have a handle on it, it must become conscious. Hence if personal or global history stutters, look for the hidden instance beneath the surface. The future of a civilisation applying to erase its historical origins is gloomy.

Westerners devote themselves to behave as processors of the local programme of the nowhere-global-machine kept beyond any understanding. To conform to their cosmogony, they project themselves as devoid of consciousness.

Dante's Ulysses call has not been heeded:

> O brothers!" I began, "who to the west Through perils without number now have reach'd,/To this the short remaining watch, that yet/Our senses have to wake, refuse not proof/Of the unpeopled world, following the track/Of Phoebus. Call to mind from whence we sprang:/Ye were not form'd to live the life of brutes/But virtue to pursue and knowledge high. (Dante 1918)

25.2.9.4 An Hypertrophied Section

This section's hypertrophy is proportionate to the overweening, pathological domination of the 1st culture over our civilisation. It was necessary to mine down to the earliest deposits of our science to become aware of the rules we abide by, without even knowing. The 1st culture can only be cracked from inside. It is totally proof against any outside attack: these are excluded from the onset, since any otherness is denied. The whole purpose to this point was to cross a level of reflexivity, to obtain a sideways view of ourselves. By looking for an origin, we found that an intrinsic impossibility of a strictly conservative thought is to account for an origin, for anything new, or created, or even happening. Which of course are intrinsic parts of our lives. The end result is that the 1st culture permanently collapses like the wave functions of quantum mechanics in the face of reality. You certainly cannot contradict from outside a system which stands the contradiction of *every* event.

Overall our all-left-brained culture harbours countless myths, and sadly, takes them seriously —as absolute, first-order, without a distance—, whereas as a matter for the 2nd culture, myths are always considered from a distanced, relative point of view. However, all these myths are fortunately built on the same pattern. We have therefore no reservations left for plunging into the 2nd culture and finding even earlier, mythical sources. If the reader has survived the journey through the 1st culture which denied his existence all the way, he can safely proceed to the 2nd, which fully accepts and even *proceeds* from his being a subject.

25.3 Tracking down the Divide, in the Internal and Mythical World of the 2nd Culture

25.3.1 Introduction

The net result of this long trip is that the sciences cover only a limited sector of the full circle. A physics which cannot account for time, in which nothing can *happen*, can only be relevant to very limited slices of our reality, necessarily immersed in a larger context relevant only to the 2nd culture.

25.3.2 The Emergence of Consciousness

25.3.2.1 The Expulsion from Eden: Not an outside, But an Internal Event

What does the Eden myth tell? Man was expelled from the heavenly garden, for having eaten from the tree of the knowledge of good and evil. The matter is evidently not any knowledge in any operative, exoteric sense, nor is it about feeling good or bad things: every living being makes such categorisation. Every hunter knows that many an animal is clever, cunning. Not *more knowledge* makes the split from animals, but *another* knowledge. Besides, the tree was there in the open for whoever to see and its fruit was also known, and somewhat frightening; if it were only to pick and eat, what could be so special about that event? Significantly the atmosphere is of intimate pondering, a personal experience; it does not picture a social group. The event did not occur anywhere outside, but inside —what else could prevent any return to the garden? The fruit borne was to cross a threshold of knowledge, causing the appearance of the image of the self within the self. The event is to know oneself, as a knowing being: consciousness. Millennia later, this would be the unique, essential message on the front of the Delphis temple, 'know thyself'. A few more millennia later, this would be repeated in the self-given species name *homo sapiens sapiens* (Linnaeus): the man who knows he knows.

25.3.2.2 Splits, Reflexiveness and Mimetism

Moreover, it is written black and white: man *saw himself*, naked —which he was no more as any animal the moment before—, and was ashamed. The event was vision of oneself, as if sideways, or from an overhanging position.

Backtracking a bit, the text is even more explicit: God created man in his own image (Genesis 1:26) —a splitting in two—, mirrors that man creates God in his own image; God and consciousness appear synchronously, and this dual nature is clear once God is identified with the human mind. Explicit again: *You shall be as Gods* (Genesis 3:5) (Fromm 1966). In the previous state, man —*homo*— is created male and female confounded —remember that in Hebrew *adamah* for Adam means soil, and in many a myth the first man is moulded out of clay; that *homo* and *humus* are from the same root. Then another split in two happens from a *side* of *homo* —not from a *rib*, as loose translations write—; *homo* wakes up one morning to realise being paired, man and woman, same and different, complementary —their eyes now open, they see they are naked. Incidentally, the root of *Eve* means life, that which is breathed into the nostrils of *adamah*.

Again we find a double, same though different, locked in a circular reference. There is a mimetic component inherent in the birth of man, even more basic and intrinsic than the treatments that René Girard gave of it (2002, 2004), which appear as echoes to this founding event. Consciousness emerges self-instituted, as a circularity, a reflexiveness, coupled with a remoteness, being out of oneself: divided from the onset.

At the same time, they become conscious of being mortal, and subject to diseases —you do not seriously believe the previous *homo*, or any animal to be immortal? The myth is all about consciousness of these facts, projecting them as fated becoming. At this awakening, man and woman *know* one another: they conceive of sexuality —where the living world has only reproduction— as the symbolic answer to mortality.

25.3.2.3 Disconnection from the Animal World and from Nature

Those who have gone through this transition know they are irrevocably different from animals. Innocence is gone. This feeling of a previous innocence is that of being all in one piece committed into action, connected to the world without a gap. A latitude to think, the freedom of the mind is consubstantial to creating a distance, a disconnection, the possibility to go adrift. The mind sinks in itself, afflicted by the very fact of his self-awareness, autonomy, freedom. The disconnection we are tracking, is precisely and intrinsically the very origin of man, in our culture's imagery. Man is not only disunited from animals, but from nature entirely: disunion is inside him.

There are many recurrent stages of successively enlarged consciousness. The process of individuation of a conscious entity, and the feeling of being unique that comes with it is repeated in every culture. It seems that many a people have invariably felt they were the true human beings, and often named themselves so — from the European colonists furrowing the sixteenth century seas to hunter-gatherers of this day or Indians of the New World. A culture spontaneously sees itself the centre of mankind, the centre of a world entirely organised around it. Just as the typical observer finds it normal to always be at the centre of the rainbow.

25.3.2.4 Kipling's Play

An interesting light can be shed from the other end of the Eurasian continent. It is obvious for everyone that Rudyard Kipling delivered in writing the fabulous world of his Indian childhood, just as Andersen, Grimm or Perrault wrote canonical versions of a treasure of tales, conveyed over centuries, and probably much longer, through Africa, Europe and the Middle East, or as Homer has sealed in the Iliad and the Odyssey stories transmitted orally for ages. In *The Crab that played with the sea*, Kipling (1902) relates the same event:

> [...] the Time of the Very Beginnings; [...] when the Eldest Magician was getting Things ready. First he got the Earth ready; then he got the Sea ready; and then he told all the Animals that they could come out and play. And the Animals said, 'O Eldest Magician, what shall we play at?' and he said, 'I will show you.' He took the Elephant —All-the-Elephant-there-was— and said,'Play at being an Elephant,' and All-the-Elephant-there-was played. [...] One by one he took all the beasts and birds and fishes and told them what to play at. But towards evening, [...] there came up the Man [...]. And the Eldest Magician said, 'Ho, Son of Adam, this is the play of the Very Beginning; but you are too wise for this play.' And the Man saluted and said, 'Yes, I am too wise for this play; but see that you make all the Animals obedient to me.'

This is it: no animal can question it, an elephant plays at being an elephant, nothing else, a cat is fully a cat, without a gap. Man however, has no assigned play, and wonders which he should choose.

25.3.2.5 Epimetheus and Prometheus

To return to Greek mythology, Epimetheus 'who thinks after', and Prometheus, 'who thinks in advance' people the earth with mortal beings. Two brothers, mirror again of a partition.

Epimetheus begins with animals, granting each one wings to fly, a shell for protection, claws, fast-running ... all arranged related to one another: a cosmos, we now call it an ecosystem, with a rôle for each one. When he comes to man, he has no talent left, no rôle for him on the stage of the cosmos. Prometheus helps by stealing fire and arts for him, igniting human creative genius, his ability to mint new worlds, and the dizzying feel of being a god that comes with it. And with the hubristic hope of escaping into these worlds. Apparently Zeus' punishment has Prometheus as its sole target. But it does hit the Promethean instance in the psyche of every man.

25.3.2.6 Aristotle, Choice and Mimetic Processes

In the *Nicomachean Ethics*, Aristotle (1926) says likewise: man is the only animal born without an *ergon*, without a task, a rôle, a function. (Indeed, Aristotle goes on with man's proper rôle being rational thought; fair perhaps, but not to the large portion of mankind which has not valued rationality, *de facto* excluded.)

Man wonders what to play at; unlike other animals, he has no assigned play, he is free to play at being all sorts of things. He can even invent —chess playing for instance. But which one is worth devoting oneself to? To escape from this dizzying freedom —a result of man being his own mirror—, he is tempted to mimic those of his kind who seem to know, look determined: man is subject to the mimetic mirror of his fellow men. Subject to believing that he has found how to play at being a man, by mimicking one. Without noticing that every man is in the same position: overlooking the subjective relativity. A desperate attempt to recover the conviction lost with the unity that was before Eden.

Here is mimetism as the essential glue that holds together human societies, but also with its dual, destructive aspect well-recognised by René Girard: Human worlds are loosely constrained, as a correlate to the emerging freedom of human behaviour. Thus in order to show coherency —to exist at all as organised entities—, they must restrain the range of the actual, in the huge space of the possible. Customs are built out of essentially arbitrary conventions, convened collective agreements that are best described as mimetic: a spontaneous tendency to behave alike. The drawback is that the slightest event can turn a fad into a fluctuation that spreads all over the society, and become de rigueur. Throughout time, travelling has always been an eye-opener —to those prepared for it, would insist Pasteur (1939)— by laying open the diversity and thus relativity of social habits —whose Latin root significantly means 'how you wear', 'appearance.' And now the reason for the vested interest of the ancient Greeks in the customs of barbarians all around their world is clear: each was to reinforce their conviction that no law or norm has extra-social origin. Which only members of their self-instituted society were aware of. The relation was thus quite asymmetrical. While other peoples were just strange to each other, all these non-selves nourished the Greek self: it fed the democratic debate on values, that could only exist in their agoras.

Hence customs are conventions, agreement to a given play. The Greek awakening comes with a feeling of vacuity that re-surfaces repeatedly in all Western history; Shakespeare mentions it in the first scene of *The Merchant of Venice:* "I hold the world but the world, Gratiano; a stage, where every man must play a part, mine a sad one."

Now René Girard (2001) has particularly analysed the consequences of having to hint on fellow men to know what is appropriate, what is to be desired. The consequence is to stage an open competition, with all individual differentiation levelled, a convergence and conflict of desires, each other individual at once confirming what is object of desire, and frustrating its reach. Girard has found that countless myths around the world recount this process culminating in a mimetic crisis: the host suddenly focusing on a single scapegoat as the ultimate obstacle to all the frustrated desires, and committing a collective murder, followed by shared relief, recovery of peace and connection in the community. Thus the murder is recalled as what bound the community together; and ritually replayed as such — for instance a goat bearing all the sins of the city is cast every year into the desert. With the seeming paradox that the initial victim, in time, becomes the founding deity of a people, the one that holds them together, as the Latin root *religere* for religion implies.

This picture of the autonomous, imaginary character of human representations can, however, be made much larger: the general issue we have seen above is that each perception is a bet, every knowledge is hypothetical, an interpretation, that every theory about the world is a speculation, no matter the amount of mathematics and experimental confirmation put in it. Another aspect is that framing a uniform space —e.g., all men decreed equal— de facto creates a competitive setting —a market. Economic markets are inherently fluctuating and unstable, subject to such mimetic crises: a speculative bubble is a *hypothetic* bubble, an imaginary collective event, disconnected from the underlying reality, and bound to finish with the bubble bursting. Tocqueville has accurately discussed the distress resulting from dismantling the Old Regime ecosystem, however inegalitarian, at the price of all men losing their assigned rôle in society. The promised freedom and equality raised great hopes when made true by the Revolution, but it amplified the distress further with exacerbated frustration and disappointment by setting a large, uniform, undifferentiated mass: the perfect, exact conditions for mimetic crises, —Tocqueville foreshadows René Girard. (Precisely, a collective crisis such as a lapidation, implies that members of the host lose all individuality, gel and function as a whole). Tocqueville (1864, p. 224) mentions increasing suicides or mental illness. With individuals losing their social envelope —what connected them to the others— individualism is born, as a natural reaction to non-differentiation (de Tocqueville 1864, p. 162).

25.3.2.7 Psychic Defunctionalisation Relative to Its Biological Substrate

To return to the intimate aspect of the emergence of man. The advent of an autonomous psyche, not totally bound to the efficiency of action of a nervous system, nor entirely engaged in survival, is what Cornelius Castoriadis (2002 p. 86) calls defunctionalisation of the psyche relative to biology. It is not enslaved by the biological substrate it grows from. It enjoys a freedom, a latitude to develop for its own sake. Our psyche speculates, freely interprets those events it meets, and has leisure to make up all sorts of wild imagining about them —and it does so liberally, as seen from Parmenides to Hawking. To put is blandly, you can safely think many more stupid things than you can *do*. To extend Aristotle's reflection, the mind does not have an assigned function. It abides by its own rules. It *must* define its own directives.

When Albert Camus wipes out 2500 years of philosophy at the first sentence of *The myth of Sisyphus* (1957), thrusting upon society that "There is but one truly serious philosophical problem and that is suicide," he points precisely to the crux of the matter, where the demands and directives of the psyche project onto itself, and it decides it should not exist, and to sap the biological being that conditions its existence, anti-biologically, anti-evolutionarily. To ban itself for purely endogenous cognitive reasons. Rules prevail over life. Cognition is *deregulated.* It demonstrates its freedom, if only that were necessary. Camus' sentence strikes again upon the difference between *knowing* something (like a 'fact') and consciousness, which involves necessarily framing it in a meaning involving all one's existence and becoming. Suicide was well-known from ancient times. But no more than Socrates' was understood, was it realised it tremendous metaphysical meaning, comparable to the rational paradoxes found early by the ancient Greeks. Just as the tree was apparent in the middle of Eden, but its meaning would appear later. Quite to the point, "what we call basic truths are simply the ones we discover after all the others," again says Camus in *The fall* (1991).

As we see it in full generality, an emerging phenomenon is the creation of a space of freedom, and is thus defunctionalised relative to its originating substrate.

25.3.2.8 Intrinsic Pain and Extrinsic Reward

In animals, the nervous system is designed for action, as Henri Laborit (1986 p. 1) states it, this is its primary biological function. Of course it is not infallible —again, nothing can be warranted about incoming events, actions are but bets. In man, it has a range of autonomous function, coupled with suspended action. The outside world and the world in the mind do interact; in a sense the first always prevails since it decides whether life continues —"two sitting intellectuals don't get as far as a walking brute" (de la Patelière, Audiard 1961). In another, the mind, while it is afforded in command by its biological purveying substrate, can decide to cease its existence by getting rid of its housing living body.

The price of consciousness —an autonomous psyche— is also the appearance of a sort of pain whose source is not an outside factor such as a wound, or hunger, but originates in the reflexivity itself. Such is the case also for shame, guilt, sin, nudity . . . —why the fateful Eden tree was not of knowledge, but knowledge *of good and evil.* The pain comes from an inconsistency peculiar to the mind. It comes from the self. Being inherent to the experience of consciousness, it explains Eden mixes blessing and punishment. Just like there is no return to the innocence prior to Eden, no wall, no armour stops this sort of pain. Victor Hugo (1859) has rendered like no other the poignancy of mistaking a reflective effect for an outside agent, in "La Conscience," —the eye of God pursuing Cain after Abel's murder. "The walls were thick as mountains. On the door they graved: "Let not God enter here." [. . .] But when he sat, so ghost-like, in his chair, And they had closed the dungeon o'er his head, The Eye was in the tomb and fixed on Cain." Another quite clear statement that God equates the human mind.

Now for the rebound that may come from the ability to forge new worlds. Speculations that burgeon in our mind away from the primary world may at times

—quite rarely actually— in return change the deal in that world. Quite rarely, but quite biased in our memories, because they are such acme, *kairos* instants in the generic *chronos* of life. The lion's metaphor illustrates what a sequel to Eden could be (Douzal 1998): The action takes place in the wooded savanna, where the lion as undisputed master, has each day his toll of man. One morning, he finds the man absorbed, not the least mindful about the lion, taken aback: —Oh! yes, I almost forgot, nothing is the same, you cannot eat me anymore. The lion inspects his claws, teeth, muscles, scans the savanna around: there is nothing new under the sun. —Yes says the man, I have got an idea. —How come an idea in *your* head would prevent me from eating you? —See the spear in the tree, over there. —*Spear*? The man cuts it out of the trunk: now you cannot *savage* me. The lion must admit that the savanna has tumbled overnight. The game is not the same. Man has escaped from the animal world.

25.3.2.9 Stuttering Genesis, Naming, Abstracting and Icarus

Scholars have long noted that Genesis stutters —creations reiterated— and have even traced back many various contributing legends that result in the present patchwork. But there is something other to be found in it. Let us first return to the deliberations under the tree in Eden, God forbidding to eat the fruit now interpreted as internal conflict, the human mind internally feared, and pondering if he should cross the gate to this huge space of many possible plays he foresees. He sees all the animals playing from an overhanging position, recalling us of the view from nowhere. He was once among them, he is now distinct (not only distinct —no species confuses its mates, and who is friend or foe or else—: consciously so) detached from this nature that gave birth to him.

On the seventh day, the world is completed and the mind rests. It does not, since the mind never rests —this has been long recognised in Buddhism—, it has all the animals unfolding before his eyes *—mentally*, who could really *do* that?—, just as in Kipling's Very Beginnings, and to the end of operating them within, he *names* them. Naming permits manipulating from a distance: abstraction. Language separates, disconnects thus frees manipulations —not all that can be said about animals can be done. It also reconnects in that it *points to* what it means —semantics. Names unfold a space which flattens everything uniformly, all laid down to be sighted all at once. Things are as if viewed from nowhere, like Icarus flying over the Labyrinth; the moment he views that its walls have only resulted from digging relentlessly along the same rules, and that many other plays are possible.

As for the first Genesis, it was but a rehearsal, a bootstrap to set the scene the mind wants to consider. A flashback, now the mind is there to talk —why he was said to be floating over the very initial undifferentiated matrix (Genesis 1:2). We have recognised the reasons for these repetitions, discussing how to build up a space: necessarily, one must start from an initial substrate which already supplies everything we intend to make. Then we do no more than to categorise it into a crisper image. This iterative process is the inevitable side-effect of language, which already stages a conservative constraint that we have seen even more strictly enforced in rational thought, which makes it impossible to account for an origin.

25.3.2.10 Escaping from the Labyrinth of Established Play

Watzlawick et al. (1974, p. 98) quote a modern account of this very process:

> Nigel Howard has presented a game-theoretical model of what he calls "existentialist axiom," and has shown that, indeed, "if a person comes to 'know' a theory about his behavior, he is no longer bound by it but becomes free to disobey it", and "... a conscious decision maker can always choose to disobey any theory predicting his behavior. We may say that he can always "transcend" such a theory."

This can be read as a comment on the expulsion of Eden, itself being our first documented discovery that we are in a play. It confirms that the key to freedom is reflexivity: at any point, an enlarged stage of consciousness. But we are still on a *stage!* Shakespeare has told us. Beneath our feet, anytime the boards can give way, opening a trap. With a glimpse of awareness tinged with panic. There are no definite bottom boards you can safely rely on —unless you follow Hawking's way of self-deception, convinced that the ultimate theory is a few steps away, together with the basic building blocks of the universe. As if it had an alphabetic floor.

Man has the distinctive potential to open up at any time another level of consciousness, and realise he plays by given rules, instantly viewing a huge space of other possible rules. In a legendary Japanese tale, the samurai Tadanobu is attacked at his mistress' without his weapons; he picks a goban and fiercely fights off his enemies. By contrast, the AlphaGo programme cannot invent the goban as a weapon. It can only play go games, unable to change rules, play something else; it is stuck forever on its alphabetic floor, forever in the same Labyrinth. Although it beat the lead player of the moment, in this regard it is devoid of interest. A machine has no alterity. It cannot step outside its rules, their span is held tight together in a block-universe, closed on itself. One more time: *a machine excludes any emergence*. This is what we expect from a machine: no uncertainty comes from inside it. It is reliable. But at heart, we are interested in new, emerging things. The gap between machine and a subject, a living being cannot by nature be bridged, and artificial intelligence proper is impossible.

History offers countless cases of being locked in rules, and inventing escapes from them. Napoléon is defeated in Spain, he would win any war but the Spaniards refuse to play war, instead they invent guerrilla warfare. By the convened rules, once the (symbolic) capital is occupied, you capitulate. Napoléon reaches Moscow on fire, he has won nothing: the Tsar of Russia plays by other rules. Whereby you defeat a dominant position, not at its own game, but by inventing a new one. In the last few decades, the computer industry has seen a series of such emergences. In the 1980s, IBM could not be displaced as number one, and was not, but microcomputers were invented, and grew so much that Microsoft became the global leader. With the web a new world burgeons in turn, so much so that new players begin to dwarf the others.

25.3.2.11 Freedom and the Torments of Choice

The expanse of a name space, of many possible rules to play, of the sight of the stormy sky crossed by the jetliner: awakening is a creation of space, of freedom; a giddy feeling, that of being God, of *one's bodily envelope* extending to the whole. It comes at a risk that the Greeks have called hubris: Daedalus warns Icarus against taking leave of his senses, as he launches him into the sky. Hubris is the delusion of being able to heave oneself up into this open space, and forget about the ground, of cutting Utopia's isthmus to the mainland. Eventually all myths are warnings against the many difficulties resulting from becoming conscious.

But freedom has another, immediate negative side: the torments of choice, since all possible choices appear equal, and not all of them can be lived. Contrary to the accepted view, cognition is not fitted to choice. Its core matter is to lay down a space of possible options. Whereas animals opt and act in one piece, cognition in man ponders and speculates, withdrawn in mental space. And languishes with its unity from before Eden, now it is fragmented in choice.

25.3.2.12 Eden and the Elusive Emergence

Despite the claims of exegetes, Eden does not conceal meanings, on the contrary, it sticks to the most down-to-earth way of expressing key issues, and they become obvious if we stick to the letter. The serpent, who is vector to the expulsion from the garden (Genesis 3), is invoked as the most clever of all animals, *reasoning, blessed with language:* of evidence it is a psychic instance, —just as the authority of God is another— and the reported argument is an inner, silent dialogue —the serpent being about the most silent of animals. That Eve is primarily involved means that *life* could not be contained, it springs and the bodily *humus* was to follow.

The serpent is also found in many earlier myths, for reasons found in the remanent child's expression "the snake biting its own tail" —the Ouroboros. By doing so, it should disappear, become nil; being here, it proves to be self-generating, it symbolises the circular, self-enacting character of an emergence. It resists any rational account, it escapes any attempt at a direct expression, because conservativity is antinomic to emergence. The written form of the myth is not for want of being explicit. It could not be more explicit.

Up to this point the tale entirely takes place within the intimacy of the personal experience of consciousness. And such experience is to be lived through again by every man. The ontogeny of the personal psyche must recapitulate the mythical stages. For another effect of emergence, or creation, appearance, being an impassable horizon of language, is that *consciousness cannot be communicated exoterically.* Its conditions for blossoming can be cultivated, but it cannot be commanded, as hinted in the classic Buddhist tale: Master, how long shall I have to wait for awakening? —Fifteen years. —And if I keep on practising hard every day? —Then thirty years. This is a good place to answer our little riddle: dance is a perfect example of a knowledge that cannot be communicated exoterically.

Once we find one, we realise there are many. By the way, it is actually more pertinent to say that ultimately no knowledge can be communicated exoterically. Read William P. Thurston (1994): he explains quite convincingly that *mathematics* are communicated very efficiently in direct presence, and become more and more unintelligible as they are more formally expressed.

Any direct attempt to reduce emergence to rationality ends up either in an infinite regress, promising to terminate where the rainbow ends or in circular arguments, both irrecoverable flaws pointed early on by ancient sceptics.

25.3.2.13 Conditions for Emergence and the Lesion Method

Let as return quickly to some past episodes. We have seen that mathematics after trying different bases for a secure foundation, ended up proving it could not find one. Eventually, mathematicians removed the foundations, and returned to their job of proving theorems that other colleagues accepted. Hence mathematics was based on nothing, *save* the cognitive subject who does them and lives the mathematical experience. Although this is seldom acknowledged in mathematical circles, the cognitive subject is *de facto* the frame of reference, the supporting substrate to which all mathematics is relative. He should never be overlooked when we talk about mathematics. About *anything*, as Descartes put it: "the proposition *I am, I exist*, is necessarily true whenever it is put forward by me or conceived in my mind." This formulation of his famous *cogito* is more of an enactment, than it is a truth. The very fact of being able to express oneself asserts first and foremost one's own existence: again the medium is the message.

It states the existence of an autonomous cognition, the miracle of consciousness, the *revolution* of Eden, to continue with the pattern started in a previous section. This mind does not exist by itself, out of nowhere, as sooner or —more likely— later one is bound to realise. Although defunctionalised, it is *relative* to a biological body, as Descartes writes to Chanut:

> Health is the greatest of all the goods which concern our bodies, but it is the one we least reflect upon and savour. The knowledge of truth is like the health of the soul: once a man possesses it, he thinks no more of it. (Cottingham et al. 1991)

Unavoidably, it seems, this is another basic truth, to be found after all the others.

This experience too must be relived by every man —or nearly every man, for few are those really struck in consciousness by illness, ageing and death; their own, and that of all living things. According to the legend, the life of Siddhartha Gautama changed dramatically on the spot. Many instead pursue immortality; the legend of king Gilgamesh in Mesopotamia c. –2500, of Qin Shi Huang (–259 to –210), the first emperor of China are exemplary. Contrast also Descartes with Socrates in Plato's *Phaedo*, eager to jettison his body, sure to release his conscious bubble into perfection.

Why does every child believe to be immortal, or so it seems? Recall how we build up a perceptive space: a phenomenon appears —thus we may begin to conceive of it—, by contrasting against the background of ambient conditions. The observer of living systems discovers the existence of functions only when confronted with the rare case where they fail. And what is more, he recognises them only in the best case, since any such exception is likely to be cognitively discarded as abnormal. In the generic situation, life goes on, nothing happens on the surface. We can only conceive that the calm surface is the result of intensely coordinated functions underneath, when we have had the experience of malfunction. This mere fact of perceptive contrast is well known in biology, even called the method of lesion in neurology. World war I offered thousands of such cases to medical curiosity.

Still, because "nature loves to hide", as Heraclitus (Héraclite 1998) wrote, we keep being deceived. For instance, the reader certainly acknowledges this 'lesion effect,' but only insofar as it concerns the *particular case* of living systems. But wait: what else but *living situations* do you cope with? Only those you set in physics classroom are devoid of life, so to speak. It takes an effort to realise that whichever way we turn, *we only find living processes*. Hence in the general case in our world, we discover things exist mostly not when they appear, but when they cease to work, otherwise they are hidden, at least to naïve eyes.

This picture seems to take us back to the opaque mythical world of the ancient Greeks. There is a difference: we know from Descartes how to investigate phenomena under the veil of reality. And the reader knows by now that we are describing a horizon of reality, and should never fool ourselves that we operate on an alphabetic floor.

Today's roar about 'ecosystem services' is completely unaware that it is condemned to list each function brought to us by ecological systems, *as we provoke their failure*. We welcome every new case, as if it were a pleasure to see nature unknitting as we dissect it. The agreement is so general that the French Constitution went completely astray in its *Charte de l'environnement*, by implying that an ultimate inventory is to come, —a generalised Euclid myth— and is the solution. Alas, the case is that *the problem is the solution* (Watzlawick et al. 1974) How many times have you heard the lament "species disappear everyday before they are even classified"? Was it a single time to criticise it? It implies that once a species is catalogued, classified, dried and stored in a collection, it is *saved*. We are relieved. We must realise here that we are blindly pursuing the task of Genesis' seventh day. It amounts to a frenzy of conversion from underlying living categories —as soon as they are identified as such—, into formal records. The extended, totalising formalisation process required by the idol of the 1st culture: forcing reality into the ontological alphabet it is claimed to be from the outset —note the paradox.

25.3.2.14 The Triumphant Monad Character of an Emergence

And now the symmetric of the method of lesion: a phenomenon emerging does so from a substrate that caters for all its needs. Otherwise it would not appear at all. It draws all its resources from that substrate, and arises as an individuation, contrasting itself against the substrate. Individuation takes place when its elements establish *more relations* with one another than with what, by contrast, becomes the outside, thus drawing a frontier. This *defines* individuation —Cyril Stanley Smith (1981 p. 369) calls it *entitation*. Inside the border, the individuality's autonomous rules apply, whose primary function is to maintain the individual. This characteristic operational closure describes well the compartmented organisation of living systems at all scales.

At the initial stage, an emergence completely *ignores* its nurturing substrate; the substrate is taken for granted, so to say, and by hypothesis, it presents no limits, no border. The emergence keeps devouring it, in the all-powerful state of whoever finds no limits. It is closed onto itself in a *monadic*, unitary state: it is the whole universe. This monad is *defunctionalised* regarding its substrate.

Concerning the emergence of the human psyche, these reflections were jotted down in a lightning inspiration by Sigmund Freud in his notebook "I am the breast," as the monadic feeling of the newborn. The infant initially does not perceive the frontier between his body and the breast, himself and his mother, bathing in self-satisfaction. His whole story will be of the concomitant discovery of the limits of self and non-self. The first frontier he meets, a frustration in psychoanalytic language, is the absence of the breast, the pain of hunger. This is the initial symmetry breaking of the psyche. Discovering that mother is other. A first realisation that there is no permanent substrate, there are always *conditions* to its presence.

All lifelong learning can be seen as a succession of such symmetry breaking, whereby something appearing initially as uniformly unconditional, appears under a deeper perspective to be relative to ever more branching conditions —as we have seen about the perspective experience of reality. In other words, *learning is only learning relativities*. Hence the fact that physics revolutions came by expressing new relativities comes as no surprise at all. This process of acquisition of structure describes learning, but also describes our science's account of more and more structures layering from the big bang to the formation of the solar system, the geologic history, and the evolution of life on earth. Our scientific cosmology from physics to biology recounts an accelerating series of emergences. A story which does not differ in principle from most creation myths evolving from indistinct chaos to an organised world.

As for the psyche, it carries an enduring limbic, archaic memory of the lost initial monadic state. This lost unity is an irrecoverable disconnection. Every emerging stage of awareness replays this split, with its ambiguous duality of awakening to enlarged horizons and losing earlier unity. Why the tale of the emergence of *homo*, and immediate split into man and woman, and *I am and I see that I am*, resonates so deeply in every individual experience.

25.3.3 *Emergence, Evolutionary Radiation and Growth*

25.3.3.1 Evolutionary Radiation: Creating Space by Rules of Generating Forms

It is now easy to sketch elements on the evolution of life in the terms we have established. The history of life shows characteristically many events of inflationary production of forms. For instance during the Cambrian explosion, most known phyla —the basic body plans of multicellular organisms— are generated over a short period: the *rules of production* for multicellular organisation are invented, or some threshold is passed that brings them into play. By the previous rules, all the possible niches of the biosphere were occupied, all the available space was completely filled with colonies of bacteria, and species blocked against one another in the famous red queen race (Van Valen 1973) (this model is a mere occurrence of what we have described as a still, living situation, hiding a wealth of internal processes confronting one another, underlying mechanisms showing only under lesion). The *monadic* ignition of the rules to build organised living forms *creates* a new, thus completely open space, that *expands as new forms furnish it* in an explosive chain propagation. Every combinatorial variation spawns a new form which finds as substrate the space of the previous biosphere to feed on, and, without restrictions or competition, carries on the inflationary process of evolutionary radiation — and exerts increasing selective pressure upon the substrate of unicellular life: the emergence is *defunctionalised*, deregulated. Thus one space is growing out of the other.

Then, as this burgeoning space fills up, species begin to interact strongly and block against one another. There is no more room for branching new forms; now accidents can only prune branches. The creation phase is closed. The species tend to *a co-evolved state* defined by *a growing, stabilising set of mutual co-determinations*: we return to a red queen race. The qualification of being *evolved* cannot apply to a single species, only to an ecosystem as a whole: it is a *relational* property. The process leads to *re-functionalise* the initial monadic state into a structured ecosystem. Stability comes with complexity —which we already knew, since explicit formal objects which in a sense are a summit of stability, being so invariant that some (Platonists) claim them absolute, emerge onto an elaborate, complex cognition.

Now we may draw the general principle: an explosion of forms implies that a substrate space is available, and a production rule has emerged to consume it and generate the expansion of a new space of forms. The principle works abstractly at any level; e.g., once many nuclides are formed and available in the universe they provide a substrate such that in appropriate conditions, news spaces of chemical reactions develop. Chemistry then has a birthdate of emergence, sometime in the history of our universe.

The simplest, most elementary biological case of this sort of process of expansion is already instructive: the exponential growth of yeast on a nutrient medium, until it bangs into the wall of the Petri box is such an explosion (without variation). A homogeneous medium is the perfect substrate for explosive growth (as is, sees the happy reader, a mimetic crisis). This is enough to demonstrate that industrial agriculture, and its large monospecific —and even nearly clonal— fields are biologically absurd: they are just huge, open Petri boxes, inviting any pest that has the right rule for consumption to rush in. Or intensely attractive potential wells for invasions. Pure and uniform, their sole merit is to appeal to rational thought, because these properties are those which ease mental manipulation.

To continue with our yeast, now collapsing by exhaustion of its substrate. It had the rule long-time ready for growing there, but why after so many millions of years of evolution, does it keep running blindly into it medium's limits? Because, in the more normal scenario of a spore, luckily airdropped onto a windfall, a rush to finish the substrate, terminated by emitting new spores has proven effective: it has cycled uninterrupted for tens of million years.

The palaeontological record shows many such explosions. Some stem from the radically new invention 'on site,' of a *space of forms*, e.g., photosynthesis, the Cambrian explosion . . . or Eden; others, that may appear to us as simply investing a pre-existing domain, e.g., continents, when water-bound fish evolved into land-going vertebrates, are nevertheless inventions of *a new process of generating forms*, from the insider point of view. Another case occurs when an exogenous disaster leads to a mass extinction —they mark all significant geological divisions. The clearings opened permit surviving species to restart an expansion. Typical is the radiation of mammals after the dinosaurs became extinct.

Quite an important remark is that an earlier generation of a space of form *can never be repeated*. For instance, no new body plan has a chance to appear today, because the space is occupied: existing forms would not allow newcomers to enter the game. The substrate space of a prior radiation of forms is irrevocably closed by other spaces silted over it. Thus the next emergence will issue when a way to use existing form spaces as a substrate is found. A condition for a form space to become a substrate is its wide, continuous extension. At any rate, there is no return to an Eden, as long as there remains a living form to keep its memory. It appears that even the most massive recorded extinctions left relict forms that kept prior spaces from burgeoning again. Among mass extinctions, probably the Great oxygenation event is the first recorded, and the Permo-Triassic one, the most extensive —but we are working hard to compete for a new record. In any case, bear in mind that erased memory cannot be recovered —that's the definition of erasing.

Thus, we may regard the Petri box as a landscape after a mass extinction has removed *all forms competing with the yeast*, leaving the spore as sole survivor, to repeat exoterically the monadic blossom of the same space —in other words, a commemorative ritual, a compulsive repetition, life rambling on the same story. Gardens have seen billions of such (nearly) identically iterated universes, each time a fallen piece of fruit is invaded by a yeast spore. Aren't these multiverses much nicer to regard than those physicists entertain?

25.3.3.2 Trophic Pyramids and Degrees of Freedom

As we see it, the notion of substrate space is to a large extent congruent with that of a nutrient medium, or in a classic view of ecology, levels of a trophic pyramid, with primary producers (mainly green plants), then herbivores, carnivores; each level being somewhat substrate to the next. This view is meaningful only for essentially human-visible macroscopic trophic chains, and moreover, under the condition of considering only growing living sizes along the chain. The perspective on the living world offered by this level of interaction with it, leaves a huge living world behind the horizon. At a closer look, the compartments of trophic networks are infinitely more intricate and interlaced, and numerous.

Anyhow, the earth provides life with two primary substrate sources, one external, the sun, the other is the capital of internal heat resulting from the initial gravitational convergence of its formation augmented by disintegrating nuclides within its volume. The internal capital is by far not secondary: it drives plates tectonics and volcanism, which renew minerals on the surface and power cycles of mountain-building and sedimentation, thus reviving living habitats that otherwise fade to a thin stream of life within a few million years. Any production of structure on earth draws from these two energies. All men's activity as living beings draws from the one and only 'trophic pyramid.' Any structure they build is nourished by that blood. Additionally, their technoscience permits to exploit all sorts of geological pools, some notably from past plant primary production (oil, coal, etc.). Therefore the overall 'power' of the biosphere is bounded, and so is that of any of its living compartments.

The resulting "power to generate structure" corresponds to *the degrees of freedom* of an emergence. With a consequence we have not yet attracted attention to: an emerging phenomenon reduces the degrees of freedom of its underlying substrates, by putting constraints on them. For instance, the cells of my body are subjected to my body, and my organism lasts because they keep agreeing to coordinate and cooperate. The cells are constrained, their freedom is reduced. A cell which releases itself from this network of constraints, sets itself free and develops *by its own rules —auto-nomous—*, is an emergence. It starts growing as *a defunctionalised bubble;* a *monad* that knows nothing but itself, ignoring its organism-substrate, although it draws all its resources from it, calling for ever more by inducing vascularisation. Eventually the expansion of this *tumour* disorganises its body host functions and both collapse together.

Any emergence which lasts too long in such a monadic, defunctionalised stage is aptly termed a cancer.

Overall the effect of an emergence is to transfer degrees of freedom, or structure-building, from substrate spaces to emerging space.

25.3.4 Our Civilisation in the Light of the Concept of Emergence

Our civilisation is characteristically in a newborn monadic expansion, without alterity, a suckling unaware of the substrate it takes the lifeblood out of. It keeps on proclaiming 'I am the breast' in all possible manners. If an emergence does not evolve to self-awareness, having its bodily envelope defined by confronting an otherness, and thus coming to be in tight mutual co-determination with its environment, its fate is that of a cancer: to collapse with its substrate for having disorganised it, which necessarily comes earlier than exhausting its resources.

What is remarkable for the present time, is that for the first time in the known universe, we are witnessing an emergence as *members* of it (not the inner experience of Eden): France for instance, has mutated during the twentieth century from a society of citizens to a society of organisations of all sorts, which are born, live and die by formal processes —they are probably close to outnumbering humans. We are in the era of the *preeminence of organisations*; practically nothing can be done which is not mediated by one —Castoriadis deems it a society of lobbies. To assess the extent of the trend, consider that quality methods are becoming the universal rule in every organisation, and their effect is to break it into as many formal components as seems fit, thus making an 'organisation of organisations,' repeating the same pattern over again. Quality is *formalisation for dummies*, you will never see someone who can do mathematics in quality. Not even Socrates would have dreamt of such an extended society-machine.

The activity in our societies shifts to always higher layers of producing emerging formalism and organisations, in a frenzy that amounts to a mass psychosis. Now every concrete act requires ten administrative acts.

To illustrate, consider all the most sincerely well-meaning *solutions* envisioned to the present integral crisis: we already mentioned the teeming ecosystem services, whereby the problem is the solution. A second belief is that all sorts of markets are necessary to "regulate natural processes" (for carbon, etc.). Quite an absurdity, since the nature of life *is* homeostasis, while on the other hand, markets are notoriously instable —speculative bubbles. For the very reason that they bear the most simple large-scale structure we can conceive of: that of a monad, or at best a homogeneous substrate. We know now we could not provide for a better medium for autonomous fluctuations, mimetic crises, emergences. Absurd also because a market connects all the many separate compartments of living systems, and makes them all fungible in one universal medium: money, thereby destroying their structure by short-circuiting them all together, pouring their contents into one melting pot. The non-structure of markets is meant to be comfortable for the mind, it is extremely attractive because it is so easy to handle, it lends itself well to ratiocination, and this is nothing else than Narcissistic seduction, but it is at the other end of the spectrum of stability, as embodied before our eyes by the complex structuring of natural systems. If only our eyes are not wide shut Schnitzler et al. (1999). Thirdly, the belief is spreading that *every* natural system (a species, a valley, a mountain, a river, a shore…) should have political representation, and a legal standing (Stone 2010); at this condition only could we weigh our relation to nature. The basic problem is *there is no end to the list of natural systems* that need formal representation in markets, politics and law. Note the extension: not only every species should be 'saved' by registering it in our systematic catalogues. Now every subset of a natural system demands a whole collection of catalogue entries. The *list of catalogues* is itself endless. The frenzy is fast and furious to exoterically convert every notion, every function freshly identified into a formal representation, to *re-instantiate* it, pure and clean, the ultimate simulation that enacts the ultimate-to-be theory. And, in passing, this policy has *already proved ineffective:* check, in one French département (county), how many associations, foundations, syndicates, and the like, have been created to protect 'natural systems' and *terroirs:* hundreds of them, ever more numerous. Each, motivated by lesions caused by changing interactions with the environment.

You want another example. Just take a look at the proceedings of the Workshop on mathematical and statistical challenges for sustainability (Rehmeyer et al. 2011), as a sample of the global babel, claims for more sensors, satellites, cameras, more data, more thorough models, more extensive simulations, what else can we do to win *the climate battle*? (Note the term. Where is the battlefield? Napoleon tried to win battles against no army, he lost.) What is the Great promise here? That once the formalisation is completed, we reach nirvana: Parmenides' dream come true. Remember: there is no such thing as *the* formalisation, there is not one and only absolute formalisation. We will not meet where the rainbow ends. There shall be no completion, and *reality is not formal* although this is precisely the overwhelming credo of the Western world, living entirely on the left side of Parmenides' bifurcation.

As this inflationary organisational formalism consumes resources, it keeps increasing the overall strain on the one and only trophic pyramid it intends to rescue: hence again, as Watzlawick et al. (1974) say, *the problem is the solution*. It boosts what it plans to curb, for want of awareness of being running through the same corridors, always digging higher walls in the labyrinth carved by Parmenides, now raised to the dimensions of a deep canyon. Indeed, we started by observing that the present crisis obviously keeps on developing despite every attempt to control it: it is *invariant* by all our range of actions. It can only be controlled by something completely other to what we are able to conceive of.

Of course, the strain put on the trophic pyramid applies in particular to men. As the primary substrate out of which all sorts of formal structures grow, they must agree to dedicate their entire energy in cultivating formal structures. They must *value* it above all. They must abide by the rules that build and operate such structures, which implies a drastic reduction of men's degrees of freedom, as for any substrate to an emergence. However, there is a singularity here: for the first time, an entity is enslaved of its own free will. The stringent injunction to any agent of Western organisations is to behave like a processor —everything is a process as regarded by quality methods.

25.4 Change

25.4.1 A Step in Self-Awareness for a Refunctionalisation

Change, for a cognitive species, means above all to acquire a new view of oneself sideways, see by the corridors of which labyrinth we abide, what sort of play we are part of —"The only true voyage of discovery, the only fountain of Eternal Youth, would be not to visit strange lands but to possess other eyes" (Proust 1929). And at this point of the journey, my Ulysses-reader has no doubts left at all that the last horizon to be driven ashore is oneself.

All three revolutions that we have considered began with the emergence of a new vision of the world, a cosmogony. Hence a new vision of oneself, a self-consciousness, since these are relative.

Although there are people around the world with evidently more viable cosmogonies, we cannot backtrack to them. The best we can do is to withdraw from engaging them in the cogs of our economic machinery, whose extensive interaction with all environments and people *decompartmentalises* cultures, and therefore living habitats, resulting in a simplified, homogeneous soup.

There is no other option than to heed Dante's Ulysses call: life at all scales proceeds by integrating experiences, thus the way out of the labyrinth is the same challenge for an individual or a civilisation, it consists in rebuilding a consistent, viable arrangement of one's historical experience: evolving it into a pleasant ecosystem of co-determining entities, stable and displaying many degrees of freedom. Ignoring or denying one's past is the worst option, because it is not forgotten, it is there in the layered sediments, only is it inaccessible. It imposes its invariant constraints, without entering into playful interaction. The past can be overcome —layered properly in remote, peaceful memory— only once assimilated, that is, when it interacts, co-evolves with the whole system (be it cognitive, social-historical, ecological).

The stake of realising that the present turn of our civilisation implies a drastic restriction of our consciousness' degrees of freedom is no small. Thus we decide whether we want to remain fully human.

25.4.2 The Other Way: Acting

We must enter into practices that relate us in an actual co-determination with the various living habitats of our world. Or return to them, since there remain many people who have weighed and cultivated their relationship to their land through many changes across millennia. Farmers leaving the industrial model exemplify such an endeavour; they proceed from values that contrast the mainstream. Not only are they vital, for change must be made in deed, they also sow the ferment of ideas that prepare metamorphoses and may someday turn creeds on their heads.

References

Aristotle (1926) Nicomachean ethics. Loeb classical library. Harvard University Press, Cambridge, MA, p 73

Bell ET (1933) Numerology, 1st edn. The Williams & Wilkins Co., Baltimore, vii+191 pp

Borges JL (2004) The Aleph and other stories. Penguin Classics, 224 pp. ISBN 978-014-24-7889. Translated and annotated from El Aleph by Andrew Hurley

Buckley P, Peat FD (eds) (1996) Glimpsing reality: ideas in physics and the link to biology, revised edn. University of Toronto Press, Toronto, 199 pp. ISBN 0-8020-0575-6

Burnet J (1920) Early Greek philosophy, 3rd edn. A & C Black Ltd., London

Camus A (1957) The myth of Sisyphus. Hamish Hamilton, London

Camus A (1991) The fall. Vintage books. ISBN 978-0679720225. Translated from La Chute, Gallimard NRF, Paris, 1957

Capart A, Capart D (1986) L'homme et les déluges. 340 pp., Hayez la longue vue, éditeur, 4 rue fin, 1080 Bruxelles, Belgique. ISBN 2-87126-003-6

Castoriadis C (1986) Domaine de l'homme: Les Carrefours du labyrinthe 2. Collection Empreintes. Éditions du Seuil, 27 rue Jacob, F-75006 Paris. ISBN 2-02-009175-5

Castoriadis C (2002) Sujet et vérité dans le monde social-historique: séminaires 1986–1987 —La création humaine 1. La couleur des idées. 500 pp., Éditions du Seuil, 25 boulevard Romain-Rolland, CS 21418, F-75993 Paris cedex 14. ISBN 978-2-02-055161-8. Texte établi, présenté et annoté par Enrique Escobar et Pascal Vernay

Castoriadis C (2005) Une société à la dérive: entretiens et débats (1974–1997). La Couleur des idées. Éditions du Seuil, 25 boulevard Romain-Rolland, CS 21418, F-75993 Paris Cedex 14. ISBN 978-2-020788533

Cohen-Tannoudji G (1992) Universal constants in physics. McGraw-Hill, New York, USA. ISBN 978-0070116511

Cottingham J, Stoohof R, Murdoch, D, Kenny A (Eds) (1991) The philosophical writings of Descartes. Cambridge University Press, the Pitt building, Trumpington street, Cambridge CB2 1RP, UK. ISBN 978-0521403238

Dante (1918) The divine comedy of Dante Alighieri, vol. 1, Inferno. Harvard University Press, Cambridge, MA. The Italian text with a translation in English blank verse and a commentary by Courtney Langdon

Davis PJ, Hersh R, Marchisotto EA (2012) The Mathematical Experience, Study Edition. Modern Birkhäuser Classics. xxv+500 pp., Birkhäuser Boston, 675 Massachusetts avenue, Cambridge, MA 02139, USA, first edition, Birkhäuser, 1981; reprint of the study edition, Birkhäuser, 1995; copyright Springer science+business media, LLC, New York, USA edn. ISBN 978-0-8176-8294-1, eISBN 978-0-8167-8295-8. doi: https://doi.org/10.1007/978-0-8176-8295-8

de Saint-Exupéry, A. (1939) Terre des hommes. nrf. 224 pp., Gallimard. ISBN 2-07-025659-6

de Tocqueville A (1856) The old regime and the revolution. Harper & Brothers, Franklin Square, New York, 344 pp. Translated by John Bonner; a missing passage in his translation has been inserted anew

de Tocqueville A (1864) De la démocratie en Amérique. OEuvres complètes d'Alexis de Tocqueville. vi+563 pp., Michel Lévy frères, libraires éditeurs, 2 bis rue Vivienne et 15 boulevard des Italiens, Paris, France, quatorzième édition edn. URL http://gallica.bnf.fr/ark:/12148/bpt6k213943r/. Publiées par Madame de Tocqueville et Gustave de Beaumont

Descartes R (Ed) (1983) OEuvres complètes, vol. 6, Discours de la méthode et essais of Bibliothèque des oeuvres philosophiques. Vrin. ISBN 978-2-7116-0196-7. Ch. Adam and P. Tannery (eds.); original anynymous edition, Leyde: Jan Maire, 1637

Douzal V (1998) Analyse de données d'évaluation sensorielle. Thèse de doctorat de l'université de Paris vi

Douzal, V. (2015) Making science relative to a cognitive subject. FQXi Trick or truth essay contest. URL http://fqxi.org/community/forum/topic/2464, http://fqxi.org/community/forum/topic/essay-download/2464/__details/douzal_ToT2.pdf. http://fqxi.org/community/forum/topic/2464, http://fqxi.org/community/forum/topic/essay-download/2464/__details/douzal_ToT2.pdf

Dreyfus HL (1992) What computers still can't do: a critique of artificial reason. MIT Press, Cambridge, MA, liii+354 pp. ISBN 978-0-262-54067-4. Revised edition of What computers can't do, 1979

Eddington AS (1920) Space, time and gravitation: an outline of the general relativity theory. Cambridge University Press, Cambridge

Edwards B (2012) Drawing on the right side of the brain. Jeremy P. Tarcher, Penguin, 375 Hudson street, New York, 10014, USA, definitive, 4th edn. ISBN 978-1-101-55839-3

Euclide (1966) Les Œuvres d'Euclide. xxviii+627 pp., Librairie scientifique et technique Albert Blanchard, Paris. « Traduites littéralement d'après un manuscrit très-ancien, resté inconnu jusqu'à nos jours, par F. Peyrard, à Paris, chez C.-F. Patris, Imprimeur-Libraire, rue de la Colombe, no4, en la Cité, 1819. » Nouveau tirage, augmenté d'une importante introduction par M. Jean Itard

Everett DL (2008) Don't sleep, there are snakes: life and language in the Amazonian jungle. Pantheon Books, a division of Random House, Inc., New york, xviii+286 pp. ISBN 978-0-375-42502-8

Fourquet J, Dubrulle J-P (2015) Ifop poll, l'attirance des français pour un gouvernement technocratique ou autoritaire. Ifop, Immeuble Le Millénaire 2, 35 rue de la Gare F-75019 Paris, France, Département Opinion et Stratégies d'Entreprise, for Atlantico.fr, http://www.ifop.fr/media/poll/3185-1-study_file.pdf. http://www.atlantico.fr/decryptage/sondage-choc-attirance-francais-pour-gouvernement-technocratique-non-elu-ou-autoritaire-html

Fromm E (1966) You shall be as gods: a radical interpretation of the Old Testament and its tradition. Holt, Rinehart and Winston, New York, first. ISBN 0030611601

Gallois LCAG, Ray A (Eds) (1840) Réimpression de l'ancien Moniteur, depuis la réunion des États-Généraux jusqu'au Consulat (mai 1789–novembre 1799), vol. 1. Bureau central, 13, qui Malaquais, Paris, France. URL http://catalogue.bnf.fr/ark:/12148/cb311819153

Girard R (2001) Celui par qui le scandale arrive. Desclée de Brouwer, 76 bis rue des Saints-Pères, F-75007 Paris, France. ISBN 2-220-05011-4

Girard R (2002) La voie méconnue du réel: une théorie des mythes archaïques et modernes. Le livre de Poche, biblio essais no 4359. 319 pp., Grasset et Fasquelle. Textes écrits en anglais par René Girard et publiés dans des revues universitaires, traduits par Bee Formentelli

Girard R (2004) Les origines de la culture: entretiens avec Pierpaolo Antonello et João Cezar de Castro Rocha. 285 pp., Desclée de Brouwer, 76 bis rue des Saints-Pères, F-75007 Paris, France. ISBN 2-220-05355-5. Texte établi avec la collaboration de Valérie Morlot-Duhoux et Camille Chaplain

Gonseth F (1949) La géométrie et le problème de l'espace. Éditions du Griffon, Neuchâtel

Hawking S (1996) A brief history of time. Bantham Books, New York, USA, updated and expanded tenth anniversary edition; first edtion, 1988 edn. ISBN 0-553-10953-7

Héraclite (1998) Fragments. 496 pp., Presses universitaires de France, collection Épiméthée, Paris. Texte établi, traduit et commenté par Marcel Conche, 1re édition, 1986, 4e édition

Hugo V (1859) La légende des siècles, vol. Première série. Hetzel. Translated by Dublin University Magazine

Innis HA (1950) Empire and communications. Oxford University Press, Amen House, London E.C. 4, England

Ismard P (2015) La démocratie contre les experts: les esclaves publics en Grèce ancienne. L'Univers historique. Éditions du Seuil, 25 boulevard Romain-Rolland, CS 21418, F-75993 Paris Cedex 14. ISBN 978-2-021123623

Kipling R (1902) Just so stories. Macmillan, London

Knuth DE (1997) Fundamental Algorithms, vol. 1 of The Art of Computer Programming. xx + 650 pp., AddisonWesley Publishing Co., Inc. (Pearson Education), Uper Saddle River, New Jersey, USA, third edn. ISBN 0-201-89683-4

Knuth DE (2011) Combinatorial algorithms Part 1, vol. 4 A of The art of computer programming. Addison Wesley Publishing Co., Inc. (Pearson Education), Upper Saddle River, first, first printing, January edition. ISBN 0-201-03804-8

Kundera M (1984) The Unbearable lightness of being. Faber & Faber, 3 Queen square London WC1N 3AU Great Britain. ISBN 0-571-13539-0. Translated from Czech Nesnesitelná lehkost bytí by Michael Henry Heim

Laborit H (1986) L'inhibition de l'action: biologie comportementale et physio-pathologie. Masson, Paris. 2e édition

Lennon J, McCartney P (1965) Nowhere man. The Beatles "Rubber Soul" album, EMI

Mannoni O (1969) Clef pour l'imaginaire ou l'Autre scène. Les Champs freudiens. Éditions du Seuil, 25 boulevard Romain-Rolland, CS 21418, F-75993 Paris cedex 14

McLuhan M (2011) The Gutenberg galaxy: the making of typographic man. University of Toronto Press, Toronto, lxii+332 pp. ISBN 978-1-4426-1269-3

McLuhan M, Fiore Q, Agel J (2008) The medium is the massage. Penguin Books, 80 Strand, London WC2R ORL, England. ISBN 978-0-141-03682-6. First edition, Bantam Book, 1967

Nagel T (1986) The view from nowhere. Oxford University Press, Inc., 198 Madison Avenue, New York, 10016-4314, USA. ISBN 978-0-19-505644-0

Nietzsche F (1973) Beyond good and evil: prelude to a philosophy of the future. Penguin Classics. Penguin Books, 80 Strand, London WC2R ORL England. ISBN 978-0-140-44923-5. Translated by R. J. Hollingdale, with an introduction by Michael Tanner

Parnas DL (2017) The real risks of artificial intelligence. Commun. ACM 60(10):27–31. ISSN 0001-0782. doi: https://doi.org/10.1145/3132724. URL http: //doi.acm.org/10.1145/3132724

Pasteur L (1939) OEuvres de Pasteur, vol. 7, Mélanges scientifiques et littéraires, index analytique. vi+667 pp., Masson et Cie, 120 boulevard Saint-Germain, Paris, France. Réunies par Louis Pasteur Vallćry-Radot

Pauli W (1948) Dialectica 2(3–4):307–422. ISSN 1746-8361. URL http://onlinelibrary.wiley.com/doi/10.1111/dltc.1948.2.issue-3-4/issuetoc

Plato (1966) Plato in twelve wolumes, vol. 1. Harvard University Press and William Heinemann Ltd., Cambridge, Massachusetts, USA, and London, England. URL http://www.perseus.tufts.edu/hopper/text?doc=Plat.+Phaedo. Translated by Harold North Fowler; introduction by W. R. M. Lamb

Poincaré H (1968) La science et l'hypothèse. 252 pp., Flammarion, collection Champs no 56, Paris. Édition originale, Paris 1902; chez Flammarion en 1968, dépôt légal, collection Champs, 1989; est fait presqu'entièrement d'articles publiés entre 1891 et 1907, remaniés et réunis par Poincaré

Popper KR (1979) Objective knowledge. Oxford University Press, Great Clarendon street, Oxford OX2 6DP, UK, revised edition, corrections and a new appendix 2; first edition, 1972 edn. ISBN 978-0198750246

Popper KR (2005) The logic of scientific discovery. Routledge Classics. Routledge (Taylor & Francis Group), 11 New Fetter Lane, London EC4P 4EE, UK. ISBN 0-203-99462-0 (ebook) 0–415–27843–0 (hardcover). First edition, Hutchinson & Co. 1959

Prévost J-L (1984) Haroun Tazieff raconte sa terre: Les colères de la terre. Société nationale de télévision française. Documentaire, 57 mn

Prigogine I (1982) Only an illusion. The Tanner lectures on human values, delivered at Jawaharlal Nehru University. URL http://tannerlectures.utah.edu/_documents/a-to-z/p/Prigogine84.pdf, 2015-12-28

Proust M (1929) Remembrance of things past, Vol. 6, The captive. Chatto & Windus. Translated from À la recherche du temps perdu, La Captive, Paris: Grasset, 1923 by Charles Kenneth Scott Moncrieff

Rehmeyer J, Cozzens M, Roberts FS (eds) (2011) Mathematical and statistical challenges for sustainability. Report on a workshop held at the DIMACS center for discrete mathematics and theoretical computer science, DIMACS Center, CoRE building, Rutgers University, 15–17 November 2010

Santayana G (1906) The life or reason: or the phases of human progress, Introduction and reason in common sense. Archibald Constable & Co. Ltd., London, ix+291 pp

Schnitzler A, Kubrick S, Raphael F (1999) Eyes wide shut. Number 10795. Pocket, 2 avenue d'Italie F-75627 Paris cedex 13, France, 1999. ISBN 2-266-09701-6. Traumnovelle, Arthur Schnitzler, Translated from German by Dominique Auclère, followed by Eyes wide shut, scenario by Stanley Kubrick and Frederic Raphael, Warner Bros, translated from English by Carole d'Yvoire, Anne and Georges Duter

Smith CS (1981) A search for structure: selected essays on science, art, and history. MIT Press, Cambridge, MA/London, x+410 pp. ISBN 0-262-19191-1

Snow CP (1961) The two cultures and the scientific revolution. Cambridge university Press, New York, iii+58 pp. first edition, 1959 edn. The Rede lecture, 1959

Speziali P (Ed) (1972) Correspondance, 1903–1955, Albert Einstein, Michele Besso. Hermann, Paris, France

Stone CD (2010) Should trees have a standing? Lay, morality, and the environment. Oxford University Press, Great Clarendon Street, Oxford OX2 6DP, England, 3rd edn. ISBN 978-0199736072

Thurston WP (1994) On proof and progress in mathematics. Bull Amer Math Soc 30(2):161–177. https://doi.org/10.1090/S0273-0979-1994-00502-6

Tufte ER (1997) Visual explanations: images and quantities, evidence and narrative. 159 pp., Graphics Press LLC, Post office box 430, Cheshire, Connecticut 06410, usa, eighth printing, with revisions, june 2007 edn. ISBN 0-9613921-2-6

Voilquin J (ed) (1941) Les penseurs grecs avant Socrate: de Thalès de Milet à Prodicos. Classiques Garnier. xvi+ pp., Librairie Garnier frères, 6 rue des Saint-Pères, Paris, France. Traduction, introduction, notice et notes par Jean Voilquin

von Glasersfeld E (2001) The radical constructivist view of science. Found Sci 6(1): 31–43. ISSN 1572-8471. doi: 10.1023/A:1011345023932. URL https://doi.org/10.1023/A:1011345023932.

von Helmholtz H (1985) Treatise on physiological optics, vol. 3, The perceptions of vision. The Classics of ophthalmology library, division of Gryphon Editions, Ltd., P.O. Box 76108, Birmingham, Alabama 35253, USA. Translated from the third German edition, 1910, edited by James P. C. Southall, published by the Optical society of America, 1925

Watzlawick P (Ed) (1980) The invented reality: how do we know what we believe we know? Contributions to constructivism. W. W. Norton & Company, New York, 352 pp. ISBN 978-0393333473

Watzlawick P, Weakland J, Fisch R (1974) Change: principles of problem formation and problem resolution, 1st edn. W. W. Norton & Company, New York. ISBN 978-0393011043

Williams JM (1990) Style: toward clarity and grace. Chicago guides to writing, editing, and publishing. The University of Chicago Press, Chicago, xviii+208 pp., 60637, USA; London, England. ISBN 0-226-89914-4. With two chapter coauthored by Gregory G. Colomb. Published by arrangement with Scott, Foresman and Company, 1981, 1985, 1990

Zinsser W (August 11 2009) writing English as a second language. https://theamericanscholar.org/writing-english-as-a-second-language/. A talk to the incoming international students at the Columbia Graduate School of Journalism

Chapter 26
Transverse Ontology Analysis: What Coviability Means

Thérèse Libourel

Contents

26.1 Introduction

Everything written in this work legitimately raises the following question: does the coviability "paradigm" have a common meaning for all researchers? This chapter will therefore be devoted to the construction and search for this common meaning.

The challenge is considerable. In order to rise to it, we will suggest an analysis from a knowledge engineering perspective (a research area of computer science) in order to later compare the results with the diverse views of thematic technicians. Beginning with an ontological approach (we shall explain the main definitions of the term ontology below), we shall present, assuming potential bias, the main concepts set out in the various chapters of this book. The analysis undertaken will suggest a first draft of the ontology of coviability that we hope will subsequently be enriched with further reflections and reasoning based on the implementation of coviability.

T. Libourel (✉)
Université Montpellier, UMR ESPACE-DEV, Montpellier, France
e-mail: therese.libourel@umontpellier.fr

O. Barrière et al. (eds.), *Coviability of Social and Ecological Systems: Reconnecting Mankind to the Biosphere in an Era of Global Change*,
https://doi.org/10.1007/978-3-319-78497-7_26

This chapter will be structured as follows: Sect. 26.2 will be devoted to the main definitions relating to ontologies, Sect. 26.3 will be given over to the knowledge engineering approach and Sect. 26.4 to the presentation of a first draft ontology which needs to be shared and fine-tuned.

26.2 About Ontologies

The term ontology is currently widely used. It is firstly necessary to provide definitions and to specify its meaning in the proposal below.

Historically ontology (from onto, taken from the Greek ὤν, ὄντος "being", is the present participle of the verb είμί "to be") is a discipline of philosophy which, although the term is more recent (introduced by Goclenius,[1] in its Latin form in the seventeenth century), dates back more than 2300 years. It refers to a very broad domain which was namely addressed by the Greek philosophers Partemides, Heraclitus, Georgia and Aristotle.

Aristotle defined it as the science of Being as a conscious being (the first philosophy prote philosophia) (Dhondt 1961). In his treatise "Categories" (Aristote 2008), Aristotle gives a list of ten categories which constitute the different descriptions associated with the manifestations of being in the world: substance (quiddity or essence), and the nine accidents (quality, quantity, relation, activity, passivity,[2] place, time, position, possession).

Among the ten categories, substance has greater importance, it constitutes the essence, which makes an entity what it is, and which thereby individualizes and differentiates one entity in relation to all of the others, ensuring a structure that remains stable throughout the continual changes of the world (Psyché et al. 2003).

However, Ontology is usually understood more as a science of "existing items" (the set of objects recognized as existing in a field) than as a science of conscious being. In other words, ontology is more interested in what exists (living beings or things) than in the principles[3] of what exists (Being).

---

[1]Rudolph Gockel Goclenius (the younger), Lexicon philosophicum dated 1613.

[2]Aristotle's name paschein for this category has traditionally been translated into English as "affection" and "passion" (also "passivity").

[3]"The word principle is one of the most widely used words in philosophical vocabulary. It comes from the Greek term archè which designates the beginning, the starting point. For Platon and Artistotle the archè most worth searching for would be an ultimate principle on which theoretical knowledge would be based. It has two main meanings, the first being objective, ontological or metaphysical, the second subjective, logical or psychological. In the first meaning, a principle is a reality, being, object or fact, on which other realities, more or less significant in number, depend and result from. In the second sense, a principle is a truth, a general proposal, on which other truths or proposals, more or less significant in number, depend or result from and on which demonstration is based (extract from http://www.cosmosvisions.com)".

"At the beginning of the fourth book of his Metaphysics, Aristotle gives a new presentation of supreme[4] science:" There is a science that studies the being as consciously being, along with the attributes belonging to it (. . .)."

"The new definition of supreme science is firstly striking in that Aristotle conceives it within the framework of his theory of the sciences as he elaborated it in the Second Analytics."

In this theory, each science is determined by its object of study, since for the Stagirite[5] the method of investigation does not give rise to differentiation. However, all scientific knowledge is a knowledge of the cause, and namely of the cause in relation to the particular focus of research; it is in this sense that Aristotle affirms that we must look for "the first cause" or "the principal cause": "to start from what is first is to start from clear principles."

A science must prove something from the characteristics that are specific to its object of study, that is to say, it only considers the attributes of its object considered "in its own right".

To go further in our epistemological approach, the science of knowledge in agreement with "the ontological question, for epistemology, the science of knowledge is first and foremost that of the reality of theoretical entities which science speaks of. In other words, the ontological question for science is firstly the question of the referent of scientific discourse: to ask what exists is to ask what is real; and to ask what is real is to ask what we are speaking about in science "(Ricoeur 2017).

The term "ontologies" (Ontology being reserved for philosophical science), which appeared about a decade ago in the field of cognitive sciences, knowledge engineering and artificial intelligence (AI), represents the "artefacts" created by conceptual modeling and is able to play the role of conceptual reference (Chaumier 2007). Taken in its broadest sense, the term ontology is more or less synonymous with theory or the conception of reality.

It is with this in mind that we give some definitions currently in use:

- "An ontology defines the basic terms and relations comprising the vocabulary of a topic area as well as the rules for combining terms and relations to define extensions to the vocabulary" (Neches et al. 1991)
- "An ontology is an explicit specification of a conceptualization." (Gruber 1993)
- "Ontologies are defined as a formal specification of a shared conceptualization" (Borst et al. 1997)
- "An ontology may take a variety of forms, but necessarily it will include a vocabulary of terms, and some specification as to their meaning. This includes

[4]The Aristotelian expression "supreme science" is not equivalent to "first philosophy", where the word "first" marks the dignity of the object of this science (Mansion 1958). The first philosophy is identical to the study of the first substance, while supreme science is synonymous with what was later called metaphysics. The identity of the first philosophy and of supreme science is precisely one of the central problems of Aristotelian metaphysics.

[5]Aristotle was born in Stagirite in 384 BC. It is for this reason that the philosopher is nicknamed "Stagirite".

definitions and an indication of how concepts are inter-related, collectively imposing a structure on the domain and restricting the possible interpretation of terms. An ontology is virtually always the manifestation of a shared understanding of a domain that is agreed between a number of agents. Such agreement facilitates the accurate and effective communication of meaning, which in turn leads to other benefits such as inter-operability, reuse and sharing." (Uschold et al. 1998)

- "Ontologies provide a vocabulary for representing knowledge about a domain and for describing specific situations in a given domain." (Fikes and Farquhar 1999)

Fundamentally, the role of ontologies is to improve communication between humans and further still, beyond humans and computers and finally between computers. Being in a multidisciplinary context, research and writings are expressed in a specialized language in the form of a specific terminology and conceptualization. The existence of such "jargons" leads to problems of comprehension and difficulties in sharing knowledge among users.

By referring to the semiotic triangle of Odgens and Richard (1923) (Fig. 26.1), every word (sign, symbol) signifies a thing, an object or an action to which it refers (referent) and is associated with a concept, intention, mental image (which it makes us think of) which is the signified and which depends on the person with whom we are conversing (Fig. 26.2).

From this observation, we can observe various forms of possibilities regarding the implementation of the ontological approach:

- The simplest form consists in building a controlled vocabulary, that is, a list of terms agreed by a community or organization.

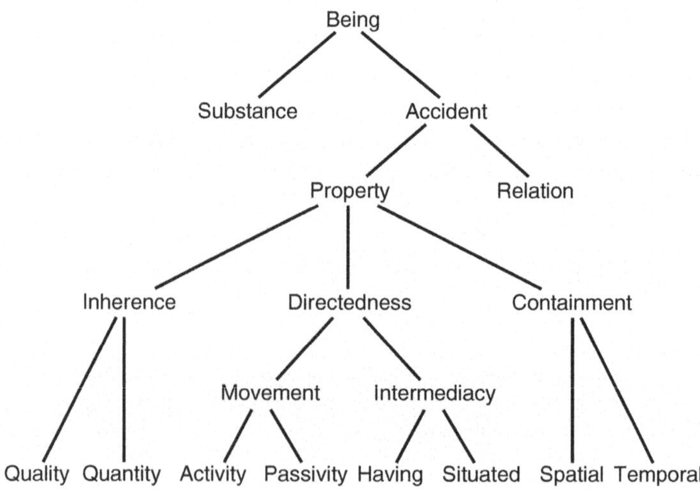

Fig. 26.1 Aristotle's categories (according to J. Sowa http://www.jfsowa.com)

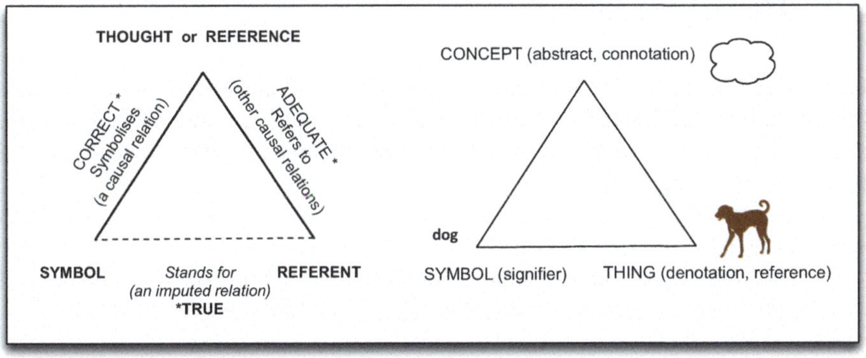

Fig. 26.2 The semiotic triangle of Odgens & Richard. (Source: Ogden and Richards 1923)

- More elaborate, a taxonomy is a controlled vocabulary organized according to a hierarchy. For example, we can have the terms "vehicle", "car" and "motorcycle". The terms "motorcycle" and "car" are specializations of the term "vehicle" because a car and a motorcycle are types of vehicles.
- A thesaurus is a taxonomy with more information about each concept signifier including alternative and preferred terms ("animals" or "animaux" in French). In addition, a thesaurus can contain relationships between terms that signify concepts: hierarchical relationships already present in the taxonomy (generalization/specialization), partitive relations (part-all relations), associative relations, instantiation relations (giving examples).
- An ontology, according to Gruber's definition: "An ontology is an explicit specification of a conceptualization [. . .]. A conceptualization is an abstract and simplified view of the world that we want to represent". It is based on two dimensions: the conceptualization of a domain, in other words, a choice as to how to describe and represent a domain; and the specification of this conceptualization, in other words, the transcription of the conceptualization into a formal and operational description.

The conceptual vision of an ontology corresponds to the reification of signifieds (in a class or as an individual entity). An individual is defined by a set of properties that can be pairs (attribute/value) or by relationships between individuals. Classes are organized according to a specialization relationship such as an "is-a" relationship. Individuals are linked to classes by the "is-instance-of" instantiation relation.

Specification, generally in the case of computing, includes a formalization stage followed by an implementation stage. Formalization calls for known languages and formalisms: description logic, semantic Web formalisms (DAML + OIL, OWL, RDF/S), conceptual graphs and so forth. The implementation is carried out in different ways (using existing or created tools) depending on the desired usage.

26.3 The Approach

Our approach is based on knowledge engineering[6] and aims to model knowledge which is either individual or collective, explicit or implicit, stable or evolutionary, expert or technical. It makes knowledge accessible in a form that depends on the context, whether operational or not.

To meet this objective and without claiming to be exhaustive, a set of texts from this book (texts composing Parts I, II, III) has been analyzed.

The analyses were elaborated using a mind map highlighting the signified derived from the signifiers presented in the various chapters of the book.

26.3.1 The Notion of Heuristic or Mind Map

As a reminder, a mind map[7] is also often called a cognitive map or a mental map. It consists in a diagram that allows the reader to visually represent his path of thought.[8] The idea is that, in a text, only a few keywords[9] called "reminder words" are necessary to understand and memorize it. We were therefore able to trace all of the maps presented in the overviews *of* Parts I, II, III of the book using Freemind[10] software.

Each heuristic map has been constructed in the following way: in the centre is the unifying term of the work on coviability. From this centre, branches radiate in all directions, bringing the main ideas of the text in the form of "headlines". In some cases, links appear to underline semantic correlations. The map is both a snapshot of the reader's feelings (with all the ambiguity of a non-specialist interpretation) and a vision of the essential terms of the statement (with specialized or correlative links connecting them).

The collection of mental maps obtained will serve as a basis to construct the ontology.

[6]This discipline is at the crossroads of several disciplines from which it borrows concepts, notably: Knowledge Engineering, Ontology (philosophy) and Linguistics.

[7]The term "heuristic" comes from ancient Greek *είρίσκω*, eurisko, meaning "I find".

[8]https://en.wikipedia.org/wiki/Mind_map

[9]The term might be more appropriate if defined as a lexical unit with a meaning given by the author.

[10]http://freemind.sourceforge.net/wiki/index.php/Main_Page

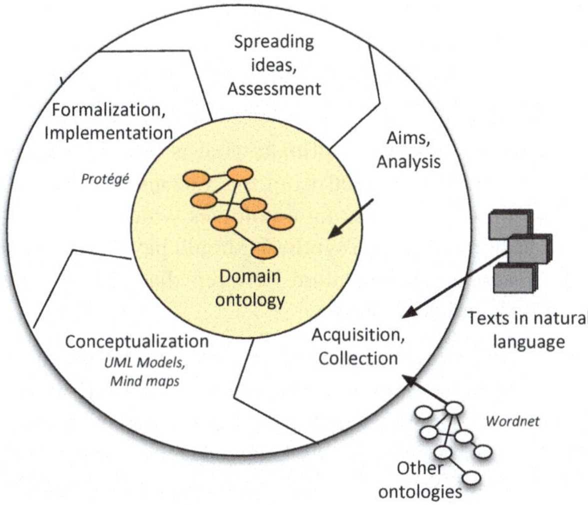

Fig. 26.3 Life cycle of an ontology. (Source: Adapted from Yang 2006)

26.3.2 Stages of Construction

The approach followed is similar to various proposals which exist in terms of software engineering or knowledge representation[11] methodologies.

In knowledge engineering in general, the different stages of the life cycle of an ontology are: analysis and acquisition, conceptualization, ontology design or formalization and implementation, publication and evaluation (Fig. 26.3).

(a) Analysis and acquisition constitute the two prior stages to construction. It is firstly necessary to understand the purpose and scope of the ontology and collect the necessary knowledge.

It is common to distinguish several types of ontology (Psyché et al. 2003) which depend on the ontology's purpose and on the scope and precision of the domain modeled:

- Top-level ontologies are universal ontologies that do not depend on a domain or a particular problem. They concern very general concepts such as the digital world, time, space, actions ... These concepts must be consensual for a very large community[12];

---

[11] A comparative and detailed study of these methods and methodologies can be found in (Fernández-López and Gómez-Pérez 2002).

[12] In this category we can mention the Cyc project http://www.cycel.com which has the objective of developing a general knowledge database permitting all forms of reasoning, the SUMO project (Standard Upper Ontology) http://www.ontologyportal.org/, the DOLCE project (Descriptive

- Core ontologies, which are less generic than the previous ones, but which are generic enough to be used across several domains;
- Domain ontologies are a set of vocabularies and concepts that describes a domain of application, based on specific knowledge of this domain.

Within the context of this book, the ultimate goal is to obtain a domain ontology based on shared knowledge extracted from the various chapters of the book and from knowledge resulting from existing ontologies which can serve as a top-level reference. These first steps towards a synthesis should pave the way for collaborative work in order to ensure communication between disciplines, going beyond the interoperability of data and applications.

(b) The *conceptualization* phase consisted in creating conceptual models to structure knowledge of the domain. For this, when reading this work, the author has developed a set of mind maps which will be presented in the overviews of Parts I, II and III. The comparison of this knowledge with that already present and analyzed within the Wordnet[13] Linguistic Ontology has permitted conceptual models regrouping significant concepts and relationships in the domain targeted to be developed. These resulting models are presented in the form of Unified Modeling Language (UML) diagrams.

(c) *The ontology creation stage.* The formalization and implementation of the knowledge base is carried out using the Protégé Editor software (protege.stanford.edu). Its representation will be achieved on the basis of the previous models and using a formal ontological representation language which will later be exploitable by machines.

(d) The *publication* and *evaluation* stage has not yet been carried out (except through some one-off exchanges). This stage is obviously essential since the construction process must remain iterative.

26.4 Results: Towards a First Ontology

26.4.1 Conceptualization

We firstly focused on grouping together the associated terms and concepts common to the majority of concepts in the work whilst positioning them in relation to the

Ontology for Linguistic and Cognitive Engineering) http://www/loa.istc.cnr.it/old/DOLCE.html and the "Basic Formal Ontology" project (BFO) http://iformis.uni-saarland.de/bfo/

[13]Wordnet is a linguistic ontology developed under the direction of George (A. Miller 1995). Nouns, verbs, adjectives and adverbs are grouped into sets of cognitive synonyms (synsets), each expressing a distinct concept. Synsets are interlinked by means of conceptual semantic and lexical relations. The resulting network of meaningfully related words and concepts can be navigated with the browser. WordNet is also freely and publicly available for download. WordNet's structure makes it a useful tool for computational linguistics and natural language processing.

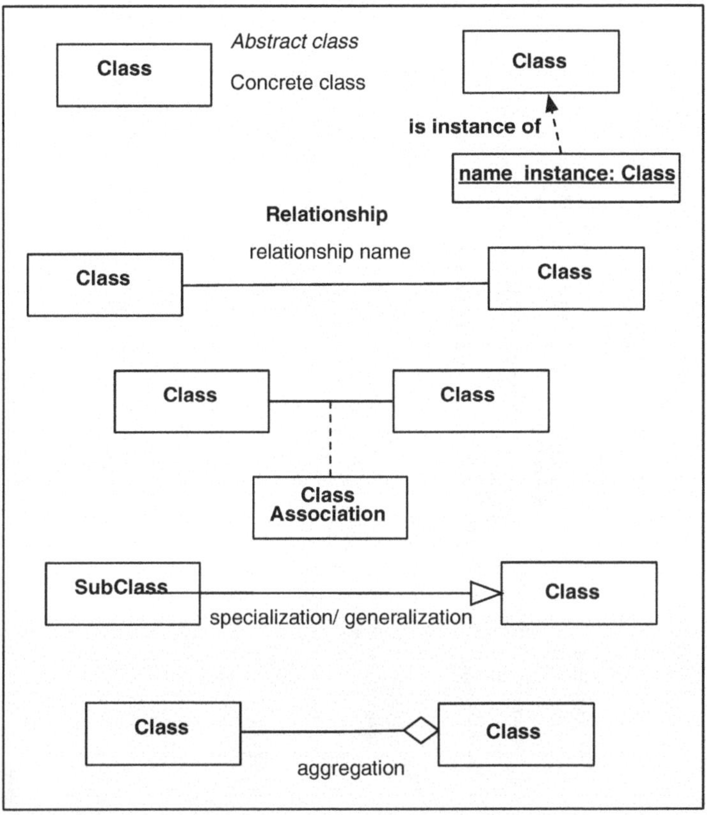

Fig. 26.4 Legend for UML symbols

WordNet reference basis within several conceptual models represented in the form of UML diagrams. The legend of these diagrams is summarized below, the classes represent what is signifieds (concepts), and various relations (simple relationships, specialization/generalization, part-of/aggregation) are used to connect concepts. Individual instances are sometimes given as examples of concepts (Fig. 26.4).

26.4.1.1 Fundamentals: The Core Model

The model (Fig. 26.5) represents what is referred to as core knowledge. The hierarchy of terminological specialization was conceived based on the classification

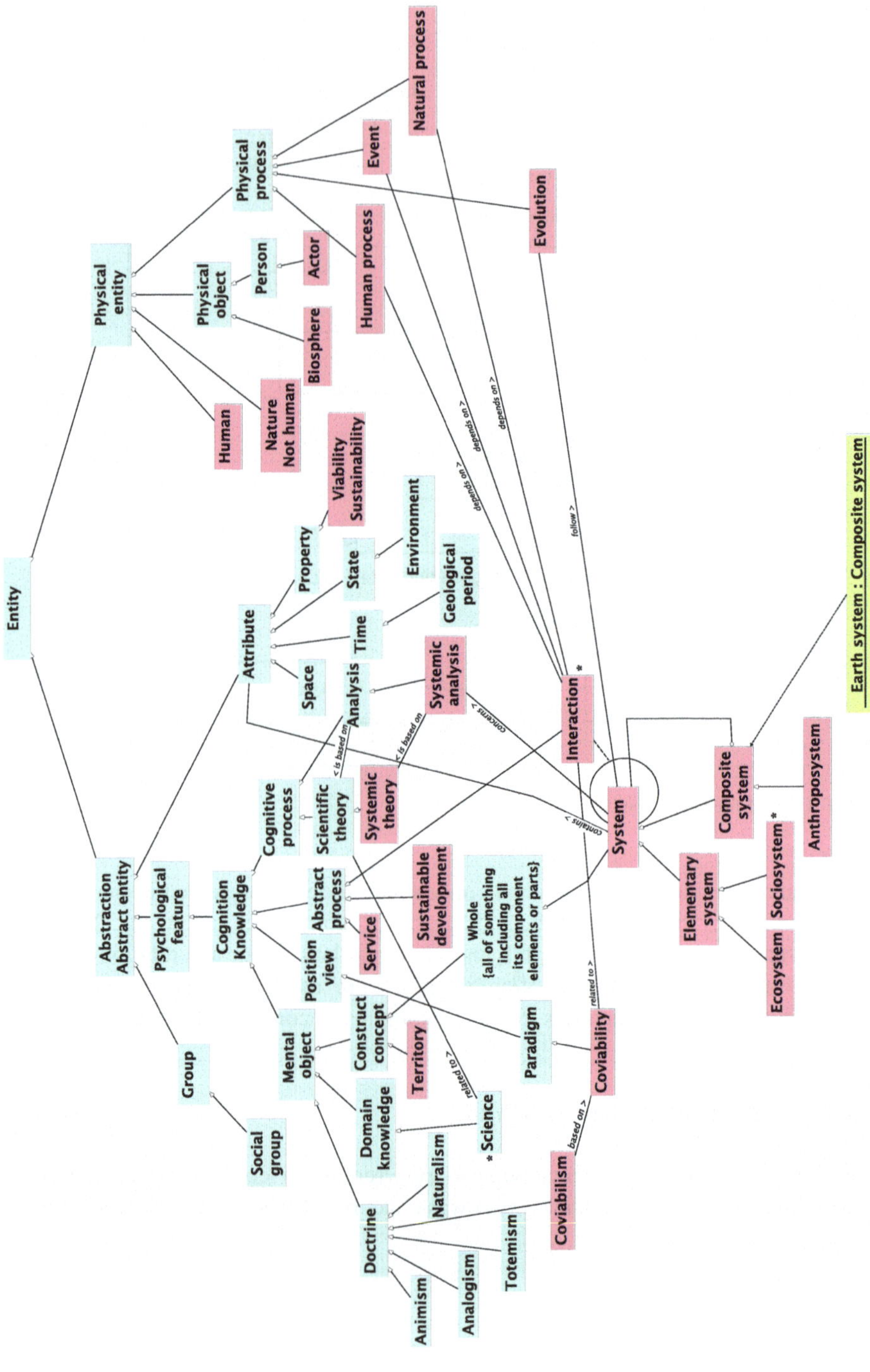

Fig. 26.5 The core conceptual model

proposed by WordNet: the main root[14] (also called Top) representing the concept of entity is available as an abstract entity and as a physical entity.

Abstract entities or abstractions are firstly specialized according to psychological features or characteristics. These are defined as knowledge or cognition (result of perception, learning, reasoning) that is subdivided into mental objects, positions or views (or posture), abstract process and cognitive process.

Mental objects are specialized according to doctrines (thinking modes), domain knowledge and constructed concepts.

Doctrines are subdivided into doctrines accepted by certain groups, such as animism, totemism, analogism, naturalism introduced by Descola (2005), and to which we propose to add coviabilism (see Chap. 6 Barrière & Behnassi).

Domain knowledge subsumes[15] all science. Among the specializations of the conceptis developed we can note the notion of "whole" which leads to the notion of system, which is one of the elements present as a common theme in this work along with the concept of territory (Elissalde 2002).

The concept of position or view relates to the way of interpreting things and expressing them in the form of paradigms linked to a way of thinking. The coviability paradigm is one of these specializations and is directly related to coviabilism.

Among the abstract processes, we can note the notion of interaction that occurs between different systems, then the notion of service, which is mentally associated with biological or societal functions, and finally, the notion of sustainable development, a development process that can reduce both social inequalities and man's impact on the environment.

As for cognitive processes, they give rise to scientific theories or more simply, to various analyses, including systemic analysis.

Finally, abstractions are firstly specialized according to groups or collectives refined into social groups and secondly, into attributes which are broken down into space, time, state and property.

The concept of system (Le Moigne 1994) makes it possible to trace the contours of various possibilities, in effect, it can be subdivided into elementary systems (which in our context can be socio and elementary ecosystems) or into composite systems. The concept of anthroposystem[16] (Lévêque et al. 2003) is a composite system. The aggregation relationship between the composite system and the system allows for an "embedded" vision of the composite system (similar to "Russian dolls").

---

[14]When represented by a computer, a hierarchy is either a tree or a lattice structure. The complete hierarchical structure is displayed from the root (or Top Summit) to the base.

[15]Subsumption designates a relationship of inclusion among concepts, events. Human and natural processes are included in physical processes.

[16]The anthroposystem, "can be defined as a structural and functional entity taking into account the interactions between societies and environments and integrating in the same space one or more natural subsystems and one or more social subsystems, all of them co-evolving over the long term", (Levêque et al. 2003).

Every system is connected to various attributes which will enable us (for some of them) to give them spatiotemporal landmarks as well as certain properties, in particular viability. The system definition is completed by the reflexive link that expresses the fact that a system interacts with others, allowing the reification of the interaction. Any interaction seen as an abstract process is related to human processes, natural processes or events. Every system follows an evolution, linked to its own dynamics.

The Earth system is an instance of a composite system. It is therefore seen as a whole and is a sophisticated aggregate of social and natural subsystems co-evolving in space and time.

Physical entities (those that have a physical existence) in our context are subdivided into categories (specialization); human, non-human, physical objects (such as the biosphere, people and social players) and physical processes, which subsume events (punctual phenomena), human processes and natural processes.

In the "core" model, light gray elements represent the elements present in the generic WorNet ontology, to which we have attached dark gray elements which represent fundamental elements related to the coviability paradigm. The elements marked with an asterisk (*) will be refined in the complementary models.

The new "coviabilism" ontology analyzes the evolution of the Earth system[17] via the interaction of its various constituting systems and the various regulations that will enable its viability. It highlights the paradigm of coviability between human and non-human elements reconnected within the biosphere.

26.4.1.2 Enriching the Science Concept

Given the multi-disciplinarity of this book, perfecting the terminology of the various scientific domains evoked has led us to produce the conceptual model in Fig. 26.6. This completes and refines[18] the core model since it stipulates all of the signifieds relating to the science concept using knowledge relating to various disciplinary fields that have put forward considerations relating to the coviability paradigm.

The science element (reproduced in relation to its position in the core concept model) is specialized according to broad categories (not exhaustive but corresponding to those mentioned in the book): mathematics, natural science, engineering, human and social sciences. Moreover, it is linked to the concept of scientific theory.

Each of these categories can be extended (thoroughly or otherwise) into a hierarchy of specializations:

Mathematics can be subdivided into applied mathematics and then statistics; Natural science into life sciences, physical sciences, geology; Engineering into computer science; Human and social sciences into philosophy, anthropology, geography, political science, economics and law.

[17] The Earth system is an instance of the Composite System.

[18] The term "refined" is used in the sense of progressively furthering knowledge.

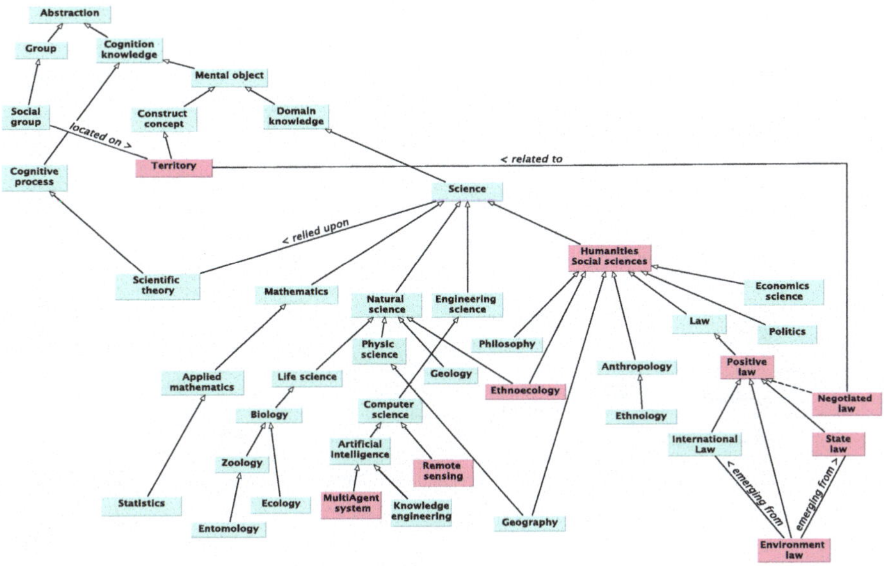

Fig. 26.6 The conceptual model: science (the elements in light gray are from WordNet and those in dark gray are those which are specific to the domain analyzed)

It should namely be noted that legal science can be subdivided into positive law, which includes international law, state law and environmental law which emerges from the two previous ones. Negotiated law carries an "environmental jurisdiction" capable of responding to environmental challenges. It is linked to the concept of territory, a specialization of the notion of constructed concept, which is itself linked to the social group.

26.4.1.3 Enriching the Sociosystem Concept

The conceptual model in Fig. 26.7 completes and refines the core model by specifying all of the signifieds relating to the notion of Sociosystem. As before, the new or reviewed elements in our context (in dark gray) are placed within the hierarchy of the core conceptual model (in light gray).

Every sociosystem is located in an anthropogenic area[19] linked to a built up territory that is linked to a culture that is itself a specialization of the knowledge present within the social groups situated within the terrority.

It is managed by a governance (specialization of abstraction) which is based on institutions that are structures and which include institutional players. This governance is both created by the legal system and a producer of this legal system.

[19]In an anthropogenic zone, several sociosystems and consequently several cultures can be found.

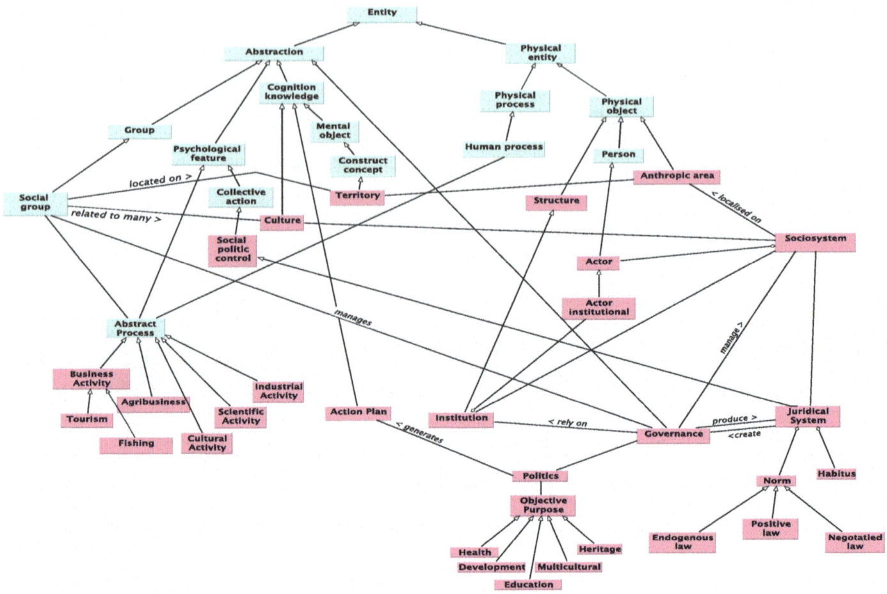

Fig. 26.7 The conceptual model: sociosystem

The legal system is an aggregate of habitus and standards which compose endogenous law, positive law and negotiated law. It permits the control and regulation of society. Policy put in place corresponds to objectives related to health, development, heritage, education, multi-culturalism ... that give rise to various action plans. The social group is linked to abstract processes (defined as functional characteristics): commercial, scientific, cultural, industrial, agricultural activities all implemented by players carrying out human processes.

26.4.1.4 Some Examples

Numerous works illustrate the general concepts enumerated above. Let us take a few examples, by introducing, in the conceptual models, instances or individuals explaining the situation described (Figs. 26.8 and 26.9).

In the first example (inspired by Chap. 12), any composite eco-socio-system which is an aggregate of a coral ecosystem and a sociosystem is subjected to a complex interaction due to various processes, the direct effects of human processes (waste water discharge, etc.), process-related events (hurricanes) or indirectly generated from interactions between human and natural processes (climate change, acidification of oceans, eutrophication). This interaction affects the state and evolution of the component systems. Via legal acts, the social system can ensure that a regulation, as a specific interaction, is established between these evolutions and that the composite system remains viable.

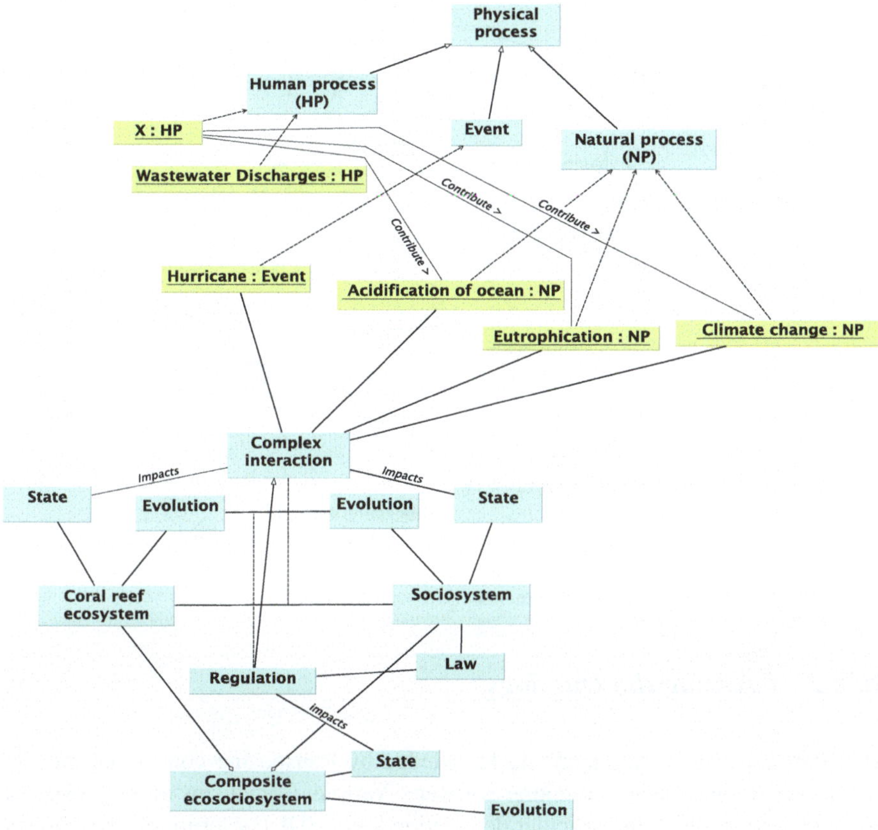

Fig. 26.8 The conceptual model: coral reefs (under the constraint of climate change)

Another example built around the governance of its legal system and its institutions (inspired by Chaps. 6 and 21). The Earth system is a composite system composed of various, interlocked sub-systems (namely socio-systems which in turn, are linked to institutions, governance, laws). On a global level, we can give examples such as the United Nations and its High Commissioner that approves international conventions such as the Universal Declaration of Human Rights (1948) or the World Charter for Nature (1982). We can give another example at a European level with the European Union and the European Commission adopting the European directive establishing a framework for community action in the water policy field (n°2000/60/CE dated 23/10/00); at a national level, each state has its own institutions and sets up specific laws. This global vision underlines the fact that local visions should be in harmony or comply with wider visions in order to put in place effective regulations.

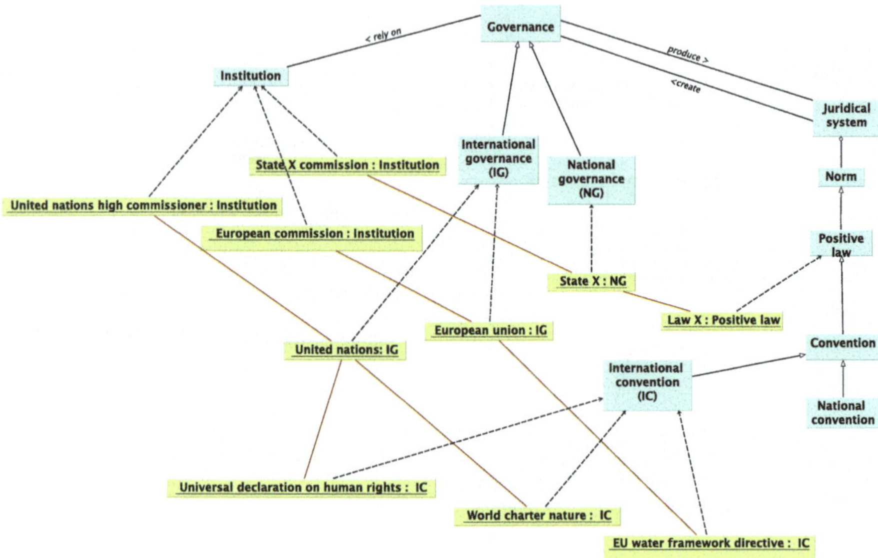

Fig. 26.9 The conceptual model: governance (European example)

26.4.2 Creating the Ontology

This step consists in systematizing knowledge by transposing conceptual models into formal models using computer software. Various formalisms already exist as previously mentioned, in particular description logics (DL) or terminological logics (Baader et al. 2010) and the formal standard languages of the World Wide Web Consortium (W3C) such as RDF and OWL based on the theoretical foundations of DL.

The name "Description Logic" associates the notion of description with that of logic: it consists in reasoning on the basis of a described knowledge.

To represent the knowledge of a domain, within what we refer to as a knowledge base, the DL needs to define elements of general knowledge, individuals, as well as the relationship between these individuals and the elements of general knowledge.

Without going into too much detail, we can retain that a knowledge base expressed in terms of DL is composed of two parts named TBox (for Terminological Box) and ABox (for Assertion Box):

- The TBox terminological level describes the general knowledge elements of a domain, in other words, the concepts and the roles. Concepts and roles have structured definitions, elaborated from numerous constructors. A semantic is associated with every concept and role description via an interpretation. A concept corresponds to a generic entity and represents a set of individuals (detailed entity), whilst a role represents a binary relationship between individuals. The subsumption relation organizes concepts and roles by level of generality:

intuitively, a concept C subsumes a concept D if C is more general that D in the sense that the set of individuals represented by C contains the set of individuals represented by D (Napoli 1997).
- The ABox factual level describes individuals by specifying contingent assertions concerning these named individuals.

Within this context, possible reasoning (in accordance with the semantics of concepts and roles) is based on the inference of individual properties, the inference of relations between individuals, the subsumption of concepts and the consistency of the knowledge base.

The OWL standard (https://www.w3.org/2001/sw/wiki/OWL) implements various description logics and allows the description of ontologies through the usage of software tools. Our choice for a first trial was "Protégé" software (http://protege.stanford.edu/).

"Protégé" is a tool, a platform and an ontological library, developed by the Stanford Center for Biomedical Informatics Research at the Standford University School of Medicine. It is free and its source code is published under a free license (the Mozilla Public License).

This tool is first and foremost an editor that allows the construction of a knowledge database constitutive of an ontology. To do this, the software environment offers several tabs: Classes or Entities dedicated to the creation of concepts, Object properties dedicated to the creation of roles, Data Properties dedicated to the creation of the characteristics particular to a class, etc (Fig. 26.10).

The above screen shot displays a part of the hierarchy of concepts and the formal representation of the individual "Earth system" which is an instance of the Composite system, as well as a part of the hierarchy of roles (ObjectProperty) displaying a formal representation of the "follow" role which enables individuals from the System and Elementary System concepts to be linked to individuals from the Evolution concept.

26.5 Conclusion

The proposal presented in this chapter is only an initial reflection inspired by the various texts in this work.

The construction of an ontology is by nature iterative, the interaction with all of the contributors to this book and further still, with the scientific community is

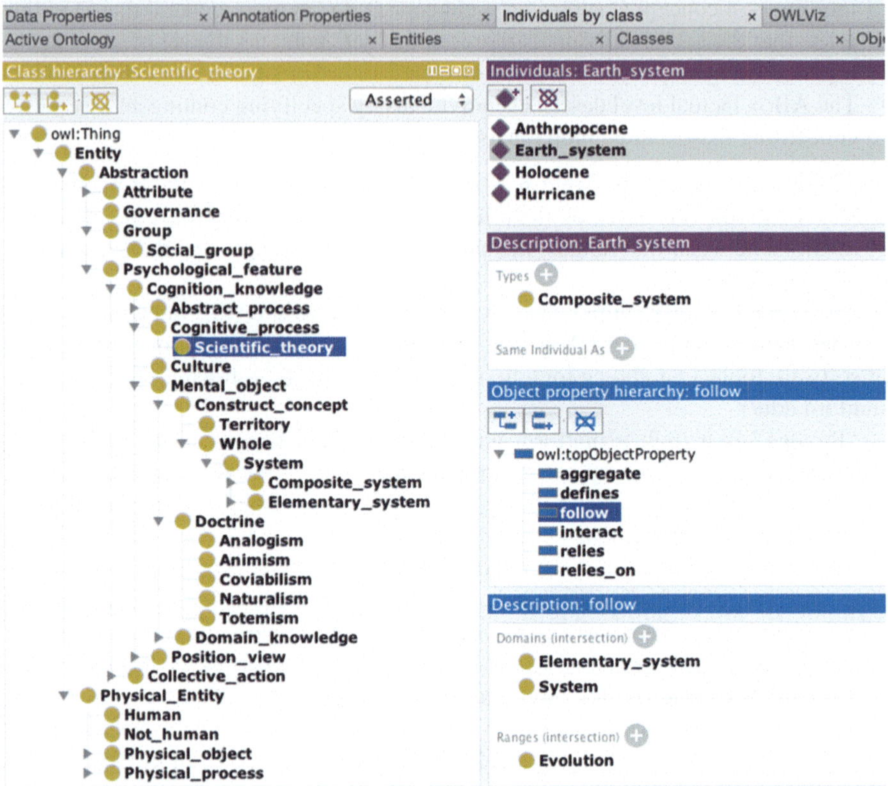

Fig. 26.10 Protégé: screen snapshot

essential[20] and should enable the validation of this prototype as well as the complete definition of the underlying terminology.

On a conceptual level, the first choice, induced by the existence of Wordnet, produced the definition and initial subdivision of entities into abstract entities and physical entities.

This choice may of course be disputed. Other aspects could have been chosen with reference to the "Basic Formal Ontology" (BFO) top level ontology for example (Arp et al. 2015). BFO classifies entities into "continuants" (also called "endurants") whose existence is continuous and which keep their identity throughout their existence. BFO also classifies entities into "occurents" (or "perdurants") referring to events, processes, changes and activities.

[20]The concepts of territory, culture, governance, jurisdiction, as the wealth of texts in this book demonstrates, are namely subject to varying levels of acceptance.

Concerning the approach, besides the fact that it has not yet been able to be very collective it could become more sophisticated if it were to base itself on an automatic analysis of the texts composing the work. This would help to pinpoint essential information.

In these conditions, a more substantial corpus could also be used. However, beyond these "categorizations", interpretation will always be subject to the creativity and the profile of analysts.

The initial objective pursued was to take account of all of the knowledge that it was possible to explain in order to better understand the complexity of the coviability paradigm and the associated ontology known as "coviability" (see Chap. 6 Barrière & Bhenassi).

A first step has consequently been taken. A follow-up is necessary. The collaborative dimension and the interdisciplinary exchanges are fundamental and constitute one of the priorities that we are setting ourselves in order to clearly explain the "coviabilism" ontology.

References

Aristote (2008) Catégories. De l'interprétation (Organon I et II), éd. Vrin, coll. "Bibliothèque des textes philosophiques", 2008, présentation et traduction de J. Tricot, pp 7–9

Arp R, Smith B, Spear AD (2015) Building ontologies with basic formal ontology. Mit Press, Cambridge

Baader F, Calvanese D, McGuinness DL, Nardi D, Patel-Schneider PF (2010) The description logic handbook: theory, implementation, and applications. Cambridge University Press, Cambridge, MA

Borst P, Akkermans H, Top J (1997) Engineering ontologies. Int J Hum Comput Stud 46(2–3):365–406

Chaumier J (2007) Les ontologies. Antécédents, aspects techniques et limites. Documentaliste-Sciences de l'Information 1/2007 44:81–83. URL: www.cairn.info/revue-documentaliste-sciences-de-l-information-2007-1-page-81.htm. https://doi.org/10.3917/docsi.441.0081

Descola P (2005) Par-delà nature et culture. Gallimard, Collection Bibliothèque des Sciences humaines

Dhondt U (1961) Science suprême et ontologie chez Aristote. In: Revue Philosophique de Louvain. Troisième série, Tome 59, N°61, pp 5–30

Elissalde B (2002) Une géographie des territoires. L'Information géographique 65:193–205

Fernández-López M, Gómez-Pérez A (2002) Overview and analysis of methodologies for building ontologies. Knowl Eng Rev 17(2):129–156

Fikes R, Farquhar A (1999) Distributed repositories of highly expressive reusable ontologies. IEEE Intell Syst 14(2):73–79

Gruber TR (1993) A translation approach to portable ontology specifications. Knowledge Acquis 5(2):199–220

Le Moigne, JL (1994) *La théorie du système général: théorie de la modélisation*. Collection Les classiques du réseau intelligence de la complexité. www.mcxapc.org

Lévêque C, Muxart T, Abbadie L, Weill A, Van der Leeuw S (2003) L'anthroposystème: entité structurelle et fonctionnelle des interactions sociétés-milieux. In Levêque Ch, van der Leeuws (éds) Quelles natures voulons-nous?. Elsevier, Paris, 110–129

Mansion A (1958) Philosophie première, philosophie seconde et métaphysique chez Aristote. Revue Philosophique de Louvain 56(50):165–221

Miller GA (1995) WordNet: a lexical database for English. Commun ACM *38*(11):39–41
Napoli A (1997) *Une introduction aux logiques de descriptions* (Doctoral dissertation, INRIA)
Neches R, Fikes RE, Finin T, Gruber T, Patil R, Senator T, Swartout WR (1991) Enabling technology for knowledge sharing. AI magazine *12*(3):36
Ogden CK, Richards IA (1923) The meaning of meaning. A Study of the Influence of Language upon Thought and of the Science of Symbolism (London 1923,% 1949), 11
Psyché V, Mendes O, Bourdeau J (2003) Apport de l'ingénierie ontologique aux environnements de formation à distance. *Sciences et Technologies de l'Information et de la Communication pour l'Éducation et la Formation*, *10*, 89–126
Ricoeur P (2017) "Ontologie", Encyclopædia Universalis (en ligne). http://www.universalis.fr/encyclopedie/ontologie/
Uschold M, King M, Moralee S, Zorgios Y (1998) The enterprise ontology. Knowl Eng Rev *13*(1):31–89
Yang K (2006) A conceptual framework for semantic web-based ecommerce, mémoire de maîtrise. Université Laval, Québec

Chapter 27
Coviability as a Scientific Paradigm for an Ecological Transition, from an Overview to a Definition

Olivier Barrière, Thérèse Libourel, Maud Loireau, Voyner Ravena-Cañete, Catherine Prost, Gilbert David, Serge Morand, Laurence Pascal, and Vincent Douzal

Contents

Humanity lives on premature loan year after year: it consumes in a few months all the resources that Earth can produce in a year. Ecosystems are increasingly overexploited, compromising their ability to regenerate. The calculation of humanity's ecological footprint has led to the determination of an "Earth Overshoot Day." In 2017, this day was August 2; it defines humans' consumption, which exceeds by 70% the available resources. Human beings thus incur a debt "because we cut down trees at a rate higher than their growth, we collect more fish in the seas than are born each year, and we release more carbon into the

O. Barrière (✉) · M. Loireau · G. David
IRD - French National Research Institute for Sustainable Development, UMR ESPACE-DEV, Montpellier, France
e-mail: olivier.barriere@ird.fr; maud.loireau@ird.fr; gilbert.david@ird.fr

T. Libourel
Université Montpellier, UMR ESPACE-DEV, Montpellier, France
e-mail: therese.libourel@umontpellier.fr

V. Ravena-Cañete
The Federal University of Pará, Belém, Brazil
e-mail: ravenacanete@gmail.com

O. Barrière et al. (eds.), *Coviability of Social and Ecological Systems: Reconnecting Mankind to the Biosphere in an Era of Global Change*,
https://doi.org/10.1007/978-3-319-78497-7_27

atmosphere than forests and oceans can absorb (. . .) we are gradually moving from scarcity to a shortage of resources" (http://www.overshootday.org).

Coviability of Social and Ecological Systems: Reconnecting Mankind to the Biosphere in an Era of Global Change is divided into two volumes. The first volume is entitled "The foundation of a new paradigm," and the second is entitled "Coviability questioned by a diversity of situations." This book defines the new concept of coviability in terms of interdisciplinarity in order to build a new paradigm[1] of scientific origin that could be an imperative referential to political and legal spheres.

The works presented in this chapter cannot as such provide a summary of the two volumes of the book. This chapter attempts rather to contextualize and define an emerging paradigm through a heterogeneity of research teams whose themes, approaches, protocols and analyses differ.

Given the global ecological challenges, should public policies focus on the issue of development and thus limit the human relationship to its environment in terms of "services" rendered (see Everard 2017)? Already in June 1995, Jacques Weber suggested rewording sustainable development to "sustainable development in the long term." For him, this rewording means "searching a lasting sustainability of ecosystems and lifestyles of which they are the pillars". Thus, introducing viability in development automatically rejects the analysis of resource-based dynamics of

[1]A paradigm is a way of seeing things, a representation of the world. It represents a whole set of beliefs, recognized values and techniques that are common to members of a given group. In social sciences, this term is used to "designate the general (explicit and implicit) theoretical structures or the ways of thinking in which research, surveys or analyzes of social phenomena take place". Thus, a paradigm is "a coherent set of hypotheses which constitutes a whole and which offers the scientist a point of view on the phenomena he/she is studying, a matrix which conditions his/her vision, a coherent representation of the world which shapes his/her way of thinking about phenomena." The concept of paradigm thus refers, at the same time, to a cognitive aspect, its content (ideas, theories, knowledge,... ontology) and to a social supporting aspect (the scientific community). (Olivier Martin, "Paradigm", Sociology [Online], The 100 Words of Sociology, posted on November 1, 2013, accessed on September 25, 2017. URL: http://sociologie.revues.org/1997 and Gilles Willett, "Paradigm, theory, model, schema: what is it?", Communication and Organization [Online], 10 | 1996, posted on March 26, 2012, accessed on September 24, 2017. URL: http://communicationorganisation.revues.org/1873; DOI: 10.4000/communicationorganisation.1873).

C. Prost
The Federal University of Bahia, Salvador, Bahia, Brazil
e-mail: cathprost@yahoo.com

S. Morand
CNRS-ISEM, University of Montpellier, Montpellier, France
e-mail: serge.morand@umontpellier.fr

L. Pascal
University of Montpellier, Montpellier, France
e-mail: laurence.pascal@umontpellier.fr

V. Douzal
IRSTEA, National Research Institute of Science and Technology for Environment and Agriculture, Montpellier, France
e-mail: vincent.douzal@teledetection.fr

inventory management (Weber 2013). Viability as a property, i.e. a quality or attribute characterizing a state, does not preclude development as a process, i.e., an ordered sequence of facts, operations, or actions. The new paradigm of socio-ecological coviability therefore does not clash with the objective of sustainable development. This paradigm rather proposes a different posture, presented in this chapter.

The combination of climate change, land and biodiversity degradation, with global population growth (from 7.6 billion in 2017 to 8.6 billion in 2030 to 9.8 billion in 2050 and 11.2 billion in 2100)[2] is leading to major changes in ecosystems and resources, in societies, and their relationship to the ecosystems and resources that mark the beginning of the Anthropocene era. The rapidity of these changes, which receive little to no supervision, calls for a view that transcends utilitarian and financial analysis. It is necessary to think of the relationship between the human and the nonhuman as that of humanity to the biosphere. It is also important to consider this relationship both in terms of the relationship uniting the human to a nonhuman "environment" and the relationship between humans.

The interactions between human and nonhuman species, between these species and the "environment" within which they live, form a system and give rise to the notion of "ecosystem." Humans are, therefore, "by nature" part of ecosystems. Ecological science, the science of interactions, completely integrates them into the biocenosis. The opportunity of the separation between the (social) human system and the ecological system is the first question we asked ourselves right at the beginning, in this book's introduction.

In an anthropocentric rationale, the paradigm of naturalist ontology[3] (defined in the first chapters of the book) leads to materializing the nonhuman through the notion of capital. The latter acknowledges only the human intrinsic value (Maris 2016, 19), a value of productivity and consumption. The human is thus no longer part of the ecosystem, the biotic community (holistic vision of Aldo Leopold, Larrère 1997). Nature becomes "capitalized" (see the general introduction of this book) and enters into a national account (Weber 1986; Comolet 1994; Richard 2012). This commodification of nature is explained: it results from a "fantasy of rational domination of the world" (Madelin 2017, 13; see Douzal, Chap. 25 in this book (Vol1)).

The step has been taken to transform ecological functions into services rendered to human societies: *The services of ecological systems and the natural capital stocks that produce them are critical to the functioning of the Earth's life-support system. They contribute to human welfare, both directly and indirectly, and therefore represent part of the total economic value of the planet* » (Costanza et al. 1997).

---

[2]The "World Population Prospects: The 2017 Review" report, published by the United Nations Department of Economic and Social Affairs (DESA), 2017.

[3]On the notion of ontology, see Chap. 26 in this book: (Thérèse Libourel: "Transverse ontology analysis") : "*Ontology is usually understood more as a science of "existing items" (the set of objects recognized as existing in a field) than as a science of conscious being. In other words, ontology is more interested in what exists (living beings or things) than in the principles of what exists (Being)*".

The ecosystem, where humans are excluded, becomes a pedestal of benefits to humans (MEA 2005). This avowed paradigm of "nature at the service of man" underlines a teleological view of the world widely discussed by many authors, including Norgaard (2010), Maris (2014), Silvertown (2015) and Madelin (2017). Their strong criticism shows that a paradigm supported by a dominant neo-liberal ideology confers a monetarized value on ecological functions,[4] but often without a market. Moreover, the benefits of ecological functions are only "*the contingency of the instrumental value of a function*," which is extremely variable (Maris 2014, 33). Finally, without developing further, the criticism underlines the overly restrictive idea of nature seen through a stock-flow, which is managed from an economic angle[5] without any real "major institutional change" (a change in the manner of consuming): "*The ecosystem services approach can be a part of a larger solution, but its dominance in our characterization of our situation and the solution is blinding us to the ecological, economic, and political complexities of the challenges we actually face*" (Norgaard 2010, 1226).[6]

This very "naturalistic" representation of the human position within the biosphere is generally adopted because it is easier to implement by the international governing community (MEA 2005). However, this representation cannot be considered universal, even if it is adopted by the international community (MEA 2005), for there is a diversity of societies and socio-cultural ontologies that are linked to a plurality of paradigms of the relationship of humans with the world, and relationships between humans.

This diversity comprises two known examples on the international scene: well-being, radical criticism of all forms of development by Andean Indians; and Ubuntu, the secular philosophy of Bantu-speaking African peoples. These two ontologies share similar ethics in community care. In this context, community is understood as a group of humans and non-humans, be they alive or not. Although the concept of living well is still under construction, Gudynas (2016) emphasizes that the notion of well-being may be reached only within a framework of deep relationships within a broad community, a perspective in which nature has rights. This implies a frank

[4]"(...) markets degrade some goods and practices by turning them into commodities. For example, the possibility that nature has an intrinsic, existential value of its own that is independent of its use to humans cannot be accommodated by the market because nature itself is not an actor in that market. Nature is devalued by monetization. All non-commercial notions are invisible to 'the one-eyed imperatives' of capital" (Silvertown 2015, 643, referring to Robertson 2006).

[5]For whom the intrinsic value of an environment only concerns natural capital and extrinsic value through its services (generally assimilated to flows generated by capital).

[6]The author distances themselves from ecosystemic services, refusing to focalize on this paradigm: "ecological economists need to resist using current dominate ways of thinking to reach short-run, partial solutions and favor both emerging and the multiplicity of less dominant ways of analyzing problems to promote a rich understanding of the complexities of society and nature. (...) While we find a stress on transaction costs and institutions in ecological economic analyses (...), this approach needs to be more integral to ecological economics and this framing needs to be spread more broadly. (...) Environmental governance can no more succeed around the metaphor of ecosystem services apart from the richness of ecological thinking than mortgage markets can succeed on the myth that housing prices will always rise. (...) as ecological economists we still too rarely argue that turning down the economic drivers and/or decoupling economic activity from environmental consequences should be the first steps toward a solution" (Norgaard 2010).

opposition to the systematic and blind increase in the consumption of material goods regardless of its integration into a balanced whole, as often prioritized in a so-called development process. For its part, the ethics defended by the Ubuntu philosophy concentrate on the care that humans must give to themselves and to others, be they human or nonhuman. Ramose (2016) points out that the emphasis put on connecting with others and prioritizing the preservation of life at the expense of economic wealth converges with the notion of "commons", a crucial element for preserving social practices of many of the world's traditional peoples.

These socio-cultural ontologies could contribute to analyzing and searching solutions to the ecological problems. They could also indicate ways to reach coviability. Thus, the mechanical design of nature conflicts with that of a living Earth "*where the human being is no longer a self-sufficient creature, overlooking a planet-object, but a being in a socio-ecological relationship with the community that constitutes and exceeds it*" (Madelin 2017, 16).

It is urgent to seek renewal. If "*Western societies have based their development on subjugating nature and are currently propagating this model on a planetary scale through a process of cultural, economic and financial globalization*" (Maris 2014, 63), the challenge of climate change emphasizes the urgency to abandon both this naturalistic enslavement and this cultural homogenization. It is time to question the very future of humanity, with common concerns notably on biodiversity (1992 Convention of Biological Diversity), desertification (1994 Convention to Combat Desertification) and the climate (2015 Paris Agreement) with "the urgent threat of climate change." Ecological urgency is becoming an imperative, expressed on an international scale: "the importance of ensuring the integrity of all ecosystems, including oceans, and the protection of biodiversity, recognized by some cultures as Mother Earth, and noting the importance for some of the concept of "climate justice", when taking action to address climate change" (2015 Paris Agreement).

Building on the Millennium Development Goals, the Actions Agenda drafted by the UN, on September 2015; "Transforming our world: the 2030 Agenda for Sustainable Development" (UN General Assembly Resolution, 25 September 2015 – A/70/L1) has set goals called "Sustainable Development Goals." These goals target the viability of both humans and the planet, which is implicitly integrated in the viability of nature ("we are responsible for the future of humanity and the planet").[7] The notion of biosphere is not referred to, preferring that of "planet," as it associates humanity to the biosphere. Already in the Preamble, the positioning of humanity is read, grouping the objectives in terms of five challenges: People, Planet,[8] Prosperity, Peace, Partnership (UN, A/69/L.85).

Resulting from a long process of international discussions between the political, scientific and citizen spheres, the seventeen Sustainable Development Goals are

---

[7] An already existing certainty which is developed by the eco-development movement (Sachs 1980).

[8] *We are determined to protect the planet from degradation, including through sustainable consumption and production, sustainably managing its natural resources and taking urgent action on climate change, so that it can support the needs of the present and future generations (UN, A/69/L.85).*

already a political framework to which countries refer. The objective is to encourage countries to negotiate with each other and develop their strategies and actions on the global, national, and local levels. Beyond the desire of bringing together all the concerns about the environment and societies discussed in particular conventions, this politically-oriented framework seeks to integrate them into a global vision to "change the world" through action. Dividing them into targets is a first step towards a collective and coherent action, with reference to an integrated and global system: that of planet Earth.

The "protection of the planet" expressed in the SDGs is a common project aiming to create commitment from all "peoples of the United Nations." Its objective is to link the future of the humans living in the present and the future generations to the 17 Goals, declared as inseparable and interconnected. From the disappearance of poverty to the protection of the planet, the SDGs are presented as a "charter for humanity and for the planet in the twenty-first century," which aims to "change the world" through action.

It is possible to extract from the SDGs categories of general objectives and to arrange them preferably in the "Sustainability Objectives" category, or in the "Growth Objectives" category. This shared vision between "sustainability" and "growth" reveals confusion, even an antagonism: would not growth objectives be the means proposed to achieve the sustainability objectives, and not the objectives themselves? In this case, are there no other means other than growth (or development) given the limits of the planet (Latouche 2012)? This question should be asked in international discussions, especially in order to take into consideration local realities.

The paradigm of socio-ecological coviability[9] is implicitly present in the "transformation of the world" program, which serves as a societal project based on values considered as universal. The relation between humanity and nature is effected by the notion of harmony, clarified in the Preamble and reiterated in the enunciation of the "Project." However, it encompasses an ambiguity: that of advocating "sustained" economic growth (strong) by the "sustainable" use (restricted) of natural resources (UNO, A/69/L.85 §9).

The prospects for the environment by the Organisation for Economic Co-operation and Development (OECD) (Organisation de Coopération et de Développement Economique (OCDE)) to 2050 (OECD 2012) emphasize the urgency, at the global level, to modify humanity's development mode: an "urgent need to change the course of our future developments" (OECD 2012). This world organization calls for immediate global action in order to avoid the costs and consequences of inaction, from both the human and economic viewpoints; as the "Continued degradation and erosion of natural environmental capital are expected to 2050, with the risk of irreversible changes that could endanger two centuries of rising living standards" (OECD 2012). This concern, expressed by the OECD,

---

[9]For the notion of socio-ecological system, see the general introductions and the Introductory Chapter of this book, see also Schoon and van der Leeuw 2015 on the evolution of this concept.

bases itself on the observed irreversible disturbances linked to climate change, the never-ending degradation of biodiversity, the growing pressure on water resources, the increasing pollution of water, and the atmospheric pollution, which is becoming the leading environmental cause of premature death globally, and so on.

The editors suggest finding avenues to break away from the sole paradigm of rational domination of the world by means of the emergence, according to a very interdisciplinary approach, of the paradigm of scientific coviability, which offers an ideological and political neutrality of the human/nonhuman relationship.

Moreover, the authors in this book have developed their analysis on the basis of their research protocol, according to their disciplinary field, their research topics and scientific experiments. This endeavor is based on achievements, particularly mathematical, that we highlighted in the General Introduction and the Introductory Chapter as well as on new disciplinary and interdisciplinary explorations of the concept.

The interest of working on socio-ecological coviability is to position ourselves:

- in the rigorous formalization of a concept that can become a paradigm firmly rooted in scientific work and as free as possible from ethnocentrism;
- in an approach that finds its foundations already established in societies, in science and in international law, even Constitutions and national legislations;
- in a universal awakening of the ecological and societal challenges of human/nonhuman relationships at both global and local scales.

This cross-disciplinary reading of the book is based on all the chapters of Volumes 1 and 2 and revolves around three cogs: (1) formalizing a universal paradigm of the human/nonhuman integrative relationship (or societies/nature), (2) the determinants of this paradigm of coviability extracted from the results presented in the chapters and (3) the highlighting of a principle reorienting the objective of "sustainable development".

27.1 From a Naturalist Ontology to an Integrative Ontology

> By becoming aware of the nature around him, Man simultaneously discovers his alterity vis-à-vis this nature: to see better, he stands outside. However, Man then develops an unfortunate tendency to perceive himself as autonomous, separated from an environment he perceives as external to him: emergence induces the rupture of the deep solidarities that bind man to his environment. (Jean-Marie Pelt 1977)

> ... this traditional dichotomy of humanity-versus-nature is false and dangerous. On the one hand, it perpetuates our destructive mishandling of the biosphere. On the other hand, it scants the self-understanding that Homo sapiens needs to settle down on our home planet, hence as a prerequisite to survival. Nature, to put the matter as succinctly as possible, is part of us, and we are part of nature. (Edward Osborne Wilson 2007)

The idea of separation formalized in the notion of an environment relates to a naturalistic ontology (as previously seen; see General Introduction/Table 27.1). Not all societies have adopted this separation between humans and other living beings. The very notion of "nature" or "environment" is difficult to translate into bozo or wayana, for example. Each has their own "nature," but the notion of "nonhuman"[10] is favored because each defines "nature" according to their own preferences and socio-cultural schemes (Descola 2011).

This naturalistic ontology is today relativized and incindentally put into question. We perceived this in December 2015 with the Paris Agreement, in the face of climate change where belonging to a whole such as the biosphere leads the international community to invent or reinvent the first steps of a form of ontology advocating the continuity of man-nature.[11] We unavoidably see in it a passage from an ontology of separation to an ontology of union or better, the integrative[12] (some speak of ecological ontology, see below). Ecological awareness modifies the dominant ontology because of scientific representations that "*open up new possibilities: not only disenchantment but also another form of knowledge of continuities and limits. From this may result an ecological ontology as technologies increasingly produce evidence of interdependence between the elements that naturalistic ontology disjoin*" (Feenberg 2013, 116).

The naturalistic ontology confronts the ethics of humanity's responsibility towards the planet and therefore towards itself. Criminalizing the destruction of ecosystems or ecocide testifies to this. The destruction of "that which is necessary for humans to exist"[13] is a violation of human rights. This analysis was born from spreading defoliant herbicide (Agent Orange) during the Vietnam War in the 1950s and 1960s (Zierler 2011). The biologist activist Arthur Galston denounced this infringement in 1970 using the notion of ecocide for the first time. The term was reused at the first Stockholm Conference on the Human Environment in 1972.

In 2010, the lawyer Polly Higgins proposed to the UN's International Law Commission the concept of ecocide as an international crime against peace, after those of genocide, a crime against humanity, a crime of aggression, and a war crime. The direct attack on humanity is underlined by the very definition of ecocide which is "*the partial or total degradation of an ecosystem, caused by man or by other causes, and which significantly diminishes the enjoyment of a territory by its inhabitants*" (Polly Higgins 2015) or "extensive damage or destruction of an ecosystem of a particular territory" (Neyret 2015).[14] For Valérie Cabanes "*ecocide (the action of destroying "home Earth") is not just another crime, added to all other human rights abuses. Ecocide is nowadays the major crime, the one that ruins the very conditions of habitability of Earth*" (Cabanes 2016). The direct violation of the

[10]Terminology emanating from the Actor-Network Theory, cf. Akrich et al. 2006.

[11]"Human beings feel connected to nature, not only in a chemical or physical manner, but in a rather vague spiritual one. . The Earth is our 'mother', our 'home', and the planet must be 'preserved from evil,' a reckless 'exploitation'" (Feenberg 2013, 116).

[12]"We may be witnessing the birth of a new Western ontology"(Feenberg 2013, 118).

[13]"Outside a legal definition, ecocide is any extensive damage or destruction of the natural landscape and disruption or loss of ecosystems of a given territory to such an extent that the survival of the inhabitants of that territory is endangered" (Lay et al. 2015, 433).

[14]On the notion of territory, see: Caron et al. 2017.

Table 27.1 Some inputs of the coviability paradigm for the International Community

The expectations of the International Community	The elements brought by Coviability as a new paradigm
Convention on Biological Diversity (1992): *intrinsic value of biodiversity, access and sharing of genetic resources, ...*	Links biodiversity to humans without biodiversity being the object of humans (capital or service): proposes a unique legal status of the human/nonhuman link to be developed, and justifies its conservation other than by its financialization
	Frees itself from the grip of the dominant paradigm in order to reason and decide differently: for example, it rethinks he relationship to land, the notion of appropriation,... to perceive the collective and the common
SDG "transforming our world" (2015): fighting against the degradation of the planet	Adopts values, public policies and regulations that establish another relationship to the biosphere, in terms of the continuity from the human to the nonhuman; the aim is to develop a status of responsible humanity towards the planet and future generations
OECD: horizon 2050: "*changing the course of our future development*"	Views the relationship between man and his environment in a non-anthropocentric way and proposes an integrative ontology.
Paris Agreement (2015) on climate change: *containing the rise* of the average temperature of the planet by 2°, *reinforcing the capacities of adaptation, promoting resilience, a development with low emissions of* greenhouse gases,...	Establishes the viability link as the foundation and stake of environmental law and human rights.
	Reasons in terms of long-term interactions, rather than just in terms of growth, to identify viability links.
	Establishes a correlation of regulation and public policies with ecological constraints
"*Living in harmony with nature*" transition (Strategic Plan for Biodiversity 2011–2020, COP Cancun 2016, UNEP interactive dialogues: "*the interdependence of humanity with Earth*"; ...)	Formalizes the goal of harmony: the viability of systems depends on their interactions
Convention to combat desertification, COP 13, Ordos-China, 2017: Targets: "(a) the amount of the land to restore or rehabilitate during the next 12 years, (b) to agree on measures that address the related emerging threats of forced migration, sand and dust storms, and (c) to consent on actions to strengthen the resilience of communities to drought	Redefines the relationship between man and the earth: establishes a link of viability through a continuity that the law must formalize through land ecology[a]

[a]For an approach on land ecology see Barrière 2017

rights of human existence thus gives birth to a crime: ecocide, which has not been fully understood yet. At the international level, ecocide has not achieved the status of imperative norm (*jus cogens*) (Lay et al. 2015).

Outside the criminal field, legal experts are working on translating into law an ethics of responsibility towards the environment, the "ecological public order" (Belaïdi 2004; Boutelet and Fritz 2005, see Chap. 15 by Belaïdi in this book). This law, still "in the making," views the environment as "*a common good whose protection, restoration and transmission to future generations are based on an ethical responsibility in order to respect its integrity and dynamics and thus to take into account the common Good and Well-being*" (Belaidi 2014). For Nadia Belaïdi, ecological public order is a tool for promoting and protecting the environment as a social value, which is essential for present and future generations. It is defined as "*a set of rules accepted and recognized by all, whose goal is to protect the ecological processes supporting all life in the perspective of ensuring sustainable development and the well-being of humanity*" (Belaïdi 2004).

With the emergence of new legal concepts, elaborating a coviability ontology is defined by its integrative nature. Why "integrative"? The reason is that, in essence, socio-ecological coviability is the result of connections, assemblages, and even fusions between the human and the nonhuman. This point is emphasized and demonstrated throughout the chapters of this book (Volumes 1 and 2).

In the face of the major issues related to global, environmental, economic, social and cultural changes, which affect the entire planet, the naturalistic paradigm domination compromises the future of humanity by pursuing a rationale of consumption in its productivist race. The nature as object, constructed by the capitalist economy, opposes the necessary connection of humanity to the biosphere (see in particular Chap. 11, Bertrand). The pursuit of the Millennium Assessment by the Sustainable Development Goals attests to this by advocating an ideal "new world," which is not only devoid of poverty, hunger, equity, and so on, but also ecologically sustainable. The OECD forecasts we have just presented (see above) underline the urgent need to act.

The integrative ontology of coviability is based on the perspective of a human/nonhuman harmony invoked by the international community. In the first instance, this involves laying down/placing (?) the UN representation with all the ambivalences and even the contradictions that characterize speeches and resolutions. Second, there is the idea of harmony promoted by the United Nations, referring to a balance, or a synchronism, until an agreement is reached. Finally, in a third step, we formalize the contours of the definition of the coviability paradigm. We do so through the presentation of an integrative ontology of coviability, which we call "coviabilism." By this, we intend to clarify the foundation of socio-ecological coviability.

27.1.1 The Issue of a "Harmony" Invoked by the International Community

In the Man-biosphere relationship, the term "harmony" has emerged as a connecting link between the human and the nonhuman at the international level. The challenge of harmony has led to recognizing the concept of "living in harmony with nature" by the United Nations General Assembly. Since 2009, it has adopted eight resolutions on this theme, the first of which declares April 22 as International Mother Earth Day. Adopting this concept engenders rethinking environmental law in terms of a biosphere law, or Earth rights, called "wild rights" by Earth Jurisprudence. These innovations originate in a search for a new universal paradigm of the relationship of man with the Planet, which unites rather than divides them. The will to leave the anthropocentric rationale behind in order to achieve an ecological transition with the objective of becoming "in harmony with nature", is gradually appearing. "The Strategic Plan for Biodiversity 2011–2020 entitled "Living in Harmony with Nature," associated with the 20 Aichi Targets (COP10, Nagoya, 2010), is carried on in the 13th Conference of the Parties to the Convention on Biological Diversity (COP13, Cancun, 2016) by creating an interactive dialogue on "Living in harmony with nature" to limit the degradation of biodiversity. In 2016, this dialogue follows a series of annual sessions from 2009 to 2015, held by the United Nations General Assembly.

In 2009, the United Nations General Assembly established the International Mother Earth Day (Resolution 63/278 adopted by the General Assembly on 22 April 2009). The groundwork of this decision lies on one consideration and one observation at least, i.e. that "*Acknowledging that the Earth and its ecosystems are our home, and convinced that in order to achieve a just balance among the economic, social, and environmental needs of present and future generations, it is necessary to promote harmony with nature and the Earth.*" The observation consists of "*Recognizing that Mother Earth is a common expression for the planet earth in a number of countries and regions, which reflects the interdependence that exists among human beings, other living species and the planet we all inhabit.*"

27.1.1.1 The UN Objective of "Living in Harmony with Nature"

The first resolution of the United Nations General Assembly on Harmony with Nature was adopted on 21 December 2009 (A/RES/64/196). In the topic on "Sustainable Development" a sub-topic entitled "Harmony with Nature" was included. This resulted in: six reports from the Secretary-General, the first dates from 19 August 2010 (A/65/314) and the last from 4 August 2015 (A/70/268). There are also seven interactive Dialogues of the General Assembly, which ran from April 2011 to April 2017. They were organized to commemorate the International Mother

Earth Day (or nurturing earth[15]). Finally, a website was created specifically for this concept: http://www.harmonywithnatureun.org. Thus a real process has been born borrowing the paradigm of harmony with nature for humanity by recalling "the interdependence that exists between the human being, other living species and the planet on which we all live."

In its resolution of December 19, 2014 (69/224), the General Assembly, "*expressing the conviction that, in order to achieve a just balance among the economic, social and environmental needs of present and future generations, it is necessary to promote harmony with nature.*" In this light, the General Assembly calls for "*an inclusive and interactive dialogue*" on the question of harmony with nature, and calls *"for holistic and integrated approaches to sustainable development that will guide humanity to live in harmony with nature and lead to efforts to restore the health and integrity of the Earth's ecosystems.*"

The latest Resolution to date, of 22 December 2015 (A/RES/70/208), establishes a "virtual dialogue" on harmony with nature. It calls for rethinking relationships with nature in a perspective of sustainable development, which requires that "*societies make radical changes in the way they produce and consume.*" States are invited to take as example the indigenous cultures "so that the protection of nature is taken into account" given their unique ontological relationship with nature. The Resolution also reminds us that the goal of sustainable development requires "*raising awareness of the fundamental links of interdependence between man and nature and to strengthen them*" by going beyond an anthropocentric approach. For this purpose, the Resolution uses the concept of "Earth Jurisprudence" as a perspective.

27.1.1.2 Earth Jurisprudence

The Interactive Dialogue of the UN General Assembly, dated 21 April 2017, endorses an "Earth Law," called "Earth Jurisprudence." It resulted with recognizing the intrinsic value of nature and the urgent need to "to reconsider how it perceives and interacts with the natural world."[16] The question of paradigm continuously recurs in the exchanges by taking reference from indigenous peoples "*Indigenous peoples' philosophies, spiritualities and traditional forms of knowledge worldwide express the understanding that human governance systems must be derived from the laws of the Earth and comply with them. Experts from around the world working in the natural and social sciences similarly recognize the need for an evolved, holistic*

[15] Mother Earth Day has been celebrated since 1970 in the United States. (cf. http://www.earthday.org).

[16] "In order to forge a balanced, healthy relationship between human activity and the Earth, there is an urgent need for society to reconsider how it perceives and interacts with the natural world".

worldview that must be rooted in respect for Nature and in the interdependence of the well-being of humankind and of the Earth."[17]

The stated idea of "Earth Jurisprudence"[18] or "wild law" carries within it the seeds of a revolution of thought. This revolution aims to abandon the naturalistic ontology in favor of a (re)union of human beings and nonhumans, forming a "broader Earth community": "*wild law or Earth Jurisprudence is an emerging theory of law and governance that seeks to evolve law in a fashion that recognises our relationship to the broader Earth community*" (Burdon 2010). Through a governance and an integrated networked law,[19] we transcend the idea of a "natural law"[20] (based on human nature) to a "wild" law (Burdon 2010;[21] Cullinan 2011; Howe 2017). Cormac Cullinan defines it as "*laws that regulate humans in a manner that creates the freedom for all members of the Earth Community to play a role in the continuing co-evolution of the planet*" (Cullinan 2011). We are nearing a sort of "ecological law" that confers a legal status to the environment (Ost 1995), with a relationship to land that leaves the realms of possession (Barrière 2011, 2017), and confers rights to nature (nature as a subject of rights: Laastad 2016; Serres 1992; Hermitte 2011). This law, which we could describe as "Earth Jurisprudence," remains outside "natural law." The latter reflects divine prescriptions (Medevielle 2010), while "*Earth Jurisprudence sets out the general norms out of which practical "wild" laws can be deduced*" (Rühs and Jone 2016).

In Florida (USA, Barry University) in 2006 a team of legal experts created a Center dedicated to "Earth Jurisprudence,"[22] starting from the observation that humanity is an integral and interdependent part of nature; that humanity has an obligation to protect the long-term health of nature[23]; and that environmental law

[17]http://www.harmonywithnatureun.org/index.php?page=view&type=12&nr=58&menu=198

[18]Berry 2000; Burdon 2010; Murray 2015: Rühs and Jone 2016.

[19]Earth Jurisprudence recognizes that the Earth is the source of natural laws that govern life. It provides a cohesive framework that underpins many disciplines, weaving them together to create a more effective, holistic governance approach, one that reflects the integrated nature of the world in which we live. Earth jurisprudence inspires citizens and societies to reconsider how they interact with the natural world". (Concept note: Interactive dialogue of the general assembly on harmony with nature in commemoration of international mother earth day, theme: earth jurisprudence, UN HQ New York, 21 April 2017, trusteeship Council).

[20]The School of Natural Law dates from the seventeenth century, but its origin comes from antiquity. For the Roman legal experts, natural law is confused with the law of peoples. However, its universal character (common to all) allows it to comply with nature and reason; "natural law is a rule suggested to us by sound reason, according to which we necessarily judge that an action is unjust or moral according to its conformity with reasonable nature, and thus God, who is the author of nature, defends one and commands the other "(Grotius – Hugo de Groo, 1583–1645- Droit de la Guerre et de la paix, see Charmont 1927, 17).

[21]"(. . .) developing law that truly reflects our place in nature and facilitates a mutually enhancing relationship with the earth".

[22]The Center has made "Advancing law, policy, and governance systems aimed to legally protect the sustainability of life and health on Earth" its mission (http://www.earthjurist.org).

[23]"We have an obligation to protect and sustain a viable Earth for current and future generations. A truly sustainable future is based on healthy ecosystems. Current environmental laws seldom

needs to refocus itself on a goal of ecosystem health. The Center proceeds by applying a legal ethic that "*seeks to develop a philosophy and practice of law that gives greater consideration to nature, by recognizing the interconnectedness of Earth's natural systems, the inherent rights and value of nature, and the dependence of humanity and all living beings on a healthy Earth*" (http://www.earthjurist.org).

27.1.1.3 A Planned, Debated, and Proclaimed Transition Toward "A Life in Harmony with Nature"

The vision of the Strategic Plan for Biodiversity 2011–2020 helps in defining ways to "live in harmony with nature": a wise usage of biodiversity (valued, conserved and restored), while "*maintaining ecosystem services, sustaining a healthy planet and delivering essential benefits for all people*" (UNEP/CBD/COP/DEC/X/2).

Recently, as its mission, the Strategic Plan proposes to "*take effective and urgent action to halt the loss of biodiversity in order to ensure that by 2020 ecosystems are resilient and continue to provide essential services, thereby securing the planet's variety of life, and contributing to human well-being, and poverty eradication. To ensure this, pressures on biodiversity are reduced, ecosystems are restored, biological resources are sustainably used and benefits arising out of utilization of genetic resources are shared in a fair and equitable manner; adequate financial resources are provided, capacities are enhanced, biodiversity issues and values mainstreamed, appropriate policies are effectively implemented, and decision-making is based on sound science and the precautionary approach*" (UNEP/CBD/COP/DEC/X/2).

Six years later, the 13th Conference of the Parties to the Convention on Biological Diversity (Cancún, Mexico, 4–17 December 2016) retained the concept of "living in harmony with nature" as an interactive dialogue.[24] The Conference defines it as a way to take into account the ecological limits of the planet and ensure the preservation of biodiversity for future generations (UNEP/CBD/COP/13/9). To this end, the Conference is part of an ecosystem approach that constitutes its main framework for action (Decision II/8). *"Living in Harmony" integrates interactions between organisms and their environment and "It recognizes that humans, with their cultural diversity, are an integral component of ecosystems*" (Decision VI/6).

The conference takes as example the manner by which local communities and indigenous peoples live in harmony with nature. They are considered "living examples of lifestyles in harmony with nature," ones that generate biocultural systems. The conference makes explicit reference to Article 10c) of the Convention.

consider the true welfare of future generations or nature, but instead routinely apply a cost-benefit analysis that uses an economic measure to prioritize short-term human interests over the undervalued long-term benefits of a healthy Earth" (http://www.earthjurist.org).

[24]"Review of the concept of 'living in harmony with nature' under the convention and related processes", UNEP/CBD/COP/13/1.

The article calls for the protection and promotion of "customary use of biological resources," and it recalls the adoption in 2010 of the global plan for action on the sustainable customary usage of the biological diversity by the former Conference of the Parties (UNEP/CBD/COP/13/9).

The Cancun Conference, in its interactive dialogue, states the interdependence of humanity with Earth: "*At the dialogues, experts from around the world working in the field of natural and social sciences have presented the need for an evolved and holistic worldview that must be rooted in respect for nature and in the interdependence of the well-being of humankind and of the Earth. Humanity is an inextricably part of the community of life on Earth, and the experts state that we cannot override the laws that maintain the homeostatic balances of the Earth System. Economic growth for some has been achieved at the expense of the natural world as well as many human populations*" (UNEP/CBD/COP/13/9).

Since 2011, the interactive dialogues of the United Nations General Assembly on "Living in harmony with Nature"[25] have questioned the idea of reconnecting humanity to the biosphere with its economic alternatives and an associated paradigm. Specifically, the terms were as follows: *How humans can better reconnect with the world around them* (first dialogue, 2011); *Different economic approaches to further a more ethical basis for the relationship between humanity and the Earth* (third dialogue, 2013); *Possible key characteristics of a new, non-anthropocentric paradigm and the further identification of strategies on how society subsequently would need to function consistent with this paradigm* (fourth dialogue, 2014).

Dialogues continue (UNEP/CBD/COP/13/9, 4 October 2016) around the central question of how to integrate into practice[26] and law[27] the concept of "living in harmony with nature." The latter leads to questions about the option of recognizing the moral personality of nature by granting it rights through "Earth Jurisprudence."[28] Is the idea of granting rights to nature in harmony with the "approach based on human rights"? Another question emerges, among others, of the contribution of scientific and local knowledge to "living in harmony with nature."[29]

[25] See: http://www.harmonywithnatureun.org/documents.html

[26] What actions can be taken at different levels to mainstream the concept of living in harmony with nature across the various sectors, including but not limited to agriculture, fisheries, forestry and tourism? (UNEP/CBD/COP/13/9, 4 October 2016).

[27] How can rights-based approaches and the customary laws of indigenous peoples and local communities promote living in harmony with nature, considering multiple and/or diverse views, and legal pluralism? (UNEP/CBD/COP/13/9, 4 October 2016); "Quelles mesures peuvent être prises à divers niveaux pour aborder les questions d'accès, d'utilisation coutumière durable, et de manière plus générale l'utilisation durable de la biodiversité, notamment au niveau local et à d'autres niveaux?" (UNEP/CBD/COP/13/9, 4 October 2016)/"What actions can be taken at different levels to address access, customary sustainable use, and more broadly sustainable use of biodiversity, including at local and other levels?" (UNEP/CBD/COP/13/9, 4 October 2016).

[28] "What are the latest developments regarding Earth jurisprudence worldwide, promoting the rights of nature?" (UNEP/CBD/COP/13/9, 4 October 2016).

[29] "How can science, together with traditional knowledge, inform approaches to living in harmony with nature?" (UNEP/CBD/COP/13/9, 4 October 2016).

The General Assembly dialogue, held in September 2016, developed the first steps towards recognizing the rights of nature adopted by the United Nations Conference on Sustainable Development, held in Rio de Janeiro, Brazil, in June 2012. In the final draft of the document adopted, entitled "The future we want,"[30] State Heads and Governments recognize that "*planet Earth and its ecosystems are our home*" and in order to achieve "*a just balance among the needs of present and future generations, it is necessary to promote Harmony with Nature*" (UNEP/CBD/COP/13/9, 4 October 2016).

Harmony with nature is also recognized in the 12th Objective (target 12.8) of the Sustainable Development Goals[31]: *ensure that people everywhere have the relevant information and awareness needed for sustainable development and lifestyles in harmony with nature.*

The resolutions of the United Nations General Assembly have increasingly recognized the importance of non-anthropocentric worldviews, promoting "living in harmony with nature." This harmony is based on "*recognizing the intrinsic value of nature, in which the relations between humanity and the planet are symbiotic, interdependent and subject to the natural laws of the Universe.*"[32] This concept of harmony will integrate the 14th Conference of the Parties of the CBD.[33]

Finally, at the occasion of the United Nations Conference on Biodiversity, the Cancun Declaration on mainstreaming the conservation and sustainable use of biodiversity for well-being (December 2–3, 2016, Cancun/Mexico) recognizes and emphasizes the importance of "living in harmony with nature": "*It is essential to live in harmony with nature, recognized by some cultures as Mother Earth, as a fundamental condition for the well-being of all life, which depends on the conservation and sustainable use of biodiversity, ...*"

The statement in this first point correlates a socio-cultural paradigm (living in harmony with nature) with a challenge of societal and ecological nature (the well-being of all life). As a result, dependency is created when the planet is (re)qualified as "nurturer,"[34] source of "symbiosis," stated in 2011 by the United Nations: "*A symbiosis between human beings and nature promotes a mutually beneficial relationship*,"[35] without clarifying this mutual relationship.

However, we remain in a Man-nature dichotomy with the concept of "living in harmony with nature." The hiatus stems from the definition of "nature." Nature is

---

[30]General Assembly resolution 66/288, annex.

[31]"*Transforming our world: the 2030 Agenda for Sustainable Development*," General Assembly resolution 70/1, annex.

[32]http://www.harmonywithnatureun.org/chronology.html

[33]"*According to the multi-year programme of work of the Conference of the Parties up to 2020 (Decision XII/31), the Conference of the Parties is due to consider approaches to living in harmony with nature at its fourteenth meeting. Thus, the dialogue may provide an opportunity to advance preliminary thinking on this concept*" (UNEP/CBD/COP/13/9, 4 October 2016).

[34]See Resolution of 21 December 2009,A/RES/64/196.

[35]United Nations General Assembly, Interactive Dialogue on Harmony with Nature, April 20, 2011, New York.

still lived as opposed to humans in the naturalistic ontology. The relationship of symbiosis forces a separation between two organisms that are, by definition, unable to survive without the other. The declared wish of the United Nations to leave behind anthropocentrism proceeds by creating an openness toward other paradigms, be they new or borrowed from the societies of the whole world, indigenous communities or traditional populations.[36]

We propose adopting this openness; we have called it "coviabilism" in a chapter in the first part of this book (Chap. 6, Barrière and Behnassi). We shall now explore this new way of being, of thinking, and of participating in the world.

27.1.2 "Coviabilism" Ontology as a New Socio-ecological Paradigm

Throughout this book (in its two volumes), the notion of coviability is expressed via more than forty chapters according to differences related to objects and research issues. It is clear that the paradigm of coviability raises questions and encourages scientific thought and analysis. The novelty of this term in most disciplines contributes in defining it as a paradigm "that takes us to another world." Depending on the situation, the need to "push boundaries," to abandon a naturalistic rationale that seems inevitable, emerges from these presentations.

We have previously seen the growing importance of "living in harmony with nature" as developed by the United Nations. Admittedly, UN reports and resolutions echo a need to break out of the paradigm of the Man-nature dichotomy. The aim is to leave behind this paradigm for another that is more universal, one which takes into account "traditional" populations (see above). This aim becomes an increasingly asserted challenge (13th Conference of the Parties of the Convention on Biological Diversity, Cancun, 2016, which explicitly calls for it in the 14th Conference).

This challenge led to suggesting an ontological outline called "coviabilism" in the categorization initiated by Philippe Descola (2005), in this book, as: "*the nature/societies interdependence that participates in the paradigm of the unity of living beings and which approaches totemism (and analogism) through the continuity of interiority and physicality*" (Chap. 6, Barrière and Behnassi).

The notion of ontology, which defines the being and which underlies that of paradigm,[37] and which also defines the way of looking and perceiving, translates

[36]From a political viewpoint, traditional populations are those that "adopt a specific way of life marked by a strong symbiosis and the relative harmony with the environment in which they live, by developing techniques of low environmental impact, and by a weak articulation with the market, a great knowledge of the biodiversity that surrounds them and a mode of production based on family labor. (...)".This term is a national or western society construct to classify other societies, minority groups "having a way of life different from the society around them (...)" (Ravena-Cañete and Ravena Cañete 2011).

[37]cf. the definitions given above, in notes 1 and 3.

a mental representation. Ontology expresses a way of being and fitting into the world. Everyone, thus, has in their disposal their "own world" since universalism is (in this context) relative. Therefore, as we have already pointed out as early as the General Introduction of this book, the need to escape a claim to the universality of the naturalist paradigm, dissociating nature and culture, requires from each of us, writers and readers, to disengage from our ethnocentric and anthropocentric envelope promoting a "naturalization" of the world.[38]

The emergence of a "coviabilism" ontology, in addition to the four ontologies defined by Philippe Descola (2005),[39] will revolve around two key inputs: the socio-ecological model and the continuity between man and nature.

27.1.2.1 A Socio-ecological Scheme

In Philippe Descola's anthropological approach (2005), the scheme defines models of relationships and behaviors that intervene in practices through psychic, sensorimotor, or emotional dispositions. The scheme is at the origin of the flow of perception (ways of being), of the organization of behaviors and practical activities (ways of doing things) and the expression of thought (ways of thinking). Thus, it will determine, on the one hand, the properties of the "existing," and on the other hand, the form of the links that they maintain between them. The scheme will allow humans and nonhumans to be identified through the differences and similarities that analogies and contrasts will infer. These latter singularize the scheme by attaching it to the duality of interiority and physicality, an internal self and a physical self. According to Descola (2005), this duality is universal; it is shared by all peoples, and has no ethnocentrism. Through combining interiority and physicality,[40] this schematic form of identification defines four main types of ontology. These are animism, totemism, analogism and naturalism (presented in Chap. 6 by Barrière and Behnassi and reused in Fig. 27.1) and are "*systems of properties of the existing, serving as anchors for contrasting forms of cosmology, as a model of social links, and as theories of identity and alterity*" (Descola 2005, 176).

[38]Here we agree with Philippe Descola who speaks of "certainties of naturalism" (Descola 2005, 241 et s.).

[39]Presented in the General Introduction and in Chap. 6 of this book, Barrière and Behnasssi, and mentioned in other chapters.

[40]Inwardness refers to the spirit, soul or consciousness with intentionality, subjectivity, reflexivity, affects, as well as the ability to signify or to dream. Physicality represents the external form, the substance, the physiological, perceptual and sensorimotor processes, in other words the morphological and physiological elements peculiar to a being (Descola 2005, 168 and 169). (Descola 2005, 168 and 169).

The challenge is to detect whether there is a continuity or not between the human being and the nonhuman being. This continuity is determined by internal homology (interiority) or external homology (physicality).[41]

This procedure of analyzing the human/nonhuman relationship provides an identity and relational schematization of the human being in the biosphere. Of the four ontologies presented by Philippe Descola, three maintain a link between humans and the biosphere through continuity and/or a relational connection. The fourth, the "naturalist" ontology, replaces this connection, as it adopts a relationship based on "subordination" and "predation."

In a time of international awareness[42] of a new geological era, the Anthropocene, should we keep to these four ontologies? Indeed, the naturalistic ontology is already ahead of this transformation: the importance given to the environment in the majority of the national Constitutions,[43] the growth in the number of protected areas in the world (nearly 15% of continental areas and 14% of the oceans and coastal areas, UNEP-WCMC and IUCN 2016), and the development of strong environmental values reflected by public ecological and energy transition policies, and by new convictions ("*no Plan B for action as there is no Planet B*"[44]). Even if still strongly present, another version of this naturalist ontology could gradually supplant it[45] (Fig. 27.1).

[41]"What is important is not to see this or that animal or plant, but to specify these 'objects' by imputing or denying them an interiority or physicality similar to those we attribute to ourselves" (Descola 2005).

[42]An awareness expressed in the five successive global planetary summits on the environment organized since 1972, the Paris Agreement on Climate Change of 2015, bringing together 194 countries, plus the European Community, and the Conferences of the Parties of the different international climate, biodiversity, desertification, etc. conventions.

[43]In 2013, 149 constitutions out of 193 in the world included environmental preservation (Boyd 2012, 2014).

[44]« no Plan B for action as there is no Planet B. » Ban Ki-moon, UN Headquarters, New York, 23 September 2014, Climate Summit 2014, "Catalyzing Action" (available on line: http://www.un.org/apps/news/story.asp?NewsID=48766#.WUfNuDPpOEJ).

[45]for example, the whole agro-ecological movement that is developing in the world. See also the solutions presented in Cyril Dion's book (2015): "Tomorrow: All Over the Globe, Solutions Already Exist » and Mélanie Laurent et Cyril Dion's film "Tomorrow": "As mankind is threatened by the collapse of the ecosystems, Cyril, Mélanie, Alexandre, Laurent, Raphäel and Antoine, all in their thirties, explore the world in search of solutions that can save their children, and with them, future generations. Using the most successful experiments in every area (agriculture, energy, habitat, economy, education, democracy...) they try to put back together the puzzle which may tell a new story of the future » (https://www.demain-lefilm.com/en). Also, Isabelle Antunes's film"Happy Rain, when floods become sustainable wealth" (2015) on the conversion of Bangladeshi farmers to fishermen and fish farmers (https://vimeo.com/100228720). "Here, the environment is not a reality independent of the human being, a kind of worshiping deity or a fatality. The real way of perceiving the environment is to admit that we are part of it, it is our way of life. We have to develop ways to observe in order to understand, reflect and identify opportunities, then take the time to build forward-looking, respectful solutions that continue to teach us about ourselves" (I. Antunes, (http://www.up-magazine.info/index.php/arts/5375-terrehappy-rain-le-film-d-une-guerriere).

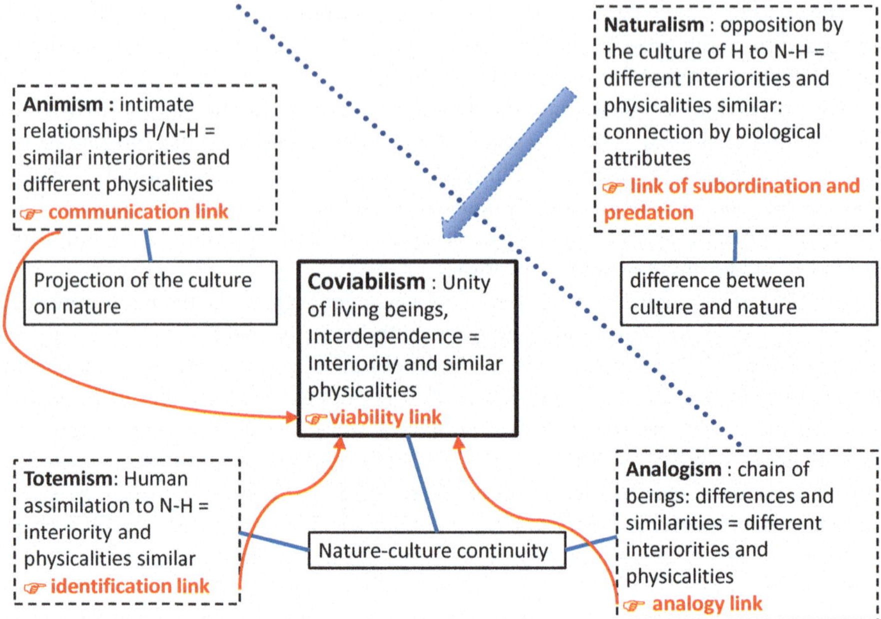

Fig. 27.1 General view of the relationships between societies and nature through the ontological prism (by taking Descola's ontologies, 2005 and Chap. 6 Barrière and Behnassi)

This new ontology is the result of a socio-ecological scheme of (re)connecting humanity to the biosphere (or the nonhuman) which, for example, the United Nations echoes by the concept of "living in harmony with nature" (which we saw earlier).

The viability link formalizing the paradigm of coviability translates the relationship through interactions necessary to ensure the viability (ability to live) of the human/nonhuman integration. This link translates a connection in order to bond with each other and sustain each other.

27.1.2.2 The Continuity Between the Human and the Nonhuman Within the Biosphere

The idea of continuity between the human and the nonhuman opposes an anthropocentric conception of the relation to the world. The notion of humanity could thus take another dimension by thinking more in terms of the "collective"[46] (Latour

[46]"There are no longer two separated attractors, one representing unity in the form of nature and the other keeping multiplicity in the form of societies. That which is collective means: the whole and not divided in two" (Latour 2004, 95). (Latour 2004, 95).

2004). Unable to be viable alone,[47] starting from the mathematical analysis, the notion of "link" allowing "bonding" emerges. If humanity forms a whole in the ecosystem, the collective is defined by this ecosystem consisting simultaneously of the human and the nonhuman.

In the ontologies presented above and in Fig. 27.1, the links of communication, identification, and analogy converge on a link of viability. The continuity that is realized by these links breaks the link of subordination and predation of naturalistic ontology. Besides, is this separation of culture/nature really justified when one increasingly questions the fact that culture is not exclusive to humans (Whiten et al. 1999; Noad et al. 2000; Luke and Whitehead 2001; Laland and Janik 2006; Kopps and Sherwin 2012), and when the United Nations explicitly recognizes this[48]?

The human-nonhuman continuity translates a (re)connection of humanity to the biosphere. By definition, continuity opposes compartmentalization or separation. The anthropological approach defines continuity by resemblance and discontinuity by difference (Descola 2005, seen in Chap. 6: Barrière and Behnassi).

Articulating the continuous and the discontinuous would therefore be a game of resemblance-dissimilarity, or of homogeneity and heterogeneity between living beings (human and nonhuman). We have previously seen that this game is related to physical materiality (physicality) and the phenomena and processes related to the mind, intelligence, affectivity and will, the psyche (interiority). Resulting from Cartesian science, the relationships defined are not objective in nature; they are rather subjective in nature, which is the logic of understanding and that of the different forms of human cognition. These are projections defining identification schemes (what we are within the living or the collective) and the relational schemes (the relationship with other living being) by means of socio-cultural paradigms, which are the ways of "seeing the world."

Here, we reach the heart of our topic: coviability. The "co" of coviability, making "a whole", marks the continuity depicted by a link of exteriority and interiority, which may be analogous or different between humans and nonhumans. The links are under five functions, each of which specifies a type of link (see Fig. 27.1):

- the assimilation of humans to nonhumans (totemism) generates a link of identification;
- intimate relationships between humans and nonhumans (animism) give rise to a communication link;
- the chain of different and identical beings confers a link of analogy (analogism);

[47]because by definition there are always interrelations between living beings (Selosse 2017) . See. the systemic approach in this book.

[48]"Recognizing that a number of socially complex mammal species, such as several species of cetaceans, great apes and elephants, show that they have a nonhuman culture (hereinafter 'culture') (...) there is evidence that the influence of culture and social complexity may be a conservation issue" UNEP / CMS / Resolution 11.23 Consequences of cetacean culture for their conservation. It was adopted by the Conference of the Parties in its 11th meeting (Quito, 4–9 November 2014).

- the opposition of humans to nonhumans is realized by a link of subordination and predation (naturalism);
- in the era of globalization and the awareness of the finitude of planet Earth, a link of viability emerges, one resulting from the human/nonhuman interdependency (coviabilism). This link of universal dimension has the vocation to supplant, or at least to limit, the relationship of opposition and subordination of naturalistic ontology and to integrate the other three ontologies.

Beyond culture and nature, the viability link formalizes the nature-culture continuity and a form of projection in each other. Thus, this link expresses integration within a whole, called collective (see *above* Latour 2004), constituted of humans and nonhumans. This "whole" leads to a "unity," participating in spiritual foundations.

This scientific vision is also present in examples from the five denominations (Christianity, Islam, Hinduism, Taoism, Buddhism) where we find the idea that everything is linked and inseparable. The relationship of societies with nature is affirmed as interwoven "*We are part of nature, included in it and thus in constant interaction with it.* (...) *It is essential to seek comprehensive solutions which consider the interactions within natural systems themselves and with social systems. We are faced not with two separate crises, one environmental and the other social, but rather with one complex crisis which is both social and environmental*" (François 2015, n°139). The Earth is seen as a cradle for humanity.[49] The metaphysical vision of creation in Islam requires that there be "*one God, one Nature and one substance*" (Wahdat al wujûd). This is existential monism (see Surah of Bees, An-Nahl). The order of the world, *dharma*, is the order of nature: a higher order to which Man subject. This order encompasses the whole in which the divine is One, called *Brahman*, the "Encompassing." Unity is the Absolute that transcends the world. The universe is a field of intertwined forces. Taoism praises nature, the "flow of life that should not be opposed," meaning "non-action." Harmony imposes itself throughout the cycles and systems of nature.

The interdependence of all beings is a fundamental law in Buddhism that "does not invite fusion with nature, but it proposes a wisdom of interdependence (*conditioned co-production, pratîtyasamutpada*)" (Brosse 2002). For the Dalai Lama, the connection with all living sentient beings generates universal responsibility (Dalai Lama and Stril-Rever 2016: 175). Interdependence, *tendrel*, constitutes a fundamental law of nature: "*not only the myriad forms of life, but also the material phenomena, at their most subtle level, are governed by the laws of interdependence. From the lands where we live to the oceans, through the clouds, forests and flowers that surround us, all these phenomena occur in dependence of subtle patterns of energy. Without their proper interaction, they dissolve and disintegrate. (...)*

[49] Koran 78–6: "Have we not given you the Earth as a cradle, the Night as cloth, sent down abundant water to grow grains of plants and luxuriant orchards?"

Ignoring interdependence has not only harmed our natural environment,[50] *but also our societies*" (Dalai Lama and Stril-Rever 2016: 53). In ancient written works, everything is connected with "the interdependence of creatures," wanted by God. "*The sun and the moon, the cedar and the little flower, the eagle and the sparrow: the spectacle of their countless diversities and inequalities tells us that no creature is self-sufficient. Creatures exist only in dependence on each other, to become whole though each other, serving each other*" (François 2015, n° 86). The relationship with the environment is expressed in terms of interactions: "*there is an interaction between ecosystems and between the various worlds of social reference, and so, once again, it appears that 'the whole is greater than the part'*" (Exhort. apost. Evangelii gaudium (24 November 2013), n. 237: AAS 105 (2013), 1116 (François 2015, n°141).

The "whole" and "interdependence" develop through a connection of interiorities and physicalities that defines solidarity (Mathevet et al. 2010, 2016; Thompson et al. 2011). In the human/nonhuman relationship, a solidarity of a socio-ecological nature generates an ecological debt in the relationships between humans, due to the consequences of a lifestyle on the Planet that causes ecological damage (Pouchain 2014; Michelot 2016a, b).

All the works presented throughout the chapters of this book (the two volumes) contribute to the emergence of the new ontology of coviabilism presented in this first stage. The second stage, which we are about to deal with, investigates what constitutes coviability. The last stage of this results chapter puts into perspective the proposal of a principle of coviability.

27.2 Reminder of the Determinants of Socio-ecological Coviability to Reach a Definition

The question is to know what does and does not lead to coviability and to define its determinants. Is his question, which refers to that of a coviability space or slider, correctly put forward? Of all the works presented in this book, the question may be understood more in terms of characterization or diagnosis rather than being "core," *stricto sensus*. Indeed, if the theory of viability posits the hypothesis of a core of viability,[51] and is organized around it, the transdisciplinary perspective, social

[50] The term "environment" does not exist in Tibetan, "because traditionally, the nonhuman world was treated as an integral part of the human world, with shared destinies" (Dalaï Lama and Stril-Rever 2016: 79).

[51] For Jean-Pierre Aubin, "there is a mathematical version of the vague viability developed by Olivier Dordan over all these years. If the core of viability was the first to be introduced chronologically (1985), it concerns only perennial, eternal evolutions. As for ephemeral evolutions, which are born and die, the adequate metaphor is that of viable capture pools (2000), which could (or should) have been called "ephemeral viability nuclei". In this case, the duration of the time window of life could be a solution and not data on the problem. Unfortunately, posing additional

and environmental sciences together, leads us to go beyond the idea of borders, of limits. The vagueness, the imprecision, the impossibility to, to predict and to set the (human and nonhuman) biological behaviors and reactions, emerge too often from a Cartesian, scientific, even political rationale. Unpredictability feeds on the complexity of socio-ecological interactions; and may first of all be linked to the socio-cultural paradigm, in that everyone looks at "that which surrounds us," more commonly, "the World". Scientific protocols and political rationales, which are by nature very ethnocentric, depend on them. By refuting any sort of determinism, we can retain the idea of threshold, of a *tipping point*.[52]

The fusion of humanity with the biosphere finds its limits by these points (*tipping point*), which create a rupture in the human to the nonhuman continuity. The breaking point constitutes a switch over threshold into a new state, which may or may not be irreversible, a non-viable state (temporarily or permanently). The issue of ecological crisis becomes in fact a socio-ecological question, defining viability breaking points and consequently those of coviability. The notion of threshold must be discussed in relation to the critical point slider of a switch over to non-coviability. However, future work, particularly on climate change, could go as far as to consider new margins of coviability, redefining breaking points. What creates a breaking point? A "switch" toward a temporary or permanent irreversibility: entry into a non-viable state such as coral death or non-pollination. The switch generates a system of "crisis," which could be, for instance, ecological in nature or pollination related. The crisis could be an "equilibrium break" or a break in a state of coviability between the human and the nonhuman. We shall only address the example of coral bleaching (Van Hooidonk, R. et al., 2016; Potts et al. 2010; Aizen et al. 2008; Goulson et al. 2015; IPBES 2017), crisis in pollination[53] (Steffan-Dewenter et al. 2005) or even the prospect of a 6th mass extinction of animals (Ceballosa et al. 2017).

As already pointed out, socio-ecological coviability is defined more by a viability orchestrated at different levels based on mutual links, interactions and dependencies.

This viability means that the classified elements exist during the time window[54] (time and duration) during which these entities live and "exchange." Each time window determines the set of viable elements and each living element belongs to a time window. They overlap. These exchanges, these interactions, constitute, to a large extent, the essence of coviability. The complexity of these exchanges is a source of coviability.

---

mathematical difficulties, they appeared later, too late, because most users speak only of nuclei" (personal communication).

[52]A Tipping Point is « the critical point in a situation, process, or system beyond which a significant and often unstoppable effect or change takes place » (Merriam-Webster.com. Merriam-Webster, n.d. Web. 25 Sept. 2017).

[53]This example echoes several works developed in the book, notably in the Introductory Chapter, then in Chap. 20 (Billet) and Chap. 43 (Rodet) which deal with the legal regulation of pollination, becoming a paying service (the price of coviability), and the effects of intensive beekeeping on bee health.

[54]According to Jean-Pierre Aubin (2010, 2013), in mathematics, time and duration (of life) define "time windows" by two numbers, their duration and either their end or their beginning. See especially: http://vimades.com/AUBIN/TempsDuréeDurance.pdf

The first convergence of the work presented in this book concerns the fact that the viability of the social system and the viability of the ecological system are interdependent and create a systemic challenge. Therefore, we are not speaking of two systems but of only one system. It is socio-ecological in nature and it corresponds to the integration of (social and ecological) subsystems, within a single system associating humans with nonhumans or "coupled Human and Natural systems"[55] (Liu et al. 2007). Thus we return to explicitly consider all the actors (called collective, see above) including the human being as an active component of the ecosystem. The duality of human Societies (social systems) and All nonhumans (ecological or natural systems) are found in works on socio-ecosystems that reason in terms of relations between systems. Through these relationships that form a link, a framework system emerges encompassing humans and nonhumans. American researchers[56] have defined it as a mode of analysis called Coupled Human and Natural Systems (CHANS) (Liu et al. 2007; Hull et al. 2015). The purpose of the CHANS framework is to understand and model interactions between natural and human systems. More specifically, they target the reactions and feedback from coupled human and natural systems, in order to respond to global ecological and human challenges.[57]

The socio-ecological systemic challenge focuses on that which constitutes a system at various scales and temporalities, still to be defined (see Lagadeuc and Chenorkian 2009). It also focuses on its systemic "envelope," that which is in the system and in its environment (Chap. 5, Fargette et al.). For example, by being open, the oasis located in a particular place does not constitute a system locally (within the territory), but on a larger scale (Chapter Fargette et al., T2). Generalized in any territory, the socio-ecological systemic challenge is both intra-territorial (coherence of relations among sub-systems of the territory), and extra-territorial (integration of intra-territorial sub-systemic parts within their encompassing system).[58] The constitution of a system requires a coherence between structure and functioning (Chap. 5, Fargette et al.). Several chapters in Volume 2 fuel this question on the basis of very different terrains, including heterogeneous contexts.

Based on the analyses carried out by all the teams contributing in this work, we deduce that the determinants and structuring elements of socio-ecological coviability are based on systemic complexity. This systemic complexity involves viability links defining coviability and three bases: (a) anthropogenic and environmental constraints, (b) regulation that is formalized by law and (c) viability, expressed by biological or social facts (Fig. 27.2).

[55]"Coupled human and natural systems are integrated systems in which oeiole –interact with natural components" (Liu et al. 2007).

[56]Jianguo Liu and Vanessa Hull, from the Center for Systems Integration and Sustainability (CSIS – Michigan State University) and Mao-Ning Tuanmu from the Department of Ecology and Evolutionary Biology (Yale University).

[57]"Such an approach is needed to tackle the increasingly complex global challenges of our time including biodiversity loss, climate change, deforestation, degradation of ecosystem services, disease spread, famine, and social unrest" (Hull et al. 2015).

[58]Chapter 32 on the example of territorialized tourist system, Dérioz et al., vol. 2.

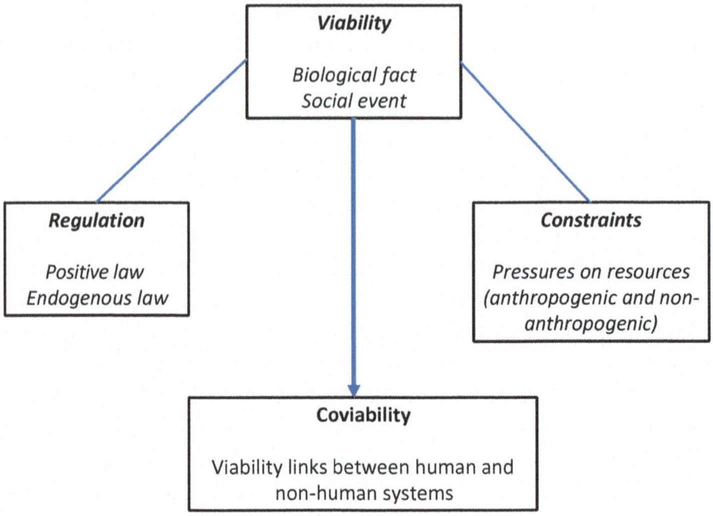

Fig. 27.2 The three elements structuring socio-ecological coviability, dependent on the links of viability between systems

The socio-ecological viability, as property (aptitude to live), results from a set of constraints (or stresses) (Chaps. 3, 4, 13, and 17 …) and the product of a regulation (almost all chapters), which allows the reproduction and the sustainability of the system. However, the state of viability remains dependent on the interactions of the system with others. Without handling these interactions, viability remains fiction. The following Fig. 27.3 completes the previous one (Fig. 27.2) by a series of determinants that keep viability in a state of fiction as long as the "co" is not implemented. Viability becomes real only through coviability, otherwise the system is not sustainable or viable. In a systemic approach, coviability requires at least two initially viable systems. By uniting and putting them in a state of interaction, they remain viable. They improve their synergies and/or minimize their competition: it is a path towards the emergence of an integrated system.

Biological and social facts become socio-biological ones within systems necessarily united, identified in the lower part of Fig. 27.3. Regulation and constraint intervene in the autopoiesis, homeostasis and resilience of systems. The "co" of viability depends on a relational structure specific to the situations studied. It shows a heterogeneity that reflects the diversity of the works, which select their own questions, methods and analysis within unique socio-ecological and geographical contexts (Fig. 27.3).

At this stage, the works presented in the chapters of this book have all enabled a definition of socio-ecological coviability to be achieved: *a property of interactive dependence between humans and nonhumans joined in a relationship that is contained by regulations and constraints. This relationship establishes a link of viability subjected to an integration threshold of the complex human/nonhuman system determining the limits of coviability elasticity, whose realization remains the coevolution in an integrated socio-ecological system.*

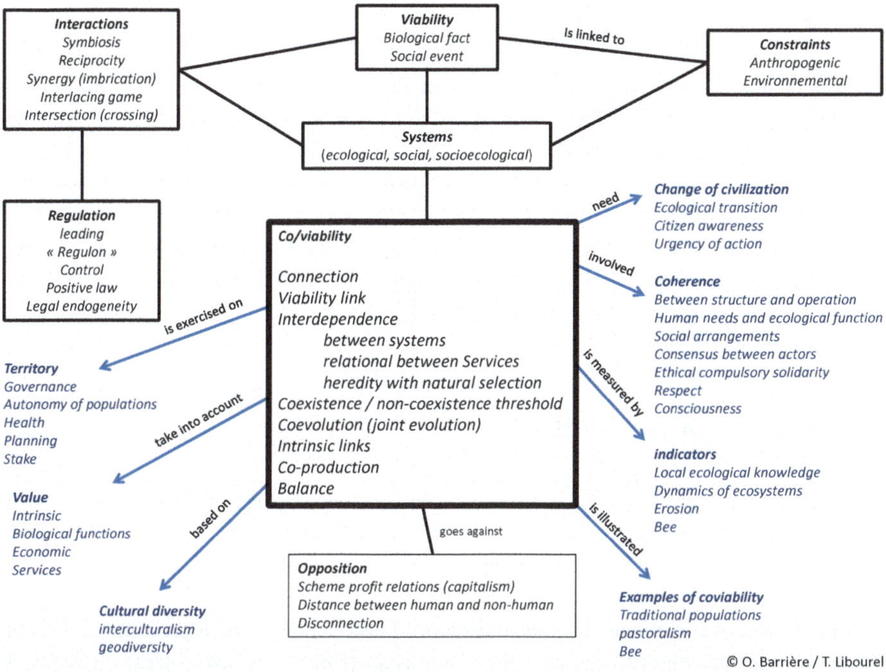

Fig. 27.3 Summary of coviability's socio-ecological determinants, based on all the works in this book (two volumes)

The concept of coviability expresses a paradigm from which emerges a universal ontology of a connection between humanity and the biosphere: coviabilism. This latter defines the unity of the living in a mental and physical continuity.

Through this definition, perspectives are deduced for building the paradigm of coviability into a legal principle in order for it to be made into a reality in the social body at different scales and for it to be reflected in public policies.

27.3 From Paradigm to a Legal Principle of Socio-ecological Coviability

Here, legal science allows a cross-disciplinary approach by "questioning" scientific and nonscientific expertise (local knowledge, inheritance, and so on) to answer questions of justice (environmental, health, and so on), of rights (human, indigenous peoples, and so on), and of responsibility (future generations, animals, nature, and so on). This cross-disciplinarity promotes the paradigm of socio-ecological coviability to a juridical principle of socio-ecological coviability.

> The question that poses itself is whether it is time to reconsider the foundations on which our social community is based in this new era of environmental crisis (Rühs and Jone 2016)

The challenge of thinking in terms of a new paradigm, be it economic, well-being, or of another relationship of the human to the biosphere, lies both at the levels of its legitimacy and its effectiveness. The ethical foundation of the paradigm is destined to be included in law in order to live. This "juridicization" is realized in the form of a normative principle (*principle of law*) which expresses unwritten rules of law. The principle, from the Latin *principium*, "that which serves as a basis for something" comes from *primo* (first) and *capere* (take). The principle takes first place, first rank; it is a beginning.

27.3.1 A "Law for Nature?

Essentially, the objective of the consecration of the paradigm of coviability would beto lay down the bases of a line of conduct, a legal norm enacting a "must-be." The resulting need is the legal recognition of the ecological limits of the Planet (Berry 2000).

The notion of limit leads to that of threshold and hence to the idea of a whole that breaks down the objectification of the nonhuman (t nature as an object) because of its instrumentalization.[59] Thus, logically, the legal personality should be extended to nonhumans ("nature," according to the authors cited).

The coviability paradigm responds to a "new 'biospheric' ethic," which contributes in maintaining the cosmic order,[60] "sustaining the earth's biosphere"[61] (Goldsmith 2003) by values of existence, an ethic generating rules of conduct, standards of behavior, ethics and morality.

Elevating natural law to a constitutional level (Rühs and Jone 2016)[62] would provide a legal framework for the implementation of "Earth Jurisprudence" in order to accomplish the necessary paradigm shift (ibid.). The recent development of "global environmental constitutionalism" reveals a new way of developing international law and environmental governance as a universal concern (Bosselmann 2015). The goal

[59]Emmenegger and Tschentscher (1994) has pointed out that giving rights to nature itself is a better solution for environmental protection than only having human duties towards the environment [14]. This is because substantive rights and protection of intrinsic value are two sides of the same coin, the one cannot be without the other. In other words, continuing to see nature as an object will result only in the protection of its instrumental value to humans (Rühs and Jone 2016).

[60]"A biospheric ethic, one compatible with the ecological view of the world we live in, (. . .) would involve human beings in helping to sustain the earth's biosphere. And unethical behaviour would be that which disrupts and destroy the biosphere" (Goldsmith 2003).

[61]« We must place ethical values in their appropriate context, that of mediating sustainable human behaviour in relationship to society, the ecosystem, the biosphere and the cosmos itself » (Goldsmith 2003).

[62]"a constitutional right of nature is needed to address the challenges that we now face globally".

behind this constitutional path is to provide stronger protection for the nonhuman, by creating substantial rights "of" nature, and not only substantial human rights "to" a healthy environment (ibid.). However, the paradigm of coviability allows us to go further by recognizing a right focusing on the link of viability, with a legal personality granted more to a property of connection (this link) than to a subject scarcely definable, or which can too often be appropriated.

Ruhs and Jones's works (2016) demonstrate that the very definition of nature constitutes a legal problem in granting rights, what is "nature"? In addition, the new Ecuadorian constitution (of 2008) is a rare case in the world (with Bolivia[63] New Zealand[64] and India[65]) of the personification of nature; the authors question this experiment, which continues to separate the human being from nature. However, one might think that defining the viability link is just as difficult. This is certainly the case if we stop at materiality, and maybe less so if this link of intangible viability is defined in a cosmological relationship identified by a river, a land, a mountain, a tree, a bird, and so on. This perspective leads to rethinking the relationship between law and the living, and more specifically to the land.

27.3.2 *Should the Legal Relationship to Land Be Rethought?*

The question on relationship to land relates to the foundations of human connection to the biosphere through this umbilical link between humanity and the land. The legal nature of this link formalizes a relational paradigm of denounced capitalist obedience: "*connection with nature —and specifically, with land—underpins any transformation of property law from an anthropocentric, individualist concept to a more ecocentric and relational one*" (Howe 2017).

The relationship stemming from the capitalist rationale, evoked in the works of this book (in particular Chap. 11, Bertrand), touches upon a core aspect of the issue of reconnecting humans to the biosphere. Recognizing the paradigm of coviability by a principle of coviability would allow a shift from this paradigm of the regime of land appropriation, reasoning more in terms of ecological integrity than in economic terms. The links of interdependency and thus connectivity to the biosphere associating humans in ecosystems cannot be part of this property regime, as it transforms the nonhuman into capital and the link of viability into a commercial link.

[63]On 21 December 2010 Bolivia adopted a "Mother Earth Law" ("Ley de Derechos de la Madre Tierre," Ley 071) which established 11 fundamental rights for nature, including the right to live and exist.

[64]In March 2017, the New Zealand Parliament donated to the Whanganui River a legal personality, granting it the status of living entity.

[65]On March 20, 2017, the Ganges River and the Yamuna River were designated as "living entities with legal status" by the High Court of the Himalayan State of Uttarakhand.

The two leading questions on the rights of nature and the rights of humans on land (in terms of *land tenure*), which we have just seen, reflect on the necessary objective rebuilding environmental law to coviability law.

27.3.3 The Input of a Fundamental Principle Resulting from the Paradigm of Coviability: A Rebuilding of Environmental Law into Coviability Law

The Global Pact for the Environment[66] project recognizes the need for action, and reaffirms the need to ensure ecosystem resilience in order to preserve biodiversity and contribute to human well-being and to poverty elimination. The Pact adopts rights, duties, and principles posited in 26 articles. The first article concerns a right to "*an ecologically sound environment adequate for their health, well-being, dignity, culture and fulfilment*" (art.1). The second is a duty to "*take care of the environment*" by contributing to the "*restoration of the integrity of the Earth's ecosystem*" (art.2). The Pact remains in the paradigm of sustainable development (the commitment to "*pursue sustainable development*," art.3).

In this first version, the Global Pact for the Environment revels in the rationale of separating the human from the nonhuman, where "environment" is the only term mentioned. Interdependence does not appear; the dichotomy is well anchored by insisting on "sustainable development." Very anthropocentric, the Pact remains settled on an ethnocentric rationale of international environmental law since 1972. The question remains on effectiveness, on the implementation of a text of principles, which, for the writers, must become binding.

The ambition of rebuilding environmental law by providing a principle of coviability has the merit of innovation. The paradigm shift proposed through the concept of coviability transforms the process of thinking and acting, sometimes already implicit.[67] Giving horizontality to the verticality in the hierarchy of norms is suggested in order to allow it to be legitimate and effective. The suggested principle

---

[66] An initiative given by 80 environmental law specialists from 40 countries; the aim of this pact-in-progress for States is to strengthen international law by 26 articles. The project is led by Laurent Fabius (President of the French Constitutional Council) and the Club des juristes. It is destined to become a binding international treaty:http://www.leclubdesjuristes.com/vers-un-pacte-mondial-pour-l-environnement-programme/

[67] The French Supreme Court is moving in the direction of coviability when it affirms the interaction and the interdependence of man and nature in the following terms: "the habit in simplifying the premises of an argument to make it easier, led to consider man, detached from his natural environment, disregarding the constant interaction of man with nature and forgetting that nature is part of man, just as he is part of it; (it) stems from this interdependence that any significant breach of the natural environment is an attack on the community of men who live in interaction with it and that this attack must be compensated (...)" V. Court of Cassation, crim. 25 sept. 2012, n° 10–82.938. Cited in Chap. 21 Treillard et al. of this book.

of coviability transforms the substance of the legal challenge: the link rather than the object (see above). More precisely, the model is no longer that of a (sustainable) development process, but that of viability, the property of a process, by conferring a status on the nonhuman, or rather on the link of viability. Thus, environmental law would be revolutionized in its definition and its function.

The input of a fundamental principle based on coviability rebuilds environmental law to one of coviability through five structuring arguments:

- It is no longer a law "of the environment" or "to the environment" but a law of the relationship between humans and the biosphere: the link of viability and interrelations generates a law of coviability. Legal regulation bases itself on the link, the connection of humans to the biosphere: there is a move from a human/nonhuman dichotomy towards a symphony in which humans are an integrated part of the ecosystem: the social human is integrated into the ecological system.
- The direction of sustainable development is based on the coviabilism ontology proposed, which is based on the perspective of viability.
- The subject of the law must not be restricted to the human only. Without necessarily personifying "nature," it is necessary to give a status to nonhumans and especially legalize the link of viability.
- Add to the vertical normative creation, a horizontal creation: a law negotiated for implementation, an effectiveness of the law of the human/nonhuman relationship.
- Recognizing legal pluralism within States: the rights of indigenous or local populations, in agreement with cultural pluralism.
- Recognizing[68] legal habitus in the definition of law, which exceeds rules and norms; moving out of a restrictive normative approach of law.

27.4 Conclusion

The 44 chapters in the two volumes of this volume have progressively contributed in defining the concept of socio-ecological coviability, with a dip into cultural coviability (see Chaps.18 Barrière Catherine, and 19 Canete Voyner). From the different analyses that have led to the definition presented above, a first attempt to summarize is shown in Fig. 27.4. Coviability results from a property of viability

[68]As principles structuring the cohesion of society. The habitus is an "immanent law, deposited in each agent by premium education, which is the condition not only of the concertation of practices but also of the practices of concertation, since the adjustments consciously made by the agents themselves presuppose mastery of a common code and that collective mobilization enterprises can not reach without a minimum of agreement between the habitus of the mobilizing agents (for instance, prophet, party leader, and so on) and the dispositions of those of whom they strive to express the aspirations. "(Bourdieu 2000, 272).

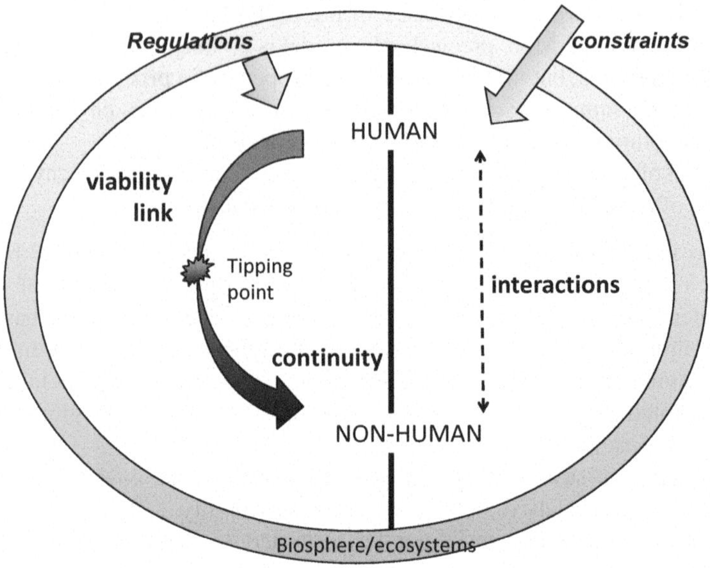

Fig. 27.4 Socio-ecological coviability: a continuity dependent on interactions and a viability link

of everything that stems from the coviability between the constitutive subsystems (see Chap. 5, Fargette et al.).

The previously defined link of viability, by a necessary connection for an aptitude to live, expresses a connection through bonding to stay together (Fig. 27.4).

The contributions of the socio-ecological coviability paradigm to the concerns of the international community, notably through the SDGs (UN, A/69/L.85), the OECD's 2050 Horizon (OECD 2012) and the various international conventions and agreements, are to reposition human evolution within an Earth system that contains it and a biosphere upon which it will always remain dependent (Table 27.1).

- Socio-ecological coviability aims at achieving sustainability of the Earth, human and nonhuman included, thanks to societal and environmental viability objectives, undivided and interconnected (human – nonhuman coviability, man's relation to the biosphere).
- Sustainable development goals will only be achieved if the interactions between humans, and between humans and nonhumans, have an evolving dynamic that allows them to adapt and the Earth system to regulate itself to maintain the viability of the biosphere.
- The means of achieving the sustainable development goals should not favor blind growth. This anthropocentric posture makes man the main preoccupation and therefore favors the human/nonhuman dichotomy. For the international community's joint project to succeed in "changing the world" through sustainable action, the means to act must remain open to other visions on the well-being of both humans and the planet.

This book comprises a great disciplinary diversity, but certain scientific fields such as economics are not present enough. Thus, the book presented should be seen more as a starting point, a commitment to the definition and perspectives of a new paradigm. The aim of the dynamics launched by the volume is to integrate a greater plurality of disciplines, with the hope of leading to a pre-configuration of transdisciplinarity.

The number of researchers and works involved in this project has enabled the stage of rethinking the ecological transition using a rationale of viability to be reached, in order to achieve socio-ecological coviability.

We have put forward a legal dimension in this chapter on how to achieve this. In this dimension, we propose the formalization of a new principle of law, one that brings about renewal rather than utopia. Climate change and overall ecological urgency, forces us to react globally through political directions and choices, especially in order to achieve a social and ecological transition.

Through this coviability approach we have entered into a complex area, that of planet Earth and sciences. As a result, the data provided in this chapter are multiple and sometimes makes one's head spin due to their specificities. The data participates in arguing the importance of promoting such a paradigm, which cannot be carried out in a summary manner. This justifies the hundred or so researchers involved. Other forms of much shorter and more brief releases will follow to render this work more accessible, especially for decision-makers.

References

Aizen MA, Garibaldi LA, Cunningham SA, Klein AM (2008) Long-term global trends in crop yield and production reveal no current pollination shortage but increasing pollinator dependency. Curr Biol 18:1572–1575. https://doi.org/10.1016/j.cub.2008.08.066

Akrich M, Callon M, Latour B (2006) Sociologie de la traduction. Presse des Mines. https://doi.org/10.4000/books.pressesmines.1181

Aubin J-P (2010) Valoriser la double nature du temps : une métaphore mathématique, Framework Programme Marie Curie Initial Training Network (FP7-PEOPLE-2010-ITN), project SADCO, Available on line : http://vimades.com/AUBIN/Fluidity.pdf

Aubin J-P (2013) Chaperoning state evolutions by variable durations. SIAM J Control Optim 51:2081–2101. https://doi.org/10.1137/120879853DOI

Barrière O (2011) Pluralisme juridique et patrimonialisation : entre paradigmes de l'« appropriation » et du « patrimoine commun ». In: Fouck SML, Hidair I (eds) La question du patrimoine en Guyane française. Diversité culturelle et patrimonialisation. Processus et dynamiques des constructions identitaires. Ibis Rouge Editions, Matoury, pp 43–75

Barrière O (2017) Human relationship to the land from a legal perspective as a human and environmental security challenge. In: Behnassi M, McGlade K (eds) Chap 14, environmental change and human security in Africa and the Middle East. Springer, pp 259–304. https://doi.org/10.1007/978-3-319-45648-5_14

Belaidi N (2004) La lutte contre les atteintes globales à l'environnement : vers une ordre public écologique ?, thèse Droit, Université de Bourgogne, 612p

Belaidi N (2014) Identité et perspectives d'un ordre public écologique. Droit et Cultures/L'Harmattan 68(2):15–49

Berry T (2000) The great work: our way into the future. Broadway Books, New York

Bosselmann K (2015) Global environmental constitutionalism: mapping the terrain. Widener Law Rev 21:1–24

Bourdieu P (2000) Esquisse d'une théorie de la pratique. Editions du seuil, Paris

Boutelet M, Fritz J-C (2005) L'ordre public écologique. Toward an ecological public order. Bruylant, Bruxelles

Boyd DR (2012) The environmental rights revolution: a global study of constitutions, human rights and the environment, Law and Society. University of British Columbia (UBC Press), Vancouver

Boyd DR (2014) The status of constitutional protection for the environment in other nations, David Suzuki Foundation Working Paper, Canada. Available online http://davidsuzuki.org/publications/2014/whitepapers/DSF%20White%20Paper%204.pdf

Brosse J (2002) Ecologie et spiritualité. Albin Michel, Paris

Burdon PD (2010) Wild law: the philosophy of earth jurisprudence. Altern Law J, 35(2), University of Adelaide Law Research Paper No. 2011-009. Available at SSRN https://ssrn.com/abstract=1636564

Cabanes V (2016) Un nouveau droit pour la terre, pour en finir avec l'écocide. Seuil, Paris

Caron P, Valette É, Wassenaar T (2017) Coppens d'Eeckenbrugge Geo, Papazian Vatché (coordinators). In: Living territories to transform the world. Quae editions

Ceballosa G, Ehrlichb PR, Dirzo R (2017) Biological annihilation via the ongoing sixth mass extinction signaled by vertebrate population losses and declines. PNAS (Proceedings of the National Academy of Sciences of the United States of America). https://doi.org/10.1073/pnas.1704949114. On line : www.pnas.org/cgi/doi/10.1073/pnas.1704949114

Charmont J (1927) La renaissance du droit naturel. Librairie de jurisprudence ancienne et moderne, Paris

Comolet A (1994) L'évaluation et la comptabilisation du patrimoine naturel, définitions, méthodes et pratiques. L'Harmattan, coll. Environnement, Paris

Costanza R et al (1997) The value of the world's ecosystem services and natural capital. Nature 387:253–260

Cullinan C (2011) Wild law: a manifesto for earth justice, 2nd edn. Green Books, Devon

Dalaï-Lama (Le) et Stril-Rever S (2016) Nouvelle réalité. L'âge de la responsabilité universelle. Les Arènes Edition

Descola Philippe (2005) Par delà nature et culture, Gallimard/In english : (2013) Beyond nature and culture, University of Chicago Press, Available on line: https://itp.nyu.edu/classes/interspecies/texts/philippe-descola-beyond-nature-and-culture_ch1.pdf

Descola P (2011) L'écologie des autres. L'anthropologie et la question de la nature. Quae, Paris

Dion C (2015) Demain, un nouveau monde en marche. Partout dans le monde des solutions existent, Actes Sud/In english: (2017) Tomorrow: All Over the Globe, Solutions Already Exist, Actes Sud

Emmenegger S, Tschentscher A (1994) Taking nature's rights seriously: the long way to biocentrism in environmental law. Georget Int Environ Law J 6:545–592

Everard M (2017) Ecosystem services key issues. Routledge, London

Feenberg A (2013) L'anthropologie et la question de la Nature. Réflexions sur L'Écologie des autres, de Philippe Descola. Revue du MAUSS 2–42:105–118. https://doi.org/10.3917/rdm.042.0105

François (2015) Lettre encyclique, *Laudato Si*, sur la sauvegarde de la maison commune Encyclical letter Laudato si' of the holy father francis on care for our common home

Goldsmith E (2003) Towards a biospheric ethic. Institute of Science and Society. Available online: http://www.i-sis.org.uk/biosphericethic.php

Goulson D, Elizabeth N, Cristina B, Ellen LR (2015) Bee declines driven by combined stress from parasites, pesticides, and lack of flowers. Science 347(6229):1255957. https://doi.org/10.1126/science.1255957

Gudynas E (2016) Bem viver. In: D'Alisa G, Demaria F, Kallis G (eds) Decrescimento. Vocabulário para um mundo novo. Editorial, Porto Alegre, pp 260–264

Hermitte M-A (2011) La nature, sujet de droit ?, Annales Histoire Sciences sociales, Editions de l'EHESS, 1, 173–212. On line: http://www.cairn.info/revue-annales-2011-1-page-173.htm

Higgins P (2015) Eradicating ecocide: laws and governance to prevent the destruction of our Earth. Shepheard-Walwyn, London

Howe HR (2017) Making wild law work—the role of 'connection with nature' and education in developing an Ecocentric property law. J Environ Law 29:19–45. https://doi.org/10.1093/jel/eqw029

Hull V, Tuanmu M-N, Liu J (2015) Synthesis of human-nature feedbacks. Ecol Soc 20(3):17. https://doi.org/10.5751/ES-07404-200317

IPBES (2017) The assessment report on pollinators, pollination and food production. IPBES, Bonn

Kopps AM, Sherwin WB (2012) Modelling the emergence and stability of a vertically transmitted cultural trait in bottlenose dolphins. Anim Behav 84(6):1347–1362. https://doi.org/10.1016/j.anbehav.2012.08.029

Laastad SG (2016) Nature as a subject of rights, a discourse analysis on Ecuador's constitutional rights of nature, master thesis in human geography. Department of Sociology and Human Geography, University of Oslo. On line Avalaible: https://www.duo.uio.no/handle/10852/51892

Lagadeuc Y, Chenorkian R (2009) Les systèmes socio-écologiques: vers une approche spatiale et temporelle. Nat Sci Soc 17(2):194–196

Laland KN, Janik VM (2006) The animal cultures debate. Trends Ecol Evol 21(10). https://doi.org/10.1016/j.tree.2006.06.005

Larrère C (1997) La communauté biotique: l'héritage d'Aldo Leopold. In: Larrère C (ed) Les philosophies de l'environnement. Presses Universitaires de France, Paris, pp 60–84

Latouche S (2012) L'âge des limites. Editions Mille et une nuits, Paris

Latour Bruno (2004) Politiques de la nature. Comment faire entrer les sciences en démocratie, La Découverte (first edition 1999)/Politics of nature: how to bring the sciences into democracy (translated by Catherine Porter), Harvard University Press

Lay B, Neyret L, Short D, Oposa A (2015) Timely and necessary: ecocide law as urgent and emerging. J Jurisprud 28:431–452

Liu J, Dietz T, Carpenter SR, Alberti M, Folke C, Moran E, Pell AN, Deadman P, Kratz T, Lubchenco J, Ostrom E, Ouyang Z, Provencher W, Redman CL, Schneider SH, Taylor WW (2007) Complexity of coupled human and natural systems. Science 317(5844):1513–1516

Luke R, Whitehead H (2001) Culture in whales and dolphins. Behav Brain Sci 24(02):309–324

Madelin P (2017) Après le capitalisme. Essai d'écologie politique. Ecosociété, Montréal

Maris V (2014) Nature à vendre. Les limites des services écosystémiques. Quae, Paris

Maris V (2016) Philosophie de la biodiversité: Petite éthique pour une nature en péril. Buchet-Chastel, Paris

Mathevet R, Thompson J, Delanoë O, Cheylan M, Gil-Fourrier C, Bonnin M (2010) La solidarité écologique: un nouveau concept pour une gestion intégrée des parcs nationaux et des territoires. Nat Sci Soc 18:424–433. https://doi.org/10.1051/nss/2011006

Mathevet R, Thompson JD, Folke C, Chapin FS (2016) Protected areas and their surrounding territory: socioecological systems in the context of ecological solidarity. Ecol Appl 26(1):5–16

Médevielle G (2010) Nature et loi naturelle comme concepts théologiques. Rech Sci Relig 2–98:245–265

Michelot A (dir.) (2016a) La dette écologique: mise en perspective de ses définitions et de ses implications. VertigO – la revue électronique en sciences de l'environnement [En ligne], Hors-série 26

Michelot A (2016b) « La dette écologique en questions : propos introductifs », VertigO – la revue électronique en sciences de l'environnement [En ligne], Hors-série 26, mis en ligne le 01 octobre 2016, consulté le 20 juin 2017. URL: http://vertigo.revues.org/17751

Millenium Ecosystem Assesment MEA (2005) Ecosystems and human Well-being: biodiversity synthesis. Washington, DC, World Resources Institute

Murray J (2015) Earth jurisprudence, wild law, emergent law: the emerging field of ecology & law. Liverpool Law Rev 36:105. https://doi.org/10.1007/s10991-015-9170-y

Neyret L (2015) Des écocrimes à l'écocide. Le droit pénal au secours de l'environnement, bruylant. Bruylant, Bruxelles

Noad MJ, Cato DH, Bryden MM, Jenner M-N, Jenner KCS (2000) Cultural revolution in whale songs. Nature 408(6812):537–537. https://doi.org/10.1038/35046199

Norgaard RB (2010) Ecosystem services: from eye-opening metaphor to complexity blinder. Ecol Econ 69:1219–1227. https://doi.org/10.1016/j.ecolecon.2009.11.009

OECD (2012) OECD environmental outlook to 2050. The consequences of inaction. OECD Publishing, Paris. https://doi.org/10.1787/9789264122246-en

Ost F (1995) La nature hors la loi. L'écologie à l'épreuve du droit. éditions la découverte, Paris

Pelt J-M (1977) L'homme renaturé. Du Seuil, Paris

Potts et al (2010) Global pollinator declines: trends, impacts and drivers. Trends Ecol Evol 25(6):345–353. https://doi.org/10.1016/j.tree.2010.01.007

Pouchain D (2014) La dette écologique: d'une notion politique à un concept philosophique ?/The ecological debt: from a political notion to a philosophical concept? Développement durable et territoires 5-1: Écologie industrielle, économie de la fonctionnalité

Ramose MB (2016) Ubuntu. In: D'Alisa G, Demaria F, Kallis G (eds) Decrescimento. Vocabulário para um mundo novo. Editorial, Porto Alegre, pp 273–275

Ravena-Cañete TM, Ravena-Cañete V (2011) Por uma sociologia do campo jurídico na/da Amazônia: as populações tradicionais amazônicas em foco. Rev Sociol Jurídica 13(1):26–54

Richard J (2012) Comptabilité et développement durable. Economica, Paris

Robertson MM (2006) The nature that capital can see: science, state, and market in the commodification of ecosystem services. Environ Plann D: Soc Space 24:367–387

Rühs N, Jone A (2016) The implementation of earth jurisprudence through substantive constitutional rights of nature. Sustainability 8(2):174. https://doi.org/10.3390/su8020174

Sachs I (1980) Stratégies de l'éco-développement. Editions ouvrières, Coll. Développement et civilisation, Paris, p 250

Schoon M, van der Leeuw S (2015) The shift toward social-ecological systems perspectives: insights into the human-nature relationship. Nat Sci Soc 23(2):166–174. https://doi.org/10.1051/nss/2015034

Selosse M-A (2017) Jamais seul, Ces microbes qui construisent les plantes, les animaux et les civilisations. Acte Sud editions

Serres M (1992) Le contrat naturel. Flammarion, Paris

Silvertown J (2015) Have ecosystem services been oversold? Trends Ecol Evol:641–648. https://doi.org/10.1016/j.tree.2015.08.007

Steffan-Dewenter I, Potts GS, Packer L (2005) Pollinator diversity and crop pollination services are at risk. Trends Ecol Evol 20(12):651–652

UNEP-WCMC and IUCN (2016) Protected planet report 2016 how protected areas contribute to achieving global targets for biodiversity

Thompson D et al (2011) Ecological solidarity as a conceptual tool for rethinking ecological and social interdependence in conservation policy for protected areas and their surrounding landscape. C. R. Biol 2011. https://doi.org/10.1016/j.crvi.2011.02.001

Van Hooidonk R et al (2016) Local-scale projections of coral reef futures and implications of the Paris Agreement. Sci Rep 6:39666. https://doi.org/10.1038/srep39666

Weber JP (1986) Le patrimoine naturel, In Les Comptes du Patrimoine Naturel, Comptes et Planification, Les collections de l'Insee (n°137–138). INSEE, Paris, pp 35–62

Weber J (2013) Gestion des ressources renouvelables: fondements théoriques d'un programme de recherche (juin 1995), in Rendre possible, Jacques weber, itinéraire d'un économiste passe-frontière, 35–52, dir. Meriem Bouemrane, Antona Martine, Robert Barbault, Marie-Christine Cormier-Salem, Quae edition

Whiten A, Goodall J, McGrew WC, Nishida T, Reynolds V, Sugiyama Y, Tutin CEG, Wrangham RW, Boesch e C (1999) Cultures in chimpanzees. Nature 399(6737):682–685. https://doi.org/10.1038/21415

Wilson EO (2007) Foreword. In: Penn DJ, Mysterud I (eds) Evolutionary perspectives on environmental problems. Aldine Transaction, New Brunswick

Zierler D (2011) The invention of ecocide, agent orange, Vietnam, and the scientists who changed the way we think about the environment. University of Georgia Press, Athens, GA

Correction to: Coviability of Social and Ecological Systems: Reconnecting Mankind to the Biosphere in an Era of Global Change

Olivier Barrière, Mohamed Behnassi, Gilbert David, Vincent Douzal, Mireille Fargette, Thérèse Libourel, Maud Loireau, Laurence Pascal, Catherine Prost, Voyner Ravena-Cañete, Frédérique Seyler, Serge Morand

Correction to:
The Frontmatter in : General Introduction
O. Barrière et al. (eds.), *Coviability of Social and Ecological Systems: Reconnecting Mankind to the Biosphere in an Era of Global Change*, https://doi.org/10.1007/978-3-319-78497-7

The original version of the book was inadvertently published with surname and first name of the authors interchanged in the Reference section of the General Introduction in the front matter. This has now been corrected.

The updated version of the book can be found at
https://doi.org/10.1007/978-3-319-78497-7

O. Barrière et al. (eds.), *Coviability of Social and Ecological Systems: Reconnecting Mankind to the Biosphere in an Era of Global Change*,
https://doi.org/10.1007/978-3-319-78497-7_28

The manufacturer's authorised representative in the EU is Springer Nature Customer Service Centre GmbH, Europaplatz 3, 69115 Heidelberg, Germany. If you have any concerns regarding our products, please contact ProductSafety@springernature.com

Printed and bound by CPI Group (UK) Ltd, Croydon, CR0 4YY
25/09/2026
02222851-0017